T0289639

Matrix, Numerical, and Optimization Methods in Science and Engineering

Address vector and matrix methods necessary in numerical methods and optimization of systems in science and engineering with this unified text. The book treats the mathematical models that describe and predict the evolution of our processes and systems, and the numerical methods required to obtain approximate solutions. It explores the dynamical systems theory used to describe and characterize system behavior, alongside the techniques used to optimize their performance. The book integrates and unifies matrix and eigenfunction methods with their applications in numerical and optimization methods. Consolidating, generalizing, and unifying these topics into a single coherent subject, this practical resource is suitable for advanced undergraduate students and graduate students in engineering, physical sciences, and applied mathematics.

Kevin W. Cassel is professor of mechanical and aerospace engineering and professor of applied mathematics at the Illinois Institute of Technology. He is also a fellow of the American Society of Mechanical Engineers and an associate fellow of the American Institute of Aeronautics and Astronautics.

Matrix, Numerical, and Optimization Methods in Science and Engineering

KEVIN W. CASSEL
Illinois Institute of Technology

CAMBRIDGE
UNIVERSITY PRESS

University Printing House, Cambridge CB2 8BS, United Kingdom

One Liberty Plaza, 20th Floor, New York, NY 10006, USA

477 Williamstown Road, Port Melbourne, VIC 3207, Australia

314–321, 3rd Floor, Plot 3, Splendor Forum, Jasola District Centre, New Delhi – 110025, India

79 Anson Road, #06–04/06, Singapore 079906

Cambridge University Press is part of the University of Cambridge.

It furthers the University's mission by disseminating knowledge in the pursuit of education, learning, and research at the highest international levels of excellence.

www.cambridge.org
Information on this title: www.cambridge.org/9781108479097
DOI: 10.1017/9781108782333

First published 2021

A catalogue record for this publication is available from the British Library.

Library of Congress Cataloging-in-Publication Data

Names: Cassel, Kevin W., 1966– author.
Title: Matrix, numerical, and optimization methods in science
and engineering / Kevin W. Cassel.
Description: Cambridge ; New York, NY : Cambridge University Press, 2021. |
Includes bibliographical references and index.
Identifiers: LCCN 2020022768 (print) | LCCN 2020022769 (ebook) |
ISBN 9781108479097 (hardback) | ISBN 9781108782333 (ebook)
Subjects: LCSH: Matrices. | Dynamics. | Numerical analysis. |
Linear systems. | Mathematical optimization. | Engineering mathematics.
Classification: LCC QA188 .C37 2021 (print) | LCC QA188 (ebook) |
DDC 512.9/434–dc23
LC record available at https://lccn.loc.gov/2020022768
LC ebook record available at https://lccn.loc.gov/2020022769

ISBN 978-1-108-47909-7 Hardback

The heavens declare the glory of God;
  the skies proclaim the work of his hands.
Day after day they pour forth speech;
  night after night they display knowledge.
There is no speech or language
  where their voice is not heard.
Their voice goes out into all the earth,
  their words to the ends of the world.
(Psalm 19:1–4)

Contents

Preface

So much of the mathematics that we know today was originally developed to treat particular problems and applications – the mathematics and applications were inseparable. As the years went by, the mathematical methods naturally were extended, unified, and formalized. This has provided a solid foundation on which to build mathematics as a standalone field and basis for extension to additional application areas, sometimes, in fact, providing the impetus for whole new fields of endeavor. Despite the fact that this evolution has served mathematics, science, and engineering immeasurably, this tendency naturally widens the gaps between pure theoretical mathematics, applied mathematics, and science and engineering applications as time progresses. As such, it becomes increasingly difficult to strike the right balance in textbooks and courses that encourages one to learn the mathematics in the context of the applications to which the scientist and engineer are ultimately interested.

Whatever the approach, the goal should be to increase the student's intellectual dexterity in research and/or practice. This requires a depth of knowledge in the fundamentals of the subject area along with the underlying mathematical techniques on which the field is based. Given the volume of theory, methods, and techniques required in the arsenal of the researcher and practitioner, I believe that this objective is best served by how the mathematical subjects are discretized into somewhat self-contained topics around which the associated methods and applications are hung.

The objective of the present text is to integrate matrix methods, dynamical systems, numerical methods, and optimization methods into a single coherent subject and extend our skill to the level that is required for graduate-level study and research in science and engineering through **consolidation**, **generalization**, and **unification** of topics. These objectives guide the choice and ordering of topics and lead to a framework that enables us to provide a unified treatment of the mathematical techniques so that the reader sees the entire picture from mathematics – without overdoing the rigor – to methods to applications in a logically arranged, and clearly articulated, manner. In this way, students retain their focus throughout on the most challenging aspects of the material to be learned – the mathematical methods. Once they have done so, it is then straightforward to see how these techniques can be extended to more complicated and disparate applications in their chosen field.

Because engineers and scientists are naturally curious about applications, we merely need to tap into this curiosity to provide motivation for mathematical developments. Consequently, the keys to learning mathematics for scientists and engineers are to (1) sufficiently motivate the need for it topic by topic, and (2) apply it right away and often. In order to leverage the reader's inherent interest in applications, therefore, the overall pattern of the text as a whole as well as within each topic is to *motivate* → *learn* → *interpret* → *apply* → *extend*. This approach more clearly uses the applications to help the student learn the fundamental mathematical techniques, and it also provides deeper insight into the applications by unifying the underlying mathematics. It encourages a deeper understanding of matrix methods and its intimate connection with numerical methods and optimization. The primary virtue of this approach is that the reader clearly sees the connections, both mathematical and physical, between a wide variety of topics.

For a subject as ubiquitous as matrix methods, great care must be exercised when selecting specific topics for inclusion in such a book lest it become an unwieldy encyclopedia. The most compelling answer to the question, "Why do scientists and engineers need a deep knowledge of matrix methods?" is that they provide the mathematical foundation for the numerical methods, optimization, and dynamical systems theory that are central to so much of modern research and practice in the physical sciences and engineering.

Perhaps the most prevalent and ubiquitous applications of matrix methods are in the numerical techniques for obtaining approximate solutions to algebraic and differential equations that govern the behavior of mechanical, electrical, chemical, and biological systems. These numerical methods pervade all areas of science and engineering and pick up where analytical methods fail us. It is becoming increasingly clear that serious researchers and practitioners must have a solid foundation in both matrix methods and numerical techniques. The most effective way to articulate such foundational material is in a unified and comprehensive manner – a token chapter on elementary numerical methods is insufficient. On the other hand, undergraduate texts in numerical methods for engineers and scientists generally only require a minimal background in linear algebra. However, a bit more formal understanding of vectors, matrices, linear systems of algebraic equations, and eigenproblems can significantly enhance the learning of such subjects. The strategy of this text clearly highlights and leverages the integral and essential nature of linear algebra in numerical methods.

The present text has been written in the same style as *Variational Methods with Applications in Science and Engineering* (Cassel, 2013) with the same emphasis on broad applications in science and engineering. Observe in Figure 0.1 how the topics from the two books complement and overlap one another. The present book consists of three parts: Part I – Matrix Methods, Part II – Numerical Methods, and Part III – Least Squares and Optimization. Of course, all of these subjects are treated within a unified framework with matrix methods providing the catalyst. After completing Part I, the reader can choose to proceed directly to Part II on numerical methods or Part III on least-squares and optimization. Part I provides all of the prerequisite material for both parts. Although there will be some numerical methods introduced in Part III, none of them depend on material in Part II.

In Part I on matrix methods, the focus is on topics that are common to numerous areas of science and engineering, not subject-specific topics. Because of our interest

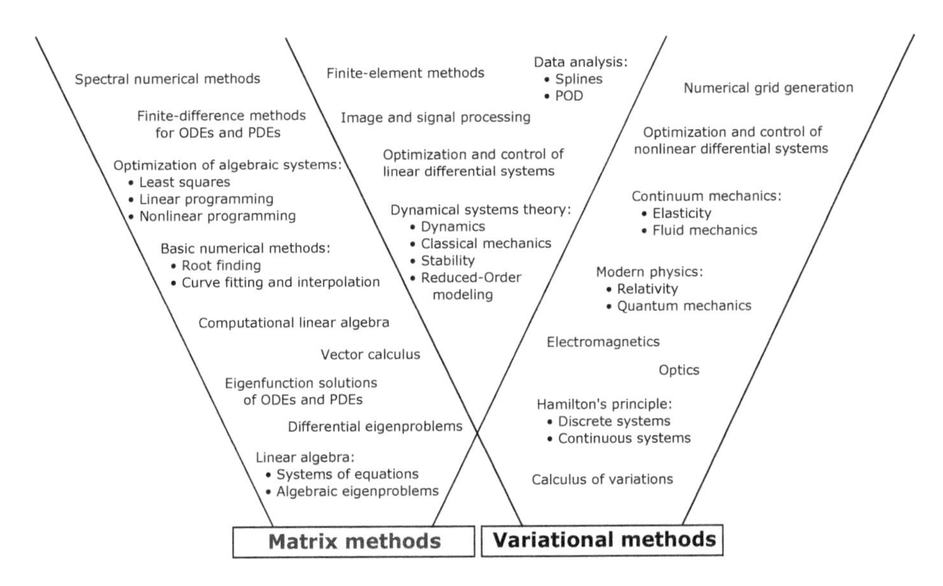

Figure 0.1 Correspondence of topics in matrix methods and variational methods with applications in science and engineering.

in continuous, as well as discrete, systems, "matrix methods" is interpreted loosely to include function spaces and eigenfunction methods. Far from being a recounting of standard methods, the material is infused with a strong dose of the applications that motivate them – including to dynamical systems theory. We progress from the basic operations on vectors and matrices in Chapter 1 in a logical fashion all the way to the singular-value decomposition at the end of Chapter 2 that is at the heart of so many modern applications to discrete and continuous systems. Integrating and juxtaposing the differential eigenproblem in Chapter 3 with its algebraic counterpart in Chapter 2 also serves to unify topics that are often separated and presented in different settings and courses. The methods articulated in Part I are appropriate for relatively small systems that can be solved exactly by hand or using symbolic mathematical software. Despite this obvious limitation, they are essential knowledge and provide a starting point for all subsequent discussion of numerical methods and optimization techniques.

Whereas Part I primarily deals with the traditional methods used for hand calculations of small systems, Part II focuses on numerical methods used to approximate solutions of very large systems on computers. In order to solve moderate to large systems of algebraic equations, for example, we must revisit linear systems of algebraic equations and the eigenproblem to see how the methods in Part I can be adapted for large systems. This is known as *computational*, or *numerical, linear algebra* and is covered in Chapter 6. The remaining chapters in Part II encompass a broad range of numerical techniques in widespread use for obtaining approximate solutions to ordinary and partial differential equations. The focus is on finite-difference methods with limited coverage of finite-element and spectral methods primarily for comparison. Unifying all of these topics under the umbrella of matrix methods provides both a broader and deeper understanding of the numerical methods themselves and the issues that arise in their deployment. In particular, it encourages a unification of topics that is not possible otherwise.

Similar to Part II, those least-squares and optimization topics included in Part III are of general interest in a wide variety of fields involving discrete and continuous systems and follow directly from material discussed in Part I. We begin with a general treatment of the least-squares problem, which is used in least-squares regression curve fitting and interpolation for data analysis and is the basis for numerous optimization and control techniques used in research and practice. Root-finding techniques for algebraic systems are integrated with optimization methods used to solve linear and nonlinear programming problems. Finally, Chapter 13 builds on the optimization foundation to address data-driven methods in reduced-order modeling featuring a clear treatment of proper-orthogonal decomposition, also called principal component analysis, and its extensions. This treatment clearly highlights the centrality of Galerkin projection in such methods.

It goes without saying that each individual reader will wish that there was more material specifically related to applications in their chosen field of study, and certain application areas have received little or no attention. Because of the broad range of applications that draw on matrix methods, we are limited in how far we may proceed along the *mathematics* → *methods* → *applications* continuum to primarily focus on methods that have broad applicability in science and engineering. For example, there is no mention of machine learning and artificial intelligence as these also depend on probability and statistics, which are beyond the scope of this text. Similarly, other than a brief mention in the context of singular-value decomposition, there is no material directly addressing image and signal processing – except that the underlying Fourier analysis methods are covered. The reader is referred to chapter 11 of Cassel (2013) for an introduction to the variational approach to such data analysis.

This text is targeted at the advanced undergraduate or graduate engineering, physical sciences, or applied mathematics student. It may also serve as a reference for researchers and practitioners in the many fields that make use of matrix, numerical, or optimization methods. The prerequisite material required is an undergraduate-level understanding of calculus, elementary complex variables, and ordinary and partial differential equations typical of engineering and physical science programs. Because of the intended audience of the book, there is little emphasis on mathematical proofs except where necessary to highlight certain essential features. Instead, the material is presented in a manner that promotes development of an intuition about the concepts and methods with an emphasis on applications to numerical and optimization methods as well as dynamical systems theory.

This book could serve as the primary text or a reference for courses in linear algebra or matrix methods, linear systems, basic numerical methods, optimization and control, and advanced numerical methods for partial differential equations. At the Illinois Institute of Technology, the material covers a portion of an engineering analysis course for first-year graduate students in various engineering disciplines (which also includes complex variables and variational methods), an undergraduate numerical methods course for mechanical and aerospace engineers (we call it computational mechanics), and a graduate course in computational methods for partial differential equations, such as computational fluid dynamics (CFD) and heat transfer. In addition, it could serve as a supplement or reference for courses in dynamical systems theory, classical mechanics, mechanical vibrations, structural mechanics, or optimization.

I would like to thank the many students who have helped shape my thinking on how best to arrange and articulate this material, and I particularly appreciate the comments and suggestions of the reviewers who greatly assisted in improving this somewhat unconventional amalgamation of topics. I can be reached at cassel@iit.edu if you have any comments on the text.

Part I

Matrix Methods

1 Vector and Matrix Algebra

Mathematics is the music of reason.

The world of ideas which [mathematics] discloses or illuminates, the contemplation of divine beauty and order which it induces, the harmonious connexion of its parts, the infinite hierarchy and absolute evidence of the truths with which it is concerned, these, and such like, are the surest grounds of the title of mathematics to human regard, and would remain unimpeached and unimpaired were the plan of the universe unrolled like a map at our feet, and the mind of man qualified to take in the whole scheme of creation at a glance. (James Joseph Sylvester)

Approximately 50 years before publication of this book, Neil Armstrong was the first human to step foot on the Moon and – thankfully – return safely to Earth. While the Apollo 11 mission that facilitated this amazing accomplishment in July of 1969 was largely symbolic from a scientific perspective, it represented the culmination of a decade of focused effort to create, develop, test, and implement a whole host of new technologies that were necessary to make such a feat possible. Many of these technologies are now fully integrated into modern life.

The march to the Moon during the 1960s was unprecedented in modern technological history for its boldness, aggressive time frame, scale, scientific and technological developments, and ultimate success. The state of the art in space travel in the early 1960s was to put a human in orbit 280 kilometers (km) (175 miles) above the Earth for several hours. It was in this context on September 12, 1962, that President Kennedy boldly – and some argued foolishly – set the United States on a course "to go to the Moon in this decade and do the other things, not because they are easy, but because they are hard." The Moon is 385,000 km (240,000 miles) from Earth – orbital flights were a mere baby step toward this aggressive goal. It was not that the technology was available and we simply needed to marshal the financial resources and political will to accomplish this bold task, it was not clear *if* the technology could be developed at all, let alone in the aggressive time frame proposed.

This problem appeared straightforward on paper. Newton's laws of motion had been articulated nearly 300 years before, and they had served remarkably well as the basis for an uncountable array of terrestrial applications and cosmological predictions. But to successfully land an object the size of a small truck on a moving body 385,000 km from its origin presented a whole host of issues. The mathematical model in the form of Newton's laws could not be fully trusted to accurately navigate such extreme distances. Nor could noisy and inaccurate measurements made on board the spacecraft as well as back on Earth be relied upon either. How could a variety of measurements, each with various levels of reliability and accuracy, be combined with mathematical predictions of the spacecraft's trajectory to provide the best estimate of its location and determine the necessary course corrections to keep it on its precise trajectory toward its target – and back?

The obstacle was not that the equations were not correct, or that they could not be accurately calculated; the problem was that there were numerous opportunities for the spacecraft to deviate slightly from its intended path. Each of these could have led to large deviations from the target trajectory over such large distances. A mechanism was required to correct the trajectory of the spacecraft along its route, but that was the easy part. How does one determine precisely where you are when so far from, well, anything? How does one adjust for the inevitable errors in making such a *state estimate*?

While so-called *filtering* techniques were available at the time, they were either not accurate enough for such a mission or too computationally intensive for on-board computers. It turned out that the methodology to address just such a predicament had been published by Kalman (1960) only 18 months before Kennedy's famous speech. In fact, a small group of researchers from NASA Ames Research Center had been in contact with Kalman in the late 1950s to discuss his new method and how it could be used for midcourse navigation correction of the Apollo spacecraft.[1] Kalman's original method applied to linear estimation problems, whereas NASA scientists were faced with a nonlinear problem. The modifications made to the original approach came to be known as the "extended Kalman filter," which remains one of the dominant approaches for treating nonlinear estimation problems today.

While on-board Kalman filtering ended up only being used as a backup to ground-based measurements for midcourse navigation between the Earth and Moon, it was found to be essential for the rendezvous problem. The three Apollo astronauts traveled together in the Command Module until the spacecraft entered lunar orbit. The Lunar Module then detached from the Command Module with Neil Armstrong and Buzz Aldrin aboard, while Michael Collins stayed in the Command Module as it remained in lunar orbit. After their historic landing on the Moon's surface, the Lunar Module then had to take off from the lunar surface and rendezvous with the Command Module in lunar orbit.[2] This was the most difficult navigational challenge of the entire mission. Because both spacecraft were moving, it presented a unique navigational challenge to facilitate their rendezvous with the necessary precision to dock the two spacecraft and reunite the three history-making astronauts. Kalman filtering was used to ingest the model predictions and on-board measurements to obtain relative state estimates

[1] NASA Technical Memorandum 86847, "Discovery of the Kalman Filter as a Practical Tool for Aerospace and Industry," by L. A. McGee and S. F. Schmidt (1985).

[2] The picture on the cover of the book was taken by Michael Collins in the Command Module just before the rendezvous was executed.

between the two orbiting spacecraft. The calculations were performed redundantly on both the Lunar and Command Modules' guidance computers for comparison.[3]

Before we can introduce state estimation in Section 11.9, there is a great deal of mathematical machinery that we need to explore. This includes material from this book on matrix methods as well as variational methods (see, for example, Cassel 2013). There are numerous additional examples of new technologies and techniques developed for the specific purpose of landing a man on the Moon – many of them depending upon the matrix, numerical, and optimization methods that are the subjects of this text.

1.1 Introduction

For most scientists and engineers, our first exposure to vectors and matrices is in the context of mechanics. Vectors are used to represent quantities, such as velocity or force, that have both a magnitude and direction in contrast to scalar quantities, such as pressure and temperature, that only have a magnitude. Likewise, matrices generally first appear when stress and strain tensors are introduced, once again in a mechanics setting. At first, such threedimensional vectors and matrices appear to be simply a convenient way to tabulate such quantities in an orderly fashion. In large part, this is true. In contrast to many areas of mathematics, for which certain operations would not be possible without it, vectors and matrices are not mathematical elements of necessity; rather they are constructs of convenience. It could be argued that there is not a single application in this book that *requires* matrix methods. However, this ubiquitous framework not only supplies a convenient way of representing large and complex data sets, it also provides a common formalism for their analysis; the topics contained in this book would be far more complex and confusing without the machinery of matrix methods.

The benefit of first being exposed to vectors and matrices through mechanics is that we naturally develop a strong geometric interpretation of them from the start. Because it appeals to our visual sensibilities, therefore, we hardly realize that we are learning the basics of *linear algebra*. In such a mechanics context, however, there is no reason to consider vectors larger than three dimensions corresponding to the three directions in our various coordinate systems that represent physical space. Unfortunately, our visual interpretation of vectors and matrices does not carry over to higher dimensions; we cannot even sketch a vector larger than three dimensions let alone impose a meaningful geometric interpretation. This is when linear algebra seems to lose its moorings in physically understandable reality and is simply a fun playground for mathematicians.

[3] MIT Report E-2411, "Apollo Navigation, Guidance, and Control: A Progress Report," by D. G. Hoag (1969). MIT Report R-649, "The Apollo Rendezvous Navigation Filter Theory, Description and Performance" (Volume 1), by E. S. Muller Jr. and P. M. Kachmar (1970).

Given its roots in algebra, it is not surprising that mathematicians have long viewed linear algebra as an essential weapon in their arsenal. For those of us who have benefited from a course in linear algebra taught by a mathematician, complementing the geometrically rich and physically practical exposure in mechanics with the formal framework of operations and methods provides a strong, and frankly necessary, foundation for research and practice in almost all areas of science and engineering. However, the mathematician's discussion of "singular matrices," "null spaces," and "vector bases" often leaves us with the notion that linear algebra, beyond our initial mechanics-driven exposure, is of very little relevance to the scientist or engineer. On the contrary, the formalism of linear algebra provides the mathematical foundation for three of the most far-reaching and widely applicable "applications" of matrix methods in science and engineering, namely dynamical systems theory, numerical methods, and optimization. Together they provide the tools necessary to solve, analyze, and optimize large-scale, complex systems of practical interest in both research and industrial practice. As we will see, even the analysis and prediction of continuous systems governed by differential equations ultimately reduces to solution of a matrix problem. This is because more often than not, numerical methods must be used, which convert the continuous governing equations into a discrete system of algebraic equations.

Although "linear" algebra is strictly speaking a special case of algebra,[4] it is in many ways a dramatic extension of the algebra that we learn in our formative years. This is certainly the case with regard to the applications that matrix methods address for scientists and engineers. Because linear methods are so well developed mathematically, with their wide-ranging set of tools, it is often tempting to reframe nonlinear problems in such a way as to allow for the utilization of linear methods. In many cases, this can be formally justified; however, one needs to be careful to do so in a manner that is faithful to the true nature of the underlying system and the information being sought. This theme will be revisited throughout the text as many of our applications exhibit nonlinear behavior.

Whereas vectors and matrices arise in a wide variety of applications and settings, the mathematics of these constructs is the same regardless of where the vectors or matrices have their origin. We will focus in Part I on the mathematics, but with little emphasis on formalism and proofs, and mention or illustrate many of the applications in science and engineering. First, however, let us motivate the need for such mathematical machinery using two simple geometric scenarios. After introducing some basic definitions and algebraic operations, we will then return in Section 1.4.1 to introduce some additional applications of matrix problems common in science and engineering.

1.1.1 Equation of a Line

We know intuitively that the shortest distance between two points is a straight line. Mathematically, this is reflected in the fact that there is a single unique straight line

[4] Although algebra can be traced back to the ancient Babylonians and Egyptians, matrix algebra was not formalized until the middle of the nineteenth century by Arthur Cayley.

that connects any two points. Consider the two points (x_1, y_1) and (x_2, y_2) in two dimensions. In order to determine the line,

$$a_0 + a_1 x = y,$$

passing through these two points, we could substitute the two points into this equation for the line as follows:

$$\begin{aligned} a_0 + a_1 x_1 &= y_1, \\ a_0 + a_1 x_2 &= y_2. \end{aligned} \tag{1.1}$$

Because the values of x and y are known for the two points, this is two equations for the two unknown constants a_0 and a_1, and we expect a unique solution. As will be shown in the next section, these coupled algebraic equations can conveniently be written in vector-matrix form as

$$\begin{bmatrix} 1 & x_1 \\ 1 & x_2 \end{bmatrix} \begin{bmatrix} a_0 \\ a_1 \end{bmatrix} = \begin{bmatrix} y_1 \\ y_2 \end{bmatrix}.$$

In order to see how this corresponds to the system of equations (1.1), matrix multiplication will need to be defined. Working from left to right, we have a matrix that includes the x values multiplied by a vector containing the coefficients in the equation of the line set equal to a vector of the y values. In the present case, the matrix and right-hand-side vector are known, and a solution is sought for the coefficients a_0 and a_1 in the equation for the line. Performing the same exercise in three dimensions would result in the need to solve three equations for three unknowns, which again could be expressed in matrix form.

What if we have more than two points? Say that instead we have $N > 2$ such points. Certainly, it would not be possible to determine a single straight line that connects all N of the points (unless they so happen to all be collinear). Expressed as before, we would have N equations for the two unknown coefficients; this is called an *overdetermined system* as there are more equations than unknowns and no unique solution exists. While there is not a single line that connects all of the points in this general case, we could imagine that there is a single line that best represents the points as illustrated in Figure 1.1. This is known as *linear least-squares regression* and illustrates that there will be times when a "solution" of the system of equations is sought even when a unique solution does not exist. Least-squares methods will be taken up in Section 10.2 after we have covered the necessary background material.

1.1.2 Linear Transformation

As a second example, consider geometric transformations. When scientists and engineers communicate their ideas, theories, and results, they must always draw attention to their *reference frame*. For example, when describing the motion of a passenger walking down the aisle of an airplane in flight, is the description from the point of view of the passenger, the airplane, a fixed point on the Earth's surface, the center of the Earth, the center of the Sun, or some other point in the universe? Obviously, the

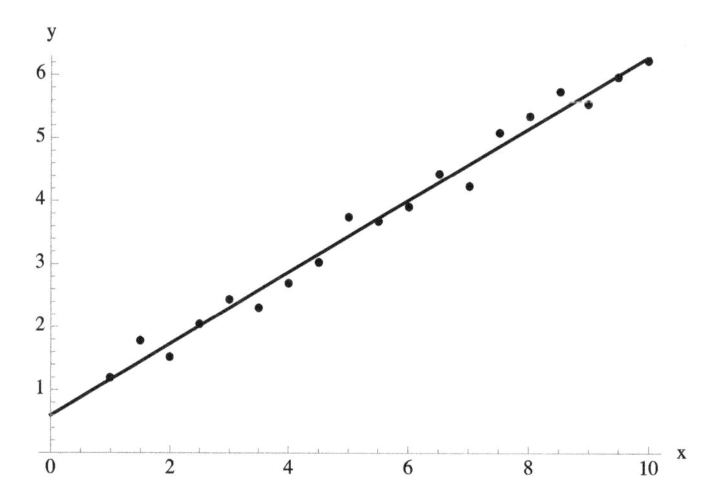

Figure 1.1 Best-fit line for $N = 19$ data points.

description will be quite different depending on which of these reference frames is used. Mathematically, we formally define the reference frame and how we will locate positions relative to it using a *coordinate system* or *vector basis*.[5] This is comprised of specifying the location of the coordinate system's origin – the location from which each coordinate and vector is measured – and each of the coordinate directions emanating from this origin.

Of course, one may specify a problem with respect to a different coordinate system than someone else, or different aspects of the problem may be best analyzed using different coordinate systems. In order to communicate information between the two coordinate systems, a *transformation* is necessary. For example, let us say that a point in the two-dimensional coordinate system X is given by $\mathbf{x}_1 = (2,5)$, and the same point with respect to another coordinate system Y is defined by $\mathbf{y}_1 = (-3,4)$. If this is a *linear* transformation, then we must be able to multiply $\mathbf{x}_1$ times something, say $\mathbf{A}$, to produce $\mathbf{y}_1$ in the form

$$\mathbf{A}\mathbf{x}_1 = \mathbf{y}_1. \tag{1.2}$$

Because both $\mathbf{x}_1$ and $\mathbf{y}_1$ involve two values, or coordinates, $\mathbf{A}$ is clearly not simply a single scalar value under general circumstances. Instead, we will need to include some combination of both coordinates to get from one coordinate system to the other. This could involve, for example, each coordinate of $\mathbf{y}_1$ being some linear combination of the two coordinates of $\mathbf{x}_1$, such that

[5] We generally call it a *coordinate system* when referring to a system with one, two, or three spatial dimensions, for which we can illustrate the coordinate system geometrically. For systems with more than three dimensions, that is, where the dimensions do not correspond to spatial coordinates, we use the more general *vector basis*, or simply *basis*, terminology. Consequently, a coordinate system is simply a vector basis for a two- or three-dimensional vector space.

$$\begin{aligned} 2a + 5b &= -3, \\ 2c + 5d &= 4. \end{aligned} \tag{1.3}$$

If we suitably define multiplication, (1.3) can be written in the compact form of (1.2) if $\mathbf{A}$ is defined as

$$\mathbf{A} = \begin{bmatrix} a & b \\ c & d \end{bmatrix},$$

which is a *transformation matrix*.[6] It just remains to determine the constants a, b, c, and d that comprise the transformation matrix. What we have in (1.3) is a system of two coupled algebraic equations for these four unknown constants. Clearly, there is no unique solution as there are many, in fact infinite, ways to transform the single point $\mathbf{x}_1$ into the point $\mathbf{y}_1$. For example, we could simply translate the coordinate system or rotate it clockwise or counterclockwise to transform one point to the other.

A unique transformation is obtained if we supply an additional pair of *image* points in the two coordinate systems, say $\mathbf{x}_2 = (-1, 6)$ and $\mathbf{y}_2 = (8, -7)$. This results in the two additional equations

$$\begin{aligned} -a + 6b &= 8, \\ -c + 6d &= -7, \end{aligned} \tag{1.4}$$

which in compact form is

$$\mathbf{A}\mathbf{x}_2 = \mathbf{y}_2. \tag{1.5}$$

Note that the transformation matrix is the same in (1.2) and (1.5). Equations (1.3) and (1.4) now provide four equations for the four unknowns, and the transformation is unique.

Is there a way to represent the *system of linear algebraic equations* given by (1.3) and (1.4) in a convenient mathematical form? How do we solve such a system? Can we be sure that this solution is unique? Can we represent *linear transformations* in a general fashion for systems having any number of coordinates? These questions are the subject of this chapter. As we will see, the points and the coordinate directions with respect to which they are defined will be conveniently represented as *vectors*. The transformation will be denoted by a *matrix* as will be the coefficients in the system of linear algebraic equations to be solved for the transformation matrix. The unknowns in the transformation matrix will be combined to form the *solution vector*.

Such changes of coordinate system (basis) and the transformation matrices that accomplish them will be a consistent theme throughout the text. We will often have occasion to transform a problem into a more desirable basis in order to facilitate interpretation, expose features, diagnose attributes, or ease solution. We will return

[6] Given their ubiquity across so many areas of mathematics, science, and engineering, one might be surprised to learn that the term "matrix" was not coined until 1850 by James Joseph Sylvester. This is more than a century after such classical fields as complex variables, differential calculus, and variational calculus had their genesis.

to this important topic in Section 1.8. Along the way, we will encounter numerous physical applications that lend themselves to similar mathematical representation as this geometric example despite their vastly different physical interpretations. As is so often the case in mathematics, this remarkable ability to unify many disparate applications within the same mathematical constructs and operations is what renders mathematics so essential to the scientist or engineer.

1.2 Definitions

Let us begin by defining vectors and matrices and several common types of matrices. One can think of a matrix as the mathematical analog of a table in a text or a spreadsheet on a computer.

Matrix: A *matrix* is an ordered arrangement of numbers, variables, or functions comprised of a rectangular grouping of elements arranged in rows and columns as follows:

$$\mathbf{A} = \begin{bmatrix} A_{11} & A_{12} & \cdots & A_{1N} \\ A_{21} & A_{22} & \cdots & A_{2N} \\ \vdots & \vdots & \ddots & \vdots \\ A_{M1} & A_{M2} & \cdots & A_{MN} \end{bmatrix} = [A_{mn}].$$

The size of the matrix is denoted by the number of rows, M, and the number of columns, N. We say that the matrix $\mathbf{A}$ is $M \times N$, which we read as "M by N." If $M = N$, then the matrix is said to be *square*. Each element A_{mn} in the matrix is uniquely identified by two subscripts, with the first, m, being its row and the second, n, being its column. Thus, $1 \leq m \leq M$ and $1 \leq n \leq N$. The elements A_{mn} may be real or complex numbers, variables, or functions.

The *main diagonal* of the matrix $\mathbf{A}$ is given by $A_{11}, A_{22}, \ldots, A_{MM}$ or A_{NN}; if the matrix is square, then $A_{MM} = A_{NN}$. Two matrices are said to be equal, that is $\mathbf{A} = \mathbf{B}$, if their sizes are the same and $A_{mn} = B_{mn}$ for all m and n.

Vector: A (column) *vector* is an $N \times 1$ matrix. For example,

$$\mathbf{u} = \begin{bmatrix} u_1 \\ u_2 \\ \vdots \\ u_N \end{bmatrix}.$$

The vector is said to be N-dimensional and can be considered a point in an N-dimensional coordinate system. By common convention, matrices are denoted by bold capital letters and vectors by bold lowercase letters.

Matrix Transpose (Adjoint): The *transpose* of matrix $\mathbf{A}$ is obtained by interchanging its rows and columns as follows:

$$\mathbf{A}^T = \begin{bmatrix} A_{11} & A_{21} & \cdots & A_{M1} \\ A_{12} & A_{22} & \cdots & A_{M2} \\ \vdots & \vdots & \ddots & \vdots \\ A_{1N} & A_{2N} & \cdots & A_{MN} \end{bmatrix} = [A_{nm}],$$

which results in an $N \times M$ matrix. If $\mathbf{A}^T = \mathbf{A}$, then $\mathbf{A}$ is said to be *symmetric* ($A_{nm} = A_{mn}$). Note that a matrix must be square to be symmetric. If instead the matrix is such that $\mathbf{A} = -\mathbf{A}^T$, it is called *skew-symmetric*. Note that for this to be true, the elements along the main diagonal of $\mathbf{A}$ must all be zero.

If the elements of $\mathbf{A}$ are complex and $\overline{\mathbf{A}}^T = \mathbf{A}$, then $\mathbf{A}$ is a *Hermitian matrix* ($\overline{A}_{nm} = A_{mn}$), where the overbar represents the *complex conjugate*, and $\overline{\mathbf{A}}^T$ is the *conjugate transpose* of $\mathbf{A}$. Note that a symmetric matrix is a special case of a Hermitian matrix.

Zero Matrix ($\mathbf{0}$): Matrix of all zeros.

Identity Matrix ($\mathbf{I}$): Square matrix with ones on the main diagonal and zeros everywhere else, for example,

$$\mathbf{I}_5 = \begin{bmatrix} 1 & 0 & 0 & 0 & 0 \\ 0 & 1 & 0 & 0 & 0 \\ 0 & 0 & 1 & 0 & 0 \\ 0 & 0 & 0 & 1 & 0 \\ 0 & 0 & 0 & 0 & 1 \end{bmatrix} = \left[\delta_{mn}\right],$$

where

$$\delta_{mn} = \begin{cases} 1, & m = n \\ 0, & m \neq n \end{cases}.$$

Triangular Matrix: All elements above (left triangular) or below (right triangular) the main diagonal are zero. For example,

$$\mathbf{L} = \begin{bmatrix} A_{11} & 0 & 0 & 0 & 0 \\ A_{21} & A_{22} & 0 & 0 & 0 \\ A_{31} & A_{32} & A_{33} & 0 & 0 \\ A_{41} & A_{42} & A_{43} & A_{44} & 0 \\ A_{51} & A_{52} & A_{53} & A_{54} & A_{55} \end{bmatrix}, \quad \mathbf{R} = \begin{bmatrix} A_{11} & A_{12} & A_{13} & A_{14} & A_{15} \\ 0 & A_{22} & A_{23} & A_{24} & A_{25} \\ 0 & 0 & A_{33} & A_{34} & A_{35} \\ 0 & 0 & 0 & A_{44} & A_{45} \\ 0 & 0 & 0 & 0 & A_{55} \end{bmatrix}.$$

Tridiagonal Matrix: All elements are zero except along the lower (first subdiagonal), main, and upper (first superdiagonal) diagonals as follows:

$$\mathbf{A} = \begin{bmatrix} A_{11} & A_{12} & 0 & 0 & 0 \\ A_{21} & A_{22} & A_{23} & 0 & 0 \\ 0 & A_{32} & A_{33} & A_{34} & 0 \\ 0 & 0 & A_{43} & A_{44} & A_{45} \\ 0 & 0 & 0 & A_{54} & A_{55} \end{bmatrix}.$$

Hessenberg Matrix: All elements are zero below the lower diagonal, that is

$$\mathbf{A} = \begin{bmatrix} A_{11} & A_{12} & A_{13} & A_{14} & A_{15} \\ A_{21} & A_{22} & A_{23} & A_{24} & A_{25} \\ 0 & A_{32} & A_{33} & A_{34} & A_{35} \\ 0 & 0 & A_{43} & A_{44} & A_{45} \\ 0 & 0 & 0 & A_{54} & A_{55} \end{bmatrix}.$$

Toeplitz Matrix: Each diagonal is a constant, such that

$$\mathbf{A} = \begin{bmatrix} A_{11} & A_{12} & A_{13} & A_{14} & A_{15} \\ A_{21} & A_{11} & A_{12} & A_{13} & A_{14} \\ A_{31} & A_{21} & A_{11} & A_{12} & A_{13} \\ A_{41} & A_{31} & A_{21} & A_{11} & A_{12} \\ A_{51} & A_{41} & A_{31} & A_{21} & A_{11} \end{bmatrix}.$$

Matrix Inverse: If a square matrix $\mathbf{A}$ is *invertible*, then its inverse $\mathbf{A}^{-1}$ is such that

$$\mathbf{A}\mathbf{A}^{-1} = \mathbf{A}^{-1}\mathbf{A} = \mathbf{I}.$$

Note: If $\mathbf{A}$ is the 2×2 matrix

$$\mathbf{A} = \begin{bmatrix} A_{11} & A_{12} \\ A_{21} & A_{22} \end{bmatrix},$$

then its inverse is

$$\mathbf{A}^{-1} = \frac{1}{|\mathbf{A}|} \begin{bmatrix} A_{22} & -A_{12} \\ -A_{21} & A_{11} \end{bmatrix},$$

where the diagonal elements A_{11} and A_{22} have been exchanged, the off-diagonal elements A_{12} and A_{21} have each switched signs, and $|\mathbf{A}| = A_{11}A_{22} - A_{12}A_{21}$ is the *determinant* of $\mathbf{A}$ (see Section 1.4.3).

Orthogonal Matrix: An $N \times N$ square matrix $\mathbf{A}$ is *orthogonal* if

$$\mathbf{A}\mathbf{A}^T = \mathbf{A}^T\mathbf{A} = \mathbf{I}.$$

It follows that

$$\mathbf{A}^T = \mathbf{A}^{-1}$$

for an orthogonal matrix. Such a matrix is called orthogonal because its column (and row) vectors are mutually orthogonal (see Sections 1.2 and 2.4).

Block Matrix: A block matrix is comprised of smaller submatrices, called *blocks*. For example, suppose that matrix $\mathbf{A}$ is a 2×2 block matrix comprised of four submatrices as follows:

$$\mathbf{A} = \begin{bmatrix} \mathbf{A}_{11} & \mathbf{A}_{12} \\ \mathbf{A}_{21} & \mathbf{A}_{22} \end{bmatrix}.$$

Table 1.1 Corresponding types of real and complex matrices.

Real	Complex
Symmetric: $\mathbf{A} = \mathbf{A}^T$	Hermitian: $\mathbf{A} = \overline{\mathbf{A}}^T$
Skew-symmetric: $\mathbf{A} = -\mathbf{A}^T$	Skew-Hermitian: $\mathbf{A} = -\overline{\mathbf{A}}^T$
Orthogonal: $\mathbf{A}^{-1} = \mathbf{A}^T$	Unitary: $\mathbf{A}^{-1} = \overline{\mathbf{A}}^T$

Although each submatrix need not be of the same size, blocks in the same row must have the same number of rows, and those in the same column must have the same number of columns. That is, $\mathbf{A}_{11}$ and $\mathbf{A}_{12}$ must have the same number of rows, $\mathbf{A}_{21}$ and $\mathbf{A}_{22}$ must have the same number of rows, $\mathbf{A}_{11}$ and $\mathbf{A}_{21}$ must have the same number of columns, and $\mathbf{A}_{12}$ and $\mathbf{A}_{22}$ must also have the same number of columns.

Generalizations of some of the preceding definitions to matrices with complex elements are given in Table 1.1.

1.3 Algebraic Operations

This being linear algebra, we must extend the algebraic operations of addition and multiplication, which are so familiar for scalars, to vectors and matrices.

Addition: For $\mathbf{A}$ and $\mathbf{B}$ having the same size $M \times N$, their sum is an $M \times N$ matrix obtained by adding the corresponding elements in $\mathbf{A}$ and $\mathbf{B}$, that is, $\mathbf{A} + \mathbf{B} = [A_{mn} + B_{mn}]$. For example:

$$\begin{bmatrix} A_{11} & A_{12} & A_{13} \\ A_{21} & A_{22} & A_{23} \\ A_{31} & A_{32} & A_{33} \end{bmatrix} + \begin{bmatrix} B_{11} & B_{12} & B_{13} \\ B_{21} & B_{22} & B_{23} \\ B_{31} & B_{32} & B_{33} \end{bmatrix} = \begin{bmatrix} A_{11}+B_{11} & A_{12}+B_{12} & A_{13}+B_{13} \\ A_{21}+B_{21} & A_{22}+B_{22} & A_{23}+B_{23} \\ A_{31}+B_{31} & A_{32}+B_{32} & A_{33}+B_{33} \end{bmatrix}.$$

Subtraction is simply a special case of addition.

Multiplication: In the case of multiplication, we must consider multiplication of a scalar and a matrix as well as a matrix and a matrix.

1. *Multiplication by a scalar*: When multiplying a scalar by a matrix, we simply multiply each element of the matrix by the scalar c. That is,

$$c\mathbf{A} = [c\, A_{mn}]\,.$$

2. *Multiplication by a matrix*: The definition of matrix multiplication is motivated by its use in linear transformations (see Sections 1.1.2 and 1.8). For the matrix product $\mathbf{AB}$ to exist, the number of columns of $\mathbf{A}$ must equal the number of rows of $\mathbf{B}$. Then

$$\underset{M\times K}{\mathbf{A}}\;\underset{K\times N}{\mathbf{B}} = \underset{M\times N}{\mathbf{C}}$$

produces an $M \times N$ matrix $\mathbf{C}$. Written out, we have

$$\begin{bmatrix} A_{11} & A_{12} & \cdots & A_{1K} \\ A_{21} & A_{22} & \cdots & A_{2K} \\ \vdots & \vdots & & \vdots \\ A_{m1} & A_{m2} & \cdots & A_{mK} \\ \vdots & \vdots & & \vdots \\ A_{M1} & A_{M2} & \cdots & A_{MK} \end{bmatrix} \begin{bmatrix} B_{11} & B_{12} & \cdots & B_{1n} & \cdots & B_{1N} \\ B_{21} & B_{22} & \cdots & B_{2n} & \cdots & B_{2N} \\ \vdots & \vdots & & \vdots & & \vdots \\ B_{K1} & B_{K2} & \cdots & B_{Kn} & \cdots & B_{KN} \end{bmatrix}$$

$$= \begin{bmatrix} C_{11} & C_{12} & \cdots & C_{1N} \\ C_{21} & C_{22} & \cdots & C_{2N} \\ \vdots & \vdots & C_{mn} & \vdots \\ C_{M1} & C_{M2} & \cdots & C_{MN} \end{bmatrix},$$

where

$$C_{mn} = A_{m1}B_{1n} + A_{m2}B_{2n} + \cdots + A_{mK}B_{Kn} = \sum_{k=1}^{K} A_{mk}B_{kn}.$$

That is, C_{mn} is the *inner product* of the mth row of **A** and the nth column of **B** (see Section 1.6 for details on the inner product).

Note that in general, $\mathbf{AB} \neq \mathbf{BA}$ (even if square), that is, matrix multiplication does not commute, and *premultiplying* **B** by **A** is not the same as *postmultiplying* **B** by **A**. Moreover, there is no such thing as division in linear algebra; taking its place is the concept of the matrix inverse defined in the previous section and discussed further in Section 1.5.2. More precisely, division in scalar algebra is a special case of the inverse in linear (matrix) algebra.

Rules: For matrices **A**, **B**, and **C**, and the identity matrix **I** of the appropriate sizes, and l and k integer:

$$\mathbf{A}(\mathbf{BC}) = (\mathbf{AB})\mathbf{C} \quad \text{(but cannot change order of } \mathbf{A}, \mathbf{B}, \text{ and } \mathbf{C}),$$

$$\mathbf{A}(\mathbf{B} + \mathbf{C}) = \mathbf{AB} + \mathbf{AC},$$

$$(\mathbf{B} + \mathbf{C})\mathbf{A} = \mathbf{BA} + \mathbf{CA},$$

$$\mathbf{AI} = \mathbf{IA} = \mathbf{A},$$

$$(\mathbf{A} + \mathbf{B})^T = \mathbf{A}^T + \mathbf{B}^T,$$

$$(\mathbf{AB})^T = \mathbf{B}^T\mathbf{A}^T \quad \text{(note reverse order)},$$

$$(\mathbf{AB})^{-1} = \mathbf{B}^{-1}\mathbf{A}^{-1} \quad \text{(if } \mathbf{A} \text{ and } \mathbf{B} \text{ invertible; note reverse order)},$$

$$\mathbf{A}^k = \mathbf{AA}\cdots\mathbf{A} \quad (k \text{ factors}),$$

$$\mathbf{A}^{-k} = \left(\mathbf{A}^{-1}\right)^k = \mathbf{A}^{-1}\mathbf{A}^{-1}\cdots\mathbf{A}^{-1} \quad (k \text{ factors}),$$

$$\mathbf{A}^k\mathbf{A}^l = \mathbf{A}^{k+l},$$

$$\left(\mathbf{A}^k\right)^l = \mathbf{A}^{kl},$$

$$\mathbf{A}^0 = \mathbf{I}.$$

Note that scalar and vector arithmetic is a special case of matrix arithmetic, as a scalar is simply a 1×1 matrix and a vector is an $N \times 1$ matrix.

We can combine vector addition and scalar multiplication to devise a very common algebraic operation in matrix methods known as a *linear combination* of vectors. Let us say that we have the M vectors $\mathbf{u}_1$, $\mathbf{u}_2$, ..., $\mathbf{u}_M$, which are each N-dimensional. A linear combination of these vectors is given by

$$\mathbf{u} = a_1\mathbf{u}_1 + a_2\mathbf{u}_2 + \cdots + a_M\mathbf{u}_M,$$

where $a_m, m = 1, \dots M$ are scalar constants. Thus, we are taking the combination of some portion of each vector through summing to form the N-dimensional vector $\mathbf{u}$. In Section 1.7, we will regard the vectors $\mathbf{u}_m, m = 1, \dots, M$ to be the *basis* of a *vector space* comprised of all possible vectors $\mathbf{u}$ given by all possible values of the coefficients $a_m, m = 1, \dots M$.

1.4 Systems of Linear Algebraic Equations – Preliminaries

The primary application of vectors and matrices is in representing and solving systems of coupled linear algebraic equations. The size of the system may be as small as 2×2, for a simple dynamical system, to $N \times N$, where N is in the millions, for systems that arise from implementation of numerical methods applied to complex physical problems. In the following, we discuss the properties of such systems and methods for determining their solution.

1.4.1 Applications

There is nothing inherently physical about the way vectors and matrices are defined and manipulated. However, they provide a convenient means of representing the mathematical models that apply in a vast array of applications that span all areas of physics, engineering, computer science, and image processing, for example. Once we become comfortable with them, vectors and matrices simply become natural extensions of the basic algebra and calculus that are so inherent to mathematics. Linear algebra, then, is not so much a separate branch of applied mathematics as it is an essential part of our mathematical repertoire.

The most common class of problems in matrix methods is in solving systems of coupled linear algebraic equations. These arise, for example, in solving for the forces in static (nonmoving) systems of discrete objects, parallel electrical circuits containing only voltage sources and resistors, and linear coordinate transformations. Perhaps the most important application is in numerical methods, which are designed to reduce continuous differential equations to – typically very large – systems of linear algebraic equations so as to make them amenable to solution via matrix methods. We address three of these applications in the following series of examples, and additional applications will be interspersed periodically in order to remind ourselves that we are developing the tools to solve real problems and also to expand our understanding of what the general mathematical constructs represent in physical and numerical contexts.

A system of linear algebraic equations may be represented in the form

$$\mathbf{A}\mathbf{u} = \mathbf{b},$$

where $\mathbf{A}$ is a given coefficient matrix, $\mathbf{b}$ is a given right-hand-side vector, and $\mathbf{u}$ is the solution vector that is sought. Solution techniques are covered in the next section.

Example 1.1 We seek to determine the currents u in the parallel electrical circuit shown in Figure 1.2 (see Jeffrey 2002).

Solution

Recall Ohm's law for resistors

$$V = uR,$$

where V is the voltage drop across the resistor, u is the current through the resistor, and R is its resistance. In addition, we have Kirchhoff's laws:

1. The sum of the currents into each junction is equal to the sum of the currents out of each junction.
2. The sum of the voltage drops around each closed circuit is zero.

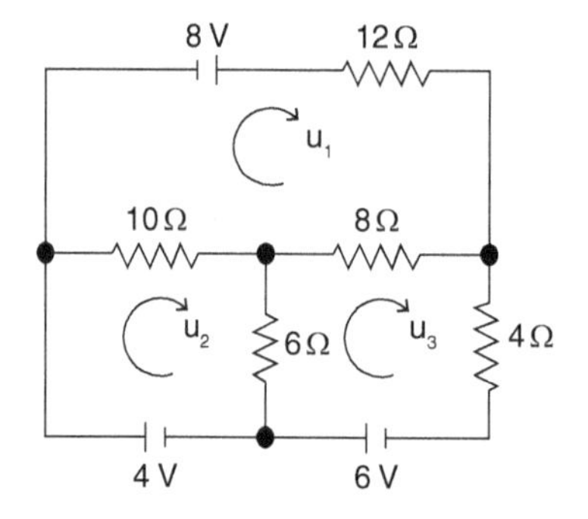

Figure 1.2 Schematic of the electrical circuit in Example 1.1.

Applying the second of Kirchhoff's laws around each of the three parallel circuits gives the three coupled equations

$$\begin{aligned}
&\text{Loop 1:} \quad 8 - 12u_1 - 8(u_1 - u_3) - 10(u_1 - u_2) = 0,\\
&\text{Loop 2:} \quad 4 - 10(u_2 - u_1) - 6(u_2 - u_3) = 0,\\
&\text{Loop 3:} \quad 6 - 6(u_3 - u_2) - 8(u_3 - u_1) - 4u_3 = 0.
\end{aligned}$$

Upon simplifying, this gives the system of linear algebraic equations

$$\begin{aligned}
30u_1 - 10u_2 - 8u_3 &= 8,\\
-10u_1 + 16u_2 - 6u_3 &= 4,\\
-8u_1 - 6u_2 + 18u_3 &= 6,
\end{aligned}$$

which in matrix form $\mathbf{Au} = \mathbf{b}$ is

$$\begin{bmatrix} 30 & -10 & -8 \\ -10 & 16 & -6 \\ -8 & -6 & 18 \end{bmatrix} \begin{bmatrix} u_1 \\ u_2 \\ u_3 \end{bmatrix} = \begin{bmatrix} 8 \\ 4 \\ 6 \end{bmatrix}.$$

Solving this system of coupled, linear, algebraic equations would provide the sought-after solution for the currents $\mathbf{u} = \begin{bmatrix} u_1 & u_2 & u_3 \end{bmatrix}^T$.

Note that if the circuit also includes capacitors and/or inductors, application of Kirchhoff's laws would produce a system of ordinary differential equations as illustrated in Example 2.9.

Example 1.2 Recall that for a structure in static equilibrium, the sum of the forces and moments are both zero, which holds for each individual member as well as the entire structure. Determine the forces in each member of the truss structure shown in Figure 1.3 (see Jeffrey 2002).

Solution

Note that all members are of length ℓ; therefore, the structure is comprised of equilateral triangles with all angles being $\pi/3$. We first obtain the reaction forces at the supports 1 and 2 by summing the forces and moments for the entire structure as follows:

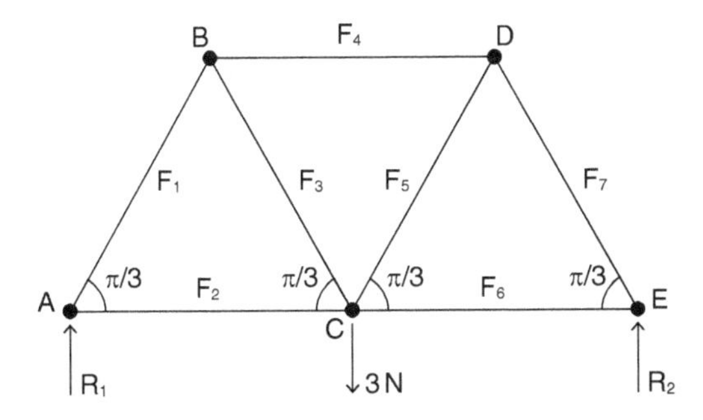

Figure 1.3 Schematic of the truss structure in Example 1.2. All members are of length ℓ.

$$\begin{aligned} \sum F_y = 0 \quad &\Rightarrow \quad R_1 + R_2 - 3 = 0, \\ \sum M_A = 0 \quad &\rightarrow \quad 2\ell R_2 - 3\ell = 0; \end{aligned}$$

therefore, reaction forces are

$$R_1 = \frac{3}{2}, \quad R_2 = \frac{3}{2}.$$

Next we draw a free-body diagram for each of the joints A, B, C, D, and E and sum the forces in the x- and y-directions to zero as follows:

$$\begin{aligned} \sum F_{A_x} = 0 \quad &\Rightarrow \quad F_2 + F_1 \cos\left(\tfrac{\pi}{3}\right) = 0, \\ \sum F_{A_y} = 0 \quad &\Rightarrow \quad R_1 + F_1 \sin\left(\tfrac{\pi}{3}\right) = 0, \\ \sum F_{B_x} = 0 \quad &\Rightarrow \quad F_4 + F_3 \cos\left(\tfrac{\pi}{3}\right) - F_1 \cos\left(\tfrac{\pi}{3}\right) = 0, \\ \sum F_{B_y} = 0 \quad &\Rightarrow \quad -F_1 \sin\left(\tfrac{\pi}{3}\right) - F_3 \sin\left(\tfrac{\pi}{3}\right) = 0, \\ &\quad \vdots \end{aligned}$$

and similarly for joints C, D, and E. Because $\cos\left(\frac{\pi}{3}\right) = \frac{1}{2}$ and $\sin\left(\frac{\pi}{3}\right) = \frac{\sqrt{3}}{2}$, the resulting system of equations in matrix form is

$$\begin{bmatrix} \frac{1}{2} & 1 & 0 & 0 & 0 & 0 & 0 \\ \frac{\sqrt{3}}{2} & 0 & 0 & 0 & 0 & 0 & 0 \\ \frac{1}{2} & 0 & -\frac{1}{2} & -1 & 0 & 0 & 0 \\ \frac{\sqrt{3}}{2} & 0 & \frac{\sqrt{3}}{2} & 0 & 0 & 0 & 0 \\ 0 & 1 & \frac{1}{2} & 0 & -\frac{1}{2} & -1 & 0 \\ 0 & 0 & \frac{\sqrt{3}}{2} & 0 & \frac{\sqrt{3}}{2} & 0 & 0 \\ 0 & 0 & 0 & 1 & \frac{1}{2} & 0 & -\frac{1}{2} \\ 0 & 0 & 0 & 0 & \frac{\sqrt{3}}{2} & 0 & \frac{\sqrt{3}}{2} \\ 0 & 0 & 0 & 0 & 0 & 1 & \frac{1}{2} \\ 0 & 0 & 0 & 0 & 0 & 0 & \frac{\sqrt{3}}{2} \end{bmatrix} \begin{bmatrix} F_1 \\ F_2 \\ F_3 \\ F_4 \\ F_5 \\ F_6 \\ F_7 \end{bmatrix} = \begin{bmatrix} 0 \\ -\frac{3}{2} \\ 0 \\ 0 \\ 0 \\ 3 \\ 0 \\ 0 \\ 0 \\ -\frac{3}{2} \end{bmatrix}.$$

Carefully examining the coefficient matrix and right-hand-side vector, one can see that the coefficient matrix contains information that only relates to the structure's geometry, while the right-hand-side vector is determined by the external loads only. Hence, different loading scenarios for the same structure can be considered by simply adjusting **b** accordingly. Observe that there are ten equations (two for each joint), but only seven unknown forces (one for each member); therefore, three of the equations must be linear combinations of the other equations in order to have a unique solution (see Section 1.6).

The previous two examples illustrate how matrices can be used to represent the governing equations of a physical system directly. There are many such examples in science and engineering. In the truss problem, for example, the equilibrium equations are applied to each *discrete* element of the structure leading to a system of linear algebraic equations for the forces in each member that must be solved simultaneously. In other words, the *discretization* follows directly from the geometry of the problem. For continuous systems, such as solids, fluids, and heat transfer problems involving continuous media, the governing equations are in the form of continuous ordinary or partial differential equations for which the discretization is less obvious. This gets us into the important and expansive topic of numerical methods. Although this is beyond the scope of our considerations at this time (see Part II), let us briefly motivate the overall approach using a simple one-dimensional example.

Example 1.3 Consider the fluid flow between two infinite parallel flat plates with the upper surface moving with constant speed U and an applied pressure gradient in the x-direction. This is known as Couette flow and is shown schematically in Figure 1.4.

Solution

The one-dimensional, fully developed flow is governed by the ordinary differential equation, which is derived from the Navier–Stokes equations enforcing conservation of momentum,

$$\frac{d^2u}{dy^2} = \frac{1}{\mu}\frac{dp}{dx} = \text{constant}, \tag{1.6}$$

where $u(y)$ is the fluid velocity in the x-direction, which we seek; $p(x)$ is the specified linear pressure distribution in the x-direction (such that dp/dx is a constant); and μ is the fluid viscosity. The no-slip boundary conditions at the lower and upper surfaces, are

$$u = 0 \quad \text{at} \quad y = 0, \tag{1.7}$$

$$u = U \quad \text{at} \quad y = H, \tag{1.8}$$

respectively. Equations (1.6) through (1.8) represent the mathematical model of the physical phenomenon, which is in the form of a differential equation.

In order to discretize the continuous problem, we first divide the continuous domain $0 \leq y \leq H$ into I equal subintervals of length $\Delta y = H/I$ as shown in Figure 1.5.

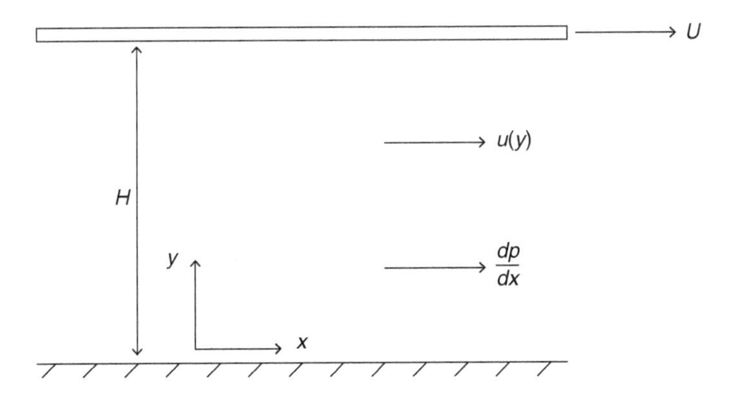

Figure 1.4 Schematic of Couette flow.

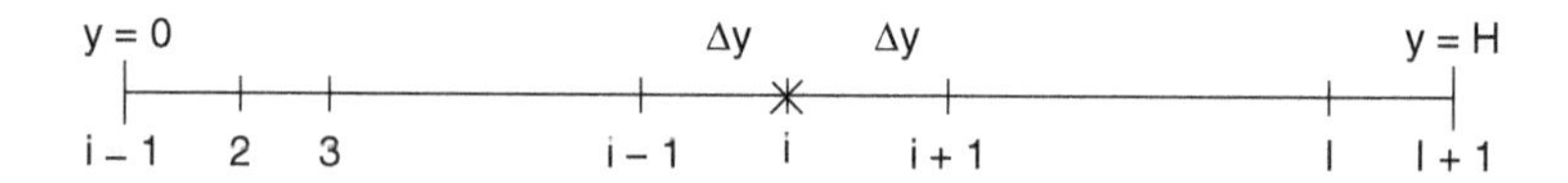

Figure 1.5 Grid used for Couette flow example. The "×" marks the location where the differential equation is approximated.

Now we must discretize the governing equation (1.6) on this *grid* or *mesh*. To see how this is done using the *finite-difference method*, recall the definition of the derivative

$$\left(\frac{du}{dy}\right)_{y=y_i} = \lim_{\Delta y \to 0} \frac{u(y_{i+1}) - u(y_i)}{\Delta y},$$

where $\Delta y = y_{i+1} - y_i$. This suggests that we can approximate the derivative by taking Δy to be small, but not in the limit as it goes all the way to zero. Thus,

$$\left(\frac{du}{dy}\right)_{y=y_i} \approx \frac{u(y_{i+1}) - u(y_i)}{\Delta y} = \frac{u_{i+1} - u_i}{\Delta y},$$

where $u_i = u(y_i)$. This is called a *forward difference*. A more accurate approximation is given by a *central difference* of the form

$$\left(\frac{du}{dy}\right)_{y=y_i} \approx \frac{u_{i+1} - u_{i-1}}{2\Delta y}. \tag{1.9}$$

We can obtain a central-difference approximation for the second-order derivative

$$\frac{d^2u}{dy^2} = \frac{d}{dy}\left(\frac{du}{dy}\right)$$

as in (1.6) by applying (1.9) midway between successive grid points as follows:

$$\left(\frac{d^2u}{dy^2}\right)_{y=y_i} \approx \frac{\left(\frac{du}{dy}\right)_{i+1/2} - \left(\frac{du}{dy}\right)_{i-1/2}}{\Delta y}.$$

Approximating du/dy as in (1.9) yields

$$\left(\frac{d^2u}{dy^2}\right)_{y=y_i} \approx \frac{\frac{u_{i+1}-u_i}{\Delta y} - \frac{u_i - u_{i-1}}{\Delta y}}{\Delta y},$$

or

$$\left(\frac{d^2u}{dy^2}\right)_{y=y_i} \approx \frac{u_{i+1} - 2u_i + u_{i-1}}{(\Delta y)^2}. \tag{1.10}$$

This is the central-difference approximation for the second-order derivative. As we will see in Chapter 7, the finite-difference approximations (1.9) and (1.10) can be formally derived using Taylor series expansions.

Given (1.10), we can discretize the governing differential (1.6) according to

$$\frac{u_{i+1} - 2u_i + u_{i-1}}{(\Delta y)^2} = \frac{1}{\mu}\frac{dp}{dx},$$

or

$$au_{i-1} + bu_i + cu_{i+1} = d, \tag{1.11}$$

where

$$a = 1, \quad b = -2, \quad c = 1, \quad d = \frac{(\Delta y)^2}{\mu}\frac{dp}{dx}.$$

The boundary conditions (1.7) and (1.8) then require that

$$u_1 = 0, \quad u_{I+1} = U. \tag{1.12}$$

Applying (1.11) at each *node point* in the grid corresponding to $i = 2,3,4,\ldots,I$ (u_1 and u_{I+1} are known) gives the system of equations

$$\begin{array}{lrcl}
i = 2: & au_1 + bu_2 + cu_3 & = & d, \\
i = 3: & au_2 + bu_3 + cu_4 & = & d, \\
i = 4: & au_3 + bu_4 + cu_5 & = & d, \\
 & & \vdots & \\
i = i: & au_{i-1} + bu_i + cu_{i+1} & = & d, \\
 & & \vdots & \\
i = I-1: & au_{I-2} + bu_{I-1} + cu_I & = & d, \\
i = I: & au_{I-1} + bu_I + cu_{I+1} & = & d.
\end{array}$$

Given the boundary conditions (1.12), the system in matrix form is

$$\begin{bmatrix}
b & c & & & & & & \\
a & b & c & & & & & \\
 & a & b & c & & & & \\
 & & \ddots & \ddots & \ddots & & & \\
 & & & a & b & c & & \\
 & & & & \ddots & \ddots & \ddots & \\
 & & & & & a & b & c \\
 & & & & & & a & b
\end{bmatrix}
\begin{bmatrix} u_1 \\ u_2 \\ u_3 \\ \vdots \\ u_i \\ \vdots \\ u_{I-1} \\ u_I \end{bmatrix}
=
\begin{bmatrix} d \\ d \\ d \\ \vdots \\ d \\ \vdots \\ d \\ d - cU \end{bmatrix},$$

where all of the empty elements in the coefficient matrix are zeros. This is called a *tridiagonal Toeplitz matrix* as only three of the diagonals have nonzero elements (tridiagonal), and the elements along each diagonal are constant (Toeplitz). This structure arises because of how the equation is discretized using central differences that involve u_{i-1}, u_i, and u_{i+1}. In general, the number of domain subdivisions I can be very large, thereby leading to a large system of algebraic equations to solve for the velocities $u_2, u_3, \ldots, u_i, \ldots, u_I$.

In all three examples, we end up with a system of linear algebraic equations to solve for the currents, forces, or discretized velocities. We will encounter numerous additional examples throughout the text.

1.4.2 General Considerations

A linear equation is one in which only polynomial terms of first degree or less appear. For example, a linear equation in N variables, $u_1, u_2, \ldots, u_N$, is of the form

$$A_1 u_1 + A_2 u_2 + \cdots + A_N u_N = b,$$

where $A_n, n = 1, \ldots, N$ and b are constants. A system of M coupled linear equations for N variables is of the form

$$\begin{aligned} A_{11}u_1 + A_{12}u_2 + \cdots + A_{1N}u_N &= b_1, \\ A_{21}u_1 + A_{22}u_2 + \cdots + A_{2N}u_N &= b_2, \\ &\vdots \\ A_{M1}u_1 + A_{M2}u_2 + \cdots + A_{MN}u_N &= b_M. \end{aligned}$$

The solution vector $\mathbf{u} = \begin{bmatrix} u_1 & u_2 & \cdots & u_N \end{bmatrix}^T$ must satisfy all M equations simultaneously. This system may be written in matrix form $\mathbf{Au} = \mathbf{b}$ as

$$\begin{bmatrix} A_{11} & A_{12} & \cdots & A_{1N} \\ A_{21} & A_{22} & \cdots & A_{2N} \\ \vdots & \vdots & \ddots & \vdots \\ A_{M1} & A_{M2} & \cdots & A_{MN} \end{bmatrix} \begin{bmatrix} u_1 \\ u_2 \\ \vdots \\ u_N \end{bmatrix} = \begin{bmatrix} b_1 \\ b_2 \\ \vdots \\ b_M \end{bmatrix}.$$

Note that the coefficient matrix $\mathbf{A}$ is $M \times N$, the solution vector $\mathbf{u}$ is $N \times 1$, and the right-hand-side vector $\mathbf{b}$ is $M \times 1$ according to the rules of matrix multiplication.

REMARKS:

1. *A system of equations may be considered a transformation, in which premultiplication by the matrix* $\mathbf{A}$ *transforms the vector* $\mathbf{u}$ *into the vector* $\mathbf{b}$ *(see Section 1.8).*
2. *A system with* $\mathbf{b} = \mathbf{0}$ *is said to be homogeneous.*
3. *The solution of* $\mathbf{Au} = \mathbf{b}$ *is the list of values* $\mathbf{u}$ *at which all of the equations intersect. For example, consider two lines* l_1 *and* l_2 *given by*

   $$A_{11}u_1 + A_{12}u_2 = b_1,$$

   and

   $$A_{21}u_1 + A_{22}u_2 = b_2,$$

   respectively. As shown in Figure 1.6, the possibilities are:
   1. l_1 *and* l_2 *have a unique intersection point at* $(u_1, u_2) \Rightarrow$ *one unique solution.*
   2. l_1 *and* l_2 *are the same line* $\Rightarrow$ *infinite solutions.*
   3. l_1 *and* l_2 *are parallel lines* $\Rightarrow$ *no solution (inconsistent).*

   The same three possibilities hold for systems of any size.
4. *As in Strang (2006), one can also interpret a system of linear algebraic equations as a linear combination of the column vectors of* $\mathbf{A}$. *If the column vectors of* $\mathbf{A}$

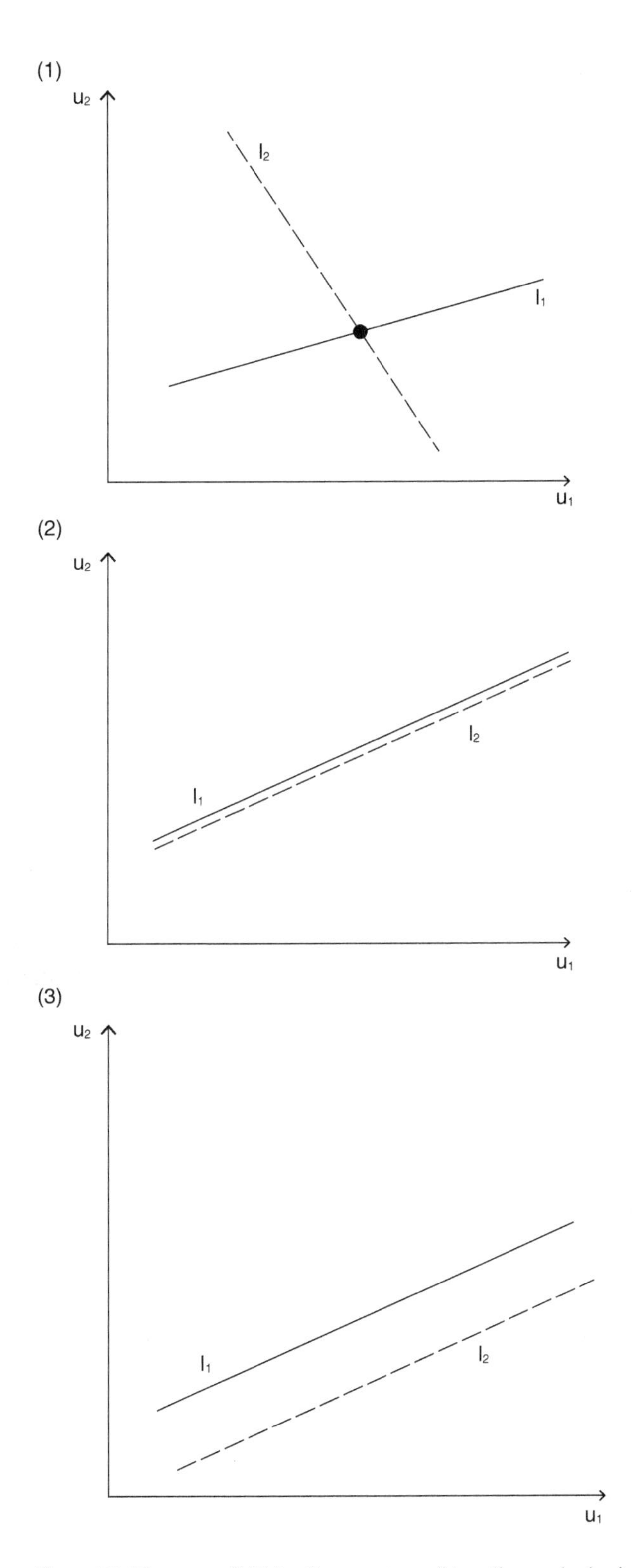

Figure 1.6 Three possibilities for a system of two linear algebraic equations.

are denoted by $\mathbf{a}_n, n = 1, \ldots N$*, then the system of equations* $\mathbf{Au} = \mathbf{b}$ *can be represented as a linear combination as follows:*

$$u_1\mathbf{a}_1 + u_2\mathbf{a}_2 + \cdots + u_N\mathbf{a}_N = \mathbf{b}.$$

In this interpretation, the solution vector $\mathbf{u}$ *contains the "amounts" of each of the column vectors required to form the right-hand-side vector* $\mathbf{b}$*, that is, the linear combination of the column vectors that produces* $\mathbf{b}$*. This interpretation emphasizes the fact that the right-hand-side vector must be in the column space of the matrix* $\mathbf{A}$*, that is, the vector space having the column vectors of* $\mathbf{A}$ *as its basis, in order to have a unique solution (see Section 1.7.4).*

1.4.3 Determinants

When solving systems of linear algebraic equations, the *determinant* of the coefficient matrix proves to be a useful diagnostic as well as being instrumental in solving such systems. For example, it allows us to determine whether the coefficient matrix is invertible or not and, thus, whether the system of equations has a unique solution, an infinity of solutions, or no solutions. It also appears in Cramer's rule for solving systems of algebraic equations in Section 1.5.3. We first motivate the determinant and why it is useful, then describe how to compute it for moderate-sized matrices, and finally describe some of its properties.

Consider the system of linear algebraic equations

$$A_{11}u_1 + A_{12}u_2 = b_1,$$
$$A_{21}u_1 + A_{22}u_2 = b_2,$$

where A_{11}, A_{12}, A_{21}, A_{22}, b_1, and b_2 are known constants, and u_1 and u_2 are the variables to be found. In order to solve for the unknown variables, multiply the first equation by A_{22}, the second by A_{12}, and then subtract. This eliminates the u_2 variable, and solving for u_1 gives

$$u_1 = \frac{A_{22}b_1 - A_{12}b_2}{A_{11}A_{22} - A_{12}A_{21}}.$$

Similarly, to obtain u_2, multiply the first equation by A_{21}, the second by A_{11}, and then subtract. This eliminates the u_1 variable, and solving for u_2 gives

$$u_2 = \frac{A_{11}b_2 - A_{21}b_1}{A_{11}A_{22} - A_{12}A_{21}}.$$

Observe that the denominators are the same in both cases. We call this denominator the *determinant* of the 2×2 coefficient matrix

$$\mathbf{A} = \begin{bmatrix} A_{11} & A_{12} \\ A_{21} & A_{22} \end{bmatrix},$$

and denote it by

$$|\mathbf{A}| = A_{11}A_{22} - A_{12}A_{21}.$$

Note that $|\mathbf{A}| \neq 0$ for there to be a solution for u_1 and u_2. If $|\mathbf{A}| = 0$, we say that $\mathbf{A}$ is *singular*.

For a 3×3 matrix

$$\mathbf{A} = \begin{bmatrix} A_{11} & A_{12} & A_{13} \\ A_{21} & A_{22} & A_{23} \\ A_{31} & A_{32} & A_{33} \end{bmatrix}, \tag{1.13}$$

the determinant is given by

$$\begin{aligned} |\mathbf{A}| = & A_{11}A_{22}A_{33} + A_{12}A_{23}A_{31} + A_{13}A_{21}A_{32} \\ & - A_{13}A_{22}A_{31} - A_{12}A_{21}A_{33} - A_{11}A_{23}A_{32}. \end{aligned} \tag{1.14}$$

Note that this diagonal multiplication method does not work for $\mathbf{A}$ larger than 3×3. Therefore, a *cofactor expansion* must be used.

For a square $N \times N$ matrix $\mathbf{A}$, M_{mn} is the *minor* and C_{mn} is the *cofactor* of A_{mn}. The minor M_{mn} is the determinant of the submatrix that remains after the mth row and nth column are removed, and the cofactor of $\mathbf{A}$ is

$$C_{mn} = (-1)^{m+n} M_{mn}.$$

We can then write the *cofactor matrix*, which is

$$\mathbf{C} = [C_{mn}] = \begin{bmatrix} +M_{11} & -M_{12} & +M_{13} & -M_{14} & \cdots \\ -M_{21} & +M_{22} & -M_{23} & +M_{24} & \\ +M_{31} & -M_{32} & +M_{33} & -M_{34} & \\ -M_{41} & +M_{42} & -M_{43} & +M_{44} & \\ \vdots & & & & \ddots \end{bmatrix},$$

where we note the sign pattern for the cofactors.

A *cofactor expansion* can be used to obtain determinants by converting the determinant of a large matrix into a sum of terms involving determinants of smaller matrices that can be easily evaluated. To illustrate, consider the determinant of 3×3 matrix $\mathbf{A}$ given by (1.13). The cofactor expansion can be formed using any row or column of matrix $\mathbf{A}$. For example, using the first row

$$\begin{aligned} |\mathbf{A}| &= A_{11}C_{11} + A_{12}C_{12} + A_{13}C_{13} \\ &= A_{11}[+M_{11}] + A_{12}[-M_{12}] + A_{13}[+M_{13}] \\ &= A_{11}\left[+\begin{vmatrix} A_{22} & A_{23} \\ A_{32} & A_{33} \end{vmatrix}\right] + A_{12}\left[-\begin{vmatrix} A_{21} & A_{23} \\ A_{31} & A_{33} \end{vmatrix}\right] + A_{13}\left[+\begin{vmatrix} A_{21} & A_{22} \\ A_{31} & A_{32} \end{vmatrix}\right] \\ &= A_{11}\left(A_{22}A_{33} - A_{23}A_{32}\right) - A_{12}\left(A_{21}A_{33} - A_{23}A_{31}\right) \\ &\quad + A_{13}\left(A_{21}A_{32} - A_{22}A_{31}\right) \\ |\mathbf{A}| &= A_{11}A_{22}A_{33} + A_{12}A_{23}A_{31} + A_{13}A_{21}A_{32} \\ &\quad - A_{11}A_{23}A_{32} - A_{12}A_{21}A_{33} - A_{13}A_{22}A_{31}, \end{aligned}$$

which is the same as (1.14).

The cofactor matrix is used in Cramer's rule in Section 1.5.3, and the cofactor expansion is used to obtain the cross product of two vectors in Section 1.6.1.

Properties of Determinants (**A** and **B** are $N \times N$):

1. If any row or column of **A** is all zeros, $|\mathbf{A}| = 0$.
2. If two rows (or columns) of **A** are interchanged, the sign of $|\mathbf{A}|$ changes.
3. If any row (or column) of **A** is a linear combination of the other rows (or columns), then $|\mathbf{A}| = 0$.
4. $|\mathbf{A}^T| = |\mathbf{A}|$.
5. In general, $|\mathbf{A} + \mathbf{B}| \neq |\mathbf{A}| + |\mathbf{B}|$.
6. $|\mathbf{AB}| = |\mathbf{BA}| = |\mathbf{A}||\mathbf{B}|$.
7. If $|\mathbf{A}| = 0$, then **A** is *singular* and not invertible. If $|\mathbf{A}| \neq 0$, then **A** is *nonsingular* and invertible, and $|\mathbf{A}^{-1}| = \dfrac{1}{|\mathbf{A}|}$.
8. The determinant of a triangular matrix is the product of the main diagonal elements.
9. The determinant of an orthogonal matrix is $|\mathbf{A}| = \pm 1$ (see Section 2.4).

1.5 Systems of Linear Algebraic Equations – Solution Methods

Let us turn our attention to basic methods for solving systems of linear algebraic equations. Methods developed in this chapter are suitable for hand calculations involving small systems of equations or using mathematics software, such as MATLAB or Mathematica,[7] for moderate-sized systems. Techniques for solving very large systems of equations computationally are discussed in Part II, and all of the methods discussed in this text for solving systems of linear algebraic equations will be summarized in Table 6.1 with reference to their respective sections.

1.5.1 Gaussian Elimination

Although not always the most efficient means of solution, *Gaussian elimination* is the most general approach, the starting point for certain specialized techniques, and the standard by which other methods are compared. Its primary virtue is that it can be applied to *any* system of equations to determine its solution(s), and indeed decipher whether it even has a solution.

The strategy in Gaussian elimination is to solve the system of linear equations $\mathbf{Au} = \mathbf{b}$ by first converting it to a new system that has the same solution **u** but is straightforward to solve. This is accomplished using *elementary row operations*, which do not change the solution vector, in a prescribed sequence. The elementary row operations correspond to actions that could be applied to a set of equations without changing their solution and are:

[7] MATLAB and Mathematica are commercial products of MathWorks and Wolfram Research, respectively.

1. Interchange two equations (rows).
2. Multiply an equation (row) by a nonzero number.
3. Add a multiple of one equation (row) to another.

But what is the objective of the elementary row operations? Observe that if our system were such that the coefficient matrix was simply an identity matrix, then the equations would be uncoupled. In this case, the solution of each equation can be written down by inspection. This is known as the *reduced row-echelon form* of the system of equations. Requiring far fewer elementary row operations, and only a bit more work to obtain the solution, we seek the *row-echelon form* of the system of equations. This consists of ones down the main diagonal with zeros below. Therefore, it is a right, or upper, triangular matrix. Thus, the objective of Gaussian elimination is to perform a sequence of elementary row operations that zero out the elements below the main diagonal and leaves leading ones in each row.

Procedure:

1. Write the *augmented matrix*:

$$[\mathbf{A}|\mathbf{b}] = \left[\begin{array}{cccc|c} A_{11} & A_{12} & \cdots & A_{1N} & b_1 \\ A_{21} & A_{22} & \cdots & A_{2N} & b_2 \\ \vdots & \vdots & & \vdots & \vdots \\ A_{M1} & A_{M2} & \cdots & A_{MN} & b_M \end{array}\right].$$

2. Move rows (equations) with leading zeros to the bottom.
3. Divide the first row by A_{11}:

$$\left[\begin{array}{cccc|c} 1 & \hat{A}_{12} & \cdots & \hat{A}_{1N} & \hat{b}_1 \\ A_{21} & A_{22} & \cdots & A_{2N} & b_2 \\ \vdots & \vdots & & \vdots & \vdots \\ A_{M1} & A_{M2} & \cdots & A_{MN} & b_M \end{array}\right].$$

4. Multiply the first row by A_{m1} and subtract from the mth row, where $m = 2, 3, \ldots, M$:

$$\left[\begin{array}{cccc|c} 1 & \hat{A}_{12} & \cdots & \hat{A}_{1N} & \hat{b}_1 \\ 0 & \hat{A}_{22} & \cdots & \hat{A}_{2N} & \hat{b}_2 \\ 0 & \hat{A}_{32} & \cdots & \hat{A}_{3N} & \hat{b}_3 \\ \vdots & \vdots & & \vdots & \vdots \\ 0 & \hat{A}_{M2} & \cdots & \hat{A}_{MN} & \hat{b}_M \end{array}\right].$$

5. Repeat steps (2) – (4), called *forward elimination*, on the lower-right $(M-1) \times (N-1)$ submatrix until the matrix is in row-echelon form, in which case the first nonzero element in each row is a one with zeros below it.
6. Obtain the solution by *backward substitution*. First, solve each equation for its leading variable (for example, $u_1 = \ldots, u_2 = \ldots, u_3 = \cdots$). Second, starting at the bottom, back substitute each equation into all of the equations above it. Finally,

the remaining $N - \text{rank}(\mathbf{A})$ variables are arbitrary, where the *rank* of matrix $\mathbf{A}$, denoted by $\text{rank}(\mathbf{A})$, is defined below.

Example 1.4 Using Gaussian elimination, solve the system of linear algebraic equations

$$\begin{aligned} 2u_1 + u_3 &= 3, \\ u_2 + 2u_3 &= 3, \\ u_1 + 2u_2 &= 3. \end{aligned}$$

Solution

In the matrix form $\mathbf{Au} = \mathbf{b}$, this system of equations is represented in the form

$$\begin{bmatrix} 2 & 0 & 1 \\ 0 & 1 & 2 \\ 1 & 2 & 0 \end{bmatrix} \begin{bmatrix} u_1 \\ u_2 \\ u_3 \end{bmatrix} = \begin{bmatrix} 3 \\ 3 \\ 3 \end{bmatrix}.$$

Note that the determinant of $\mathbf{A}$ is $|\mathbf{A}| = -9$; therefore, the matrix is nonsingular, and we expect a unique solution.

In order to apply Gaussian elimination, let us write the augmented matrix

$$\left[\begin{array}{ccc|c} 2 & 0 & 1 & 3 \\ 0 & 1 & 2 & 3 \\ 1 & 2 & 0 & 3 \end{array} \right],$$

and perform elementary row operations to produce the row-echelon form via forward elimination. In order to have a leading one in the first column of the first row, dividing through by two produces

$$\left[\begin{array}{ccc|c} 1 & 0 & \frac{1}{2} & \frac{3}{2} \\ 0 & 1 & 2 & 3 \\ 1 & 2 & 0 & 3 \end{array} \right].$$

We first seek to eliminate the elements below the leading one in the first column of the first row. The first column of the second row already contains a zero, so we focus on the third row. To eliminate the leading one, subtract the third row from the first row to produce

$$\left[\begin{array}{ccc|c} 1 & 0 & \frac{1}{2} & \frac{3}{2} \\ 0 & 1 & 2 & 3 \\ 0 & -2 & \frac{1}{2} & -\frac{3}{2} \end{array} \right].$$

Now the submatrix that results from eliminating the first row and first column is considered. There is already a leading one in the second column of the second row, so we seek to eliminate the element directly beneath it. To do so, multiply the second row by two and add to the third row to yield

$$\left[\begin{array}{ccc|c} 1 & 0 & \frac{1}{2} & \frac{3}{2} \\ 0 & 1 & 2 & 3 \\ 0 & 0 & \frac{9}{2} & \frac{9}{2} \end{array}\right].$$

To obtain row-echelon form, divide the third row by 9/2 to give

$$\left[\begin{array}{ccc|c} 1 & 0 & \frac{1}{2} & \frac{3}{2} \\ 0 & 1 & 2 & 3 \\ 0 & 0 & 1 & 1 \end{array}\right].$$

Having obtained the row-echelon form, we now perform the backward substitution to obtain the solution for $\mathbf{u}$. Beginning with the third row, we see that

$$u_3 = 1.$$

Then from the second row

$$u_2 + 2u_3 = 3,$$

or substituting $u_3 = 1$, we have

$$u_2 = 1.$$

Similarly, from the first row,

$$u_1 + \frac{1}{2}u_3 = \frac{3}{2},$$

or again substituting $u_3 = 1$ leads to

$$u_1 = 1.$$

Therefore, the solution is

$$\begin{bmatrix} u_1 \\ u_2 \\ u_3 \end{bmatrix} = \begin{bmatrix} 1 \\ 1 \\ 1 \end{bmatrix}.$$

Substitution will reveal that this solution satisfies all three of the original equations.

The preceding example consisted of three equations for the three unknowns, and a unique solution was obtained through Gaussian elimination. Recall, however, that a system of equations also may produce an infinity of solutions or no solution at all. Once again, Gaussian elimination will reveal either of these three possibilities. However, there are additional diagnostics that will help determine the outcome based on evaluating the ranks of the coefficient and augmented matrices and comparing with the number of unknowns.

The *rank* of matrix $\mathbf{A}$, denoted by rank($\mathbf{A}$), is given by the size of the largest square submatrix of $\mathbf{A}$ with a nonzero determinant. Equivalently, the rank of $\mathbf{A}$ is the number of nonzero rows when reduced to row-echelon form $\hat{\mathbf{A}}$. This is because elementary row operations do not change the rank of a matrix. If the rank of $\mathbf{A}$ and the rank of the augmented matrix $[\mathbf{A}|\mathbf{b}]$, denoted by rank($\mathbf{A}_{aug}$), are equal, the system is said to be *consistent* and the system has one or more solutions. If they are not the same, then the system is *inconsistent*, and it has no solutions.

With M being the number of equations (rows), N being the number of unknowns (columns), rank($\mathbf{A}$) being the rank of the coefficient matrix $\mathbf{A}$, and rank($\mathbf{A}_{aug}$) being the rank of the augmented matrix $\mathbf{A}_{aug}$, the three possibilities are as follows:

1. If $\text{rank}(\mathbf{A}) = \text{rank}(\mathbf{A}_{aug}) = N$, the system is consistent with as many independent equations as unknowns, such that there is a unique solution.
2. If $\text{rank}(\mathbf{A}) = \text{rank}(\mathbf{A}_{aug}) < N$, the system is consistent but with fewer independent equations than unknowns. Then there is an $[N - \text{rank}(\mathbf{A})]$-parameter family of solutions, such that $[N - \text{rank}(\mathbf{A})]$ of the variables are arbitrary and there are *infinite solutions*. Because there are fewer equations than required for a unique solution, the system is said to be *underdetermined*. In such cases, the *least-squares method* can be employed to obtain the solution that satisfies an additional criterion (see Section 10.2.2).
3. If $\text{rank}(\mathbf{A}) < \text{rank}(\mathbf{A}_{aug})$,[8] then the $\hat{A}_{ij}$ elements for the ith row $[i > \text{rank}(\mathbf{A})]$ are all zero, but at least one of the corresponding $\hat{b}_i$ is nonzero, and the system is *inconsistent* and *no solution* exists. Because there are more equations than required for a unique solution, the system is said to be *overdetermined*. In such cases, the *least-squares method* can be employed to obtain the solution that is "closest" to satisfying the equations (see Section 10.2.1).

Next we consider a series of three closely related examples that illustrate each of the three possible outcomes when seeking a solution of a system of linear algebraic equations as summarized in Figure 1.6.

Example 1.5 Using Gaussian elimination, solve the system of linear algebraic equations

$$\begin{aligned} u_1 + 2u_2 + 3u_3 &= 1, \\ 2u_1 + 3u_2 + 4u_3 &= 2, \\ 3u_1 + 4u_2 + 5u_3 &= 3. \end{aligned}$$

Solution

In the matrix form $\mathbf{Au} = \mathbf{b}$, this system of equations is represented in the form

$$\begin{bmatrix} 1 & 2 & 3 \\ 2 & 3 & 4 \\ 3 & 4 & 5 \end{bmatrix} \begin{bmatrix} u_1 \\ u_2 \\ u_3 \end{bmatrix} = \begin{bmatrix} 1 \\ 2 \\ 3 \end{bmatrix}.$$

Despite the clear pattern in the coefficient matrix and right-hand-side vector, there is nothing particularly special about this matrix problem. Note that the determinant of $\mathbf{A}$ is $|\mathbf{A}| = 0$; therefore, the matrix is singular, in which case we either have no solution or an infinite number of solutions. That is, there is no unique solution.

[8] Note that it is not possible for $\text{rank}(\mathbf{A}) > \text{rank}(\mathbf{A}_{aug})$.

In order to apply Gaussian elimination, let us write the augmented matrix

$$\left[\begin{array}{ccc|c} 1 & 2 & 3 & 1 \\ 2 & 3 & 4 & 2 \\ 3 & 4 & 5 & 3 \end{array}\right],$$

and perform elementary row operations to produce the row-echelon form via forward elimination. Because there is already a leading one in the first column of the first row, we proceed to eliminating the elements in rows two and three below the leading one in the first row. Multiplying two times the first row and subtracting the second row, as well as multiplying three times the first row and subtracting the third row yields

$$\left[\begin{array}{ccc|c} 1 & 2 & 3 & 1 \\ 0 & 1 & 2 & 0 \\ 0 & 2 & 4 & 0 \end{array}\right].$$

There is already a leading one in the second row, so we proceed to eliminate the element below it in the third row. To do so, multiply two times the second row and subtract the third row as follows:

$$\left[\begin{array}{ccc|c} 1 & 2 & 3 & 1 \\ 0 & 1 & 2 & 0 \\ 0 & 0 & 0 & 0 \end{array}\right],$$

which is in row-echelon form.

Observe that while we have three unknowns ($N = 3$), the rank of the coefficient matrix $\mathbf{A}$ is rank($\mathbf{A}$) $= 2$. Recall that the rank of the coefficient matrix is equal to the number of nonzero rows after row reduction. The rank of the augmented matrix is rank($\mathbf{A}_{aug}$) $= 2$ as well. Because the ranks of the coefficient and augmented matrices are the same, the system of equations is consistent, but because rank($\mathbf{A}$) $< N$, we have an $[N - \text{rank}(\mathbf{A})]$-parameter family of solutions. That is, there are an infinite number of solutions. Specifically, we have the two equations for the three unknowns

$$\begin{aligned} u_1 + 2u_2 + 3u_3 &= 1, \\ u_2 + 2u_3 &= 0. \end{aligned}$$

The second equation requires that $u_2 = -2u_3$, which when substituted into the first equation leads to $u_1 = u_3 + 1$. Because u_3 is left to be arbitrary, we have a one-parameter family (infinity) of solutions.

In the next example, we alter the right-hand-side vector $\mathbf{b}$ from the previous example but leave the coefficient matrix $\mathbf{A}$ unaltered.

Example 1.6 Using Gaussian elimination, solve the system of linear algebraic equations

$$u_1 + 2u_2 + 3u_3 = 3,$$
$$2u_1 + 3u_2 + 4u_3 = 1,$$
$$3u_1 + 4u_2 + 5u_3 = 2.$$

Solution

In the matrix form $\mathbf{Au} = \mathbf{b}$, this system of equations is represented in the form

$$\begin{bmatrix} 1 & 2 & 3 \\ 2 & 3 & 4 \\ 3 & 4 & 5 \end{bmatrix} \begin{bmatrix} u_1 \\ u_2 \\ u_3 \end{bmatrix} = \begin{bmatrix} 3 \\ 1 \\ 2 \end{bmatrix}.$$

Because the coefficient matrix is unchanged from the previous example, the determinant is still zero, and the same series of elementary row operations leads to the row-echelon form.

Writing the augmented matrix gives

$$\left[\begin{array}{ccc|c} 1 & 2 & 3 & 3 \\ 2 & 3 & 4 & 1 \\ 3 & 4 & 5 & 2 \end{array}\right].$$

Multiplying two times the first row and subtracting the second row, as well as multiplying three times the first row and subtracting the third row, yields

$$\left[\begin{array}{ccc|c} 1 & 2 & 3 & 3 \\ 0 & 1 & 2 & 5 \\ 0 & 2 & 4 & 7 \end{array}\right].$$

Multiplying two times the second row and subtracting the third row yields the row-echelon form

$$\left[\begin{array}{ccc|c} 1 & 2 & 3 & 3 \\ 0 & 1 & 2 & 5 \\ 0 & 0 & 0 & 3 \end{array}\right].$$

Note that while the rank of the coefficient matrix is still $\text{rank}(\mathbf{A}) = 2$, the rank of the augmented matrix is now $\text{rank}(\mathbf{A}_{aug}) = 3$. Because the ranks of the coefficient and augmented matrices are no longer the same, the system of equations is inconsistent, and there is no solution. Observe that the third equation requires that

$$0u_1 + 0u_2 + 0u_3 = 3,$$

which of course is not satisfied by any solution $\mathbf{u}$. The fact that it is not possible to satisfy this equation reflects the inconsistency of the system of equations.

Again, let us change the right-hand-side vector $\mathbf{b}$ and also one element in the coefficient matrix $\mathbf{A}$.

Example 1.7 Using Gaussian elimination, solve the system of linear algebraic equations

$$\begin{aligned} u_1 + 2u_2 + 3u_3 &= 1, \\ 2u_1 + 3u_2 + 4u_3 &= 1, \\ 3u_1 + 4u_2 + 6u_3 &= 1. \end{aligned}$$

Solution

In the matrix form $\mathbf{Au} = \mathbf{b}$, this system of equations is represented in the form

$$\begin{bmatrix} 1 & 2 & 3 \\ 2 & 3 & 4 \\ 3 & 4 & 6 \end{bmatrix} \begin{bmatrix} u_1 \\ u_2 \\ u_3 \end{bmatrix} = \begin{bmatrix} 1 \\ 1 \\ 1 \end{bmatrix}.$$

The determinant of $\mathbf{A}$ is now $|\mathbf{A}| = -1$; therefore, the matrix is nonsingular and invertible, and the system of equations has a unique solution. Again, we perform the same sequence of elementary row operations as only the last element has been changed in the coefficient matrix.

Writing the augmented matrix gives

$$\left[\begin{array}{ccc|c} 1 & 2 & 3 & 1 \\ 2 & 3 & 4 & 1 \\ 3 & 4 & 6 & 1 \end{array}\right].$$

Multiplying two times the first row and subtracting the second row, as well as multiplying three times the first row and subtracting the third row, yields

$$\left[\begin{array}{ccc|c} 1 & 2 & 3 & 1 \\ 0 & 1 & 2 & 1 \\ 0 & 2 & 3 & 2 \end{array}\right].$$

Multiplying two times the second row and subtracting the third row yields the row-echelon form

$$\left[\begin{array}{ccc|c} 1 & 2 & 3 & 1 \\ 0 & 1 & 2 & 1 \\ 0 & 0 & 1 & 0 \end{array}\right].$$

In this case, the ranks of the coefficient and augmented matrices are the same and equal the number of unknowns, such that $N = \text{rank}(\mathbf{A}) = \text{rank}(\mathbf{A}_{aug}) = 3$, which reflects the fact that the system is consistent and has a unique solution.

This unique solution is obtained using backward substitution as follows. The third equation indicates that

$$u_3 = 0,$$

which when substituted into the second equation requires that

$$u_2 = 1.$$

Finally, substituting both of these results into the first equation yields

$$u_1 = -1.$$

Therefore, the solution vector is

$$\begin{bmatrix} u_1 \\ u_2 \\ u_3 \end{bmatrix} = \begin{bmatrix} -1 \\ 1 \\ 0 \end{bmatrix}.$$

Recall that if $\mathbf{A}$ is an $N \times N$ triangular matrix, then

$$|\mathbf{A}| = A_{11}A_{22} \cdots A_{NN},$$

in which case the determinant is the product of the elements on the main diagonal. Therefore, we may row reduce the matrix $\mathbf{A}$ to triangular form in order to simplify the determinant calculation. In doing so, however, we must take into account the influence of each elementary row operation on the determinant as follows:

- Interchange rows $\Rightarrow |\hat{\mathbf{A}}| = -|\mathbf{A}|$.
- Multiply the row by k $\Rightarrow |\hat{\mathbf{A}}| = k|\mathbf{A}|$.
- Add a multiple of one row to another $\Rightarrow |\hat{\mathbf{A}}| = |\mathbf{A}|$.

1.5.2 Matrix Inverse

Given a system of equations $\mathbf{Au} = \mathbf{b}$, with an invertible coefficient matrix, having $|\mathbf{A}| \neq 0$, an alternative to Gaussian elimination for solving the system is based on using the matrix inverse. Premultiplying by $\mathbf{A}^{-1}$ gives

$$\mathbf{A}^{-1}\mathbf{Au} = \mathbf{A}^{-1}\mathbf{b}.$$

Because $\mathbf{A}^{-1}\mathbf{A} = \mathbf{I}$, the solution vector may be obtained from

$$\mathbf{u} = \mathbf{A}^{-1}\mathbf{b}.$$

Note that one can easily change $\mathbf{b}$ and premultiply by $\mathbf{A}^{-1}$ to obtain a solution for a different right-hand-side vector. If $\mathbf{b} = \mathbf{0}$, and the system is *homogeneous*, the only solution is the trivial solution $\mathbf{u} = \mathbf{0}$.[9]

One technique for obtaining the matrix inverse is to use Gaussian elimination. This is based on the fact that the elementary row operations that reduce $\mathbf{A}_N$ to $\mathbf{I}_N$ also reduce $\mathbf{I}_N$ to $\mathbf{A}_N^{-1}$. Therefore, we augment the coefficient and identity matrices according to

$$\left[\begin{array}{c|c} \mathbf{A}_N & \mathbf{I}_N \end{array} \right],$$

and perform the elementary row operations that reduce this to

$$\left[\begin{array}{c|c} \mathbf{I}_N & \mathbf{A}_N^{-1} \end{array} \right].$$

Note that $\mathbf{A}$ is not invertible if the row operations do not produce $\mathbf{I}_N$.

[9] Note that while the terms *homogeneous* and *trivial* both have to do with something being zero, homogeneous refers to the right-hand side of an equation, while trivial refers to the solution.

Example 1.8 As in Example 1.4 using Gaussian elimination, consider the system of linear algebraic equations

$$2u_1 + u_3 = 3,$$
$$u_2 + 2u_3 = 3,$$
$$u_1 + 2u_2 = 3.$$

Solution

In matrix form $\mathbf{Au} = \mathbf{b}$, this is

$$\begin{bmatrix} 2 & 0 & 1 \\ 0 & 1 & 2 \\ 1 & 2 & 0 \end{bmatrix} \begin{bmatrix} u_1 \\ u_2 \\ u_3 \end{bmatrix} = \begin{bmatrix} 3 \\ 3 \\ 3 \end{bmatrix}.$$

Recall that the determinant of $\mathbf{A}$ is $|\mathbf{A}| = -9$; therefore, the matrix is nonsingular and invertible.

Augmenting the coefficient matrix $\mathbf{A}$ with the 3×3 identify matrix $\mathbf{I}$ gives

$$\left[\begin{array}{ccc|ccc} 2 & 0 & 1 & 1 & 0 & 0 \\ 0 & 1 & 2 & 0 & 1 & 0 \\ 1 & 2 & 0 & 0 & 0 & 1 \end{array}\right].$$

Performing the same Gaussian elimination steps as in Example 1.4 to obtain the row-echelon form produces

$$\left[\begin{array}{ccc|ccc} 1 & 0 & \frac{1}{2} & \frac{1}{2} & 0 & 0 \\ 0 & 1 & 2 & 0 & 1 & 0 \\ 0 & 0 & 1 & \frac{1}{9} & \frac{4}{9} & -\frac{2}{9} \end{array}\right].$$

To obtain the inverse of $\mathbf{A}$ on the right, we must have the identify matrix on the left, that is, we must have the left side in reduced row-echelon form. To eliminate the two in the third column of the second row, multiply the third row by -2 and add to the second row. Similarly, to eliminate the one-half in the third column of the first row, multiply the third row by $-1/2$ and add to the first row. This leads to

$$\left[\begin{array}{ccc|ccc} 1 & 0 & 0 & \frac{4}{9} & -\frac{2}{9} & \frac{1}{9} \\ 0 & 1 & 0 & -\frac{2}{9} & \frac{1}{9} & \frac{4}{9} \\ 0 & 0 & 1 & \frac{1}{9} & \frac{4}{9} & -\frac{2}{9} \end{array}\right].$$

Therefore, the inverse of $\mathbf{A}$ is

$$\mathbf{A}^{-1} = \frac{1}{9} \begin{bmatrix} 4 & -2 & 1 \\ -2 & 1 & 4 \\ 1 & 4 & -2 \end{bmatrix},$$

and the solution vector can then be obtained as follows:

$$\mathbf{u} = \mathbf{A}^{-1}\mathbf{b} = \frac{1}{9} \begin{bmatrix} 4 & -2 & 1 \\ -2 & 1 & 4 \\ 1 & 4 & -2 \end{bmatrix} \begin{bmatrix} 3 \\ 3 \\ 3 \end{bmatrix} = \frac{1}{9} \begin{bmatrix} 9 \\ 9 \\ 9 \end{bmatrix} = \begin{bmatrix} 1 \\ 1 \\ 1 \end{bmatrix},$$

which is the same solution as obtained using Gaussian elimination directly in Example 1.4.

See Chapter 6 for additional methods for determining the inverse of a matrix that are suitable for computer algorithms, including LU decomposition and Cholesky decomposition.

1.5.3 Cramer's Rule

Cramer's rule is an alternative method for obtaining the unique solution of a system $\mathbf{Au} = \mathbf{b}$, where $\mathbf{A}$ is $N \times N$ and nonsingular ($|\mathbf{A}| \neq 0$). Recall that the cofactor matrix for an $N \times N$ matrix $\mathbf{A}$ is

$$\mathbf{C} = \begin{bmatrix} C_{11} & C_{12} & \cdots & C_{1N} \\ C_{21} & C_{22} & \cdots & C_{2N} \\ \vdots & \vdots & & \vdots \\ C_{N1} & C_{N2} & \cdots & C_{NN} \end{bmatrix},$$

where C_{mn} is the cofactor of A_{mn}. We define the *adjugate* of $\mathbf{A}$ to be the transpose of the cofactor matrix as follows:

$$\text{Adj}(\mathbf{A}) = \mathbf{C}^T.$$

The adjugate matrix is often referred to as the *adjoint* matrix; however, this can lead to confusion with the term *adjoint* used in other settings.

Evaluating the product of matrix $\mathbf{A}$ with its adjugate, it can be shown using the rules of cofactor expansions that

$$\mathbf{A}\,\text{Adj}(\mathbf{A}) = |\mathbf{A}|\,\mathbf{I}.$$

Therefore, because $\mathbf{AA}^{-1} = \mathbf{I}$ (if $\mathbf{A}$ is invertible), we can write

$$\mathbf{A}^{-1} = \frac{1}{|\mathbf{A}|}\text{Adj}(\mathbf{A}).$$

This leads, for example, to the familiar result for the inverse of a 2×2 matrix given in Section 1.2.

More generally, this result leads to Cramer's rule for the solution of the system $\mathbf{Au} = \mathbf{b}$ by recognizing that the nth element of the vector $\text{Adj}(\mathbf{A})\mathbf{b}$ is equal to $|\mathbf{A}_n|$, where $\mathbf{A}_n$, $n = 1, 2, \ldots, N$, is obtained by replacing the nth column of $\mathbf{A}$ by the right-hand-side vector

$$\mathbf{b} = \begin{bmatrix} b_1 \\ b_2 \\ \vdots \\ b_N \end{bmatrix}.$$

Then the system has a unique solution, which is

$$u_1 = \frac{|\mathbf{A}_1|}{|\mathbf{A}|}, \quad u_2 = \frac{|\mathbf{A}_2|}{|\mathbf{A}|}, \quad \ldots, \quad u_N = \frac{|\mathbf{A}_N|}{|\mathbf{A}|}.$$

REMARKS:

1. *If the system is homogeneous, in which case* $\mathbf{b} = \mathbf{0}$, *then* $|\mathbf{A}_j| = 0$, *and the unique solution is the trivial solution* $\mathbf{u} = \mathbf{0}$ *as expected.*
2. *Although Cramer's rule applies for any size nonsingular matrix, it is only efficient for small systems* ($N \leq 3$) *owing to the ease of calculating determinants of* 3×3 *and smaller systems. For large systems, using Gaussian elimination is typically more practical.*

Example 1.9 Let us once again consider the system of linear algebraic equations from Examples 1.4 and 1.8, which is

$$2u_1 + u_3 = 3,$$
$$u_2 + 2u_3 = 3,$$
$$u_1 + 2u_2 = 3.$$

Solution

In matrix form $\mathbf{Au} = \mathbf{b}$, the system is

$$\begin{bmatrix} 2 & 0 & 1 \\ 0 & 1 & 2 \\ 1 & 2 & 0 \end{bmatrix} \begin{bmatrix} u_1 \\ u_2 \\ u_3 \end{bmatrix} = \begin{bmatrix} 3 \\ 3 \\ 3 \end{bmatrix}.$$

Again, the determinant of $\mathbf{A}$ is $|\mathbf{A}| = -9$. By Cramer's rule, the solution is

$$u_1 = \frac{|\mathbf{A}_1|}{|\mathbf{A}|} = \frac{1}{-9} \begin{vmatrix} 3 & 0 & 1 \\ 3 & 1 & 2 \\ 3 & 2 & 0 \end{vmatrix} = \frac{-9}{-9} = 1,$$

$$u_2 = \frac{|\mathbf{A}_2|}{|\mathbf{A}|} = \frac{1}{-9} \begin{vmatrix} 2 & 3 & 1 \\ 0 & 3 & 2 \\ 1 & 3 & 0 \end{vmatrix} = \frac{-9}{-9} = 1,$$

$$u_3 = \frac{|\mathbf{A}_3|}{|\mathbf{A}|} = \frac{1}{-9} \begin{vmatrix} 2 & 0 & 3 \\ 0 & 1 & 3 \\ 1 & 2 & 3 \end{vmatrix} = \frac{-9}{-9} = 1,$$

which is the same as obtained using the inverse and Gaussian elimination in the earlier examples.

1.6 Vector Operations

Vectors, which are simply $N \times 1$ matrices, can serve many purposes. In mechanics, vectors represent the magnitude and direction of quantities, such as forces, moments, and velocities; in electromagnetics, they represent the magnitude and direction of the

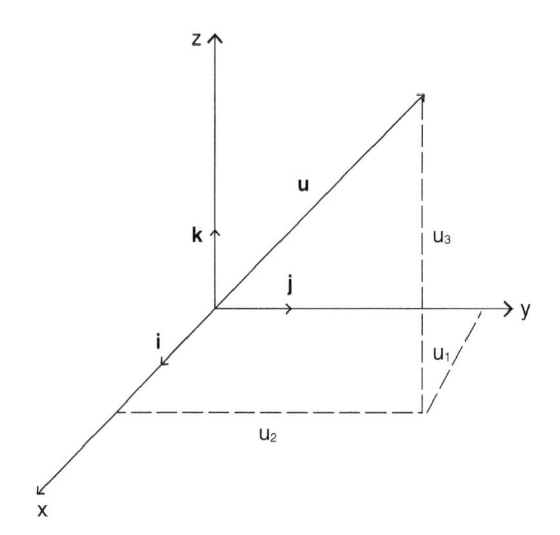

Figure 1.7 Three-dimensional vector **u** in Cartesian coordinates.

electric and magnetic fields. We first describe some operations and properties unique to vectors followed by a brief discussion of the use of vectors in forming bases for vector spaces in the next section. In order to appeal to straightforward geometric interpretations, let us first review vectors in three dimensions before extending to higher dimensions.

1.6.1 Three-Dimensional Vectors

Recall that a *scalar* is a quantity specified only by a magnitude, such as mass, length, temperature, pressure, energy, et cetera, whereas a *vector* is a quantity specified by both a magnitude and direction, such as position, velocity, force, moment, momentum, et cetera. In three dimensions, a vector **u** is given by

$$\mathbf{u} = \begin{bmatrix} u_1 \\ u_2 \\ u_3 \end{bmatrix},$$

which is shown for Cartesian coordinates in Figure 1.7, in which all three basis vectors are mutually orthogonal. The unit vectors **i**, **j**, and **k** are in each of the coordinate directions. Finally, a *tensor* is a quantity specified by multiple magnitudes and directions. In three-dimensional applications, it is given by nine quantities expressed as a 3×3 matrix, such as stress, strain, moment of inertia, et cetera. See Section 4.2 for more on tensors.

Vector addition is simply a special case of matrix addition. For example, adding the two three-dimensional vectors **u** and **v** leads to

$$\mathbf{u} + \mathbf{v} = \begin{bmatrix} u_1 \\ u_2 \\ u_3 \end{bmatrix} + \begin{bmatrix} v_1 \\ v_2 \\ v_3 \end{bmatrix} = \begin{bmatrix} u_1 + v_1 \\ u_2 + v_2 \\ u_3 + v_3 \end{bmatrix}.$$

Observe that adding two vectors results in another vector called the *resultant vector*.

Multiplying a scalar times a vector simply scales the length of the vector accordingly without changing its direction. It is a special case of matrix-scalar multiplication. For example,

$$k\mathbf{u} = k \begin{bmatrix} u_1 \\ u_2 \\ u_3 \end{bmatrix} = \begin{bmatrix} ku_1 \\ ku_2 \\ ku_3 \end{bmatrix},$$

which is a vector.

Matrix-matrix multiplication applied to vectors is known as the *inner product*.[10] The inner product of **u** and **v** is denoted by

$$\langle \mathbf{u}, \mathbf{v} \rangle = \mathbf{u}^T \mathbf{v} = \begin{bmatrix} u_1 & u_2 & u_3 \end{bmatrix} \begin{bmatrix} v_1 \\ v_2 \\ v_3 \end{bmatrix} = u_1 v_1 + u_2 v_2 + u_3 v_3 = \mathbf{v}^T \mathbf{u} = \langle \mathbf{v}, \mathbf{u} \rangle,$$

which is a scalar.[11] If $\langle \mathbf{u}, \mathbf{v} \rangle = 0$, we say that the vectors **u** and **v** are *orthogonal*. In two and three dimensions, this means that they are geometrically perpendicular.

The inner product can be used to determine the length of a vector, which is known as its *norm*, and given by

$$\|\mathbf{u}\| = \langle \mathbf{u}, \mathbf{u} \rangle^{1/2} = \sqrt{u_1^2 + u_2^2 + u_3^2}.$$

A *unit vector* is one having length (norm) equal to one. The inner product and norm can also be used to determine the angle between two vectors. From the Pythagoras theorem, the angle θ between two vectors, as illustrated in Figure 1.8, is such that

$$\cos\theta = \frac{\langle \mathbf{u}, \mathbf{v} \rangle}{\|\mathbf{u}\| \, \|\mathbf{v}\|}. \tag{1.15}$$

If $\langle \mathbf{u}, \mathbf{v} \rangle = 0$, then $\theta = \frac{\pi}{2}$, and the vectors are orthogonal. For more on vector norms, see Section 1.9.

A vector operation that is unique to the two- and three-dimensional cases is the *cross product*.[12] Recall that the inner (dot or scalar) product of two vectors produces a scalar; the cross product of two vectors produces a vector.

In Cartesian coordinates, where the vectors $\mathbf{u} = u_1\mathbf{i} + u_2\mathbf{j} + u_3\mathbf{k}$ and $\mathbf{v} = v_1\mathbf{i} + v_2\mathbf{j} + v_3\mathbf{k}$, the cross product is

$$\mathbf{w} = \mathbf{u} \times \mathbf{v} = \begin{vmatrix} \mathbf{i} & \mathbf{j} & \mathbf{k} \\ u_1 & u_2 & u_3 \\ v_1 & v_2 & v_3 \end{vmatrix} = (u_2 v_3 - u_3 v_2)\,\mathbf{i} - (u_1 v_3 - u_3 v_1)\,\mathbf{j} + (u_1 v_2 - u_2 v_1)\,\mathbf{k}.$$

Note how the cofactor expansion about the first row is used to obtain each component of the cross product. The cross product produces a vector **w** that is perpendicular to the plane in which the vectors **u** and **v** lie and points in the direction according to

[10] The inner product also goes by the terms *dot product* and *scalar product*.

[11] Different authors use various notation for the inner product, including $\langle \mathbf{u}, \mathbf{v} \rangle = \mathbf{u} \cdot \mathbf{v} = \mathbf{u}^T \mathbf{v} = (\mathbf{u}, \mathbf{v})$.

[12] The cross product is sometimes referred to as a *vector product* because it produces a vector in contrast to the inner (scalar) product, which produces a scalar.

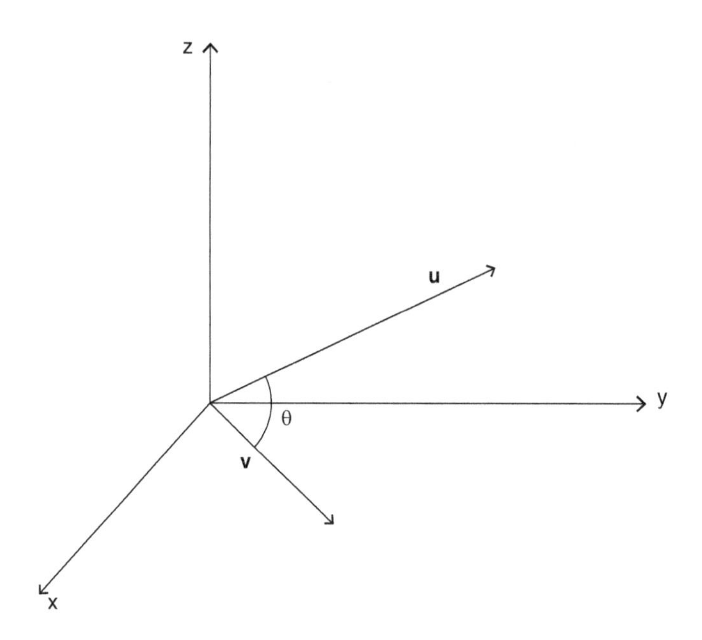

Figure 1.8 Angle between two vectors.

the right-hand rule in sweeping from $\mathbf{u}$ to $\mathbf{v}$. See Section 4.1 for the cross product in cylindrical and spherical coordinates.

REMARKS:

1. *Observe that the order of the vectors matters when taking the cross product according to the right-hand rule, whereas it does not for the inner product. That is, while* $\langle \mathbf{u}, \mathbf{v} \rangle = \langle \mathbf{v}, \mathbf{u} \rangle$ *produces a scalar of the same magnitude,* $\mathbf{u} \times \mathbf{v} = -\mathbf{v} \times \mathbf{u}$, *that is, the two cross products each produce a vector of the same magnitude but of opposite sign.*
2. *If either* $\mathbf{u}$ *or* $\mathbf{v}$ *is zero, or if* $\mathbf{u}$ *and* $\mathbf{v}$ *are parallel, then the cross product is the zero vector* $\mathbf{w} = \mathbf{0}$.
3. *If the two vectors* $\mathbf{u}$ *and* $\mathbf{v}$ *are taken to be the two adjacent sides of a parallelogram, then the norm (length) of the cross product* $\|\mathbf{w}\| = \|\mathbf{u} \times \mathbf{v}\|$ *is the area of the parallelogram.*
4. *From* (1.15), *observe that for the inner product*

$$\langle \mathbf{u}, \mathbf{v} \rangle = \|\mathbf{u}\| \, \|\mathbf{v}\| \cos \theta,$$

which is the projection of one vector onto the other. Similarly, for the cross product

$$\mathbf{u} \times \mathbf{v} = \|\mathbf{u}\| \, \|\mathbf{v}\| \sin \theta \, \mathbf{n},$$

where $\mathbf{n}$ *is a unit vector normal to the plane containing* $\mathbf{u}$ *and* $\mathbf{v}$. *The cross product is a measure of rotation of some sort. For example, it is typically first encountered in physics and mechanics when computing moments or torques produced by forces acting about an axis.*

1.6.2 Vectors in Mechanics – Caution

In our mathematics courses, we learn that vectors can be moved freely throughout a coordinate system as long as they maintain their same magnitude and direction. This is why we learned to add vectors by placing the "tail" of one vector at the "tip" of the other as illustrated in Figure 1.9 to give the resultant vector $\mathbf{u} + \mathbf{v}$.

In mechanics, however, where we deal with forces and moments expressed as vectors, we must be more careful. In statics and dynamics of *rigid* – non-deformable – bodies, forces and moments must be applied along their "line of action," which is the infinitely long line on which the vector resides. As illustrated in Figure 1.10, pushing and pulling on a *rigid* body are the same. This is known as the *principle of transmissibility*.

In the mechanics of deformable bodies, such as strength of materials, force and moment vectors must be applied at their "point of action." That is, they can no longer be moved freely along their line of action. In other words, pushing and pulling on a deformable body, such as a string, are not the same as illustrated in Figure 1.11. In summary, mathematically, vectors can be moved freely in three-dimensional space as long as they maintain the same magnitude and direction. In rigid-body mechanics, they must maintain the same magnitude, direction, and *line* of action. In deformable-body mechanics, they must maintain the same magnitude, direction, and *point* of action.

1.6.3 *N*-Dimensional Vectors

The inner product and norm operations, but not the cross product, extend naturally to N-dimensional vectors; for example, the inner product of two N-dimensional vectors

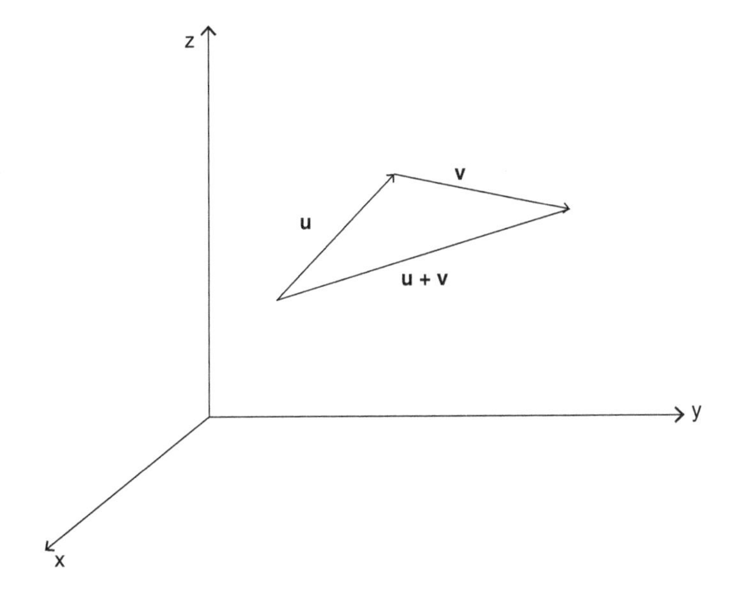

Figure 1.9 Vector addition.

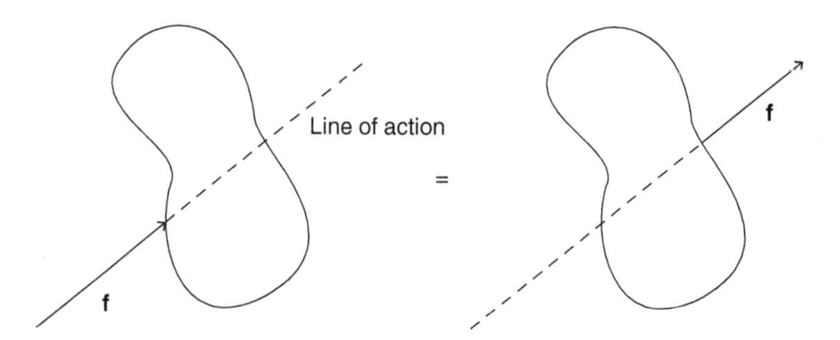

Figure 1.10 Force vector must act along its line of action in rigid-body statics and dynamics.

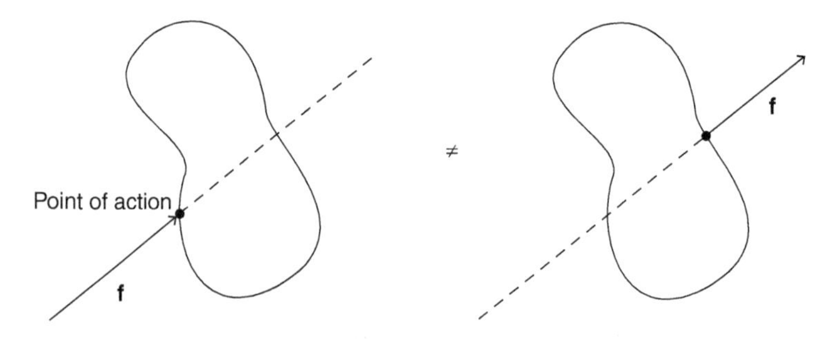

Figure 1.11 Force vector must act at its point of action in deformable-body mechanics of solids.

is given by

$$\langle \mathbf{u}, \mathbf{v} \rangle = u_1 v_1 + u_2 v_2 + \cdots + u_N v_N,$$

and the norm[13] is

$$\|\mathbf{u}\| = \langle \mathbf{u}, \mathbf{u} \rangle^{1/2} = \sqrt{u_1^2 + u_2^2 + \cdots + u_N^2};$$

however, graphical interpretations do not.

Given the term "inner" product, one may wonder if there is an "outer" product. Indeed there is; however, we will not find much practical use for it (see Sections 2.8 and 4.2 for two exceptions). Whereas the inner product of $\mathbf{u}$ and $\mathbf{v}$ is $\mathbf{u}^T\mathbf{v}$, which produces a scalar (1×1 matrix), the outer product is $\mathbf{u}\mathbf{v}^T$, which produces an $M \times N$ matrix, where $\mathbf{u}$ is $M \times 1$ and $\mathbf{v}$ is $N \times 1$ ($M = N$ for there to be an inner product). The outer product is also called a *dyad*.

1.7 Vector Spaces, Bases, and Orthogonalization

Mathematically, the primary use of vectors is in providing a *basis* – a generalization of a coordinate system – for a *vector space*. For example, the orthogonal unit vectors

[13] Note that $\mathbf{u}^2$ is taken to mean the inner product of $\mathbf{u}$ with itself, that is, $\langle \mathbf{u}, \mathbf{u} \rangle$; however, one must be careful using this notation as $\sqrt{\mathbf{u}^2} \neq \mathbf{u}$ because $\mathbf{u}^2$ is a scalar.

i, **j**, and **k** provide a basis for three-dimensional Cartesian coordinates, such that any three-dimensional vector can be written as a linear combination of the basis vectors.

1.7.1 Linear Independence of Vectors

In order to form the *basis* for a vector space, the basis vectors must be *linearly independent*. Essentially, this means that none of the vectors can be written as linear combinations of the others.

In two dimensions, if two vectors $\mathbf{u}_1$ and $\mathbf{u}_2$ are such that constants a_1 and a_2 exist for which

$$a_1\mathbf{u}_1 + a_2\mathbf{u}_2 = \mathbf{0},$$

where a_1 and a_2 are not both zero, then (1) $\mathbf{u}_1$ is a scalar multiple of $\mathbf{u}_2$, (2) $\mathbf{u}_1$ and $\mathbf{u}_2$ are parallel, and (3) $\mathbf{u}_1$ and $\mathbf{u}_2$ are *linearly dependent*. If no such a_1 and a_2 exist, and the preceding expression is only satisfied if $a_1 = a_2 = 0$, then $\mathbf{u}_1$ and $\mathbf{u}_2$ are *linearly independent*.

Similarly, in N dimensions, a set of M vectors $\mathbf{u}_1, \mathbf{u}_2, \ldots, \mathbf{u}_M$ are *linearly independent* if their linear combination is zero, in which case

$$a_1\mathbf{u}_1 + a_2\mathbf{u}_2 + \cdots + a_M\mathbf{u}_M = \mathbf{0} \tag{1.16}$$

only if $a_1 = a_2 = \cdots = a_M = 0$. In other words, none of the vectors may be expressed as a linear combination of the others. If the a_m coefficients are not all zero, then the $\mathbf{u}_m$ vectors are *linearly dependent*.

To obtain a criterion for the existence of the a_m coefficients, take the inner product of each vector $\mathbf{u}_m, m = 1, 2, \ldots, M$ with (1.16), which produces the following system of equations for the coefficients $\mathbf{a}_m, m = 1, 2, \ldots, M$:

$$\begin{aligned}
a_1\mathbf{u}_1^2 + a_2\langle\mathbf{u}_1,\mathbf{u}_2\rangle + \cdots + a_M\langle\mathbf{u}_1,\mathbf{u}_M\rangle &= 0,\\
a_1\langle\mathbf{u}_2,\mathbf{u}_1\rangle + a_2\mathbf{u}_2^2 + \cdots + a_M\langle\mathbf{u}_2,\mathbf{u}_M\rangle &= 0,\\
&\vdots\\
a_1\langle\mathbf{u}_M,\mathbf{u}_1\rangle + a_2\langle\mathbf{u}_M,\mathbf{u}_2\rangle + \cdots + a_M\mathbf{u}_M^2 &= 0,
\end{aligned}$$

or in matrix form

$$\begin{bmatrix} \mathbf{u}_1^2 & \langle\mathbf{u}_1,\mathbf{u}_2\rangle & \cdots & \langle\mathbf{u}_1,\mathbf{u}_M\rangle \\ \langle\mathbf{u}_2,\mathbf{u}_1\rangle & \mathbf{u}_2^2 & \cdots & \langle\mathbf{u}_2,\mathbf{u}_M\rangle \\ \vdots & \vdots & \ddots & \vdots \\ \langle\mathbf{u}_M,\mathbf{u}_1\rangle & \langle\mathbf{u}_M,\mathbf{u}_2\rangle & \cdots & \mathbf{u}_M^2 \end{bmatrix} \begin{bmatrix} a_1 \\ a_2 \\ \vdots \\ a_M \end{bmatrix} = \begin{bmatrix} 0 \\ 0 \\ \vdots \\ 0 \end{bmatrix}.$$

The coefficient matrix $\mathbf{G}$ is called the *Gram*, or *Gramian, matrix*. If $G = |\mathbf{G}| \neq 0$, where G is the *Gram determinant*, $\mathbf{G}$ is nonsingular and invertible, and the only solution of the homogeneous system is $\mathbf{a} = \mathbf{0}$, which is the trivial solution. In this case, the vectors $\mathbf{u}_1, \ldots, \mathbf{u}_M$ are linearly independent. If $G = |\mathbf{G}| = 0$, an infinity of solutions exist for the coefficients, that is, there is no unique solution, and the vectors $\mathbf{u}_1, \ldots, \mathbf{u}_M$ are linearly dependent.

REMARKS:

1. *The Gram matrix* $\mathbf{G}$ *is symmetric owing to the properties of inner products.*
2. *If the matrix* $\mathbf{A}$ *is formed by placing the vectors* $\mathbf{u}_m, m = 1, \ldots, M$ *as its columns, then* $\mathbf{G} = \mathbf{A}^T\mathbf{A}$, *which will feature prominently in Sections 2.4 and 2.8.*
3. *Taking the rows (or columns) of a matrix as a set of vectors, the number of linearly independent row (or column) vectors equals the rank of the matrix.*

Example 1.10 Consider the case when $\mathbf{u}_1, \ldots, \mathbf{u}_M$ are all mutually orthogonal, and determine if the vectors are linearly independent.

Solution

Because the vectors are orthogonal, their inner products are all zero except along the main diagonal, in which case

$$\begin{bmatrix} \|\mathbf{u}_1\|^2 & 0 & \cdots & 0 \\ 0 & \|\mathbf{u}_2\|^2 & \cdots & 0 \\ \vdots & \vdots & \ddots & \vdots \\ 0 & 0 & \cdots & \|\mathbf{u}_M\|^2 \end{bmatrix} \begin{bmatrix} a_1 \\ a_2 \\ \vdots \\ a_M \end{bmatrix} = \begin{bmatrix} 0 \\ 0 \\ \vdots \\ 0 \end{bmatrix}.$$

Thus, $G \neq 0$ and the only solution is the trivial solution $\mathbf{a} = \mathbf{0}$, and $\mathbf{u}_1, \ldots, \mathbf{u}_M$ are linearly independent. As one would expect, orthogonal vectors are linearly independent.

1.7.2 Basis of a Vector Space

Consider the linearly independent, N-dimensional vectors $\mathbf{u}_1, \mathbf{u}_2, \ldots, \mathbf{u}_R$. All of the vectors $\mathbf{v}$ that are linear combinations of these vectors $\mathbf{u}_m$, such that

$$\mathbf{v} = a_1\mathbf{u}_1 + a_2\mathbf{u}_2 + \cdots + a_R\mathbf{u}_R,$$

where the a_m coefficients are arbitrary, form a *vector space* V with *dimension* R. We say that (1) V is a subspace of N-dimensional space; (2) the vectors $\mathbf{u}_1, \ldots, \mathbf{u}_R$ *span* V and form a *basis* for it; (3) if $\mathbf{u}_m$ are unit vectors, each a_m is the *component* of $\mathbf{v}$ along $\mathbf{u}_m$; (4) R is the rank of the Gram matrix $\mathbf{G}$ with defect equal to $N - R$; (5) R is also the rank of the matrix with the $\mathbf{u}_m$ vectors as its rows or columns; and (6) the *range* of a matrix $\mathbf{A}$, denoted by range($\mathbf{A}$), is the vector space spanned by the columns of $\mathbf{A}$.

Example 1.11 Determine whether the vectors

$$\mathbf{u}_1 = \begin{bmatrix} 1 \\ 1 \\ 2 \end{bmatrix}, \quad \mathbf{u}_2 = \begin{bmatrix} 1 \\ 0 \\ 1 \end{bmatrix}, \quad \mathbf{u}_3 = \begin{bmatrix} 2 \\ 1 \\ 3 \end{bmatrix}$$

span three-dimensional space.

Solution

This is the same as asking if an arbitrary three-dimensional vector $\mathbf{v}$ can be expressed as a linear combination of $\mathbf{u}_1$, $\mathbf{u}_2$, and $\mathbf{u}_3$? That is,

$$\mathbf{v} = a_1\mathbf{u}_1 + a_2\mathbf{u}_2 + a_3\mathbf{u}_3,$$

or

$$\begin{bmatrix} v_1 \\ v_2 \\ v_3 \end{bmatrix} = a_1 \begin{bmatrix} 1 \\ 1 \\ 2 \end{bmatrix} + a_2 \begin{bmatrix} 1 \\ 0 \\ 1 \end{bmatrix} + a_3 \begin{bmatrix} 2 \\ 1 \\ 3 \end{bmatrix},$$

or

$$\begin{aligned} v_1 &= a_1 + a_2 + 2a_3, \\ v_2 &= a_1 + a_3, \\ v_3 &= 2a_1 + a_2 + 3a_3, \end{aligned}$$

or

$$\underbrace{\begin{bmatrix} 1 & 1 & 2 \\ 1 & 0 & 1 \\ 2 & 1 & 3 \end{bmatrix}}_{\mathbf{A}} \begin{bmatrix} a_1 \\ a_2 \\ a_3 \end{bmatrix} = \begin{bmatrix} v_1 \\ v_2 \\ v_3 \end{bmatrix}.$$

Note that the columns of $\mathbf{A}$ are $\mathbf{u}_1$, $\mathbf{u}_2$, and $\mathbf{u}_3$.

Is this linear system *consistent* for all $\mathbf{v}$? To find out, evaluate the determinant (only possible if $\mathbf{A}$ is square)

$$|\mathbf{A}| = 0 + 2 + 2 - 0 - 3 - 1 = 0.$$

Because the determinant is zero, the system is inconsistent, and no unique solution exists for $\mathbf{a}$ with any $\mathbf{v}$. Hence, the vectors $\mathbf{u}_1$, $\mathbf{u}_2$, and $\mathbf{u}_3$ do not span all of three-dimensional space.

Alternatively, we could determine the Gram matrix $\mathbf{G}$ and evaluate the Gram determinant as follows:

$$\mathbf{G} = \begin{bmatrix} \mathbf{u}_1^2 & \langle \mathbf{u}_1, \mathbf{u}_2 \rangle & \langle \mathbf{u}_1, \mathbf{u}_3 \rangle \\ \langle \mathbf{u}_2, \mathbf{u}_1 \rangle & \mathbf{u}_2^2 & \langle \mathbf{u}_2, \mathbf{u}_3 \rangle \\ \langle \mathbf{u}_3, \mathbf{u}_1 \rangle & \langle \mathbf{u}_3, \mathbf{u}_2 \rangle & \mathbf{u}_3^2 \end{bmatrix},$$

which in this case is

$$\mathbf{G} = \begin{bmatrix} 6 & 3 & 9 \\ 3 & 2 & 5 \\ 9 & 5 & 14 \end{bmatrix}.$$

Observe that $\mathbf{G}$ is symmetric as expected. The Gram determinant is $G = |\mathbf{G}| = 0$; therefore, $\mathbf{u}_1$, $\mathbf{u}_2$, and $\mathbf{u}_3$ are not linearly independent and cannot span all of three-dimensional space.

Alternatively, we could use Gaussian elimination to row reduce the system to determine if $\text{rank}(\mathbf{A}) = N$. In this example, $\text{rank}(\mathbf{A}) < N$ because the third equation is the sum of the first and second equations. Observe that it is easiest to do the first method (if $\mathbf{A}$ is square), but that either of the other two methods is required for nonsquare $\mathbf{A}$.

In contrast to the preceding example, if the three three-dimensional basis vectors are linearly independent, they span all of three-dimensional vector space. As such, any three-dimensional vector can be written as a linear combination of these three basis vectors. This holds in general as well. That is, an N-dimensional vector space is spanned by *any* set of N linearly independent, N-dimensional vectors.

Summarizing: If $\mathbf{A}$ is an $N \times N$ matrix, the following statements are equivalent:

1. $|\mathbf{A}| \neq 0$.
2. $\mathbf{A}$ is *nonsingular*.
3. $\mathbf{A}$ is *invertible*.
4. $\mathbf{A}$ has *rank* $\text{rank}(\mathbf{A}) = N$.
5. The row (and column) vectors of $\mathbf{A}$ are *linearly independent*.
6. The row (and column) vectors of $\mathbf{A}$ *span* N-dimensional space and form a *basis* for it.
7. $\mathbf{Au} = \mathbf{0}$ has only the *trivial* solution, which is $\mathbf{u} = \mathbf{A}^{-1}\mathbf{0} = \mathbf{0}$.
8. $\mathbf{Au} = \mathbf{b}$ is *consistent* for every $\mathbf{b}$, with $\mathbf{u} = \mathbf{A}^{-1}\mathbf{b}$, and there is a unique solution for each $\mathbf{b}$.

Note that the opposite of (1) through (7), but not (8), is also equivalent, in which case the system of equations would either be *inconsistent* – having no solution – or the system would have an infinity of solutions.

1.7.3 Gram–Schmidt Orthogonalization

Although any set of N linearly independent, N-dimensional vectors can form a basis for an N-dimensional vector space, it is often convenient to work with basis vectors that are mutually orthogonal and of unit length. We say that they are *orthonormal* as they are orthogonal vectors normalized to length one. This can be accomplished using *Gram–Schmidt orthogonalization*, which produces an orthogonal set of unit vectors $(\mathbf{q}_1, \mathbf{q}_2, \ldots, \mathbf{q}_S)$ from S linearly independent vectors $(\mathbf{u}_1, \mathbf{u}_2, \ldots, \mathbf{u}_S)$. Then any N-dimensional vector $\mathbf{v}$ in the vector space V spanned by $\mathbf{q}_1, \mathbf{q}_2, \ldots, \mathbf{q}_S$ can be written as a linear combination of the orthonormal basis vectors as follows:

$$\mathbf{v} = \langle \mathbf{v}, \mathbf{q}_1 \rangle \mathbf{q}_1 + \langle \mathbf{v}, \mathbf{q}_2 \rangle \mathbf{q}_2 + \cdots + \langle \mathbf{v}, \mathbf{q}_S \rangle \mathbf{q}_S,$$

where the inner product $\langle \mathbf{v}, \mathbf{q}_i \rangle$ is the component of $\mathbf{v}$ in the $\mathbf{q}_i$ direction.

Procedure (using $\mathbf{u}_1$ as the reference vector):

1. Normalize $\mathbf{u}_1$ according to

$$\mathbf{q}_1 = \frac{\mathbf{u}_1}{\|\mathbf{u}_1\|},$$

such that $\mathbf{q}_1$ is a unit vector.

2. Determine the component of $\mathbf{u}_2$ that is orthogonal to $\mathbf{q}_1$. This is accomplished by subtracting the component of $\mathbf{u}_2$ in the $\mathbf{q}_1$ direction from $\mathbf{u}_2$ to obtain

$$\hat{\mathbf{u}}_2 = \mathbf{u}_2 - \langle \mathbf{u}_2, \mathbf{q}_1 \rangle \mathbf{q}_1,$$

which is orthogonal to $\mathbf{q}_1$ as illustrated in Figure 1.12. Normalizing $\hat{\mathbf{u}}_2$ gives

$$\mathbf{q}_2 = \frac{\hat{\mathbf{u}}_2}{\|\hat{\mathbf{u}}_2\|}.$$

3. Determine the component of $\mathbf{u}_3$ that is orthogonal to both $\mathbf{q}_1$ *and* $\mathbf{q}_2$. This is accomplished by subtracting the components of $\mathbf{u}_3$ in the $\mathbf{q}_1$ and $\mathbf{q}_2$ directions from $\mathbf{u}_3$ as follows:

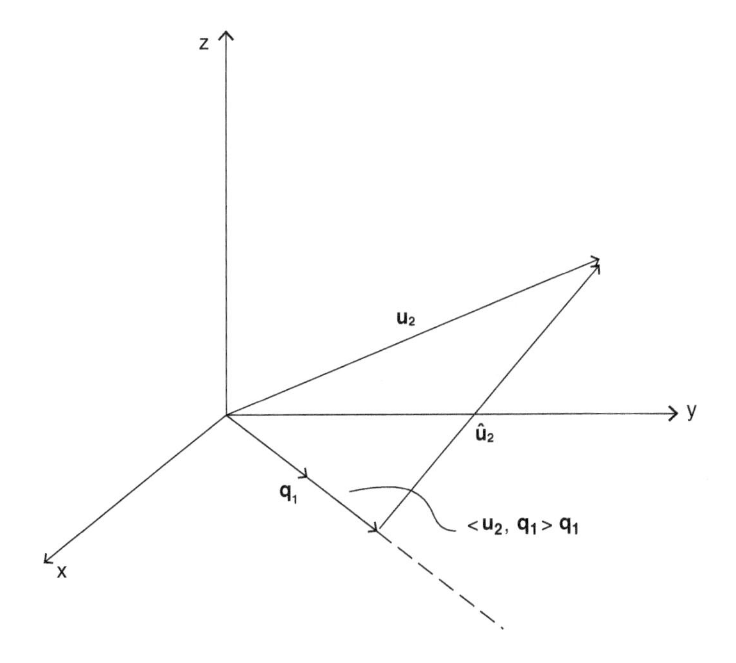

Figure 1.12 Step 2 in Gram–Schmidt orthogonalization.

$$\hat{\mathbf{u}}_3 = \mathbf{u}_3 - \langle \mathbf{u}_3, \mathbf{q}_1 \rangle \mathbf{q}_1 - \langle \mathbf{u}_3, \mathbf{q}_2 \rangle \mathbf{q}_2,$$

thereby producing a vector that is orthogonal to both $\mathbf{q}_1$ and $\mathbf{q}_2$. Normalizing $\hat{\mathbf{u}}_3$ produces

$$\mathbf{q}_3 = \frac{\hat{\mathbf{u}}_3}{\|\hat{\mathbf{u}}_3\|}.$$

4. Continue for $4 \leq i \leq S$:

$$\hat{\mathbf{u}}_i = \mathbf{u}_i - \sum_{k=1}^{i-1} \langle \mathbf{u}_i, \mathbf{q}_k \rangle \mathbf{q}_k, \quad \mathbf{q}_i = \frac{\hat{\mathbf{u}}_i}{\|\hat{\mathbf{u}}_i\|}. \tag{1.17}$$

Let us confirm that Gram–Schmidt orthogonalization in the form of (1.17) does indeed produce a vector $\hat{\mathbf{u}}_i$ that is orthogonal to each of the previously obtained orthonormal vectors $\mathbf{q}_j, j = 1, \ldots, i-1$. To do so, we show that the inner products of $\hat{\mathbf{u}}_i$ and $\mathbf{q}_j$ are all zero. This inner product is

$$\langle \hat{\mathbf{u}}_i, \mathbf{q}_j \rangle = \langle \mathbf{u}_i, \mathbf{q}_j \rangle - \sum_{k=1}^{i-1} \langle \mathbf{u}_i, \mathbf{q}_k \rangle \langle \mathbf{q}_k, \mathbf{q}_j \rangle, \quad j = 1, \ldots, i-1.$$

But $\langle \mathbf{q}_k, \mathbf{q}_j \rangle = 0$ for $j \neq k$, and $\langle \mathbf{q}_k, \mathbf{q}_j \rangle = 1$ when $j = k$, thereby producing

$$\langle \hat{\mathbf{u}}_i, \mathbf{q}_j \rangle = \langle \mathbf{u}_i, \mathbf{q}_j \rangle - \langle \mathbf{u}_i, \mathbf{q}_j \rangle = 0.$$

Consequently, $\hat{\mathbf{u}}_i$ is orthogonal to each of the previously obtained vectors $\mathbf{q}_j, j = 1, \ldots, i-1$.

REMARKS:

1. *The Gram–Schmidt orthogonalization procedure always results in orthonormal basis vectors if the* $\mathbf{u}_i$ *vectors are linearly independent, but changing the order of the linearly independent vectors* $\mathbf{u}_i$ *will alter the resulting orthonormal vectors* $\mathbf{q}_i$. *In other words, Gram–Schmidt orthogonalization is not unique.*
2. *It will be found that this orthogonalization procedure bears a close relationship to QR decomposition (see Section 2.10) and least-squares methods (see Section 10.2).*

1.7.4 Row, Column, and Null Spaces of a Matrix

Because the rows and columns of a matrix form a set of vectors, it is possible, and often useful, to consider the vector space spanned by the row or column vectors of a matrix, that is, the vector space for which the row or column vectors form a basis. The *row space* of matrix $\mathbf{A}$ is the vector space spanned by its row vectors, and the *column space*, or *range*, is that spanned by its column vectors.

Let us consider the $M \times N$ matrix $\mathbf{A}$ of the form

$$\mathbf{A} = \begin{bmatrix} A_{11} & A_{12} & \cdots & A_{1N} \\ A_{21} & A_{22} & \cdots & A_{2N} \\ \vdots & \vdots & \ddots & \vdots \\ A_{M1} & A_{M2} & \cdots & A_{MN} \end{bmatrix},$$

where $\mathbf{u}_m, m = 1, \ldots, M$, are the row vectors of $\mathbf{A}$, and $\mathbf{v}_n, n = 1, \ldots, N$, are the column vectors of $\mathbf{A}$. The row representation of $\mathbf{A}$ is then

$$\mathbf{A} = \begin{bmatrix} \mathbf{u}_1^T \\ \mathbf{u}_2^T \\ \vdots \\ \mathbf{u}_M^T \end{bmatrix}, \quad \text{where} \quad \mathbf{u}_m = \begin{bmatrix} A_{m1} \\ A_{m2} \\ \vdots \\ A_{mN} \end{bmatrix}, m = 1, \ldots, M. \tag{1.18}$$

Similarly, the column representation of $\mathbf{A}$ is

$$\mathbf{A} = \begin{bmatrix} \mathbf{v}_1 & \mathbf{v}_2 & \cdots & \mathbf{v}_N \end{bmatrix}, \tag{1.19}$$

where

$$\mathbf{v}_n = \begin{bmatrix} A_{1n} \\ A_{2n} \\ \vdots \\ A_{Mn} \end{bmatrix}, \quad n = 1, \ldots, N.$$

Let us consider the system of linear algebraic equations

$$\mathbf{A}\mathbf{u} = \mathbf{b}. \tag{1.20}$$

Suppose $\mathbf{u}_H$ is a solution of the homogeneous system

$$\mathbf{A}\mathbf{u}_H = \mathbf{0}, \tag{1.21}$$

and $\mathbf{u}_P$ is a particular solution of the nonhomogeneous system

$$\mathbf{A}\mathbf{u}_P = \mathbf{b}. \tag{1.22}$$

Because this is a linear system of algebraic equations, the general solution is of the form

$$\mathbf{u} = \mathbf{u}_P + k\mathbf{u}_H,$$

where k is an arbitrary constant. In order to show this, observe that

$$\begin{aligned} \mathbf{A}(\mathbf{u}_P + k\mathbf{u}_H) &= \mathbf{b} \\ \mathbf{A}\mathbf{u}_P + k\mathbf{A}\mathbf{u}_H &= \mathbf{b} \\ \mathbf{b} + k\mathbf{0} &= \mathbf{b}, \end{aligned}$$

which is of course true for all $\mathbf{b}$. We say that the homogeneous solution $\mathbf{u}_H$ is in the *null space*, and $\mathbf{b}$ is in the *column space*, of the matrix $\mathbf{A}$ as further clarified later in this section.

Let us first consider the homogeneous equation (1.21) using the row representation as follows:

$$\begin{bmatrix} \cdots & \mathbf{u}_1^T & \cdots \\ \cdots & \mathbf{u}_2^T & \cdots \\ & \vdots & \\ \cdots & \mathbf{u}_M^T & \cdots \end{bmatrix} \begin{bmatrix} \vdots \\ \mathbf{u}_H \\ \vdots \end{bmatrix} = \begin{bmatrix} \vdots \\ \mathbf{0} \\ \vdots \end{bmatrix}.$$

This requires that

$$\langle \mathbf{u}_1, \mathbf{u}_H \rangle = 0, \quad \langle \mathbf{u}_2, \mathbf{u}_H \rangle = 0, \quad \ldots, \quad \langle \mathbf{u}_M, \mathbf{u}_H \rangle = 0;$$

that is, $\mathbf{u}_H$ is orthogonal to each of the row vectors of $\mathbf{A}$. The following possibilities then exist:

1. If the rank of $\mathbf{A}$ is N [rank($\mathbf{A}$) $= N$], then the row vectors $\mathbf{u}_m, m = 1, \ldots, M$, are linearly independent, and they span the N-dimensional row space. In this case, $\mathbf{u}_H = \mathbf{0}$ is the only solution of $\mathbf{A}\mathbf{u}_H = \mathbf{0}$, and the null space is the zero vector.
2. If rank($\mathbf{A}$) $< N$, in which case only rank($\mathbf{A}$) of the $\mathbf{u}_m$, $m = 1, \ldots, M$ vectors are linearly independent, then the rank($\mathbf{A}$)-dimensional space spanned by the rank($\mathbf{A}$) linearly independent row vectors is the *row space*. There are then [N − rank($\mathbf{A}$)] arbitrary vectors that satisfy $\mathbf{A}\mathbf{u}_H = \mathbf{0}$ and are thus orthogonal to the row space. The [N − rank($\mathbf{A}$)]-dimensional space where these $\mathbf{x}_H$ vectors exist is called the *orthogonal complement* of the row space, or the *null space*, of $\mathbf{A}$. It follows that a nonzero vector $\mathbf{u}_H$ satisfies $\mathbf{A}\mathbf{u}_H = \mathbf{0}$ *if and only if* it is in the null space of matrix $\mathbf{A}$.

Now consider the nonhomogeneous equation (1.20) using the column representation as follows:

$$\begin{bmatrix} \vdots & \vdots & & \vdots \\ \mathbf{v}_1 & \mathbf{v}_2 & \cdots & \mathbf{v}_N \\ \vdots & \vdots & & \vdots \end{bmatrix} \begin{bmatrix} u_1 \\ u_2 \\ \vdots \\ u_N \end{bmatrix} = \begin{bmatrix} \vdots \\ \mathbf{b} \\ \vdots \end{bmatrix}.$$

This requires that vector $\mathbf{b}$ be a linear combination of the column vectors $\mathbf{v}_n, n = 1, \ldots, N$, in which case

$$u_1 \mathbf{v}_1 + u_2 \mathbf{v}_2 + \cdots + u_N \mathbf{v}_N = \mathbf{b}.$$

Then $\mathbf{b}$ is in the column space of $\mathbf{A}$, and the column space defines all possible vectors $\mathbf{b}$. The following possibilities then exist:

1. If $\mathbf{b}$ is in the column space of $\mathbf{A}$, then the system $\mathbf{A}\mathbf{u} = \mathbf{b}$ is consistent and has a unique solution if rank($\mathbf{A}$) $= N$ or an infinity of solutions if rank($\mathbf{A}$) $< N$.
2. If $\mathbf{b}$ is not in the column space of $\mathbf{A}$, then the system $\mathbf{A}\mathbf{u} = \mathbf{b}$ is inconsistent.

REMARKS:

1. *The dimensions of the row and column space are the same as one another and equal the rank of matrix* $\mathbf{A}$, *that is,* $\text{rank}(\mathbf{A}) = \text{rank}(\mathbf{A}^T)$.
2. *For an* $M \times N$ *matrix, the dimension of the null space is equal to* $N - \text{rank}(\mathbf{A})$.
3. *If one considers the matrix* $\mathbf{A}$ *to be a linear transformation from an* N*-dimensional vector space to an* M*-dimensional one, then the column space of the matrix equals the image of this linear transformation.*
4. *Elementary row operations do not change the row or column space of a matrix.*

Our first exposure to linear algebra is typically in the context of solving systems of linear algebraic equations for which a unique solution is sought. In such cases, the coefficient matrix is typically square and the rank equals the number of unknowns. Indeed, only square matrices have a determinant, can be invertible and have eigenvalues and eigenvectors (see Chapter 2). As a result, it is tempting to dismiss discussion of rectangular matrices, or more precisely those having $\text{rank}(\mathbf{A}) \neq N$, as only being of interest to mathematicians. Being comfortable with vector spaces and the particulars of matrices that are not square will pay dividends when we get to applications, such as least-squares methods, singular-value decomposition, and proper-orthogonal decomposition, for example. Rest assured that we would not include these topics if not relevant to applications of interest to scientists and engineers.

1.8 Linear Transformations

Thus far, we have viewed the matrix problem $\mathbf{Au} = \mathbf{b}$ as a system of linear algebraic equations, in which we seek the solution vector $\mathbf{u}$ for a given coefficient matrix $\mathbf{A}$ and right-hand-side vector $\mathbf{b}$. Alternatively, the matrix problem

$$\underset{M\times N}{\mathbf{A}}\ \underset{N\times 1}{\mathbf{x}} = \underset{M\times 1}{\mathbf{y}},$$

where $\mathbf{A}$ is specified and $\mathbf{x}$ and $\mathbf{y}$ are arbitrary vectors, may be viewed as a *linear transformation*, or *mapping*, from an N-dimensional vector space to an M-dimensional vector space, which are the sizes of $\mathbf{x}$ and $\mathbf{y}$, respectively. We say that the matrix $\mathbf{A}$ operates on the vector $\mathbf{x}$ to transform it into the vector $\mathbf{y}$. In this context, $\mathbf{A}$ transforms vectors from an N-space *domain* to an M-space *range* as shown in Figure 1.13. Such transformation matrices will be encountered in numerous settings throughout the remainder of the text, and thinking of matrices as linear operators in this manner will prove helpful in understanding many of the methods to come.

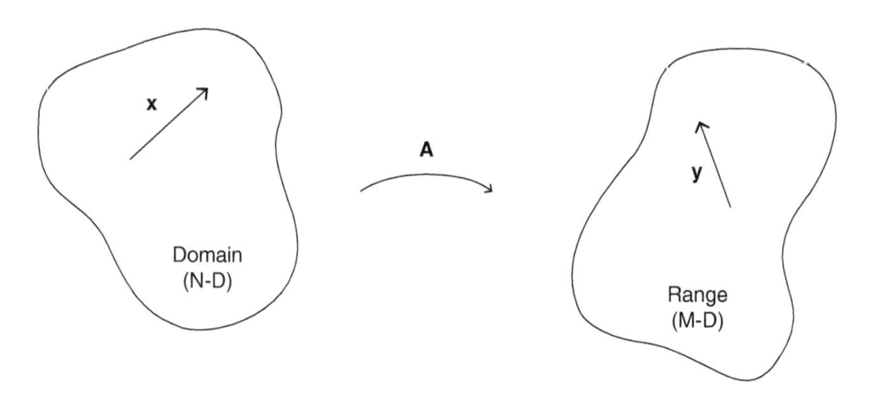

Figure 1.13 Linear transformation from $\mathbf{x}$ in the domain to $\mathbf{y}$ in the range of $\mathbf{A}$.

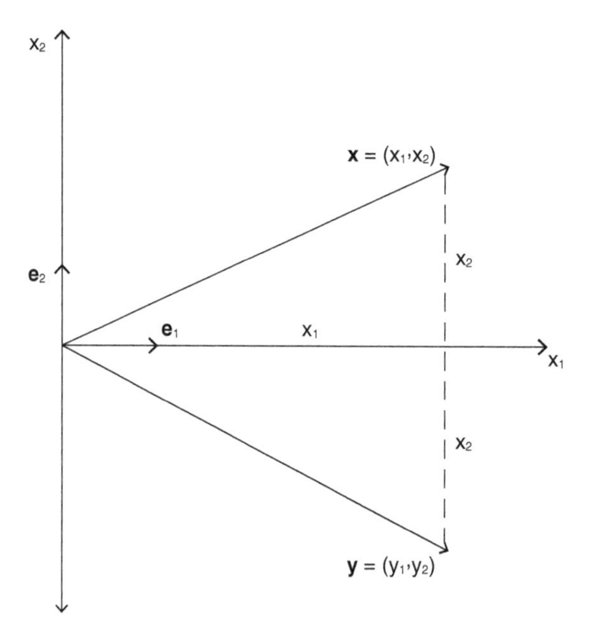

Figure 1.14 Reflection of vector $\mathbf{x}$ about the x_1-axis to form vector $\mathbf{y}$ in Example 1.12

Example 1.12 Determine the transformation matrix $\mathbf{A}$ that transforms the vector $\mathbf{x}$ to the vector $\mathbf{y}$ by reflecting it about the x_1-axis as illustrated in Figure 1.14.

Solution

The vector $\mathbf{y}$ is given by

$$\mathbf{y} = y_1\mathbf{e}_1 + y_2\mathbf{e}_2,$$

where $\mathbf{e}_1$ and $\mathbf{e}_2$ are unit vectors in the x_1 and x_2 coordinate directions, respectively. As can be observed graphically, however, the components of $\mathbf{y} = \begin{bmatrix} y_1 & y_2 \end{bmatrix}^T$ are related to the components of $\mathbf{x} = \begin{bmatrix} x_1 & x_2 \end{bmatrix}^T$ according to

$$y_1 = x_1, \quad y_2 = -x_2,$$

for such a reflection. These two relationships can be expressed as a matrix transformation of the form

$$\begin{bmatrix} y_1 \\ y_2 \end{bmatrix} = \begin{bmatrix} 1 & 0 \\ 0 & -1 \end{bmatrix} \begin{bmatrix} x_1 \\ x_2 \end{bmatrix}.$$

Example 1.13 Determine the transformation matrix $\mathbf{A}$ that transforms the vector $\mathbf{x}$ to the vector $\mathbf{y}$ by rotating it counterclockwise about the origin through an angle θ as illustrated in Figure 1.15. Both vectors are of length r.

Solution
The vector $\mathbf{x}$ is given by

$$\mathbf{x} = x_1\mathbf{e}_1 + x_2\mathbf{e}_2 = r\cos\alpha\ \mathbf{e}_1 + r\sin\alpha\ \mathbf{e}_2,$$

where $\mathbf{e}_1$ and $\mathbf{e}_2$ are unit vectors in the x_1 and x_2 coordinate directions, respectively. Thus,

$$x_1 = r\cos\alpha, \quad x_2 = r\sin\alpha. \tag{1.23}$$

Similarly, the vector $\mathbf{y}$ is given by

$$\begin{aligned} \mathbf{y} &= y_1\mathbf{e}_1 + y_2\mathbf{e}_2, \\ &= r\cos(\alpha+\theta)\mathbf{e}_1 + r\sin(\alpha+\theta)\mathbf{e}_2, \\ \mathbf{y} &= r\,(\cos\alpha\cos\theta - \sin\alpha\sin\theta)\,\mathbf{e}_1 + r\,(\sin\alpha\cos\theta + \cos\alpha\sin\theta)\,\mathbf{e}_2. \end{aligned}$$

From (1.23), however, this can be written in terms of x_1 and x_2 as follows:

$$\mathbf{y} = (x_1\cos\theta - x_2\sin\theta)\,\mathbf{e}_1 + (x_2\cos\theta + x_1\sin\theta)\,\mathbf{e}_2.$$

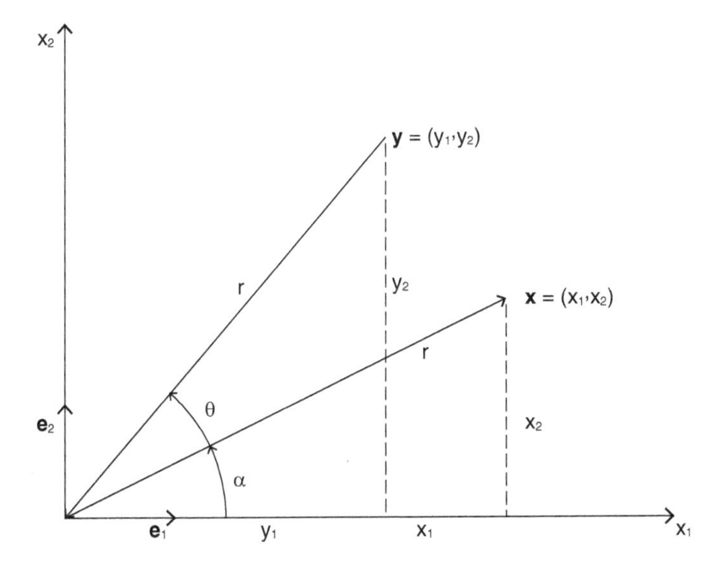

Figure 1.15 Rotation of vector $\mathbf{x}$ through an angle θ to form vector $\mathbf{y}$ in Example 1.13.

Therefore,

$$y_1 = x_1 \cos\theta - x_2 \sin\theta, \quad y_2 = x_1 \sin\theta + x_2 \cos\theta,$$

which in matrix form is

$$\begin{bmatrix} y_1 \\ y_2 \end{bmatrix} = \begin{bmatrix} \cos\theta & -\sin\theta \\ \sin\theta & \cos\theta \end{bmatrix} \begin{bmatrix} x_1 \\ x_2 \end{bmatrix}.$$

This transformation matrix rotates a vector counterclockwise about the origin through an angle θ without altering its length.

REMARKS:

1. *Because the transformations are linear, they can be superimposed. For example,*

$$\mathbf{Ax} + \mathbf{Bx} = \mathbf{y}$$

is equivalent to

$$(\mathbf{A} + \mathbf{B})\,\mathbf{x} = \mathbf{y}.$$

2. *Successively applying a series of transformations to* $\mathbf{x}$*, such as*

$$\mathbf{x}_1 = \mathbf{A}_1\mathbf{x}, \quad \mathbf{x}_2 = \mathbf{A}_2\mathbf{x}_1, \quad \ldots, \quad \mathbf{x}_K = \mathbf{A}_K\mathbf{x}_{K-1},$$

is equivalent to applying the transformation

$$\mathbf{x}_K = \mathbf{Ax},$$

where $\mathbf{A} = \mathbf{A}_K \cdots \mathbf{A}_2\mathbf{A}_1$ *(note the reverse order).*

3. *If the transformation matrix* $\mathbf{A}$ *is invertible, then* $\mathbf{A}^{-1}\mathbf{y}$ *transforms* $\mathbf{y}$ *back to* $\mathbf{x}$*, which is the inverse transformation.*
4. *If* $\mathbf{y} = \mathbf{Ax}$*, then* $\mathbf{y}$ *is a linear combination of the columns* $(\mathbf{a}_1, \mathbf{a}_2, \ldots, \mathbf{a}_N)$ *of* $\mathbf{A}$ *in the form*

$$\mathbf{y} = \mathbf{Ax} = x_1\mathbf{a}_1 + x_2\mathbf{a}_2 + \cdots + x_N\mathbf{a}_N.$$

That is, $\mathbf{y}$ *is in the column space of matrix* $\mathbf{A}$*.*

5. *In the case when a matrix* $\mathbf{A}$ *transforms a nonzero vector* $\mathbf{x}$ *to the zero vector* $\mathbf{0}$*, that is,* $\mathbf{Ax} = \mathbf{0}$*, the vector* $\mathbf{x}$ *is said to be in the null space of* $\mathbf{A}$*. The null space is a subspace of that defined by the row or column vectors of* $\mathbf{A}$ *(see Section 1.7.4).*
6. *If the transformation matrix* $\mathbf{A}$ *is orthogonal, it preserves the length of the vector, that is,* $\|\mathbf{x}\| = \|\mathbf{y}\|$*. It is shown in Section 2.4 that the rotation matrix in Example 1.13 is orthogonal.*
7. *For more on the mathematics of linear algebra, see Strang (2006), Strang (2016), Lay, Lay, and McDonald (2016), and Meckes and Meckes (2018).*

1.9 Note on Norms

Throughout this chapter, we have made use of the norm as a measure of the size or length of a vector. It essentially replaces the absolute value of real numbers or the modulus of complex numbers. Thus far, the norm has been defined for an N-dimensional vector $\mathbf{u}$ as

$$\|\mathbf{u}\| = \langle \mathbf{u}, \mathbf{u} \rangle^{1/2} = \sqrt{u_1^2 + u_2^2 + \cdots + u_N^2} = \left(\sum_{n=1}^{N} u_n^2 \right)^{1/2},$$

which is the square root of the inner product of the vector with itself. In two and three dimensions, this definition of the norm corresponds to the Euclidian length of the vector – the straight-line distance from the origin to the point (u_1, u_2, u_3). It turns out that while this is the most common definition of the norm, there are others that are encountered as well.

In general, we define the so-called L_p-norm, or simply the p-norm, of the vector $\mathbf{u}$ as follows:

$$\|\mathbf{u}\|_p = \left(\sum_{n=1}^{N} |u_n|^p \right)^{1/p}, \tag{1.24}$$

where $1 \le p < \infty$. Therefore, our previous definition of the norm corresponds to the L_2-norm, or simply the "2-norm." As a result, many authors indicate this norm by $\|\mathbf{u}\|_2$. We will use this notation when necessary to prevent any confusion; however, if no subscript is given, then it is assumed to be the L_2-norm.

Other common norms correspond to $p = 1$ and $p = \infty$. The L_1-norm is the sum of the absolute values of each component of $\mathbf{u}$ according to

$$\|\mathbf{u}\|_1 = \sum_{n=1}^{N} |u_n|.$$

This corresponds to only being able to travel from the origin to the point along lines parallel to the coordinate axes. The L_∞-norm, or ∞-norm, is given by

$$\|\mathbf{u}\|_\infty = \max_{1 \le n \le N} |u_n|,$$

which is simply the largest component of $\mathbf{u}$ by absolute value.

We will also have occasion to compute norms of matrices in order to quantify their "size." The L_1-norm of a matrix $\mathbf{A}$ is the largest L_1-norm of the column vectors of the $M \times N$ matrix $\mathbf{A}$ as follows:

$$\|\mathbf{A}\|_1 = \max_{1 \le n \le N} \|\mathbf{a}_n\|_1,$$

where $\mathbf{a}_n$ denotes the nth column of $\mathbf{A}$. The L_∞-norm of a matrix $\mathbf{A}$ is the largest L_1-norm of the row vectors of $\mathbf{A}$, that is,

$$\|\mathbf{A}\|_\infty = \max_{1 \le m \le M} \|\mathbf{a}_m\|_1,$$

where $\mathbf{a}_m$ denotes the mth row of $\mathbf{A}$. Therefore, the 1-norm is the maximum column sum, and the ∞-norm is the maximum row sum of the matrix $\mathbf{A}$. Finally, the L_2-norm of a matrix is given by

$$\|\mathbf{A}\|_2 = \left[\rho\left(\mathbf{A}^T\mathbf{A}\right)\right]^{1/2} = \sigma_1,$$

where $\rho\left(\mathbf{A}^T\mathbf{A}\right)$ is the spectral radius, that is, the largest eigenvalue by magnitude, of $\mathbf{A}^T\mathbf{A}$ (see Section 2.2), and σ_1 is the maximum singular value of $\mathbf{A}$ (see Section 2.8). Hence, it is also sometimes referred to as the *spectral norm*. If $\mathbf{A}$ is symmetric, then the L_2-norm is the spectral radius of $\mathbf{A}$, that is, $\|\mathbf{A}\|_2 = \rho(\mathbf{A})$.

To more directly mimic the L_2-norm for vectors, we also define the *Frobenius norm* of a matrix given by

$$\|\mathbf{A}\|_F = \left(\sum_{m=1}^{M}\sum_{n=1}^{N}|A_{mn}|^2\right)^{1/2}.$$

This is simply the square root of the sum of the squares of all the elements of the matrix $\mathbf{A}$. Alternatively, it can also be expressed in terms of $\mathbf{A}\mathbf{A}^T$ as

$$\|\mathbf{A}\|_F = \sqrt{\operatorname{tr}(\mathbf{A}\mathbf{A}^T)} = \left(\sum_{m=1}^{M}\lambda_m\right)^{1/2} = \left(\sum_{m=1}^{M}\sigma_m^2\right)^{1/2},$$

where $\operatorname{tr}(\mathbf{A}\mathbf{A}^T)$ is the trace of $\mathbf{A}\mathbf{A}^T$, λ_m are the eigenvalues of $\mathbf{A}\mathbf{A}^T$, and σ_m are the singular values of $\mathbf{A}$ (see Chapter 2).

REMARKS:

1. *When both vectors and matrices are present, we use the same norm for both when performing operations.*
2. *In one dimension, the L_2-norm is simply the absolute value.*
3. *Unless indicated otherwise, we will use the L_2-norm for both vectors and matrices. Its popularity stems in part from the fact that it can often be interpreted in terms of energies in certain applications. In addition, it extends naturally to functions as in Chapter 3.*
4. *Although the L_2-norm is most commonly used, the L_1-norm and L_∞-norm are typically more convenient to compute.*
5. *The Schwarz inequality for vectors is*

$$|\langle\mathbf{u},\mathbf{v}\rangle| \leq \|\mathbf{u}\|_2\|\mathbf{v}\|_2.$$

More generally, it can be proven that

$$\|\mathbf{A}\mathbf{B}\| \leq \|\mathbf{A}\|\,\|\mathbf{B}\|.$$

These inequalities prove useful in determining bounds on various quantities of interest involving vectors and matrices.

6. *Vector and matrix norms will play a prominent role in defining the condition number in Chapter 6 that is so central to computational methods.*

7. *For more on the mathematical properties of norms, see Golub and Van Loan (2013) and Horn and Johnson (2013).*

1.10 Briefly on Bases

Let us end this chapter where we started, with the choice of coordinate systems and vector bases. Early in our training as scientists and engineers, we learn that the key to solving many problems is a judicious choice of coordinate system, or vector basis, with respect to which to represent the problem. Do we employ a Cartesian, cylindrical, or spherical coordinate system? Where is the origin of the coordinate system logically placed? How are the coordinate directions oriented? These choices are somewhat arbitrary and are simply a matter of convenience in many cases, while in other situations it is the difference between being able to obtain a solution or not. We will see over and over again that this important lesson holds true in more advanced pursuits as well.

The reason why this focus on bases is so helpful is that more than half the battle of solving a problem is typically knowing what form, or basis, the solution can most naturally be written with respect to. In other words, if we know what vectors or functions the solution can be written in terms of, then we just need to determine how much of each basis is required for the particular solution at hand. The simplest example of this is the basis vectors of a coordinate system. For example, we know that any vector in three-dimensional Cartesian coordinates can be written as a linear combination of the unit vectors **i**, **j**, and **k** in the three coordinate directions. For a given vector, it then remains to simply determine how much of each unit vector is required to represent that particular vector – that is, to determine the components u_x, u_y, and u_z in the vector

$$\mathbf{u} = u_x\mathbf{i} + u_y\mathbf{j} + u_z\mathbf{k}.$$

We espoused the virtues of using orthonormal basis vectors to provide a convenient representation of a problem and ease many of the operations, such as projecting vectors onto coordinate axes and transforming vectors. The Gram–Schmidt orthogonalization procedure was introduced as a means to obtain such a basis from a set of linearly independent basis vectors.

It will also prove helpful to view solutions of various types of problems in terms of linear combinations of basis vectors or functions. For example, we know that the solution of a linear, constant-coefficient ordinary differential equation is given in terms of the functions e^{mx}, where there are as many m's as the order of the differential equation. The solution of such a differential equation then involves simply determining the m's and how much of each e^{mx} term is required. For a second-order linear ordinary differential equation with constant coefficients, therefore, we seek the constants m_1, m_2, c_1, c_2 in the function

$$u(x) = c_1e^{m_1x} + c_2e^{m_2x}.$$

In all such scenarios, then, the two-step process involves answering the following questions:

1. How do we choose or determine the set of basis vectors or functions to use for a given problem?
2. Given the set of basis vectors or functions, how do we determine the coefficients for each?

As we develop a variety of techniques for solving and analyzing the various types of matrix and differential problems and their solutions, it is helpful to take note of the following questions as they relate to bases:

1. For the given form of coefficient or transformation matrix $\mathbf{A}$, what can we say about the form of the solution vectors or eigenvectors? Can we represent the solution in terms of a particular set of basis vectors? For example, can we use orthonormal basis vectors or must we employ a more general set of linearly independent, but not orthogonal, basis vectors?
2. Can we modify the basis vectors in order to facilitate obtaining the solution or ease interpretation?

Such questions related to bases will be a common theme throughout the text as they greatly help in understanding the motivation as well as relative advantages and disadvantages of various solution techniques and numerical methods. This being the case, the reader is strongly encouraged to become comfortable with the terminology and techniques surrounding basis vectors (and eventually functions). To aid in this process, we will close each of the first four chapters in Part I with a "Briefly on Bases" section that highlights what that chapter has contributed to the preceding discussion. In so doing, the reader will have a solid foundation on which to build in Part III, where these ideas are central to methods for analyzing and understanding large-scale numerical solutions and experimental data sets.

Finally, the notational conventions used throughout the text are summarized in Table 1.2.

Exercises

The following exercises are to be completed using hand calculations. When appropriate, the results can be checked using built-in functions in Python, MATLAB, or Mathematica for the vector and matrix operations.

1.1 Consider the matrix

$$\mathbf{A} = \begin{bmatrix} 1 & 2 & 1 \\ 2 & 1 & 0 \\ -1 & 0 & 1 \end{bmatrix}.$$

Determine $\mathbf{A}^T$, $\mathbf{A}^2$, $|\mathbf{A}|$, and $\mathbf{A}^{-1}$.

Table 1.2 Notational conventions.

Quantity/operation	Convention	Examples
Vectors	Bold lowercase letters and square brackets	$\mathbf{a} = [\cdot]$
Matrices	Bold uppercase letters and square brackets	$\mathbf{A} = [\cdot]$
Zero vector or matrix	Bold zero	$\mathbf{0}$
Identity matrix	Bold I	$\mathbf{I}$
Matrix/array size	Uppercase letters	I, J, K, M, N
Matrix/array index	Lowercase letters and subscripts	i, j, k, m, n
Trace of matrix		$\text{tr}(\mathbf{A})$
Determinant		$\vert\mathbf{A}\vert$
Condition number		$\text{cond}(\mathbf{A})$, $\kappa(\mathbf{A})$
Spectral radius		$\rho(\mathbf{A})$
Matrix rank		$\text{rank}\,(\mathbf{A})$
Vector or matrix norm		$\Vert\mathbf{a}\Vert$, $\Vert\mathbf{A}\Vert$
Matrix transpose	Superscript T	$\mathbf{A}^T$
Matrix inverse	Superscript -1	$\mathbf{A}^{-1}$
Matrix pseudo-inverse	Superscript $+$	$\mathbf{A}^+$
Scalar/dot/inner product	Angle brackets	$\langle\mathbf{a}, \mathbf{b}\rangle = \mathbf{a} \cdot \mathbf{b} = \mathbf{a}^T\mathbf{b}$
Outer product		$\mathbf{a}\mathbf{b}^T$
Inner product of functions	Angle brackets	$\langle g, h\rangle = \int_a^b g(x)h(x)dx$
Functions	Lowercase with independent variables in parentheses	$f(x,t)$
Functionals	Uppercase with dependent functions in square brackets	$I[u(x), v(x)]$, $J[u(t)]$
Spatial ordinary derivative	Prime, superscript in parentheses	$u'(x) = du/dx$, $u^{(n)}(x) = d^n u/dx^n$
Temporal ordinary derivative	Dot	$\dot{u}(t) = du/dt, \ddot{u}(t) = d^2u/dt^2$
Partial derivatives	Subscripts	$u_{xx} = \partial^2 u/\partial x^2$
Eigenvalues	Lowercase lambda	λ
Lagrange multipliers	Uppercase lambda	Λ
Gradient operator	Del (nabla)	$\boldsymbol{\nabla}$
Laplacian operator	Del (nabla) squared	$\boldsymbol{\nabla}^2 = \boldsymbol{\nabla} \cdot \boldsymbol{\nabla}$
General differential operator		$\mathcal{L}$
Adjoint differential operator		$\mathcal{L}^*$
Imaginary number	San serif i	i
Real and imaginary parts		$\text{Re}(u)$, $\text{Im}(u)$
Complex conjugate	Overbar	$\overline{\mathbf{A}}$
Numerical time step		Δt
Numerical grid spacing		$\Delta x, \Delta y$
Numerical grid point	Index subscript(s)	$u_{i,j}$
Numerical iteration number	Superscript in parentheses	$u^{(n)}$

1.2 Consider the two vectors and one matrix

$$\mathbf{u} = \begin{bmatrix} -3 \\ 2 \\ 1 \\ -1 \end{bmatrix}, \quad \mathbf{v} = \begin{bmatrix} 2 \\ -2 \\ 1 \\ 3 \end{bmatrix}, \quad \mathbf{A} = \begin{bmatrix} 1 & 5 & 4 & -2 \\ -7 & 2 & 1 & 4 \\ 3 & 2 & -1 & 1 \\ 1 & 3 & -4 & 2 \end{bmatrix}.$$

Determine $\mathbf{u} + \mathbf{v}$, $\langle \mathbf{u}, \mathbf{v} \rangle$, $\mathbf{A}^T$, $|\mathbf{A}|$, $\mathbf{A}^{-1}$, and $\mathbf{Au}$.

1.3 Show that

$$|\mathbf{AB}| = |\mathbf{A}|\,|\mathbf{B}|$$

if $\mathbf{A}$ and $\mathbf{B}$ are general 2×2 matrices.

1.4 For a general 3×3 matrix $\mathbf{A}$, write out the cofactor expansion of the determinant of $\mathbf{A}$ in the form

$$|\mathbf{A}| = A_{11}C_{11} + A_{12}C_{12} + A_{13}C_{13},$$

and reorder the terms to show that

$$|\mathbf{A}| = \left|\mathbf{A}^T\right|.$$

1.5 The *law of cosines* for a triangle with sides of length a, b, and c, in which the angle opposite the side of length c is C, takes the form

$$c^2 = a^2 + b^2 - 2ab\cos C.$$

Prove this by taking vectors $\mathbf{a}$, $\mathbf{b}$, and $\mathbf{c}$ along each side, such that $\mathbf{c} = \mathbf{a} - \mathbf{b}$, and considering the inner product $\langle \mathbf{c}, \mathbf{c} \rangle = \langle \mathbf{a} - \mathbf{b}, \mathbf{a} - \mathbf{b} \rangle$.

1.6 Let $\mathbf{x}$ and $\mathbf{y}$ be N-dimensional vectors and λ be a scalar. Prove that

$$\|\mathbf{x} + \lambda\mathbf{y}\|^2 + \|\mathbf{x} - \lambda\mathbf{y}\|^2 = 2\left(\|\mathbf{x}\|^2 + \lambda^2\|\mathbf{y}\|^2\right).$$

1.7 Let $\mathbf{x}$ and $\mathbf{y}$ be N-dimensional orthogonal vectors. Prove that the *Pythagoras theorem* takes the form

$$\|\mathbf{x} + \mathbf{y}\|^2 = \|\mathbf{x}\|^2 + \|\mathbf{y}\|^2.$$

1.8 Consider the determinant

$$P(\lambda) = \begin{vmatrix} 4-\lambda & 0 & 1 \\ 1 & -\lambda & 1 \\ -1 & -2 & 2-\lambda \end{vmatrix},$$

where λ is an unknown parameter. By expanding the determinant, show that $P(\lambda)$ is a polynomial for λ. Determine the value(s) of λ for which $P(\lambda) = 0$.

1.9 Determine the values of λ for which the following system of equations has a non-trivial solution:

$$\begin{aligned} 3u_1 + u_2 - \lambda u_3 &= 0, \\ 4u_1 - 2u_2 - 3u_3 &= 0, \\ 2\lambda u_1 + 4u_2 + \lambda u_3 &= 0. \end{aligned}$$

For each possible value of λ, determine the most general solution for $\mathbf{u}$.

1.10 Determine the rank of the matrix

$$\mathbf{A} = \begin{bmatrix} -1 & 1 & 0 \\ 0 & -1 & 1 \\ -2 & 1 & 1 \end{bmatrix}.$$

1.11 For the matrices

$$\mathbf{A} = \begin{bmatrix} 3 & -1 & 1 \\ 1 & 4 & 0 \\ 2 & 1 & -3 \end{bmatrix}, \quad \mathbf{B} = \begin{bmatrix} 1 & -3 & 1 \\ 2 & 0 & 5 \\ 3 & 1 & 2 \end{bmatrix},$$

verify that

$$(\mathbf{AB})^{-1} = \mathbf{B}^{-1}\mathbf{A}^{-1}.$$

1.12 Prove that $(\mathbf{AB})^{-1} = \mathbf{B}^{-1}\mathbf{A}^{-1}$ if $\mathbf{A}$ and $\mathbf{B}$ are invertible and of the same size.

1.13 Consider the following system of equations expressed in matrix form:

$$\begin{bmatrix} 1 & 1 & 1 & 1 \\ 0 & 1 & 1 & 1 \\ 0 & 0 & 1 & 1 \\ 0 & 0 & 0 & 0 \end{bmatrix} \begin{bmatrix} u_1 \\ u_2 \\ u_3 \\ u_4 \end{bmatrix} = \begin{bmatrix} 0 \\ 1 \\ 2 \\ 3 \end{bmatrix}.$$

Determine whether this system has no solution, one unique solution, or an infinity of solutions. Justify your answer.

1.14 Consider the following system of equations:

$$\begin{aligned} u_1 + 3u_2 + 2u_3 &= 0, \\ u_1 + u_2 + u_3 &= 1, \\ 2u_2 + u_3 &= 2. \end{aligned}$$

Determine whether this system has no solution, one unique solution, or an infinity of solutions. Justify your answer.

1.15 Given the system of equations

$$\begin{aligned} u_1 + 5u_2 &= 3, \\ 7u_1 - 3u_2 &= -1, \end{aligned}$$

determine the solution using Cramer's rule.

1.16 Given the system of equations

$$\begin{aligned} 2u_1 + u_3 &= 9, \\ 2u_2 + u_3 &= 3, \\ u_1 + 2u_3 &= 3, \end{aligned}$$

determine the solution using Cramer's rule.

1.17 Consider the matrix

$$\mathbf{A} = \begin{bmatrix} 1 & 4 & 1 & 0 \\ 2 & 1 & 3 & 1 \\ 5 & 6 & 7 & 2 \\ 2 & 1 & 0 & 1 \end{bmatrix}.$$

Determine the solution of the homogeneous linear system of equations $\mathbf{Au} = \mathbf{0}$.

1.18 Consider the system of linear algebraic equations $\mathbf{Au} = \mathbf{b}$, where

$$\mathbf{A} = \begin{bmatrix} 1 & 2 & 1 \\ 1 & 1 & 2 \\ 2 & 1 & 1 \\ 0 & 3 & 5 \end{bmatrix}, \quad \mathbf{b} = \begin{bmatrix} 4 \\ 0 \\ 4 \\ 1 \end{bmatrix}.$$

Determine whether the system is consistent or inconsistent. If it is consistent, obtain its solution(s) for $\mathbf{u}$.

1.19 Consider the system of linear algebraic equations

$$\begin{aligned} 2u_1 - u_2 - u_3 &= 2, \\ u_1 + 2u_2 + u_3 &= 2, \\ 4u_1 - 7u_2 - 5u_3 &= 2. \end{aligned}$$

Show that this system of equations has a one-parameter family of solutions by considering the ranks of relevant matrices. Determine this solution.

1.20 For the system of linear algebraic equations $\mathbf{Au} = \mathbf{b}$, where

$$\mathbf{A} = \begin{bmatrix} 2 & -4 & 1 \\ 6 & 2 & -1 \\ -2 & 6 & -2 \end{bmatrix}, \quad \mathbf{b} = \begin{bmatrix} 4 \\ 10 \\ -6 \end{bmatrix},$$

obtain the solution $\mathbf{u}$ using (1) Gaussian elimination, (2) matrix inverse, and (3) Cramer's rule.

1.21 In Example 1.1, Ohm's and Kirchhoff's laws were applied to a parallel electric circuit involving voltage sources and resistors to obtain the system of linear algebraic equations $\mathbf{Au} = \mathbf{b}$ of the form

$$\begin{bmatrix} 30 & -10 & -8 \\ -10 & 16 & -6 \\ -8 & -6 & 18 \end{bmatrix} \begin{bmatrix} u_1 \\ u_2 \\ u_3 \end{bmatrix} = \begin{bmatrix} 8 \\ 4 \\ 6 \end{bmatrix}.$$

Solve the system of equations for the three currents using (1) Gaussian elimination, (2) matrix inverse, and (3) Cramer's rule.

1.22 Consider the matrix

$$\mathbf{A} = \begin{bmatrix} 1 & -1 & 1 \\ -1 & 1 & -1 \\ 1 & 1 & 1 \end{bmatrix}.$$

Is the matrix $\mathbf{A}$ singular or nonsingular? Does the system of equations $\mathbf{Au} = \mathbf{0}$ have a nontrivial solution? Are the column vectors of $\mathbf{A}$ linearly dependent?

1.23 Are the vectors $\mathbf{u}_1 = \begin{bmatrix} 2 & 0 & 1 \end{bmatrix}^T$ and $\mathbf{u}_2 = \begin{bmatrix} 1 & 1 & 0 \end{bmatrix}^T$ linearly independent? Are they orthogonal? Explain your answers.

1.24 Determine whether the vectors $\mathbf{u}_1 = \begin{bmatrix} 1 & 2 & 1 \end{bmatrix}^T$, $\mathbf{u}_2 = \begin{bmatrix} 0 & 1 & 1 \end{bmatrix}^T$, and $\mathbf{u}_3 = \begin{bmatrix} 1 & 3 & 2 \end{bmatrix}^T$ span three-dimensional space.

1.25 Determine whether the vectors $\mathbf{u}_1 = \begin{bmatrix} -1 & 0 & -2 \end{bmatrix}^T$, $\mathbf{u}_2 = \begin{bmatrix} 1 & -1 & 1 \end{bmatrix}^T$ and $\mathbf{u}_3 = \begin{bmatrix} 0 & 1 & 1 \end{bmatrix}^T$ are linearly independent.

1.26 Determine if the vector $\mathbf{v} = \begin{bmatrix} 6 & 1 & -6 & 2 \end{bmatrix}^T$ is in the vector space spanned by the basis vectors $\mathbf{u}_1 = \begin{bmatrix} 1 & 1 & -1 & 1 \end{bmatrix}^T$, $\mathbf{u}_2 = \begin{bmatrix} -1 & 0 & 1 & 1 \end{bmatrix}^T$, and $\mathbf{u}_3 = \begin{bmatrix} 1 & -1 & -1 & 0 \end{bmatrix}^T$.

1.27 Consider the following vectors in three-dimensional Cartesian coordinates:

$$\mathbf{u}_1 = -\mathbf{i} + \mathbf{k},$$
$$\mathbf{u}_2 = 2\mathbf{j} + \mathbf{k},$$
$$\mathbf{u}_3 = \mathbf{i} + \mathbf{j} + \mathbf{k}.$$

Use Gram–Schmidt orthogonalization to obtain a set of orthonormal basis vectors for three-dimensional vector space.

1.28 Using Gram–Schmidt orthogonalization, construct a set of three mutually orthogonal unit vectors from the set of linearly independent vectors $\mathbf{u}_1 = \begin{bmatrix} -1 & 2 & 0 \end{bmatrix}^T$, $\mathbf{u}_2 = \begin{bmatrix} 1 & 1 & -1 \end{bmatrix}^T$, and $\mathbf{u}_3 = \begin{bmatrix} 1 & -1 & 1 \end{bmatrix}^T$.

1.29 Using Gram–Schmidt orthogonalization, construct a set of three mutually orthogonal unit vectors from the set of linearly independent vectors $\mathbf{u}_1 = \begin{bmatrix} 1 & 0 & 2 & 2 \end{bmatrix}^T$, $\mathbf{u}_2 = \begin{bmatrix} 1 & 1 & 0 & 1 \end{bmatrix}^T$, and $\mathbf{u}_3 = \begin{bmatrix} 1 & 1 & 0 & 0 \end{bmatrix}^T$.

1.30 Show that the vector $\mathbf{v} = \begin{bmatrix} 2 & 1 & 2 & 0 \end{bmatrix}^T$ is in the vector space spanned by the three vectors given in Exercise 1.29, and express $\mathbf{v}$ as a linear combination of the orthonormal vectors $\mathbf{q}_1$, $\mathbf{q}_2$, and $\mathbf{q}_3$ that result from Gram–Schmidt orthogonalization.

2 Algebraic Eigenproblems and Their Applications

The knowledge we have of mathematical truths is not only certain, but real knowledge; and not the bare empty vision of vain, insignificant chimeras of the brain.
(John Locke)

Recall from Section 1.8 that we may view a matrix $\mathbf{A}$ as a linear transformation from a vector $\mathbf{x}$ to a vector $\mathbf{y}$ in the form $\mathbf{Ax} = \mathbf{y}$. In general, such a linear transformation may translate, rotate, reflect, and/or scale (stretch) a vector, and any transformation can be decomposed into a series of these simple operations. In the context of such transformations, it is often useful to characterize the transformation matrix $\mathbf{A}$ in terms of its eigenvalues and eigenvectors. The eigenvectors are those special vectors that are *only* stretched by the transformation matrix, and the eigenvalues are their scale factors. Far from being a mathematical novelty that only occurs for certain special matrices, it turns out that all square matrices have a set of eigenvalue–eigenvector pairs. Even more interesting for us, these eigenpairs are found to have a wide variety of applications in nearly all areas of engineering and the physical sciences.

2.1 Applications of Eigenproblems

2.1.1 Geometric Example

We used a simple two-dimensional geometric example in Section 1.1.2 to introduce and motivate the need for vectors and matrices. Recall that we specified two image points in each of two coordinate systems and considered a route to determining the transformation matrix between them. Let us similarly motivate the *eigenproblem.* In particular, consider the case when the transformation from one two-dimensional coordinate system (basis) to another only stretches the original vectors by a specified amount without any additional translations, rotations, or other operations.

Returning to our two coordinate systems X and Y, consider the two vectors $\mathbf{x}_1 = \begin{bmatrix} 1 & -2 \end{bmatrix}^T$ and $\mathbf{x}_2 = \begin{bmatrix} 3 & -1 \end{bmatrix}^T$ in X.[1] We seek the transformation matrix $\mathbf{A}$ that only stretches these two vectors by prescribed stretch factors $\lambda_1 = 2$ and $\lambda_2 = \frac{1}{3}$.

[1] Note that we are now being more precise in defining the points as vectors than was the case in the introduction to Chapter 1, where they were simply thought of as pairs of coordinates.

Therefore, the transformation $\mathbf{A}\mathbf{x} = \mathbf{y}$ is such that

$$\mathbf{A}\mathbf{x}_1 = \lambda_1 \mathbf{x}_1, \quad \mathbf{A}\mathbf{x}_2 = \lambda_2 \mathbf{x}_2,$$

in which case $\mathbf{y}_1 = \lambda_1 \mathbf{x}_1$ and $\mathbf{y}_2 = \lambda_2 \mathbf{x}_2$ are simply scalar multiples of the original vectors. Written out, these are

$$\begin{bmatrix} A_{11} & A_{12} \\ A_{21} & A_{22} \end{bmatrix} \begin{bmatrix} 1 \\ -2 \end{bmatrix} = 2 \begin{bmatrix} 1 \\ -2 \end{bmatrix}, \quad \begin{bmatrix} A_{11} & A_{12} \\ A_{21} & A_{22} \end{bmatrix} \begin{bmatrix} 3 \\ -1 \end{bmatrix} = \frac{1}{3} \begin{bmatrix} 3 \\ -1 \end{bmatrix}.$$

This produces four equations for the four unknown elements of the transformation matrix $\mathbf{A}$ that give rise to this unique geometric property.

It is important to note that the transformation that we would obtain from this exercise would only stretch the two vectors that we specified – they are the eigenvectors of $\mathbf{A}$; all other vectors would undergo additional translations, rotations, et cetera as well. That is why these eigenvectors are so special. Under normal circumstances, we will be seeking the values λ_n and vectors $\mathbf{x}_n$ for a given matrix $\mathbf{A}$ rather than the other way around. This is the *eigenproblem*[2] for matrix $\mathbf{A}$, where the possible values of λ_n are called the *eigenvalues* and the corresponding vectors $\mathbf{x}_n$ are called the *eigenvectors*.

2.1.2 Principal Stresses

Let us consider a physical example – the stresses acting on an infinitesimally small tetrahedral element of a solid or fluid as illustrated in Figure 2.1. Note that A is the area of triangle B-C-D. The stress field is given by the *stress tensor*, which is a 3×3 matrix of the form[3]

$$\boldsymbol{\tau} = \begin{bmatrix} \tau_{xx} & \tau_{xy} & \tau_{xz} \\ \tau_{xy} & \tau_{yy} & \tau_{yz} \\ \tau_{xz} & \tau_{yz} & \tau_{zz} \end{bmatrix},$$

where the diagonal elements τ_{xx}, τ_{yy}, and τ_{zz} are the *normal stresses*, and the off-diagonal elements $\tau_{xy} = \tau_{yx}$, $\tau_{xz} = \tau_{zx}$, and $\tau_{yz} = \tau_{zy}$ are the *shear stresses*. Stresses are defined such that the first subscript indicates the outward normal to the surface, and the second subscript indicates the direction of the stress on that face as shown in Figure 2.2. Also recall that $\tau_{ij} = \tau_{ji}$; therefore, the stress tensor is always symmetric, such that $\boldsymbol{\tau} = \boldsymbol{\tau}^T$.

Let us take $\mathbf{n}$ to be the normal unit vector, having length one, to the inclined face B-C-D on which only a normal stress τ_n acts, then

$$\mathbf{n} = n_x \mathbf{i} + n_y \mathbf{j} + n_z \mathbf{k}.$$

Enforcing static equilibrium of the element in the x-direction, which requires that $\sum F_x = 0$, yields

$$\langle \tau_n \mathbf{n}, \mathbf{i} \rangle A = \tau_{xx} \langle A\mathbf{n}, \mathbf{i} \rangle + \tau_{xy} \langle A\mathbf{n}, \mathbf{j} \rangle + \tau_{xz} \langle A\mathbf{n}, \mathbf{k} \rangle,$$

[2] The German word "eigen" means *proper* or *characteristic*.

[3] See Section 4.2 for more on tensors.

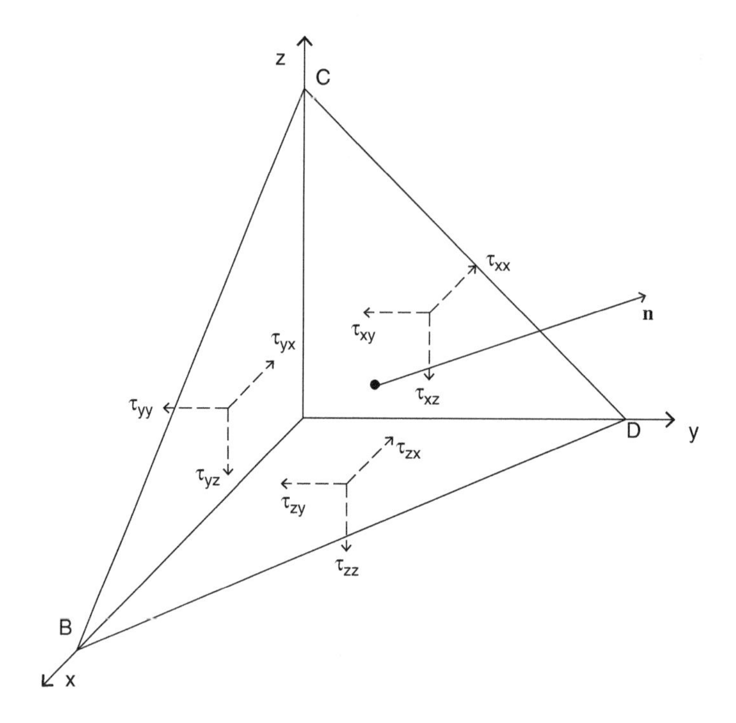

Figure 2.1 Stresses acting on a tetrahedral element.

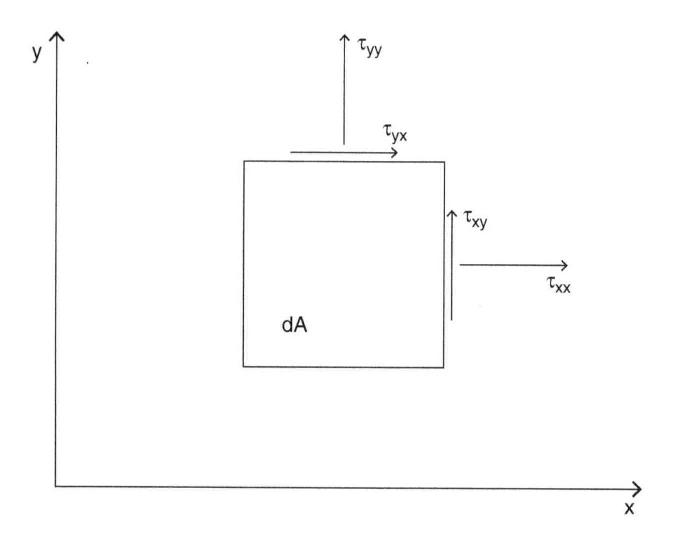

Figure 2.2 Stresses acting on an infinitesimally small area.

where the inner product $\langle \cdot \rangle$ on the left-hand side is the component of the stress in the x-direction that is acting on the B-C-D surface, and those in the terms on the right-hand side are the areas of the x-, y-, and z-faces, respectively. Remember that the inner (dot) product with a unit vector gives the component in that direction. Simplifying yields

$$\tau_n n_x = \tau_{xx} n_x + \tau_{xy} n_y + \tau_{xz} n_z.$$

Similarly, in the y- and z-directions, we have

$$\sum F_y = 0: \quad \tau_n n_y = \tau_{xy} n_x + \tau_{yy} n_y + \tau_{yz} n_z,$$
$$\sum F_z = 0: \quad \tau_n n_z = \tau_{xz} n_x + \tau_{yz} n_y + \tau_{zz} n_z.$$

In matrix form, this system of equations is given by

$$\begin{bmatrix} \tau_{xx} & \tau_{xy} & \tau_{xz} \\ \tau_{xy} & \tau_{yy} & \tau_{yz} \\ \tau_{xz} & \tau_{yz} & \tau_{zz} \end{bmatrix} \begin{bmatrix} n_x \\ n_y \\ n_z \end{bmatrix} = \tau_n \begin{bmatrix} n_x \\ n_y \\ n_z \end{bmatrix},$$

or

$$\boldsymbol{\tau} \mathbf{n}_n = \tau_n \mathbf{n}_n,$$

where $\boldsymbol{\tau}$ is the stress tensor. The *principal axes* $\mathbf{n}_n$ are the coordinate axes with respect to which only normal stresses act, such that there are no shear stresses. These are called *principal stresses*. Thus, the three eigenvalues τ_n of the stress tensor are the principal stresses, and the three corresponding eigenvectors $\mathbf{n}_n$ are the principal axes on which they each act. We see that the answer to the physical question, "What stress state only has normal stress?" is answered mathematically by asking, "What are the eigenvalues and eigenvectors of the stress tensor?" Geometrically, this is encapsulated by Mohr's circle, and the principal stresses are the minimum and maximum normal stresses acting on any orientation. See Example 2.1 for an illustration of determining the principal axes and stresses.

In a similar manner, the *principal moments of inertia* are the eigenvalues of the symmetric moment of inertia tensor, which is given by

$$\begin{bmatrix} I_{xx} & I_{xy} & I_{xz} \\ I_{xy} & I_{yy} & I_{yz} \\ I_{xz} & I_{yz} & I_{zz} \end{bmatrix},$$

where I_{xx}, I_{yy}, and I_{zz} are the *moments of inertia*, and I_{xy}, I_{xz}, and I_{yz} are the *products of inertia* of a body about the corresponding axes. The moment of inertia tensor accounts for how the mass is distributed throughout a rigid body. The eigenvectors are the corresponding coordinate directions – principal axes – about which the principal moments of inertia act and there are no products of inertia.

2.1.3 Systems of Linear Ordinary Differential Equations

The application that will occupy much of our attention in this chapter is obtaining solutions of coupled systems of linear ordinary differential equations. *Linear systems*, or more generally *dynamical systems*, are largely synonymous with matrix methods. The governing equations of any time-dependent, discrete, linear system, whether electrical, mechanical, biological, or chemical, can be expressed as a system of first-order, linear ordinary differential equations of the form

$$\dot{\mathbf{u}}(t) = \mathbf{A}\mathbf{u}(t) + \mathbf{f}(t),$$

where the dot represents a time derivative, $\mathbf{A}$ is a known coefficient matrix, $\mathbf{f}(t)$ is a known vector of specified functions of time, and $\mathbf{u}(t)$ is the solution (state) vector of dependent variables that is sought. Expressed in this first-order form, it is often referred to as the *state-space representation*. The diagonalization procedure used to solve such systems is discussed in Section 2.5 and has widespread application in a variety of fields of science and engineering. It involves transforming the system of equations into an alternative one that is much easier to solve, where the new system is based on a different set of variables found using a modal matrix consisting of the eigenvectors of the coefficient matrix $\mathbf{A}$.

2.1.4 Additional Applications

There are many applications in which determining the vectors that are only stretched by a particular matrix (eigenvectors) and the corresponding stretch factors (eigenvalues) hold great significance physically; they can literally mean the difference between life and death![4] In addition to those highlighted above, some common applications where eigenvalues and eigenvectors are encountered are:

- The natural frequencies of vibration of mechanical or electrical dynamical systems correspond to eigenvalues (see Section 2.6.3).
- Stability of dynamical systems subject to small disturbances is determined from a consideration of eigenvalues (see Section 5.3).
- There are several powerful matrix decompositions that are related to, based on, or used to solve the eigenproblem (see Sections 2.7–2.10).
- Optimization using *quadratic programming* reduces to solving generalized eigenproblems (see Section 4.3.3).
- The eigenproblem has important applications in numerical methods, such as determining if an iterative method will converge toward the exact solution (see Section 6.4.2).
- Extraction of dominant features and reduction of large data sets using proper-orthogonal decomposition reduces to solving a large eigenproblem (see Chapter 13).

In many applications, the eigenvectors can be thought of as providing a vector basis for representing the system that is better in some sense than the original basis. For example, the principal axes provide a basis with respect to which only principal (normal) stresses and principal moments of inertia act, transforming a quadratic to its canonical form via diagonalization using the eigenvectors corresponds to a change of basis to one in which the axes of the quadratic are aligned with the basis vectors (see Section 2.3.3), and we will solve systems of ordinary differential equations by transforming them to a basis through diagonalization that uncouples the modes and eases solution (see Section 2.6). In addition, several advanced techniques to be discussed in Part III are based on the idea of determining an "optimal" basis for a particular problem or set of data.

[4] In the context of structural or aerodynamic stability, for example.

2.2 Eigenvalues and Eigenvectors

For a given $N \times N$ matrix $\mathbf{A}$, the general form of the algebraic eigenproblem is

$$\mathbf{A}\mathbf{u}_n = \lambda_n \mathbf{u}_n, \tag{2.1}$$

where $\mathbf{A}$ is the known matrix, λ_n are the scalar *eigenvalues* representing the stretch factors, and $\mathbf{u}_n$ are the *eigenvectors* and are the vectors that are transformed into themselves. The eigenvalues and eigenvectors are sought for a given matrix $\mathbf{A}$.[5]

For the majority of values of λ_n with a given matrix $\mathbf{A}$, the system (2.1) only has the trivial solution $\mathbf{u}_n = \mathbf{0}$. We are interested in the values of λ_n that produce nontrivial solutions; these are the eigenvalues or characteristic values. Each eigenvalue $\lambda_n, n = 1, \ldots, N$ has an associated eigenvector or characteristic vector $\mathbf{u}_n$. Graphically, recall that in general, multiplying $\mathbf{A}\mathbf{u}$ may translate, rotate, and/or scale the vector $\mathbf{u}$. If $\mathbf{u}_n$ is an eigenvector of $\mathbf{A}$, however, we may interpret the linear transformation $\mathbf{A}\mathbf{u}_n$ as simply scaling the vector $\mathbf{u}_n$ by a factor λ_n as described earlier and illustrated in Figure 2.3.

Note that the eigenvalues and eigenvectors are of the matrix $\mathbf{A}$; however, the eigenproblem (2.1) is a system of algebraic equations.[6] In order to determine the eigenvalues and eigenvectors, (2.1) may be written in the form

$$\mathbf{A}\mathbf{u}_n = \lambda_n \mathbf{I}\mathbf{u}_n,$$

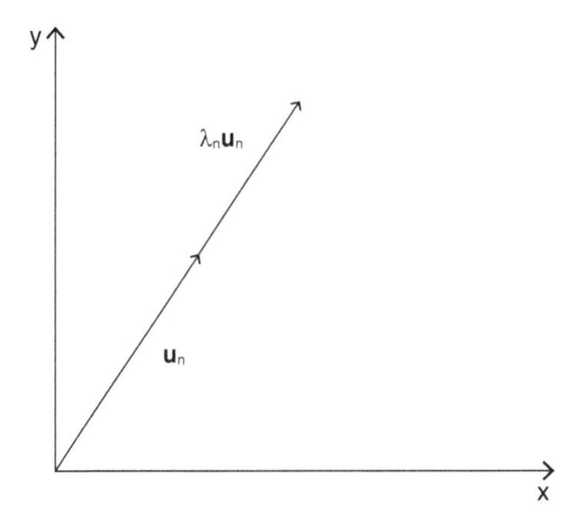

Figure 2.3 Graphical interpretation of eigenvalues λ_n and eigenvectors $\mathbf{u}_n$.

[5] The notation $\mathbf{u}_n$ is used for the eigenvectors to clearly distinguish them from the dependent variable $\mathbf{u}$, to emphasis that there are $n = 1, \ldots, N$ eigenvalues and eigenvectors, and to indicate that they come in pairs $(\lambda_n, \mathbf{u}_n), n = 1, \ldots, N$.

[6] It is important to distinguish between operations on matrices versus operations on entire systems of equations. For example, the determinant and inverse are operations on a matrix, while Gaussian elimination is an operation on a system of equations. More subtly, eigenvalues and eigenvectors are of a matrix, but the eigenproblem itself is a system of algebraic equations.

with the identity matrix inserted so that the right-hand side is of the same "shape" as the left-hand side. Rearranging gives

$$(\mathbf{A} - \lambda_n \mathbf{I})\,\mathbf{u}_n = \mathbf{0}, \tag{2.2}$$

where the right-hand side is the zero vector. Because this system of equations is homogeneous, a nontrivial solution will only exist if

$$|\mathbf{A} - \lambda_n \mathbf{I}| = 0,$$

such that $\mathbf{A} - \lambda_n \mathbf{I}$ is singular. If $|\mathbf{A} - \lambda_n \mathbf{I}| \neq 0$, then $\mathbf{A} - \lambda_n \mathbf{I}$ is nonsingular and $\mathbf{u}_n = \mathbf{0}$, which is the trivial solution, is the only solution. Writing out the determinant yields

$$\begin{vmatrix} A_{11} - \lambda_n & A_{12} & \cdots & A_{1N} \\ A_{21} & A_{22} - \lambda_n & \cdots & A_{2N} \\ \vdots & \vdots & \ddots & \vdots \\ A_{N1} & A_{N2} & \cdots & A_{NN} - \lambda_n \end{vmatrix} = 0.$$

Setting this determinant equal to zero results in a polynomial equation of degree N for λ_n, which is called the *characteristic equation*. The N solutions to the characteristic equation are the eigenvalues $\lambda_1, \lambda_2, \ldots, \lambda_N$, which may be real or complex.

For each $\lambda_n, n = 1, \ldots, N$, there is a nontrivial solution $\mathbf{u}_n, n = 1, \ldots, N$; these are the eigenvectors. For example, if $N = 3$, the characteristic equation is a cubic polynomial of the form

$$c_1 \lambda_n^3 + c_2 \lambda_n^2 + c_3 \lambda_n + c_4 = 0;$$

therefore, there are three eigenvalues λ_1, λ_2, and λ_3 and three corresponding eigenvectors $\mathbf{u}_1$, $\mathbf{u}_2$, and $\mathbf{u}_3$. Let us illustrate how the eigenvalues and eigenvectors are determined by returning to the problem of determining the principal stresses of a stress tensor.

Example 2.1 Obtain the principal axes and principal stresses for the two-dimensional stress distribution given by the stress tensor

$$\boldsymbol{\tau} = \begin{bmatrix} \tau_{xx} & \tau_{xy} \\ \tau_{yx} & \tau_{yy} \end{bmatrix} = \begin{bmatrix} 3 & -1 \\ -1 & 3 \end{bmatrix}.$$

Solution

The normal stresses are $\tau_{xx} = 3$ and $\tau_{yy} = 3$, and $\tau_{xy} = -1$ and $\tau_{yx} = -1$ are the shear stresses acting on the body. The principal axes of the stress tensor $\boldsymbol{\tau}$ correspond to the orientations of the axes for which only normal stresses act on the body, that is, there are no shear stresses. The principal axes are the eigenvectors of the stress tensor $\boldsymbol{\tau}$. The eigenvalues of the stress tensor are the magnitudes of the principal (normal) stresses. To determine the eigenvalues and eigenvectors of $\boldsymbol{\tau}$, we write $(\boldsymbol{\tau} - \tau_n \mathbf{I})\mathbf{n}_n = \mathbf{0}$, which is

$$\begin{bmatrix} 3-\tau_n & -1 \\ -1 & 3-\tau_n \end{bmatrix} \mathbf{n}_n = \mathbf{0}. \tag{2.3}$$

For a nontrivial solution to exist

$$\begin{vmatrix} 3-\tau_n & -1 \\ -1 & 3-\tau_n \end{vmatrix} = 0$$

$$(3-\tau_n)(3-\tau_n) - (-1)(-1) = 0$$

$$\tau_n^2 - 6\tau_n + 8 = 0$$

$$(\tau_n - 2)(\tau_n - 4) = 0$$

$$\therefore \tau_1 = 2, \quad \tau_2 = 4.$$

Thus, the minimum normal stress for any orientation is $\tau_1 = 2$, and the maximum normal stress is $\tau_2 = 4$. To determine the eigenvector $\mathbf{n}_1$ corresponding to $\tau_1 = 2$, substitute $\tau_n = 2$ into (2.3), which produces

$$\begin{bmatrix} 3-2 & -1 \\ -1 & 3-2 \end{bmatrix} \mathbf{n}_n = \mathbf{0},$$

or

$$n_1 - n_2 = 0,$$
$$-n_1 + n_2 = 0.$$

Observe that these equations are the same; therefore, we have one equation for two unknowns. Let $n_1 = c_1$, in which case $n_2 = c_1$ from the preceding equation, and the eigenvector is

$$\mathbf{n}_1 = c_1 \begin{bmatrix} 1 \\ 1 \end{bmatrix},$$

where c_1 is arbitrary. This is the eigenvector corresponding to $\tau_1 = 2$ determined to a scalar multiple; hence, any scalar multiple of an eigenvector will satisfy (2.1).

Obtain the eigenvector $\mathbf{n}_2$ corresponding to $\tau_2 = 4$:

$$\begin{bmatrix} 3-4 & -1 \\ -1 & 3-4 \end{bmatrix} \mathbf{n}_n = \mathbf{0},$$

or

$$-n_1 - n_2 = 0,$$
$$-n_1 - n_2 = 0,$$

which again is one equation for two unknowns; therefore, we have one arbitrary constant. Let $n_2 = c_2$, in which case $n_1 = -c_2$ from the preceding equation, and the eigenvector is

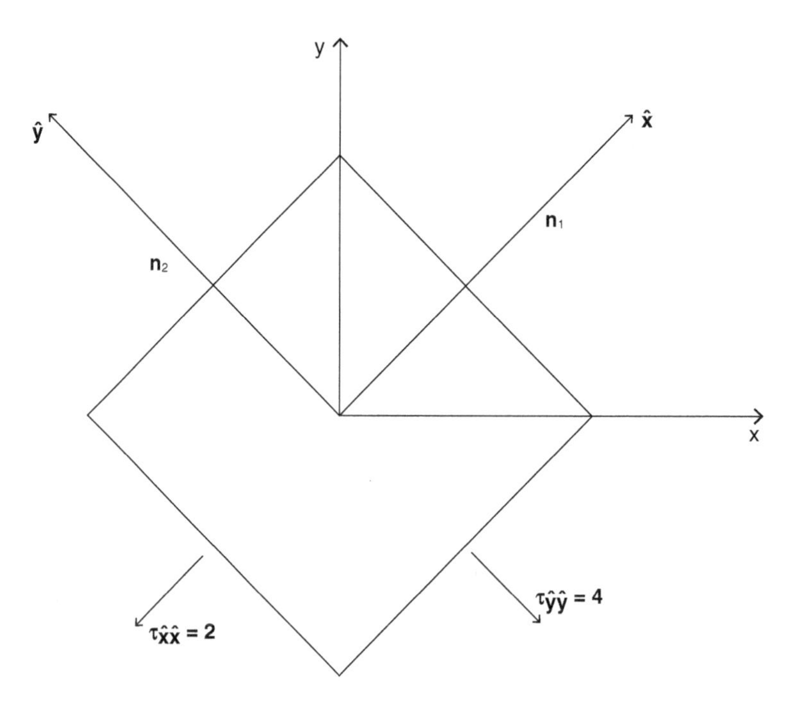

Figure 2.4 Principal stresses and principal axes for Example 2.1.

$$\mathbf{n}_2 = c_2 \begin{bmatrix} -1 \\ 1 \end{bmatrix},$$

which is the eigenvector corresponding to $\tau_2 = 4$ determined to a scalar multiple.

Therefore, $\mathbf{n}_1$ and $\mathbf{n}_2$ are the principal axes, and τ_1 and τ_2 are the principal (normal) stresses in the $\mathbf{n}_1$ and $\mathbf{n}_2$ directions, respectively, with no shear stresses as shown in Figure 2.4.

REMARKS:

1. *The physical fact that such principal stresses and axes exist for any stress field is a consequence of the mathematical properties of real symmetric matrices. Specifically, they are always diagonalizable, such that in this case*

$$\hat{\tau} = \begin{bmatrix} \tau_{\hat{x}\hat{x}} & 0 \\ 0 & \tau_{\hat{y}\hat{y}} \end{bmatrix} = \begin{bmatrix} 2 & 0 \\ 0 & 4 \end{bmatrix}.$$

   See Sections 2.3.3 and 2.5 for more on diagonalization.
2. *Observe that the principal axes (eigenvectors) are orthogonal, in which case* $\langle \mathbf{n}_1, \mathbf{n}_2 \rangle = 0$. *As shown in Section 2.3.1, this is always true for real symmetric matrices with distinct eigenvalues.*

REMARKS:

1. *Mathematically, the eigenvalues and eigenvectors can be real or complex. Depending on the application, however, they may be necessarily one or the other. For example, when determining the principal stresses as in the previous example, the eigenvalues are always real.*
2. *As in the previous example, we always lose at least one equation when determining eigenvectors. That is, for $N \times N$ matrix $\mathbf{A}$, the rank of $\mathbf{A} - \lambda_n \mathbf{I}$ is always less than N, such that $\mathbf{A} - \lambda_n \mathbf{I}$ is singular in order to obtain a nontrivial solution of the homogeneous system* (2.2). *Therefore, there is always at least one arbitrary constant reflecting the fact that there are an infinity of solutions.*
3. *In the preceding example, the eigenvalues are distinct, with $\lambda_1 \neq \lambda_2$. In some cases, however, eigenvalues may be repeated as in the next example.*

Example 2.2 Determine the eigenvalues and eigenvectors of the 4×4 matrix

$$\mathbf{A} = \begin{bmatrix} 0 & 0 & 1 & 1 \\ -1 & 2 & 0 & 1 \\ -1 & 0 & 2 & 1 \\ 1 & 0 & -1 & 0 \end{bmatrix}.$$

Solution

In order to determine the eigenvalues, we take the determinant

$$|\mathbf{A} - \lambda_n \mathbf{I}| = 0,$$

or

$$\begin{vmatrix} -\lambda_n & 0 & 1 & 1 \\ -1 & 2-\lambda_n & 0 & 1 \\ -1 & 0 & 2-\lambda_n & 1 \\ 1 & 0 & -1 & -\lambda_n \end{vmatrix} = 0.$$

To evaluate the determinant of a 4×4 matrix requires forming a cofactor expansion (see Section 1.4.2). Observing that the second column of our determinant has all zeros except for one element, we take the cofactor expansion about this column, which is

$$+(2-\lambda_n) \begin{vmatrix} -\lambda_n & 1 & 1 \\ -1 & 2-\lambda_n & 1 \\ 1 & -1 & -\lambda_n \end{vmatrix} = 0.$$

Evaluating the determinant of the 3×3 minor matrix leads to

$$(2-\lambda_n)\left[\lambda_n^2(2-\lambda_n) + 1 + 1 - (2-\lambda_n) - \lambda_n - \lambda_n\right] = 0$$

$$(2-\lambda_n)\left[\lambda_n^2(2-\lambda_n) - \lambda_n\right] = 0$$

$$\lambda_n(\lambda_n - 2)\left(\lambda_n^2 - 2\lambda_n + 1\right) = 0$$

$$\lambda_n(\lambda_n - 2)(\lambda_n - 1)^2 = 0.$$

As here, when taking determinants of matrices that are larger than 2×2, look for opportunities to factor out $(\lambda_n - \lambda_i)$ factors rather than forming the Nth-degree polynomial in the standard form, which is typically difficult to factor. The four eigenvalues are

$$\lambda_1 = 0, \quad \lambda_2 = 1, \quad \lambda_3 = 1, \quad \lambda_4 = 2.$$

Thus, there are two distinct eigenvalues, and two of the eigenvalues are repeated. We say that the eigenvalue $\lambda_2 = \lambda_3 = 1$ has *multiplicity* two.

Let us first consider the eigenvectors corresponding to the two distinct eigenvalues. For $\lambda_1 = 0$, we have the system

$$\begin{bmatrix} 0 & 0 & 1 & 1 \\ -1 & 2 & 0 & 1 \\ -1 & 0 & 2 & 1 \\ 1 & 0 & -1 & 0 \end{bmatrix} \mathbf{u}_1 = \mathbf{0}.$$

Although it is not obvious by observation that we lose one equation, note that the sum of the third and fourth equations equals the first equation.[7] Therefore, the system reduces to the final three equations for the four unknowns according to

$$\begin{aligned} -u_1 + 2u_2 + u_4 &= 0, \\ -u_1 + 2u_3 + u_4 &= 0, \\ u_1 - u_3 &= 0. \end{aligned}$$

Because we have only three equations for four unknowns, let $u_1 = c_1$. Then back substituting into the preceding equations leads to $u_3 = c_1$, $u_4 = -c_1$, and $u_2 = c_1$, respectively. Therefore, the eigenvector corresponding to the eigenvalue $\lambda_1 = 0$ is

$$\mathbf{u}_1 = c_1 \begin{bmatrix} 1 \\ 1 \\ 1 \\ -1 \end{bmatrix}.$$

Similarly, for the eigenvalue $\lambda_4 = 2$, we have the system

$$\begin{bmatrix} -2 & 0 & 1 & 1 \\ -1 & 0 & 0 & 1 \\ -1 & 0 & 0 & 1 \\ 1 & 0 & -1 & -2 \end{bmatrix} \mathbf{u}_4 = \mathbf{0}.$$

The second and third equations are the same; therefore, the remaining equations are

$$\begin{aligned} -2u_1 + u_3 + u_4 &= 0, \\ -u_1 + u_4 &= 0, \\ u_1 - u_3 - 2u_4 &= 0. \end{aligned}$$

[7] If you do not see this, use Gaussian elimination.

In order to solve this system of three equations for four unknowns, we can use Gaussian elimination to reduce the system to

$$\begin{aligned} u_1 - u_3 - 2u_4 &= 0, \\ u_3 + 3u_4 &= 0, \\ 2u_4 &= 0. \end{aligned}$$

Backward substitution leads to the eigenvector corresponding to the eigenvalue $\lambda_4 = 2$ being

$$\mathbf{u}_4 = c_4 \begin{bmatrix} 0 \\ 1 \\ 0 \\ 0 \end{bmatrix},$$

where we note that x_2 is arbitrary as it does not appear in any of the equations.

For the repeated eigenvalue $\lambda_2 = \lambda_3 = 1$, we attempt to find two linearly independent eigenvectors. Setting $\lambda_n = 1$ in $(\mathbf{A} - \lambda_n \mathbf{I})\mathbf{u}_n = \mathbf{0}$ gives

$$\begin{bmatrix} -1 & 0 & 1 & 1 \\ -1 & 1 & 0 & 1 \\ -1 & 0 & 1 & 1 \\ 1 & 0 & -1 & -1 \end{bmatrix} \mathbf{u}_{2,3} = \mathbf{0}.$$

The first, third, and fourth rows (equations) are the same, so u_1, u_2, u_3, and u_4 are determined from the two equations

$$\begin{aligned} -u_1 + u_3 + u_4 &= 0, \\ -u_1 + u_2 + u_4 &= 0. \end{aligned}$$

We have two equations for four unknowns requiring two arbitrary constants, so let $u_1 = c_2$ and $u_2 = c_3$; thus,

$$\begin{aligned} u_1 &= c_2, \\ u_2 &= c_3, \\ u_4 &= c_2 - c_3, \\ u_3 &= c_2 - u_4 = c_2 - (c_2 - c_3) = c_3. \end{aligned}$$

Therefore, the two eigenvectors must satisfy

$$\mathbf{u}_{2,3} = \begin{bmatrix} c_2 \\ c_3 \\ c_3 \\ c_2 - c_3 \end{bmatrix}.$$

The parameters c_2 and c_3 are arbitrary, so we may choose any two unique pairs of values that result in linearly independent eigenvectors. Choosing, for example, $c_2 = 1$ and $c_3 = 1$ for $\mathbf{u}_2$ and $c_2 = 1$ and $c_3 = 0$ for $\mathbf{u}_3$ gives the additional two eigenvectors (along with $\mathbf{u}_1$ and $\mathbf{u}_4$)

$$\mathbf{u}_2 = \begin{bmatrix} 1 \\ 1 \\ 1 \\ 0 \end{bmatrix}, \quad \mathbf{u}_3 = \begin{bmatrix} 1 \\ 0 \\ 0 \\ 1 \end{bmatrix}.$$

Note that the arbitrary constant is always implied, even if it is not shown explicitly. In this case, the four eigenvectors are linearly independent; therefore, they provide a basis for the four-dimensional vector space associated with matrix **A**. We will see in later sections how this is useful.

In the case of triangular matrices, the eigenvalues are found along the main diagonal. For the diagonal case,[8] the eigenvectors and inverse are also especially straightforward to obtain (nearly) by inspection. The eigenvalues of an $N \times N$ diagonal matrix are on the main diagonal as follows:

$$\mathbf{A} = \begin{bmatrix} \lambda_1 & 0 & \cdots & 0 \\ 0 & \lambda_2 & \cdots & 0 \\ \vdots & \vdots & \ddots & \vdots \\ 0 & 0 & \cdots & \lambda_N \end{bmatrix},$$

and the corresponding eigenvectors are

$$\mathbf{u}_1 = \begin{bmatrix} 1 \\ 0 \\ \vdots \\ 0 \end{bmatrix}, \quad \mathbf{u}_2 = \begin{bmatrix} 0 \\ 1 \\ \vdots \\ 0 \end{bmatrix}, \quad \ldots, \quad \mathbf{u}_N = \begin{bmatrix} 0 \\ 0 \\ \vdots \\ 1 \end{bmatrix},$$

which are mutually orthogonal. If $|\mathbf{A}| \neq 0$, such that **A** is not singular, then the inverse of **A** is

$$\mathbf{A}^{-1} = \begin{bmatrix} \frac{1}{\lambda_1} & 0 & \cdots & 0 \\ 0 & \frac{1}{\lambda_2} & \cdots & 0 \\ \vdots & \vdots & \ddots & \vdots \\ 0 & 0 & \cdots & \frac{1}{\lambda_N} \end{bmatrix}.$$

Finally, let us determine the eigenvalues and eigenvectors of the integer power of a matrix in terms of the eigenvalues and eigenvectors of the matrix itself. Consider the eigenproblem $\mathbf{A}\mathbf{u}_n = \lambda_n \mathbf{u}_n$. Premultiplying by **A** gives

$$\mathbf{A}\left(\mathbf{A}\mathbf{u}_n\right) = \lambda_n \mathbf{A}\mathbf{u}_n,$$

or from the original eigenproblem

$$\mathbf{A}^2 \mathbf{u}_n = \lambda_n^2 \mathbf{u}_n.$$

[8] Of course, a diagonal matrix is just a special case of a triangular matrix.

Doing so again yields

$$\mathbf{A}(\mathbf{A}^2\mathbf{u}_n) = \lambda_n^2\mathbf{A}\mathbf{u}_n,$$

or from the original eigenproblem

$$\mathbf{A}^3\mathbf{u}_n = \lambda_n^3\mathbf{u}_n.$$

Generalizing, we obtain the important result that

$$\mathbf{A}^k\mathbf{u}_n = \lambda_n^k\mathbf{u}_n, \quad k = 1, 2, 3, \ldots.$$

That is, if λ_n are the eigenvalues of $\mathbf{A}$, then the eigenvalues of $\mathbf{A}^k$ are λ_n^k, and the eigenvectors of $\mathbf{A}$ and $\mathbf{A}^k$ are the same. We will take advantage of this result several times throughout the text.

REMARKS:

1. *The eigenvectors obtained using the procedure in the previous two examples will be referred to as* regular eigenvectors.
2. *The set of all eigenvalues of* $\mathbf{A}$, $\lambda_1, \lambda_2, \ldots, \lambda_N$, *is the* spectrum *of* $\mathbf{A}$.
3. *The* spectral radius *of* $\mathbf{A}$ *is* $\rho = \max(|\lambda_1|, |\lambda_2|, \ldots, |\lambda_N|)$.
4. *Because the determinant of a matrix and its transpose are equal,*

$$|\mathbf{A} - \lambda\mathbf{I}| = \left|(\mathbf{A} - \lambda\mathbf{I})^T\right| = \left|\mathbf{A}^T - \lambda\mathbf{I}\right|;$$

   therefore, the characteristic polynomials and eigenvalues for $\mathbf{A}$ *and* $\mathbf{A}^T$ *are the same.*
5. *The* N *eigenvalues of an* $N \times N$ *matrix* $\mathbf{A}$ *are such that*

$$\lambda_1 + \lambda_2 + \cdots + \lambda_N = A_{11} + A_{22} + \cdots + A_{NN} = \text{tr}(\mathbf{A}),$$

   where tr($\mathbf{A}$) *is the* trace *of* $\mathbf{A}$, *which is the sum of the elements on the main diagonal.*
6. *It can be shown that* $|\mathbf{A}| = \lambda_1\lambda_2\cdots\lambda_N$. *Therefore,* $\mathbf{A}$ *is a singular matrix* if and only if *at least one of the eigenvalues is zero.*
7. *The eigenvectors corresponding to distinct eigenvalues are always linearly independent.*
8. Cayley–Hamilton theorem: *If* $P_N(\lambda) = 0$ *is the* N*th-degree characteristic polynomial of an* $N \times N$ *matrix* $\mathbf{A}$, *then* $\mathbf{A}$ *satisfies*

$$P_N(\mathbf{A}) = \mathbf{0}.$$

   That is, $\mathbf{A}$ *satisfies its own characteristic polynomial. See Section 2.7 for a proof of the Cayley–Hamilton theorem.*
9. *In some cases with nonsymmetric matrices having repeated eigenvalues, there are fewer regular eigenvectors than eigenvalues. For example, an eigenvalue with multiplicity two may only have one corresponding* regular eigenvector. *When an* $N \times N$ *matrix has fewer than* N *linearly independent regular eigenvectors, we*

say the matrix is defective. *A procedure for obtaining* generalized eigenvectors *in such cases will be provided in Section 2.5.2.*

10. *Some applications, such as quadratic programming treated in Section 4.3.3, result in the so-called* generalized eigenproblem

$$\mathbf{A}\mathbf{u}_n = \lambda_n \mathbf{B}\mathbf{u}_n,$$

such that the regular eigenproblem (2.1) *corresponds to* $\mathbf{B} = \mathbf{I}$. *Both types of eigenproblems can be treated using very similar techniques. Do not confuse the generalized eigenproblem with generalized eigenvectors; they are not related.*

11. *Computer algorithms for calculating the eigenvalues and eigenvectors of large matrices numerically are typically based on QR decomposition, which is introduced in Section 2.10. Such algorithms, and their variants, are described in Section 6.5.*

2.3 Real Symmetric Matrices

In many applications, the square matrix $\mathbf{A}$ is *real* and *symmetric*, in which case $\mathbf{A} = \mathbf{A}^T$. This is the case in mechanics, for example, for the stress, strain, and moment of inertia tensors (matrices). The equality of mixed partial derivatives, for example, $\partial^2 u/\partial x \partial y = \partial^2 u/\partial y \partial x$ when $u(x, y)$ is smooth, in vector calculus leads to a symmetric Hessian matrix (see Chapter 4). Many numerical methods produce symmetric matrices, as we will see in Part II. In addition, the covariance matrix, which will feature prominently in proper-orthogonal decomposition in Chapter 13, is symmetric. Let us consider some properties and applications of this special case, and we will introduce some important operations, such as diagonalization, along the way that have broader applicability.

2.3.1 Properties of Eigenvectors

Consider two *distinct* eigenvalues λ_1 and λ_2 of $\mathbf{A}$, such that $\lambda_1 \neq \lambda_2$, and their associated eigenvectors $\mathbf{u}_1$ and $\mathbf{u}_2$, in which case

$$\mathbf{A}\mathbf{u}_1 = \lambda_1 \mathbf{u}_1, \quad \mathbf{A}\mathbf{u}_2 = \lambda_2 \mathbf{u}_2.$$

Taking the transpose of the first eigenrelation

$$(\mathbf{A}\mathbf{u}_1)^T = \lambda_1 \mathbf{u}_1^T,$$

and recalling that $(\mathbf{A}\mathbf{u}_1)^T = \mathbf{u}_1^T \mathbf{A}^T$ and postmultiplying by $\mathbf{u}_2$ gives

$$\mathbf{u}_1^T \mathbf{A}^T \mathbf{u}_2 = \lambda_1 \mathbf{u}_1^T \mathbf{u}_2. \tag{2.4}$$

Premultiplying the second eigenrelation by $\mathbf{u}_1^T$ yields

$$\mathbf{u}_1^T \mathbf{A}\mathbf{u}_2 = \lambda_2 \mathbf{u}_1^T \mathbf{u}_2. \tag{2.5}$$

Subtracting (2.4) from (2.5) with $\mathbf{A} = \mathbf{A}^T$ gives

$$(\lambda_2 - \lambda_1)\mathbf{u}_1^T\mathbf{u}_2 = 0,$$

or

$$(\lambda_2 - \lambda_1)\langle \mathbf{u}_1, \mathbf{u}_2 \rangle = 0.$$

Because the eigenvalues are distinct, such that $\lambda_2 \neq \lambda_1$, this result requires that

$$\langle \mathbf{u}_1, \mathbf{u}_2 \rangle = 0.$$

Therefore, the two eigenvectors corresponding to two distinct eigenvalues of a real symmetric matrix are orthogonal.

This result explains why the principal axes of a stress tensor, which is symmetric, are mutually orthogonal as in Example 2.1. It also can be proven that all of the eigenvalues of a real symmetric matrix are real. Recall that a Hermitian matrix is such that $\mathbf{A} = \overline{\mathbf{A}}^T$. As with real symmetric matrices, the eigenvalues of a Hermitian matrix are real, and the eigenvectors are mutually orthogonal.

For a symmetric matrix with repeated eigenvalues, the eigenvectors are at least linearly independent, but can be *made* to be mutually orthogonal.[9] This will be illustrated in the context of Example 2.8 in order to clarify this point.

These results have implications for vector bases. If all of the eigenvalues of a real symmetric matrix are distinct, then the eigenvectors are all mutually orthogonal. Therefore, a full Gram–Schmidt orthogonalization is not necessary. In this case, we could simply normalize the eigenvectors according to

$$\mathbf{q}_n = \frac{\mathbf{u}_n}{\|\mathbf{u}_n\|}, \quad n = 1, 2, \ldots, N$$

in order to obtain an orthonormal basis for N-space.[10] If instead an eigenvalue is repeated $S \leq N$ times, there are S corresponding eigenvectors that are linearly independent, but not necessarily orthogonal; the remaining $N - S$ eigenvectors corresponding to distinct eigenvalues are mutually orthogonal. Because the eigenvectors are all linearly independent (even if not orthogonal), they still form a basis for an N-dimensional vector space.

The following sections consider two applications involving real symmetric matrices. For an additional application to convergence of iterative numerical algorithms, see Section 6.4.2.

2.3.2 Linear Systems of Equations

In Chapter 1, we considered three methods for solving linear systems of algebraic equations $\mathbf{Au} = \mathbf{b}$, namely (1) Gaussian elimination, (2) using the inverse to evaluate

[9] Unfortunately, there is some confusion concerning this case. It is often stated that the eigenvectors for all symmetric matrices are mutually orthogonal. While the constants can be chosen such that this is the case, it is not inherently true for all symmetric matrices – only those with distinct eigenvalues.

[10] We will use $\mathbf{u}_n$ to indicate general eigenvectors and $\mathbf{q}_n$ for orthonormal eigenvectors throughout.

$\mathbf{u} = \mathbf{A}^{-1}\mathbf{b}$, and (3) Cramer's rule. Here, we obtain a particularly straightforward and efficient method for solving such systems when $\mathbf{A}$ is real and symmetric. Using the preceding properties, this can be done in the form of a linear combination of the eigenvectors of $\mathbf{A}$.

Consider the system

$$\mathbf{A}\mathbf{u} = \mathbf{b},$$

where $\mathbf{A}$ is a real and symmetric $N \times N$ matrix that is nonsingular, $\mathbf{b}$ is known, and $\mathbf{u}$ is the solution vector to be determined. Recall that if the eigenvalues of $\mathbf{A}$ are distinct, the eigenvectors are mutually orthogonal. If the eigenvalues are repeated, the eigenvectors can be chosen to be orthogonal.

Because the right-hand-side vector $\mathbf{b}$ is in N-space, it may be written in terms of a linear combination of the orthonormal eigenvectors, $\mathbf{q}_1, \dots, \mathbf{q}_N$, of $\mathbf{A}$ as follows:

$$\mathbf{b} = \hat{a}_1\mathbf{q}_1 + \hat{a}_2\mathbf{q}_2 + \cdots + \hat{a}_N\mathbf{q}_N. \tag{2.6}$$

That is, the vectors $\mathbf{q}_1, \dots, \mathbf{q}_N$ form a basis for N-space. Taking the inner product of $\mathbf{q}_1$ with $\mathbf{b}$, which is equivalent to premultiplying (2.6) by $\mathbf{q}_1^T$, gives

$$\langle \mathbf{q}_1, \mathbf{b} \rangle = \mathbf{q}_1^T\mathbf{b} = \hat{a}_1\mathbf{q}_1^T\mathbf{q}_1 + \hat{a}_2\mathbf{q}_1^T\mathbf{q}_2 + \cdots + \hat{a}_N\mathbf{q}_1^T\mathbf{q}_N,$$

but the eigenvectors are mutually orthogonal (and normalized); therefore, $\langle \mathbf{q}_1, \mathbf{q}_i \rangle = 0$ for $n = 2, \dots, N$, leaving

$$\hat{a}_1 = \langle \mathbf{b}, \mathbf{q}_1 \rangle .$$

Generalizing, we have

$$\hat{a}_n = \langle \mathbf{b}, \mathbf{q}_n \rangle , \quad n = 1, 2, \dots, N, \tag{2.7}$$

which can be evaluated to give the constants in the linear combination (2.6).

The solution vector $\mathbf{u}$ is also in N-space; consequently, it can be expressed in the form

$$\mathbf{u} = a_1\mathbf{q}_1 + a_2\mathbf{q}_2 + \cdots + a_N\mathbf{q}_N, \tag{2.8}$$

where we seek the coefficients $a_n, n = 1, 2, \dots, N$.

Substituting (2.8) and (2.6) into $\mathbf{A}\mathbf{u} = \mathbf{b}$ leads to

$$\mathbf{A}\left(a_1\mathbf{q}_1 + a_2\mathbf{q}_2 + \cdots + a_N\mathbf{q}_N\right) = \hat{a}_1\mathbf{q}_1 + \hat{a}_2\mathbf{q}_2 + \cdots + \hat{a}_N\mathbf{q}_N$$

$$a_1\mathbf{A}\mathbf{q}_1 + a_2\mathbf{A}\mathbf{q}_2 + \cdots + a_N\mathbf{A}\mathbf{q}_N = \hat{a}_1\mathbf{q}_1 + \hat{a}_2\mathbf{q}_2 + \cdots + \hat{a}_N\mathbf{q}_N$$

$$a_1\lambda_1\mathbf{q}_1 + a_2\lambda_2\mathbf{q}_2 + \cdots + a_N\lambda_N\mathbf{q}_N = \hat{a}_1\mathbf{q}_1 + \hat{a}_2\mathbf{q}_2 + \cdots + \hat{a}_N\mathbf{q}_N.$$

Note that we could seemingly have used any linearly independent basis vectors in the linear combinations (2.6) and (2.8), but using the orthonormal eigenvectors of $\mathbf{A}$ as basis vectors allows us to perform the last step in the preceding.

Because the $\mathbf{q}_i$ vectors are linearly independent, each of their coefficients must be equal according to

$$a_1 = \frac{\hat{a}_1}{\lambda_1}, \quad a_2 = \frac{\hat{a}_2}{\lambda_2}, \quad \ldots, \quad a_N = \frac{\hat{a}_N}{\lambda_N},$$

or from (2.7)

$$a_n = \frac{\langle \mathbf{b}, \mathbf{q}_n \rangle}{\lambda_n}, \quad n = 1, 2, \ldots, N.$$

Then from (2.8), the solution vector $\mathbf{u}$ for a system with a *real symmetric* coefficient matrix is

$$\mathbf{u} = \sum_{n=1}^{N} \frac{\langle \mathbf{b}, \mathbf{q}_n \rangle}{\lambda_n} \mathbf{q}_n. \tag{2.9}$$

In other words, we can conveniently form the solution of the system of linear equations $\mathbf{Au} = \mathbf{b}$ in the form of a linear combination of the orthonormal eigenvectors of the symmetric matrix $\mathbf{A}$; that is, the eigenvectors form a useful basis for expressing the solution of the system of equations. Observe that once we have obtained the eigenvalues and eigenvectors of $\mathbf{A}$, we can easily determine the solution for various right-hand-side vectors $\mathbf{b}$ from (2.9).

The notation used here will be used throughout Chapter 2. That is, $\mathbf{u}$ is the solution vector for the system of algebraic equations $\mathbf{Au} = \mathbf{b}$. The eigenvectors of $\mathbf{A}$ are denoted by $\mathbf{u}_n$, $n = 1, 2, \ldots, N$, and the orthonormal eigenvectors of $\mathbf{A}$ – if available – are denoted by $\mathbf{q}_n$, $n = 1, 2, \ldots, N$.

Let us summarize the previous procedure in the following four steps. This same procedure will be adapted to allow us to solve differential equations using eigenfunction expansions in Section 3.2.2. Given the system of algebraic equations

$$\mathbf{Au} = \mathbf{b}, \tag{2.10}$$

where $\mathbf{A}$ is an $N \times N$ real symmetric matrix, the solution can be obtained using the following procedure:

1. Extract the matrix $\mathbf{A}$ from the system of algebraic equations and form its eigenproblem

$$\mathbf{A}\mathbf{q}_n = \lambda_n \mathbf{q}_n. \tag{2.11}$$

   Determine the eigenvalues λ_n and orthonormal eigenvectors $\mathbf{q}_n$ of matrix $\mathbf{A}$.
2. Expand the known right-hand-side vector $\mathbf{b}$ as a linear combination of the eigenvectors as follows:

$$\mathbf{b} = \sum_{n=1}^{N} \hat{a}_n \mathbf{q}_n. \tag{2.12}$$

Determine the $\hat{a}_n$ coefficients by taking the inner product of each of the eigenvectors $\mathbf{q}_n$ with (2.12) leading to

$$\hat{a}_n = \langle \mathbf{b}, \mathbf{q}_n \rangle, \quad n = 1, 2, \ldots, N. \tag{2.13}$$

3. Expand the unknown solution vector $\mathbf{u}$ as a linear combination of the eigenvectors according to

$$\mathbf{u} = \sum_{n=1}^{N} a_n \mathbf{q}_n. \tag{2.14}$$

To determine the a_n coefficients, consider the original system of algebraic equations

$$\begin{aligned} \mathbf{A}\mathbf{u} &= \mathbf{b}, \\ \mathbf{A}\left(\sum_{n=1}^{N} a_n \mathbf{q}_n\right) &= \sum_{n=1}^{N} \hat{a}_n \mathbf{q}_n, \\ \sum_{n=1}^{N} a_n \mathbf{A}\mathbf{q}_n &= \sum_{n=1}^{N} \hat{a}_n \mathbf{q}_n, \\ \sum_{n=1}^{N} a_n \lambda_n \mathbf{q}_n &= \sum_{n=1}^{N} \hat{a}_n \mathbf{q}_n, \\ \therefore a_n = \frac{\hat{a}_n}{\lambda_n} &= \frac{\langle \mathbf{b}, \mathbf{q}_n \rangle}{\lambda_n}, \quad n = 1, 2, \ldots, N. \end{aligned}$$

4. Form the solution from the eigenvalues and eigenvectors using

$$\mathbf{u} = \sum_{n=1}^{N} \frac{\langle \mathbf{b}, \mathbf{q}_n \rangle}{\lambda_n} \mathbf{q}_n. \tag{2.15}$$

Example 2.3 Let us once again consider the system of linear algebraic equations from the examples in Sections 1.5.1 through 1.5.3, which is

$$\begin{aligned} 2u_1 + u_3 &= 3, \\ u_2 + 2u_3 &= 3, \\ u_1 + 2u_2 &= 3. \end{aligned}$$

Solution

In matrix form $\mathbf{A}\mathbf{u} = \mathbf{b}$, the system is

$$\begin{bmatrix} 2 & 0 & 1 \\ 0 & 1 & 2 \\ 1 & 2 & 0 \end{bmatrix} \begin{bmatrix} u_1 \\ u_2 \\ u_3 \end{bmatrix} = \begin{bmatrix} 3 \\ 3 \\ 3 \end{bmatrix},$$

where we observe that $\mathbf{A}$ is real and symmetric.

In order to apply the result (2.9), it is necessary to obtain the eigenvalues and eigenvectors of the matrix $\mathbf{A}$ (step 1). The eigenvalues are the values of λ_n that satisfy $|\mathbf{A} - \lambda_n \mathbf{I}| = 0$, that is,

$$\begin{vmatrix} 2-\lambda_n & 0 & 1 \\ 0 & 1-\lambda_n & 2 \\ 1 & 2 & -\lambda_n \end{vmatrix} = 0.$$

Evaluating the determinant yields

$$(2-\lambda_n)(1-\lambda_n)(-\lambda_n) - (1-\lambda_n) - 4(2-\lambda_n) = 0,$$

where we note that there are no common $(\lambda_n - \lambda_i)$ factors. Therefore, we simplify to obtain the characteristic polynomial

$$\lambda_n^3 - 3\lambda_n^2 - 3\lambda_n + 9 = 0.$$

Although there is an exact expression for determining the roots of a cubic polynomial (analogous to the quadratic formula), it is not widely known and not worth memorizing. Instead, we can make use of mathematical software, such as MATLAB or Mathematica, to do the heavy lifting for us. Doing so results in the eigenvalues

$$\lambda_1 = 3, \quad \lambda_2 = -\sqrt{3}, \quad \lambda_3 = \sqrt{3},$$

which are distinct. The corresponding orthonormal eigenvectors are

$$\mathbf{q}_1 = \frac{1}{\sqrt{3}}\begin{bmatrix} 1 \\ 1 \\ 1 \end{bmatrix}, \ \mathbf{q}_2 = \frac{1}{\sqrt{6(2-\sqrt{3})}}\begin{bmatrix} -2+\sqrt{3} \\ 1-\sqrt{3} \\ 1 \end{bmatrix}, \ \mathbf{q}_3 = \frac{1}{\sqrt{6(2+\sqrt{3})}}\begin{bmatrix} -2-\sqrt{3} \\ 1+\sqrt{3} \\ 1 \end{bmatrix}.$$

Surprisingly enough, evaluating

$$\mathbf{u} = \sum_{n=1}^{N} \frac{\langle \mathbf{q}_n, \mathbf{b} \rangle}{\lambda_n} \mathbf{q}_n = \frac{\langle \mathbf{q}_1, \mathbf{b} \rangle}{\lambda_1} \mathbf{q}_1 + \frac{\langle \mathbf{q}_2, \mathbf{b} \rangle}{\lambda_2} \mathbf{q}_2 + \frac{\langle \mathbf{q}_3, \mathbf{b} \rangle}{\lambda_3} \mathbf{q}_3$$

produces the correct solution

$$\begin{bmatrix} u_1 \\ u_2 \\ u_3 \end{bmatrix} = \begin{bmatrix} 1 \\ 1 \\ 1 \end{bmatrix}.$$

REMARKS:

1. *This example illustrates that determining eigenvalues and eigenvectors is not always neat and tidy even though* $\mathbf{A}$ *is only* 3×3 *and rather innocuous looking in this case (at least they were not complex!). Thankfully, the wide availability of tools such as MATLAB and Mathematica allow us to tackle such problems without difficulty.*
2. *Observe that the majority of our effort in using this approach to determining the solution of a system of algebraic equations entails obtaining the eigenvalues and eigenvectors. Consequently, it is particularly efficient if we must solve the eigenproblem for some other reason – and the matrix is real symmetric – and/or we seek the solutions for multiple right-hand-side vectors* $\mathbf{b}$.

2.3.3 Quadratic Forms – Diagonalization

The focus of matrix methods is on *linear* systems, that is, those that can be expressed as polynomials of the first degree. However, there are some limited scenarios in which these methods can be extended to *nonlinear* systems. In particular, this is the case for quadratic expressions involving polynomials of degree two. More importantly, quadratic forms are useful at this point as a means to introduce matrix diagonalization.

A second-degree expression – with no first-degree or constant terms – in $x_1, x_2, \ldots, x_N$ may be written

$$\begin{aligned} \mathcal{A} &= \sum_{m=1}^{N}\sum_{n=1}^{N} A_{mn}x_m x_n \\ &= A_{11}x_1^2 + A_{22}x_2^2 + \cdots + A_{NN}x_N^2 \\ &\quad +2A_{12}x_1x_2 + 2A_{13}x_1x_3 + \cdots + 2A_{N-1,N}x_{N-1}x_N, \end{aligned} \tag{2.16}$$

where A_{mn} and x_m are real. This is known as a *quadratic form*, and the matrix of coefficients are symmetric, such that $A_{mn} = A_{nm}$. Note that the "2" appears in the mixed product terms because of the symmetry of A_{mn}; for example, there are both x_1x_2 and x_2x_1 terms that have the same coefficient $A_{12} = A_{21}$. Observe that quadratics can be written in matrix form as

$$\mathcal{A} = \mathbf{x}^T\mathbf{A}\mathbf{x},$$

where $\mathbf{A} = [A_{mn}]$ is symmetric.

Some applications of quadratic forms include:

- In two dimensions (x_1, x_2): $\mathcal{A} = \mathbf{x}^T\mathbf{A}\mathbf{x}$ represents a *conic curve*, for example, ellipse, hyperbola, or parabola.
- In three dimensions (x_1, x_2, x_3): $\mathcal{A} = \mathbf{x}^T\mathbf{A}\mathbf{x}$ represents a *quadric surface*, for example, ellipsoid, hyperboloid, or paraboloid.
- In mechanics, the moments of inertia, angular momentum of a rotating body, and kinetic energy of a system of moving particles can be represented by quadratic forms. In addition, the potential (strain) energy of a linear spring is expressed as a quadratic.
- The least-squares method of curve fitting leads to quadratics (see Section 11.1).
- In electro-optics, the *index ellipsoid* of a crystal in the presence of an applied electric field is given by a quadratic form (see section 1.7 of Yariv and Yeh, 2007, for example).
- Objective and cost functions and functionals in optimization and control problems are often expressed in terms of quadratic forms as discussed in parts III of Cassel (2013) and the present text.

Reduction to Matrix Form

What do quadratics, which are nonlinear, have to do with linear systems of equations? We can convert the single quadratic (2.16) in N variables into a system of N linear equations by differentiating the quadratic with respect to each variable as follows:

$$f_n(\mathbf{x}) = \frac{1}{2}\frac{\partial}{\partial x_n}(\mathbf{x}^T\mathbf{A}\mathbf{x}), \quad n = 1, 2, \ldots, N.$$

This leads to the system of linear equations

$$\begin{aligned} A_{11}x_1 + A_{12}x_2 + \cdots + A_{1N}x_N &= f_1, \\ A_{12}x_1 + A_{22}x_2 + \cdots + A_{2N}x_N &= f_2, \\ &\vdots \\ A_{1N}x_1 + A_{2N}x_2 + \cdots + A_{NN}x_N &= f_N, \end{aligned} \tag{2.17}$$

or

$$\mathbf{A}\mathbf{x} = \mathbf{f},$$

where $\mathbf{A}$ is an $N \times N$ real symmetric matrix with $\mathbf{f}$ defined in this way. To see this, differentiate (2.16) with respect to each variable x_n, $n = 1, 2, \ldots, N$. Note that quadratics can be written in the equivalent forms

$$\mathcal{A} = \mathbf{x}^T\mathbf{A}\mathbf{x} = \mathbf{x}^T\mathbf{f} = \langle \mathbf{x}, \mathbf{f} \rangle. \tag{2.18}$$

Example 2.4 Determine the matrix $\mathbf{A}$ in the quadratic form for the ellipse defined by

$$3x_1^2 - 2x_1x_2 + 3x_2^2 = 8.$$

Solution

From (2.16), we have

$$A_{11} = 3, \quad A_{12} = -1, \quad A_{22} = 3.$$

Thus, in matrix form

$$\mathbf{A} = \begin{bmatrix} A_{11} & A_{12} \\ A_{12} & A_{22} \end{bmatrix} = \begin{bmatrix} 3 & -1 \\ -1 & 3 \end{bmatrix},$$

which is symmetric as expected. Checking (2.18)

$$\begin{aligned} \underset{1\times1}{\mathcal{A}} &= \underset{1\times2}{\mathbf{x}^T}\ \underset{2\times2}{\mathbf{A}}\ \underset{2\times1}{\mathbf{x}} \\ &= \begin{bmatrix} x_1 & x_2 \end{bmatrix} \begin{bmatrix} 3 & -1 \\ -1 & 3 \end{bmatrix} \begin{bmatrix} x_1 \\ x_2 \end{bmatrix} \\ &= \begin{bmatrix} x_1 & x_2 \end{bmatrix} \begin{bmatrix} 3x_1 - x_2 \\ -x_1 + 3x_2 \end{bmatrix} \\ &= x_1(3x_1 - x_2) + x_2(-x_1 + 3x_2) \\ \mathcal{A} &= 3x_1^2 - 2x_1x_2 + 3x_2^2. \end{aligned}$$

Setting $\mathcal{A} = 8$ gives the particular ellipse under consideration.

If the quadratic $\mathbf{x}^T\mathbf{A}\mathbf{x} > 0$ for all $\mathbf{x} \neq 0$, then the quadratic is *positive definite*; in other words, it is "definitely positive," and the real symmetric matrix $\mathbf{A}$ is said to be positive definite. In this case,

- The sign of the quadratic does not depend on $\mathbf{x}$ as it is always positive.
- A real symmetric matrix $\mathbf{A}$ is positive definite *if and only if* all of its eigenvalues are positive; note that $\mathbf{A}$ is nonsingular and has full rank in such a case. If the eigenvalues are all nonnegative, so they are all positive or zero, then the matrix is said to be *positive semidefinite.*
- See Section 6.3.2 for the Cholesky decomposition, which applies to positive definite, real symmetric matrices.
- In mechanics, for example, the elasticity tensor is symmetric positive definite.

If there are no mixed terms in the quadratic, in which case it is of the form

$$\mathcal{A} = A_{11}x_1^2 + A_{22}x_2^2 + \cdots + A_{NN}x_N^2,$$

the quadratic is said to be in *canonical form.*[11] In such cases,

- $\mathbf{A}$ is a diagonal matrix with nonzero elements only along the main diagonal.
- In two and three dimensions, the conic curve or quadric surface in *canonical form* is such that the major and minor axes are aligned with the coordinate axes x_1, x_2, and x_3.
- In the context of mechanics, the x_1, x_2, and x_3 directions in the canonical form are called the *principal axes*, and the following diagonalization procedure gives rise to the *parallel axis theorem*, which allows one to transform information from one coordinate system to another.

Diagonalization of Symmetric Matrices with Distinct Eigenvalues

Given a general quadratic in terms of the vector $\mathbf{x} = \begin{bmatrix} x_1 & x_2 & \cdots & x_N \end{bmatrix}^T$ with respect to an N-dimensional coordinate system X, we can use a *coordinate transformation*, say $X \to Y$, to put the quadratic in canonical form in terms of the vector $\mathbf{y} = \begin{bmatrix} y_1 & y_2 & \cdots & y_N \end{bmatrix}^T$ relative to the coordinate system Y. That is, transforming from the basis X to that given by Y places the quadratic in its canonical form.

Suppose that vectors in the two bases are related by the linear transformation

$$\mathbf{x} = \mathbf{Q}\mathbf{y},$$

where the $N \times N$ transformation matrix $\mathbf{Q}$ rotates the basis such that the quadratic is in canonical form with respect to $\mathbf{y}$. Substituting into (2.18) gives

$$\begin{aligned} \mathcal{A} &= \mathbf{x}^T\mathbf{A}\mathbf{x} \\ &= (\mathbf{Q}\mathbf{y})^T\,\mathbf{A}\,(\mathbf{Q}\mathbf{y}) \\ &= \mathbf{y}^T(\mathbf{Q}^T\mathbf{A}\mathbf{Q})\mathbf{y} \\ \mathcal{A} &= \mathbf{y}^T\mathbf{D}\mathbf{y}, \end{aligned}$$

where $\mathbf{D} = \mathbf{Q}^T\mathbf{A}\mathbf{Q}$ must be a diagonal matrix in order to produce the canonical form of the quadratic with respect to $\mathbf{y}$. We say that $\mathbf{Q}$ *diagonalizes* $\mathbf{A}$, such that premultiplying $\mathbf{A}$ by $\mathbf{Q}^T$ and postmultiplying by $\mathbf{Q}$ gives a matrix $\mathbf{D}$ that must be diagonal. We call $\mathbf{Q}$ the *orthogonal modal matrix.*

[11] "Canonical" means "standard."

The procedure to determine the modal matrix for real symmetric matrices is as follows:

1. Determine the eigenvalues, $\lambda_1, \lambda_2, \ldots, \lambda_N$, of the $N \times N$ real symmetric matrix $\mathbf{A}$.
2. Determine the orthonormal eigenvectors, $\mathbf{q}_1, \mathbf{q}_2, \ldots, \mathbf{q}_N$. Note that they are already orthogonal, and they simply need to be normalized. Then

$$\mathbf{A}\mathbf{q}_1 = \lambda_1 \mathbf{q}_1, \quad \mathbf{A}\mathbf{q}_2 = \lambda_2 \mathbf{q}_2, \quad \ldots, \quad \mathbf{A}\mathbf{q}_N = \lambda_N \mathbf{q}_N. \tag{2.19}$$

3. Construct the *orthogonal modal matrix* $\mathbf{Q}$ with columns given by the orthonormal eigenvectors $\mathbf{q}_1, \mathbf{q}_2, \ldots, \mathbf{q}_N$ as follows:

$$\mathbf{Q} = \begin{bmatrix} \vdots & \vdots & & \vdots \\ \mathbf{q}_1 & \mathbf{q}_2 & \cdots & \mathbf{q}_N \\ \vdots & \vdots & & \vdots \end{bmatrix}.$$

   Note that the order of the columns does not matter but corresponds to the order in which the eigenvalues appear along the diagonal of $\mathbf{D}$.

To show that $\mathbf{Q}^T\mathbf{A}\mathbf{Q}$ diagonalizes $\mathbf{A}$, premultiply $\mathbf{Q}$ by $\mathbf{A}$ and use equations (2.19)

$$\begin{aligned}
\mathbf{A}\mathbf{Q} &= \begin{bmatrix} \cdots & \mathbf{a}_1 & \cdots \\ \cdots & \mathbf{a}_2 & \cdots \\ & \vdots & \\ \cdots & \mathbf{a}_N & \cdots \end{bmatrix} \begin{bmatrix} \vdots & \vdots & & \vdots \\ \mathbf{q}_1 & \mathbf{q}_2 & \cdots & \mathbf{q}_N \\ \vdots & \vdots & & \vdots \end{bmatrix} \\
&= \begin{bmatrix} \langle \mathbf{a}_1, \mathbf{q}_1 \rangle & \langle \mathbf{a}_1, \mathbf{q}_2 \rangle & \cdots & \langle \mathbf{a}_1, \mathbf{q}_N \rangle \\ \langle \mathbf{a}_2, \mathbf{q}_1 \rangle & \langle \mathbf{a}_2, \mathbf{q}_2 \rangle & \cdots & \langle \mathbf{a}_2, \mathbf{q}_N \rangle \\ \vdots & \vdots & \ddots & \vdots \\ \langle \mathbf{a}_N, \mathbf{q}_1 \rangle & \langle \mathbf{a}_N, \mathbf{q}_2 \rangle & \cdots & \langle \mathbf{a}_N, \mathbf{q}_N \rangle \end{bmatrix} \\
&= \begin{bmatrix} \vdots & \vdots & & \vdots \\ \mathbf{A}\mathbf{q}_1 & \mathbf{A}\mathbf{q}_2 & \cdots & \mathbf{A}\mathbf{q}_N \\ \vdots & \vdots & & \vdots \end{bmatrix} \\
\mathbf{A}\mathbf{Q} &= \begin{bmatrix} \vdots & \vdots & & \vdots \\ \lambda_1\mathbf{q}_1 & \lambda_2\mathbf{q}_2 & \cdots & \lambda_N\mathbf{q}_N \\ \vdots & \vdots & & \vdots \end{bmatrix}.
\end{aligned}$$

Then premultiply by $\mathbf{Q}^T$

$$\begin{aligned}
\mathbf{D} &= \mathbf{Q}^T\mathbf{A}\mathbf{Q} \\
&= \begin{bmatrix} \cdots & \mathbf{q}_1 & \cdots \\ \cdots & \mathbf{q}_2 & \cdots \\ & \vdots & \\ \cdots & \mathbf{q}_N & \cdots \end{bmatrix} \begin{bmatrix} \vdots & \vdots & & \vdots \\ \lambda_1\mathbf{q}_1 & \lambda_2\mathbf{q}_2 & \cdots & \lambda_N\mathbf{q}_N \\ \vdots & \vdots & & \vdots \end{bmatrix}
\end{aligned}$$

$$\mathbf{D} = \begin{bmatrix} \lambda_1 \langle \mathbf{q}_1, \mathbf{q}_1 \rangle & \lambda_2 \langle \mathbf{q}_1, \mathbf{q}_2 \rangle & \cdots & \lambda_N \langle \mathbf{q}_1, \mathbf{q}_N \rangle \\ \lambda_1 \langle \mathbf{q}_2, \mathbf{q}_1 \rangle & \lambda_2 \langle \mathbf{q}_2, \mathbf{q}_2 \rangle & \cdots & \lambda_N \langle \mathbf{q}_2, \mathbf{q}_N \rangle \\ \vdots & \vdots & \ddots & \vdots \\ \lambda_1 \langle \mathbf{q}_N, \mathbf{q}_1 \rangle & \lambda_2 \langle \mathbf{q}_N, \mathbf{q}_2 \rangle & \cdots & \lambda_N \langle \mathbf{q}_N, \mathbf{q}_N \rangle \end{bmatrix}.$$

Because all vectors are of unit length and are mutually orthogonal, such that $\langle \mathbf{q}_i, \mathbf{q}_j \rangle = 0$ for $i \neq j$ and $\langle \mathbf{q}_i, \mathbf{q}_j \rangle = 1$ for $i = j$, $\mathbf{D}$ is the diagonal matrix

$$\mathbf{D} = \begin{bmatrix} \lambda_1 & 0 & \cdots & 0 \\ 0 & \lambda_2 & \cdots & 0 \\ \vdots & \vdots & \ddots & \vdots \\ 0 & 0 & \cdots & \lambda_N \end{bmatrix}.$$

Thus,

$$\mathcal{A} = \mathbf{y}^T \mathbf{D} \mathbf{y} = \lambda_1 y_1^2 + \lambda_2 y_2^2 + \cdots + \lambda_N y_N^2,$$

such that the quadratic is in canonical form with respect to the new basis Y determined by the eigenvectors.

REMARKS:

1. *Not only is* $\mathbf{D}$ *diagonal, we know what it is once we have the eigenvalues. That is, it is not actually necessary to evaluate* $\mathbf{Q}^T\mathbf{A}\mathbf{Q}$.
2. *The order in which the eigenvalues* λ_n *appear in* $\mathbf{D}$ *corresponds to the order in which the corresponding eigenvectors* $\mathbf{q}_n$ *are placed in the orthogonal modal matrix* $\mathbf{Q}$.
3. *Recall that the columns (and rows) of an orthogonal matrix form an orthonormal set of vectors. For quadratic forms, therefore,* $\mathbf{Q}^T = \mathbf{Q}^{-1}$, *and we have an* orthogonal modal matrix.
4. *Not only is transforming quadratic forms to canonical form an application of diagonalization, it provides us with a geometric interpretation of what the diagonalization procedure is designed to accomplish in other settings as well.*
5. *The diagonalization procedure introduced here for* $\mathbf{A}$ *symmetric and* $\mathbf{Q}$ *orthogonal is a special case of the general diagonalization procedure presented in Section 2.5 for symmetric and nonsymmetric* $\mathbf{A}$.

2.3.4 Linear Systems of Equations Revisited

Having derived the diagonalization procedure for real symmetric matrix $\mathbf{A}$, let us revisit the procedure from Section 2.3.2 for obtaining the solution of a system of linear algebraic equations

$$\mathbf{A}\mathbf{u} = \mathbf{b}$$

using the eigenvalues and eigenvectors of $\mathbf{A}$. Here, we repeat the derivation, but using the more compact matrix formulation.

For $\mathbf{A}$ real and symmetric, the orthonormal eigenvectors $\mathbf{q}_n$ comprise the columns of the $N \times N$ orthogonal matrix $\mathbf{Q}$. Equation (2.14) gives the solution in terms of the linear combination of the known eigenvectors, which in matrix form is

$$\mathbf{u} = \mathbf{Q}\mathbf{a}, \tag{2.20}$$

where

$$\mathbf{a} = \begin{bmatrix} a_1 & a_2 & \cdots & a_n & \cdots & a_N \end{bmatrix}^T.$$

Substituting this into the system of equations yields

$$\mathbf{A}\mathbf{Q}\mathbf{a} = \mathbf{b}.$$

Premultiplying both sides with $\mathbf{Q}^T$ gives

$$\mathbf{Q}^T\mathbf{A}\mathbf{Q}\mathbf{a} = \mathbf{Q}^T\mathbf{b}.$$

But $\mathbf{D} = \mathbf{Q}^T\mathbf{A}\mathbf{Q}$ is a diagonal matrix with the eigenvalues of the coefficient matrix $\mathbf{A}$ on the diagonal; therefore,

$$\mathbf{D}\mathbf{a} = \mathbf{Q}^T\mathbf{b}.$$

Premultiplying the inverse of $\mathbf{D}$ on both sides then gives the result that the coefficients that we seek are

$$\mathbf{a} = \mathbf{D}^{-1}\mathbf{Q}^T\mathbf{b}. \tag{2.21}$$

Let us show how (2.21) is the same result as obtained in Section 2.3.2. Recall that the inverse of a diagonal matrix is a diagonal matrix with the reciprocals of the diagonal elements, here the eigenvalues of $\mathbf{A}$, on the diagonal. In addition, evaluating $\mathbf{Q}^T\mathbf{b}$ leads to the result that

$$\mathbf{a} = \begin{bmatrix} \frac{1}{\lambda_1} & 0 & \cdots & 0 \\ 0 & \frac{1}{\lambda_2} & \cdots & 0 \\ \vdots & \vdots & \ddots & \vdots \\ 0 & 0 & \cdots & \frac{1}{\lambda_N} \end{bmatrix} \begin{bmatrix} \langle \mathbf{q}_1, \mathbf{b} \rangle \\ \langle \mathbf{q}_2, \mathbf{b} \rangle \\ \vdots \\ \langle \mathbf{q}_N, \mathbf{b} \rangle \end{bmatrix} = \begin{bmatrix} \frac{\langle \mathbf{q}_1, \mathbf{b} \rangle}{\lambda_1} \\ \frac{\langle \mathbf{q}_2, \mathbf{b} \rangle}{\lambda_2} \\ \vdots \\ \frac{\langle \mathbf{q}_N, \mathbf{b} \rangle}{\lambda_N} \end{bmatrix}.$$

This is the same result as found in Section 2.3.2 for the coefficients in the expansion for the solution. The solution of the system of algebraic equations $\mathbf{A}\mathbf{u} = \mathbf{b}$ is then given by $\mathbf{u} = \mathbf{Q}\mathbf{a}$, which is the matrix form of (2.15).

Example 2.5 The solution of a system of equations $\mathbf{A}\mathbf{u} = \mathbf{b}$ is sought, where

$$\mathbf{A} = \begin{bmatrix} -2 & 1 & 1 & 1 & 2 \\ 1 & 2 & 1 & -1 & 1 \\ 1 & 1 & -2 & 1 & 1 \\ 1 & -1 & 1 & 2 & 1 \\ 2 & 1 & 1 & 1 & -2 \end{bmatrix}, \quad \mathbf{b} = \begin{bmatrix} -1 \\ 2 \\ 1 \\ -2 \\ 1 \end{bmatrix}.$$

Observe that $\mathbf{A}$ is real and symmetric. Obtain the solution in terms of a linear combination of the eigenvectors of $\mathbf{A}$.

Solution

The diagonal matrix with the eigenvalues of $\mathbf{A}$ on the diagonal is given by

$$\mathbf{D} = \begin{bmatrix} -4. & 0. & 0. & 0. & 0. \\ 0. & 3.3059 & 0. & 0. & 0. \\ 0. & 0. & 3. & 0. & 0. \\ 0. & 0. & 0. & -2.78567 & 0. \\ 0. & 0. & 0. & 0. & -1.52023 \end{bmatrix}.$$

The orthogonal matrix with its columns being the eigenvectors of real symmetric $\mathbf{A}$ is

$$\mathbf{Q} = \begin{bmatrix} -0.707107 & 0.417217 & 0. & 0.251386 & 0.512577 \\ 0. & 0.514111 & -0.707107 & 0.110606 & -0.472711 \\ 0. & 0.351054 & 0. & -0.921489 & 0.166187 \\ 0. & 0.514111 & 0.707107 & 0.110606 & -0.472711 \\ 0.707107 & 0.417217 & 0. & 0.251386 & 0.512577 \end{bmatrix}.$$

The solution of the system of linear algebraic equations $\mathbf{Au} = \mathbf{b}$ is then constructed using the columns of $\mathbf{Q}$ as the basis vectors according to (2.20) and (2.21) as follows:

$$\mathbf{u} = \mathbf{Qa} = \mathbf{QD}^{-1}\mathbf{Q}^T\mathbf{b} = \begin{bmatrix} 0.321429 \\ 0.809524 \\ -0.285714 \\ -0.52381 \\ -0.178571 \end{bmatrix}, \tag{2.22}$$

where the coefficients in the linear combination

$$\mathbf{u} = a_1\mathbf{q}_1 + a_2\mathbf{q}_2 + a_3\mathbf{q}_3 + a_4\mathbf{q}_4 + a_5\mathbf{q}_5$$

are given by

$$\mathbf{a} = \mathbf{D}^{-1}\mathbf{Q}^T\mathbf{b} = \begin{bmatrix} -0.353553 \\ 0.10619 \\ -0.942809 \\ 0.330797 \\ -0.109317 \end{bmatrix}.$$

Observe that none of the coefficients $\mathbf{a}$ in the linear combination using eigenvectors of $\mathbf{A}$ are particularly small. Therefore, all five eigenvectors must be included to obtain an accurate representation of the solution. We will revisit this example in Chapter 13 using proper-orthogonal decomposition to show how we can reconstruct an accurate approximation of the solution using only one basis vector by choosing them optimally.

2.4 Normal and Orthogonal Matrices

We focus our attention in Section 2.3 on real symmetric matrices because of their prevalence in applications. The two primary results encountered for real symmetric $\mathbf{A}$ are:

1. The eigenvectors of $\mathbf{A}$ corresponding to distinct eigenvalues are mutually orthogonal.
2. The matrix $\mathbf{A}$ is diagonalizable using the orthogonal modal matrix $\mathbf{Q}$ having the orthonormal eigenvectors of $\mathbf{A}$ as its columns according to

$$\mathbf{D} = \mathbf{Q}^T\mathbf{A}\mathbf{Q},$$

where $\mathbf{D}$ is a diagonal matrix containing the eigenvalues of $\mathbf{A}$. This is referred to as a *similarity transformation* because $\mathbf{A}$ and $\mathbf{D}$ have the same eigenvalues.[12]

This is known as the *spectral theorem*. One may wonder whether a real symmetric matrix is the most general matrix for which these two powerful results are true. It turns out that it is not.

The most general matrix for which the spectral theorem holds is a *normal matrix*. A normal[13] matrix is such that it commutes with its conjugate transpose so that

$$\overline{\mathbf{A}}^T\mathbf{A} = \mathbf{A}\overline{\mathbf{A}}^T,$$

or

$$\mathbf{A}^T\mathbf{A} = \mathbf{A}\mathbf{A}^T,$$

if $\mathbf{A}$ is real.

Clearly symmetric and Hermitian matrices are normal. In addition, all orthogonal, skew-symmetric, unitary, and skew-Hermitian matrices are normal as well. However, not all normal matrices are one of these forms.

Given the rather special properties of normal matrices, one may wonder whether the products $\mathbf{A}\mathbf{A}^T$ and $\mathbf{A}^T\mathbf{A}$ themselves have any special properties as well. It turns out that for any matrix $\mathbf{A}$ of size $M \times N$, both products have the following properties:

1. They are always square; $\mathbf{A}\mathbf{A}^T$ is $M \times M$, and $\mathbf{A}^T\mathbf{A}$ is $N \times N$.
2. They are symmetric (for example, $\mathbf{A}\mathbf{A}^T = (\mathbf{A}\mathbf{A}^T)^T$).
3. $\text{rank}(\mathbf{A}) = \text{rank}(\mathbf{A}\mathbf{A}^T) = \text{rank}(\mathbf{A}^T\mathbf{A})$.
4. Both matrix products have the same nonzero eigenvalues. All additional eigenvalues are zero.
5. Both matrix products are positive semidefinite, that is, all eigenvalues are nonnegative.
6. Because both matrix products are symmetric, the eigenvectors of both are – or can be made to be – mutually orthogonal.

[12] Two matrices are said to be *similar* if they have the same eigenvalues. Recall that for two matrices to be *equivalent*, each of their corresponding elements must be the same.

[13] In mathematics, the term *normal* is synonymous with "perpendicular" or "orthogonal," so a normal matrix is one that produces normal, that is, orthogonal, eigenvectors. When we say that a vector has been *normalized*, however, this simply means that it has been made to be of unit length, that is, length one.

We will encounter these products in singular-value decomposition (see Section 2.8), polar decomposition (see Section 2.9), least-squares methods (see Section 10.2), and proper-orthogonal decomposition (see Chapter 13).

Example 2.6 Determine the requirements for a real 2×2 matrix $\mathbf{A}$ to be normal.

Solution

For a real 2×2 matrix to be normal

$$\mathbf{A}^T\mathbf{A} - \mathbf{A}\mathbf{A}^T = \mathbf{0}$$

$$\begin{bmatrix} A_{11} & A_{21} \\ A_{12} & A_{22} \end{bmatrix}\begin{bmatrix} A_{11} & A_{12} \\ A_{21} & A_{22} \end{bmatrix} - \begin{bmatrix} A_{11} & A_{12} \\ A_{21} & A_{22} \end{bmatrix}\begin{bmatrix} A_{11} & A_{21} \\ A_{12} & A_{22} \end{bmatrix} = \mathbf{0}$$

$$\begin{bmatrix} A_{11}^2 + A_{21}^2 & A_{11}A_{12} + A_{21}A_{22} \\ A_{11}A_{12} + A_{21}A_{22} & A_{12}^2 + A_{22}^2 \end{bmatrix} - \begin{bmatrix} A_{11}^2 + A_{12}^2 & A_{11}A_{21} + A_{12}A_{22} \\ A_{11}A_{21} + A_{12}A_{22} & A_{21}^2 + A_{22}^2 \end{bmatrix} = \mathbf{0}$$

$$\begin{bmatrix} A_{21}^2 - A_{12}^2 & A_{11}A_{12} + A_{21}A_{22} - A_{11}A_{21} - A_{12}A_{22} \\ A_{11}A_{12} + A_{21}A_{22} - A_{11}A_{21} - A_{12}A_{22} & A_{12}^2 - A_{21}^2 \end{bmatrix} = \mathbf{0}$$

$$(A_{12} - A_{21})\begin{bmatrix} -A_{12} - A_{21} & A_{11} - A_{22} \\ A_{11} - A_{22} & A_{12} + A_{21} \end{bmatrix} = \mathbf{0}.$$

Thus, $\mathbf{A}$ is normal if

$$A_{12} = A_{21},$$

in which case $\mathbf{A}$ is symmetric, OR

$$A_{11} = A_{22}, \quad A_{21} = -A_{12},$$

in which case $\mathbf{A}$ is of the form

$$\mathbf{A} = \begin{bmatrix} a & b \\ -b & a \end{bmatrix},$$

which is skew symmetric. Thus, for a 2×2 real matrix to be normal, it must be either symmetric or skew symmetric.

REMARKS:

1. *Recall that a symmetric matrix has – or can be made to have – orthogonal eigenvectors and real eigenvalues. While a normal matrix also has orthogonal eigenvectors, the eigenvalues may be complex.*
2. *The result from Section 2.3.2 applies for* $\mathbf{A}$ *normal. Specifically, the solution of*

$$\mathbf{Au} = \mathbf{b},$$

where $\mathbf{A}$ *is an* $N \times N$ *normal matrix, is given by*

$$\mathbf{u} = \sum_{n=1}^{N} \frac{\langle \mathbf{q}_n, \mathbf{b} \rangle}{\lambda_n} \mathbf{q}_n,$$

where λ_n *and* $\mathbf{q}_n, n = 1, \ldots, N$ *are the eigenvalues and orthonormal eigenvectors of* $\mathbf{A}$*, respectively.*

3. *Whether a matrix is normal or not has important consequences for stability of systems governed by such matrices. Specifically, the stability of systems governed by normal matrices is determined solely by the nature of the eigenvalues of* $\mathbf{A}$*. On the other hand, for systems governed by nonnormal matrices, the eigenvalues only determine the asymptotic stability of the system as* $t \to \infty$*, and the stability when* $t = O(1)$ *may exhibit a different behavior known as* transient growth *(see Section 5.3.3).*[14]
4. *For more on nonnormal matrices and their consequences, see Trefethen and Embree (2005).*

Because of their prevalence in applications and unique mathematical properties, it is worthwhile to further consider some of the attributes of orthogonal matrices. Recall that a square orthogonal matrix is one in which its column (and row) vectors are all mutually orthogonal and of length one. Let us begin by establishing their primary property highlighted thus far, which is that $\mathbf{Q}^{-1} = \mathbf{Q}^T$. To do so, consider the $N \times N$ orthogonal matrix

$$\mathbf{Q} = \begin{bmatrix} \vdots & \vdots & & \vdots \\ \mathbf{q}_1 & \mathbf{q}_2 & \cdots & \mathbf{q}_N \\ \vdots & \vdots & & \vdots \end{bmatrix},$$

where the $\mathbf{q}_i$ column vectors are mutually orthonormal. From Section 1.7.1, recall that the Gram matrix of an orthogonal matrix is defined by

$$\mathbf{Q}^T\mathbf{Q} = \begin{bmatrix} \cdots & \mathbf{q}_1 & \cdots \\ \cdots & \mathbf{q}_2 & \cdots \\ & \vdots & \\ \cdots & \mathbf{q}_N & \cdots \end{bmatrix} \begin{bmatrix} \vdots & \vdots & & \vdots \\ \mathbf{q}_1 & \mathbf{q}_2 & \cdots & \mathbf{q}_N \\ \vdots & \vdots & & \vdots \end{bmatrix} = \mathbf{I}.$$

Because of orthonormality of the vectors $\mathbf{q}_i$, therefore, matrix multiplication produces an $N \times N$ identity matrix. Similarly, because the row vectors of a square orthogonal matrix are also orthonormal, $\mathbf{Q}\mathbf{Q}^T = \mathbf{I}$ as well. Consequently, $\mathbf{Q}^T\mathbf{Q} = \mathbf{Q}\mathbf{Q}^T = \mathbf{I}$, which confirms that $\mathbf{Q}^{-1} = \mathbf{Q}^T$ for square orthogonal matrices and also confirms that orthogonal matrices are normal matrices as well.

Clearly, the determinant of an identity matrix is unity, but

$$\mathbf{I} = \mathbf{Q}^T\mathbf{Q}.$$

[14] Also see section 6.5 of Cassel (2013) for a variational perspective on transient growth.

Therefore, taking the determinant of both sides yields

$$|\mathbf{I}| = |\mathbf{Q}^T\mathbf{Q}| = |\mathbf{Q}^T||\mathbf{Q}| = |\mathbf{Q}|^2.$$

Taking the square root of $|\mathbf{Q}|^2 = 1$ then gives the determinant of an orthogonal matrix to be

$$|\mathbf{Q}| = \pm 1.$$

Next, recall from Section 1.8 that a rotation matrix in two dimensions is given by

$$\mathbf{A} = \begin{bmatrix} \cos\theta & -\sin\theta \\ \sin\theta & \cos\theta \end{bmatrix},$$

which leads to a rotation in the counterclockwise direction of angle θ. Recalling that $\cos^2\theta + \sin^2\theta = 1$, let us evaluate $\mathbf{A}^T\mathbf{A}$. This leads to

$$\begin{aligned} \mathbf{A}^T\mathbf{A} &= \begin{bmatrix} \cos\theta & \sin\theta \\ -\sin\theta & \cos\theta \end{bmatrix}\begin{bmatrix} \cos\theta & -\sin\theta \\ \sin\theta & \cos\theta \end{bmatrix} \\ &= \begin{bmatrix} \cos^2\theta + \sin^2\theta & -\sin\theta\cos\theta + \sin\theta\cos\theta \\ -\sin\theta\cos\theta + \sin\theta\cos\theta & \cos^2\theta + \sin^2\theta \end{bmatrix} \\ &= \begin{bmatrix} 1 & 0 \\ 0 & 1 \end{bmatrix} \\ \mathbf{A}^T\mathbf{A} &= \mathbf{I}. \end{aligned}$$

As a result, one can see that rotation matrices are orthogonal ($\mathbf{A} = \mathbf{Q}$), in which case taking $\mathbf{Qx}$ only rotates the vector $\mathbf{x}$ and does not scale or stretch it.

Finally, we prove that the eigenvalues of an orthogonal matrix are $\lambda_n = \pm 1$. Forming the eigenproblem for an orthogonal matrix, we have

$$\mathbf{Q}\mathbf{u}_n = \lambda_n\mathbf{u}_n.$$

Taking the norm of both sides and squaring leads to

$$\|\mathbf{Q}\mathbf{u}_n\|^2 = \lambda_n^2\|\mathbf{u}_n\|^2.$$

Alternatively, writing in terms of inner products, this is

$$\langle\mathbf{Q}\mathbf{u}_n, \mathbf{Q}\mathbf{u}_n\rangle = \lambda_n^2\langle\mathbf{u}_n, \mathbf{u}_n\rangle,$$

or

$$(\mathbf{Q}\mathbf{u}_n)^T\mathbf{Q}\mathbf{u}_n = \lambda_n^2\mathbf{u}_n^T\mathbf{u}_n.$$

Evaluating the transpose of the product on the left-hand side yields

$$\mathbf{u}_n^T\mathbf{Q}^T\mathbf{Q}\mathbf{u}_n = \lambda_n^2\mathbf{u}_n^T\mathbf{u}_n,$$

but $\mathbf{Q}^T\mathbf{Q} = \mathbf{I}$; therefore,

$$\mathbf{u}_n^T\mathbf{u}_n = \lambda_n^2\mathbf{u}_n^T\mathbf{u}_n,$$

or

$$\|\mathbf{u}_n\|^2 = \lambda_n^2 \|\mathbf{u}_n\|^2.$$

Because the eigenvectors of an orthogonal matrix have nonzero length, the norms cancel, and we are left with the result that

$$\lambda_n^2 = 1,$$

or

$$\lambda_n = \pm 1.$$

Similarly, because $\mathbf{Q}^T\mathbf{Q} = \mathbf{Q}\mathbf{Q}^T = \mathbf{I}$, the singular values of an orthogonal matrix are all $\sigma_n = 1$, in which case it is perfectly conditioned (see Section 2.8).

Far from being mathematical novelties, these properties of orthogonal matrices have far-reaching implications for how they are used in applications. Because the determinant of an orthogonal matrix is $|\mathbf{Q}| = \pm 1$ and the eigenvalues of an orthogonal matrix are $\lambda_n = \pm 1$ (and singular values are all unity), operations with orthogonal matrices are especially well behaved. For example, any measure of the "size" of the matrix and its column (or row) vectors are all unity. Consequently, mathematical and/or numerical operations with orthogonal matrices and its column (or row) vectors, in addition to being convenient, are also numerically stable as they do not cause algorithms or methods to produce results that are too large or too small (more on this in Section 6.2).

2.5 Diagonalization

In Section 2.3.3, we introduced the diagonalization procedure for real symmetric matrices having distinct eigenvalues. Here, we generalize this procedure for symmetric matrices with repeated eigenvalues and nonsymmetric matrices. Diagonalization consists of performing a *similarity transformation*, in which a new matrix $\mathbf{D}$ is produced that has the same eigenvalues as the original matrix $\mathbf{A}$, such that $\mathbf{A}$ and $\mathbf{D}$ are *similar*. The new matrix $\mathbf{D}$ is diagonal with the eigenvalues of $\mathbf{A}$ residing on its diagonal.

The outcome of the diagonalization process is determined by the nature of the eigenvalues of $\mathbf{A}$ as summarized in Table 2.1. The case considered in Section 2.3.3 is such that matrix $\mathbf{A}$ is real and symmetric with distinct eigenvalues, in which case the

Table 2.1 Summary of the eigenvectors of a matrix based on the type of matrix and the nature of its eigenvalues.

	Distinct eigenvalues	Repeated eigenvalues
Symmetric $\mathbf{A}$	Mutually orthogonal eigenvectors	Linearly independent eigenvectors
Nonsymmetric $\mathbf{A}$	Linearly independent eigenvectors	?

eigenvectors are mutually orthogonal, and the modal matrix can be made to be orthogonal (this also applies to other normal matrices as described in Section 2.4). For a nonsymmetric matrix with distinct eigenvalues, the eigenvectors are mutually linearly independent, but not necessarily orthogonal. As we will see, cases with nonsymmetric matrices with distinct eigenvalues or symmetric matrices allow for the matrix to be fully diagonalized. For a nonsymmetric matrix with repeated eigenvalues, however, the regular eigenvectors may or may not be linearly independent, in which case it may not be fully diagonalized. A very similar procedure, however, will produce the so-called *Jordan canonical form*, which is nearly diagonalized. The Jordan canonical form will be discussed further in Section 2.5.2.

We first encountered diagonalization in the context of transforming quadratics to their canonical form. Shortly, diagonalization will become central to solving systems of linear ordinary differential equations by decoupling them in a fashion that makes their solution very straightforward. It will then be shown to be the foundation for singular-value decomposition, which is at the heart of very powerful, and widely applicable, data reduction techniques, such as proper-orthogonal decomposition. Finally, diagonalization will prove useful in the development of the QR algorithm for numerically determining the eigenvalues of large systems of equations and solving least-squares problems.

In all of its applications, diagonalization can be viewed as a change of basis – or coordinate system – to one in which the quadratic form, solution of a system of linear ordinary differential equations, or a discrete data set are in some sense more effectively represented. We will return to this theme as a tool to motivate and interpret these important operations.

2.5.1 Matrices with Linearly Independent Eigenvectors

The diagonalization procedure described in Section 2.3.3 for real symmetric matrices with distinct eigenvalues is a special case of the general procedure given here. If the eigenvectors are linearly independent, including those that are also orthogonal, then the modal matrix[15] $\mathbf{U}$, whose columns are the eigenvectors of $\mathbf{A}$, produce the similarity transformation

$$\mathbf{D} = \mathbf{U}^{-1}\mathbf{A}\mathbf{U} = \begin{bmatrix} \lambda_1 & 0 & \cdots & 0 \\ 0 & \lambda_2 & \cdots & 0 \\ \vdots & \vdots & \ddots & \vdots \\ 0 & 0 & \cdots & \lambda_N \end{bmatrix},$$

where $\mathbf{D}$ is a diagonal matrix with the eigenvalues of $\mathbf{A}$ along the diagonal as shown. Note the use of the inverse in place of the transpose for this general case.

To prove that $\mathbf{U}^{-1}\mathbf{A}\mathbf{U}$ diagonalizes $\mathbf{A}$, let us establish that $\mathbf{A}$ and $\mathbf{D}$ are similar, thereby having the same eigenvalues. This is the case if they both have the same characteristic equation; therefore, consider the following:

[15] We reserve use of $\mathbf{Q}$ to indicate orthogonal modal matrices.

$$\begin{aligned}
|\mathbf{D} - \lambda\mathbf{I}| &= |\mathbf{U}^{-1}\mathbf{A}\mathbf{U} - \lambda\mathbf{I}| \\
&= \left|\mathbf{U}^{-1}\mathbf{A}\mathbf{U} - \lambda\mathbf{U}^{-1}\mathbf{I}\mathbf{U}\right| \\
&= |\mathbf{U}^{-1}(\mathbf{A} - \lambda\mathbf{I})\,\mathbf{U}| \\
&= |\mathbf{U}|\,|\mathbf{A} - \lambda\mathbf{I}|\,|\mathbf{U}^{-1}| \\
&= |\mathbf{U}|\,|\mathbf{A} - \lambda\mathbf{I}|\,\frac{1}{|\mathbf{U}|} \\
|\mathbf{D} - \lambda\mathbf{I}| &= |\mathbf{A} - \lambda\mathbf{I}|\,.
\end{aligned}$$

Consequently, $\mathbf{D}$ and $\mathbf{A}$ have the same characteristic equation and eigenvalues, and $\mathbf{U}^{-1}\mathbf{A}\mathbf{U}$ does indeed diagonalize $\mathbf{A}$.

REMARKS:

1. *This general diagonalization procedure requires premultiplying by the inverse of the modal matrix. When the modal matrix is orthogonal, as for symmetric* $\mathbf{A}$, *then* $\mathbf{U}^{-1} = \mathbf{U}^T = \mathbf{Q}^T$. *Note that whereas it is necessary for the orthogonal modal matrix* $\mathbf{Q}$ *to be formed from the normalized eigenvectors, it is not necessary to normalize the eigenvectors when forming* $\mathbf{U}$*; they simply need to be linearly independent.*
2. *An* $N \times N$ *matrix* $\mathbf{A}$ *can be diagonalized if there are N regular eigenvectors that are linearly independent (see O'Neil, 2012, for example, for a proof). Consequently, all symmetric matrices can be diagonalized, and if the matrix is not symmetric, the eigenvalues must be distinct.*
3. *It is not necessary to evaluate* $\mathbf{U}^{-1}\mathbf{A}\mathbf{U}$, *because we know the result* $\mathbf{D}$ *if the eigenvectors are linearly independent.*
4. *The term* modal matrix *arises from its application in dynamical systems in which the diagonalization, or decoupling, procedure leads to isolation of the natural frequencies – or* modes *– of vibration of the system, which correspond to the eigenvalues. The general motion of the system is then a superposition (linear combination) of these modes (see Section 2.6.3).*

2.5.2 Jordan Canonical Form

If an $N \times N$ matrix $\mathbf{A}$ has fewer than N regular eigenvectors, as may be the case for a nonsymmetric matrix with repeated eigenvalues, it is called *defective*. Additional *generalized eigenvectors* may be obtained such that they are linearly independent with the regular eigenvectors in order to form the modal matrix $\mathbf{U}$. In this case, $\mathbf{U}^{-1}\mathbf{A}\mathbf{U}$ results in the *Jordan canonical form*, which is not completely diagonalized.

For example, if $\mathbf{A}$ is 5×5 with two repeated eigenvalues λ_1 (multiplicity two) and three repeated eigenvalues λ_2 (multiplicity three), then the Jordan canonical form is

$$\mathbf{J} = \mathbf{U}^{-1}\mathbf{A}\mathbf{U} = \begin{bmatrix} \lambda_1 & a_1 & 0 & 0 & 0 \\ 0 & \lambda_1 & 0 & 0 & 0 \\ 0 & 0 & \lambda_2 & a_2 & 0 \\ 0 & 0 & 0 & \lambda_2 & a_3 \\ 0 & 0 & 0 & 0 & \lambda_2 \end{bmatrix},$$

where a_1, a_2, and a_3 above the repeated eigenvalues are 0 or 1. Unfortunately, the only way to determine the Jordan canonical matrix $\mathbf{J}$ is to actually evaluate $\mathbf{U}^{-1}\mathbf{A}\mathbf{U}$ just to determine the elements above the repeated eigenvalues.

In order to form the modal matrix $\mathbf{U}$ when there are fewer than N regular eigenvectors, we need a procedure for obtaining the necessary generalized eigenvectors from the regular ones. Recall that regular eigenvectors $\mathbf{u}_n$ satisfy the eigenproblem

$$(\mathbf{A} - \lambda_n \mathbf{I})\, \mathbf{u}_n = \mathbf{0}.$$

If for a given eigenvalue, say λ_1, with multiplicity K, we only obtain one regular eigenvector using the normal procedure, then we need to obtain $K - 1$ generalized eigenvectors. These generalized eigenvectors satisfy the sequence of eigenproblems

$$(\mathbf{A} - \lambda_1 \mathbf{I})^k\, \mathbf{u}_k = \mathbf{0}, \quad k = 2, 3, \dots, K.$$

Note that the regular eigenvector $\mathbf{u}_1$ results from the case with $k = 1$.

Rather than taking successive integer powers of $(\mathbf{A} - \lambda_1 \mathbf{I})$ and obtaining the corresponding eigenvectors, observe the following. If

$$(\mathbf{A} - \lambda_1 \mathbf{I})\, \mathbf{u}_1 = \mathbf{0} \tag{2.23}$$

produces the regular eigenvector $\mathbf{u}_1$ corresponding to the repeated eigenvalue λ_1, then the generalized eigenvector $\mathbf{u}_2$ is obtained from

$$(\mathbf{A} - \lambda_1 \mathbf{I})^2\, \mathbf{u}_2 = \mathbf{0}. \tag{2.24}$$

However, this is equivalent to writing the matrix problem

$$(\mathbf{A} - \lambda_1 \mathbf{I})\, \mathbf{u}_2 = \mathbf{u}_1.$$

To see this, multiply both sides by $(\mathbf{A} - \lambda_1 \mathbf{I})$ to produce

$$(\mathbf{A} - \lambda_1 \mathbf{I})^2\, \mathbf{u}_2 = (\mathbf{A} - \lambda_1 \mathbf{I})\, \mathbf{u}_1,$$

but from (2.23), the right-hand side is zero and we have (2.24). Thus, the generalized eigenvectors for the repeated eigenvalues can be obtained by successively solving the systems of equations

$$(\mathbf{A} - \lambda_1 \mathbf{I})\, \mathbf{u}_k = \mathbf{u}_{k-1}, \quad k = 2, 3, \dots, K, \tag{2.25}$$

starting with the regular eigenvector $\mathbf{u}_1$ on the right-hand side for $k = 2$. In this manner, the previous regular or generalized eigenvector $\mathbf{u}_{k-1}$ becomes the known right-hand side in the system of equations to be solved for the next generalized eigenvector $\mathbf{u}_k$.

REMARKS:

1. *Because* $(\mathbf{A} - \lambda_1 \mathbf{I})$ *is singular for the eigenproblem, we do not obtain a unique solution for each successive generalized eigenvector. This is consistent with the fact that the eigenvectors always include an arbitrary constant multiplier.*
2. *The resulting regular and generalized eigenvectors form a mutually linearly independent set of* N *vectors to form the modal matrix* $\mathbf{U}$.

Example 2.7 Determine the regular and generalized eigenvectors for the nonsymmetric matrix

$$\mathbf{A}=\begin{bmatrix}2 & -1 & 2 & 0\\ 0 & 3 & -1 & 0\\ 0 & 1 & 1 & 0\\ 0 & 1 & -3 & 5\end{bmatrix}.$$

Then obtain the modal matrix $\mathbf{U}$ that reduces $\mathbf{A}$ to the Jordan canonical form, and determine the Jordan canonical form.

Solution

The eigenvalues of $\mathbf{A}$ are

$$\lambda_1=\lambda_2=\lambda_3=2, \quad \lambda_4=5.$$

Corresponding to $\lambda_2=2$, we obtain the single regular eigenvector

$$\mathbf{u}_1=\begin{bmatrix}1\\ 0\\ 0\\ 0\end{bmatrix},$$

and corresponding to $\lambda_4=5$ we obtain the regular eigenvector

$$\mathbf{u}_4=\begin{bmatrix}0\\ 0\\ 0\\ 1\end{bmatrix}.$$

Therefore, we only have two regular eigenvectors.

In order to determine the two generalized eigenvectors, $\mathbf{u}_2$ and $\mathbf{u}_3$, corresponding to the repeated eigenvalue $\lambda_1=2$, we solve the system of equations

$$(\mathbf{A}-\lambda_1\mathbf{I})\,\mathbf{u}_2=\mathbf{u}_1, \tag{2.26}$$

to obtain $\mathbf{u}_2$ from the known $\mathbf{u}_1$, and then we solve

$$(\mathbf{A}-\lambda_1\mathbf{I})\,\mathbf{u}_3=\mathbf{u}_2, \tag{2.27}$$

to obtain $\mathbf{u}_3$ from $\mathbf{u}_2$. Using Gaussian elimination[16] to solve the two systems of equations in succession, we obtain

$$\mathbf{u}_2=\begin{bmatrix}c_1\\ 1\\ 1\\ \frac{2}{3}\end{bmatrix}, \quad \mathbf{u}_3=\begin{bmatrix}c_2\\ c_1+2\\ c_1+1\\ \frac{2}{3}c_1+\frac{5}{9}\end{bmatrix},$$

[16] Because the systems of equations are singular, they can only be solved using Gaussian elimination.

where c_1 and c_2 are arbitrary and arise because (2.26) and (2.27) do not have unique solutions owing to the fact that $|\mathbf{A} - \lambda_1\mathbf{I}| = 0$. Choosing $c_1 = c_2 = 0$, we have the generalized modal matrix

$$\mathbf{U} = \begin{bmatrix} 1 & 0 & 0 & 0 \\ 0 & 1 & 2 & 0 \\ 0 & 1 & 1 & 0 \\ 0 & \frac{2}{3} & \frac{5}{9} & 1 \end{bmatrix}.$$

This procedure produces four linearly independent regular and generalized eigenvectors. Pseudodiagonalizing then gives the Jordan canonical form

$$\mathbf{J} = \mathbf{U}^{-1}\mathbf{A}\mathbf{U} = \begin{bmatrix} 2 & 1 & 0 & 0 \\ 0 & 2 & 1 & 0 \\ 0 & 0 & 2 & 0 \\ 0 & 0 & 0 & 5 \end{bmatrix}.$$

Note that this requires that we actually invert $\mathbf{U}$ and evaluate $\mathbf{J} = \mathbf{U}^{-1}\mathbf{A}\mathbf{U}$, unlike cases for which $\mathbf{D} = \mathbf{U}^{-1}\mathbf{A}\mathbf{U}$ is a diagonal matrix. Observe that the eigenvalues are on the diagonal with 0 or 1 (1 in this case) above the repeated eigenvalues as expected.

Note that the generalized eigenvector(s), and the regular eigenvector(s) from which they are obtained, must be placed in the order in which they are determined to form the modal matrix $\mathbf{U}$.

2.6 Systems of Ordinary Differential Equations

One of the most important uses of the diagonalization procedure outlined in the previous sections is in solving systems of first-order linear ordinary differential equations. We focus our discussion here on so-called *initial-value problems* that evolve in time. However, these same techniques can be applied to *boundary-value problems* as well.

2.6.1 General Approach for First-Order Systems

Recall that the general solution of the first-order linear ordinary differential equation

$$\dot{u}(t) = \frac{du}{dt} = au(t),$$

where a is a constant, is

$$u(t) = ce^{at},$$

where c is a constant of integration. A dot denotes differentiation with respect to the independent variable t.[17]

[17] We follow the convention of indicating ordinary differentiation with respect to time using dots and primes for spatial independent variables.

Now consider a system of N *coupled* first-order linear ordinary differential equations

$$\begin{aligned}
\dot{u}_1(t) &= A_{11}u_1(t) + A_{12}u_2(t) + \cdots + A_{1N}u_N(t) + f_1(t), \\
\dot{u}_2(t) &= A_{21}u_1(t) + A_{22}u_2(t) + \cdots + A_{2N}u_N(t) + f_2(t), \\
&\vdots \\
\dot{u}_N(t) &= A_{N1}u_1(t) + A_{N2}u_2(t) + \cdots + A_{NN}u_N(t) + f_N(t),
\end{aligned}$$

where the A_{mn} coefficients and $f_n(t)$ functions are known, and the dependent variables $u_1(t), u_2(t), \ldots, u_N(t)$ are to be determined. Observe that each equation may involve all of the dependent variables in a coupled, interdependent fashion. If $f_n(t) = 0, n = 1, \ldots, N$, the system is said to be *homogeneous*. This system may be written in matrix form as

$$\dot{\mathbf{u}}(t) = \mathbf{A}\mathbf{u}(t) + \mathbf{f}(t), \tag{2.28}$$

where

$$\mathbf{u}(t) = \begin{bmatrix} u_1(t) \\ u_2(t) \\ \vdots \\ u_N(t) \end{bmatrix}, \quad \mathbf{A} = \begin{bmatrix} A_{11} & A_{12} & \cdots & A_{1N} \\ A_{21} & A_{22} & \cdots & A_{2N} \\ \vdots & \vdots & \ddots & \vdots \\ A_{N1} & A_{N2} & \cdots & A_{NN} \end{bmatrix}, \quad \mathbf{f}(t) = \begin{bmatrix} f_1(t) \\ f_2(t) \\ \vdots \\ f_N(t) \end{bmatrix}.$$

In order to solve this coupled system, we transform the solution vector $\mathbf{u}(t)$ into a new vector of dependent variables $\mathbf{v}(t)$ for which the equations are easily solved. In particular, we diagonalize the coefficient matrix $\mathbf{A}$ such that with respect to the new variables $\mathbf{v}(t)$, the system is uncoupled and straightforward to solve.

Once again, it is helpful to think of the diagonalization in terms of a change of basis to accomplish the uncoupling of the differential equations. This is achieved using

$$\mathbf{u}(t) = \mathbf{U}\mathbf{v}(t), \tag{2.29}$$

where the modal matrix is

$$\mathbf{U} = \begin{bmatrix} \vdots & \vdots & & \vdots \\ \mathbf{u}_1 & \mathbf{u}_2 & \cdots & \mathbf{u}_N \\ \vdots & \vdots & & \vdots \end{bmatrix}.$$

The $\mathbf{u}_n$ vectors are linearly independent eigenvectors of $\mathbf{A}$. Remember that there is no need to normalize the eigenvectors when forming the modal matrix $\mathbf{U}$. We will show that with respect to these new variables, the system of first-order ordinary differential equations becomes uncoupled. The elements of $\mathbf{U}$ are constants; therefore, differentiating (2.29) gives

$$\dot{\mathbf{u}}(t) = \mathbf{U}\dot{\mathbf{v}}(t), \tag{2.30}$$

and substituting (2.29) and (2.30) into (2.28) leads to

$$\mathbf{U}\dot{\mathbf{v}}(t) = \mathbf{A}\mathbf{U}\mathbf{v}(t) + \mathbf{f}(t).$$

Premultiplying by $\mathbf{U}^{-1}$ gives

$$\dot{\mathbf{v}}(t) = \mathbf{U}^{-1}\mathbf{A}\mathbf{U}\mathbf{v}(t) + \mathbf{U}^{-1}\mathbf{f}(t).$$

Recall that unless $\mathbf{A}$ is nonsymmetric with repeated eigenvalues, $\mathbf{D} = \mathbf{U}^{-1}\mathbf{A}\mathbf{U}$ is a diagonal matrix with the eigenvalues of $\mathbf{A}$ on the main diagonal. For example, for a homogeneous system having $\mathbf{f}(t) = 0$, we have

$$\begin{bmatrix} \dot{v}_1(t) \\ \dot{v}_2(t) \\ \vdots \\ \dot{v}_N(t) \end{bmatrix} = \begin{bmatrix} \lambda_1 & 0 & \cdots & 0 \\ 0 & \lambda_2 & \cdots & 0 \\ \vdots & \vdots & \ddots & \vdots \\ 0 & 0 & \cdots & \lambda_N \end{bmatrix} \begin{bmatrix} v_1(t) \\ v_2(t) \\ \vdots \\ v_N(t) \end{bmatrix},$$

or

$$\begin{aligned} \dot{v}_1(t) &= \lambda_1 v_1(t), \\ \dot{v}_2(t) &= \lambda_2 v_2(t), \\ &\vdots \\ \dot{v}_N(t) &= \lambda_N v_N(t). \end{aligned}$$

Therefore, the differential equations have been *uncoupled*, and the solutions in terms of the transformed variables are simply

$$v_1(t) = c_1 e^{\lambda_1 t}, \quad v_2(t) = c_2 e^{\lambda_2 t}, \quad \ldots, \quad v_N(t) = c_N e^{\lambda_N t},$$

or in vector form

$$\mathbf{v}(t) = \begin{bmatrix} c_1 e^{\lambda_1 t} \\ c_2 e^{\lambda_2 t} \\ \vdots \\ c_N e^{\lambda_N t} \end{bmatrix}.$$

Now transform back to determine the solution in terms of the original variables $\mathbf{u}(t)$ using

$$\mathbf{u}(t) = \mathbf{U}\mathbf{v}(t).$$

The constants of integration, $c_i, i = 1, \ldots, N$, are determined using the initial conditions for the differential equations.

REMARKS:

1. *For this procedure to fully uncouple the equations requires linearly independent eigenvectors; therefore, it works if* $\mathbf{A}$ *is symmetric or if* $\mathbf{A}$ *is nonsymmetric with distinct eigenvalues.*

2. *Note that if the eigenvectors of* $\mathbf{A}$ *are all mutually orthogonal, they only need to be normalized, in which case* $\mathbf{U}^{-1} = \mathbf{U}^T$, *and* $\mathbf{U} = \mathbf{Q}$ *is an orthogonal modal matrix.*
3. *Although we can write down the uncoupled solution in terms of* $\mathbf{v}(t)$ *having only the eigenvalues (there is no need to actually evaluate* $\mathbf{U}^{-1}\mathbf{AU}$*), we need the eigenvectors to form the modal matrix* $\mathbf{U}$ *in order to transform back to the original variables* $\mathbf{u}(t)$ *according to* (2.29).
4. *If* $\mathbf{A}$ *is nonsymmetric with repeated eigenvalues, in which case a Jordan canonical matrix results from the diagonalization procedure, then the system of equations in* $\mathbf{v}$ *can still be solved, although they are not fully uncoupled, via backward substitution.*
5. *If the system is nonhomogeneous, determine the* homogeneous *and* particular *solutions of the uncoupled equations and sum to obtain the general solution.*
6. *Note that there is no requirement that* $\mathbf{A}$ *be nonsingular for diagonalization.*
7. *Although the diagonalization procedure only applies to systems of first-order, linear ordinary differential equations having linearly independent eigenvectors, it can be generalized using Galerkin projection as discussed in Chapter 13 to nonlinear ordinary and partial differential equations.*

In order to more clearly see that the solution $\mathbf{u}$ of the homogeneous system of first-order, linear ordinary differential equations

$$\dot{\mathbf{u}}(t) = \mathbf{A}\mathbf{u}(t) \tag{2.31}$$

is the sum of the individual modes in the new basis defined by the eigenvectors, consider the following alternative, but equivalent, formulation. For linear equations with constant coefficients, we expect N solutions of the form

$$\mathbf{x}_n(t) = \mathbf{u}_n e^{\lambda_n t}, \quad n = 1, 2, \dots, N, \tag{2.32}$$

where the elements of the vectors $\mathbf{u}_n$ are constants. Then differentiating gives

$$\dot{\mathbf{x}}_n(t) = \lambda_n \mathbf{u}_n e^{\lambda_n t},$$

and substituting into (2.31) leads to

$$\lambda_n \mathbf{u}_n e^{\lambda_n t} = \mathbf{A}\mathbf{u}_n e^{\lambda_n t};$$

therefore, canceling the exponentials yields

$$\mathbf{A}\mathbf{u}_n = \lambda_n \mathbf{u}_n.$$

Consequently, we have an eigenproblem for the coefficient matrix $\mathbf{A}$, with the eigenvalues $\lambda_1, \lambda_2, \dots, \lambda_N$ and eigenvectors $\mathbf{u}_1, \mathbf{u}_2, \dots, \mathbf{u}_N$ comprising the solutions (2.32). The general solution is the linear combination of these vector functions

$$\mathbf{u}(t) = a_1 \mathbf{x}_1 + a_2 \mathbf{x}_2 + \cdots + a_N \mathbf{x}_N,$$

or

$$\mathbf{u}(t) = a_1\mathbf{u}_1 e^{\lambda_1 t} + a_2\mathbf{u}_2 e^{\lambda_2 t} + \cdots + a_N\mathbf{u}_N e^{\lambda_N t} = \mathbf{U}\mathbf{v},$$

where $\mathbf{U}$ and $\mathbf{v}$ are as defined previously. In this way, the solution can more directly be seen to be a sum of the individual modes in the new basis defined by the eigenvectors.

Following is a series of examples of solving systems of first-order differential equations using diagonalization. Each one introduces the approach to handling an additional aspect of these problems. We begin with a homogeneous system of first-order equations.

Example 2.8 Solve the homogeneous system of first-order equations

$$\frac{du_1}{dt} = u_2 + u_3,$$
$$\frac{du_2}{dt} = u_1 + u_3,$$
$$\frac{du_3}{dt} = u_1 + u_2.$$

Solution

We write this system in matrix form $\dot{\mathbf{u}} = \mathbf{A}\mathbf{u}$, or

$$\begin{bmatrix} \dot{u}_1(t) \\ \dot{u}_2(t) \\ \dot{u}_3(t) \end{bmatrix} = \begin{bmatrix} 0 & 1 & 1 \\ 1 & 0 & 1 \\ 1 & 1 & 0 \end{bmatrix} \begin{bmatrix} u_1(t) \\ u_2(t) \\ u_3(t) \end{bmatrix}.$$

Note that the coefficient matrix is symmetric; therefore, the eigenvectors are linearly independent (even if the eigenvalues are not distinct), and $\mathbf{A}$ is fully diagonalizable. The characteristic equation for the coefficient matrix is

$$\lambda_n^3 - 3\lambda_n - 2 = 0,$$

which gives the eigenvalues

$$\lambda_1 = -1, \quad \lambda_2 = -1, \quad \lambda_3 = 2.$$

Consider the repeated eigenvalue $\lambda_1 = \lambda_2 = -1$. Substituting into $(\mathbf{A} - \lambda_n\mathbf{I})\mathbf{u}_n = \mathbf{0}$ gives the single equation

$$u_1 + u_2 + u_3 = 0.$$

Let $u_1 = d_1$ and $u_2 = d_2$, then $u_3 = -d_1 - d_2$, and we have

$$\mathbf{u}_{1,2} = \begin{bmatrix} d_1 \\ d_2 \\ -d_1 - d_2 \end{bmatrix}. \tag{2.33}$$

To obtain $\mathbf{u}_1$, we choose $d_1 = -1$ and $d_2 = 0$, resulting in

$$\mathbf{u}_1 = \begin{bmatrix} -1 \\ 0 \\ 1 \end{bmatrix},$$

and for $\mathbf{u}_2$ we choose $d_1 = -1$ and $d_2 = 1$ to give

$$\mathbf{u}_2 = \begin{bmatrix} -1 \\ 1 \\ 0 \end{bmatrix}.$$

Be sure to confirm that $\mathbf{u}_1$ and $\mathbf{u}_2$ are linearly independent so that $\mathbf{u}_2$ cannot be written as a constant multiple of $\mathbf{u}_1$.

Substituting $\lambda_3 = 2$ gives the two equations (after some row reduction)

$$\begin{aligned} -2u_1 + u_2 + u_3 &= 0, \\ -u_2 + u_3 &= 0. \end{aligned}$$

If we let $u_3 = d_1$, then the final eigenvector is

$$\mathbf{u}_3 = \begin{bmatrix} 1 \\ 1 \\ 1 \end{bmatrix}.$$

Now we form the modal matrix

$$\mathbf{U} = \begin{bmatrix} -1 & -1 & 1 \\ 0 & 1 & 1 \\ 1 & 0 & 1 \end{bmatrix}.$$

Transforming to a new variable $\mathbf{u}(t) = \mathbf{U}\mathbf{v}(t)$, the uncoupled solutions in terms of $\mathbf{v}(t)$ are

$$v_1(t) = c_1 e^{\lambda_1 t} = c_1 e^{-t}, \quad v_2(t) = c_2 e^{\lambda_2 t} = c_2 e^{-t}, \quad v_3(t) = c_3 e^{\lambda_3 t} = c_3 e^{2t}.$$

Transforming this set of solutions back to the original variable $\mathbf{u} = \mathbf{U}\mathbf{v}$ gives

$$\begin{bmatrix} u_1(t) \\ u_2(t) \\ u_3(t) \end{bmatrix} = \begin{bmatrix} -1 & -1 & 1 \\ 0 & 1 & 1 \\ 1 & 0 & 1 \end{bmatrix} \begin{bmatrix} c_1 e^{-t} \\ c_2 e^{-t} \\ c_3 e^{2t} \end{bmatrix}.$$

Thus, the general solution of the system of first-order differential equations is

$$\begin{aligned} u_1(t) &= -c_1 e^{-t} - c_2 e^{-t} + c_3 e^{2t}, \\ u_2(t) &= c_2 e^{-t} + c_3 e^{2t}, \\ u_3(t) &= c_1 e^{-t} + c_3 e^{2t}. \end{aligned}$$

The coefficients c_1, c_2, and c_3 would be determined using the three required initial conditions.

Observe that the coefficient matrix in the previous example is real and symmetric, but with one of the eigenvalues repeated twice – having multiplicity two. In seeking the corresponding eigenvectors, note the following. Each of the two eigenvectors $\mathbf{u}_1$ and $\mathbf{u}_2$ obtained for the repeated eigenvalue $\lambda_{1,2} = -1$ are not orthogonal to each other but are orthogonal to the single eigenvector $\mathbf{u}_3$ corresponding to the eigenvalue $\lambda_3 = 2$. Owing to the symmetry of $\mathbf{A}$, the two eigenvectors $\mathbf{u}_1$ and $\mathbf{u}_2$ corresponding to the repeated eigenvalue for any choice of the arbitrary constants d_1 and d_2 in (2.33) can only be guaranteed to be linearly independent, as is the case for the constants chosen in the solution. However, the constants d_1 and d_2 can be chosen such that the corresponding eigenvectors are *made* to be orthogonal with one another – and with $\mathbf{u}_3$. For example, choosing $d_1 = -1$ and $d_2 = 0$ (same as in the example) along with $d_1 = 1$ and $d_2 = -2$ produces the eigenvectors

$$\mathbf{u}_1 = \begin{bmatrix} -1 \\ 0 \\ 1 \end{bmatrix}, \quad \mathbf{u}_2 = \begin{bmatrix} 1 \\ -2 \\ 1 \end{bmatrix},$$

which are mutually orthogonal. Consequently, while symmetric matrices with repeated eigenvalues produce (at least) linearly independent eigenvectors, they can be made to be mutually orthogonal through a suitable choice of the constants, so that the full set of eigenvectors are orthogonal to one another.

In the next example, the system of equations that results from applying the relevant physical principle is not of the usual form.

Example 2.9 Obtain the differential equations governing the parallel electric circuit shown in Figure 2.5.

Solution

In circuit analysis, we define the current u, resistance R, voltage V, inductance L, and capacitance C, which are related as follows. Ohm's law gives the voltage drop across

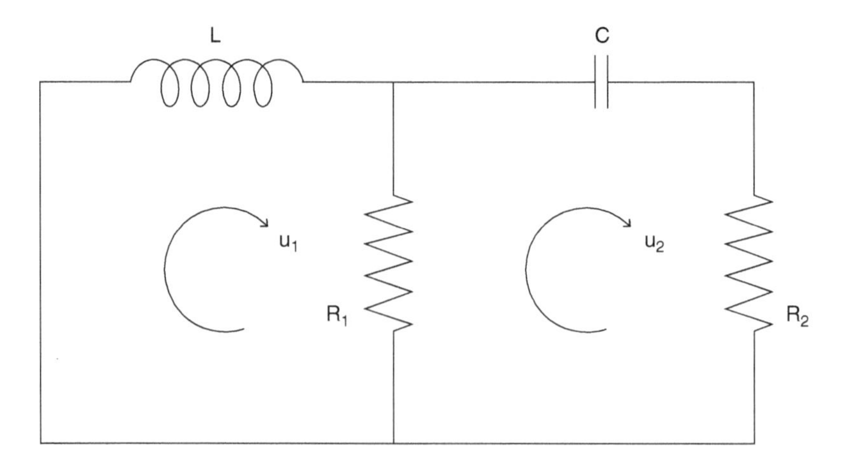

Figure 2.5 Schematic of the parallel electric circuit in Example 2.9.

a resistor as

$$V = uR,$$

and the voltage drop across an inductor is

$$V = L\frac{du}{dt},$$

and a capacitor is such that

$$u = C\frac{dV}{dt},$$

in which case

$$V = \frac{1}{C}\int u dt.$$

Parallel circuits must obey Kirchhoff's laws, which require that

1. $\sum u_{in} = \sum u_{out}$ at each junction,
2. $\sum V = 0$ around each closed circuit.

Applying Kirchhoff's second law around each loop:
Loop 1:

$$V_L + V_{R_1} = 0 \quad \Rightarrow \quad L\frac{du_1}{dt} + (u_1 - u_2)R_1 = 0.$$

Loop 2:

$$V_C + V_{R_2} + V_{R_1} = 0 \quad \Rightarrow \quad \frac{1}{C}\int u_2 dt + u_2 R_2 + (u_2 - u_1)R_1 = 0.$$

Differentiating the second equation with respect to t (to remove the integral) leads to

$$u_2 + CR_2\frac{du_2}{dt} + CR_1\left(\frac{du_2}{dt} - \frac{du_1}{dt}\right) = 0.$$

Therefore, we have the system of two ordinary differential equations for the currents $u_1(t)$ and $u_2(t)$ given by

$$L\frac{du_1}{dt} = -R_1 u_1 + R_1 u_2,$$

$$-CR_1\frac{du_1}{dt} + C(R_1 + R_2)\frac{du_2}{dt} = -u_2,$$

or in matrix form

$$\begin{bmatrix} L & 0 \\ -CR_1 & C(R_1 + R_2) \end{bmatrix} \begin{bmatrix} \dot{u}_1(t) \\ \dot{u}_2(t) \end{bmatrix} = \begin{bmatrix} -R_1 & R_1 \\ 0 & -1 \end{bmatrix}, \begin{bmatrix} u_1(t) \\ u_2(t) \end{bmatrix},$$

or

$$\mathbf{A}_1\dot{\mathbf{u}}(t) = \mathbf{A}_2\mathbf{u}(t).$$

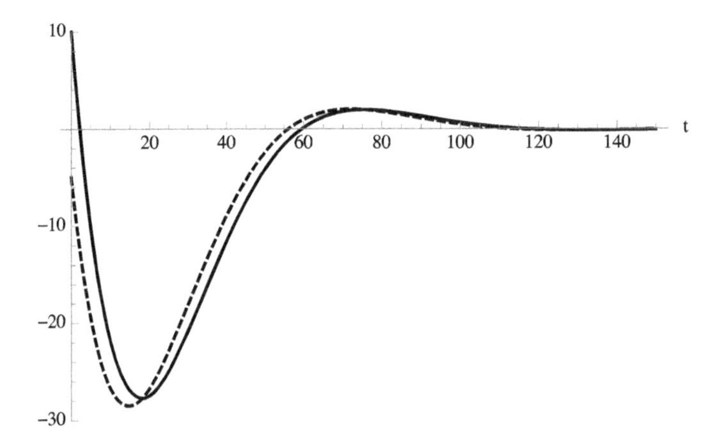

Figure 2.6 Sample solution for Example 2.9 ($u_1(t)$ – solid line, $u_2(t)$ – dashed line).

To obtain the usual form $\dot{\mathbf{u}} = \mathbf{A}\mathbf{u}$, premultiply $\mathbf{A}_1^{-1}$ on both sides to give

$$\dot{\mathbf{u}}(t) = \mathbf{A}_1^{-1}\mathbf{A}_2\mathbf{u}(t);$$

therefore,

$$\mathbf{A} = \mathbf{A}_1^{-1}\mathbf{A}_2.$$

We would then diagonalize $\mathbf{A}$ as usual in order to solve for $u_1(t)$ and $u_2(t)$.

A sample solution with $R_1 = 10$, $R_2 = 3$, $L = 30$, $C = 5$, and initial conditions $u_1(0) = 10$, and $u_2(0) = -5$ is shown in Figure 2.6. Note that because there is no voltage source, the current decays with time via a damped oscillation.

Next we consider an example consisting of a nonhomogeneous system of equations having $\mathbf{f}(t) \neq 0$.

Example 2.10 Solve the system of equations

$$\begin{aligned}\dot{u}_1(t) &= -2u_2 + \sin t,\\ \dot{u}_2(t) &= -2u_1 - t.\end{aligned}$$

Solution

In matrix form, the system of first-order linear differential equations is

$$\dot{\mathbf{u}}(t) = \mathbf{A}\mathbf{u}(t) + \mathbf{f}(t),$$

where

$$\mathbf{A} = \begin{bmatrix} 0 & -2 \\ -2 & 0 \end{bmatrix}, \quad \mathbf{f}(t) = \begin{bmatrix} \sin t \\ -t \end{bmatrix}.$$

The eigenvalues and eigenvectors of $\mathbf{A}$ are

$$\lambda_1 = 2, \qquad \mathbf{u}_1 = \begin{bmatrix} 1 \\ -1 \end{bmatrix},$$
$$\lambda_2 = -2, \qquad \mathbf{u}_2 = \begin{bmatrix} 1 \\ 1 \end{bmatrix}.$$

Note that $\mathbf{A}$ is symmetric with distinct eigenvalues; therefore, the eigenvectors are mutually orthogonal.

Forming the modal matrix and taking its inverse, we have

$$\mathbf{U} = \begin{bmatrix} 1 & 1 \\ -1 & 1 \end{bmatrix}, \quad \mathbf{U}^{-1} = \frac{1}{2}\begin{bmatrix} 1 & -1 \\ 1 & 1 \end{bmatrix}.$$

We transform to the new variables $\mathbf{v}(t)$ using

$$\mathbf{u}(t) = \mathbf{U}\mathbf{v}(t),$$

with respect to which our system of equations becomes

$$\dot{\mathbf{v}}(t) = \mathbf{U}^{-1}\mathbf{A}\mathbf{U}\mathbf{v}(t) + \mathbf{U}^{-1}\mathbf{f}(t).$$

Recalling that $\mathbf{D} = \mathbf{U}^{-1}\mathbf{A}\mathbf{U}$ has the eigenvalues along its main diagonal and evaluating $\mathbf{U}^{-1}\mathbf{f}(t)$, this becomes

$$\begin{bmatrix} \dot{v}_1(t) \\ \dot{v}_2(t) \end{bmatrix} = \begin{bmatrix} 2 & 0 \\ 0 & -2 \end{bmatrix}\begin{bmatrix} v_1(t) \\ v_2(t) \end{bmatrix} + \frac{1}{2}\begin{bmatrix} \sin t + t \\ \sin t - t \end{bmatrix};$$

therefore, we have the two uncoupled equations

$$\dot{v}_1(t) = 2v_1(t) + \frac{1}{2}\sin t + \frac{1}{2}t,$$
$$\dot{v}_2(t) = -2v_2(t) + \frac{1}{2}\sin t - \frac{1}{2}t.$$

Because the equations are *nonhomogeneous*, the solutions are of the form

$$v_1(t) = v_{1c}(t) + v_{1p}(t),$$
$$v_2(t) = v_{2c}(t) + v_{2p}(t),$$

where subscript c represents the *complementary*, or *homogeneous*, *solution*, and subscript p represents the *particular solution*. The complementary solutions are

$$v_{1c}(t) = c_1 e^{\lambda_1 t} = c_1 e^{2t},$$
$$v_{2c}(t) = c_2 e^{\lambda_2 t} = c_2 e^{-2t}.$$

We determine the particular solutions using the *method of undetermined coefficients*, which applies for right-hand sides involving polynomials, exponential functions, and trigonometric functions.

Consider the particular solution for $y_1(t)$, which is of the general form

$$v_{1p}(t) = A\sin t + B\cos t + Ct + D,$$

and differentiating gives

$$\dot{v}_{1p}(t) = A\cos t - B\sin t + C.$$

Substituting into the uncoupled equation for $v_1(t)$ yields

$$A\cos t - B\sin t + C = 2(A\sin t + B\cos t + Ct + D) + \frac{1}{2}\sin t + \frac{1}{2}t.$$

Equating like terms leads to four equations for the four unknown constants as follows:

$$\begin{aligned}
&\cos t: && A = 2B && \Rightarrow \quad A = -\frac{1}{5}, \\
&\sin t: && -B = 2A + \frac{1}{2} && \Rightarrow \quad -B = 4B + \frac{1}{2} \Rightarrow B = -\frac{1}{10}, \\
&t: && 0 = 2C + \frac{1}{2} && \Rightarrow \quad C = -\frac{1}{4}, \\
&\text{constant}: && C = 2D && \Rightarrow \quad D = -\frac{1}{8}.
\end{aligned}$$

In general, this could lead to a 4×4 system of equations for the "undetermined coefficients" A, B, C, and D. Thus, the particular solution is

$$v_{1p}(t) = -\frac{1}{5}\sin t - \frac{1}{10}\cos t - \frac{1}{4}t - \frac{1}{8},$$

and the general solution is

$$v_1(t) = v_{1c}(t) + v_{1p}(t) = c_1e^{2t} - \frac{1}{5}\sin t - \frac{1}{10}\cos t - \frac{1}{4}t - \frac{1}{8}.$$

Similarly, considering the equation for $v_2(t)$ leads to

$$v_2(t) = c_2e^{-2t} + \frac{1}{5}\sin t - \frac{1}{10}\cos t - \frac{1}{4}t + \frac{1}{8}.$$

To obtain the solution in terms of the original variable $\mathbf{u}(t)$, evaluate

$$\mathbf{u}(t) = \mathbf{U}\mathbf{v}(t),$$

or

$$\begin{bmatrix} u_1(t) \\ u_2(t) \end{bmatrix} = \begin{bmatrix} 1 & 1 \\ -1 & 1 \end{bmatrix}\begin{bmatrix} v_1(t) \\ v_2(t) \end{bmatrix} = \begin{bmatrix} v_1 + v_2 \\ -v_1 + v_2 \end{bmatrix}.$$

Thus, the general solution is

$$\begin{aligned}
u_1(t) &= c_1e^{2t} + c_2e^{-2t} - \frac{1}{5}\cos t - \frac{1}{2}t, \\
u_2(t) &= -c_1e^{2t} + c_2e^{-2t} + \frac{2}{5}\sin t + \frac{1}{4}.
\end{aligned}$$

The integration constants c_1 and c_2 would be obtained using initial conditions if given.

The method of undetermined coefficients used here to obtain the particular solution is very straightforward as the basic form of the particular solution is known for polynomial, exponential, and trigonometric right-hand sides; we simply need to determine the coefficients on each term. For more on the method of undetermined coefficients, see, for example, Jeffrey (2002) or Kreyszig (2011). In particular, be sure that $v_p(t)$ is not already part of the complementary solution $v_c(t)$; in the previous example, for instance, this would occur if $v_p(t) = Ae^{2t} + \cdots$. If this were the case, include a t factor in the particular solution, such that $v_p = Ate^{2t} + \cdots$ as discussed in the preceding references. For situations when this method does not apply, *variation of parameters* can be used (see, for example, Kreyszig, 2011).

There are two additional eventualities that must be addressed. They are (1) when the eigenvalues and/or eigenvectors are complex and (2) when we have higher-order linear differential equations as is common in applications. These are addressed in the next section.

2.6.2 Higher-Order Equations

We can extend the diagonalization approach for systems of first-order, linear differential equations to *any* linear ordinary differential equation as follows. Consider the Nth-order linear differential equation

$$u^{(N)} = F\left(t, u, \dot{u}, \ldots, u^{(N-1)}\right),$$

where superscripts in parentheses indicate the order of the derivative. This Nth-order differential equation can be converted to a system of N first-order differential equations by the substitutions

$$\begin{aligned}
u_1(t) &= u(t), \\
u_2(t) &= \dot{u}(t), \\
u_3(t) &= \ddot{u}(t), \\
&\vdots \\
u_{N-1}(t) &= u^{(N-2)}(t), \\
u_N(t) &= u^{(N-1)}(t).
\end{aligned}$$

Differentiating the preceding substitutions and writing exclusively in terms of the new variables results in a system of first-order equations

$$\begin{aligned}
\dot{u}_1(t) &= \dot{u}(t) &&= u_2(t), \\
\dot{u}_2(t) &= \ddot{u}(t) &&= u_3(t), \\
\dot{u}_3(t) &= \dddot{u}(t) &&= u_4(t), \\
&\vdots \\
\dot{u}_{N-1}(t) &= u^{(N-1)}(t) &&= u_N(t), \\
\dot{u}_N(t) &= u^{(N)}(t) &&= F(t, u_1, u_2, \ldots, u_N),
\end{aligned}$$

with the last equation following from the original differential equation.

REMARKS:

1. *Be sure that the transformed system is only in terms of the new variables* $u_1(t), u_2(t), \ldots, u_N(t)$.

2. *This approach can be used to convert any system of higher-order linear differential equations to a system of first-order linear equations. For example, three coupled second-order equations could be converted to six first-order equations.*
3. *For dynamical systems considered in Chapter 5, this first-order form is called the* state-space *representation.*

In the next example, we consider a single second-order linear differential equation. Because this is simply a linear ordinary differential equation with constant coefficients, it can be solved more directly using standard methods (in fact, you may know the solution of this equation without even solving it owing to its commonality). However, we seek to use this example to illustrate the diagonalization procedure as applied to higher-order differential equations. It also illustrates how to deal with systems of equations having complex eigenvalues and/or eigenvectors.

Example 2.11 Obtain the solution of the second-order initial-value problem

$$\frac{d^2u}{dt^2} + u = 0, \tag{2.34}$$

with the initial conditions

$$u(0) = 1, \quad \dot{u}(0) = \left.\frac{du}{dt}\right|_{t=0} = 2. \tag{2.35}$$

Solution

In order to convert this second-order equation to a system of two first-order differential equations, we make the following substitutions:

$$\begin{aligned} u_1(t) &= u(t), \\ u_2(t) &= \dot{u}(t). \end{aligned} \tag{2.36}$$

Differentiating the substitutions and transforming to $u_1(t)$ and $u_2(t)$, we have the following system of two first-order equations

$$\begin{aligned} \dot{u}_1(t) &= \dot{u}(t) = u_2(t), \\ \dot{u}_2(t) &= \ddot{u}(t) = -u(t) = -u_1(t), \end{aligned}$$

for the two unknowns $u_1(t)$ and $u_2(t)$. Note that the original second-order equation (2.34) has been used in the final equation ($\ddot{u} = -u$). Written in matrix form, $\dot{\mathbf{u}}(t) = \mathbf{A}\mathbf{u}(t)$, we have

$$\begin{bmatrix} \dot{u}_1(t) \\ \dot{u}_2(t) \end{bmatrix} = \begin{bmatrix} 0 & 1 \\ -1 & 0 \end{bmatrix} \begin{bmatrix} u_1(t) \\ u_2(t) \end{bmatrix},$$

where $\mathbf{A}$ is not symmetric.

To obtain the eigenvalues, we evaluate $|\mathbf{A} - \lambda_n \mathbf{I}| = 0$, or

$$\begin{vmatrix} -\lambda_n & 1 \\ -1 & -\lambda_n \end{vmatrix} = 0,$$

which yields the characteristic equation

$$\lambda_n^2 + 1 = 0.$$

Factoring gives the distinct eigenvalues

$$\lambda_1 = -\mathrm{i}, \quad \lambda_2 = \mathrm{i},$$

which is a complex conjugate pair.[18] Having complex eigenvalues requires a minor modification to the preceding outlined procedure, but for now we proceed as before. The corresponding eigenvectors are also complex and given by

$$\mathbf{u}_1 = \begin{bmatrix} \mathrm{i} \\ 1 \end{bmatrix}, \quad \mathbf{u}_2 = \begin{bmatrix} -\mathrm{i} \\ 1 \end{bmatrix}.$$

Consequently, forming the modal matrix, we have

$$\mathbf{U} = \begin{bmatrix} \mathrm{i} & -\mathrm{i} \\ 1 & 1 \end{bmatrix}.$$

Note that because matrix $\mathbf{A}$ is not symmetric but has distinct eigenvalues, we have linearly independent eigenvectors, and the system can be fully diagonalized. In order to uncouple the system of differential equations, we transform the problem according to

$$\mathbf{u}(t) = \mathbf{U}\mathbf{v}(t).$$

With respect to $\mathbf{v}(t)$, the solution is

$$v_1(t) = c_1 e^{-\mathrm{i}t}, \quad v_2(t) = c_2 e^{\mathrm{i}t}.$$

Transforming back using $\mathbf{u}(t) = \mathbf{U}\mathbf{v}(t)$ gives the solution of the system of first-order equations in terms of $\mathbf{u}(t)$. From the substitutions (2.36), we obtain the solution with respect to the original variable as follows:

$$u(t) = u_1(t) = c_1 \mathrm{i} e^{-\mathrm{i}t} - c_2 \mathrm{i} e^{\mathrm{i}t}.$$

Other than applying the initial conditions to obtain the integration constants c_1 and c_2, we would normally be finished at this point. Because the solution is complex, however, we must do a bit more work to obtain the real solution. It can be shown that for linear equations, both the real and imaginary parts of a complex solution are by themselves solutions of the differential equations, and that a linear combination of the real and imaginary parts, which are both real, is also a solution of the linear equations. We can extract the real and imaginary parts of exponentials by applying *Euler's formula*, which is

$$e^{a\mathrm{i}t} = \cos(at) + \mathrm{i}\sin(at).$$

[18] The imaginary number is denoted by $\mathrm{i} = \sqrt{-1}$; i is reserved for use as an index in numerical methods.

Applying the Euler formula to our solution yields

$$\begin{aligned} u(t) &= c_1 \mathrm{i} e^{-\mathrm{i}t} - c_2 \mathrm{i} e^{\mathrm{i}t} \\ &= c_1 \mathrm{i} \left[\cos(-t) + \mathrm{i}\sin(-t)\right] - c_2 \mathrm{i} \left(\cos t + \mathrm{i}\sin t\right) \\ u(t) &= c_1 \mathrm{i} \left(\cos t - \mathrm{i}\sin t\right) - c_2 \mathrm{i} \left(\cos t + \mathrm{i}\sin t\right); \end{aligned}$$

therefore, the real and imaginary parts are

$$\mathrm{Re}(u) = (c_1 + c_2)\sin t,$$
$$\mathrm{Im}(u) = (c_1 - c_2)\cos t.$$

To construct the general solution for $u(t)$, we then superimpose the real and imaginary parts to obtain the general form of the solution to the original second-order differential equation as follows:

$$u(t) = A\sin t + B\cos t. \tag{2.37}$$

The constants A and B are obtained by applying the initial conditions (2.35), which lead to $A = 2$ and $B = 1$; therefore, the final solution of (2.34) subject to the initial conditions (2.35) is

$$u(t) = 2\sin t + \cos t.$$

REMARKS:

1. *While real eigenvalues correspond to an exponential solution as in Example 2.8, observe that the imaginary eigenvalues here correspond to an oscillatory solution.*
2. *You may recognize that the second-order differential equation (2.34) considered in this example governs the motion of an undamped harmonic oscillator. In this context, the initial conditions* (2.35) *represent initial position and velocity values of the oscillator.*

REMARKS:

1. *The approach used in Example 2.11 to handle complex eigenvalues and eigenvectors holds for linear systems for which superposition of solutions is valid.*
2. *For additional examples of systems of linear ordinary differential equations and application to discrete dynamical systems, see the next section and Chapter 5.*
3. *For illustrations of the types of solutions possible when solving systems of* non-linear *equations, see Sections 5.5 and 5.6 for the forced Duffing equation and the Saltzman–Lorenz model, respectively.*

2.6.3 Illustrative Example from Dynamics

Let us consider a physical example from dynamics. It comes from Example 5.5 of Cassel (2013), where the governing state equations are obtained using Hamilton's principle. Here, they are obtained using Newton's second law and solved using matrix methods and diagonalization.

Example 2.12 Consider the spring–mass system shown in Figure 2.7. The system is such that there is no force in any of the springs when the masses are located at $x_1 = 0, x_2 = 0$. Obtain the equations of motion for the system, the natural frequencies of the system, and general solutions for the motion of the masses.

Solution

This spring–mass system has two *degrees of freedom*, which is the minimum number of dependent variables ($x_1(t)$ and $x_2(t)$ in this case) required to fully specify the state of the system. Physically, this means that the system has two *natural frequencies*, or *modes, of vibration*, and the general motion of the system is a superposition of these two modes.

In order to obtain the governing equations of motion for this system, recall Newton's second law: $\mathbf{f} = m\mathbf{a} = m\ddot{\mathbf{x}}$. Applying Newton's second law to the free-body diagram for each mass:

(1)

$2kx_1$ ← [m] → $k(x_2 - x_1)$

$$m\frac{d^2x_1}{dt^2} = k(x_2 - x_1) - 2kx_1 = -3kx_1 + kx_2, \tag{2.38}$$

(2)

$k(x_2 - x_1)$ ← [2m] → $-kx_2$

$$2m\frac{d^2x_2}{dt^2} = -kx_2 - k(x_2 - x_1) = kx_1 - 2kx_2. \tag{2.39}$$

Thus, we have a system of two second-order ordinary differential equations for $x_1(t)$ and $x_2(t)$.

It is common to write such equations of motion for dynamical systems in the matrix form

$$\mathbf{M}\ddot{\mathbf{x}} + \mathbf{C}\dot{\mathbf{x}} + \mathbf{K}\mathbf{x} = \mathbf{f}(t),$$

where $\mathbf{M}$ is the *mass matrix*, $\mathbf{C}$ is the *damping matrix*, $\mathbf{K}$ is the *stiffness matrix*, $\ddot{\mathbf{x}}(t)$ is the *acceleration vector*, $\dot{\mathbf{x}}(t)$ is the *velocity vector*, $\mathbf{x}(t)$ is the *displacement vector*, and $\mathbf{f}(t)$ is the *force vector*. In the present case

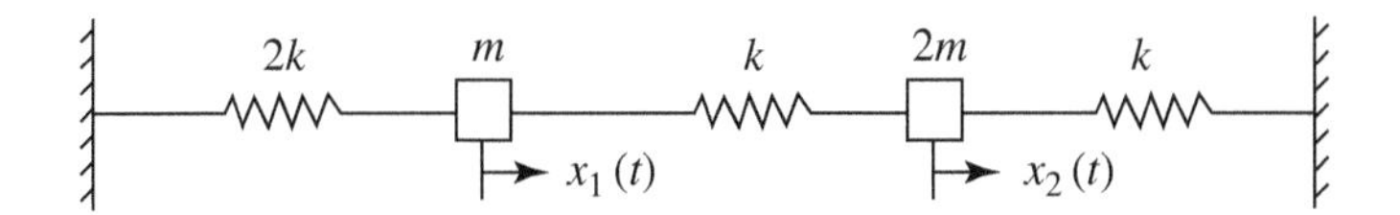

Figure 2.7 Schematic of the spring-mass system in Example 2.12.

$$\mathbf{M} = \begin{bmatrix} m & 0 \\ 0 & 2m \end{bmatrix}, \quad \mathbf{C} = \begin{bmatrix} 0 & 0 \\ 0 & 0 \end{bmatrix}, \quad \mathbf{K} = \begin{bmatrix} 3k & -k \\ -k & 2k \end{bmatrix}, \quad \mathbf{f} = \begin{bmatrix} 0 \\ 0 \end{bmatrix}, \quad \mathbf{x} = \begin{bmatrix} x_1 \\ x_2 \end{bmatrix}.$$

Before seeking the general solution of this system of equations using diagonalization, let us first obtain the *natural frequencies* (*normal modes*) of the system. A natural frequency is a solution in which all of the parts of the system oscillate with the same frequency, such that the motion of the entire system is periodic. In the present case, these will occur when the two masses oscillate with the same frequency; therefore, assume periodic motion of the form[19]

$$x_1(t) = A_1 e^{\mathrm{i}\omega t}, \quad x_2(t) = A_2 e^{\mathrm{i}\omega t},$$

where ω is the natural frequency, and A_1 and A_2 are the amplitudes of masses 1 and 2, respectively. It is understood that we are taking the real part of the final expressions for $x_1(t)$ and $x_2(t)$. For example, using Euler's formula

$$x_1(t) = Re\left[A_1 e^{\mathrm{i}\omega t}\right] = Re\left[A_1 \cos(\omega t) + A_1 \mathrm{i} \sin(\omega t)\right] = A_1 \cos(\omega t).$$

Evaluating the derivatives gives

$$\frac{d^2 x_1}{dt^2} = -A_1 \omega^2 e^{\mathrm{i}\omega t}, \quad \frac{d^2 x_2}{dt^2} = -A_2 \omega^2 e^{\mathrm{i}\omega t}.$$

Substituting into (2.38) and (2.39) and canceling $e^{\mathrm{i}\omega t}$ gives the two linear algebraic equations

$$\begin{aligned} -m\omega^2 A_1 &= -3kA_1 + kA_2, \\ -2m\omega^2 A_2 &= kA_1 - 2kA_2, \end{aligned}$$

or upon rearranging

$$\begin{aligned} 3A_1 - A_2 &= \lambda A_1, \\ -\tfrac{1}{2}A_1 + A_2 &= \lambda A_2, \end{aligned}$$

where $\lambda = m\omega^2/k$. Thus, we have two linear algebraic equations for the two unknown amplitudes A_1 and A_2. In matrix form, this is

$$\begin{bmatrix} 3 & -1 \\ -\frac{1}{2} & 1 \end{bmatrix} \begin{bmatrix} A_1 \\ A_2 \end{bmatrix} = \lambda \begin{bmatrix} A_1 \\ A_2 \end{bmatrix}.$$

Consequently, we have an eigenproblem in which the eigenvalues λ are related to the natural frequencies ω of the system, which is typical of dynamical systems. The corresponding eigenvectors are the amplitudes of oscillation of the two masses, which we do not actually care about here. Solving, we find that the eigenvalues are

$$\lambda_1 = \frac{m\omega_1^2}{k} = 2 - \frac{\sqrt{6}}{2}, \quad \lambda_2 = \frac{m\omega_2^2}{k} = 2 + \frac{\sqrt{6}}{2}.$$

[19] This form is better than setting $x_1(t) = A_1 \cos(\omega t)$ directly, for example, which does not work if we have odd-order derivatives.

Therefore, the two *natural frequencies*, or *modes*, of the system, corresponding to the two degrees of freedom, are

$$\omega_1 = 0.8805\sqrt{\frac{k}{m}}, \quad \omega_2 = 1.7958\sqrt{\frac{k}{m}}. \tag{2.40}$$

The smallest of the natural frequencies, ω_1, is known as the *fundamental mode*. As we will show, the general motion of the system is a superposition of these two modes of oscillation for systems governed by linear equations of motion, as is the case here.

Now we return to the general solution of the system of two second-order differential equations (2.38) and (2.39) – forgetting that we know the natural frequencies. In order to convert these to a system of first-order differential equations, we make the following substitutions

$$\begin{aligned} u_1(t) &= x_1(t), \\ u_2(t) &= x_2(t), \\ u_3(t) &= \dot{x}_1(t), \\ u_4(t) &= \dot{x}_2(t). \end{aligned} \tag{2.41}$$

Note that two second-order ordinary differential equations transform to four first-order equations. Differentiating the substitutions and transforming to the new variables, we have the following system of four equations

$$\begin{aligned} \dot{u}_1(t) &= \dot{x}_1(t) = u_3(t), \\ \dot{u}_2(t) &= \dot{x}_2(t) = u_4(t), \\ \dot{u}_3(t) &= \ddot{x}_1(t) = -3Kx_1(t) + Kx_2(t) = -3Ku_1(t) + Ku_2(t), \\ \dot{u}_4(t) &= \ddot{x}_2(t) = \tfrac{1}{2}Kx_1(t) - Kx_2(t) = \tfrac{1}{2}Ku_1(t) - Ku_2, \end{aligned}$$

for the four unknowns $u_1(t)$, $u_2(t)$, $u_3(t)$, and $u_4(t)$, where $K = k/m$. Written in matrix form, $\dot{\mathbf{u}}(t) = \mathbf{A}\mathbf{u}(t)$, we have the homogeneous system of equations

$$\begin{bmatrix} \dot{u}_1(t) \\ \dot{u}_2(t) \\ \dot{u}_3(t) \\ \dot{u}_4(t) \end{bmatrix} = \begin{bmatrix} 0 & 0 & 1 & 0 \\ 0 & 0 & 0 & 1 \\ -3K & K & 0 & 0 \\ \frac{1}{2}K & -K & 0 & 0 \end{bmatrix} \begin{bmatrix} u_1(t) \\ u_2(t) \\ u_3(t) \\ u_4(t) \end{bmatrix}.$$

For simplicity, let us take $K = k/m = 1$. Then the eigenvalues of $\mathbf{A}$, which can be obtained using MATLAB or Mathematica, are

$$\lambda_1 = 1.7958\mathrm{i}, \quad \lambda_2 = -1.7958\mathrm{i}, \quad \lambda_3 = 0.8805\mathrm{i}, \quad \lambda_4 = -0.8805\mathrm{i},$$

which are complex conjugate pairs. The corresponding eigenvectors are also complex:

$$\mathbf{u}_1 = \begin{bmatrix} -0.4747\mathrm{i} \\ 0.1067\mathrm{i} \\ 0.8524 \\ -0.1916 \end{bmatrix}, \quad \mathbf{u}_2 = \begin{bmatrix} 0.4747\mathrm{i} \\ -0.1067\mathrm{i} \\ 0.8524 \\ -0.1916 \end{bmatrix}, \quad \mathbf{u}_3 = \begin{bmatrix} 0.3077 \\ 0.6846 \\ 0.2709\mathrm{i} \\ 0.6027\mathrm{i} \end{bmatrix}, \quad \mathbf{u}_4 = \begin{bmatrix} 0.3077 \\ 0.6846 \\ -0.2709\mathrm{i} \\ -0.6027\mathrm{i} \end{bmatrix}.$$

Thus, forming the modal matrix, we have

$$\mathbf{U} = \begin{bmatrix} -0.4747\mathrm{i} & 0.4747\mathrm{i} & 0.3077 & 0.3077 \\ 0.1067\mathrm{i} & -0.1067\mathrm{i} & 0.6846 & 0.6846 \\ 0.8524 & 0.8524 & 0.2709\mathrm{i} & -0.2709\mathrm{i} \\ -0.1916 & -0.1916 & 0.6027\mathrm{i} & -0.6027\mathrm{i} \end{bmatrix}.$$

In order to uncouple the system of differential equations, we transform the problem according to

$$\mathbf{u}(t) = \mathbf{U}\mathbf{v}(t).$$

Recalling that $\mathbf{A}$ is nonsymmetric with distinct eigenvalues, the system is uncoupled with respect to $\mathbf{v}(t)$, and the solution is

$$v_1(t) = c_1 e^{1.7958\mathrm{i}t}, \quad v_2(t) = c_2 e^{-1.7958\mathrm{i}t}, \quad v_3(t) = c_3 e^{0.8805\mathrm{i}t}, \quad v_4(t) = c_4 e^{-0.8805\mathrm{i}t}.$$

Transforming back using $\mathbf{u}(t) = \mathbf{U}\mathbf{v}(t)$ gives the solution of the system of first-order equations in terms of $\mathbf{u}(t)$. From the substitutions (2.41), we obtain the solution with respect to the original variables as follows:

$$x_1(t) = u_1(t) = -0.4747\mathrm{i}v_1(t) + 0.4747\mathrm{i}v_2(t) + 0.3077v_3(t) + 0.3077v_4(t),$$
$$x_2(t) = u_2(t) = 0.1067\mathrm{i}v_1(t) + -0.1067\mathrm{i}v_2(t) + 0.6846v_3(t) + 0.6846v_4(t).$$

As in Example 2.11, we can extract the real and imaginary parts by applying Euler's formula. Doing this, the real and imaginary parts of $x_1(t)$ and $x_2(t)$ are

$$\mathrm{Re}(x_1) = 0.4747(c_1 + c_2)\sin(1.7958t) + 0.3077(c_3 + c_4)\cos(0.8805t),$$
$$\mathrm{Im}(x_1) = -0.4747(c_1 - c_2)\cos(1.7958t) + 0.3077(c_3 - c_4)\sin(0.8805t),$$

$$\mathrm{Re}(x_2) = -0.1067(c_1 + c_2)\sin(1.7958t) + 0.6846(c_3 + c_4)\cos(0.8805t),$$
$$\mathrm{Im}(x_2) = 0.1067(c_1 - c_2)\cos(1.7958t) + 0.6846(c_3 - c_4)\sin(0.8805t).$$

To obtain the general solution for $x_1(t)$ and $x_2(t)$, we then superimpose the real and imaginary parts, such that $x_i(t) = A\mathrm{Re}(x_i) + B\mathrm{Im}(x_i)$, where A and B are constants, giving

$$\begin{aligned} x_1(t) &= 0.4747a_1\sin(1.7958t) + 0.3077a_2\cos(0.8805t) \\ &\quad -0.4747a_3\cos(1.7958t) + 0.3077a_4\sin(0.8805t), \\ x_2(t) &= -0.1067a_1\sin(1.7958t) + 0.6846a_2\cos(0.8805t) \\ &\quad +0.1067a_3\cos(1.7958t) + 0.6846a_4\sin(0.8805t), \end{aligned} \tag{2.42}$$

where the constants are

$$a_1 = A(c_1 + c_2), \quad a_2 = A(c_3 + c_4), \quad a_3 = B(c_1 - c_2), \quad a_4 = B(c_3 - c_4).$$

Note that the general solution consists of a superposition of the two natural frequencies (2.40) determined by the previous natural modes calculation. Also observe that the imaginary eigenvalues correspond to oscillatory behavior as expected for the spring–mass system.

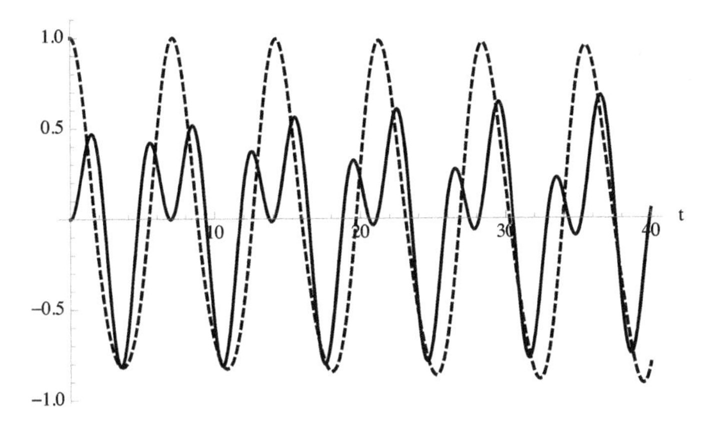

Figure 2.8 Solution for $x_1(t)$ (solid line) and $x_2(t)$ (dashed line) for Example 2.12.

To obtain the four integration constants, we require four initial conditions. We will specify the positions and velocities of the two masses as follows:

$$
\begin{aligned}
x_1 = 0, \quad \dot{x}_1 = 0 \quad \text{at} \quad t = 0, \\
x_2 = 1, \quad \dot{x}_2 = 0 \quad \text{at} \quad t = 0.
\end{aligned}
$$

Applying these initial conditions to (2.42), we obtain a system of four equations for the four unknown constants

$$
\begin{bmatrix} 0 & 0.3077 & -0.4747 & 0 \\ 0.8525 & 0 & 0 & 0.2709 \\ 0 & 0.6846 & 0.1067 & 0 \\ -0.1916 & 0 & 0 & 0.6028 \end{bmatrix} \begin{bmatrix} a_1 \\ a_2 \\ a_3 \\ a_4 \end{bmatrix} = \begin{bmatrix} 0 \\ 0 \\ 1 \\ 0 \end{bmatrix}.
$$

Observe that the initial conditions appear in the right-hand-side vector. Solving this system gives

$$
a_1 = 0, \quad a_2 = 1.3267, \quad a_3 = 0.8600, \quad a_4 = 0.
$$

Substituting these constants into the general solutions (2.42) for $x_1(t)$ and $x_2(t)$ yields the solution shown in Figure 2.8.

Recall that all general solutions, including the preceding one, are linear combinations of the two natural frequencies. Thus, we would expect that a set of initial conditions can be found that excites only one of the natural frequencies. For example, choosing initial conditions such that only ω_1 is excited is shown in Figure 2.9, and that which only excites ω_2 is shown in Figure 2.10.

REMARKS:

1. *In this example, the matrix* **A** *is comprised entirely of constants and does not depend on time. Such dynamical systems are called* autonomous.
2. *Observe that transforming a second-order system to an equivalent first-order one, which doubles the number of equations, does not introduce additional frequencies.*

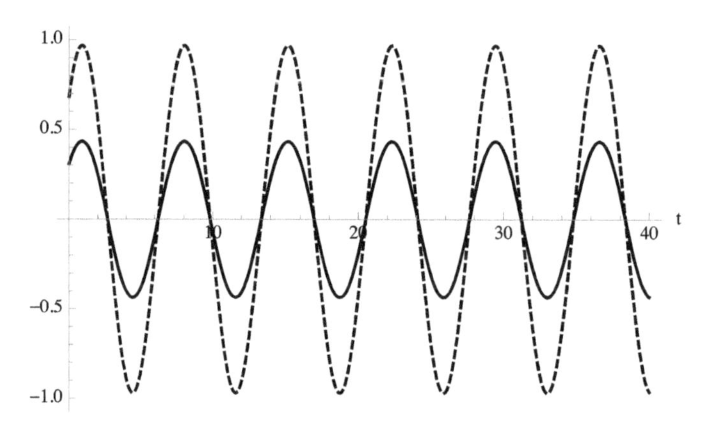

Figure 2.9 Solution for $x_1(t)$ (solid line) and $x_2(t)$ (dashed line) for Example 2.12 when only ω_1 is excited.

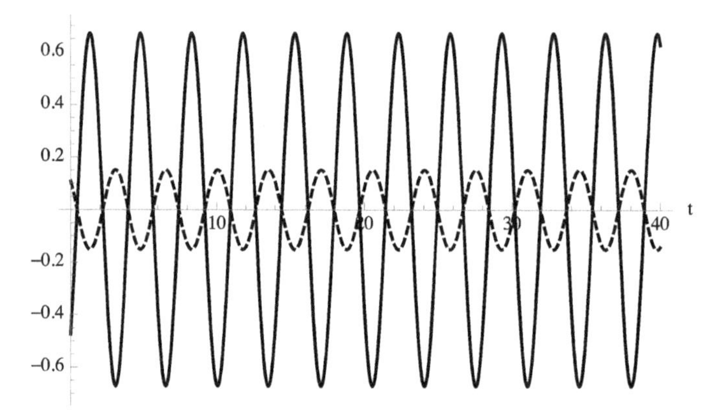

Figure 2.10 Solution for $x_1(t)$ (solid line) and $x_2(t)$ (dashed line) for Example 2.12 when only ω_2 is excited.

In the preceding example, for instance, there are still only two natural frequencies in the uncoupled solution $\mathbf{v}(t)$.

3. *The diagonalization procedure is equivalent to determining an alternative basis* $\mathbf{v}(t)$, *sometimes called* principal coordinates, *with respect to which the motions of the masses in the system are uncoupled. Although this may seem physically counterintuitive for a coupled mechanical system, such as our spring–mass example, the mathematics tells us that it must be the case – except for systems that result in nonsymmetric matrices with repeated eigenvalues, which result in the Jordan canonical form. This is similar to the change of basis through diagonalization to put quadratics in their canonical form.*
4. *See Chapter 5 for much more on how matrix methods can be used to solve, analyze, characterize, and interpret solutions of linear and nonlinear dynamical systems.*

2.7 Schur Decomposition

We now turn our attention from diagonalization of square matrices to another central pillar of matrix methods – matrix decomposition. Matrix *decomposition*, or *factorization*, is a method of taking a matrix $\mathbf{A}$ and factoring it into the product of two or more matrices that have particular properties, often for further analysis or computation. For example, the diagonalization procedure that has served us so well can be thought of as a decomposition. Namely, for a square matrix $\mathbf{A}$, we have the following *eigen-decompositions*:

- For real symmetric matrices with distinct eigenvalues, which have orthogonal eigenvectors, the matrix $\mathbf{A}$ can be decomposed as follows:
$$\mathbf{A} = \mathbf{Q}\mathbf{D}\mathbf{Q}^T, \tag{2.43}$$
where $\mathbf{Q}$ is the modal matrix comprised of the orthonormal eigenvectors of $\mathbf{A}$, and $\mathbf{D}$ is a diagonal matrix having the eigenvalues of $\mathbf{A}$ along its diagonal.
- For matrices with linearly independent eigenvectors, the matrix $\mathbf{A}$ can be decomposed as follows:
$$\mathbf{A} = \mathbf{U}\mathbf{D}\mathbf{U}^{-1},$$
where $\mathbf{U}$ is the modal matrix containing the linearly independent eigenvectors of $\mathbf{A}$.
- For nonsymmetric matrices with repeated eigenvalues, the matrix $\mathbf{A}$ can be decomposed as follows:
$$\mathbf{A} = \mathbf{U}\mathbf{J}\mathbf{U}^{-1},$$
where $\mathbf{J}$ is the nearly diagonal Jordan canonical form.

Not only is this state of affairs somewhat confusing, with different eigen-decompositions corresponding to matrices with various properties (not all of which are known ahead of time), it is also quite restrictive. In particular, only square matrices have such eigen-decompositions.

After a brief mention of the Schur decomposition, we discuss two additional decompositions, singular-value decomposition and its special case polar decomposition, that are based on solving eigenproblems. Singular-value decomposition addresses this confusion and restriction, and it is particularly useful as it has numerous applications. Finally, we return to the Gram–Schmidt orthogonalization procedure to motivate the QR decomposition, which is the basis for a numerical method for approximating the eigenvalues and eigenvectors of large matrices on a computer.

The decomposition (2.43) for real symmetric $\mathbf{A}$ is a special case of the *Schur decomposition*

$$\mathbf{A} = \mathbf{Q}\mathbf{R}\mathbf{Q}^T,$$

where $\mathbf{Q}$ is orthogonal, and $\mathbf{R}$ is a right (upper) triangular matrix that is similar to $\mathbf{A}$, that is, having the eigenvalues of $\mathbf{A}$ along the diagonal of $\mathbf{R}$. The Schur decomposition exists for any square matrix $\mathbf{A}$, even if it is defective with less than N linearly independent, regular eigenvectors. One can think of this as *triangularization* as opposed to diagonalization. The diagonal form (2.43) results if $\mathbf{A}$ is a normal matrix as discussed in Section 2.4.

Whereas diagonalization requires determination of the entire spectrum of eigenvalues and eigenvectors of $\mathbf{A}$, including generalized eigenvectors when necessary, the Schur decomposition only requires obtaining one such eigenvalue and its associated eigenvector. The remaining orthonormal vectors used to form the modal matrix $\mathbf{Q}$ need not be eigenvectors and can be obtained using Gram–Schmidt orthogonalization, for example. The remaining eigenvalues will then be found along the diagonal of the right triangular matrix $\mathbf{R}$. Because one could choose different eigenpairs to seed the orthogonalization process, which itself is not unique, the Schur decomposition is not unique. In practice, QR decomposition (see Section 2.10) is typically used to obtain the Schur decomposition, which is a step in certain algorithms for solving systems of linear algebraic equations.

As an application of the Schur decomposition, let us show how it can be used to prove the Cayley–Hamilton theorem. Doing so illustrates that decompositions often allow one to take advantage of the properties of the matrices in a decomposition to make powerful general statements about the original matrix.

The characteristic polynomial of any $N \times N$ square matrix $\mathbf{A}$ is given by

$$(\lambda - \lambda_1)(\lambda - \lambda_2)\cdots(\lambda - \lambda_N) = 0.$$

Let us replace λ with $\mathbf{A}$ in the characteristic equation as follows:

$$(\mathbf{A} - \lambda_1\mathbf{I})(\mathbf{A} - \lambda_2\mathbf{I})\cdots(\mathbf{A} - \lambda_N\mathbf{I}) = \mathbf{B},$$

and show that $\mathbf{B} = \mathbf{0}$ to confirm the Cayley–Hamilton theorem. Recall that the Schur decomposition

$$\mathbf{A} = \mathbf{QRQ}^T$$

exists for any square matrix, where $\mathbf{Q}$ is orthogonal, and $\mathbf{R}$ is right (upper) triangular. Because $\mathbf{Q}$ is orthogonal, $\mathbf{QQ}^T = \mathbf{I}$; therefore, using the Schur decomposition, the preceding equation can be written in the form

$$\left(\mathbf{QRQ}^T - \lambda_1\mathbf{QQ}^T\right)\left(\mathbf{QRQ}^T - \lambda_2\mathbf{QQ}^T\right)\cdots\left(\mathbf{QRQ}^T - \lambda_N\mathbf{QQ}^T\right) = \mathbf{B},$$

which is equivalent to

$$\mathbf{Q}(\mathbf{R} - \lambda_1\mathbf{I})\mathbf{Q}^T\mathbf{Q}(\mathbf{R} - \lambda_2\mathbf{I})\mathbf{Q}^T\cdots\mathbf{Q}(\mathbf{R} - \lambda_N\mathbf{I})\mathbf{Q}^T = \mathbf{B}.$$

Once again, however, $\mathbf{Q}^T\mathbf{Q} = \mathbf{I}$; therefore, this simplifies to

$$\mathbf{Q}(\mathbf{R} - \lambda_1\mathbf{I})(\mathbf{R} - \lambda_2\mathbf{I})\cdots(\mathbf{R} - \lambda_N\mathbf{I})\mathbf{Q}^T = \mathbf{B}.$$

Premultiplying by $\mathbf{Q}^T$ and postmultiplying by $\mathbf{Q}$ yields

$$(\mathbf{R} - \lambda_1\mathbf{I})(\mathbf{R} - \lambda_2\mathbf{I})\cdots(\mathbf{R} - \lambda_N\mathbf{I}) = \mathbf{Q}^T\mathbf{BQ}.$$

Now recall that the eigenvalues of a triangular matrix are found on its diagonal. Consequently, in each of the factors $(\mathbf{R} - \lambda_n\mathbf{I})$, one of the diagonal values vanishes after subtracting off the corresponding eigenvalue λ_n. Therefore, the first column is zero in the first factor, multiplying the first two factors will result in the first and second columns being zero, and so on. As a result, the left-hand side is the zero matrix, and $\mathbf{Q}^T\mathbf{B}\mathbf{Q} = \mathbf{0}$, which can only be true if $\mathbf{B} = \mathbf{0}$, proving the Cayley–Hamilton theorem. Observe how this proof takes maximum advantage of the properties of orthogonal and triangular matrices – this is the power of matrix decomposition.

2.8 Singular-Value Decomposition

A particularly important and useful decomposition is *singular-value decomposition* (SVD).[20] It allows us to extend the ideas of eigenvalues, eigenvectors, diagonalization, and the eigen-decomposition, which only exist for square matrices, to any – including rectangular – matrices. It can serve both as a diagnostic tool and a solution technique in a number of important applications as discussed at the end of the section and encountered throughout the text.

2.8.1 Framework for SVD

If we could be so bold as to state a wish list for a general diagonalization procedure, it would provide an eigenlike decomposition for *any* matrix $\mathbf{A}$, including rectangular matrices. In addition, at the risk of sounding too greedy, it would be awfully convenient if it always produced diagonal and orthonormal matrices as in the eigen-decomposition (2.43). Let us first tackle the generalization to rectangular matrices; the second wish will turn out to be a natural consequence of the first.

Obviously, rectangular matrices do not have eigenvalues and eigenvectors in the strict sense, so an alternative formulation is required rather than that based on the eigen-decomposition discussed thus far. The key is to recall from Section 2.4 that for any matrix $\mathbf{A}$ of size $M \times N$, the products $\mathbf{A}\mathbf{A}^T$ and $\mathbf{A}^T\mathbf{A}$ have the following properties:

1. Both products are always square; $\mathbf{A}\mathbf{A}^T$ is $M \times M$, and $\mathbf{A}^T\mathbf{A}$ is $N \times N$.
2. Both products produce symmetric matrices.
3. Both products have the same nonzero eigenvalues. All additional eigenvalues are zero.[21]
4. Both matrix products are positive semidefinite, in which all eigenvalues are nonnegative.
5. Because both matrix products are symmetric, the eigenvectors of both are mutually orthogonal.

[20] In the context of matrices, the term "singular" suggests a matrix with a zero determinant, that is, one having a zero eigenvalue. We will see later in what sense this may be true in SVD. More broadly, however, one can think of the use of the term here as indicating that singular-value decomposition applies *even* to, but not exclusively to, singular matrices.

[21] Because the larger of the two matrix products has at least one zero eigenvalue (if $M \neq N$), the matrix product is singular, which gives rise to the name of the decomposition.

The obvious question is, "How does consideration of $\mathbf{AA}^T$ and/or $\mathbf{A}^T\mathbf{A}$ help us obtain a useful decomposition of the matrix $\mathbf{A}$?" Let us start by forming the eigenproblems for these two square matrix products as follows:

$$\mathbf{AA}^T\mathbf{u}_n = \sigma_n^2\mathbf{u}_n, \quad \mathbf{A}^T\mathbf{A}\mathbf{v}_n = \sigma_n^2\mathbf{v}_n, \tag{2.44}$$

where σ_n^2 are the eigenvalues, and $\mathbf{u}_n$ and $\mathbf{v}_n$ are the respective sets of eigenvectors. These eigenproblems reflect the fact stated earlier that the nonzero eigenvalues of both products are the same. Also recall that both sets of eigenvectors are – or can be made to be – mutually orthogonal owing to symmetry of the matrix products. Analogous to the eigen-decomposition (2.43) for symmetric matrices, we have the eigen-decompositions

$$\mathbf{AA}^T = \mathbf{U}\boldsymbol{\Sigma}\boldsymbol{\Sigma}^T\mathbf{U}^T, \quad \mathbf{A}^T\mathbf{A} = \mathbf{V}\boldsymbol{\Sigma}^T\boldsymbol{\Sigma}\mathbf{V}^T, \tag{2.45}$$

where the modal matrices $\mathbf{U}$ and $\mathbf{V}$ contain the orthogonal eigenvectors $\mathbf{u}_n$ and $\mathbf{v}_n$ of $\mathbf{AA}^T$ and $\mathbf{A}^T\mathbf{A}$ as their columns, respectively, and $\boldsymbol{\Sigma}$ is an $M \times N$ diagonal matrix with σ_n along the diagonal such that $\boldsymbol{\Sigma}\boldsymbol{\Sigma}^T$ is an $M \times M$ matrix with σ_n^2 along its diagonal, and $\boldsymbol{\Sigma}^T\boldsymbol{\Sigma}$ is an $N \times N$ matrix with σ_n^2 along its diagonal.

All that we have done so far is show that we can construct the eigen-decomposition and diagonalize the square and symmetric matrix products $\mathbf{AA}^T$ and $\mathbf{A}^T\mathbf{A}$. It still remains to see how this provides a means to diagonalize matrix $\mathbf{A}$ itself. To this end, let us consider the first of (2.45):

$$\begin{aligned}
\mathbf{AA}^T &= \mathbf{U}\boldsymbol{\Sigma}\boldsymbol{\Sigma}^T\mathbf{U}^T \\
&= \mathbf{U}\boldsymbol{\Sigma}\left(\mathbf{V}^T\mathbf{V}\right)\boldsymbol{\Sigma}^T\mathbf{U}^T \\
&= \left(\mathbf{U}\boldsymbol{\Sigma}\mathbf{V}^T\right)\left(\mathbf{V}\boldsymbol{\Sigma}^T\mathbf{U}^T\right) \\
\mathbf{AA}^T &= \left(\mathbf{U}\boldsymbol{\Sigma}\mathbf{V}^T\right)\left(\mathbf{U}\boldsymbol{\Sigma}\mathbf{V}^T\right)^T,
\end{aligned}$$

where we have used the fact that $\mathbf{V}^T\mathbf{V} = \mathbf{I}$ because $\mathbf{V}$ is orthogonal. Similarly, the second of equations (2.45) yields:

$$\begin{aligned}
\mathbf{A}^T\mathbf{A} &= \mathbf{V}\boldsymbol{\Sigma}^T\boldsymbol{\Sigma}\mathbf{V}^T \\
&= \mathbf{V}\boldsymbol{\Sigma}^T\left(\mathbf{U}^T\mathbf{U}\right)\boldsymbol{\Sigma}\mathbf{V}^T \\
&= \left(\mathbf{V}\boldsymbol{\Sigma}^T\mathbf{U}^T\right)\left(\mathbf{U}\boldsymbol{\Sigma}\mathbf{V}^T\right) \\
\mathbf{A}^T\mathbf{A} &= \left(\mathbf{U}\boldsymbol{\Sigma}\mathbf{V}^T\right)^T\left(\mathbf{U}\boldsymbol{\Sigma}\mathbf{V}^T\right),
\end{aligned}$$

where we have used the fact that $\mathbf{U}^T\mathbf{U} = \mathbf{I}$ because $\mathbf{U}$ is orthogonal. In both cases, this suggests that we can factor the original matrix $\mathbf{A}$ according to

$$\mathbf{A} = \mathbf{U}\boldsymbol{\Sigma}\mathbf{V}^T, \tag{2.46}$$

where $\boldsymbol{\Sigma}$ is an $M \times N$ diagonal matrix, $\mathbf{U}$ is an $M \times M$ orthogonal matrix,[22] and $\mathbf{V}$ is an $N \times N$ orthogonal matrix. Observe the similarity to diagonalization of symmetric

[22] Because $\mathbf{U}$ is an orthogonal matrix, we would normally designate it as $\mathbf{Q}$. However, we will use the nearly universal notation for SVD in the literature, in which $\mathbf{U}$ is the matrix containing the left singular vectors, and $\mathbf{V}$ is the matrix containing the right singular vectors. This also distinguishes $\mathbf{U}$ from the orthogonal matrix $\mathbf{Q}$ in the Schur decomposition and diagonalization of symmetric matrices.

matrices in (2.43) except that we now have two orthogonal modal matrices $\mathbf{U}$ and $\mathbf{V}$ rather than just one. More importantly, this decomposition applies to *any* matrix, including rectangular ones. This is known as the SVD of the matrix $\mathbf{A}$, which can be considered to be a generalization of the eigen-decomposition of square matrices.

While the nonzero eigenvalues of both of the matrix products are the same, is there any relationship between the eigenvectors $\mathbf{u}_n$ of $\mathbf{A}\mathbf{A}^T$ and the eigenvectors $\mathbf{v}_n$ of $\mathbf{A}^T\mathbf{A}$? Indeed there is! To see it, postmultiply $\mathbf{V}$ on both sides of (2.46) to give

$$\mathbf{A}\mathbf{V} = \mathbf{U}\mathbf{\Sigma}.$$

Likewise, take the transpose of (2.46) and postmultiply $\mathbf{U}$ on both sides to yield

$$\mathbf{A}^T\mathbf{U} = \mathbf{V}\mathbf{\Sigma}^T.$$

Therefore, the eigenvectors $\mathbf{u}_n$, which are the columns of $\mathbf{U}$, and $\mathbf{v}_n$, which are the columns of $\mathbf{V}$, are related through the relationships

$$\mathbf{A}\mathbf{v}_n = \sigma_n\mathbf{u}_n, \quad \mathbf{A}^T\mathbf{u}_n = \sigma_n\mathbf{v}_n. \tag{2.47}$$

The values of σ_n are called the *singular values*, the vectors $\mathbf{u}_n$ are the *left singular vectors*, and the vectors $\mathbf{v}_n$ are the *right singular vectors* of the matrix $\mathbf{A}$.[23] Although $\mathbf{u}_n$ and $\mathbf{v}_n$ are the eigenvectors of different matrices according to equations (2.44), we see that there is a special relationship between these two sets of basis vectors as given by (2.47). To close the loop and confirm that our suggested singular-value decomposition in (2.46) is correct, the relationships (2.47) can be confirmed directly by substitution into (2.44). For example, substituting the first of (2.47) into the second of equations (2.44) and canceling one σ_n from both sides yields the second of equations (2.47). Similarly, substituting the second of equations (2.47) into the first of equations (2.44) and canceling one σ_n from both sides yields the first of equations (2.47).

As advertised, SVD allows us to extend the ideas of eigenvalues, eigenvectors, and diagonalization to any rectangular matrix. In particular, we have determined that the problem of determining the SVD of a matrix $\mathbf{A}$ is equivalent to performing the eigen-decompositions of the symmetric matrices $\mathbf{A}\mathbf{A}^T$ and $\mathbf{A}^T\mathbf{A}$.

We can interpret the components of SVD graphically as follows. Recall that when we multiply a square $N \times N$ matrix $\mathbf{A}$ by one of its $N \times 1$ eigenvectors, it returns the same vector stretched by a scalar value equal to the corresponding eigenvalue. According to the first of equations (2.47), when we multiply a rectangular $M \times N$ matrix $\mathbf{A}$ by one of its $N \times 1$ right singular vectors, it returns the corresponding $M \times 1$ left singular vector stretched by a scalar value equal to the corresponding singular value. The N-dimensional right singular vectors $\mathbf{v}_n$ form an orthonormal basis for the *domain*, while the M-dimensional left singular vectors $\mathbf{u}_n$ form an orthonormal basis for the *range* of matrix $\mathbf{A}$ (see Section 1.8). The applications of SVD go well beyond diagonalization of rectangular matrices and will be mentioned at the end of the section and highlighted throughout the text.

[23] The "left" and "right" designations simply result from the ordering of the $\mathbf{U}$ and $\mathbf{V}$ matrices in the SVD (2.46). In addition, saying that a vector is "singular" does not have any mathematical implications in the same way that saying a matrix is "singular"; it is just a name.

2.8.2 Method for SVD

Based on the discussion in the previous section, the following method can be used to obtain the components $\mathbf{U}$, $\mathbf{V}$, and $\boldsymbol{\Sigma}$ in the SVD of a matrix $\mathbf{A}$. Any $M \times N$ matrix $\mathbf{A}$, even singular or nearly singular matrices, can be decomposed as follows:

$$\mathbf{A} = \mathbf{U}\boldsymbol{\Sigma}\mathbf{V}^T,$$

where $\mathbf{U}$ is an $M \times M$ orthogonal matrix, $\mathbf{V}$ is an $N \times N$ orthogonal matrix, and $\boldsymbol{\Sigma}$ is an $M \times N$ diagonal matrix of the form

$$\boldsymbol{\Sigma} = \begin{bmatrix} \sigma_1 & 0 & \cdots & 0 & 0 & \cdots & 0 \\ 0 & \sigma_2 & \cdots & 0 & 0 & \cdots & 0 \\ \vdots & \vdots & \ddots & \vdots & \vdots & & \vdots \\ 0 & 0 & \cdots & \sigma_P & 0 & \cdots & 0 \end{bmatrix},$$

where $P = \min(M, N)$ and $\sigma_1 \geq \sigma_2 \geq \cdots \geq \sigma_P \geq 0$. The σ_n values are the nonnegative square roots of the eigenvalues of $\mathbf{A}^T\mathbf{A}$ or $\mathbf{A}\mathbf{A}^T$, whichever is smaller, and are called the *singular values* of $\mathbf{A}$. The fact that SVD orders the singular values from largest to smallest is one of its primary virtues and will prove essential in a number of its applications. It is helpful to remember that for $M \times N$ matrix $\mathbf{A}$, $\text{rank}(\mathbf{A}) = \text{rank}(\mathbf{A}^T\mathbf{A}) = \text{rank}(\mathbf{A}\mathbf{A}^T)$ is the number of *nonzero* singular values of $\mathbf{A}$. If $\text{rank}(\mathbf{A}) < P = \min(M, N)$, then $\mathbf{A}$ is singular.

According to (2.47), the columns $\mathbf{u}_n$ of matrix $\mathbf{U}$ and $\mathbf{v}_n$ of matrix $\mathbf{V}$ satisfy

$$\mathbf{A}\mathbf{v}_n = \sigma_n \mathbf{u}_n, \quad \mathbf{A}^T\mathbf{u}_n = \sigma_n \mathbf{v}_n,$$

and $\|\mathbf{u}_n\| = 1$, $\|\mathbf{v}_n\| = 1$, such that they are normalized. Premultiplying the second equation by $\mathbf{A}$ and using the first equation gives

$$\mathbf{A}\mathbf{A}^T\mathbf{u}_n = \sigma_n \mathbf{A}\mathbf{v}_n = \sigma_n^2 \mathbf{u}_n; \tag{2.48}$$

therefore, the $\mathbf{u}_n$ vectors are the eigenvectors of $\mathbf{A}\mathbf{A}^T$, with the σ_n^2 values being the eigenvalues. Similarly, premultiplying the first equation by $\mathbf{A}^T$ and using the second equation gives

$$\mathbf{A}^T\mathbf{A}\mathbf{v}_n = \sigma_n \mathbf{A}^T\mathbf{u}_n = \sigma_n^2 \mathbf{v}_n; \tag{2.49}$$

therefore, the $\mathbf{v}_n$ vectors are the eigenvectors of $\mathbf{A}^T\mathbf{A}$, with the values σ_n^2 being the eigenvalues.

The orthonormal vectors $\mathbf{u}_n$, which are the columns of $\mathbf{U}$, are called the *left singular vectors* of $\mathbf{A}$, and the orthonormal vectors $\mathbf{v}_n$, which are the columns of $\mathbf{V}$, are called the *right singular vectors* of $\mathbf{A}$. We only need to solve one of the eigenproblems (2.48) or (2.49). Once we have $\mathbf{v}_n$, for example, then we obtain $\mathbf{u}_n$ from $\mathbf{A}\mathbf{v}_n = \sigma_n \mathbf{u}_n$. Whereas, if we have $\mathbf{u}_n$, then we obtain $\mathbf{v}_n$ from $\mathbf{A}^T\mathbf{u}_n = \sigma_n \mathbf{v}_n$. We compare M and N to determine which eigenproblem will be easier (typically the smaller one).

Example 2.13 Use singular-value decomposition to factor the 2×3 matrix

$$\mathbf{A} = \begin{bmatrix} 1 & 0 & 1 \\ 1 & 1 & 0 \end{bmatrix}.$$

Solution

The transpose of $\mathbf{A}$ is

$$\mathbf{A}^T = \begin{bmatrix} 1 & 1 \\ 0 & 1 \\ 1 & 0 \end{bmatrix};$$

therefore,

$$\mathbf{A}\mathbf{A}^T = \begin{bmatrix} 2 & 1 \\ 1 & 2 \end{bmatrix}, \quad \mathbf{A}^T\mathbf{A} = \begin{bmatrix} 2 & 1 & 1 \\ 1 & 1 & 0 \\ 1 & 0 & 1 \end{bmatrix}.$$

As expected, the matrix products are both square and symmetric. Let us determine the eigenvalues of $\mathbf{A}\mathbf{A}^T\mathbf{u}_n = \sigma_n^2\mathbf{u}_n$, which is the smaller of the two. The characteristic equation for the eigenvalues σ_n^2 is

$$\sigma_n^4 - 4\sigma_n^2 + 3 = 0,$$

which when factored yields

$$\sigma_1^2 = 3, \quad \sigma_2^2 = 1,$$

noting that $\sigma_1^2 \geq \sigma_2^2$. Next we obtain the corresponding eigenvectors of $\mathbf{A}\mathbf{A}^T\mathbf{u}_n = \sigma_n^2\mathbf{u}_n$, which are the left singular vectors of $\mathbf{A}$, as follows:

$$\sigma_1^2 = 3: \quad \begin{bmatrix} -1 & 1 \\ 1 & -1 \end{bmatrix}\mathbf{u}_1 = \mathbf{0} \quad \Rightarrow \quad \mathbf{u}_1 = \frac{1}{\sqrt{2}}\begin{bmatrix} 1 \\ 1 \end{bmatrix},$$

$$\sigma_2^2 = 1: \quad \begin{bmatrix} 1 & 1 \\ 1 & 1 \end{bmatrix}\mathbf{u}_2 = \mathbf{0} \quad \Rightarrow \quad \mathbf{u}_2 = \frac{1}{\sqrt{2}}\begin{bmatrix} -1 \\ 1 \end{bmatrix}.$$

Note that $\mathbf{u}_1$ and $\mathbf{u}_2$ are orthogonal and have been normalized. Now we can form the orthogonal matrix $\mathbf{U}$ using the left singular vectors

$$\mathbf{U} = \begin{bmatrix} \mathbf{u}_1 & \mathbf{u}_2 \end{bmatrix} = \frac{1}{\sqrt{2}}\begin{bmatrix} 1 & -1 \\ 1 & 1 \end{bmatrix}.$$

To obtain $\mathbf{v}_n$, recall that $\mathbf{A}^T\mathbf{u}_n = \sigma_n\mathbf{v}_n$; thus,

$$\mathbf{v}_1 = \frac{1}{\sigma_1}\mathbf{A}^T\mathbf{u}_1 = \frac{1}{\sqrt{3}}\begin{bmatrix} 1 & 1 \\ 0 & 1 \\ 1 & 0 \end{bmatrix}\frac{1}{\sqrt{2}}\begin{bmatrix} 1 \\ 1 \end{bmatrix} = \frac{1}{\sqrt{6}}\begin{bmatrix} 2 \\ 1 \\ 1 \end{bmatrix},$$

which is normalized. Similarly,

$$\mathbf{v}_2 = \frac{1}{\sigma_2}\mathbf{A}^T\mathbf{u}_2 = \frac{1}{\sqrt{1}}\begin{bmatrix} 1 & 1 \\ 0 & 1 \\ 1 & 0 \end{bmatrix}\frac{1}{\sqrt{2}}\begin{bmatrix} -1 \\ 1 \end{bmatrix} = \frac{1}{\sqrt{2}}\begin{bmatrix} 0 \\ 1 \\ -1 \end{bmatrix}.$$

Again, note that $\mathbf{v}_1$ and $\mathbf{v}_2$ are orthonormal. The third right singular vector, $\mathbf{v}_3$, also must be orthogonal to $\mathbf{v}_1$ and $\mathbf{v}_2$; thus, letting $\mathbf{v}_3 = \begin{bmatrix} a & b & c \end{bmatrix}^T$ and evaluating $\langle \mathbf{v}_3, \mathbf{v}_1 \rangle = 0$ and $\langle \mathbf{v}_3, \mathbf{v}_2 \rangle = 0$ determines two of the three constants, and normalizing to determine the third gives[24]

$$\mathbf{v}_3 = \frac{1}{\sqrt{3}}\begin{bmatrix} 1 \\ -1 \\ -1 \end{bmatrix}.$$

Therefore, the orthogonal matrix $\mathbf{V}$ containing the right singular vectors is given by

$$\mathbf{V} = \begin{bmatrix} \mathbf{v}_1 & \mathbf{v}_2 & \mathbf{v}_3 \end{bmatrix} = \begin{bmatrix} \frac{2}{\sqrt{6}} & 0 & \frac{1}{\sqrt{3}} \\ \frac{1}{\sqrt{6}} & \frac{1}{\sqrt{2}} & -\frac{1}{\sqrt{3}} \\ \frac{1}{\sqrt{6}} & -\frac{1}{\sqrt{2}} & -\frac{1}{\sqrt{3}} \end{bmatrix},$$

and the diagonal matrix of singular values is

$$\mathbf{\Sigma} = \begin{bmatrix} \sqrt{3} & 0 & 0 \\ 0 & \sqrt{1} & 0 \end{bmatrix},$$

where the singular values of $\mathbf{A}$ are $\sigma_1 = \sqrt{3}$ and $\sigma_2 = \sqrt{1} = 1$. One can check the result of the decomposition by evaluating

$$\mathbf{A} = \mathbf{U}\mathbf{\Sigma}\mathbf{V}^T$$

to be sure it returns the original matrix.

Given the matrix form (2.46) of SVD, it is of interest to write it in vector form to provide an alternative interpretation and for future use. For illustrative purposes, let us consider the case when $P = M < N$, for which

$$\begin{aligned}
\mathbf{A} &= \mathbf{U}\mathbf{\Sigma}\mathbf{V}^T \\
&= \begin{bmatrix} \vdots & \vdots & & \vdots \\ \mathbf{u}_1 & \mathbf{u}_2 & \cdots & \mathbf{u}_M \\ \vdots & \vdots & & \vdots \end{bmatrix} \begin{bmatrix} \sigma_1 & 0 & \cdots & 0 & 0 & \cdots & 0 \\ 0 & \sigma_2 & \cdots & 0 & 0 & \cdots & 0 \\ \vdots & \vdots & \ddots & \vdots & \vdots & & \vdots \\ 0 & 0 & \cdots & \sigma_M & 0 & \cdots & 0 \end{bmatrix} \begin{bmatrix} \cdots & \mathbf{v}_1 & \cdots \\ \cdots & \mathbf{v}_2 & \cdots \\ & \vdots & \\ \cdots & \mathbf{v}_N & \cdots \end{bmatrix} \\
&= \begin{bmatrix} \vdots & \vdots & & \vdots & \vdots & & \vdots \\ \sigma_1\mathbf{u}_1 & \sigma_2\mathbf{u}_2 & \cdots & \sigma_M\mathbf{u}_M & \mathbf{0} & \cdots & \mathbf{0} \\ \vdots & \vdots & & \vdots & \vdots & & \vdots \end{bmatrix} \begin{bmatrix} \cdots & \mathbf{v}_1 & \cdots \\ \cdots & \mathbf{v}_2 & \cdots \\ & \vdots & \\ \cdots & \mathbf{v}_N & \cdots \end{bmatrix} \\
\mathbf{A} &= \sigma_1\mathbf{u}_1\mathbf{v}_1^T + \sigma_2\mathbf{u}_2\mathbf{v}_2^T + \cdots + \sigma_P\mathbf{u}_P\mathbf{v}_P^T,
\end{aligned}$$

[24] Alternatively, we could use the cross (vector) product in this case as $\mathbf{v}_3 = \mathbf{v}_1 \times \mathbf{v}_2$ is orthogonal to $\mathbf{v}_1$ and $\mathbf{v}_2$.

or using summation notation

$$\mathbf{A} = \sum_{n=1}^{P} \sigma_n \mathbf{u}_n \mathbf{v}_n^T, \tag{2.50}$$

where again $P = \min(M, N)$. Note that these vector products are examples of *outer products* (see Section 1.6.3). Because $\mathbf{u}_n$ is $M \times 1$ and $\mathbf{v}_n$ is $N \times 1$, the outer product $\mathbf{u}_n \mathbf{v}_n^T$ is a matrix of size $M \times N$, which is the same size as matrix $\mathbf{A}$. In this way, we can reconstruct the matrix $\mathbf{A}$ directly given its singular values and the left and right singular vectors as a sum of the P modes corresponding to each nonzero singular value. Because the singular values are ordered by decreasing magnitude, we could approximate matrix $\mathbf{A}$ by simply truncating the summation at fewer than P terms, eliminating those terms having small singular values, for example. This is the basis of proper-orthogonal decomposition, which will be taken up in Chapter 13, and other data-reduction techniques that take advantage of the fact that the most important information about a data set corresponds to the largest singular values, and that corresponding to the smallest singular values can be ignored without much loss of important data.

Example 2.14 Let us illustrate this SVD matrix reconstruction process using the Hilbert matrix. The Hilbert matrix is defined as

$$H[i, j] = \frac{1}{i + j - 1}.$$

For example, the 5×5 Hilbert matrix is given by

$$\mathbf{H}_5 = \begin{bmatrix} 1 & \frac{1}{2} & \frac{1}{3} & \frac{1}{4} & \frac{1}{5} \\ \frac{1}{2} & \frac{1}{3} & \frac{1}{4} & \frac{1}{5} & \frac{1}{6} \\ \frac{1}{3} & \frac{1}{4} & \frac{1}{5} & \frac{1}{6} & \frac{1}{7} \\ \frac{1}{4} & \frac{1}{5} & \frac{1}{6} & \frac{1}{7} & \frac{1}{8} \\ \frac{1}{5} & \frac{1}{6} & \frac{1}{7} & \frac{1}{8} & \frac{1}{9} \end{bmatrix}.$$

Let us consider a 40×40 Hilbert matrix. Performing a SVD of $\mathbf{H}_{40}$ produces the $\mathbf{U}$ matrix of left singular vectors, the $\mathbf{V}$ matrix of right singular vectors, and the diagonal matrix $\mathbf{\Sigma}$ containing the singular values. These singular values are plotted in Figure 2.11. Observe that there are only 15 nonzero singular values; the remaining singular values from 16 through 40 are all zero. In addition, you will note that they decrease in magnitude very rapidly. This also can be illustrated by plotting the cumulative sum of the singular values as shown in Figure 2.12.

Let us use the SVD to reconstruct the original matrix using

$$\mathbf{H} = \mathbf{U}\mathbf{\Sigma}\mathbf{V}^T$$

or (2.50) with the $P = K < 40$ modes having the largest singular values. In this case, $\mathbf{U}$ is $40 \times K$, $\mathbf{\Sigma}$ is $K \times K$, and $\mathbf{V}$ is $40 \times K$. Because only $n = 15$ of the singular values are nonzero, a reconstruction with $K = 15$ would exactly reproduce the original 40×40 Hilbert matrix. Furthermore, most of the nonzero singular values

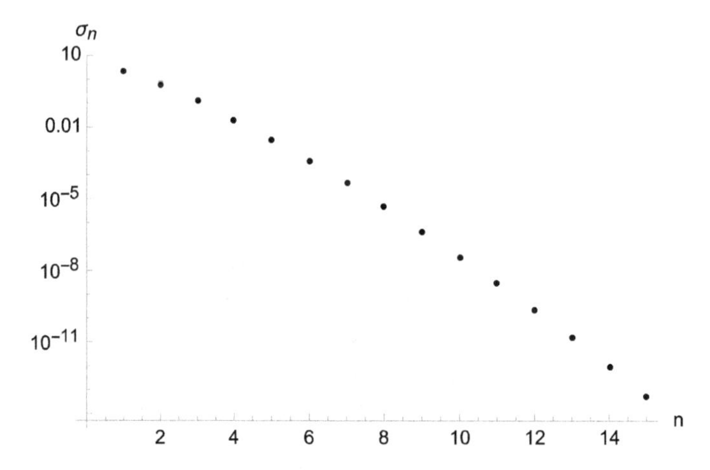

Figure 2.11 Singular values of the 40×40 Hilbert matrix on a log plot.

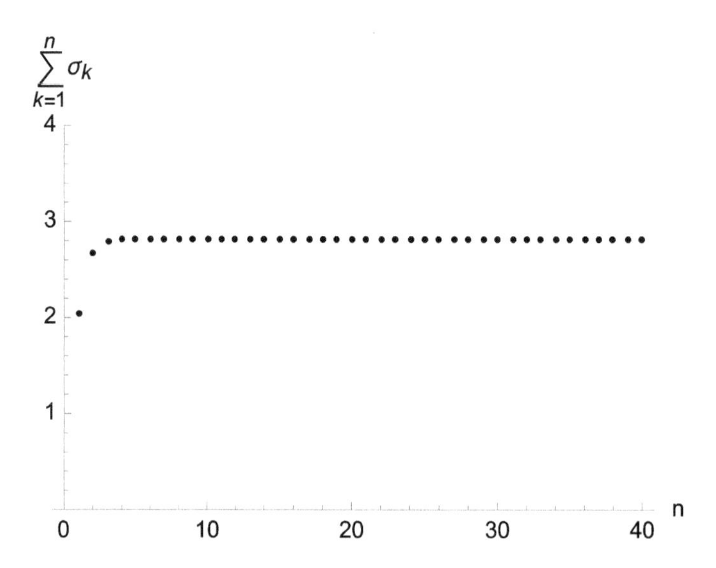

Figure 2.12 Cumulative sum of the singular values of the 40×40 Hilbert matrix.

are very small, and just the first few contain all of the important information. For example, a reconstruction with $K = 5$ reproduces the 40×40 Hilbert matrix to five significant figures. Try it!

Note that we have not made any assumptions about $\mathbf{A}$; therefore, the SVD can be obtained for *any* real or complex rectangular matrix, even if it is singular. It is rather remarkable, therefore, that any matrix can be decomposed into the product of two orthogonal matrices and one diagonal matrix. This is in marked contrast to all of our previous operations, in which various conditions must be met; a matrix only has an inverse if it is square and its determinant is nonzero, a matrix can only be diagonalized if it is square and has linearly independent eigenvectors, and so forth. Because SVD can be applied to any matrix, it opens the door to numerous applications.

Let us first consider the matrix inverse. We know that for a matrix to have an inverse, it must be square and of full rank, that is, rank($\mathbf{A}$) $= N$. In this case, $\mathbf{A}$ does not have any zero singular values, and we can take advantage of the orthogonality of $\mathbf{U}$ and $\mathbf{V}$ to compute the inverse of the matrix $\mathbf{A}$ as follows:

$$\mathbf{A}^{+} = \left(\mathbf{U}\boldsymbol{\Sigma}\mathbf{V}^{T}\right)^{-1} = \mathbf{V}\boldsymbol{\Sigma}^{-1}\mathbf{U}^{T}, \tag{2.51}$$

where the inverse of the diagonal matrix $\boldsymbol{\Sigma}$ is simply a diagonal matrix containing the reciprocals of the singular values of $\mathbf{A}$. Far more than simply providing yet another means to obtain the inverse of a matrix, this provides the definition for the *pseudo-inverse*, or *Moore–Penrose inverse*, of *any* rectangular matrix.[25] The notation $\mathbf{A}^{+}$ is used to distinguish it from the regular inverse of a square matrix, but they are the same when a regular inverse exists. In addition, if $\mathbf{A}$ is square and invertible (nonsingular), the singular values of $\mathbf{A}^{-1}$ are the reciprocals of the singular values of $\mathbf{A}$. These are very powerful results as they allow us to "solve" systems of algebraic equations that do not have a unique solution. In Section 10.2.1, for example, it will be shown how the pseudo-inverse can be used to obtain a solution of overdetermined systems of equations in a least-squares context.[26]

Finally, in the special case that $\mathbf{A}$ is an $N \times N$ real symmetric matrix, in which case $\mathbf{A}^{T}\mathbf{A} = \mathbf{A}\mathbf{A}^{T} = \mathbf{A}^{2}$, the N singular values of $\mathbf{A}$ are the absolute values of its eigenvalues, that is, $\sigma_n = |\lambda_n|$, $\mathbf{U}$ is the modal matrix $\mathbf{Q}$, and $\mathbf{v}_n = \text{sign}(\lambda_n)\mathbf{u}_n$. What about the case when matrix $\mathbf{A}$ is square, but not symmetric? Unlike the symmetric case, there is no direct relationship between the eigenvalues and singular values of $\mathbf{A}$ or the eigenvectors and the left and right singular vectors of $\mathbf{A}$ in this case.

REMARKS:

1. *If* $\mathbf{A}$ *is complex, then the singular-value decomposition of an* $M \times N$ *matrix* $\mathbf{A}$ *is*

$$\mathbf{A} = \mathbf{U}\boldsymbol{\Sigma}\overline{\mathbf{V}}^{T},$$

where $\mathbf{U}$ *is an* $M \times M$ *unitary matrix,* $\mathbf{V}$ *is an* $N \times N$ *unitary matrix, and the singular values found along the diagonal of* $\boldsymbol{\Sigma}$ *are real and nonnegative. In addition, all previous instances of the transpose become the conjugate transpose for complex* $\mathbf{A}$.

2. *Because the matrix products* $\mathbf{A}^{T}\mathbf{A}$ *and* $\mathbf{A}\mathbf{A}^{T}$ *are both symmetric, their eigenvectors are orthogonal if the eigenvalues are distinct. If some of the eigenvalues are repeated, although the associated eigenvectors can only be assured of being linearly independent, they can be chosen to be mutually orthogonal. This is necessary here in order for the left and right singular vectors to be orthogonal. For example, if* M *or* N *is at least two greater than the other, then the larger of the two matrix products will have zero as a repeated eigenvalue. Because the matrix product is*

[25] Because $\boldsymbol{\Sigma}$ is rectangular ($M \times N$), $\boldsymbol{\Sigma}^{-1}$ is interpreted as the $N \times M$ matrix having the reciprocals of the nonzero singular values along the diagonal and all other elements remaining zero.

[26] Recall that an overdetermined system has more equations than unknowns.

symmetric, however, its eigenvectors can still be chosen to be mutually orthogonal to form the complete set of left or right singular vectors.

3. *Recall from Section 1.9 that the L_2-norm of matrix $\mathbf{A}$ is given by its largest singular value*

$$\|\mathbf{A}\|_2 = \sigma_1,$$

and the Frobenius norm is

$$\|\mathbf{A}\|_F = \sqrt{\sigma_1^2 + \sigma_2^2 + \cdots + \sigma_R^2},$$

where R is the rank of $\mathbf{A}$.

4. *Recall from Chapter 1 that mathematically we are only concerned with whether a matrix is singular or nonsingular as determined by the determinant. When carrying out numerical operations with large matrices, however, we are concerned with how close to being singular a matrix is. The most common measure of how close a matrix is to being singular is the* condition number. *Matrices having very large condition numbers are said to be* ill conditioned *and are close to being singular, in which case numerical operations may produce large errors. See Section 6.2.2 for a definition of the condition number in terms of singular values and how it can be used to estimate errors in numerical algorithms.*
5. *See section 2.6 of Press et al. (2007) for more details on the numerical implementation of SVD. In particular, see the method for determining the "best" solution of singular or ill-conditioned systems of equations (yes, that is right, singular systems can be "solved").*
6. *Additional applications of SVD include least-squares curve fitting, obtaining orthonormal basis vectors without the round-off errors that can build up in the Gram–Schmidt orthogonalization procedure, image and signal compression (see next section), proper-orthogonal decomposition (POD) – also called principal-component analysis – for reduced-order modeling, pseudo-inverse for solving ill-conditioned systems numerically, and optimal control and transient growth of linear systems.*
7. *In probability and statistics, the product $\mathbf{A}\mathbf{A}^T$ is closely related to the* covariance matrix *and quantifies the degree to which multiple data values are related to one another. We will return to the covariance matrix when we discuss proper-orthogonal decomposition in Chapter 13.*

2.8.3 Image Compression with SVD

Until the early seventeenth century, our sight was limited to those things that could be observed by the naked eye. The invention of the lens facilitated development of both the microscope and the telescope that allowed us to directly observe the very small and the very large – but distant. This allowed us to extend our sight in both directions on the spatial scale from the $O(1)$ world of our everyday experience. In addition, the discovery of the electromagnetic spectrum and invention of devices to view and utilize it extended our "sight" in both directions on the electromagnetic spectrum from the visible range.

As I write this in July 2015, the New Horizons probe has just passed Pluto after nine and one-half years traveling through the solar system. There are seven instruments on the probe that are recording pictures and data that are then beamed billions of miles back to Earth for recording and processing. Three of the instruments are recording data in the visible, infrared, and ultraviolet ranges of the electromagnetic spectrum, and the data from all of the instruments are transmitted back to Earth as radio waves. In addition to interplanetary exploration, medical imaging, scientific investigation, and surveillance produce massive quantities of image data. Indeed, much of modern life relies on myriad signals transmitting our communication, television, GPS, Internet, and encryption data. More and more of these data are digital, rather than analog, and require rapid processing of large quantities of images and signals that each contain increasing amounts of detailed data. All of these developments have served to dramatically increase the volume, type, and complexity of images and signals requiring processing.

Signals and images are produced, harvested, and processed by all manner of electronic devices that utilize the entire electromagnetic spectrum. The now traditional approach to image and signal processing is based on Fourier analysis of both discrete and continuous data. More recently, advances in variational image processing have supplemented these techniques as well (see chapter 11 of Cassel, 2013, for an introduction and references). As the volume of experimental and computational data rapidly proliferates, there is a growing need for creative techniques to characterize and consolidate the data such that the essential features can be readily extracted. Here, we illustrate how SVD can be used for image compression.

A grayscale image can be represented as an $M \times N$ matrix $\mathbf{A}$ of pixels, where each element in the matrix is a number between zero and one indicating the level of white through black of each pixel.[27] Figure 2.13a shows an image of the ceiling of Bath Cathedral in England consisting of $M = 768$ by $N = 1{,}024$ pixels.

Taking the SVD of the image matrix of grayscale values allows one to fully reconstruct the original image using

$$\mathbf{A} = \mathbf{U}\boldsymbol{\Sigma}\mathbf{V}^T,$$

where $\mathbf{U}$ is $M \times M$, $\boldsymbol{\Sigma}$ is $M \times N$, and $\mathbf{V}$ is $N \times N$. Alternatively, if only the first $K < M$ modes are retained corresponding to the K largest singular values, the same reconstruction holds, except that $\mathbf{U}$ is now $M \times K$, $\boldsymbol{\Sigma}$ is $K \times K$, and $\mathbf{V}$ is $N \times K$. Alternatively, the vector form of the SVD given by (2.50) could be used with P replaced by K. Sample reconstructions are shown in Figure 2.13b–f for varying numbers of modes. As you can see, a reconstruction with $K = 100 < 768$ modes appears nearly identical to the original image, with only a very minor loss of detail.

[27] Color images would require three times as much storage for the levels of red, green, and blue for each pixel.

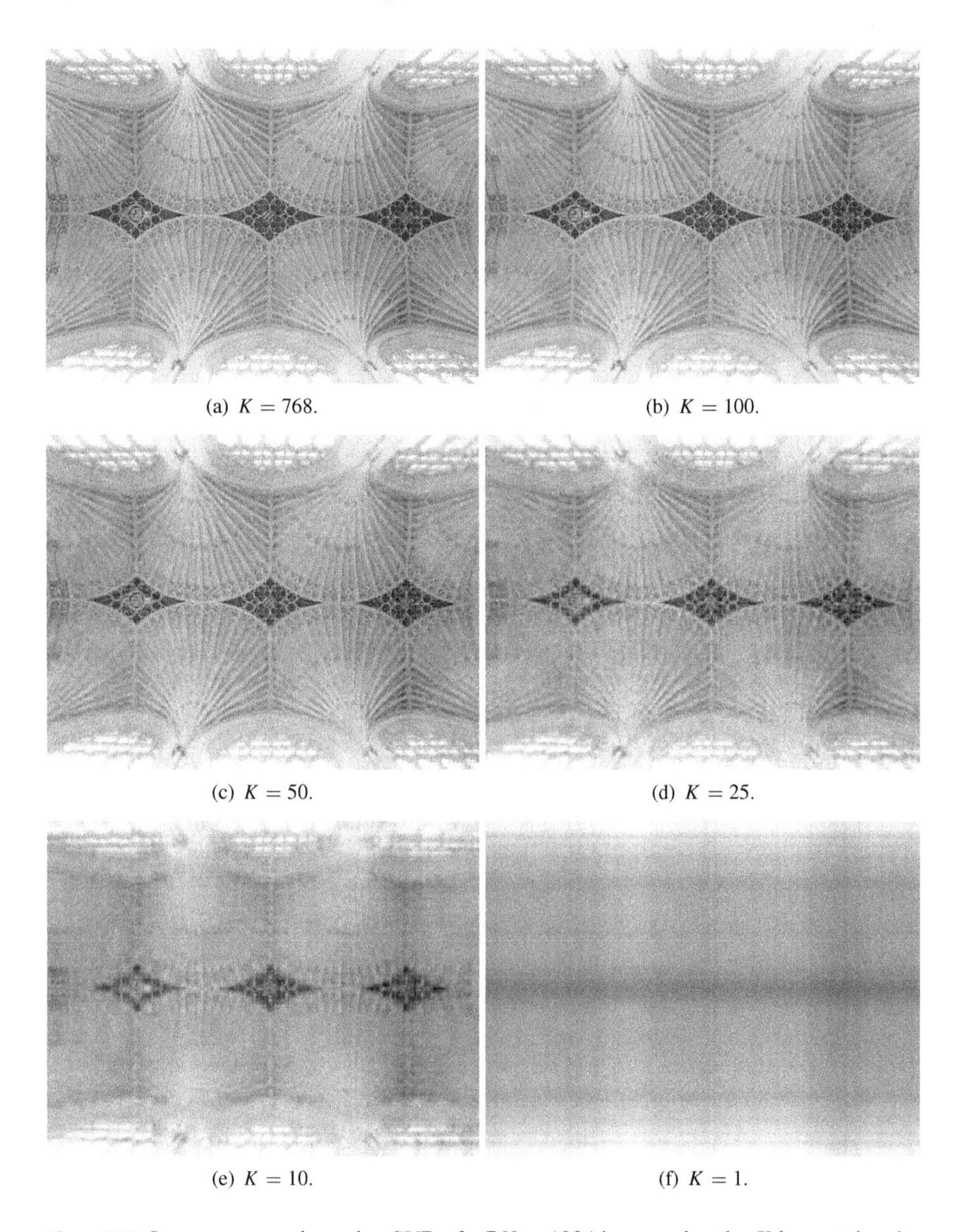

Figure 2.13 Image compression using SVD of a 768×1024 image using the K largest singular values.

As the number of modes is further reduced to $K = 50, 25, 10$, and 1, additional detail is lost in the reconstructed image.

As the number of SVD modes K is decreased, the image data are increasingly compressed – we need less data to reconstruct the image. Therefore, let us consider the corresponding reduction in storage requirements. For the sake of simplicity, let us assume the original image is square and of size $N \times N$, then the matrix representation of the original image would require N^2 values to be stored. An SVD representation of the full image would actually require more storage for the $\mathbf{U}$ and $\mathbf{V}$ matrices and the N singular values, totaling $2N^2 + N$ values. Therefore, $K \ll N$ in order for the SVD representation to require substantially less storage than the original image. If SVD is applied with $K \ll N$ singular values used for the image reconstruction, the two matrices $\mathbf{U}$ and $\mathbf{V}$ are both $N \times K$, and there are K singular values. Consequently, the total number of values to be stored is only $K(2N + 1)$, which is substantially less than $N(2N + 1)$ for $K \ll N$.

2.9 Polar Decomposition

A special case of singular-value decomposition that is useful in *continuum mechanics* occurs when the matrix $\mathbf{A}$ is square. Recall that the singular-value decomposition of matrix $\mathbf{A}$ is given by

$$\mathbf{A} = \mathbf{U}\boldsymbol{\Sigma}\mathbf{V}^T,$$

where $\mathbf{U}$ and $\mathbf{V}$ are orthogonal matrices, and $\boldsymbol{\Sigma}$ is a diagonal matrix. We now consider the special case when $\mathbf{A}$ is a real, square $N \times N$ matrix, in which case $\mathbf{U}$, $\mathbf{V}$, and $\boldsymbol{\Sigma}$ are also $N \times N$. Because $\mathbf{V}$ is an orthogonal matrix, $\mathbf{V}^T\mathbf{V} = \mathbf{I}$. Owing to the uniform sizes of each of the matrices, we can then insert $\mathbf{V}^T\mathbf{V}$ into the singular-value decomposition as follows:

$$\mathbf{A} = \mathbf{U}\mathbf{V}^T\mathbf{V}\boldsymbol{\Sigma}\mathbf{V}^T.$$

Let us break the right-hand side up into the product of two matrices[28]

$$\mathbf{A} = \mathbf{W}\mathbf{R}, \tag{2.52}$$

where

$$\mathbf{W} = \mathbf{U}\mathbf{V}^T, \quad \mathbf{R} = \mathbf{V}\boldsymbol{\Sigma}\mathbf{V}^T. \tag{2.53}$$

As shown later, $\mathbf{W}$ is an orthogonal matrix, and $\mathbf{R}$ is a symmetric matrix.

Recall that for a matrix to be orthogonal, the product with its transpose must equal the identity matrix. Because $\mathbf{U}$ and $\mathbf{V}$ are both orthogonal, $\mathbf{U}\mathbf{U}^T = \mathbf{U}^T\mathbf{U} = \mathbf{I}$ and $\mathbf{V}\mathbf{V}^T = \mathbf{V}^T\mathbf{V} = \mathbf{I}$. Therefore, it follows that

$$\mathbf{W}\mathbf{W}^T = (\mathbf{U}\mathbf{V}^T)(\mathbf{U}\mathbf{V}^T)^T = \mathbf{U}\mathbf{V}^T\mathbf{V}\mathbf{U}^T = \mathbf{U}\mathbf{I}\mathbf{U}^T = \mathbf{U}\mathbf{U}^T = \mathbf{I}.$$

Thus, the product $\mathbf{W}$ of two orthogonal matrices $\mathbf{U}$ and $\mathbf{V}^T$ is also orthogonal.

For a matrix to be symmetric, it must equal its transpose. In this case, this requires that

$$\mathbf{R} = \mathbf{R}^T,$$

[28] The use of matrices $\mathbf{W}$ and $\mathbf{R}$ here follows the usual convection in continuum mechanics literature. Note, however, that $\mathbf{R}$ is not right triangular as is our usual convention in this text.

or

$$\mathbf{V}\boldsymbol{\Sigma}\mathbf{V}^T = \left(\mathbf{V}\boldsymbol{\Sigma}\mathbf{V}^T\right)^T.$$

Evaluating the transpose on the right-hand side gives

$$\mathbf{V}\boldsymbol{\Sigma}\mathbf{V}^T = \mathbf{V}\boldsymbol{\Sigma}^T\mathbf{V}^T.$$

Premultiplying $\mathbf{V}^T$ on both sides and postmultiplying $\mathbf{V}$ on both sides leads to

$$\boldsymbol{\Sigma} = \boldsymbol{\Sigma}^T,$$

which is satisfied because $\boldsymbol{\Sigma}$ is a square diagonal matrix. In addition to being symmetric, it turns out that $\mathbf{R}$ is also positive definite, such that all of its eigenvalues are positive. Because every matrix $\mathbf{A}$ has a singular-value decomposition, it also has a polar decomposition (2.52) if it is square, where the two decomposed matrices are given by (2.53) in terms of the singular-value decomposition.

In lieu of carrying out the full singular-value decomposition to determine $\mathbf{W}$ and $\mathbf{R}$ in the polar decomposition, we can obtain these two matrices directly as follows. Taking the transpose of (2.52) yields

$$\mathbf{A}^T = \mathbf{R}\mathbf{W}^T,$$

where we recall that $\mathbf{R}$ is symmetric. Postmultiplying $\mathbf{A} = \mathbf{W}\mathbf{R}$, where $\mathbf{W}^T\mathbf{W} = \mathbf{I}$ because $\mathbf{W}$ is orthogonal, gives

$$\mathbf{A}^T\mathbf{A} = \mathbf{R}\mathbf{W}^T\mathbf{W}\mathbf{R} = \mathbf{R}^2. \tag{2.54}$$

Recall from Section 2.2 that the eigenvectors of $\mathbf{R}$ and $\mathbf{R}^2$ are the same, while the eigenvalues of $\mathbf{R}^2$ are λ_i^2, where λ_i are the eigenvalues of $\mathbf{R}$. Therefore, we determine the eigenvalues and eigenvectors of the symmetric matrix $\mathbf{R}^2 = \mathbf{A}^T\mathbf{A}$, and form the orthogonal modal matrix $\mathbf{Q}$. Then to diagonalize the symmetric matrix $\mathbf{R}^2$, we take

$$\mathbf{Q}^T\mathbf{R}^2\mathbf{Q} = \begin{bmatrix} \lambda_1^2 & 0 & \cdots & 0 \\ 0 & \lambda_2^2 & \cdots & 0 \\ \vdots & \vdots & \ddots & \vdots \\ 0 & 0 & \cdots & \lambda_N^2 \end{bmatrix}.$$

Similarly, we can diagonalize $\mathbf{R}$ according to

$$\mathbf{Q}^T\mathbf{R}\mathbf{Q} = \begin{bmatrix} \lambda_1 & 0 & \cdots & 0 \\ 0 & \lambda_2 & \cdots & 0 \\ \vdots & \vdots & \ddots & \vdots \\ 0 & 0 & \cdots & \lambda_N \end{bmatrix}.$$

Premultiplying by $\mathbf{Q}$ and postmultiplying by $\mathbf{Q}^T$ gives the symmetric matrix $\mathbf{R}$

$$\mathbf{R} = \mathbf{Q} \begin{bmatrix} \lambda_1 & 0 & \cdots & 0 \\ 0 & \lambda_2 & \cdots & 0 \\ \vdots & \vdots & \ddots & \vdots \\ 0 & 0 & \cdots & \lambda_N \end{bmatrix} \mathbf{Q}^T.$$

This is a simple calculation once we have the eigenvalues and eigenvectors of $\mathbf{A}^T\mathbf{A}$. Then from (2.52), the orthogonal matrix $\mathbf{W}$ is obtained from

$$\mathbf{W} = \mathbf{A}\mathbf{R}^{-1}.$$

Alternatively, the polar decomposition may be defined in the form

$$\mathbf{A} = \hat{\mathbf{R}}\mathbf{W}, \tag{2.55}$$

rather than (2.52), where $\hat{\mathbf{R}}$ is symmetric and positive definite. In this case, taking the transpose gives

$$\mathbf{A}^T = \mathbf{W}^T\hat{\mathbf{R}},$$

and premultiplying by $\mathbf{A} = \hat{\mathbf{R}}\mathbf{W}$ gives

$$\mathbf{A}\mathbf{A}^T = \hat{\mathbf{R}}\mathbf{W}\mathbf{W}^T\hat{\mathbf{R}} = \hat{\mathbf{R}}^2 \tag{2.56}$$

in place of (2.54). The matrix $\hat{\mathbf{R}}$ may then be obtained by solving the eigenproblem for $\mathbf{A}\mathbf{A}^T$ similar to the preceding, or if $\mathbf{W}$ is already known, then the symmetric matrix $\hat{\mathbf{R}}$ is given by

$$\hat{\mathbf{R}} = \mathbf{A}\mathbf{W}^T.$$

REMARKS:

1. *Decomposing a general square matrix* $\mathbf{A}$ *in this way allows us to take advantage of the properties of orthogonal and symmetric matrices.*
2. *If* $\mathbf{A}$ *is complex, then* $\mathbf{R}$ *and* $\hat{\mathbf{R}}$ *are positive-definite Hermitian matrices, and* $\mathbf{W}$ *is a unitary matrix.*
3. *In continuum mechanics, the polar decomposition of the deformation gradient tensor (matrix)* $\mathbf{F} = \mathbf{W}\mathbf{R} = \hat{\mathbf{R}}\mathbf{W}$ *is such that* $\mathbf{W}$ *is the* rotation matrix, $\mathbf{R}$ *is the positive-definite symmetric* right-stretch matrix, *and* $\hat{\mathbf{R}}$ *is the positive-definite symmetric* left-stretch matrix. *See, for example, Nair (2009) and Reddy (2013). Essentially, the polar decomposition separates out the expansion (stretch) and rotation components of the stress tensor acting on an object. The fact that there are two such polar decompositions, one with the rotation performed before the stretch and the other with it performed after the stretch, reflects this property of such geometric decompositions. This is analogous to polar coordinates, which consists of knowing the radial distance of a point from the origin and its angle from a reference line.*

2.10 QR Decomposition

There is one more very important matrix decomposition – QR decomposition – that will prove to be very useful. Although it does not involve the eigenvalues or eigenvectors of the matrix that we are decomposing, as in the previous decompositions, instead it is used to actually determine those eigenvalues and eigenvectors. As such, QR decomposition is the basis for numerical algorithms designed to determine the eigenvalues and eigenvectors of large matrices, solve least-squares problems, and perform orthogonalization.

In order to introduce the QR decomposition, let us revisit the Gram–Schmidt orthogonalization procedure introduced in Section 1.7.3. Recall that it is a procedure for obtaining an orthonormal set of basis vectors from a set of linearly independent vectors. The primary virtue of this method is that it enables us to develop a clear intuition as to how each vector is adjusted in order to be orthogonal to each of the previous – already orthogonal – vectors. For large numbers of large-dimensional vectors, however, it is extremely inefficient and subject to significant errors building up as we apply Gram–Schmidt to each successive vector one by one. Not only is each vector subject to the errors built up in orthogonalizing the previous vectors, orthogonalization of each successive vector requires a growing number of operations. An alternative method, which largely mitigates these issues, is based on QR decomposition.

Recall that the objective of Gram–Schmidt orthogonalization is to transform a set of S linearly independent $N \times 1$ vectors $(\mathbf{u}_1, \mathbf{u}_2, \ldots, \mathbf{u}_S)$ into a set of S orthonormal vectors $(\mathbf{q}_1, \mathbf{q}_2, \ldots, \mathbf{q}_S)$, such that they are all mutually orthogonal and normalized to unit length. Each successive vector $\mathbf{u}_i$ is made to be orthogonal to all of the previously orthonormalized vectors $(\mathbf{q}_1, \mathbf{q}_2, \ldots, \mathbf{q}_{i-1})$ using

$$\hat{\mathbf{u}}_i = \mathbf{u}_i - \sum_{k=1}^{i-1} \langle \mathbf{u}_i, \mathbf{q}_k \rangle \mathbf{q}_k, \quad i = 2, 3, \ldots, S. \tag{2.57}$$

The orthogonalized vectors $\hat{\mathbf{u}}_i$ are then normalized to length one using

$$\mathbf{q}_i = \frac{\hat{\mathbf{u}}_i}{\|\hat{\mathbf{u}}_i\|},$$

which produces the final set of orthonormal vectors.

To see how the Gram–Schmidt orthogonalization procedure can be framed as a matrix decomposition, let us solve (2.57) for $\mathbf{u}_i$ and write $\hat{\mathbf{u}}_i = \|\hat{\mathbf{u}}_i\| \mathbf{q}_i$ as follows:

$$\mathbf{u}_i = \|\hat{\mathbf{u}}_i\| \mathbf{q}_i + \sum_{k=1}^{i-1} \langle \mathbf{u}_i, \mathbf{q}_k \rangle \mathbf{q}_k. \tag{2.58}$$

Just as $\left\langle \hat{\mathbf{u}}_i, \mathbf{q}_j \right\rangle$ is the projection of the vector $\hat{\mathbf{u}}_i$ in the direction of the unit vector $\mathbf{q}_j$, $\left\langle \hat{\mathbf{u}}_i, \mathbf{q}_i \right\rangle$ is the projection of $\hat{\mathbf{u}}_i$ in the direction of the unit vector $\mathbf{q}_i$. Because $\hat{\mathbf{u}}_i$ is in the same direction as $\mathbf{q}_i$, however, this is the same as the norm (length) of $\hat{\mathbf{u}}_i$, in which case

$$\|\hat{\mathbf{u}}_i\| = \left\langle \hat{\mathbf{u}}_i, \mathbf{q}_i \right\rangle.$$

To show this in an alternative way, recall from earlier that

$$\hat{\mathbf{u}}_i = \|\hat{\mathbf{u}}_i\| \mathbf{q}_i.$$

Taking the inner product of $\mathbf{q}_i$ with both sides yields

$$\left\langle \hat{\mathbf{u}}_i, \mathbf{q}_i \right\rangle = \|\hat{\mathbf{u}}_i\| \left\langle \mathbf{q}_i, \mathbf{q}_i \right\rangle,$$

but $\langle \mathbf{q}_i, \mathbf{q}_i \rangle = 1$; therefore,

$$\|\hat{\mathbf{u}}_i\| = \left\langle \hat{\mathbf{u}}_i, \mathbf{q}_i \right\rangle.$$

Substituting (2.57) for $\hat{\mathbf{u}}_i$ into this result then gives

$$\begin{aligned}
\|\hat{\mathbf{u}}_i\| &= \left\langle \hat{\mathbf{u}}_i, \mathbf{q}_i \right\rangle \\
&= \langle \mathbf{u}_i, \mathbf{q}_i \rangle - \sum_{k=1}^{i-1} \langle \mathbf{u}_i, \mathbf{q}_k \rangle \langle \mathbf{q}_i, \mathbf{q}_k \rangle \\
\|\hat{\mathbf{u}}_i\| &= \langle \mathbf{u}_i, \mathbf{q}_i \rangle,
\end{aligned}$$

because $k \neq i$ and thus $\langle \mathbf{q}_i, \mathbf{q}_k \rangle = 0$ owing to orthogonality of the $\mathbf{q}$ vectors. Consequently, (2.58) can be written in the form

$$\mathbf{u}_i = \langle \mathbf{u}_i, \mathbf{q}_i \rangle \mathbf{q}_i + \sum_{k=1}^{i-1} \langle \mathbf{u}_i, \mathbf{q}_k \rangle \mathbf{q}_k = \sum_{k=1}^{i} \langle \mathbf{u}_i, \mathbf{q}_k \rangle \mathbf{q}_k, \tag{2.59}$$

where the first term has been subsumed in the summation by changing the upper limit from $i - 1$ to i.

Can (2.59) be expressed in a matrix form? Let us form an $N \times S$ matrix $\mathbf{A}$ having the original vectors $\mathbf{u}_i$ as its columns according to $\mathbf{A} = \begin{bmatrix} \mathbf{u}_1 & \mathbf{u}_2 & \cdots & \mathbf{u}_S \end{bmatrix}$ and another $N \times S$ matrix $\mathbf{Q}$ having the corresponding orthonormal vectors $\mathbf{q}_i$ as its columns according to $\mathbf{Q} = \begin{bmatrix} \mathbf{q}_1 & \mathbf{q}_2 & \cdots & \mathbf{q}_S \end{bmatrix}$.

Let us write out the first several expressions from (2.59) in order to observe the pattern:

$$\begin{aligned}
i = 1: &\quad \mathbf{u}_1 = \langle \mathbf{u}_1, \mathbf{q}_1 \rangle \mathbf{q}_1, \\
i = 2: &\quad \mathbf{u}_2 = \langle \mathbf{u}_2, \mathbf{q}_1 \rangle \mathbf{q}_1 + \langle \mathbf{u}_2, \mathbf{q}_2 \rangle \mathbf{q}_2, \\
i = 3: &\quad \mathbf{u}_3 = \langle \mathbf{u}_3, \mathbf{q}_1 \rangle \mathbf{q}_1 + \langle \mathbf{u}_3, \mathbf{q}_2 \rangle \mathbf{q}_2 + \langle \mathbf{u}_3, \mathbf{q}_3 \rangle \mathbf{q}_3, \\
&\vdots
\end{aligned}$$

Thus, (2.59) can be written in the matrix form

$$\mathbf{A} = \mathbf{QR}, \tag{2.60}$$

where $\mathbf{Q}$ is an $N \times S$ orthogonal matrix containing the orthonormal basis vectors, and $\mathbf{R}$ is the $S \times S$ right (upper) triangular matrix of coefficients

$$\mathbf{R} = \begin{bmatrix} \langle \mathbf{u}_1, \mathbf{q}_1 \rangle & \langle \mathbf{u}_2, \mathbf{q}_1 \rangle & \langle \mathbf{u}_3, \mathbf{q}_1 \rangle & \cdots \\ 0 & \langle \mathbf{u}_2, \mathbf{q}_2 \rangle & \langle \mathbf{u}_3, \mathbf{q}_2 \rangle & \cdots \\ 0 & 0 & \langle \mathbf{u}_3, \mathbf{q}_3 \rangle & \cdots \\ \vdots & \vdots & \vdots & \ddots \end{bmatrix}. \tag{2.61}$$

This is known as the *QR decomposition* of the $N \times S$ matrix $\mathbf{A}$.

Therefore, one way to obtain the QR decomposition is to first perform a Gram–Schmidt orthogonalization as in Section 1.7.3, from which the $\mathbf{Q}$ matrix is formed from the orthonormal vectors $\mathbf{q}_i$, and the $\mathbf{R}$ matrix is determined by carrying out the required inner products between the original vectors $\mathbf{u}_i$ and the resulting orthonormal vectors $\mathbf{q}_i$ according to (2.61). For large matrices, however, this method is inefficient and prone to numerical errors as discussed earlier. In fact, an often preferred technique for performing orthogonalization is to first obtain the QR decomposition and then extract the orthonormal vectors from $\mathbf{Q}$. In this way, the $\mathbf{Q}$ matrix in the QR decomposition provides an orthogonal basis for the vector space spanned by the column vectors of $\mathbf{A}$. The technique for actually determining the $\mathbf{Q}$ and $\mathbf{R}$ matrices in the QR decomposition is provided in Section 6.5.

In addition to constructing orthogonal basis vectors, QR decomposition can be used to determine the solution of the $N \times N$ system of equations $\mathbf{Au} = \mathbf{b}$. Given the QR decomposition of $\mathbf{A}$, the system of equations becomes

$$\mathbf{QRu} = \mathbf{b},$$

or premultiplying both sides by $\mathbf{Q}^{-1} = \mathbf{Q}^T$ yields the system

$$\mathbf{Ru} = \hat{\mathbf{b}},$$

where $\hat{\mathbf{b}} = \mathbf{Q}^T\mathbf{b}$. Because $\mathbf{R}$ is right triangular, this system of equations can be solved via simple backward substitution. This approach is rarely used, however, as it takes twice as long as LU decomposition (see Section 6.3.1) owing to the need to first obtain the QR decomposition of $\mathbf{A}$. Despite its apparent lack of usefulness at this point, the QR decomposition will feature prominently in both Parts II and III. In Section 6.5, it will be shown to be at the core of a numerical method to approximate the eigenvalues and eigenvectors of a matrix, and in Section 10.2, it will be shown how QR decomposition can be used to solve least-squares problems.

Although not covered thus far, we have referenced two additional decompositions that will be covered in Section 6.3, namely LU decomposition and Cholesky decomposition. They are used to calculate the solution of a system of linear algebraic equations and the inverse of large matrices.

2.11 Briefly on Bases

The central feature of this chapter has been the algebraic eigenproblem and its generalization, the SVD. The utility of these two decompositions is unmatched in a wide variety of physical, numerical, and data-analysis settings. In many applications, the eigenvectors or singular vectors can be thought of as providing a vector basis for representing the system that is better in some sense than the original basis. For example, in Section 2.1.2, we were reminded that the eigenvectors of a stress tensor correspond to the principal axes and provide a basis with respect to which only principal (normal) stresses act. In Section 2.3.2, we saw that in seeking the solution of a linear system of algebraic equations $\mathbf{A}\mathbf{u} = \mathbf{b}$, if the coefficient matrix $\mathbf{A}$ is symmetric, the solution vector $\mathbf{u}$ can be written as a linear combination of the mutually orthogonal eigenvectors of the symmetric matrix $\mathbf{A}$. That is, the eigenvectors provide the ideal basis for representing the solution for any right-hand side vector $\mathbf{b}$. We will see a completely analogous result in the next chapter for differential equations as well. While this specific application appears to be limited to a narrow class of problems, we will see in Part III that a similar approach can be extended to determining the optimal basis vectors for representation of *any* set of experimental or numerical data through POD. This extension is facilitated by the SVD highlighted in this chapter, so much so that some think of SVD and POD as essentially synonymous.

It has been emphasized throughout this chapter that the diagonalization procedure is a powerful technique for determining a more convenient means of representing a quadratic function or, more generally, system of linear, first-order differential equations. The eigenvectors of the coefficient matrix in each case provide the ideal basis vectors for representing the quadratic in its canonical form or uncoupling the system of first-order differential equations in order to render their solution as straightforward as possible. Therefore, the diagonalization procedure emphasizes the importance of the second of the two main questions when considering the basis to use in a particular problem:

1. What vector or function basis is most appropriate for representing the solution of a system of linear algebraic equations or a differential equation?
2. Is there a change of basis that leads to a "better" representation of the solution, whether to facilitate its solution, such as in diagonalization, or to more naturally extract useful information about solutions of the system, such as in POD?

Finally, we found that the eigenvalues of a dynamic coefficient matrix correspond to the natural frequencies of vibration of the corresponding dynamical system, which in turn form the basis functions for the general solution of the dynamical system. This will be explored more fully in Chapter 5.

2.12 Reader's Choice

At this point in the text, the reader has a choice to make. If after reading Chapters 1 and 2, the reader is in urgent need of more focus on applications in order to see how these methods are applied in science and engineering contexts, proceed to Chapter 5, which highlights numerous applications of the material in the first two chapters in dynamical systems theory. This is a collection of techniques that provide a powerful arsenal of methods for analyzing and characterizing mechanical, electrical, chemical, and biological systems that evolve with time. Alternatively, if the reader would prefer to maintain continuity of topics and consider the differential eigenproblem immediately after covering the algebraic eigenproblem, then proceed to Chapter 3. We will extend the ideas and methods used for the algebraic eigenproblem to the differential eigenproblem. This will open up a whole new set of applications. The reader is encouraged to read through Chapter 5, whether now or later, as it will furnish numerous additional and interesting examples and equip readers for analyzing a wide variety of linear and nonlinear systems.

Exercises

The following exercises are to be completed using hand calculations. When appropriate, the results can be checked using built-in functions in Python, MATLAB, or Mathematica for the vector and matrix operations.

2.1 Determine the eigenvalues and eigenvectors of the matrix

$$\mathbf{A} = \begin{bmatrix} 2 & 2 & 0 \\ 2 & 5 & 0 \\ 0 & 0 & 3 \end{bmatrix}.$$

Are the eigenvectors linearly independent? Are they mutually orthogonal? Explain your answers.

2.2 Determine the eigenvalues and eigenvectors of the matrix

$$\mathbf{A} = \begin{bmatrix} 1 & 1 & 3 \\ 1 & 1 & -3 \\ 3 & -3 & -3 \end{bmatrix}.$$

Are the eigenvectors linearly independent? Are they mutually orthogonal? Explain your answers.

2.3 Determine the eigenvalues and eigenvectors of the matrix

$$\mathbf{A} = \begin{bmatrix} 0 & 0 & 2 \\ -1 & 1 & 2 \\ -1 & 0 & 3 \end{bmatrix}.$$

Are the eigenvectors linearly independent? Are they mutually orthogonal? Explain your answers.

2.4 The stress tensor for a three-dimensional state of stress at a point is given by

$$\tau = \begin{bmatrix} 30 & 15 & 20 \\ 15 & 22 & 26 \\ 20 & 26 & 40 \end{bmatrix}.$$

Obtain the principal stresses and principal axes along which they act using built-in functions within Python, MATLAB, or Mathematica.

2.5 Determine the eigenvalues and eigenvectors (including generalized eigenvectors if necessary) of the matrix

$$\mathbf{A} = \begin{bmatrix} 0 & 0 & 3 \\ 0 & 3 & 0 \\ 3 & 1 & 0 \end{bmatrix}.$$

2.6 Determine the eigenvalues and eigenvectors (including generalized eigenvectors if necessary) of the matrix

$$\mathbf{A} = \begin{bmatrix} 1 & 2 & 0 \\ 1 & 1 & 2 \\ 0 & -1 & 1 \end{bmatrix}.$$

2.7 Determine the eigenvalues and eigenvectors (including generalized eigenvectors if necessary) of the matrix

$$\mathbf{A} = \begin{bmatrix} 5 & 1 & -4 \\ 4 & 3 & -5 \\ 3 & 1 & -2 \end{bmatrix}.$$

2.8 Determine the eigenvalues and eigenvectors (including generalized eigenvectors if necessary) of the matrix

$$\mathbf{A} = \begin{bmatrix} 2 & 2 & 0 & 0 \\ 0 & 2 & 0 & 0 \\ 0 & 0 & 3 & 3 \\ 0 & 0 & 0 & 3 \end{bmatrix}.$$

2.9 Determine the eigenvalues and eigenvectors (including generalized eigenvectors if necessary) of the matrix

$$\mathbf{A} = \begin{bmatrix} 5 & 1 & 3 & 2 \\ 0 & 5 & 0 & -3 \\ 0 & 0 & 5 & 1 \\ 0 & 0 & 0 & 5 \end{bmatrix}.$$

2.10 Consider the following matrix:

$$\mathbf{A} = \begin{bmatrix} 2 & 2 & 2 \\ 0 & 2 & 2 \\ 0 & 0 & 2 \end{bmatrix}.$$

(a) Obtain the eigenvalues and eigenvectors (including generalized eigenvectors if necessary) of the matrix $\mathbf{A}$.
(b) Determine the matrix $\mathbf{B} = \mathbf{U}^{-1}\mathbf{A}\mathbf{U}$ that results from attempting to diagonalize the matrix $\mathbf{A}$. Is this result what you expected from part (a)? Why or why not?

2.11 Consider the following matrix:

$$\mathbf{A} = \begin{bmatrix} 0 & 0 & 4 \\ 0 & 2 & 0 \\ -1 & 0 & 4 \end{bmatrix}.$$

(a) Obtain the eigenvalues and eigenvectors (including generalized eigenvectors if necessary) of the matrix $\mathbf{A}$.
(b) Determine the matrix $\mathbf{B} = \mathbf{U}^{-1}\mathbf{A}\mathbf{U}$ that results from attempting to diagonalize the matrix $\mathbf{A}$. Is this result what you expected from part (a)? Why or why not?

2.12 Consider the following matrix:

$$\mathbf{A} = \begin{bmatrix} 3 & -1 & 0 \\ 1 & 1 & 0 \\ 0 & 0 & 2 \end{bmatrix}.$$

(a) Obtain the eigenvalues and eigenvectors (including generalized eigenvectors if necessary) of the matrix $\mathbf{A}$.
(b) Determine the matrix $\mathbf{B} = \mathbf{U}^{-1}\mathbf{A}\mathbf{U}$ that results from attempting to diagonalize the matrix $\mathbf{A}$. Is this result what you expected from part (a)? Why or why not?

2.13 Verify the Cayley–Hamilton theorem for the matrix

$$\mathbf{A} = \begin{bmatrix} 1 & 2 \\ 2 & 1 \end{bmatrix}.$$

2.14 Verify the Cayley–Hamilton theorem for the matrix

$$\mathbf{A} = \begin{bmatrix} 0 & -1 \\ \pi^2 & -2\pi \end{bmatrix}.$$

2.15 Verify the Cayley–Hamilton theorem for the matrix

$$\mathbf{A} = \begin{bmatrix} -2 & -3 & -1 \\ 1 & 2 & 1 \\ 3 & 3 & 2 \end{bmatrix}.$$

2.16 Let

$$\mathbf{A} = \begin{bmatrix} 2 & 1 \\ 1 & 2 \end{bmatrix} \quad \text{and} \quad \mathbf{B} = \mathbf{A}^5 - 3\mathbf{A}^4 + 2\mathbf{A} - \mathbf{I}.$$

(a) Determine the eigenvalues and corresponding eigenvectors of $\mathbf{B}$.
(b) Determine whether $\mathbf{B}$ is positive definite.

2.17 Let

$$\mathbf{A} = \begin{bmatrix} 2 & 0 & 1 \\ 0 & 3 & 0 \\ 1 & 0 & 2 \end{bmatrix} \quad \text{and} \quad \mathbf{B} = \mathbf{A}^3 - 7\mathbf{A}^2 + 15\mathbf{A} - 9\mathbf{I}.$$

(a) Determine the eigenvalues and corresponding eigenvectors of $\mathbf{B}$.
(b) Determine whether $\mathbf{B}$ is positive definite.

2.18 For the matrix

$$\mathbf{A} = \begin{bmatrix} 5 & 1 \\ 3 & -2 \end{bmatrix},$$

determine its inverse using

$$\mathbf{A}^{-1} = \frac{-1}{c_n}\left(\mathbf{A}^{n-1} + c_1\mathbf{A}^{n-2} + \cdots + c_{n-1}\mathbf{I}\right).$$

Recall that the characteristic equation may be written in the form

$$P(\lambda) = (-1)^n[\lambda^n + c_1\lambda^{n-1} + \cdots + c_{n-1}\lambda + c_n] = 0.$$

Check the result by showing that $\mathbf{A}\mathbf{A}^{-1} = \mathbf{I}$.

2.19 The matrix

$$\mathbf{A} = \begin{bmatrix} 1 & 0 & 1 \\ 0 & -1 & 0 \\ 1 & 0 & 1 \end{bmatrix}$$

has the eigenvalues

$$\lambda_1 = -1, \quad \lambda_2 = 0, \quad \lambda_3 = 2.$$

Are the eigenvectors of **A** linearly independent? Are the eigenvectors of **A** mutually orthogonal? Is **A** invertible? Is **A** positive definite? Is **A** diagonalizable? Explain your answers.

2.20 A symmetric matrix **A** has the following spectrum

$$\lambda_1 = 0, \quad \lambda_2 = \frac{1}{2}, \quad \lambda_3 = 2, \quad \lambda_4 = \frac{3}{4}.$$

Is the matrix **A** singular? Is it positive definite? Explain why or why not in each case.

2.21 Given that the eigenvalues and normalized eigenvectors of the matrix

$$\mathbf{A} = \begin{bmatrix} 2 & 1 & 0 \\ 1 & 2 & 0 \\ 0 & 0 & 2 \end{bmatrix}$$

are

$$\lambda_1 = 1, \quad \lambda_2 = 2, \quad \lambda_3 = 3,$$

and

$$\mathbf{q}_1 = \frac{1}{\sqrt{2}}\begin{bmatrix} 1 \\ -1 \\ 0 \end{bmatrix}, \quad \mathbf{q}_2 = \begin{bmatrix} 0 \\ 0 \\ 1 \end{bmatrix}, \quad \mathbf{q}_3 = \frac{1}{\sqrt{2}}\begin{bmatrix} 1 \\ 1 \\ 0 \end{bmatrix},$$

respectively, use the eigenvalues and eigenvectors to determine the solution of the system of algebraic equations

$$\mathbf{A}\mathbf{u} = \mathbf{b},$$

where

$$\mathbf{b} = \begin{bmatrix} 3 \\ 0 \\ 6 \end{bmatrix}.$$

2.22 Obtain the solution of the system of algebraic equations $\mathbf{A}\mathbf{u} = \mathbf{b}$, where

$$\mathbf{A} = \begin{bmatrix} 2 & 1 & 0 \\ 1 & 2 & 0 \\ 0 & 0 & 2 \end{bmatrix}, \quad \mathbf{b} = \begin{bmatrix} 3 \\ 0 \\ 6 \end{bmatrix},$$

by expanding the solution vector **u** as a linear combination of the eigenvectors of **A**.

2.23 Obtain the solution of the system of algebraic equations $\mathbf{Au} - \Lambda\mathbf{u} = \mathbf{b}$, where

$$\mathbf{A} = \begin{bmatrix} 1 & 0 & 1 \\ 0 & 1 & 0 \\ 1 & 0 & 1 \end{bmatrix}, \quad \mathbf{c} = \begin{bmatrix} 1 \\ 2 \\ 1 \end{bmatrix}, \quad \Lambda = -1,$$

by expanding the solution vector $\mathbf{u}$ as a linear combination of the eigenvectors of $\mathbf{A}$.

2.24 Given a quadratic of the form $\mathcal{A} = \mathbf{x}^T\mathbf{A}\mathbf{x}$, where

$$\mathbf{A} = \begin{bmatrix} 2 & 1 & 0 \\ 1 & 1 & 0 \\ 0 & 0 & 3 \end{bmatrix},$$

is the quadratic in canonical form? Explain why, or why not.

2.25 Consider the quadratic form given by

$$5x_1^2 - 4x_1x_2 + 8x_2^2 = 36.$$

Determine the matrix that transforms this conic section into its canonical form, and write down that canonical form.

2.26 Consider the quadratic form given by

$$2x_1^2 + 4x_2^2 + 2x_3^2 - 2x_1x_2 + 4x_2x_3 - 2x_1x_3 = 8.$$

Determine the matrix that transforms this quadratic into its canonical form, and write down that canonical form.

2.27 Consider the quadratic form given by

$$\mathcal{A} = x_1^2 + 2x_1x_2 + 2x_2x_3 + x_3^2.$$

Determine the matrix that transforms this quadratic into its canonical form, and write down that canonical form.

2.28 Consider the quadratic form given by

$$\mathcal{A} = x_1^2 + 2x_1x_2 + 4x_2^2 + 3x_3^2.$$

Determine the matrix that transforms this quadratic into its canonical form, and write down that canonical form.

2.29 Consider the quadratic form given by

$$\mathcal{A} = \frac{3}{2}x_1^2 - x_1x_3 + x_2^2 + \frac{3}{2}x_3^2.$$

Determine the matrix that transforms this quadratic into its canonical form, and write down that canonical form. Classify the type of quadratic based on its canonical form?

2.30 Determine for what values of a the following quadratic is such that

$$\mathcal{A} = x_1^2 + ax_2^2 + 4x_1x_2 > 0.$$

2.31 Consider the matrix $\mathbf{A}$ and vector $\mathbf{b}$ given by

$$\mathbf{A} = \begin{bmatrix} 2 & 0 & 3 \\ 0 & 1 & 0 \\ 3 & 0 & 2 \end{bmatrix}, \quad \mathbf{b} = \begin{bmatrix} 5 \\ 1 \\ 5 \end{bmatrix}.$$

(a) Obtain the solution of the system of algebraic equations

$$\mathbf{A}\mathbf{x} = \mathbf{b}$$

by expanding the solution vector $\mathbf{x}$ as a linear combination of the eigenvectors of $\mathbf{A}$.

(b) Write down the quadratic

$$\mathbf{y}^T \mathbf{A}\mathbf{y} = 4$$

in expanded form, and determine its canonical form.

(c) Obtain the solution of the following system of differential equations using matrix operations

$$\dot{\mathbf{u}} = \mathbf{A}\mathbf{u},$$

subject to the initial conditions

$$\mathbf{u}(0) = \mathbf{b}.$$

2.32 Consider the matrix $\mathbf{A}$ and vector $\mathbf{b}$ given by

$$\mathbf{A} = \begin{bmatrix} 0 & 1 & 1 \\ 1 & 1 & 0 \\ 1 & 0 & 1 \end{bmatrix}, \quad \mathbf{b} = \begin{bmatrix} 1 \\ 0 \\ 1 \end{bmatrix}.$$

(a) Obtain the solution of the system of algebraic equations

$$\mathbf{A}\mathbf{x} = \mathbf{b}$$

by expanding the solution vector $\mathbf{x}$ as a linear combination of the eigenvectors of $\mathbf{A}$.

(b) Write down the quadratic

$$\mathbf{y}^T \mathbf{A}\mathbf{y} = 4$$

in expanded form, and determine its canonical form.

(c) Obtain the solution of the following system of differential equations using matrix operations

$$\dot{\mathbf{u}} = \mathbf{A}\mathbf{u},$$

subject to the initial conditions

$$\mathbf{u}(0) = \mathbf{b}.$$

2.33 Consider the matrix $\mathbf{A}$ and vector $\mathbf{b}$ given by

$$\mathbf{A} = \begin{bmatrix} 0 & 1 & 0 \\ 1 & 1 & 1 \\ 0 & 1 & 0 \end{bmatrix}, \quad \mathbf{b} = \begin{bmatrix} 1 \\ 2 \\ 1 \end{bmatrix}.$$

(a) Given that $\mathbf{A}$ is real and symmetric, why is it not possible to obtain the solution of the system of algebraic equations

$$\mathbf{A}\mathbf{x} = \mathbf{b}$$

by expanding the solution vector $\mathbf{x}$ as a linear combination of the eigenvectors of $\mathbf{A}$? Determine the solution using some other appropriate means.

(b) Write down the quadratic

$$\mathbf{y}^T \mathbf{A} \mathbf{y} = 1$$

in expanded form, and determine its canonical form. Given the nature of $\mathbf{A}$, what does this mean for the shape of the canonical form?

(c) Obtain the solution of the following system of differential equations using matrix operations

$$\dot{\mathbf{u}} = \mathbf{A}\mathbf{u},$$

subject to the initial conditions

$$\mathbf{u}(0) = \mathbf{b}.$$

2.34 We seek an alternative method for determining the mth power of a matrix $\mathbf{A}$ that avoids performing the matrix multiplication m times. Recall that $\mathbf{D} = \mathbf{U}^{-1}\mathbf{A}\mathbf{U}$, where $\mathbf{D}$ is a diagonal matrix with the eigenvalues of $\mathbf{A}$ along the diagonal that is produced using the modal matrix $\mathbf{U}$ of $\mathbf{A}$.

(a) For a general matrix $\mathbf{A}$, show that $\mathbf{A}^m = \mathbf{U}\mathbf{D}^m\mathbf{U}^{-1}, m = 1, 2, 3, \ldots$ (hint: first consider $\mathbf{A}^2$ and generalize).

(b) Use this general result to evaluate $\mathbf{A}^3$ for the matrix $\mathbf{A}$ given by

$$\mathbf{A} = \begin{bmatrix} 2 & 5 \\ 5 & 2 \end{bmatrix}.$$

Note: By recognizing that the integer power of a diagonal matrix $\mathbf{D}^m$ is simply a diagonal matrix with the elements being given by the mth power of each diagonal element of $\mathbf{D}$, this gives an easier method for obtaining powers of $\mathbf{A}$ with large m than performing the matrix multiplication m times.

2.35 The structure of a carbon dioxide molecule may be idealized as three masses connected by two springs, where the masses are the carbon and oxygen atoms, and the springs represent the chemical bonds between the atoms.[29] The equations of motion for the three atoms, that is, masses, are given by

$$m_O \frac{d^2 x_1}{dt^2} = -kx_1 + kx_2,$$

$$m_C \frac{d^2 x_2}{dt^2} = kx_1 - 2kx_2 + kx_3,$$

$$m_O \frac{d^2 x_3}{dt^2} = kx_2 - kx_3.$$

(a) By substituting the oscillatory solutions

$$x_1(t) = A_1 e^{i\omega t} \quad [\text{or } x_1(t) = A_1 \cos(\omega t)],$$
$$x_2(t) = A_2 e^{i\omega t} \quad [\text{or } x_2(t) = A_2 \cos(\omega t)],$$
$$x_3(t) = A_3 e^{i\omega t} \quad [\text{or } x_3(t) = A_3 \cos(\omega t)],$$

[29] Adapted from Gilat and Subramaniam (2014).

into the differential equations, determine the system of algebraic equations for the natural frequencies of vibration ω of the system of atoms from these equations of motion.

(b) For $k = 1{,}420\,kg/s^2$, $m_O = 2.6568 \times 10^{-26}\,kg$, and $m_C = 1.9926 \times 10^{-26}\,kg$, use built-in functions within Python, MATLAB, or Mathematica for the vector and matrix operation(s) required to compute the natural frequencies.

2.36 Obtain the general solution of the following system of differential equations using matrix operations:

$$\frac{du_1}{dt} = -u_2,$$

$$\frac{du_2}{dt} = u_1.$$

2.37 Obtain the general solution of the following system of differential equations using matrix operations:

$$\frac{du_1}{dt} = 2u_1 - u_2,$$

$$\frac{du_2}{dt} = u_1.$$

2.38 Obtain the general solution of the following system of differential equations using matrix operations:

$$\frac{du_1}{dt} = u_1 + 4u_2,$$

$$\frac{du_2}{dt} = -u_1 - 3u_2.$$

2.39 Obtain the solution of the following system of differential equations using matrix operations:

$$\frac{du_1}{dt} = u_1 + u_2,$$

$$\frac{du_2}{dt} = -u_1 + 3u_2,$$

subject to the initial conditions

$$u_1(0) = 1, \quad u_2(0) = 2.$$

2.40 Obtain the solution of the following system of differential equations using matrix operations:

$$\frac{du_1}{dt} = 2u_1 + 2u_2,$$

$$\frac{du_2}{dt} = 2u_1 + 2u_2,$$

subject to the initial conditions

$$u_1(0) = 1, \quad u_2(0) = 2.$$

2.41 Obtain the solution of the following system of differential equations using matrix operations:

$$\frac{du_1}{dt} = u_1 + 2u_2,$$
$$\frac{du_2}{dt} = 2u_1 + u_2,$$

subject to the initial conditions

$$u_1(0) = 1, \quad u_2(0) = 3.$$

2.42 Obtain the solution of the following system of differential equations using matrix operations:

$$\frac{du_1}{dt} = u_1 + u_2 + 2u_3,$$
$$\frac{du_2}{dt} = 2u_2 + u_3,$$
$$\frac{du_3}{dt} = 3u_3,$$

subject to the initial conditions

$$u_1(0) = 5, \quad u_2(0) = 3, \quad u_3(0) = 2.$$

2.43 Obtain the general solution of the following system of differential equations using matrix operations:

$$\frac{du_1}{dt} = 2u_1 + 2u_2 + 2u_3,$$
$$\frac{du_2}{dt} = 2u_2 + 2u_3,$$
$$\frac{du_3}{dt} = 2u_3.$$

2.44 Obtain the general solution of the following system of differential equations using matrix operations:

$$\frac{du_1}{dt} = 3u_1 + u_3,$$
$$\frac{du_2}{dt} = 3u_2,$$
$$\frac{du_3}{dt} = 3u_3.$$

2.45 Obtain the solution of the following system of differential equations using matrix operations:

$$\frac{du_1}{dt} = 2u_1 + u_3,$$
$$\frac{du_2}{dt} = u_2,$$
$$\frac{du_3}{dt} = u_2 + 3u_3,$$

subject to the initial conditions

$$u_1(0) = -1, \quad u_2(0) = 2, \quad u_3(0) = 1.$$

2.46 Obtain the solution of the following system of differential equations using matrix operations:

$$\frac{du_1}{dt} = u_1 + u_3,$$
$$\frac{du_2}{dt} = 2u_2,$$
$$\frac{du_3}{dt} = -4u_1 - 3u_3,$$

subject to the initial conditions

$$u_1(0) = 1, \quad u_2(0) = -1, \quad u_3(0) = 0.$$

2.47 Obtain the general solution of the following system of differential equations using matrix operations:

$$\frac{du_1}{dt} = 2u_1,$$
$$\frac{du_2}{dt} = 3u_1 - 2u_2,$$
$$\frac{du_3}{dt} = 2u_2 + 3u_3.$$

2.48 Obtain the solution of the following system of differential equations using matrix operations:

$$\frac{du_1}{dt} = 2u_1 + 2u_2 + 1,$$
$$\frac{du_2}{dt} = 2u_1 + 2u_2 + 1,$$

subject to the initial conditions

$$u_1(0) = 0, \quad u_2(0) = 0.$$

2.49 Obtain the general solution of the following system of differential equations using matrix operations:

$$\frac{du_1}{dt} = -10u_1 - 18u_2 + t,$$
$$\frac{du_2}{dt} = 6u_1 + 11u_2 + 3.$$

2.50 Obtain the general solution of the following system of differential equations using matrix operations:

$$\frac{du_1}{dt} = u_1 + 2u_2 + e^t,$$

$$\frac{du_2}{dt} = 2u_2 + t^2.$$

2.51 Obtain the general solution of the following system of differential equations using matrix operations:

$$\frac{du_1}{dt} = 2u_1 + u_2,$$

$$\frac{du_2}{dt} = u_1 + 2u_2 + e^{2t}.$$

2.52 Obtain the general solution of the following system of differential equations using matrix operations:

$$\frac{du_1}{dt} = -2u_1 + 2u_2 + 2u_3 + \sin t,$$

$$\frac{du_2}{dt} = -u_2 + 3,$$

$$\frac{du_3}{dt} = -2u_1 + 4u_2 + 3u_3.$$

2.53 Obtain the general solution of the following system of differential equations using matrix operations:

$$\frac{du_1}{dt} + 2\frac{du_2}{dt} = 3u_2,$$

$$2\frac{du_1}{dt} + \frac{du_2}{dt} = 3u_1.$$

2.54 Obtain the general solution of the following differential equation using matrix operations:

$$\frac{d^2u}{dt^2} + \frac{1}{2}\frac{du}{dt} - \frac{1}{2}u = \frac{1}{2}.$$

2.55 Consider the differential equation

$$\frac{d^2u}{dx^2} + u = x,$$

subject to the boundary conditions

$$u(0) = 0, \quad u(1) = 0.$$

Obtain the solution using matrix operations. Note that this is the same problem as solved in Example 3.4 using eigenfunctions expansions.

2.56 Consider the differential equation

$$\frac{d^2u}{dx^2} - 4u = 8,$$

subject to the boundary conditions

$$u(0) = 0, \quad u(1) = 0.$$

Obtain the solution of the differential equation using matrix operations. Note that this is the same problem as solved in Exercise 3.25 using eigenfunctions expansions.

2.57 Consider the differential equation

$$\frac{d^2u}{dx^2} + \frac{du}{dx} = 1,$$

subject to the boundary conditions

$$u(0) = 0, \quad u'(1) = 0.$$

Obtain the solution using matrix operations.

2.58 Consider the differential equation

$$\frac{d^2u}{dt^2} + 3\frac{du}{dt} + 2u = e^{2t},$$

subject to the initial conditions

$$u(0) = 0, \quad \dot{u}(0) = 0.$$

Obtain the solution using matrix operations.

2.59 Consider the differential equation

$$\frac{d^2u}{dt^2} + 3\frac{du}{dt} + 2u = e^{2t}\cos t,$$

subject to the initial conditions

$$u(0) = 0, \quad \dot{u}(0) = 0.$$

Obtain the solution using matrix operations.

2.60 Obtain the general solution of the following differential equation using matrix operations:

$$\frac{d^3u}{dx^3} - \frac{d^2u}{dx^2} - 2\frac{du}{dx} = 0.$$

2.61 Determine the general solution of the following differential equation using matrix operations:

$$\frac{d^3u}{dx^3} - 2\frac{d^2u}{dx^2} - \frac{du}{dx} + 2u = 0.$$

2.62 Determine the singular values of the matrix

$$\mathbf{A} = \begin{bmatrix} 1 & 0 & 1 \\ 0 & 2 & 0 \end{bmatrix}.$$

2.63 Determine the singular-value decomposition of the matrix

$$\mathbf{A} = \begin{bmatrix} 1 \\ 0 \\ 1 \end{bmatrix}.$$

2.64 Determine the singular-value decomposition of the matrix

$$\mathbf{A} = \begin{bmatrix} \sqrt{2} & 0 & \sqrt{2} \end{bmatrix}.$$

2.65 Determine the singular-value decomposition of the matrix

$$\mathbf{A} = \begin{bmatrix} 1 & 2 \\ 2 & 1 \end{bmatrix}.$$

2.66 Determine the singular-value decomposition of the matrix

$$\mathbf{A} = \begin{bmatrix} 1 & 2 \\ 2 & 2 \\ 2 & 1 \end{bmatrix}.$$

2.67 Determine the singular-value decomposition of the matrix

$$\mathbf{A} = \begin{bmatrix} 3 & 2 & 2 \\ 2 & 3 & -2 \end{bmatrix}.$$

2.68 Determine the singular-value decomposition of the matrix

$$\mathbf{A} = \begin{bmatrix} 1 & -1 \\ -2 & 2 \\ -1 & -1 \end{bmatrix}.$$

2.69 Determine the singular-value decomposition of the matrix

$$\mathbf{A} = \begin{bmatrix} 1 & 0 & 1 & 0 \\ 0 & 1 & 0 & 1 \end{bmatrix}.$$

2.70 Determine the singular-value decomposition of the matrix

$$\mathbf{A} = \begin{bmatrix} 0 & 1 & 1 \\ \sqrt{2} & 2 & 0 \\ 0 & 1 & 1 \end{bmatrix}.$$

Compare the singular values and singular vectors of $\mathbf{A}$ to its eigenvalues and eigenvectors.

2.71 Show that the polar decomposition of an orthogonal matrix $\mathbf{A}$ is given by $\mathbf{U} = \mathbf{V} = \mathbf{I}$ and $\mathbf{R} = \mathbf{A}$, where $\mathbf{I}$ is the identity matrix.

2.72 Determine the polar decomposition of the matrix

$$\mathbf{A} = \begin{bmatrix} 11 & 10 \\ -2 & 5 \end{bmatrix}.$$

2.73 Determine the polar decomposition of the matrix

$$\mathbf{A} = \frac{1}{5}\begin{bmatrix} 26 & -7 \\ 14 & 2 \end{bmatrix}.$$

2.74 If $\mathbf{A}$ is an $N \times N$ matrix and $\mathbf{Q}$ is an $N \times N$ orthogonal matrix, show that

$$\exp\left(\mathbf{Q}^T\mathbf{A}\mathbf{Q}\right) = \mathbf{Q}^T \exp(\mathbf{A})\mathbf{Q}.$$

(Hint: write the exponential on the left-hand side in terms of a Taylor series.)

3 Differential Eigenproblems and Their Applications

> The mathematician's patterns, like the painter's or the poet's, must be beautiful; the ideas, like the colors or the words, must fit together in a harmonious way. Beauty is the first test; there is no permanent place in the world for ugly mathematics. (G. H. Hardy)

The primary application of this chapter is developing methods for solving ordinary and partial differential equations, so let us briefly survey the broad landscape of differential equations before proceeding. This will allow us to put the diagonalization procedure established in Chapter 2 and those in this chapter in perspective along with the numerical methods to be covered in Part II.

Differential equations can be ordinary or partial, initial- or boundary-value, and linear or nonlinear. The distinction between ordinary and partial differential equations simply hinges on the number of independent variables – if only one, it is ordinary; if more than one, it is partial. Of course, the additional complexity of partial versus ordinary differential equations has consequences for our solution techniques, their complexity, and the nature of the possible solutions. The most general and fundamental distinction is whether a differential equation is an initial-value problem or a boundary-value problem. This speaks to the nature of the physical phenomena being modeled and has important implications for numerical solution techniques.

Initial-value problems evolve in a single coordinate direction – usually time. Therefore, they model dynamic processes and require initial conditions at the start time of the problem. Boundary-value problems, on the other hand, typically model steady – or time invariant – processes that only depend on a spatial coordinate(s) and not time. They require boundary conditions at all spatial boundaries of the domain. In the case of partial differential equations, initial-value problems correspond to parabolic or hyperbolic equations, while boundary-value problems correspond to elliptic equations as defined in Section 7.5.

Both initial- and boundary-value problems can be linear or nonlinear, which dramatically influences the solution techniques that are available for their solution as well as the nature of the solutions that may result. In general, the techniques developed in Chapters 2 and 3 result in exact solutions, but they are restricted to certain classes of linear equations. Therefore, these methods are powerful, but limited in scope. For more complicated linear, as well as nonlinear, equations, numerical methods from Part II must be employed to obtain approximate solutions. Numerical methods for boundary-value problems are covered in Chapter 8, and numerical methods for initial-value problems are covered in Chapter 9. All of the methods discussed in this text for exactly or numerically solving ordinary and partial differential equations will be summarized in Table 6.3 with reference to their respective sections.

The diagonalization procedure introduced in Chapter 2 applies to systems of linear ordinary differential equations with constant coefficients. A look back at the examples will reveal that all are initial-value problems in time. While there is no fundamental reason why diagonalization could not be applied to boundary-value problems, certain boundary conditions are not naturally accommodated within such a framework. Therefore, it is typically classified as a method primarily for use with initial-value problems.

The eigenfunction expansion approach at the heart of this chapter can be applied to certain linear ordinary and partial differential equations with constant or variable coefficients. It applies to both initial- and boundary-value problems and can be used to determine exact solutions of certain linear ordinary and partial differential equations and as the foundation for spectral numerical methods that can approximate the solution of more general equations, including nonlinear ones.

Note that there is some overlap between the cases that can be treated using diagonalization and eigenfunction expansions. When the same problem can be solved using both techniques, the same solution will be obtained, but they will be in very different forms. However, the eigenfunction expansion approach is more general and applies to a broader class of problems. In particular, whereas ordinary differential equations typically govern the physics of *discrete systems* comprised of discrete objects, such as springs and point masses, the extension to partial differential equations is a very powerful one as it allows us to treat *continuous systems*, such as solid and fluid mechanics, for which the mass is distributed throughout the domain. The solutions obtained in this chapter are series expansions of solutions in terms of eigenfunctions of differential operators. As such, vectors in Chapter 2 become functions in Chapter 3; likewise, matrices become differential operators.

This analogy between linear algebraic and differential systems suggests that the method used in Section 2.3.2 to solve systems of linear algebraic equations with symmetric coefficient matrices may lead to an approach for solving differential equations. In particular, we used the orthonormal eigenvectors of the symmetric coefficient matrix in the linear system of algebraic equations as the basis vectors for forming the solution of system $\mathbf{Au} = \mathbf{b}$. Might an analogous approach be used in which the eigenfunctions of a differential operator are used as the basis functions for the solution of a differential equation? That is, just as the solution of a system of linear algebraic equations can be written as a linear combination of the eigenvectors of the coefficient matrix, can the solution to a differential equation be written as the linear combination of the eigenfunctions of the differential operator? Indeed, this is the case for self-adjoint differential operators, which have properties analogous to symmetric matrices, namely that the eigenfunctions corresponding to distinct eigenvalues are mutually orthogonal.

Before we begin, let us address the suggested analogies between vector and function spaces as well as eigenvectors and eigenfunctions. As engineers and scientists, we view matrices and differential operators as two distinct entities, and we motivate similarities by analogy. Although we are often under the impression that the formal mathematical approach – replete with its theorems and proofs – is unnecessarily cumbersome, even though we appreciate that someone has put such methods on a firm mathematical foundation, this is an example where the more mathematical approach reveals its true value. Although from the applications point of view, matrices and differential operators seem very different and occupy different topical domains, the mathematician recognizes that matrices and certain differential operators are both *linear operators* that have particular properties, thus unifying these two constructs. Therefore, the mathematician is not surprised by the relationship between operations that apply to both; in fact, it couldn't be any other way!

3.1 Function Spaces, Bases, and Orthogonalization

You will observe that we have introduced some new, but familiar sounding, terminology. We need to establish in what way eigenfunctions of differential operators are similar – and different – from eigenvectors of matrices. In addition, what relationship, if any, does a *function space* have to a vector space? This is where we start. We begin by defining a function space by analogy to vector spaces that allows us to consider linear independence and orthogonality of functions. We then consider eigenvalues and eigenfunctions of differential operators, which allows us to develop powerful techniques for solving ordinary and partial differential equations.

3.1.1 Definitions

Just as we often use linearly independent – or orthogonal – *vectors* as a convenient basis for representing a vector space, we will often desire a set of linearly independent – or orthogonal – basis *functions* that can be used to represent a function space. Such a function space may contain the solution of a differential equation, for example, just as a vector space may contain the solution of a system of algebraic equations as in Section 2.3.2.

In order to define a *function space*, let us draw an analogy to vector spaces considered in Chapter 1.

Vector Space:

1. A set of *vectors* $\mathbf{u}_m, m = 1, \ldots, M$, are *linearly independent* if their linear combination is zero, in which case we can only write

$$a_1\mathbf{u}_1 + a_2\mathbf{u}_2 + \cdots + a_M\mathbf{u}_M = \mathbf{0}$$

   if the a_m coefficients are all zero. Thus, no vector $\mathbf{u}_m$ can be written as a linear combination of the others.
2. Any N-dimensional *vector* may be expressed as a linear combination of N linearly independent, N-dimensional *basis vectors*. Although it is not necessary, we often

prefer mutually orthogonal basis vectors, in which case

$$\langle \mathbf{u}_m, \mathbf{u}_n \rangle = 0, \quad m \neq n.$$

Function Space:

1. A set of *functions* $u_m(x), m = 1, \dots, M$, are *linearly independent* over an interval $a \leq x \leq b$ if their linear combination is zero, in which case we can only write

$$a_1 u_1(x) + a_2 u_2(x) + \cdots + a_M u_M(x) = 0$$

if the a_m coefficients are all zero. Thus, no function $u_m(x)$ can be written as a linear combination of the others over the interval $a \leq x \leq b$.

2. Any *piecewise continuous function* in the interval $a \leq x \leq b$ may be expressed as a linear combination of an infinite number of linearly independent *basis functions*. Although it is not necessary, we often prefer mutually orthogonal basis functions.

Observe that the primary difference between vector and functions spaces is their dimensionality; vector spaces are finite dimensional, while function spaces are infinite dimensional. In addition, note that linear independence and orthogonality of functions is defined with respect to a particular spatial interval.

3.1.2 Linear Independence of Functions

To determine if a set of functions are linearly independent over some interval $a \leq x \leq b$, we form the *Wronskian*, which is a determinant of the functions and their derivatives. Say we have M functions, $u_1(x), u_2(x), \dots, u_M(x)$, for which the first $M-1$ derivatives are continuous. The Wronskian is then defined by

$$W\left[u_1(x), u_2(x), \dots, u_M(x)\right] = \begin{vmatrix} u_1(x) & u_2(x) & \cdots & u_M(x) \\ u_1'(x) & u_2'(x) & \cdots & u_M'(x) \\ \vdots & \vdots & \ddots & \vdots \\ u_1^{(M-1)}(x) & u_2^{(M-1)}(x) & \cdots & u_M^{(M-1)}(x) \end{vmatrix}.$$

If the Wronskian is nonzero at some point in the interval, such that $W \neq 0$ for some x in $a \leq x \leq b$, then the functions $u_1(x)$, $u_2(x)$, ..., $u_M(x)$ are linearly independent over that interval.

In practice, it is often the case that the functions are known to be linearly independent – or even orthogonal – owing to the setting, and it is rarely necessary to actually evaluate the Wronskian. For example, this is the case for solutions of the Sturm–Liouville equations, which have self-adjoint differential operators, to be considered in Section 3.3.

3.1.3 Basis Functions

As mentioned previously, any piecewise continuous function in an interval may be expressed as a linear combination of an infinite number of linearly independent *basis functions*. Consider a function $u(x)$ over the interval $a \leq x \leq b$. If $u_1(x), u_2(x), \ldots$ are linearly independent basis functions over the interval $a \leq x \leq b$, then $u(x)$ can be expressed as

$$u(x) = a_1 u_1(x) + a_2 u_2(x) + \cdots = \sum_{n=1}^{\infty} a_n u_n(x),$$

where not all a_n coefficients are zero. Because this requires an infinite set of basis functions, we say that the function $u(x)$ is *infinite dimensional*.

Although it is not necessary, we prefer mutually orthogonal basis functions such that the inner product of two functions is zero according to

$$\langle u_m(x), u_n(x) \rangle = \int_a^b w(x) u_m(x) u_n(x) dx = 0, \quad m \neq n, \tag{3.1}$$

that is, $u_m(x)$ and $u_n(x)$ are *orthogonal*[1] over the interval $a \leq x \leq b$ with respect to the *weight function* $w(x)$. Note that $w(x) = 1$ unless stated otherwise.

Analogous to norms of vectors, we define the norm of a function, which gives a measure of a function's "size." Recall that the norm of a vector[2] is defined by

$$\|\mathbf{u}\| = \langle \mathbf{u}, \mathbf{u} \rangle^{1/2} = \left(\sum_{n=1}^{N} u_n^2 \right)^{1/2}.$$

Similarly, the norm of a function is defined by

$$\|u(x)\| = \langle u(x), u(x) \rangle^{1/2} = \left[\int_a^b w(x) u^2(x) dx \right]^{1/2},$$

where the discrete sum is replaced by a continuous integral in both the inner product and norm.

With these definitions of inner products and norms, one could apply the Gram–Schmidt orthogonalization procedure from Section 1.7.3 to obtain an *orthonormal* set of basis functions, such that

$$\langle u_m(x), u_n(x) \rangle = \int_a^b w(x) u_m(x) u_n(x) dx = \delta_{mn} = \begin{cases} 0, & m \neq n \\ 1, & m = n \end{cases},$$

where δ_{mn} is the Kronecker delta.

Example 3.1 Consider the polynomial functions $u_0(x) = x^0 = 1, u_1(x) = x^1 = x$, $u_2(x) = x^2, \ldots, u_n(x) = x^n, \ldots$ in the interval $-1 \leq x \leq 1$. Using these functions, which are linearly independent over the interval, obtain an orthonormal set of basis functions using the Gram–Schmidt orthogonalization procedure. Use the

[1] Whereas orthogonality of vectors is associated with them being geometrically perpendicular, there is no such connotation for orthogonal functions. They simply have the mathematical property that their inner product as defined here is zero.

[2] Recall from Section 1.9 that we use the L_2-norm unless indicated otherwise.

weight function

$$w_n = \frac{2n+1}{2}.$$

Solution

Begin by normalizing $u_0(x)$ to length one. The norm of $u_0(x) = 1$ is $(n = 0)$

$$\|u_0\| = \left(\int_{-1}^{1} \frac{2n+1}{2} (1)^2 dx \right)^{1/2} = \left(\frac{1}{2} x \Big|_{-1}^{1} \right)^{1/2} = 1;$$

thus, normalizing gives

$$\tilde{u}_0(x) = \frac{u_0}{\|u_0\|} = \frac{1}{1} = 1.$$

Consider $u_1(x) = x$. First, evaluate the inner product of $u_1(x)$ with the previous function $\tilde{u}_0(x)$ $(n = 0)$

$$\langle u_1, \tilde{u}_0 \rangle = \int_{-1}^{1} \frac{2n+1}{2} x(1) dx = \frac{1}{2} \frac{x^2}{2} \Big|_{-1}^{1} = 0;$$

therefore, they are already orthogonal, and we simply need to normalize $u_1(x) = x$ as before $(n = 1)$:

$$\|u_1\| = \left(\int_{-1}^{1} \frac{2n+1}{2} x^2 dx \right)^{1/2} = \left(\frac{3}{2} \frac{x^3}{3} \Big|_{-1}^{1} \right)^{1/2} = 1;$$

$$\therefore \tilde{u}_1(x) = \frac{u_1}{\|u_1\|} = x.$$

Recall that *even* functions are such that $f(-x) = f(x)$, and *odd* functions are such that $f(-x) = -f(x)$. Therefore, all of the odd functions $\left(x, x^3, \ldots\right)$ are orthogonal to all of the even functions $\left(x^0, x^2, x^4, \ldots\right)$ in the interval $-1 \leq x \leq 1$.

Now considering $u_2(x) = x^2$, the function that is mutually orthogonal to the previous ones is given by

$$\hat{u}_2 = u_2 - \langle u_2, \tilde{u}_1 \rangle \tilde{u}_1 - \langle u_2, \tilde{u}_0 \rangle \tilde{u}_0,$$

but $\langle u_2, \tilde{u}_1 \rangle = 0$, and $(n = 0)$

$$\langle u_2, \tilde{u}_0 \rangle = \int_{-1}^{1} \frac{2n+1}{2} x^2 (1) dx = \frac{1}{2} \frac{x^3}{3} \Big|_{-1}^{1} = \frac{1}{3}.$$

Note that just like vectors, the inner product of two functions results in a scalar. Then,

$$\hat{u}_2 = x^2 - \frac{1}{3}(1) = x^2 - \frac{1}{3}.$$

Normalizing gives ($n = 2$)

$$\|\hat{u}_2\| = \left[\int_{-1}^{1} \frac{2n+1}{2}\left(x^2 - \frac{1}{3}\right)^2 dx\right]^{1/2} = \frac{2}{3};$$

$$\therefore \tilde{u}_2(x) = \frac{\hat{u}_2}{\|u_2\|} = \frac{3}{2}\left(x^2 - \frac{1}{3}\right) = \frac{3x^2 - 1}{2}.$$

Repeating for u_3 leads to

$$\tilde{u}_3(x) = \frac{5x^3 - 3x}{2}.$$

This produces *Legendre polynomials*, which are denoted by $P_n(x)$ and are mutually orthogonal over the interval $-1 \le x \le 1$ (see Section 3.3.3). Any continuous – or piecewise continuous – function $f(x)$ in the interval $-1 \le x \le 1$ can be written as a linear combination of the mutually orthogonal Legendre polynomials according to

$$f(x) = \sum_{n=0}^{\infty} \hat{a}_n P_n(x) = \sum_{n=0}^{\infty} \langle f(x), P_n(x)\rangle\, P_n(x),$$

where $\hat{a}_n = \langle f(x), P_n(x)\rangle$ because $P_n(x)$ are orthonormal (recall that the same holds for vectors).

Example 3.2 Consider the trigonometric functions in the interval $0 \le x \le 2\pi$ (note that $\cos(0x) = 1$ and $\sin(0x) = 0$):

$$\begin{array}{ll} 1, \cos(x), \cos(2x), \ldots, \cos(nx), \ldots, & n = 0, 1, 2, \ldots \\ \sin(x), \sin(2x), \ldots, \sin(mx), \ldots, & m = 1, 2, 3, \ldots \end{array}$$

Using these functions, which are linearly independent over the interval, obtain an orthonormal set of basis functions using the Gram–Schmidt orthogonalization procedure.

Solution

One can confirm that all of these functions are mutually orthogonal over the interval $0 \le x \le 2\pi$; therefore, it is only necessary to normalize them according to

$$\begin{array}{lll} 1 & \rightarrow & \dfrac{1}{\sqrt{2\pi}} \\ \cos(nx), n = 1, 2, 3, \ldots & \rightarrow & \dfrac{1}{\sqrt{\pi}} \\ \sin(mx), m = 1, 2, 3, \ldots & \rightarrow & \dfrac{1}{\sqrt{\pi}} \end{array}$$

Thus, any piecewise continuous function in $0 \le x \le 2\pi$ can be written as

$$f(x) = \frac{a_0}{\sqrt{2\pi}} + \sum_{n=1}^{\infty} a_n \frac{\cos(nx)}{\sqrt{\pi}} + \sum_{m=1}^{\infty} b_m \frac{\sin(mx)}{\sqrt{\pi}},$$

where

$$
\begin{aligned}
a_0 &= \left\langle f(x), \frac{1}{\sqrt{2\pi}} \right\rangle = \frac{1}{\sqrt{2\pi}} \int_0^{2\pi} f(x)dx, \\
a_n &= \left\langle f(x), \frac{\cos(nx)}{\sqrt{\pi}} \right\rangle = \frac{1}{\sqrt{\pi}} \int_0^{2\pi} f(x)\cos(nx)dx, \\
b_m &= \left\langle f(x), \frac{\sin(mx)}{\sqrt{\pi}} \right\rangle = \frac{1}{\sqrt{\pi}} \int_0^{2\pi} f(x)\sin(mx)dx.
\end{aligned}
$$

This produces the *Fourier series* (see Section 3.3.3). Once again, any piecewise continuous function $f(x)$ in the interval $0 \leq x \leq 2\pi$ can be expressed as a Fourier series.

REMARKS:

1. *An infinite set of orthogonal functions* $u_0(x), u_1(x), u_2(x), \ldots$ *is said to be* complete *if any piecewise continuous function in the interval can be expanded in terms of* $u_0(x), u_1(x), u_2(x), \ldots$. *Note that whereas an N-dimensional* vector space *is spanned by N mutually orthogonal vectors, an infinite-dimensional* function space *requires an infinite number of mutually orthogonal functions to span.*
2. *Analogous definitions and procedures can be developed for functions of more than one variable; for example, two two-dimensional functions* $u(x, y)$ *and* $v(x, y)$ *are orthogonal in the region A with respect to the weight function* $w(x, y)$ *if*

$$
\langle u(x, y), v(x, y) \rangle = \iint_A w(x, y)u(x, y)v(x, y)dxdy = 0.
$$

3. *The orthogonality of Legendre polynomials and Fourier series, along with other sets of functions, will prove very useful and is one reason why they are referred to as* special functions.

3.2 Eigenfunctions of Differential Operators

In the previous section, it has been shown that certain sets of functions have particularly attractive properties, namely that they are mutually orthogonal according to the definition (3.1). Here, we will explore sets of basis functions obtained as the eigenfunctions of a differential operator and show how they can be used to obtain solutions of ordinary differential equations. Similar to Section 2.3.2, where we expressed the solution of a system of linear algebraic equations in terms of a linear combination of the eigenvectors of the coefficient matrix, we will see how using the eigenfunctions of a differential operator provides a desirable basis for expressing the solution of a differential equation. First, of course, we must establish what an eigenfunction is.

3.2.1 Definitions

Let us further build on the analogy between vector and function spaces in which differential operators perform a role similar to the transformation matrix **A**.

Vector Space:

Consider the linear transformation (recall Section 1.8)

$$\underset{M\times N}{\mathbf{A}}\underset{N\times 1}{\mathbf{x}} = \underset{M\times 1}{\mathbf{y}},$$

such that the matrix **A** transforms **x** from the N-dimensional *domain* of **A** into **y**, which is in the M-dimensional *range* of **A**. This is illustrated in Figure 3.1.

Function Space:

Consider the general differential operator

$$\mathcal{L} = a_0 + a_1\frac{d}{dx} + a_2\frac{d^2}{dx^2} + \cdots + a_N\frac{d^N}{dx^N},$$

which is an Nth-order linear differential operator with constant coefficients. In the differential equation

$$\mathcal{L}u(x) = f(x),$$

the differential operator $\mathcal{L}$ transforms $u(x)$ into $f(x)$ as illustrated in Figure 3.2. Note that in this case, both the domain and range of $\mathcal{L}$ are infinite dimensional.

Just as an algebraic eigenproblem can be formulated for a matrix **A** in the form

$$\mathbf{A}\mathbf{u}_n = \lambda_n\mathbf{u}_n,$$

a differential eigenproblem can be formulated for a differential operator in the form

$$\mathcal{L}u_n(x) = \lambda_n u_n(x),$$

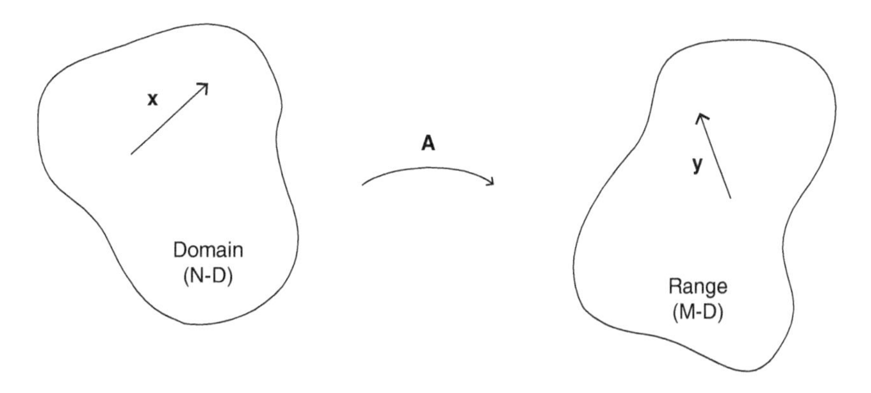

Figure 3.1 Transformation from **x** to **y** via the linear transformation matrix **A**.

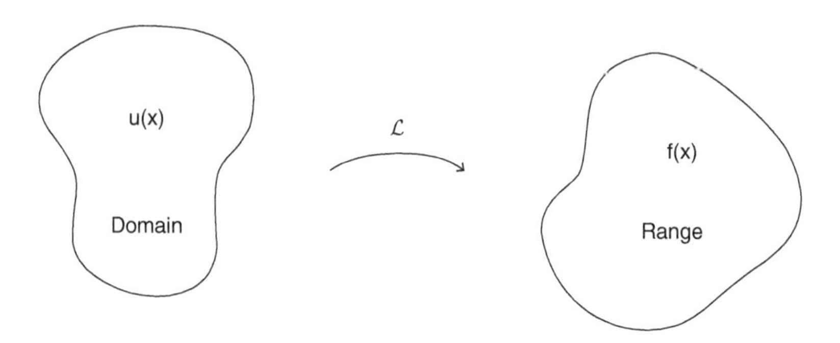

Figure 3.2 Transformation from $u(x)$ to $f(x)$ via the linear transformation $\mathcal{L}$.

where any value of λ_n that corresponds to a nontrivial solution $u_n(x)$ is an *eigenvalue* of $\mathcal{L}$, and the corresponding solution $u_n(x)$ is an *eigenfunction* of $\mathcal{L}$.[3] Similar to the way that $\mathbf{A}$ transforms $\mathbf{u}_n$ into a constant multiple of itself, $\mathcal{L}$ transforms $u_n(x)$ into a constant multiple of itself.

Observe that just as we consider eigenvalues and eigenvectors for *matrices*, not systems of algebraic equations, we consider eigenvalues and eigenfunctions of *differential operators*, not entire differential equations. Note, however, that just as the eigenproblem for a matrix is itself a system of algebraic equations, the eigenproblem for a differential operator also is a differential equation.

In differential eigenproblems, we require boundary conditions, which must be homogeneous. If $a \leq x \leq b$, then we may specify the value of the solution at the boundaries

$$u(a) = 0, \quad u(b) = 0,$$

or the derivative of the solution

$$u'(a) = 0, \quad u'(b) = 0,$$

or a linear combination of the two

$$c_1 u'(a) + c_2 u(a) = 0, \quad c_3 u'(b) + c_4 u(b) = 0,$$

or combinations of the preceding, for example,

$$u(a) = 0, \quad u'(b) = 0.$$

3.2.2 Eigenfunction Solutions of Ordinary Differential Equations

Recall from Section 2.3.2 that the solution of a system of linear algebraic equations $\mathbf{Au} = \mathbf{b}$, where $\mathbf{A}$ is real and symmetric, can be written in terms of the eigenvalues and orthonormal eigenvectors of the coefficient matrix $\mathbf{A}$. In particular, the eigenvectors of

[3] As in the eigenproblem for matrices, we use the subscript n to indicate eigenvalues and eigenfunctions in order to distinguish them from the dependent variable $u(x)$, to emphasize that there are multiple eigenvalues and eigenfunctions, and that they come in pairs $(\lambda_n, u_n(x))$.

$\mathbf{A}$ are used as a vector basis for expressing the unknown solution vector $\mathbf{u}$ and the known right-hand-side vector $\mathbf{b}$.

Given the upgrade from vectors and matrices in Chapter 2 to functions and differential operators here, we follow precisely the same four steps as in Section 2.3.2 in order to obtain the solution of linear differential equations. Given the differential equation

$$\mathcal{L}u(x) = f(x), \tag{3.2}$$

where the differential operator $\mathcal{L}$ is self-adjoint (to be defined in Section 3.3.2), the solution can be obtained using the following procedure:

1. Extract the differential operator $\mathcal{L}$ from the differential equation and form its eigenproblem

$$\mathcal{L}u_n(x) = \lambda_n u_n(x). \tag{3.3}$$

   Determine the eigenvalues λ_n and orthonormal eigenfunctions $u_n(x)$ of the differential operator $\mathcal{L}$.
2. Expand the known right-hand-side function $f(x)$ as a linear combination of the eigenfunctions as follows:

$$f(x) = \sum_{n=1}^{\infty} \hat{a}_n u_n(x). \tag{3.4}$$

   Determine the $\hat{a}_n$ coefficients by taking the inner product of each of the eigenfunctions $u_n(x)$ with (3.4) leading to

$$\hat{a}_n = \langle f(x), u_n(x) \rangle, \quad n = 1, 2, \dots. \tag{3.5}$$

3. Expand the unknown solution $u(x)$ as a linear combination of the eigenfunctions according to

$$u(x) = \sum_{n=1}^{\infty} a_n u_n(x). \tag{3.6}$$

   To determine the a_n coefficients, consider the original differential equation

$$\begin{aligned}
\mathcal{L}u(x) &= f(x), \\
\mathcal{L}\left(\sum_{n=1}^{\infty} a_n u_n(x)\right) &= \sum_{n=1}^{\infty} \hat{a}_n u_n(x), \\
\sum_{n=1}^{\infty} a_n \mathcal{L}u_n(x) &= \sum_{n=1}^{\infty} \hat{a}_n u_n(x), \\
\sum_{n=1}^{\infty} a_n \lambda_n u_n(x) &= \sum_{n=1}^{\infty} \hat{a}_n u_n(x), \\
\therefore a_n = \frac{\hat{a}_n}{\lambda_n} &= \frac{\langle f(x), u_n(x) \rangle}{\lambda_n}, \quad n = 1, 2, \dots.
\end{aligned}$$

4. Form the solution from the eigenvalues and eigenfunctions using

$$u(x) = \sum_{n=1}^{\infty} \frac{\langle f(x), u_n(x) \rangle}{\lambda_n} u_n(x). \tag{3.7}$$

As in the algebraic case, the bulk of the work goes into obtaining the eigenvalues and eigenfunctions in step 1. The eigenfunctions of the differential operator are used as basis functions for expressing the right-hand-side function $f(x)$ and the solution $u(x)$. In principle, any complete set of basis functions could be used to represent the function $f(x)$ and solution $u(x)$. By using the eigenfunctions of the differential operator, however, a particularly straightforward expression is obtained in step 3 for the a_n coefficients in the solution expansion.

We will illustrate the details of solving the eigenproblem for differential operators along with how they are used to express the solution of a differential equation through an example.

Example 3.3 Consider the simple differential equation

$$\frac{d^2u}{dx^2} = f(x), \tag{3.8}$$

over the range $0 \leq x \leq 1$ with the homogeneous boundary conditions

$$u(0) = 0, \quad u(1) = 0.$$

Solve for $u(x)$ in terms of the eigenfunctions of the differential operator in (3.8).[4]

Solution

Step 1: The differential operator in (3.8) is

$$\mathcal{L} = \frac{d^2}{dx^2}, \tag{3.9}$$

and the associated differential eigenproblem is given by

$$\mathcal{L}u_n(x) = \lambda_n u_n(x). \tag{3.10}$$

Observe once again that the original differential equation (3.8) is not an eigenproblem, but that the eigenproblem (3.10) is a differential equation.

We solve the differential eigenproblem using the techniques for solving linear ordinary differential equations to determine the eigenvalues λ_n for which nontrivial solutions $u_n(x)$ exist. In general, λ_n may be positive, negative, or zero. Let us first consider the case with $\lambda_n = 0$. From the differential eigenproblem (3.10) with the differential operator (3.9) and $\lambda_n = 0$, the eigenproblem is $u_n'' = 0$, which has the solution

$$u_n(x) = c_1 x + c_2.$$

[4] Clearly, for "simple" $f(x)$, we can simply integrate two times to obtain $u(x)$. However, what about more complicated $f(x)$ or differential operators?

Applying the boundary conditions

$$u_n(0) = 0 \quad \Rightarrow \quad 0 = c_1(0) + c_2 \quad \Rightarrow \quad c_2 = 0,$$
$$u_n(1) = 0 \quad \Rightarrow \quad 0 = c_1(1) + 0 \quad \Rightarrow \quad c_1 = 0.$$

Therefore, we only get the trivial solution $u_n(x) = 0$.

We are seeking nontrivial solutions, so try letting $\lambda_n = +\mu_n^2 > 0$ giving

$$u_n'' - \mu_n^2 u_n = 0.$$

The solution for constant coefficient, linear differential equations of this form is $u_n(x) = e^{rx}$, where r is a constant. Upon substitution, this leads to the requirement that r must satisfy

$$r^2 - \mu_n^2 = 0,$$

or

$$(r + \mu_n)(r - \mu_n) = 0.$$

Therefore, $r = \pm\mu_n$, and the solution is

$$u_n(x) = \hat{c}_3 e^{\mu_n x} + \hat{c}_4 e^{-\mu_n x},$$

or equivalently

$$u_n(x) = c_3 \cosh(\mu_n x) + c_4 \sinh(\mu_n x). \tag{3.11}$$

Applying the boundary conditions to determine the constants c_3 and c_4 in (3.11) requires that

$$u_n(0) = 0 \quad \Rightarrow \quad 0 = c_3(1) + c_4(0) \quad \Rightarrow \quad c_3 = 0,$$
$$u_n(1) = 0 \quad \Rightarrow \quad 0 = \cancelto{0}{c_3} \cosh \mu_n + c_4 \sinh \mu_n.$$

From the second equation with $c_3 = 0$, we see that $\sinh \mu_n$ only equals zero for $\mu_n = 0$, but we have already ruled out the $\lambda_n = \mu_n^2 = 0$ case; therefore, $c_4 = 0$, and we once again have the trivial solution. More generally, we have two equations for the two unknown constants, which may be written in matrix form as follows:

$$\begin{bmatrix} 1 & 0 \\ \cosh \mu_n & \sinh \mu_n \end{bmatrix} \begin{bmatrix} c_3 \\ c_4 \end{bmatrix} = \begin{bmatrix} 0 \\ 0 \end{bmatrix}.$$

For a nontrivial solution to exist for this homogeneous system

$$\begin{vmatrix} 1 & 0 \\ \cosh \mu_n & \sinh \mu_n \end{vmatrix} = 0,$$

in which case

$$\sinh \mu_n = 0.$$

Again, this is only satisfied if $\mu_n = 0$, for which the trivial solution results.

Finally, let $\lambda_n = -\mu_n^2 < 0$, in which case the differential eigenproblem becomes

$$u_n'' + \mu_n^2 u_n = 0.$$

Again, considering a solution of the form $u_n(x) = e^{rx}$, r must satisfy

$$r^2 + \mu_n^2 = 0,$$

or

$$(r + \mathrm{i}\mu_n)(r - \mathrm{i}\mu_n) = 0.$$

The solution is

$$u_n(x) = \hat{c}_5 e^{-\mathrm{i}\mu_n x} + \hat{c}_6 e^{\mathrm{i}\mu_n x},$$

or equivalently[5]

$$u_n(x) = c_5 \cos(\mu_n x) + c_6 \sin(\mu_n x). \tag{3.12}$$

Applying the boundary conditions to determine the constants c_5 and c_6 in (3.12) gives

$$\begin{aligned} u_n(0) = 0 \quad &\Rightarrow \quad 0 = c_5(1) + c_6(0) \qquad \Rightarrow \quad c_5 = 0, \\ u_n(1) = 0 \quad &\Rightarrow \quad 0 = \cancelto{0}{c_5} \cos \mu_n + c_6 \sin \mu_n. \end{aligned}$$

From the second equation with $c_5 = 0$, we see that $\sin \mu_n = 0$. More generally, we have two equations for the two unknown constants, which may be written in matrix form as follows:

$$\begin{bmatrix} 1 & 0 \\ \cos \mu_n & \sin \mu_n \end{bmatrix} \begin{bmatrix} c_5 \\ c_6 \end{bmatrix} = \begin{bmatrix} 0 \\ 0 \end{bmatrix}.$$

For a nontrivial solution to exist for this homogeneous system

$$\begin{vmatrix} 1 & 0 \\ \cos \mu_n & \sin \mu_n \end{vmatrix} = 0,$$

in which case

$$\sin \mu_n = 0.$$

This is the *characteristic equation*. Note that the characteristic equation is *transcendental*[6] rather than the *polynomial* characteristic equations encountered for matrices, and that we have an infinite number of eigenvalues and corresponding eigenfunctions. The zeros (roots) of the characteristic equation are

$$\mu_n = \pm\pi, \pm 2\pi, \ldots, \pm n\pi, \ldots, \quad n = 1, 2, \ldots.$$

Thus, the roots that give nontrivial solutions are

$$\mu_n = n\pi, \quad n = 1, 2, 3, \ldots,$$

[5] We typically prefer trigonometric functions for finite domains and exponential functions for infinite or semi-infinite domains owing to the relative ease of applying the boundary conditions.

[6] Transcendental functions are simply those that cannot be expressed as polynomials. That is, they "transcend" algebra.

and the eigenvalues are

$$\lambda_n = -\mu_n^2 = -n^2\pi^2, \quad n = 1, 2, 3, \ldots.$$

The corresponding eigenfunctions (3.12), with $c_5 = 0$, are

$$u_n(x) = c_6 \sin(n\pi x), \quad n = 1, 2, 3, \ldots.$$

The subscript n on λ_n, μ_n, and $u_n(x)$ indicates that there are an infinity of eigenvalues and eigenfunctions, one for each n. Note that it is not necessary in this case to consider negative n because we obtain the same eigenfunctions as for positive n as follows:

$$\begin{aligned} u_n(x) &= c_6 \sin(n\pi x) \\ u_{-n}(x) &= c_6 \sin(-n\pi x) = -c_6 \sin(n\pi x) = -u_n(x), \end{aligned}$$

and both represent the same eigenfunction given that c_6 is arbitrary.

The solution of the eigenproblem gives the *eigenfunctions* of the differential operator. Although the constant c_6 is arbitrary, it is often convenient to choose c_6 by normalizing the eigenfunctions, such that

$$\|u_n\| = 1,$$

or equivalently

$$\|u_n\|^2 = 1.$$

Then

$$\int_0^1 u_n^2(x)dx = 1$$

$$\int_0^1 c_6^2 \sin^2(n\pi x)dx = 1$$

$$c_6^2 \frac{1}{2} = 1$$

$$\therefore c_6 = \sqrt{2}.$$

Finally, the *normalized eigenfunctions* of the differential operator (3.9) are

$$u_n(x) = \sqrt{2}\sin(n\pi x), \quad n = 1, 2, 3, \ldots,$$

which is the *Fourier sine series*. Recall that the corresponding eigenvalues are

$$\lambda_n = -\mu_n^2 = -n^2\pi^2, \quad n = 1, 2, 3, \ldots.$$

Observe that a solution of the differential eigenproblem (3.10) exists for any value of λ_n; however, not all of these solutions satisfy the boundary conditions. Those *nontrivial* solutions that do satisfy both (3.10) and the boundary conditions are the eigenfunctions, and the corresponding values of λ_n are the eigenvalues, of the differential operator with its boundary conditions.

REMARKS:

1. *Changing the differential operator and/or the boundary conditions, which include the domain, will change the eigenvalues and eigenfunctions.*
2. *If more than one case ($\lambda_n < 0, \lambda_n = 0, \lambda_n > 0$) produces nontrivial solutions, superimpose the corresponding eigenvalues and eigenfunctions.*
3. *In what follows, we will take advantage of the orthogonality of the eigenfunctions $u_n(x)$. See Section 3.3 for why it is not necessary to explicitly check the orthogonality of the eigenfunctions in this case.*

Step 2: Having obtained the eigenvalues and eigenfunctions of the differential operator with its associated boundary conditions, we can now obtain a series solution of the original differential equation (3.8), which is repeated here

$$\frac{d^2u}{dx^2} = f(x), \quad u(0) = 0, \quad u(1) = 0. \tag{3.13}$$

Note that because the eigenfunctions, which are a Fourier sine series, are all mutually orthogonal, they provide an especially convenient set of basis functions for the function space spanned by *any* piecewise continuous function $f(x)$ in the interval $0 \leq x \leq 1$. Hence, the known right-hand side of the differential equation can be expressed as follows:

$$f(x) = \sum_{n=1}^{\infty} \hat{a}_n u_n(x), \quad 0 \leq x \leq 1.$$

In order to obtain the coefficients $\hat{a}_n$ for a given $f(x)$, take the inner product of the eigenfunctions $u_m(x)$ with both sides:

$$\langle f(x), u_m(x) \rangle = \sum_{n=1}^{\infty} \hat{a}_n \langle u_n(x), u_m(x) \rangle .$$

Because the basis functions $u_n(x)$ are orthogonal, such an inner product is called an *orthogonal projection* of $f(x)$ onto the basis functions. Because the eigenfunctions are all orthonormal, the terms on the right-hand side vanish except for that corresponding to $m = n$, for which $\langle u_n(x), u_m(x) \rangle = 1$. Thus,

$$\hat{a}_n = \langle f(x), u_n(x) \rangle , \quad n = 1, 2, 3, \ldots .$$

Therefore, the function $f(x)$ can be expanded in terms of the eigenfunctions as follows:

$$f(x) = \sum_{n=1}^{\infty} \sqrt{2} \left\langle f(x), \sqrt{2}\sin(n\pi x) \right\rangle \sin(n\pi x), \quad 0 \leq x \leq 1. \tag{3.14}$$

Step 3: We may also expand the unknown solution itself in terms of the orthonormal eigenfunctions – basis functions – according to

$$u(x) = \sum_{n=1}^{\infty} a_n u_n(x),$$

where the coefficients a_n are to be determined. As shown earlier, substitution into the differential equation, and use of the corresponding eigenproblem, leads to the result that

$$a_n = \frac{\langle f(x), u_n(x)\rangle}{\lambda_n}, \qquad n = 1, 2, 3, \ldots,$$

because we use the eigenfunctions of $\mathcal{L}$ as the basis for expressing the solution and right-hand-side functions.

Step 4: Thus, the solution of the differential equation (3.13) is given by the Fourier sine series

$$u(x) = \sum_{n=1}^{\infty} \frac{\langle f(x), u_n(x)\rangle}{\lambda_n} u_n(x) = \sum_{n=1}^{\infty} \frac{-\sqrt{2}}{n^2\pi^2} \left\langle f(x), \sqrt{2}\sin(n\pi x)\right\rangle \sin(n\pi x) \quad (3.15)$$

over the interval $0 \leq x \leq 1$.

Let us illustrate our expansions for the function $f(x)$ from (3.14) and the solution $u(x)$ from the expansion (3.15) when $f(x) = x^5$. First, let us plot the one-, five-, and 50-term expansions for $f(x)$ along with the actual function. The actual $f(x)$ is given by the solid line, and the eigenfunction expansions are given by the dashed lines in Figure 3.3. As expected, the eigenfunction representation of $f(x)$ improves as we retain additional terms. Note that while the eigenfunction expansion of $f(x)$ satisfies the same homogeneous end conditions as the differential equation, which are $u(0) = 0$ and $u(1) = 0$, it is not necessary for the actual function $f(x)$ to do so. If it does not, however, then the expansion does not apply directly at the endpoint.

Now let us plot the one-, three-, and five-term eigenfunction expansions for the solution $u(x)$ and compare with the exact solution. The exact solution is given by the solid line, and the eigenfunction expansions of the solution are given by the dashed lines in Figure 3.4. Note that although a five-term expansion for $f(x)$ is clearly

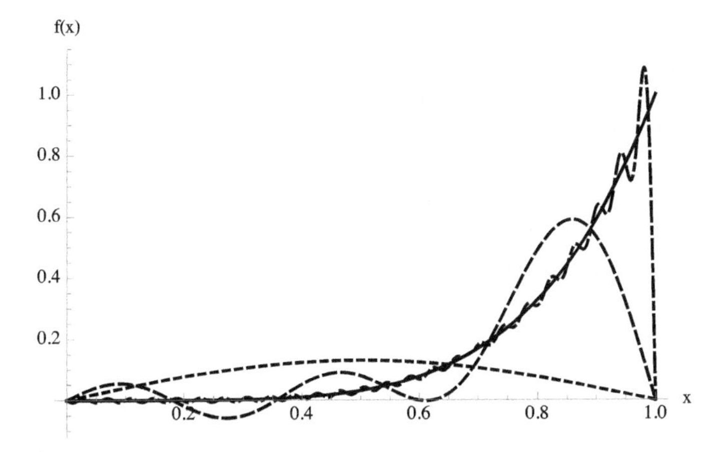

Figure 3.3 Plots of $f(x)$. Solid line is actual function and dashed lines are one-, five-, and 50-term expansions (identifiable from the number of oscillations in each) from (3.14).

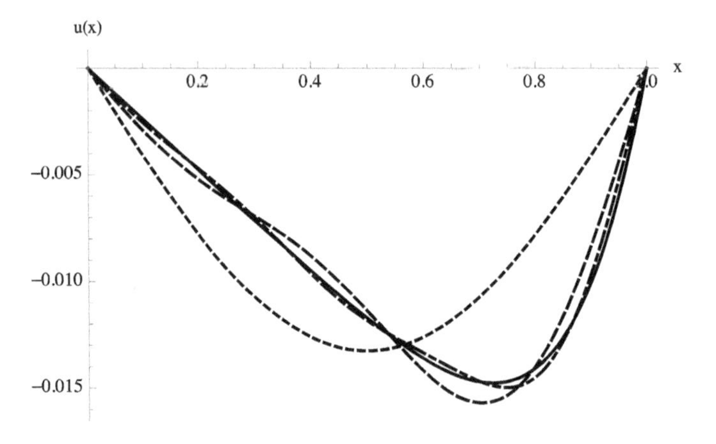

Figure 3.4 Plots of $u(x)$. Solid line is exact solution and dashed lines are one-, three-, and five-term expansions (identifiable from the number of oscillations in each) from (3.15).

inadequate, the five-term expansion for the solution $u(x)$ is comparatively close to the exact solution.

Obviously, solving $u'' = x^5$ exactly is straightforward simply by integrating twice, so what is the advantage of the eigenfunction solution approach?

1. In general, $\mathcal{L}u = f(x)$ can only be solved exactly for certain $\mathcal{L}$ and $f(x)$, whereas the eigenfunction expansion approach may be applied for general $\mathcal{L}$ and – even piecewise continuous – $f(x)$.
2. Solutions for various $f(x)$ may be obtained with minimal effort once the eigenvalues and eigenfunctions of the differential operator are obtained (recall changing the right-hand-side vector **b** in Section 2.3.2).
3. Eigenfunction expansions, such as Fourier series, may be applied to discrete data, such as from experiments. A popular approach is called *proper-orthogonal decomposition* (POD), which provides an alternative means of expressing large experimental or numerical data sets (see Chapter 13).
4. The eigenfunction approach provides the motivation for spectral numerical methods (see Section 7.3). Specifically, the eigenfunctions that arise from various differential operators prove to be an attractive source of basis functions for solving a wide variety of problems beyond those that yield to exact solutions as considered here.
5. More generally, after reading the previous section, the reader may wonder why there is so much emphasis on using the eigenfunctions of the differential operator to express the solution of the differential equation. Why not simply use some convenient – and complete – set of linearly independent functions, such as a general Fourier series, to represent the solution?

- While it is *possible* to express a solution in terms of any set of linearly independent and complete set of functions, such functions may not provide the *best* representation.
- In today's terminology, we would make the following distinction. If our desire is to represent a set of measured experimental data, we can choose any convenient set of functions, such as Fourier series, to represent the data. This is referred to as a "data-driven" scenario, in which no mathematical model is available, only raw data. As is the case here, however, we have the mathematical model; therefore, there are advantages to using the actual eigenfunctions of the differential operator in the problem of interest. As in the previous example, using the actual eigenfunctions of the differential operator allow for a particularly straightforward solution methodology that provides an exact solution of the original problem.

Example 3.4 Determine the solution of the nonhomogeneous differential equation

$$\frac{d^2u}{dx^2} + u = x, \quad u(0) = 0, \quad u(1) = 0, \tag{3.16}$$

in terms of the eigenvalues and eigenfunctions of the differential operator.

Solution

The differential operator is now

$$\mathcal{L} = \frac{d^2}{dx^2} + 1, \tag{3.17}$$

and the associated eigenproblem $\mathcal{L}u_n(x) = \lambda_n u_n(x)$ is given by

$$\frac{d^2u}{dx^2} + (1 - \lambda_n)u = 0, \quad u(0) = 0, \quad u(1) = 0. \tag{3.18}$$

The reader should confirm that cases with $\mu_n^2 = 1 - \lambda_n = 0$ and $\mu_n^2 = 1 - \lambda_n < 0$ permit only trivial solutions. Therefore, let us consider the case when $\mu_n^2 = 1 - \lambda_n > 0$. As in the previous example, the general solution in this case is given by

$$u_n(x) = c_1 \cos(\mu_n x) + c_2 \sin(\mu_n x). \tag{3.19}$$

Enforcing the boundary condition $u(0) = 0$ requires that $c_1 = 0$, and enforcing the boundary condition $u(1) = 0$ leads to the characteristic equation

$$\sin \mu_n = 0.$$

The values of μ_n that satisfy the characteristic equation, and thus give nontrivial solutions, are given by

$$\mu_n = \pm n\pi, \quad n = 1, 2, 3, \ldots.$$

Thus, the eigenvalues are

$$\lambda_n = 1 - \mu_n^2 = 1 - n^2\pi^2, \quad n = 1, 2, 3, \ldots, \tag{3.20}$$

and the eigenfunctions are

$$u_n(x) = \sqrt{2}\sin(n\pi x), \quad n = 1, 2, 3, \ldots, \tag{3.21}$$

where the $c_2 = \sqrt{2}$ factor is included to normalize the eigenfunctions to unity.

Next we expand the right-hand-side function $f(x)$ and the solution $u(x)$ of the original differential (3.16) in terms of the orthonormal eigenfunctions $u_n(x)$ according to

$$f(x) = \sum_{n=1}^{\infty} \hat{a}_n u_n(x) = \sum_{n=1}^{\infty} \langle f(x), u_n(x)\rangle\, u_n(x), \quad u(x) = \sum_{n=1}^{\infty} a_n u_n(x),$$

where the coefficients a_n are to be determined. Substituting into the differential equation (3.16) yields

$$\begin{aligned} \sum a_n u_n'' + \sum a_n u_n &= \sum \langle f, u_n\rangle\, u_n, \\ \sum a_n \left(u_n'' + u_n\right) &= \sum \langle f, u_n\rangle\, u_n, \\ \sum a_n \lambda_n u_n &= \sum \langle f, u_n\rangle\, u_n. \end{aligned}$$

For the coefficients of each eigenfunction to be equal in each term of the expansion requires that

$$a_n \lambda_n = \langle f, u_n\rangle;$$

therefore,

$$\begin{aligned} a_n &= \frac{\langle f(x), u_n(x)\rangle}{\lambda_n} \\ &= \frac{\sqrt{2}}{1 - n^2\pi^2}\int_0^1 x\sin(n\pi x)dx \\ &= \frac{\sqrt{2}}{1 - n^2\pi^2}\left[\frac{\sin(n\pi x)}{n^2\pi^2} - \frac{x\cos(n\pi x)}{n\pi}\right]_0^1 \\ &= \frac{\sqrt{2}}{1 - n^2\pi^2}\left[-\frac{\cos(n\pi)}{n\pi}\right] \\ a_n &= \frac{\sqrt{2}(-1)^n}{n\pi(n^2\pi^2 - 1)}. \end{aligned}$$

Consequently, the solution is

$$u(x) = \sum_{n=1}^{\infty} \frac{2(-1)^n}{n\pi(n^2\pi^2 - 1)}\sin(n\pi x). \tag{3.22}$$

3.2.3 Stability of Continuous Systems – Beam-Column Buckling

Recall from Chapter 2 that the dynamics of *discrete systems*, in which the mass is concentrated in discrete objects, are comprised of springs, masses, pendulums, et

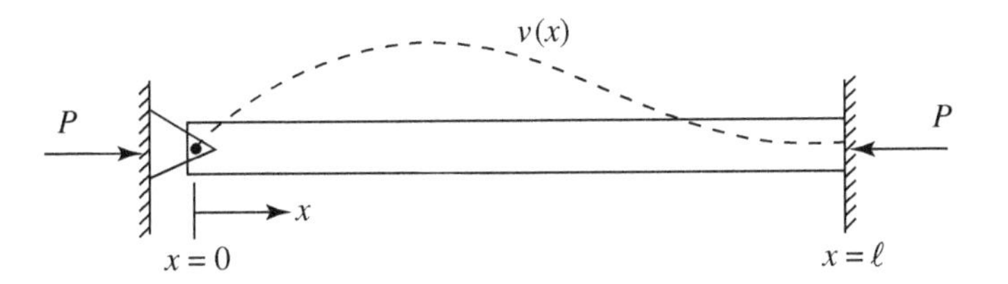

Figure 3.5 Schematic of the beam-column buckling problem.

cetera. For instance, recall the spring–mass example in Section 2.6.3. Discrete systems have a finite number of degrees of freedom, and the equations of motion are ordinary differential equations in time. Stability of a discrete system results in an *algebraic eigenproblem* of the form

$$\mathbf{A}\mathbf{u}_n = \lambda_n \mathbf{u}_n.$$

The dynamics of *continuous systems*, in which the mass is distributed throughout the system, involve strings, beams, membranes, et cetera. Continuous systems have an infinite number of degrees of freedom, and the equations of motion are partial differential equations in space and time. Stability of a continuous system results in a *differential eigenproblem* of the form

$$\mathcal{L}u_n = \lambda_n u_n.$$

For example, let us consider stability of the beam-column buckling problem.

Example 3.5 The buckling equation governing the lateral *stability* of a beam-column under axial load is the fourth-order differential equation[7]

$$\frac{d^2}{dx^2}\left(EI\frac{d^2v}{dx^2}\right) + P\frac{d^2v}{dx^2} = 0, \quad 0 \le x \le \ell.$$

Here, x is the axial coordinate along the beam-column, $v(x)$ is its lateral deflection, and P is the axial compressive force applied to the ends of the beam-column as shown in Figure 3.5. The properties of the beam-column are the Young's modulus E and the moment of inertia I of the cross-section. The product EI represents its *stiffness*.

If the end at $x = 0$ is hinged, in which case it cannot sustain a moment, and the end at $x = \ell = 1$ is fixed, then the boundary conditions are

$$v(0) = 0, \quad v''(0) = 0,$$
$$v(1) = 0, \quad v'(1) = 0.$$

Determine the buckling mode shapes and the corresponding critical buckling loads.

Solution

For a constant cross-section beam-column with uniform properties, E and I are con-

[7] See section 6.6 of Cassel (2013) for a derivation.

stant; therefore, we may define

$$\lambda_n = \frac{P_n}{EI},$$

and write the equation in the form

$$\frac{d^4 v_n}{dx^4} + \lambda_n \frac{d^2 v_n}{dx^2} = 0, \tag{3.23}$$

which may be regarded as a *generalized eigenproblem* of the form $\mathcal{L}_1 v_n = \lambda_n \mathcal{L}_2 v_n$.[8] In this case, the given differential equation governing stability of the beam-column *is* the eigenproblem. Note that this may more readily be recognized as an eigenproblem of the usual form if the substitution $u_n = d^2 v_n/dx^2$ is made resulting in $u_n'' + \lambda_n u_n = 0$; however, we will consider the fourth-order equation for the purposes of this example. Trivial solutions with $v_n(x) = 0$, corresponding to no lateral deflection, will occur for most $\lambda_n = P_n/EI$. However, certain values of the parameter λ_n will produce nontrivial solutions corresponding to buckling; these values of λ_n are the eigenvalues.

The fourth-order linear ordinary differential equation with constant coefficients (3.23) has solutions of the form $v_n(x) = e^{rx}$. Letting $\lambda_n = +\mu_n^2 > 0$ ($\lambda_n = 0$ and $\lambda_n < 0$ produce only trivial solutions for the given boundary conditions[9]), then r must satisfy

$$r^4 + \mu_n^2 r^2 = 0$$

$$r^2\left(r^2 + \mu_n^2\right) = 0$$

$$r^2\left(r + \mathrm{i}\mu_n\right)\left(r - \mathrm{i}\mu_n\right) = 0$$

$$\therefore r = 0, 0, \pm\, \mathrm{i}\mu_n.$$

Taking into account that $r = 0$ is a double root, the general solution of (3.23) is

$$v_n(x) = c_1 e^{0x} + c_2 x e^{0x} + \hat{c}_3 e^{\mathrm{i}\mu_n x} + \hat{c}_4 e^{-\mathrm{i}\mu_n x},$$

or

$$v_n(x) = c_1 + c_2 x + c_3 \sin(\mu_n x) + c_4 \cos(\mu_n x). \tag{3.24}$$

Applying the boundary conditions at $x = 0$, we find that

$$v_n(0) = 0 \quad \Rightarrow \quad 0 = c_1 + c_4 \quad \Rightarrow \quad c_1 = -c_4,$$
$$v_n''(0) = 0 \quad \Rightarrow \quad 0 = -\mu_n^2 c_4 \quad \Rightarrow \quad c_4 = 0.$$

[8] Recall that the generalized algebraic eigenproblem is given by $\mathbf{A}\mathbf{u}_n = \lambda_n \mathbf{B}\mathbf{u}_n$.

[9] This makes sense physically because $\lambda_n = P_n/EI = 0$ means there is no load, and $P_n < 0$ means there is a tensile load, in which case there is no buckling.

Thus, $c_1 = c_4 = 0$. Applying the boundary conditions at $x = 1$ leads to

$$\begin{aligned} v_n(1) = 0 \quad &\Rightarrow \quad 0 = c_2 + c_3 \sin \mu_n, \\ v_n'(1) = 0 \quad &\Rightarrow \quad 0 = c_2 + c_3 \mu_n \cos \mu_n. \end{aligned} \tag{3.25}$$

In order to have a nontrivial solution for c_2 and c_3, we must have

$$\begin{vmatrix} 1 & \sin \mu_n \\ 1 & \mu_n \cos \mu_n \end{vmatrix} = 0,$$

which gives the characteristic equation

$$\tan \mu_n = \mu_n.$$

Plotting $\tan \mu_n$ and μ_n in Figure 3.6, the points of intersection are the roots. The roots of the characteristic equation are ($\mu_0 = 0$ gives the trivial solution)

$$\mu_1 = 1.43\pi, \quad \mu_2 = 2.46\pi, \quad \mu_3 = 3.47\pi, \quad \ldots,$$

which have been obtained numerically. Thus, with $\lambda_n = \mu_n^2$, the eigenvalues are

$$\lambda_1 = 2.05\pi^2, \quad \lambda_2 = 6.05\pi^2, \quad \lambda_3 = 12.05\pi^2, \quad \ldots.$$

From the first relationship in (3.25), we see that

$$c_3 = -\frac{c_2}{\sin \mu_n};$$

therefore, from the solution (3.24), the eigenfunctions are

$$v_n(x) = c_2 \left[x - \frac{\sin(\mu_n x)}{\sin \mu_n} \right], \quad n = 1, 2, 3, \ldots. \tag{3.26}$$

As before, c_2 is an arbitrary constant that may be chosen such that the eigenfunctions are normalized if desired.

Any differential equation with the same differential operator and boundary conditions as in (3.23), including nonhomogeneous forms, can be solved by expanding

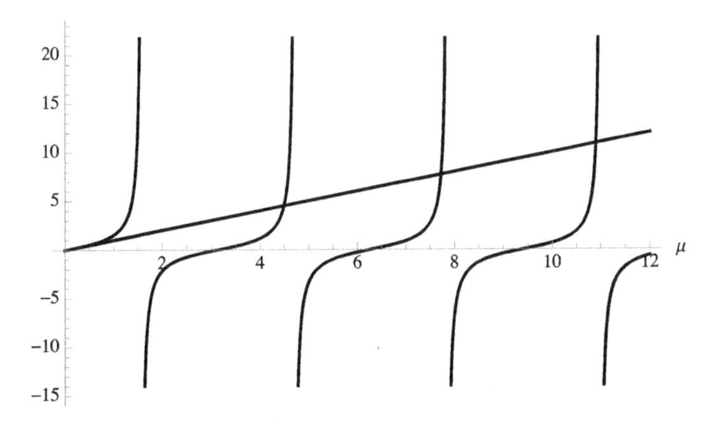

Figure 3.6 Plots of $\tan \mu$ and μ.

the solution – and the right-hand side – in terms of the eigenfunctions (3.26). A nonhomogeneous equation in this context would correspond to application of lateral loads on the column.

For the homogeneous case considered here, however, solutions for different values of n correspond to various possible modes of the solution for the lateral deflection. In particular, with respect to the eigenvalues, we have

$$P_n = \lambda_n EI, \quad n = 1, 2, 3, \ldots,$$

which are called the *critical buckling loads* of the column. So for $P < P_1$, the only solution is the trivial solution, and the column does not deflect laterally. When

$$P = P_1 = 2.05\pi^2 EI,$$

which is called the *Euler buckling load*, the column deflects laterally according to the following mode shape:

$$v_1(x) = x - \frac{\sin(1.43\pi x)}{\sin(1.43\pi)},$$

corresponding to a single "bulge" in the column as shown in Figure 3.7. Recall that the column is hinged at $x = 0$ and fixed at $x = 1$.

REMARKS:

1. *Achieving the higher critical loads and corresponding deflection modes for $n = 2, 3, \ldots$ would require appropriately placed constraints that restrict the lateral movement of the column and suppression of the lower modes.*
2. *Observe the correspondence between the mathematical and physical aspects of the beam-column buckling problem shown in Table 3.1.*

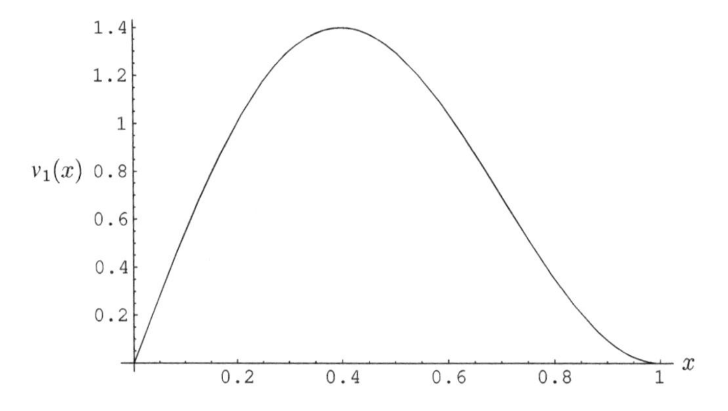

Figure 3.7 Mode shape for the Euler buckling load P_1.

Table 3.1 Corresponding mathematical and physical aspects of the beam-column buckling problem.

Mathematical	Physical
Eigenproblem	Governing (buckling) equation
Eigenvalues	Critical buckling loads
Trivial solution	No buckling (stable)
Eigenfunction (nontrivial)	Buckling mode shapes
Nonhomogeneous equation	Lateral loads

3.3 Adjoint and Self-Adjoint Differential Operators

In the following sections, we want to build further on the analogy between eigenvectors and eigenfunctions to determine for which differential operators the eigenfunctions are mutually orthogonal. We also desire a means to confirm from the differential operator itself that the eigenfunctions in the previous examples are mutually orthogonal, which has been assumed. Recall that in the matrix case, the eigenvectors are orthogonal if the matrix is real and symmetric with distinct eigenvalues. Is there an analogy in differential operators to a matrix being symmetric?

3.3.1 Adjoint of a Differential Operator

Consider an $N \times N$ matrix $\mathbf{A}$ and *arbitrary* $N \times 1$ vectors $\mathbf{u}$ and $\mathbf{v}$. Recall that the definition of the inner product of two vectors is

$$\langle \mathbf{v}, \mathbf{u} \rangle = \mathbf{v}^T \mathbf{u}.$$

Likewise, then

$$\langle \mathbf{v}, \mathbf{A}\mathbf{u} \rangle = \mathbf{v}^T \mathbf{A}\mathbf{u},$$

but the right-hand side can be rewritten in the form

$$\langle \mathbf{v}, \mathbf{A}\mathbf{u} \rangle = \left(\mathbf{A}^T \mathbf{v}\right)^T \mathbf{u}.$$

In terms of the inner product, this requires that

$$\langle \mathbf{v}, \mathbf{A}\mathbf{u} \rangle = \left\langle \mathbf{u}, \mathbf{A}^T \mathbf{v} \right\rangle.$$

Noting that the roles of the vectors $\mathbf{u}$ and $\mathbf{v}$ have been switched in the inner products on each side of this equation, we can think of this as a means – albeit a somewhat complicated one – to define the transpose of a matrix $\mathbf{A}^T$ in terms of inner products of vectors.

By analogy, a differential operator $\mathcal{L}$ has an *adjoint* operator $\mathcal{L}^*$ that satisfies

$$\langle v, \mathcal{L}u \rangle = \left\langle u, \mathcal{L}^* v \right\rangle, \tag{3.27}$$

where $u(x)$ and $v(x)$ are *arbitrary* functions with homogeneous boundary conditions. In this way, the adjoint of a differential operator is analogous to the transpose of a matrix.[10]

In order to illustrate the approach for determining the adjoint of a differential operator, consider the general second-order linear differential equation with variable coefficients

$$\mathcal{L}u = \frac{1}{w(x)}\left[a_0(x)u''(x) + a_1(x)u'(x) + a_2(x)u(x)\right] = 0, \quad a \leq x \leq b, \tag{3.28}$$

where $w(x)$ is a weight function. To obtain the adjoint operator, consider an arbitrary function $v(x)$, and take the inner product with $\mathcal{L}u$ to obtain the left-hand side of (3.27) as follows:

$$\langle v, \mathcal{L}u \rangle = \int_a^b w(x)v(x)\left\{\frac{1}{w(x)}\left[a_0(x)u''(x) + a_1(x)u'(x) + a_2(x)u(x)\right]\right\} dx, \tag{3.29}$$

where the inner product is taken with respect to the weight function $w(x)$. We want to switch the roles of $u(x)$ and $v(x)$ in the inner product by interchanging derivatives on $u(x)$ for derivatives on $v(x)$ in order to obtain the right-hand side of (3.27). This is accomplished using integration by parts.[11] The third term only involves multiplication; therefore, there is no problem switching the $u(x)$ and $v(x)$ functions. Integrating the second term by parts gives

$$\int_a^b a_1 v u' dx = a_1 v u \Big|_a^b - \int_a^b u(a_1 v)' dx,$$

where

$$\int p dq = pq - \int q dp$$

with

$$\begin{aligned} p &= a_1 v, & q &= u, \\ dp &= (a_1 v)' dx, & dq &= u' dx. \end{aligned}$$

Similarly, integrating the first term of (3.29) by parts twice results in

$$\int_a^b a_0 v u'' dx = \underbrace{a_0 v u' \Big|_a^b - \int_a^b u'(a_0 v)' dx}_{(1)} = \underbrace{\Big[a_0 v u' - (a_0 v)' u\Big]_a^b + \int_a^b u(a_0 v)'' dx}_{(2)},$$

(1)

$$\begin{aligned} p &= a_0 v, & q &= u', \\ dp &= (a_0 v)' dx, & dq &= u'' dx, \end{aligned}$$

[10] In an unfortunate twist of terminology, the term *adjoint* has different meanings in the context of matrices. Recall from Section 1.4.3 that the adjoint of a matrix is related to its cofactor matrix, while from Section 1.2, it is also another term for taking the transpose of a matrix.

[11] See section 1.6 of Cassel (2013) for a review of integration by parts.

(2)

$$\hat{p} = (a_0 v)', \qquad \hat{q} = u,$$
$$d\hat{p} = (a_0 v)'' dx, \quad d\hat{q} = u' dx.$$

Substituting into (3.29) gives

$$\begin{aligned} \langle v, \mathcal{L}u \rangle &= \Big[a_0 v u' - (a_0 v)' u + a_1 v u \Big]_a^b \\ &\quad + \int_a^b w(x) u(x) \left\{ \frac{1}{w(x)} \left[(a_0 v)'' - (a_1 v)' + a_2 v \right] \right\} dx, \end{aligned}$$

where the integral is $\langle u, \mathcal{L}^* v \rangle$, and the expression in {} is $\mathcal{L}^* v$. Therefore, if the boundary conditions on $u(x)$ and $v(x)$ are homogeneous, in which case the terms evaluated at $x = a$ and $x = b$ vanish, we have

$$\langle v, \mathcal{L}u \rangle = \langle u, \mathcal{L}^* v \rangle,$$

where the *adjoint* operator $\mathcal{L}^*$ of the differential operator $\mathcal{L}$ is

$$\mathcal{L}^* v = \frac{1}{w(x)} \left\{ [a_0(x) v]'' - [a_1(x) v]' + a_2(x) v \right\}. \tag{3.30}$$

Note that the variable coefficients move inside the derivatives, and the odd-order derivative changes sign as compared to $\mathcal{L}u$. This is the case for higher-order derivatives as well.

Example 3.6 Determine the adjoint operator for

$$\mathcal{L} = \frac{d^2}{dx^2} + x \frac{d}{dx}, \quad 0 \le x \le 1,$$

with homogeneous boundary conditions.

Solution

From (3.28), we have

$$a_0(x) = 1, \quad a_1(x) = x, \quad a_2(x) = 0, \quad w(x) = 1.$$

Then from (3.30), the adjoint operator is

$$\begin{aligned} \mathcal{L}^* v &= [a_0 v]'' - [a_1 v]' + a_2 v \\ &= v'' - (xv)' \\ \mathcal{L}^* v &= v'' - xv' - v. \end{aligned}$$

Thus, the adjoint operator of $\mathcal{L}$ is

$$\mathcal{L}^* = \frac{d^2}{dx^2} - x \frac{d}{dx} - 1.$$

Note that $\mathcal{L}^* \ne \mathcal{L}$ in this example.

3.3.2 Requirement for an Operator to Be Self-Adjoint

Recall that a matrix that is equal to its transpose is said to be *symmetric*, or *Hermitian* for its conjugate transpose, and has the property that its eigenvectors are mutually orthogonal if its eigenvalues are distinct as proven in Section 2.3.1. Likewise, if the differential operator and its adjoint are the same, in which case $\mathcal{L} = \mathcal{L}^*$, then the differential operator $\mathcal{L}$ is said to be *self-adjoint*, or *Hermitian*, and distinct eigenvalues produce orthogonal eigenfunctions.

To prove this, consider the case when $\mathcal{L} = \mathcal{L}^*$ and let $u(x)$ and $v(x)$ now be two eigenfunctions of the differential operator, such that $\mathcal{L}u = \lambda_1 u, \mathcal{L}v = \lambda_2 v$. Evaluating (3.27) yields

$$\begin{aligned}\langle u, \mathcal{L}v\rangle &= \langle v, \mathcal{L}u\rangle\\ \langle u, \lambda_2 v\rangle &= \langle v, \lambda_1 u\rangle\\ (\lambda_1 - \lambda_2)\langle u, v\rangle &= 0.\end{aligned}$$

Consequently, if $\lambda_1 \neq \lambda_2$, the corresponding eigenfunctions must be orthogonal in order for their inner product to be zero.

As illustrated in the previous example, not all differential equations of the form (3.28) have differential operators that are self-adjoint, that is, for arbitrary $a_0(x), a_1(x)$, and $a_2(x)$ coefficients. Let us determine the subset of such equations that are self-adjoint.

Recall from (3.30) that the adjoint operator of a general second-order, linear differential operator with variable coefficients is given by

$$\mathcal{L}^* v = \frac{1}{w(x)}\left\{[a_0 v]'' - [a_1 v]' + a_2 v\right\}.$$

We rewrite the adjoint operator $\mathcal{L}^*$ of $\mathcal{L}$ by carrying out the differentiations via the product rule and collecting terms as follows:

$$\mathcal{L}^* v = \frac{1}{w(x)}\left\{a_0 v'' + \left[2a_0' - a_1\right] v' + \left[a_0'' - a_1' + a_2\right] v\right\}. \tag{3.31}$$

For $\mathcal{L}$ to be self-adjoint, the operators $\mathcal{L}$ and $\mathcal{L}^*$ in equations (3.28) and (3.31), respectively, must be the same. Given that the first term is already identical, consider the second and third terms. Equivalence of the second terms requires that

$$a_1(x) = 2a_0'(x) - a_1(x),$$

or

$$a_1(x) = a_0'(x). \tag{3.32}$$

Equivalence of the third terms requires that

$$a_2(x) = a_0''(x) - a_1'(x) + a_2(x),$$

but this is always the case if (3.32) is satisfied, such that $a_1'(x) = a_0''(x)$. Therefore, substitution of the condition (3.32) into (3.28) requires that

$$\mathcal{L}u = \frac{1}{w(x)} \left\{ a_0 u'' + a_0' u' + a_2 u \right\} = 0.$$

The differential operator in the preceding expression may be written in the form

$$\mathcal{L} = \frac{1}{w(x)} \left\{ \frac{d}{dx} \left[a_0(x) \frac{d}{dx} \right] + a_2(x) \right\},$$

which is called the *Sturm–Liouville differential operator.*

Therefore, a second-order linear differential operator of the general form (3.28) is self-adjoint if and only if it is of the Sturm–Liouville form

$$\mathcal{L} = \frac{1}{w(x)} \left\{ \frac{d}{dx} \left[p(x) \frac{d}{dx} \right] + q(x) \right\}, \tag{3.33}$$

where $p(x) > 0$ and $w(x) > 0$ in $a \leq x \leq b$, and the boundary conditions are homogeneous. It follows that the corresponding eigenfunctions of the Sturm–Liouville differential operator are orthogonal with respect to the weight function $w(x)$. As we will see, any second-order, linear ordinary differential equation can be converted to Sturm–Liouville form. This is useful for the many physical phenomena governed by such equations.

Similarly, consider the fourth-order Sturm–Liouville differential operator

$$\mathcal{L} = \frac{1}{w(x)} \left\{ \frac{d^2}{dx^2} \left[s(x) \frac{d^2}{dx^2} \right] + \frac{d}{dx} \left[p(x) \frac{d}{dx} \right] + q(x) \right\}.$$

This operator also can be shown to be self-adjoint if the boundary conditions are homogeneous of the form

$$u = 0, \quad u' = 0,$$

or

$$u = 0, \quad s(x)u'' = 0,$$

or

$$u' = 0, \quad \left[s(x)u'' \right]' = 0,$$

at $x = a$ and $x = b$. Then the corresponding eigenfunctions are orthogonal on the interval $a \leq x \leq b$ with respect to $w(x)$.

3.3.3 Eigenfunctions of Sturm–Liouville Operators

Let us consider the eigenproblem $w(x)\mathcal{L}u_n(x) = -\lambda_n w(x)u_n(x)$ for the Sturm–Liouville differential operator (3.33), which is

$$\frac{d}{dx}\left[p(x)\frac{du_n}{dx}\right] + \left[q(x) + \lambda_n w(x)\right]u_n = 0, \quad a \le x \le b. \tag{3.34}$$

Recall that $p(x) > 0$ and $w(x) > 0$ in $a \le x \le b$. Note inclusion of the weight function $w(x)$ in the differential eigenproblem. The Sturm–Liouville equation is a linear, homogeneous, second-order ordinary differential equation, and it is the eigenproblem for the Sturm–Liouville differential operator. The Sturm–Liouville equation is a boundary-value problem and requires homogeneous boundary conditions of the form

$$c_1 u_n(a) + c_2 u_n'(a) = 0, \quad c_3 u_n(b) + c_4 u_n'(b) = 0,$$

where c_1 and c_2 are not both zero, and c_3 and c_4 are not both zero. That is, a linear combination of $u_n(x)$ and $u_n'(x)$ is zero at the endpoints.

In summary, for the Sturm–Liouville differential operator:

1. The eigenvalues are *distinct* and *nonnegative* (not proven here).
2. The eigenfunctions $u_n(x)$ are *orthogonal* with respect to the weight function $w(x)$, such that

$$\langle u_n, u_m \rangle = \int_a^b w(x)u_n(x)u_m(x)dx = 0, \quad m \ne n.$$

Recall that the norm of $u_n(x)$ is also defined with respect to the weight function according to

$$\|u_n\|^2 = \int_a^b w(x)u_n^2(x)dx.$$

Solutions to the eigenproblems associated with the Sturm–Liouville differential operator with various $p(x)$, $q(x)$, and $w(x)$ and having appropriate boundary conditions produce several common sets of basis functions, for example Fourier series, Legendre polynomials, Bessel functions, and Chebyshev polynomials[12] (see, for example, Greenberg 1998, Jeffrey 2002, and Asmar 2005).

Fourier Series

We have already found in Section 3.2 that for eigenproblems of the form

$$\frac{d^2u_n}{dx^2} + \lambda_n u_n = 0, \quad u_n(0) = 0, \quad u_n(1) = 0,$$

the eigenfunctions of the differential operator produce a *Fourier sine series*

$$u_n(x) = a_n \sin(n\pi x), \quad n = 1, 2, 3, \ldots.$$

[12] These functions (and others) are so special that they are often referred to as *special functions*.

Similarly, the boundary-value problem

$$\frac{d^2u_n}{dx^2} + \lambda_n u_n = 0, \quad u_n'(0) = 0, \quad u_n'(1) = 0$$

produces eigenfunctions that form the *Fourier cosine series*, for which

$$u_n(x) = a_n \cos(n\pi x), \quad n = 0, 1, 2, \ldots.$$

Observe that the only change from the previous case is the boundary conditions.

Bessel Functions

The differential operator associated with the eigenproblem

$$\frac{d}{dx}\left(x\frac{du_n}{dx}\right) + \left(-\frac{\nu^2}{x} + \mu_n^2 x\right) u_n = 0, \quad 0 \leq x \leq 1,$$

has as its eigenfunctions the *Bessel functions* $J_\nu(\mu_n x), J_{-\nu}(\mu_n x)$ if ν is not an integer, and $J_\nu(\mu_n x), Y_\nu(\mu_n x)$ if ν is an integer. $J_{\pm\nu}$ are the Bessel functions of the first kind, and Y_ν are the Bessel functions of the second kind. For example, see Figure 3.8 for plots of $J_0(x), J_1(x), Y_0(x)$, and $Y_1(x)$.

Note that the preceding differential equation is a Sturm–Liouville equation with

$$p(x) = x, \quad q(x) = -\frac{\nu^2}{x}, \quad w(x) = x, \quad \lambda_n = \mu_n^2.$$

Bessel functions are orthogonal over the interval $0 \leq x \leq 1$ with respect to the weight function $w(x) = x$, and the Bessel equation arises when solving partial differential equations involving the Laplacian operator ∇^2 in cylindrical coordinates (see Section 3.4.3).

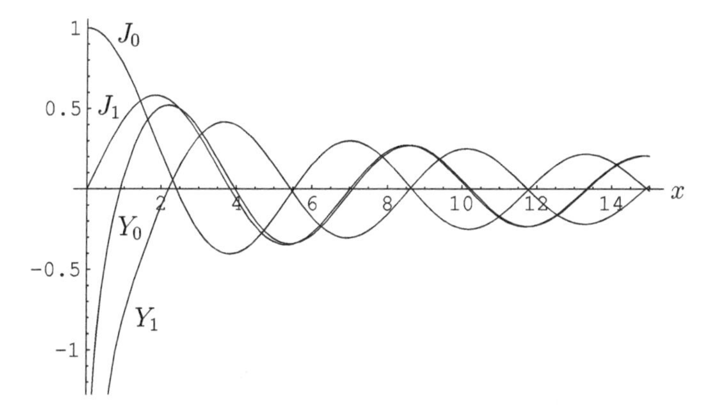

Figure 3.8 Bessel functions $J_0(x), J_1(x), Y_0(x)$, and $Y_1(x)$.

Legendre Polynomials

Legendre polynomials arise as eigenfunctions of the differential operator associated with the eigenproblem

$$\frac{d}{dx}\left[\left(1-x^2\right)\frac{du_n}{dx}\right]+\nu(\nu+1)u_n=0, \quad -1\leq x\leq 1,$$

where in the Sturm–Liouville equation

$$p(x)=1-x^2, \quad q(x)=0, \quad w(x)=1, \quad \lambda_n=\nu(\nu+1).$$

The Legendre polynomials are

$$P_0(x)=1, \quad P_1(x)=x, \quad P_2(x)=\frac{1}{2}(3x^2-1), \quad \ldots$$

with the recursion relation

$$(\nu+1)P_{\nu+1}(x)-(2\nu+1)xP_\nu(x)+\nu P_{\nu-1}(x)=0.$$

Given any two successive Legendre polynomials, the remaining sequence may be obtained using the recursion relation. We have already shown that Legendre polynomials are orthogonal over the interval $-1\leq x\leq 1$ (see Section 3.1). Note that the Legendre equation arises when solving Laplace's equation $\nabla^2 u=0$ in spherical coordinates.

Chebyshev Polynomials

The differential operator from the following eigenproblem produces Chebyshev polynomials as its eigenfunctions:

$$\frac{d}{dx}\left[\left(1-x^2\right)^{1/2}\frac{du_n}{dx}\right]+\nu^2\left(1-x^2\right)^{-1/2}u_n=0, \quad -1\leq x\leq 1, \tag{3.35}$$

where in the Sturm–Liouville equation

$$p(x)=\left(1-x^2\right)^{1/2}, \quad q(x)=0, \quad w(x)=\left(1-x^2\right)^{-1/2}, \quad \lambda_n=\nu^2.$$

The Chebyshev polynomials $T_\nu(x)$ of degree ν are

$$T_0(x)=1, \quad T_1(x)=x, \quad T_2(x)=2x^2-1, \quad \ldots$$

with the recursion relation

$$T_{\nu+1}(x)-2xT_\nu(x)+T_{\nu-1}(x)=0.$$

Chebyshev polynomials are orthogonal over the interval $-1\leq x\leq 1$.

Converting to Self-Adjoint Form

The Sturm–Liouville forms of the eigenproblems given earlier are all self-adjoint. However, alternative forms of these same equations may not be. For example, the Chebyshev equation written in the form

$$\left(1-x^2\right)\frac{d^2u_n}{dx^2}-x\frac{du_n}{dx}+\nu^2u_n=0$$

is not self-adjoint. The self-adjoint Sturm–Liouville form (3.35) is obtained by multiplying by $\left(1-x^2\right)^{-1/2}$.

In general, any second-order linear ordinary differential eigenproblem of the form

$$a_0(x)\frac{d^2u_n}{dx^2}+a_1(x)\frac{du_n}{dx}+\left[a_2(x)+\lambda_n a_3(x)\right]u_n=0 \tag{3.36}$$

can be reformulated as a Sturm–Liouville equation. To show how, let us write (3.36) in the form

$$\frac{d^2u_n}{dx^2}+\frac{a_1(x)}{a_0(x)}\frac{du_n}{dx}+\left[\frac{a_2(x)}{a_0(x)}+\lambda_n\frac{a_3(x)}{a_0(x)}\right]u_n=0.$$

Consider the Sturm–Liouville equation (3.34), which may be written in the form

$$p(x)\frac{d^2u_n}{dx^2}+\frac{dp}{dx}\frac{du_n}{dx}+\left[q(x)+\lambda_n w(x)\right]u_n=0,$$

or

$$\frac{d^2u_n}{dx^2}+\frac{1}{p(x)}\frac{dp}{dx}\frac{du_n}{dx}+\left[\frac{q(x)}{p(x)}+\lambda_n\frac{w(x)}{p(x)}\right]u_n=0.$$

Comparing these two equations, we see that in order to be self-adjoint, we must have

$$\frac{p'}{p}=\frac{a_1}{a_0},\quad \frac{q}{p}=\frac{a_2}{a_0},\quad \frac{w}{p}=\frac{a_3}{a_0}.$$

Integrating the first of these yields

$$\ln p=\int\frac{a_1}{a_0}dx;$$

thus,

$$p(x)=\exp\left[\int\frac{a_1(x)}{a_0(x)}dx\right], \tag{3.37}$$

and the other two relationships lead to

$$q(x)=\frac{a_2(x)}{a_0(x)}p(x),\quad w(x)=\frac{a_3(x)}{a_0(x)}p(x). \tag{3.38}$$

Note that $a_3(x)\neq 0$ for differential eigenproblems. This is how we determine the weight function $w(x)$ for a given equation. Using equations (3.37) and (3.38), we can convert any second-order linear eigenproblem of the form (3.36) into the self-adjoint Sturm–Liouville form (3.34).

Nonhomogeneous Equations

Just as in Section 3.2, solutions to nonhomogeneous forms of differential equations with the preceding operators are obtained by expanding both the solution and the right-hand side in terms of the eigenfunctions (basis functions) of the differential operators and determining the coefficients in the expansions.

We expand the right-hand side $f(x)$ in terms of eigenfunctions $u_n(x), n = 0, 1, 2, \ldots$ according to

$$f(x) = \sum_{n=0}^{\infty} \hat{a}_n u_n(x) = \hat{a}_0 u_0(x) + \hat{a}_1 u_1(x) + \cdots .$$

To determine the coefficients $\hat{a}_n$, evaluate the inner product $\langle u_m(x), f(x) \rangle$, which is equivalent to multiplying the preceding expression by $w(x)u_m(x)$ and integrating over the interval $a \leq x \leq b$. Once again, we are projecting the function $f(x)$ onto the orthogonal basis functions $u_n(x)$. If the eigenfunctions are orthogonal, then all of the terms in the expansion with $n \neq m$ vanish, leaving that with $n = m$ providing

$$\int_a^b w(x)u_m(x)f(x)dx = \hat{a}_m \int_a^b w(x)\left[u_m(x)\right]^2 dx,$$

or

$$\langle u_m(x), f(x) \rangle = \hat{a}_m \left\| u_m(x) \right\|^2.$$

Thus, the coefficients in the expansion for $f(x)$ are

$$\hat{a}_n = \frac{\langle u_n(x), f(x) \rangle}{\| u_n(x) \|^2}.$$

Note that if the eigenfunctions are normalized with respect to the weight function $w(x)$, then $\|u_n(x)\|^2 = 1$ in the denominator. Once again, it is the orthogonality of the eigenfunctions that makes this approach possible. Observe the similarity between this result and (2.7) for linear systems of algebraic equations.

Nonhomogeneous Boundary Conditions

Recall that the boundary conditions must be homogeneous in order for the boundary terms to vanish in the integration by parts. For a linear ordinary differential equation having nonhomogeneous boundary conditions, a simple transformation will convert it to a nonhomogeneous equation with homogeneous boundary conditions. Let us consider an example to illustrate how this is done.

Consider the ordinary differential equation

$$\frac{d^2u}{dx^2} + (1+x)\frac{du}{dx} + u = 0, \quad u(0) = 0, \quad u(1) = 1. \tag{3.39}$$

In order to shift the nonhomogeneity from the boundary condition at $x = 1$ to the equation, let us transform according to

$$u(x) = \phi_0(x) + U(x), \tag{3.40}$$

where $\phi_0(x)$ is specified to satisfy the boundary conditions. Then the new boundary conditions in terms of $U(x)$ are homogeneous. The simplest form of $\phi_0(x)$ that satisfies both boundary conditions is

$$\phi_0(x) = x. \tag{3.41}$$

Applying the transformation (3.40) with (3.41) to the differential equation (3.39) produces the transformed problem

$$\frac{d^2U}{dx^2} + (1+x)\frac{dU}{dx} + U = -(1+2x), \quad U(0) = 0, \quad U(1) = 0. \tag{3.42}$$

Observe that the differential operator and the type of the boundary conditions remain unchanged by the transformation. However, the nonhomogeneity is moved from the boundary conditions to the equation itself. Note that although the boundary conditions are now homogeneous, the equation is not self-adjoint; it could be converted to a self-adjoint Sturm–Liouville form using the previously outlined procedure.

3.4 Partial Differential Equations – Separation of Variables

Thus far, we have considered eigenfunction expansions for cases governed by ordinary differential equations, having one independent variable x. In some cases, we can extend the eigenfunction expansion approach using the *method of separation of variables* to solve partial differential equations as well.

3.4.1 Laplace Equation

Let us begin by illustrating the method of separation of variables using the Laplace equation, which governs steady heat conduction and electrostatic fields, for example. The Laplace equation for $u(x, y)$ in two-dimensional Cartesian coordinates is

$$\frac{\partial^2 u}{\partial x^2} + \frac{\partial^2 u}{\partial y^2} = 0, \tag{3.43}$$

with boundary conditions u or $\partial u/\partial n$ specified on each boundary, where n represents the normal to the boundary (see Figure 3.9). According to Chapter 7, Laplace's equation is representative of *elliptic partial differential equations* and governs diffusive processes.

It is supposed that the solution $u(x, y)$ can be written as the product of two functions, one a function of x only and one a function of y only, thereby *separating the variables*, as follows:

$$u(x, y) = \phi(x)\psi(y). \tag{3.44}$$

In general, there is no reason to expect that a solution $u(x, y)$ is separable in this manner, but that does not mean we cannot try! Substituting into the Laplace equation (3.43) gives

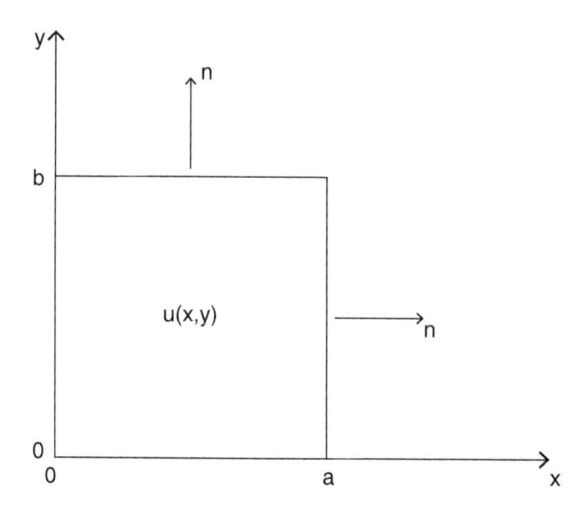

Figure 3.9 Domain for the Laplace equation with n indicating outward facing normals to the boundary.

$$\psi \frac{d^2\phi}{dx^2} + \phi \frac{d^2\psi}{dy^2} = 0,$$

where partial derivatives become ordinary because $\phi(x)$ and $\psi(y)$ only depend on one variable. Upon separating the dependence on x and y to opposite sides of the equation, we have

$$\frac{1}{\phi(x)} \frac{d^2\phi}{dx^2} = -\frac{1}{\psi(y)} \frac{d^2\psi}{dy^2} = \lambda.$$

Because the left-hand side is a function of x only, and the right-hand side is a function of y only, the equation must be equal to a constant, say λ, as x and y may be varied independently of each other. The constant λ is called the *separation constant*. Thus, we have two ordinary differential equations

$$\frac{1}{\phi(x)} \frac{d^2\phi}{dx^2} = \lambda \quad \Rightarrow \quad \frac{d^2\phi}{dx^2} - \lambda\phi(x) = 0, \tag{3.45}$$

and

$$-\frac{1}{\psi(y)} \frac{d^2\psi}{dy^2} = \lambda \quad \Rightarrow \quad \frac{d^2\psi}{dy^2} + \lambda\psi(y) = 0. \tag{3.46}$$

Clearly, not all partial differential equations can be written in the separable form (3.44). When they can, however, the method of separation of variables allows one to convert a partial differential equation into a set of ordinary differential equations. For the Laplace equation considered here, the second-order partial differential equation in two dimensions is converted into two second-order ordinary differential equations – one in each independent variable. For one of these ordinary differential equations to be an eigenproblem, both of its boundary conditions must be homogeneous; this equation will be considered first to obtain the eigenvalues and eigenfunctions of its differential operator. These eigenvalues are then used to complete the second of the two ordinary differential equations, which can be solved if one of its two boundary conditions is homogeneous. Therefore, three of the four boundary conditions for the Laplace equation must be homogeneous.

Example 3.7 Consider the temperature distribution $u(x, y)$ owing to steady conduction in the rectangular domain given in Figure 3.10. Steady heat conduction is governed by Laplace's equation (3.43), and the boundary conditions are

$$u = 0 \quad \text{at} \quad x = 0, x = a, y = 0, \qquad u = f(x) \quad \text{at} \quad y = b. \tag{3.47}$$

That is, the temperature is zero on three boundaries and some specified distribution $f(x)$ on the fourth boundary.

Solution

Separating variables as in (3.44) leads to the ordinary differential equations (3.45) and (3.46). We first consider the eigenproblem, which is the equation having two homogeneous boundary conditions. Thus, consider (3.45)

$$\frac{d^2\phi_n}{dx^2} - \lambda_n\phi_n = 0,$$

which for $\lambda_n = -\mu_n^2 < 0$ ($\lambda_n = 0$ and $\lambda_n = +\mu_n^2 > 0$ produce trivial solutions) has the solution (see Example 3.3)

$$\phi_n(x) = c_1\cos(\mu_n x) + c_2\sin(\mu_n x). \tag{3.48}$$

The boundary condition $u(0, y) = 0$ requires that $\phi_n(0) = 0$; therefore, $c_1 = 0$. The boundary condition $u(a, y) = 0$ requires that $\phi_n(a) = 0$; therefore, the characteristic equation is

$$\sin\left(\mu_n a\right) = 0,$$

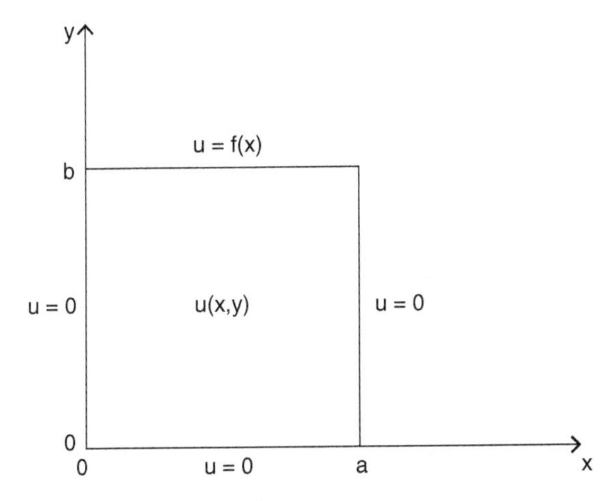

Figure 3.10 Domain for heat conduction Example 3.7.

which is satisfied when

$$\mu_n = \frac{n\pi}{a}, \quad n = 1, 2, \ldots. \tag{3.49}$$

The eigenfunctions are then

$$\phi_n(x) = c_2 \sin\left(\frac{n\pi}{a}x\right), \quad n = 1, 2, \ldots, \tag{3.50}$$

where c_2 is arbitrary and set equal to one for convenience (we will normalize the eigenfunctions later). Now consider (3.46), recalling that we have the eigenvalues $\lambda_n = -\mu_n^2 = -\left(\frac{n\pi}{a}\right)^2$,

$$\frac{d^2\psi}{dy^2} - \mu_n^2 \psi = 0,$$

which has the solution

$$\psi_n(y) = c_3 \cosh\left(\frac{n\pi}{a}y\right) + c_4 \sinh\left(\frac{n\pi}{a}y\right).$$

However, $u(x,0) = 0$; therefore, $\psi_n(0) = 0$ requiring that $c_3 = 0$, and

$$\psi_n(y) = c_4 \sinh\left(\frac{n\pi}{a}y\right). \tag{3.51}$$

Then the solution is ($a_n = c_4$)

$$u(x,y) = \sum_{n=1}^{\infty} u_n(x,y) = \sum_{n=1}^{\infty} \phi_n(x)\psi_n(y) = \sum_{n=1}^{\infty} a_n \sin\left(\frac{n\pi}{a}x\right) \sinh\left(\frac{n\pi}{a}y\right), \tag{3.52}$$

which is essentially an eigenfunction expansion in $\phi_n(x)$ with variable coefficients $\psi_n(y)$. We determine the a_n coefficients by applying the remaining nonhomogeneous boundary condition (3.47) at $y = b$ as follows:

$$u(x,b) = \sum_{n=1}^{\infty} \phi_n(x)\psi_n(b) = f(x).$$

Recognizing that the $\psi_n(b)$ are constants, taking the inner product of the eigenfunctions $\phi_m(x)$ with both sides gives

$$\|\phi_n(x)\|^2 \psi_n(b) = \langle f(x), \phi_n(x) \rangle,$$

where all the terms in the summation on the left-hand side vanish owing to orthogonality of eigenfunctions except when $m = n$ (recall that we did not normalize the eigenfunctions $\phi_n(x)$). Then with $\|\phi_n\|^2 = a/2$, solving for the constants $\psi_n(b)$ yields

$$\psi_n(b) = \frac{\langle f(x), \phi_n(x) \rangle}{\|\phi_n(x)\|^2} = \frac{2}{a} \int_0^a f(x) \sin\left(\frac{n\pi}{a}x\right) dx, \quad n = 1, 2, \ldots, \tag{3.53}$$

which are the Fourier sine coefficients of $f(x)$. As before, if we had chosen the c_2 coefficients in order to normalize the eigenfunctions (3.50), then $\|\phi_n(x)\|^2 = 1$ in the denominator. From (3.51) at $y = b$, the constants a_n in the solution (3.52) are

$$a_n = \frac{\psi_n(b)}{\sinh\left(\frac{n\pi b}{a}\right)};$$

therefore, the solution (3.52) becomes

$$u(x,y) = \sum_{n=1}^{\infty} u_n(x,y) = \sum_{n=1}^{\infty} \phi_n(x)\psi_n(y) = \sum_{n=1}^{\infty} \psi_n(b) \sin\left(\frac{n\pi}{a}x\right) \frac{\sinh\left(\frac{n\pi}{a}y\right)}{\sinh\left(\frac{n\pi b}{a}\right)}, \tag{3.54}$$

where $\psi_n(b)$ are the Fourier sine coefficients of $f(x)$ obtained from (3.53).

For example, consider the case with $a = b = 1$, and $f(x) = 1$. Then (3.53) becomes

$$\psi_n(1) = 2\int_0^1 \sin(n\pi x)\, dx = \frac{2}{n\pi}[1 - \cos(n\pi)], \quad n = 1, 2, \ldots,$$

which is zero for n even, and $\frac{4}{n\pi}$ for n odd. Therefore, let us define a new index according to $n = 2m + 1, m = 0, 1, 2, 3, \ldots$, in which case

$$\psi_m(1) = \frac{4}{\pi(2m+1)}.$$

The solution (3.54) is then

$$u(x,y) = \frac{4}{\pi} \sum_{m=0}^{\infty} \frac{\sin[(2m+1)\pi x] \sinh[(2m+1)\pi y]}{(2m+1)\sinh[(2m+1)\pi]}.$$

A contour plot of this solution is shown in Figure 3.11, where the contours represent constant temperature isotherms in heat conduction.

REMARKS:

1. *The preceding approach works when three of the four boundaries have homogeneous boundary conditions. Because the equations are linear, more general cases may be treated using superposition. For example, see Figure 3.12 for an example having two nonhomogeneous boundary conditions.*
2. *When we obtain the solutions*

$$u_n(x,y) = \phi_n(x)\psi_n(y),$$

they each satisfy the Laplace equation individually for $n = 1, 2, 3, \ldots$. Because the Laplace equation is linear, we obtain the most general solution by superimposing these solutions according to

$$u(x,y) = \sum u_n(x,y) = \sum \phi_n(x)\psi_n(y).$$

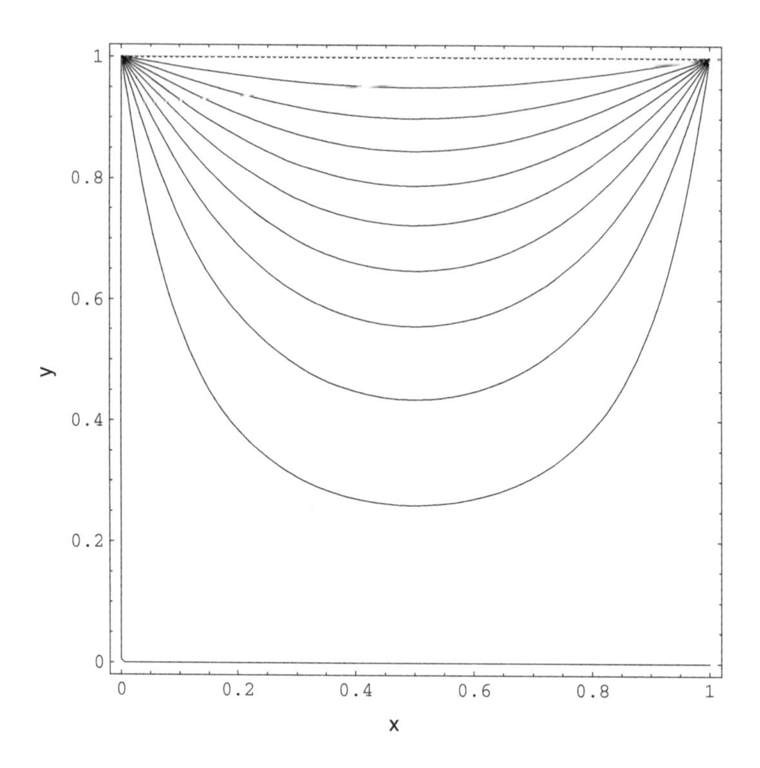

Figure 3.11 Isotherms for Example 3.7 in increments of $u = 0.1$.

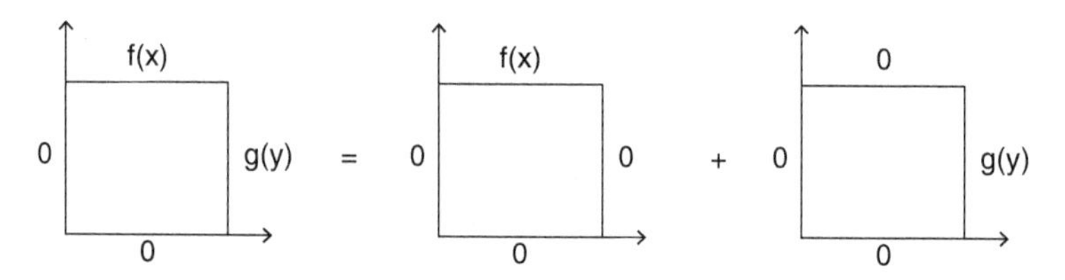

Figure 3.12 Superposition of eigenfunction solutions for a problem with two nonhomogeneous boundary conditions.

3.4.2 Unsteady Diffusion Equation

Consider the one-dimensional, unsteady diffusion equation

$$\frac{\partial u}{\partial t} = \alpha \frac{\partial^2 u}{\partial x^2}, \tag{3.55}$$

where $\alpha > 0$ is the diffusivity. It governs the temperature distribution $u(x,t)$ owing to unsteady conduction in the one-dimensional domain $0 \leq x \leq \ell$, for example. The initial condition is

$$u(x,0) = f(x), \tag{3.56}$$

and the homogeneous boundary conditions are

$$u(0,t) = 0, \quad u(\ell,t) = 0. \tag{3.57}$$

According to Chapter 7, the unsteady diffusion equation is an example of a *parabolic partial differential equation*. Note that the second-order derivatives on the right-hand side represent diffusion in the same way that the second-order terms model diffusion in the Laplace equation. The unsteady diffusion equation may also be solved using the method of separation of variables in certain cases. As with the Laplace equation, we seek a solution comprised of the product of two functions, one a function of time only and the other a function of space only according to

$$u(x,t) = \phi(x)\psi(t). \tag{3.58}$$

Substituting into (3.55) gives

$$\phi \frac{d\psi}{dt} = \alpha \psi \frac{d^2\phi}{dx^2},$$

which after separating variables leads to

$$\frac{1}{\alpha\psi}\frac{d\psi}{dt} = \frac{1}{\phi}\frac{d^2\phi}{dx^2} = \lambda = -\mu^2,$$

where we note that $\lambda = 0$ and $\lambda > 0$ lead to trivial solutions for the eigenproblem with the given boundary conditions. Consequently, the single partial differential equation (3.55) becomes the two ordinary differential equations

$$\frac{d^2\phi}{dx^2} + \mu^2\phi = 0, \tag{3.59}$$

$$\frac{d\psi}{dt} + \alpha\mu^2\psi = 0. \tag{3.60}$$

Equation (3.59) along with the homogeneous boundary conditions (3.57) represents the eigenproblem for $\phi(x)$. The general solution of (3.59) is

$$\phi_n(x) = c_1 \cos(\mu_n x) + c_2 \sin(\mu_n x). \tag{3.61}$$

From the boundary condition $u(0,t) = 0$, we must have $\phi_n(0) = 0$, which requires that $c_1 = 0$. Then from $u(\ell,t) = 0$, we must have $\phi_n(\ell) = 0$, which gives the characteristic equation

$$\sin(\mu_n \ell) = 0;$$

therefore,

$$\mu_n = \frac{n\pi}{\ell}, \quad n = 1,2,3,\ldots, \tag{3.62}$$

from which we obtain the eigenvalues

$$\lambda_n = -\mu_n^2 = -\left(\frac{n\pi}{\ell}\right)^2, \quad n = 1,2,3,\ldots. \tag{3.63}$$

If we let $c_2 = 1$ for convenience, the eigenfunctions are

$$\phi_n(x) = \sin\left(\frac{n\pi}{\ell}x\right), \quad n = 1,2,3,\ldots. \tag{3.64}$$

Let us now consider (3.60) with the result (3.62). The solution of this first-order ordinary differential equation is

$$\psi_n(t) = a_n \exp(-\alpha \mu_n^2 t), \quad n = 1,2,3,\dots. \tag{3.65}$$

From (3.58), with (3.64) and (3.65), the solution is

$$u(x,t) = \sum_{n=1}^{\infty} \phi_n(x)\psi_n(t) = \sum_{n=1}^{\infty} a_n \exp\left(-\frac{\alpha n^2\pi^2}{\ell^2}t\right) \sin\left(\frac{n\pi}{\ell}x\right). \tag{3.66}$$

The a_n coefficients are determined through application of the nonhomogeneous initial condition (3.56) applied at $t = 0$, which requires that

$$u(x,0) = \sum_{n=1}^{\infty} a_n \sin\left(\frac{n\pi}{\ell}x\right) = f(x).$$

Alternatively, we may write this as

$$\sum_{n=1}^{\infty} a_n\phi_n(x) = f(x).$$

Taking the inner product of the eigenfunctions $\phi_m(x)$ with both sides, the only non-vanishing term occurs when $m = n$ owing to orthogonality of the eigenfunctions. Therefore,

$$a_n\|\phi_n(x)\|^2 = \langle f(x), \phi_n(x)\rangle,$$

which gives the coefficients as

$$a_n = \frac{\langle f(x), \phi_n(x)\rangle}{\|\phi_n(x)\|^2} = \frac{2}{\ell}\int_0^{\ell} f(x)\sin\left(\frac{n\pi}{\ell}x\right)dx, \quad n = 1,2,3,\dots. \tag{3.67}$$

These are the Fourier sine coefficients of the initial condition $f(x)$. Thus, the eigenfunction solution is given by (3.66) with the coefficients (3.67) for a given $f(x)$ initial condition.

We will repeat this problem using a far more general, but closely related, technique based on Galerkin projection in Chapter 13. The advantage of Galerkin projection is that it can be applied using any set of basis functions, not just the eigenfunctions of the differential operator, and can be extended naturally to non-self-adjoint and nonlinear partial differential equations as well.

3.4.3 Wave Equation

The wave equation is the canonical hyperbolic partial differential equation (see Section 7.5). Using the method of separation of variables, we obtain solutions in terms of eigenfunction expansions for the one- and two-dimensional wave equations as illustrated here.

Consider a general hyperbolic partial differential equation of the form

$$\mathcal{M}\ddot{u} + \mathcal{K}u = 0, \tag{3.68}$$

where $\mathcal{M}$ and $\mathcal{K}$ are linear differential operators in space, dots denote differentiation in time, and $u(x,y,z,t)$ is the unsteady displacement of a vibrating string, membrane, beam, et cetera.

In order to convert the partial differential equation (3.68) into ordinary differential equations, we use the *method of separation of variables*. As for the one-dimensional, unsteady diffusion equation, we write $u(x,y,z,t)$ as the product of two functions, one that accounts for the spatial dependence and one for the temporal dependence, as follows:

$$u(x,y,z,t) = \phi(x,y,z)\psi(t). \tag{3.69}$$

The spatial modes $\phi_n(x,y,z)$ are the basis functions. Substituting into the governing equation (3.68) gives

$$\ddot{\psi}\mathcal{M}\phi + \psi\mathcal{K}\phi = 0,$$

or separating variables leads to

$$\frac{\mathcal{K}\phi(x,y,z)}{\mathcal{M}\phi(x,y,z)} = -\frac{\ddot{\psi}(t)}{\psi(t)} = \lambda.$$

Because the left-hand side is a function of x, y, and z only, and the right-hand side of t only, the equation must be equal to a separation constant, say λ. Then

$$\mathcal{K}\phi(x,y,z) - \lambda\mathcal{M}\phi(x,y,z) = 0, \tag{3.70}$$

$$\ddot{\psi}(t) + \lambda\psi(t) = 0, \tag{3.71}$$

where (3.70), which is a partial differential equation, is an eigenproblem for the spatial vibrational displacement modes $\phi(x,y,z)$ of the system, and (3.71) is an ordinary differential equation in time.

For a continuous system, there are an infinity of eigenvalues λ representing the natural frequencies of the system and corresponding eigenfunctions $\phi(x,y,z)$ representing the vibrational mode shapes. The general, time-dependent motion of the system is a superposition of these eigenmodes, which is determined from a solution of the ordinary differential equation (3.71) in time – with λ being the eigenvalues of (3.70).

Example 3.8 Consider the one-dimensional wave equation

$$\frac{\partial^2 u}{\partial t^2} = c^2 \frac{\partial^2 u}{\partial x^2}, \tag{3.72}$$

where c is the wave speed in the material, and $u(x,t)$ is the displacement. The wave equation[13] governs, for example:

[13] See section 5.4 of Cassel (2013) for a derivation.

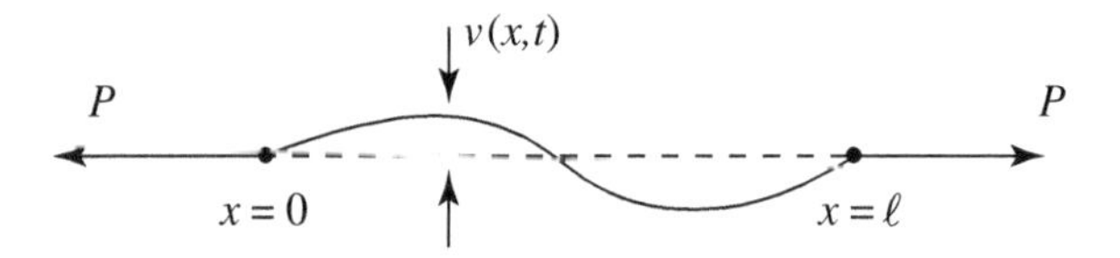

Figure 3.13 Vibration of an elastic string.

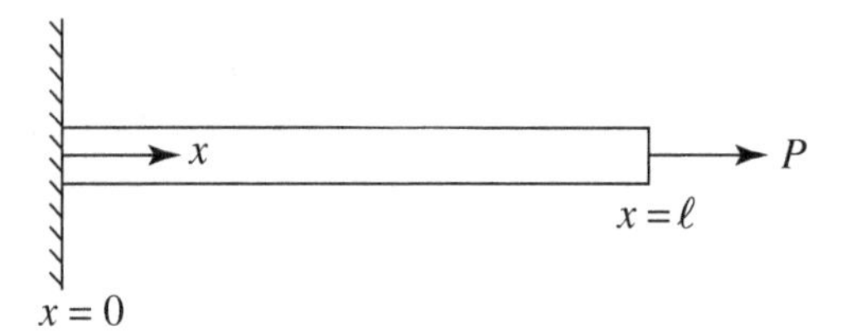

Figure 3.14 Vibration of a longitudinal rod.

1. Lateral vibration of a string, with $c^2 = P/\rho$, where P is the tension in the string, and ρ is the mass per unit length of the string, as shown in Figure 3.13.
2. Longitudinal vibration of a rod, with $c^2 = E/\rho$, where E is the Young's modulus of the rod, and ρ is the mass per unit volume of the rod, as shown in Figure 3.14.

Solve the wave equation using an eigenfunction expansion.

Solution

Writing the wave equation (3.72) in the form of (3.68), we have the spatial differential operators

$$\mathcal{M} = -\frac{1}{c^2}, \quad \mathcal{K} = \frac{\partial^2}{\partial x^2}.$$

From equations (3.70) and (3.71), and letting $\lambda = \omega^2 > 0$, we have

$$\frac{d^2\phi}{dx^2} + \frac{\omega^2}{c^2}\phi = 0,$$

$$\frac{d^2\psi}{dt^2} + \omega^2\psi = 0,$$

which are two ordinary differential equations in x and t, respectively, from one partial differential equation. From Example 3.3 with $\mu_n = \omega_n/c$ and $\mu_n = \omega_n$, the solutions to these equations are

$$\phi_n(x) = c_1 \cos\left(\frac{\omega_n}{c}x\right) + c_2 \sin\left(\frac{\omega_n}{c}x\right), \tag{3.73}$$

$$\psi_n(t) = c_3 \cos(\omega_n t) + c_4 \sin(\omega_n t). \tag{3.74}$$

Recall that (3.73) are the spatial eigenfunctions $\phi_n(x)$, that is, the vibrational modes or basis functions, and ω_n^2/c^2 are the eigenvalues, which are related to the natural frequencies of vibration.

To obtain the four constants, we need boundary and initial conditions. For a vibrating string, for example, consider the homogeneous boundary conditions

$$u(0,t) = 0, \quad u(\ell,t) = 0,$$

corresponding to zero displacement at both ends. Noting that the boundary conditions in x on $u(x,t) = \phi(x)\psi(t)$ must be satisfied by $\phi(x)$, we require

$$u(0,t) = 0 \quad \Rightarrow \quad \phi_n(0) = 0 \quad \Rightarrow \quad c_1 = 0.$$

Similarly, from

$$u(\ell,t) = 0 \quad \Rightarrow \quad \phi_n(\ell) = 0,$$

and we have

$$\sin\left(\frac{\omega_n \ell}{c}\right) = 0, \tag{3.75}$$

which is the *characteristic equation* for the natural frequencies of vibration. The eigenvalues satisfy

$$\frac{\omega_n \ell}{c} = n\pi, \quad n = 1,2,\ldots;$$

therefore, the natural frequencies are

$$\omega_n = \frac{n\pi c}{\ell}, \quad n = 1,2,\ldots. \tag{3.76}$$

The eigenfunctions are

$$\phi_n(x) = c_2 \sin\left(\frac{n\pi}{\ell}x\right), \quad n = 1,2,\ldots, \tag{3.77}$$

where c_2 is arbitrary and may be taken as $c_2 = 1$. Thus, the eigenfunctions – vibration mode shapes or basis functions – are a Fourier sine series. Note that because it is a continuous system, the string or rod has infinite degrees of freedom corresponding to the infinity of natural frequencies (as compared to the spring–mass example in Section 2.6.3, which has two natural frequencies).

The general, time-dependent motion of the string is a superposition of the infinite eigenfunctions, that is, natural modes of vibration. The actual motion is determined from the temporal solution $\psi_n(t)$ and initial conditions, which are as follows:

$$\text{Initial displacement:} \quad u(x,0) = f(x),$$

$$\text{Initial velocity:} \quad \frac{\partial u}{\partial t}(x,0) = g(x).$$

Note that two initial conditions are required as the equation is second order in time. From (3.74), the solution is

$$\psi_n(t) = c_3 \cos(\omega_n t) + c_4 \sin(\omega_n t), \tag{3.78}$$

where $\omega_n = n\pi c/\ell$, and differentiating in time gives

$$\dot{\psi}_n(t) = -c_3\omega_n \sin(\omega_n t) + c_4\omega_n \cos(\omega_n t). \tag{3.79}$$

But from equations (3.78) and (3.79)

$$\psi_n(0) = c_3, \quad \dot{\psi}_n(0) = c_4\omega_n;$$

therefore, substituting into (3.78) leads to

$$\psi_n(t) = \psi_n(0)\cos(\omega_n t) + \frac{1}{\omega_n}\dot{\psi}_n(0)\sin(\omega_n t). \tag{3.80}$$

From the initial conditions

$$\begin{aligned} u(x,0) = f(x) \quad &\Rightarrow \quad \sum \phi_n(x)\psi_n(0) = f(x), \\ \frac{\partial u}{\partial t}(x,0) = g(x) \quad &\Rightarrow \quad \sum \phi_n(x)\dot{\psi}_n(0) = g(x). \end{aligned} \tag{3.81}$$

Taking the inner product of $\phi_m(x)$ with both sides of the first equation in (3.81) gives

$$\|\phi_n(x)\|^2\psi_n(0) = \langle f(x), \phi_n(x)\rangle,$$

where all terms are zero owing to orthogonality except when $m = n$. Then

$$\psi_n(0) = \frac{\langle f(x), \phi_n(x)\rangle}{\|\phi_n(x)\|^2} = \frac{2}{\ell}\int_0^\ell f(x)\sin\left(\frac{n\pi}{\ell}x\right)dx.$$

Similarly, from the second equation in (3.81)

$$\|\phi_n(x)\|^2\dot{\psi}_n(0) = \langle g(x), \phi_n(x)\rangle,$$

$$\therefore \dot{\psi}_n(0) = \frac{\langle g(x), \phi_n(x)\rangle}{\|\phi_n(x)\|^2} = \frac{2}{\ell}\int_0^\ell g(x)\sin\left(\frac{n\pi}{\ell}x\right)dx.$$

Both $\psi_n(0)$ and $\dot{\psi}_n(0)$ are Fourier sine coefficients of $f(x)$ and $g(x)$, respectively. Finally, the solution is

$$u(x,t) = \sum_{n=1}^{\infty}\phi_n(x)\psi_n(t) = \sum_{n=1}^{\infty}\sin\left(\frac{n\pi}{\ell}x\right)\left[\psi_n(0)\cos(\omega_n t) + \frac{1}{\omega_n}\dot{\psi}_n(0)\sin(\omega_n t)\right], \tag{3.82}$$

where from (3.76), the frequencies of vibration for each of the modes are

$$\omega_n = \frac{n\pi c}{\ell}, \quad n = 1, 2, \ldots.$$

For example, consider the initial conditions

$$u(x,0) = f(x) = ae^{-b(x-1/2)^2}, \quad \dot{u}(x,0) = g(x) = 0,$$

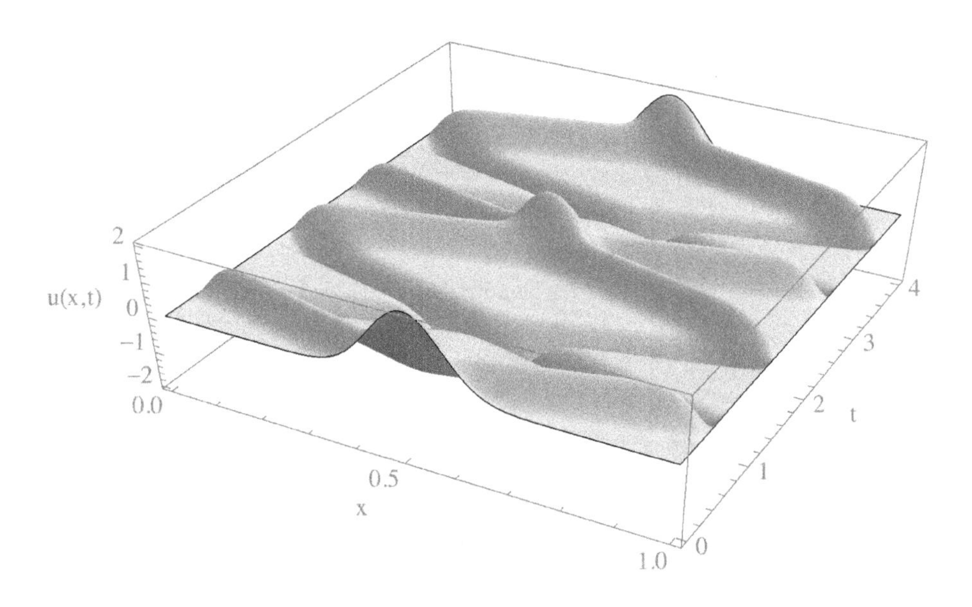

Figure 3.15 Solution of the wave equation with $\ell = 1, a = 1, b = 100$.

which corresponds to a Gaussian displacement distribution centered at $x = 1/2$ at $t = 0$. Then we obtain

$$\psi_n(0) = \frac{\langle f(x), \phi_n(x) \rangle}{\|\phi_n(x)\|^2} = \frac{2}{\ell} \int_0^{\ell} a e^{-b(x-1/2)^2} \sin\left(\frac{n\pi}{\ell} x\right) dx = \ldots,$$

$$\dot{\psi}_n(0) = \frac{\langle g(x), \phi_n(x) \rangle}{\|\phi_n(x)\|^2} = 0.$$

Plotting the solution (3.82) for $u(x,t)$ with $\ell = 1, a = 1, b = 100$ gives the solution shown in Figure 3.15. Observe that the initial Gaussian distribution at $t = 0$ leads to the propagation of right- and left-moving waves that reflect off the boundaries at $x = 0$ and $x = \ell = 1$.

Thus far our examples have involved Fourier series for the eigenfunctions. In the following example, we encounter Bessel functions as the eigenfunctions.

Example 3.9 As an additional application to vibrations problems, consider the vibration of a circular membrane of radius one.

Solution

The governing equation for the lateral displacement $u(r, \theta, t)$ is the two-dimensional wave equation in cylindrical coordinates

$$\frac{\partial^2 u}{\partial t^2} = c^2 \left(\frac{\partial^2 u}{\partial r^2} + \frac{1}{r} \frac{\partial u}{\partial r} + \frac{1}{r^2} \frac{\partial^2 u}{\partial \theta^2} \right) \quad \left(= c^2 \nabla^2 u \right), \tag{3.83}$$

where in (3.68)

$$\mathcal{M} = -\frac{1}{c^2}, \quad \mathcal{K} = \frac{\partial^2}{\partial r^2} + \frac{1}{r}\frac{\partial}{\partial r} + \frac{1}{r^2}\frac{\partial^2}{\partial \theta^2}.$$

The boundary condition is

$$u(1, \theta) = 0,$$

which indicates that the membrane is clamped along its circular perimeter. Upon separating the spatial and temporal variables according to $u(r, \theta, t) = \phi(r, \theta)\psi(t)$, equation (3.70) for the vibrational modes becomes

$$\frac{\partial^2 \phi}{\partial r^2} + \frac{1}{r}\frac{\partial \phi}{\partial r} + \frac{1}{r^2}\frac{\partial^2 \phi}{\partial \theta^2} + \frac{\lambda}{c^2}\phi = 0, \tag{3.84}$$

which is known as the Helmholtz partial differential equation $\nabla^2\phi + \frac{\lambda}{c^2}\phi = 0$.

Because (3.84) is a partial differential equation, let us again separate variables according to

$$\phi(r, \theta) = G(r)H(\theta),$$

which leads to

$$\frac{r^2}{G}\left(\frac{d^2G}{dr^2} + \frac{1}{r}\frac{dG}{dr} + \frac{\lambda}{c^2}G\right) = -\frac{1}{H}\frac{d^2H}{d\theta^2} = \sigma, \tag{3.85}$$

where σ is the separation constant. Then we have the two ordinary differential eigenproblems

$$r^2\frac{d^2G}{dr^2} + r\frac{dG}{dr} + \left(\frac{\lambda}{c^2}r^2 - \sigma\right)G = 0, \tag{3.86}$$

$$\frac{d^2H}{d\theta^2} + \sigma H = 0. \tag{3.87}$$

The general solution of (3.87) (with $\sigma > 0$) is of the form

$$H(\theta) = c_1 \cos\left(\sqrt{\sigma}\theta\right) + c_2 \sin\left(\sqrt{\sigma}\theta\right). \tag{3.88}$$

To be single-valued in the circular domain, however, the solution must be 2π-periodic in θ, requiring that

$$\phi(r, \theta + 2\pi) = \phi(r, \theta).$$

For the cosine term to be 2π-periodic,

$$\cos\left[\sqrt{\sigma}(\theta + 2\pi)\right] = \cos\left(\sqrt{\sigma}\theta\right),$$

or

$$\cos\left(\sqrt{\sigma}\theta\right)\cos\left(2\pi\sqrt{\sigma}\right) - \sin\left(\sqrt{\sigma}\theta\right)\sin\left(2\pi\sqrt{\sigma}\right) = \cos\left(\sqrt{\sigma}\theta\right).$$

Matching coefficients of the $\cos\left(\sqrt{\sigma}\theta\right)$ and $\sin\left(\sqrt{\sigma}\theta\right)$ terms requires that

$$\cos\left(2\pi\sqrt{\sigma}\right) = 1 \quad \Rightarrow \quad \sqrt{\sigma} = 0, 1, 2, 3, \ldots,$$

and

$$\sin\left(2\pi\sqrt{\sigma}\right) = 0 \quad \Rightarrow \quad \sqrt{\sigma} = 0, \frac{1}{2}, 1, \frac{3}{2}, 2, \ldots.$$

Because both conditions must be satisfied, 2π-periodicity of the $\cos\left(\sqrt{\sigma}\theta\right)$ term requires that the eigenvalues be

$$\sqrt{\sigma} = n = 0, 1, 2, 3, \ldots. \tag{3.89}$$

Consideration of 2π-periodicity of the $\sin\left(\sqrt{\sigma}\theta\right)$ term in (3.88) produces the same requirement. Therefore, (3.88) becomes

$$H_n(\theta) = c_1 \cos(n\theta) + c_2 \sin(n\theta), \quad n = 0, 1, 2, 3, \ldots. \tag{3.90}$$

Equation (3.86) may be written in the form

$$\frac{d}{dr}\left(r\frac{dG_n}{dr}\right) + \left(-\frac{n^2}{r} + \frac{\lambda}{c^2}r\right)G_n = 0, \tag{3.91}$$

which is Bessel's equation. As stated in Section 3.3.3, we expect to arrive at the Bessel equation when solving an equation with the Laplacian operator ∇^2 in cylindrical coordinates as is the case here. The general solution is

$$G_n(r) = c_3 J_n\left(\frac{\sqrt{\lambda}}{c}r\right) + c_4 Y_n\left(\frac{\sqrt{\lambda}}{c}r\right), \tag{3.92}$$

where J_n and Y_n are Bessel functions of the first and second kinds, respectively. Recall that Y_n are unbounded at $r = 0$; therefore, we must set $c_4 = 0$. Taking $c_3 = 1$, we have

$$\phi_n(r,\theta) = G_n(r)H_n(\theta) = J_n\left(\frac{\sqrt{\lambda}}{c}r\right)\left[c_1\cos(n\theta) + c_2\sin(n\theta)\right], \quad n = 0, 1, 2, \ldots,$$

where in order to satisfy the boundary condition at $r = 1$, we have the characteristic equation $J_n(\sqrt{\lambda}/c) = 0$. That is, $\lambda_{m,n}$ is chosen such that $\sqrt{\lambda_{m,n}}/c$ are zeros of the Bessel functions J_n (there are an infinity of zeros, $m = 1, 2, 3, \ldots$, for each Bessel function J_n). For example, four modes are shown in Figure 3.16.

Observe that $n + 1$ corresponds to the number of "lobes" in the azimuthal θ-direction, and m corresponds to the number of regions having positive or negative sign in the radial r-direction. As in the previous example, the solutions for $\psi(t)$ along with initial conditions would determine how the spatial modes evolve in time to form the full solution

$$u(r,\theta,t) = \sum_{n=0}^{\infty} \phi_n(r,\theta)\psi_n(t).$$

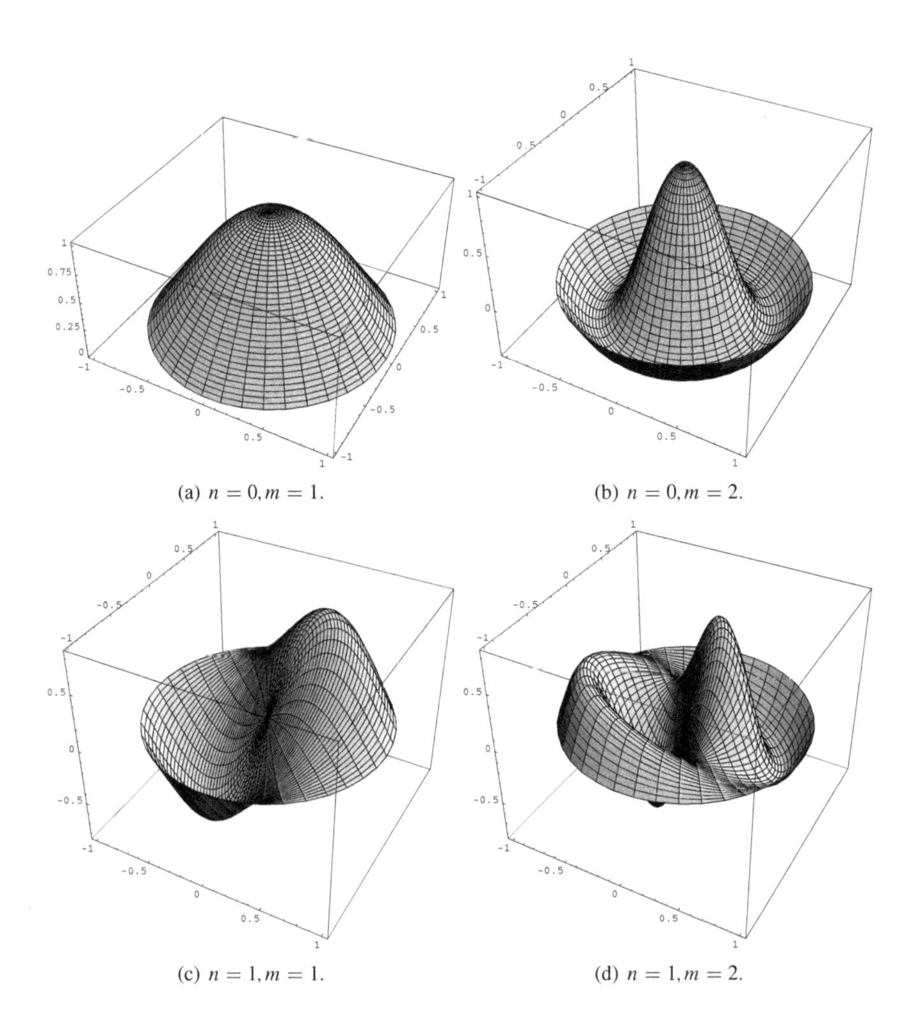

(a) $n = 0, m = 1$.

(b) $n = 0, m = 2$.

(c) $n = 1, m = 1$.

(d) $n = 1, m = 2$.

Figure 3.16 Various modes (basis functions) in the solution of the wave equation in cylindrical coordinates.

REMARKS:

1. *Whereas the eigenvalues and eigenfunctions have physical meaning in the vibrations context, where they are the natural frequencies and modes of vibration, respectively, in other contexts, such as heat conduction and electromagnetics, they do not and are merely a mathematical device by which to obtain the solution via an eigenfunction expansion in terms of appropriate basis functions.*
2. *It is rather remarkable that we have been able to extend and generalize methods developed to solve algebraic systems of equations to solve ordinary differential equations and now partial differential equations. This approach will be further extended to become the foundation for spectral numerical methods (Section 7.3) and reduced-order modeling (Chapter 13). This is the remarkable power of mathematics!*

3. *Observe that all of the partial differential equations considered here have been on* bounded *domains, in which case separation of variables with eigenfunction expansions provides a solution in some cases. For partial differential equations on* unbounded *domains, we can sometimes use integral transforms, such as Fourier, Laplace, or Hankel (see Asmar, 2005).*

3.4.4 Electromagnetics

In chapter 7 of Cassel (2013), it is shown that the Maxwell's equations for electromagnetics reduce to wave equations governing the electric and magnetic potentials as well as the electric and magnetic fields if no electric charges or currents are present. If the fields are steady and do not change with time, the wave equations for the electric and magnetic potentials and fields all become Laplace equations. Therefore, separation of variables can be used as in Sections 3.4.1 and 3.4.3 to solve electromagnetics problems in finite domains.

3.4.5 Schrödinger Equation

The Schrödinger equation of quantum mechanics is derived using variational methods in section 8.2.1 of Cassel (2013). It governs the probability *wave function* that allows us to determine the probability of a particle being in a certain location. The wave function $\psi(x, y, z, t)$ provides the continuous distribution of the intensity of the electron wave. In one spatial dimension, for example, once we have solved for the wave functions $\psi(x,t)$ and its complex conjugate $\bar{\psi}(x,t)$, the probability P of an electron being located in the range $a \leq x \leq b$ at time t can be calculated. This probability is then the square of the magnitude of the wave function, or equivalently the product of the wave function ψ and its complex conjugate $\bar{\psi}$, evaluated at that position according to

$$P = \int_a^b |\psi(x,t)|^2 dx = \int_a^b \psi(x,t)\bar{\psi}(x,t)dx.$$

Because the total probability for all locations must be unity, we have that the norm of the wave function is unity according to

$$\int_{-\infty}^{\infty} \psi(x,t)\bar{\psi}(x,t)dx = 1.$$

As we will see, the wave function comes in the form of eigenfunctions, where the eigenvalues provide the discrete frequencies to which the energy of the electron is proportional, where Planck's universal constant $\hbar$ provides the constant of proportionality. The wave function is complex, and because the Schrödinger equation is linear, the real and imaginary parts are each solutions of the equation independently.

The unsteady Schrödinger equation is given in terms of the Laplacian by

$$\mathrm{i}\hbar\frac{\partial\psi}{\partial t} = -\frac{\hbar^2}{2\mu}\nabla^2\psi + (V - E)\,\psi,$$

where μ is the *reduced mass* of the system, which is the equivalent mass of a point located at the center of gravity of the system and is constant in the non-relativistic case considered here, and $V(x)$ is the potential energy. The first term of Schrödinger's equation results from the kinetic energy. Observe that we have an unsteady eigenproblem in which the Lagrange multiplier E is the eigenvalue and $\psi(x,t)$ – or $\bar{\psi}(x,t)$ – is the eigenfunction.

In order to illustrate solutions of the Schrödinger equation, let us consider the case for which it was originally developed – the hydrogen atom. Because there is only a single electron orbiting a single proton, the hydrogen atom can be modeled using the Coulomb potential

$$V(r) = -\frac{e^2}{4\pi\epsilon_0 r}, \tag{3.93}$$

where e is the charge of an electron, ϵ_0 is the permittivity of a vacuum, and r is the distance from the proton to the electron orbiting it.

Because of the spherical geometry, we write the Schrödinger equation in spherical coordinates (r,θ,ϕ) with the proton located at the origin. Substituting the potential energy from (3.93) gives the Schrödinger equation for the hydrogen atom as

$$\mathrm{i}\hbar\frac{\partial\psi}{\partial t} = -\frac{\hbar^2}{2\mu}\nabla^2\psi - \frac{e^2}{4\pi\epsilon_0 r}\psi - E\psi. \tag{3.94}$$

Begin by separating the variables according to

$$\psi(r,\theta,\phi,t) = u(r,\theta,\phi)T(t), \tag{3.95}$$

such that u contains the spatial dependence, while T contains the temporal dependence. Substituting into (3.94) and separating variables produces the two equations

$$\frac{dT}{dt} = -\frac{\mathrm{i}}{\hbar}ET \tag{3.96}$$

for the temporal behavior, and

$$-\frac{\hbar^2}{2\mu}\nabla^2 u - \frac{e^2}{4\pi\epsilon_0 r}u = Eu \tag{3.97}$$

for the spatial behavior. The solution of the ordinary differential equation (3.96) then gives the temporal behavior as

$$T_n(t) = c_0 e^{-\mathrm{i}E_n t/\hbar}, \tag{3.98}$$

which will be complete once the eigenvalues E_n are determined.

Now consider (3.97) for the spatial behavior. Observe that this is a differential eigenproblem for the eigenvalues E, which are the Lagrange multiplier and separation constant, and the eigenfunctions $u(r,\theta,\phi)$ for the spatial behavior of the wave function. The Laplacian operator in spherical coordinates is (see Section 4.1.2)

$$\nabla^2 = \frac{1}{r^2}\frac{\partial}{\partial r}\left(r^2\frac{\partial}{\partial r}\right) + \frac{1}{r^2\sin\theta}\frac{\partial}{\partial\theta}\left(\sin\theta\frac{\partial}{\partial\theta}\right) + \frac{1}{r^2\sin^2\theta}\frac{\partial^2}{\partial\phi^2}$$

$$= \frac{\partial^2}{\partial r^2} + \frac{2}{r}\frac{\partial}{\partial r} + \frac{1}{r^2}\left(\frac{\partial^2}{\partial\theta^2} + \cot\theta\frac{\partial}{\partial\theta} + \csc^2\theta\frac{\partial^2}{\partial\phi^2}\right).$$

Therefore, (3.97) becomes

$$-\frac{\hbar^2}{2\mu}\left[\frac{\partial^2 u}{\partial r^2} + \frac{2}{r}\frac{\partial u}{\partial r} + \frac{1}{r^2}\left(\frac{\partial^2 u}{\partial\theta^2} + \cot\theta\frac{\partial u}{\partial\theta} + \csc^2\theta\frac{\partial^2 u}{\partial\phi^2}\right)\right] - \frac{e^2}{4\pi\epsilon_0 r}u = Eu. \tag{3.99}$$

Separating variables once again according to

$$u(r,\theta,\phi) = R(r)Y(\theta,\phi) \tag{3.100}$$

leads to two equations. The first is an ordinary differential equation for the radial component

$$-\frac{\hbar^2}{2\mu}\left(\frac{d^2R}{dr^2} + \frac{2}{r}\frac{dR}{dr} - \frac{\eta}{r^2}R\right) - \frac{e^2}{4\pi\epsilon_0 r}R = ER, \tag{3.101}$$

where η is the new separation constant (along with E). The second is a partial differential equation

$$\frac{\partial^2 Y}{\partial\theta^2} + \cot\theta\frac{\partial Y}{\partial\theta} + \csc^2\theta\frac{\partial^2 Y}{\partial\phi^2} + \eta Y = 0. \tag{3.102}$$

This latter equation is that for the spherical harmonics, which only has nontrivial solutions when

$$\eta = \eta_l = l(l+1), \quad l = 0,1,2,\ldots.$$

For each η_l, there are $2l+1$ *spherical harmonics*, which are given by

$$Y_l^m(\theta,\phi) = \sqrt{\frac{2l+1}{4\pi}\frac{(l-m)!}{(l+m)!}}P_l^m(\cos\theta)e^{im\phi}, \quad m = -l, -l+1, \ldots, l-1, l,$$

where we make use of the following relationship between *associated Legendre functions* $P_l^m(x)$ and Legendre polynomials $P_l(x)$:

$$P_l^m(x) = (-1)^m\left(1-x^2\right)^{m/2}\frac{d^m P_l(x)}{dx^m}.$$

For more on spherical harmonics, see Asmar (2005).

Equation (3.101) for the radial behavior $R(r)$, with the separation constant, is now

$$-\frac{\hbar^2}{2\mu}\left(\frac{d^2R}{dr^2}+\frac{2}{r}\frac{dR}{dr}-\frac{l(l+1)}{r^2}R\right)-\frac{e^2}{4\pi\epsilon_0 r}R=ER. \tag{3.103}$$

This equation can be converted into a differential equation with known eigenfunctions through two substitutions. First, let us make the substitution

$$r=\frac{s}{\alpha},\quad R(r)=w(s),$$

where

$$\alpha=\sqrt{-\frac{8\mu E}{\hbar^2}}.$$

Equation (3.103) becomes

$$\frac{d^2w}{ds^2}+\frac{2}{s}\frac{dw}{ds}-\frac{l(l+1)}{s^2}w-\frac{1}{4}w+\frac{n}{x}w=0, \tag{3.104}$$

where

$$n=\frac{2\mu e^2}{4\pi\epsilon_0\alpha\hbar^2}=\frac{e^2}{4\pi\epsilon_0\hbar}\sqrt{\frac{\mu}{-2E}}.$$

Note that $E<0$ owing to quantum mechanical considerations. The second substitution is given by

$$w(s)=s^l e^{-s/2}y(s),$$

which leads to the equation

$$s\frac{d^2y}{ds^2}+[2(l+1)-s]\frac{dy}{ds}+(n-l-1)y=0. \tag{3.105}$$

If n is a positive integer, such that $n-l-1$ is a nonnegative integer, then equation (3.105) is a *generalized Laguerre differential equation* having polynomial solutions with degree $n-l-1$ given by the *generalized Laguerre polynomials*

$$y_{nl}(s)=L_{n-l-1}^{2l+1}(s).$$

Unraveling the two substitutions, the solution for the radial component of the solution is given by

$$R_{nl}(r)=w_{nl}(\alpha_n r)=(\alpha_n r)^l e^{-\alpha_n r/2}y_{nl}(\alpha_n r)=(\alpha_n r)^l e^{-\alpha_n r/2}L_{n-l-1}^{2l+1}(\alpha_n r), \tag{3.106}$$

where

$$\alpha_n=\frac{\mu e^2}{\pi\epsilon_0 n\hbar^2}.$$

Again, see Asmar (2005) for more on generalized Laguerre polynomials.

The wave function for the spherical hydrogen atom is then given by

$$\psi_{lmn}(r,\theta,\phi,t)=R_{nl}(r)Y_l^m(\theta,\phi)T_n(t).$$

We find that the state of the hydrogen atom is determined by three integer quantum numbers known as the *principal*, *orbital* or *angular momentum*, and *magnetic quantum numbers*. The corresponding eigenfunctions ψ_n that result from the method of separation of variables provide the probability density functions from which the probability P is obtained for the location of the electron. The permissible discrete energy levels of the hydrogen atom E_n correspond to eigenvalues and only depend upon the principal quantum number n and are given by

$$E_n = \frac{1}{n^2} E_1, \quad n = 1, 2, \ldots, \infty,$$

where

$$E_1 = -\frac{\mu}{32} \left(\frac{e^2}{\pi \epsilon_0 \hbar} \right)^2 = -13.598 \, eV,$$

where the units are electron volts. When the electron moves from an energy level E_l to E_m, a photon of energy $E = E_l - E_m$ is given off. The lowest energy E_1 is known as the *ionization energy* or *ground energy*, and the theoretical value provided earlier has been confirmed by experiments. Observe that the energy levels are all negative, and the difference between respective energy levels decreases until reaching $E_\infty = 0$. For more complex atoms, the energy levels depend upon all three quantum numbers, not just the principal one as for the hydrogen atom.

3.5 Briefly on Bases

The parallels between the general approaches to solving differential equations in this chapter and how the eigenvectors were used as the basis for expressing the solution of the system of linear algebraic equations having a symmetric coefficient matrix in Section 2.3.2 have been emphasized throughout the chapter. In particular, for differential equations with a self-adjoint differential operator, the eigenfunctions are mutually orthogonal and provide the basis functions with respect to which the solution of the differential equation can be conveniently expressed as an eigenfunction expansion. This holds for ordinary differential equations as well as partial differential equations through separation of variables. The primary difference is that now there are an infinity of basis functions required for an exact solution.

Once again, it is remarkable – but not unexpected mathematically – that a completely analogous approach can be applied to systems of linear algebraic equations, ordinary differential equations, and partial differential equations. As we will see, these basic approaches can be extended to become the foundation for the spectral numerical method for solving more general differential equations (see Section 7.3) and proper-orthogonal decomposition for determining the dominant features in a numerical or experimental data set (see Chapter 13). Data reduction techniques, such as proper-orthogonal decomposition, are essentially methods for obtaining an optimal set of basis functions for accurate representation of the data set using the minimal amount of information. In addition, it will be shown in Chapter 13 that if the eigenfunctions of the differential operator are used as the basis functions in a Galerkin projection to obtain a reduced-order model, it is found to be perfectly consistent with the eigenfunction solution of the differential equation.

Exercises

3.1 Determine whether the following two functions are orthogonal over the interval $-1 \leq x \leq 1$ with respect to the weight function $w(x) = 1$:

$$u_1 = 3x^2 - 1, \quad u_2 = x(5x^2 - 3).$$

3.2 Verify that the set of functions

$$u_n(x) = \sin\left(\frac{n\pi x}{\ell}\right), \quad n = 1, 2, \ldots$$

are orthogonal for all n over the interval $0 \leq x \leq \ell$ with respect to the weight function $w(x) = 1$, and obtain their norms.

3.3 Verify that the set of functions

$$u_n(x) = \cos\left(\frac{n\pi x}{\ell}\right), \quad n = 1, 2, \ldots$$

are orthogonal for all n over the interval $0 \leq x \leq \ell$ with respect to the weight function $w(x) = 1$, and obtain their norms.

3.4 Expand the polynomial $3x^4 - 4x^2 - x$ in terms of a linear combination of the Chebyshev polynomials $T_0(x)$, $T_1(x)$, $T_2(x)$, $T_3(x)$, and $T_4(x)$.

3.5 Perform the integration by parts necessary to determine the adjoint operator of

$$\mathcal{L}u = \frac{d^2u}{dx^2} + u,$$

with homogeneous boundary conditions. Check your result using (3.30). Is the differential operator $\mathcal{L}$ self-adjoint?

3.6 Perform the integration by parts necessary to determine the adjoint operator of

$$\mathcal{L}u = \frac{d^2u}{dx^2} - 2\frac{du}{dx} + u,$$

with homogeneous boundary conditions. Check your result using (3.30). Is the differential operator $\mathcal{L}$ self-adjoint?

3.7 Perform the integration by parts necessary to determine the adjoint operator of

$$\mathcal{L}u = x^2\frac{d^2u}{dx^2} + \frac{du}{dx} + 2u,$$

with homogeneous boundary conditions. Check your result using (3.30). Is the differential operator $\mathcal{L}$ self-adjoint?

3.8 Show that Bessel's equation of order ν, which is given by

$$x^2\frac{d^2u}{dx^2} + x\frac{du}{dx} + (x^2 - \nu^2)u = 0,$$

is not self-adjoint.

3.9 Determine the value of a that makes the following equation self-adjoint:

$$(a \sin x)\frac{d^2u}{dx^2} + (\cos x)\frac{du}{dx} + 2u = 0.$$

3.10 Is the differential operator

$$\mathcal{L} = \frac{d^2}{dx^2} - x\frac{d}{dx} + x^2$$

with homogeneous boundary conditions self-adjoint?

3.11 Is the differential operator

$$\mathcal{L} = x^2\frac{d^2}{dx^2} + x\frac{d}{dx} + 1$$

with homogeneous boundary conditions self-adjoint?

3.12 Is the differential operator

$$\mathcal{L} = \frac{d^2}{dx^2} + \frac{d}{dx} + 1$$

with homogeneous boundary conditions self-adjoint?

3.13 Obtain the self-adjoint form of the differential operator

$$\mathcal{L} = \frac{d^2}{dx^2} + x\frac{d}{dx}.$$

3.14 Reduce the differential eigenproblem

$$(1 - x^2)\frac{d^2u}{dx^2} - x\frac{du}{dx} = \lambda u$$

to Sturm–Liouville form.

3.15 Consider the differential operator

$$\mathcal{L} = \frac{d^2}{dx^2} + \frac{1}{2}x^{-1/2}\frac{d}{dx} + x^2.$$

Obtain the adjoint operator of $\mathcal{L}$. Is $\mathcal{L}$ self-adjoint? If not, obtain the differential operator $\mathcal{L}$ in self-adjoint Sturm–Liouville form.

3.16 Consider the differential equation

$$\mathcal{L}u = \frac{d^2u}{dx^2} + u = \sin x,$$

with the nonhomogeneous boundary conditions

$$u(1) = 1, \quad u(2) = 3.$$

Determine the values of A and B in the transformation

$$u(x) = U(x) + Ax + B,$$

for which the boundary conditions are made to be homogeneous, that is, $U(1) = 0, U(2) = 0$. Write down the transformed problem in terms of $U(x)$.

3.17 Consider the differential equation

$$\mathcal{L}u = x^2\frac{d^2u}{dx^2} - 2x\frac{du}{dx} + 2u = 0.$$

(a) Obtain the adjoint operator of $\mathcal{L}$ assuming homogeneous boundary conditions. Is $\mathcal{L}$ self-adjoint?

(b) Now consider the differential equation subject to the following nonhomogeneous boundary conditions:

$$u(-1) = -2, \quad u(1) = 0.$$

Show that the boundary conditions may be made homogeneous, that is, $U(-1) = 0, U(1) = 0$, using the transformation

$$u(x) = U(x) + x - 1.$$

In addition, through multiplication by a suitable factor, show that the differential operator $\mathcal{L}$ may be converted to the Sturm–Liouville form

$$\mathcal{L}_{SL} = \frac{d}{dx}\left[p(x)\frac{d}{dx}\right] + q(x),$$

where $p(x) = 1/x^2$ and $q(x) = 2/x^4$.

(c) Obtain the exact solution of the original differential equation $\mathcal{L}u = 0$, with the nonhomogeneous boundary conditions, and compare it to the solution of the Sturm–Liouville form with homogeneous boundary conditions. (Hint: this is a Cauchy–Euler equation.)

3.18 Determine the eigenvalues and eigenfunctions of the differential eigenproblem

$$\frac{d^2u}{dx^2} = \lambda u, \quad u'(0) = 0, \quad u(1) = 0.$$

3.19 Obtain the eigenvalues and eigenfunctions of the differential eigenproblem

$$\frac{d^2u}{dx^2} = \lambda u, \quad u(0) + u'(0) = 0, \quad u(1) + u'(1) = 0.$$

List all of the roots of the characteristic equation for $\lambda = -\mu^2$.

3.20 Consider the eigenproblem

$$\frac{d^2u}{dx^2} = \lambda u, \quad u'(0) = 0, \quad u'(1) = 0.$$

Obtain the eigenvalues and eigenfunctions for this differential operator.

3.21 Consider the eigenproblem

$$\frac{d^2u}{dx^2} = \lambda u, \quad u(0) + u'(0) = 0, \quad u(1) - u'(1) = 0.$$

Obtain the characteristic equation and the eigenfunctions for this differential operator if $\lambda = -\mu^2$. Be sure to determine whether $\mu = 0$ is an eigenvalue.

3.22 Consider the differential eigenproblem

$$\frac{d^2u}{dx^2} = \lambda u, \quad u(0) + u'(0) = 0, \quad u(1) = 0.$$

Obtain the characteristic equation and the eigenfunctions for this differential operator if $\lambda = -\mu^2$.

3.23 Consider the differential eigenproblem

$$\frac{d^2u}{dx^2} = \lambda u, \quad u(0) - u'(0) = 0, \quad u(1) = 0.$$

(a) Obtain the characteristic equation and the eigenfunctions for this differential operator if $\lambda = -\mu^2$ (do not normalize the eigenfunctions).

(b) Using the results from (a), show how you would solve the differential equation

$$\frac{d^2u}{dx^2} + \Lambda u = f(x),$$

where Λ is distinct from any eigenvalue of the operator. The boundary conditions are the same as before. Show the steps, but do not perform any integration (inner products or norms).

3.24 Consider the differential eigenproblem

$$\frac{d^2u}{dx^2} = \lambda u, \quad u(0) - u'(0) = 0, \quad u(1) - 2u'(1) = 0.$$

(a) Obtain the characteristic equation and the eigenfunctions for this differential operator if $\lambda = -\mu^2$.

(b) Using the eigenfunctions from (a), solve the differential equation

$$\frac{d^2u}{dx^2} - 4u = 1 + x,$$

with the given boundary conditions.

3.25 Consider the differential equation

$$\frac{d^2u}{dx^2} - 4u = 8, \quad u(0) = 0, \quad u(1) = 0.$$

Obtain the solution of the differential equation in terms of an eigenfunction expansion. Note that this is the same problem as solved in Exercise 2.56 using diagonalization.

3.26 Obtain the characteristic equation and the eigenfunctions of the eigenproblem

$$\frac{d^2u}{dx^2} - 2\frac{du}{dx} = \lambda u, \quad u(0) = 0, \quad u(1) - u'(1) = 0.$$

3.27 Recall that the Sturm–Liouville differential operator is

$$\mathcal{L} = \frac{1}{w(x)}\left\{\frac{d}{dx}\left[p(x)\frac{d}{dx}\right] + q(x)\right\}.$$

Consider the eigenproblem $\mathcal{L}u(x) = \lambda u(x)$ for the case with $p(x) = 1$, $q(x) = 1$, and $w(x) = 1$ and boundary conditions

$$u(0) = 0, \quad u(1) = 0.$$

(a) Determine the eigenvalues and eigenfunctions of the differential operator.

(b) Using the eigenfunctions obtained in (a), obtain the eigenfunction expansion of the solution of the differential equation

$$\mathcal{L}u(x) = f(x),$$

where $f(x) = x$.

3.28 Determine the eigenvalues and eigenfunctions of the differential eigenproblem ($\lambda = \mu^2$):

$$x^2\frac{d^2u}{dx^2} + x\frac{du}{dx} = \lambda u, \quad u(1) = 0, \quad u(4) = 0.$$

Hint: this is a Cauchy–Euler equation.

3.29 Consider the differential eigenproblem

$$\frac{d^2u}{dx^2} - 2\frac{du}{dx} + u = \lambda u, \quad u(0) = 0, \quad u(1) - u'(1) = 0.$$

(a) Obtain the eigenvalues and eigenfunctions for this eigenproblem.
(b) Determine the self-adjoint Sturm–Liouville form of the differential eigenproblem.

3.30 Consider the differential operator

$$\mathcal{L}u = \frac{d^2u}{dx^2} - 2\frac{du}{dx} + u, \quad u(0) = 0, \quad u'(1) = 0.$$

(a) Obtain the adjoint operator of $\mathcal{L}$. Is $\mathcal{L}$ self-adjoint?
(b) Obtain the the characteristic equation and the eigenfunctions of the operator $\mathcal{L}$.

3.31 Consider the differential operator

$$\mathcal{L}u = x^2\frac{d^2u}{dx^2} + x\frac{du}{dx} + 4u, \quad u(1) = 0, \quad u(2) = 0.$$

Obtain the characteristic equation and the eigenfunctions for the differential operator $\mathcal{L}$ for *either* the positive ($\lambda = \mu^2 > 0$) *or* negative ($\lambda = -\mu^2 < 0$) case.

3.32 Consider the differential eigenproblem

$$\frac{d^2u}{dx^2} + 4\frac{du}{dx} + (3 + \lambda)u = 0, \quad u(0) = 0, \quad u(1) = 0.$$

Obtain the eigenvalues and eigenfunctions for this differential operator.

3.33 Determine whether or not the partial differential equation

$$y\frac{\partial^2 u}{\partial x^2} + \frac{1}{x^2}\frac{\partial^2 u}{\partial y^2} = 0$$

can be solved using the method of separation of variables (do not attempt to obtain the solution).

3.34 Determine whether or not the partial differential equation

$$\frac{\partial^2 u}{\partial x^2} + \frac{\partial u}{\partial x} + \frac{\partial^2 u}{\partial y^2} = 0$$

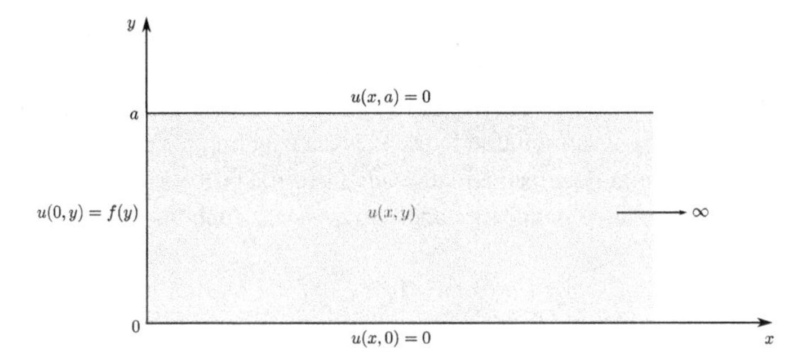

Figure 3.17 Schematic for Exercise 3.36.

can be solved using the method of separation of variables (do not attempt to obtain the solution).

3.35 Determine whether or not the partial differential equation

$$\frac{\partial u}{\partial t} = \frac{\partial u}{\partial x} + 3\frac{\partial^2 u}{\partial x^2}$$

can be solved using the method of separation of variables (do not attempt to obtain the solution).

3.36 Consider steady-state heat conduction in the semi-infinite slab shown in Figure 3.17, which is governed by

$$\frac{\partial^2 u}{\partial x^2} + \frac{\partial^2 u}{\partial y^2} = 0.$$

The boundary conditions are

$$u(x,0) = u(x,a) = 0, \quad \text{for} \quad 0 < x < \infty,$$

and

$$u(0,y) = f(y).$$

Using the method of separation of variables, obtain the solution for $u(x, y)$ in terms of an eigenfunction expansion.

3.37 We seek the solution of the Poisson equation, which is a nonhomogeneous version of the Laplace equation, in the unit square $0 \leq x \leq 1, 0 \leq y \leq 1$. The Poisson equation is

$$\frac{\partial^2 u}{\partial x^2} + \frac{\partial^2 u}{\partial y^2} = f(x,y),$$

and we impose the derivative (Neumann) boundary conditions

$$\frac{\partial u}{\partial x} = 0 \quad \text{at} \quad x = 0, x = 1,$$

and

$$\frac{\partial u}{\partial y} = 0 \quad \text{at} \quad y = 0, y = 1.$$

(a) Determine the eigenvalues and eigenfunctions of the differential operator in the homogeneous form of the Poisson equation for the given boundary conditions.

(b) Briefly outline how you would use the eigenfunctions from (a) to obtain the general form of the solution to the Poisson equation for a given $f(x, y)$.

3.38 Reconsider the one-dimensional, unsteady diffusion problem from Section 3.4.2 but now with an insulated boundary condition at $x = \ell$, such that

$$u(0,t) = 0, \quad \frac{\partial u}{\partial x}(\ell,t) = 0,$$

and the initial temperature distribution is

$$u(x,0) = f(x) = u_0\left(1 + \frac{x}{\ell}\right),$$

where u_0 is a constant.

3.39 Reconsider the one-dimensional, unsteady diffusion problem from Section 3.4.2 but now with a convection boundary condition at $x = 0$, such that

$$u(0,t) - \frac{\partial u}{\partial x}(0,t) = 0, \quad u(\ell,t) = 0.$$

3.40 The steady-state temperature distribution is sought in a semicircular domain with $y > 0$. The semicircle is centered at the origin and has radius R, and the straight portion is aligned with the x-axis. The boundary temperature along the straight portion is given by $u = 0$, and that along the semicircular boundary is given by $u = u_0\theta(\pi - \theta)$, where u_0 is a constant. The Laplace equation governing steady heat conduction in polar coordinates is given by

$$\frac{\partial^2 u}{\partial r^2} + \frac{1}{r}\frac{\partial u}{\partial r} + \frac{1}{r^2}\frac{\partial^2 u}{\partial \theta^2} = 0.$$

3.41 A membrane in the rectangular domain $0 \le x \le a$ and $0 \le y \le b$ has its edges clamped. The governing equation is the two-dimensional wave equation in Cartesian coordinates

$$\frac{\partial^2 u}{\partial t^2} = c^2\left(\frac{\partial^2 u}{\partial x^2} + \frac{\partial^2 u}{\partial y^2}\right),$$

where c is the wave speed, and $u(x, y, t)$ is the vibrational amplitude. Show that the natural frequencies of vibration (related to the eigenvalues λ_{mn}) are given by

$$\omega_{mn} = c\sqrt{\left(\frac{n\pi}{a}\right)^2 + \left(\frac{m\pi}{b}\right)^2},$$

and that the corresponding eigenfunctions determining the modes of vibration are

$$\phi_{mn}(x,y) = \sin\left(\frac{n\pi x}{a}\right)\sin\left(\frac{m\pi y}{b}\right).$$

3.42 Consider the transverse vibration of a beam of length $0 \le x \le \pi$ governed by

$$\frac{\partial^2 u}{\partial t^2} + \frac{\partial^4 u}{dx^4} = 0.$$

Determine the vibrational mode shapes $\phi(x)$ for the following two cases:

(a) Both ends are free, for which the boundary conditions are given by

$$\phi''(0) = \phi'''(0) = 0, \quad \phi''(\pi) = \phi'''(\pi) = 0.$$

(b) Both ends are clamped, for which the boundary conditions are given by

$$\phi(0) = \phi'(0) = 0, \quad \phi(\pi) = \phi'(\pi) = 0.$$

4 Vector and Matrix Calculus

One geometry cannot be more true than another; it can only be more convenient. (Jules Henri Poincaré)

In Chapters 1 and 2, we have focused on *algebra* involving vectors and matrices; here, we treat *calculus* of these important and ubiquitous constructs. These operations will prove essential in various applications throughout the remainder of the text. For example, there are several important operators that arise involving differentiation (gradient, divergence, and curl) and integration (divergence and Stokes' theorems) in two- and three-dimensional settings in continuum mechanics, electromagnetics, modern physics, and other fields. These operations are defined for three-dimensional vectors and extended to the tensors that are used in these areas to represent quantities that have a magnitude and two directions. Simplistically, a tensor is a 3×3 matrix; however, the way they are used in various physics-based applications dictates that they have additional properties beyond those of generic vectors and matrices. In particular, just as vectors in mechanics consist of the three components of a velocity or force, for example, in the coordinate directions of a three-dimensional coordinate system, tensors encapsulate quantities, such as stress and strain, that require nine quantities to fully specify at a point in three-dimensional space.

Vector calculus also will be essential to converting equations between their differential and integral forms, converting differential equations from one coordinate system to another, and leads to a compact representation of partial differential equations. Once we understand these vector and integral forms and their physical implications, they become a powerful tool in representing various physical phenomena in a compact and illuminating form.

Finally, determining extrema of algebraic functions requires evaluating first and second derivatives. This segues into least-squares methods and optimization, which are treated briefly here and in more detail in Chapters 10 and 12. The reader is referred to more detailed treatments of vector calculus in Greenberg (1998), Jeffrey (2002), Kreyszig (2011), and Hildebrand (1976), for example; here we focus on those topics of importance in later developments, such as identifying extrema with and without constraints.

4.1 Vector Calculus

4.1.1 Cartesian and Cylindrical Coordinate Systems

Before proceeding to the derivative operators, let us first consider the two most commonly used coordinate systems, namely Cartesian and cylindrical. In the Cartesian coordinate system (see Figure 4.1), the unit vectors in the x, y, and z coordinate directions are **i**, **j**, and **k**, respectively. Therefore, a vector in three-dimensional Cartesian coordinates is given by

$$\mathbf{a} = a_x\mathbf{i} + a_y\mathbf{j} + a_z\mathbf{k},$$

where, for example, a_x is the component of the vector **a** in the x-direction, which is given by the inner (dot) product $a_x = \langle \mathbf{a}, \mathbf{i} \rangle = \mathbf{a} \cdot \mathbf{i}$. Note that in the Cartesian coordinate system, the unit vectors **i**, **j**, and **k** are independent of changes in x, y, and z. Therefore, derivatives of the unit vectors with respect to each of the coordinate directions vanish. For example,

$$\begin{aligned}
\frac{\partial \mathbf{a}}{\partial x} &= \frac{\partial}{\partial x}\left(a_x\mathbf{i} + a_y\mathbf{j} + a_z\mathbf{k}\right), \\
&= \frac{\partial a_x}{\partial x}\mathbf{i} + a_x\cancelto{0}{\frac{\partial \mathbf{i}}{\partial x}} + \frac{\partial a_y}{\partial x}\mathbf{j} + a_y\cancelto{0}{\frac{\partial \mathbf{j}}{\partial x}} + \frac{\partial a_z}{\partial x}\mathbf{k} + a_z\cancelto{0}{\frac{\partial \mathbf{k}}{\partial x}}, \\
\frac{\partial \mathbf{a}}{\partial x} &= \frac{\partial a_x}{\partial x}\mathbf{i} + \frac{\partial a_y}{\partial x}\mathbf{j} + \frac{\partial a_z}{\partial x}\mathbf{k}.
\end{aligned}$$

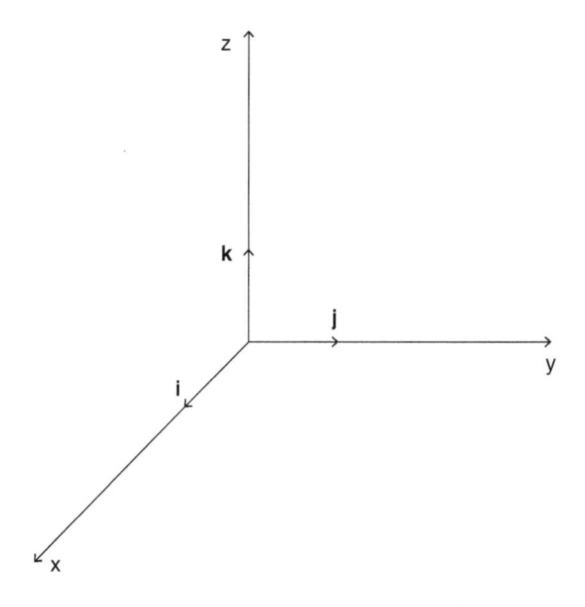

Figure 4.1 Cartesian coordinate system.

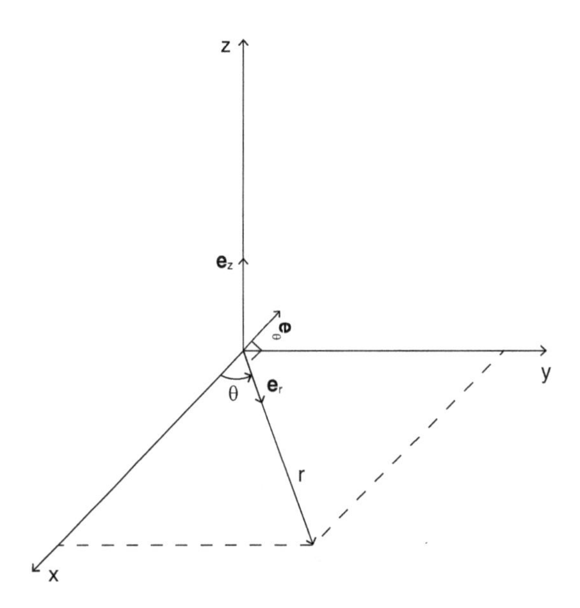

Figure 4.2 Cylindrical coordinate system.

Note that this is the same as simply taking the partial derivative with respect to x of each of the components of $\mathbf{a}$.

Cylindrical coordinates, which are shown in Figure 4.2, require a bit more care. A vector in three-dimensional cylindrical coordinates is given by

$$\mathbf{a} = a_r \mathbf{e}_r + a_\theta \mathbf{e}_\theta + a_z \mathbf{e}_z.$$

In terms of Cartesian coordinates, the unit vectors are

$$\mathbf{e}_r = \cos\theta \mathbf{i} + \sin\theta \mathbf{j}, \quad \mathbf{e}_\theta = -\sin\theta \mathbf{i} + \cos\theta \mathbf{j}, \quad \mathbf{e}_z = \mathbf{k}.$$

Note that the unit vector $\mathbf{e}_z$ is independent of r, θ, and z, and the unit vectors $\mathbf{e}_r$ and $\mathbf{e}_\theta$ are independent of r and z; however, they are dependent on changes in θ. For example,

$$\frac{\partial \mathbf{e}_r}{\partial \theta} = \frac{\partial}{\partial \theta}(\cos\theta \mathbf{i} + \sin\theta \mathbf{j}) = -\sin\theta \mathbf{i} + \cos\theta \mathbf{j} = \mathbf{e}_\theta.$$

Similarly,

$$\frac{\partial \mathbf{e}_\theta}{\partial \theta} = \frac{\partial}{\partial \theta}(-\sin\theta \mathbf{i} + \cos\theta \mathbf{j}) = -\cos\theta \mathbf{i} - \sin\theta \mathbf{j} = -\mathbf{e}_r.$$

Summarizing, we have that

$$\frac{\partial \mathbf{e}_r}{\partial r} = 0, \quad \frac{\partial \mathbf{e}_\theta}{\partial r} = 0, \quad \frac{\partial \mathbf{e}_z}{\partial r} = 0,$$

$$\frac{\partial \mathbf{e}_r}{\partial \theta} = \mathbf{e}_\theta, \quad \frac{\partial \mathbf{e}_\theta}{\partial \theta} = -\mathbf{e}_r, \quad \frac{\partial \mathbf{e}_z}{\partial \theta} = 0,$$

$$\frac{\partial \mathbf{e}_r}{\partial z} = 0, \quad \frac{\partial \mathbf{e}_\theta}{\partial z} = 0, \quad \frac{\partial \mathbf{e}_z}{\partial z} = 0.$$

The key in cylindrical coordinates is to remember that one must first take the derivative before performing vector operations. This will be illustrated in Example 4.1.

4.1.2 Derivative Operators

The fundamental building block of vector calculus is the vector differential operator $\boldsymbol{\nabla}$, called "del," which is given in Cartesian coordinates by[1]

$$\boldsymbol{\nabla} = \frac{\partial}{\partial x}\mathbf{i} + \frac{\partial}{\partial y}\mathbf{j} + \frac{\partial}{\partial z}\mathbf{k}. \tag{4.1}$$

It is a vector operator that operates on scalar functions, such as $\phi(x, y, z)$, and vector functions, such as $\mathbf{f}(x, y, z)$, in three distinct ways. The first is denoted by $\boldsymbol{\nabla}\phi$, and is called the *gradient*, which is the spatial rate of change of the scalar field $\phi(x, y, z)$; it is the three-dimensional version of the derivative (slope). The second is denoted by $\boldsymbol{\nabla} \cdot \mathbf{f}$, read "del dot ef," and is called the *divergence*. It is the spatial rate of change of the vector field $\mathbf{f}(x, y, z)$ indicating expansion or contraction of the vector field. The third is denoted by $\boldsymbol{\nabla} \times \mathbf{f}$, read "del cross ef," and is called the *curl*. It is the spatial rate of change of the vector field $\mathbf{f}(x, y, z)$ indicating rotation of the vector field. We first focus on these operations in Cartesian coordinates and then generalize to cylindrical and spherical coordinates.

Gradient (grad ϕ) and Directional Derivative

In Cartesian coordinates, the gradient of the scalar field $\phi(x, y, z)$ is

$$\boldsymbol{\nabla}\phi = \left(\frac{\partial}{\partial x}\mathbf{i} + \frac{\partial}{\partial y}\mathbf{j} + \frac{\partial}{\partial z}\mathbf{k}\right)\phi = \frac{\partial \phi}{\partial x}\mathbf{i} + \frac{\partial \phi}{\partial y}\mathbf{j} + \frac{\partial \phi}{\partial z}\mathbf{k}. \tag{4.2}$$

The gradient of a scalar function is a vector field that represents the direction and magnitude of the greatest spatial rate of change of $\phi(x, y, z)$ at each point. That is, it indicates the steepest slope of the scalar field at each point as illustrated in Figure 4.3. The component of the gradient vector in the direction of some unit vector $\mathbf{e}$ is known as the *directional derivative* and is given by

$$D_e\phi = \mathbf{e} \cdot \boldsymbol{\nabla}\phi.$$

[1] It is important to realize that when writing vectors containing derivatives in this manner, we do not mean to imply that in the first term, for example, that $\partial/\partial x$ is operating on the unit vector $\mathbf{i}$; it is merely the x-component of the vector. This is why some authors place the unit vector before the respective components to prevent any such confusion.

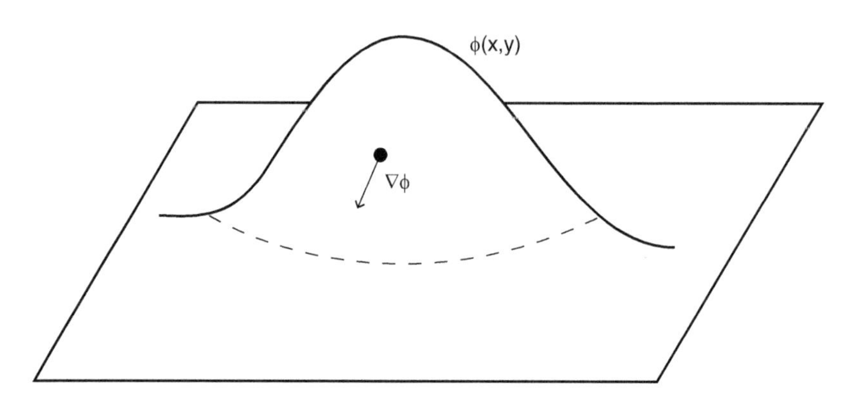

Figure 4.3 The gradient of a scalar field.

While the usual partial derivative gives the rate of change of the function with respect to the coordinate directions, the directional derivative represents the rate of change of ϕ in the $\mathbf{e}$ direction, which can be any direction and not limited to the coordinate directions.

Divergence (div f)

In Cartesian coordinates, the divergence of the vector field $\mathbf{f}(x, y, z)$ is the inner (dot) product of the del vector and the vector $\mathbf{f}$, as follows:

$$\nabla \cdot \mathbf{f} = \left(\frac{\partial}{\partial x}\mathbf{i} + \frac{\partial}{\partial y}\mathbf{j} + \frac{\partial}{\partial z}\mathbf{k}\right) \cdot (f_x\mathbf{i} + f_y\mathbf{j} + f_z\mathbf{k}) = \frac{\partial f_x}{\partial x} + \frac{\partial f_y}{\partial y} + \frac{\partial f_z}{\partial z}. \tag{4.3}$$

The divergence of a vector function is a scalar field that represents the net normal component of the vector field $\mathbf{f}$ passing out through the surface of an infinitesimal volume surrounding each point. It indicates the expansion of a vector field.

Curl (curl f)

In Cartesian coordinates, the curl of the vector field $\mathbf{f}(x, y, z)$ is the cross product of the del vector and the vector $\mathbf{f}$, as follows (see Section 1.6.1):

$$\nabla \times \mathbf{f} = \begin{vmatrix} \mathbf{i} & \mathbf{j} & \mathbf{k} \\ \dfrac{\partial}{\partial x} & \dfrac{\partial}{\partial y} & \dfrac{\partial}{\partial z} \\ f_x & f_y & f_z \end{vmatrix} = \left(\frac{\partial f_z}{\partial y} - \frac{\partial f_y}{\partial z}\right)\mathbf{i} - \left(\frac{\partial f_z}{\partial x} - \frac{\partial f_x}{\partial z}\right)\mathbf{j} + \left(\frac{\partial f_y}{\partial x} - \frac{\partial f_x}{\partial y}\right)\mathbf{k}. \tag{4.4}$$

Recall that we use a cofactor expansion about the first row to evaluate the determinant. The curl is a vector field that represents the net tangential component of the vector field $\mathbf{f}$ along the surface of an infinitesimal volume surrounding each point. It indicates the rotational characteristics of a vector field. For example, the curl of a force field is the moment or torque, and the curl of a velocity field yields the rate of rotation of the particles.

Laplacian

In Cartesian coordinates, the Laplacian operator is

$$\nabla^2 = \nabla \cdot \nabla = \left(\frac{\partial}{\partial x}\mathbf{i} + \frac{\partial}{\partial y}\mathbf{j} + \frac{\partial}{\partial z}\mathbf{k}\right) \cdot \left(\frac{\partial}{\partial x}\mathbf{i} + \frac{\partial}{\partial y}\mathbf{j} + \frac{\partial}{\partial z}\mathbf{k}\right) = \frac{\partial^2}{\partial x^2} + \frac{\partial^2}{\partial y^2} + \frac{\partial^2}{\partial z^2}, \tag{4.5}$$

where we note that it is the dot product of del with itself, which is denoted by ∇^2. The Laplacian can operate on both scalar and vector functions to produce a scalar or vector expression, respectively. For example, $\nabla^2 \phi = 0$ is the Laplace equation, which we have encountered several times already.

Gradient, Divergence, Curl, and Laplacian in Curvilinear Orthogonal Coordinates

There are three commonly used curvilinear orthogonal coordinate systems: Cartesian, cylindrical, and spherical coordinates. We can unify the writing of the gradient, divergence, curl, and Laplacian in each of these coordinate systems as follows.

If ϕ is a scalar function and $\mathbf{f} = f_1\mathbf{e}_1 + f_2\mathbf{e}_2 + f_3\mathbf{e}_3$ is a vector function of orthogonal curvilinear coordinates q_1, q_2, and q_3, with $\mathbf{e}_1$, $\mathbf{e}_2$, and $\mathbf{e}_3$ being unit vectors in the direction of increasing q_1, q_2, and q_3, respectively, then the gradient, divergence, curl, and Laplacian are defined as follows:
Gradient:

$$\nabla \phi = \frac{1}{h_1}\frac{\partial \phi}{\partial q_1}\mathbf{e}_1 + \frac{1}{h_2}\frac{\partial \phi}{\partial q_2}\mathbf{e}_2 + \frac{1}{h_3}\frac{\partial \phi}{\partial q_3}\mathbf{e}_3. \tag{4.6}$$

Divergence:

$$\nabla \cdot \mathbf{f} = \frac{1}{h_1 h_2 h_3}\left[\frac{\partial}{\partial q_1}(h_2 h_3 f_1) + \frac{\partial}{\partial q_2}(h_3 h_1 f_2) + \frac{\partial}{\partial q_3}(h_1 h_2 f_3)\right]. \tag{4.7}$$

Curl:

$$\nabla \times \mathbf{f} = \frac{1}{h_1 h_2 h_3}\begin{vmatrix} h_1\mathbf{e}_1 & h_2\mathbf{e}_2 & h_3\mathbf{e}_3 \\ \dfrac{\partial}{\partial q_1} & \dfrac{\partial}{\partial q_2} & \dfrac{\partial}{\partial q_3} \\ h_1 f_1 & h_2 f_2 & h_3 f_3 \end{vmatrix}. \tag{4.8}$$

Laplacian:

$$\nabla^2 \phi = \frac{1}{h_1 h_2 h_3}\left[\frac{\partial}{\partial q_1}\left(\frac{h_2 h_3}{h_1}\frac{\partial \phi}{\partial q_1}\right) + \frac{\partial}{\partial q_2}\left(\frac{h_3 h_1}{h_2}\frac{\partial \phi}{\partial q_2}\right) + \frac{\partial}{\partial q_3}\left(\frac{h_1 h_2}{h_3}\frac{\partial \phi}{\partial q_3}\right)\right]. \tag{4.9}$$

In the preceding expressions, h_1, h_2, and h_3 are the *scale factors*, which are given by

$$h_1 = \left|\frac{\partial \mathbf{r}}{\partial q_1}\right|, \quad h_2 = \left|\frac{\partial \mathbf{r}}{\partial q_2}\right|, \quad h_3 = \left|\frac{\partial \mathbf{r}}{\partial q_3}\right|,$$

where $\mathbf{r}$ represents the position vector of a point in space. In Cartesian coordinates, for example, $\mathbf{r} = x\mathbf{i} + y\mathbf{j} + z\mathbf{k}$.

In the following, we provide the q_i, h_i, and $\mathbf{e}_i$ for Cartesian, cylindrical, and spherical coordinates for use in (4.6) through (4.9).

Cartesian Coordinates: (x, y, z)

$$\begin{aligned} q_1 &= x, & h_1 &= 1, & \mathbf{e}_1 &= \mathbf{i}, \\ q_2 &= y, & h_2 &= 1, & \mathbf{e}_2 &= \mathbf{j}, \\ q_3 &= z, & h_3 &= 1, & \mathbf{e}_3 &= \mathbf{k}. \end{aligned}$$

Cylindrical Coordinates: (r, θ, z)

$$\begin{aligned} q_1 &= r, & h_1 &= 1, & \mathbf{e}_1 &= \mathbf{e}_r, \\ q_2 &= \theta, & h_2 &= r, & \mathbf{e}_2 &= \mathbf{e}_\theta, \\ q_3 &= z, & h_3 &= 1, & \mathbf{e}_3 &= \mathbf{e}_z. \end{aligned}$$

Spherical Coordinates: (R, ϕ, θ)

$$\begin{aligned} q_1 &= R, & h_1 &= 1, & \mathbf{e}_1 &= \mathbf{e}_R, \\ q_2 &= \phi, & h_2 &= R, & \mathbf{e}_2 &= \mathbf{e}_\phi, \\ q_3 &= \theta, & h_3 &= R\sin\phi, & \mathbf{e}_3 &= \mathbf{e}_\theta. \end{aligned}$$

Example 4.1 Evaluate the divergence of the vector

$$\mathbf{f} = f_r \mathbf{e}_r$$

in cylindrical coordinates, where f_r is a constant.

Solution

Remembering that one must first take the derivative before performing vector operations, evaluating the divergence in cylindrical coordinates yields (keeping in mind that f_r is a constant)

$$\begin{aligned} \boldsymbol{\nabla}\cdot\mathbf{f} &= \left(\frac{\partial}{\partial r}\mathbf{e}_r + \frac{1}{r}\frac{\partial}{\partial\theta}\mathbf{e}_\theta + \frac{\partial}{\partial z}\mathbf{e}_z\right)\cdot(f_r\mathbf{e}_r) \\ &= \mathbf{e}_r\cdot\frac{\partial}{\partial r}(f_r\mathbf{e}_r) + \mathbf{e}_\theta\cdot\frac{1}{r}\frac{\partial}{\partial\theta}(f_r\mathbf{e}_r) + \mathbf{e}_z\cdot\frac{\partial}{\partial z}(f_r\mathbf{e}_r) \\ &= \mathbf{e}_r\cdot f_r\overset{0}{\cancel{\frac{\partial\mathbf{e}_r}{\partial r}}} + \mathbf{e}_\theta\cdot\frac{f_r}{r}\frac{\partial\mathbf{e}_r}{\partial\theta} + \mathbf{e}_z\cdot f_r\overset{0}{\cancel{\frac{\partial\mathbf{e}_r}{\partial z}}} \\ &= \mathbf{e}_\theta\cdot\frac{f_r}{r}\mathbf{e}_\theta \\ \boldsymbol{\nabla}\cdot\mathbf{f} &= \frac{f_r}{r}, \end{aligned}$$

where we have used the fact that $\partial\mathbf{e}_r/\partial\theta = \mathbf{e}_\theta$ obtained earlier. If we would have first carried out the dot products before differentiating, we would have obtained $\boldsymbol{\nabla}\cdot\mathbf{f} = \partial f_r/\partial r = 0$, which is incorrect.

Gradient, Divergence, and Curl Vector Identities

The following properties can be established for the gradient, divergence, and curl of scalar and vector functions. The curl of the gradient of any scalar vanishes, that is,

$$\nabla \times \nabla\phi = \mathbf{0}.$$

The divergence of the curl of any vector vanishes according to

$$\nabla \cdot (\nabla \times \mathbf{f}) = 0.$$

Note also that the divergence and curl of the product of a scalar and vector obey the product rule. For example,

$$\nabla \cdot (\phi\mathbf{f}) = \mathbf{f} \cdot \nabla\phi + \phi\nabla \cdot \mathbf{f}, \quad \nabla \times (\phi\mathbf{f}) = \phi\nabla \times \mathbf{f} + \nabla\phi \times \mathbf{f}.$$

In these relationships, recall that the gradient of a scalar is a vector, the curl of a vector is a vector, and the divergence of a vector is a scalar.

4.1.3 Integral Theorems

In addition to differentiation, calculus involves its inverse – integration. Because of the different forms of differentiation, there are corresponding integral theorems.

Divergence Theorem

Let $\mathbf{f}$ be any continuously differentiable vector field in a volume $\bar{V}$ surrounded by a surface S, then

$$\oint_S \mathbf{f} \cdot d\mathbf{A} = \int_{\bar{V}} \nabla \cdot \mathbf{f} d\bar{V}, \tag{4.10}$$

where a differential area on surface S is $d\mathbf{A} = dA\,\mathbf{n}$, with $\mathbf{n}$ being the outward facing normal at $d\mathbf{A}$ as shown in Figure 4.4. The divergence theorem transforms a surface integral over S to a volume integral over $\bar{V}$ and states that the integral of the normal component of $\mathbf{f}$ over the bounding surface is equal to the volume integral of the divergence of $\mathbf{f}$ throughout the volume $\bar{V}$. Recalling that the divergence of a vector field measures its expansion properties, the divergence theorem takes into account that the expansion of each differential volume within $\bar{V}$ is "absorbed" by the neighboring differential volumes until the surface is reached, such that the net – integrated – result is that the total expansion of the volume is given by the expansion at its bounding surface.

There are additional forms of the divergence theorem corresponding to the other types of derivatives as well. For the gradient, we have

$$\oint_S \phi d\mathbf{A} = \int_{\bar{V}} \nabla\phi d\bar{V}, \tag{4.11}$$

and for the curl

$$\oint_S \mathbf{f} \times d\mathbf{A} = -\int_{\bar{V}} \nabla \times \mathbf{f} d\bar{V}. \tag{4.12}$$

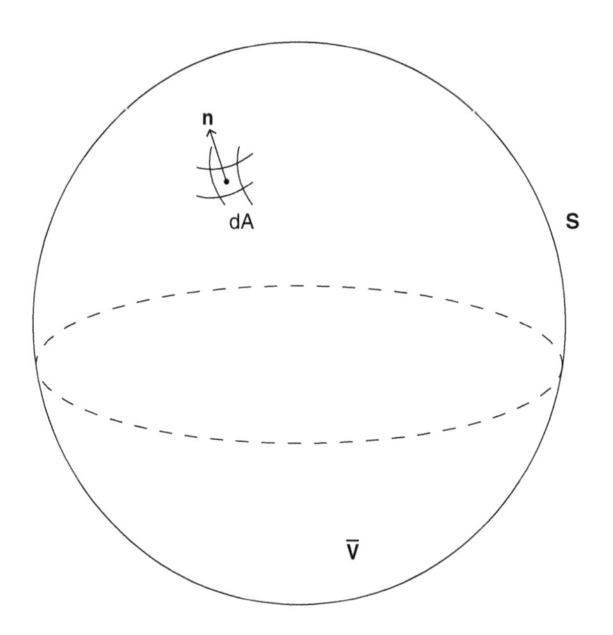

Figure 4.4 Schematic for the divergence theorem.

The latter case is also known as *Gauss' theorem*. This states that the integral – sum – of the tangential components of **f** over the bounding surface equals the volume integral of the curl of **f** throughout the entire volume. Once again, observe how the curl, which indicates rotation, of each differential element in the volume cancels with that of its neighboring elements until the surface S is reached, thereby leading to the left-hand side, which is only evaluated at the surface.

Stokes' and Green's Theorems

In two dimensions, we have a similar relationship between the integral around a curve C bounding an area domain A as illustrated in Figure 4.5. Here, $d\mathbf{s}$ is a unit vector tangent to C. Stokes' theorem states that

$$\oint_C \mathbf{f} \cdot d\mathbf{s} = \int_A (\nabla \times \mathbf{f}) \cdot d\mathbf{A}. \tag{4.13}$$

Stokes' theorem transforms a line integral over the bounding curve to an area integral and relates the line integral of the tangential component of **f** over the closed curve C to the integral of the normal component of the curl of **f** over the area A.

In two-dimensional Cartesian coordinates,

$$\mathbf{f} = P(x, y)\mathbf{i} + Q(x, y)\mathbf{j},$$

and

$$d\mathbf{s} = dx\mathbf{i} + dy\mathbf{j},$$

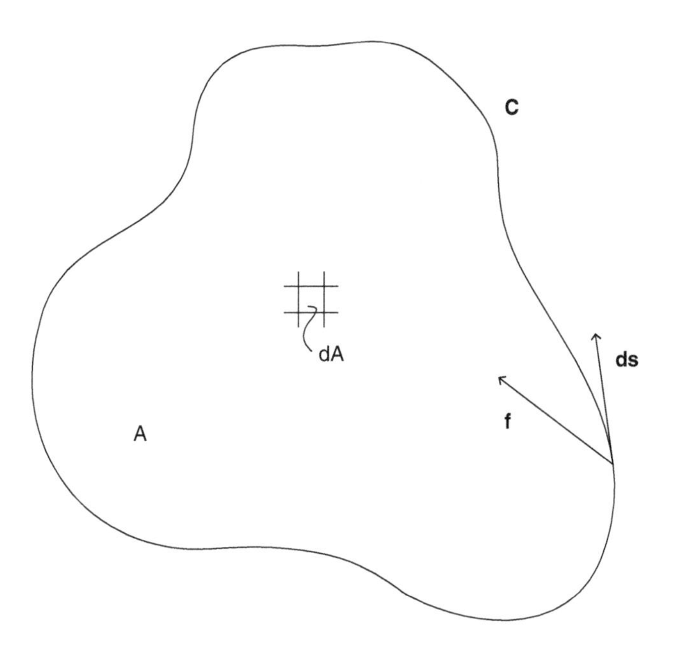

Figure 4.5 Schematic for Stokes' theorem.

then

$$\nabla \times \mathbf{f} = \left(\frac{\partial Q}{\partial x} - \frac{\partial P}{\partial y} \right) \mathbf{k}.$$

In this case, Stokes' theorem becomes

$$\oint_C (Pdx + Qdy) = \int_A \left(\frac{\partial Q}{\partial x} - \frac{\partial P}{\partial y} \right) dA. \tag{4.14}$$

This particular form of Stokes' theorem is also known as *Green's theorem.*

4.1.4 Coordinate Transformations

There will often be occasion to transform an expression or differential equation from one coordinate system to another. We consider two-dimensional coordinates for brevity; extension to three dimensions is straightforward, but tedious. Let us consider transformations of the form

$$(x, y) \Longleftrightarrow (\xi, \eta),$$

where (x, y) are the original coordinates, and (ξ, η) are the new coordinates being transformed to. There are two aspects in carrying out such coordinate transformation: (1) what criterion must be satisfied in order for the transformation to be one-to-one, and (2) how are derivatives in one coordinate system transformed to the other. The former requirement means that each point in one coordinate system will have a single unique image point in the other coordinate system.

The criterion for a one-to-one transformation is obtained by determining the requirements for a unique solution of the differentials of one set of variables in terms of the other. For

$$\xi = \xi(x, y), \quad \eta = \eta(x, y),$$

the total differentials are

$$d\xi = \frac{\partial \xi}{\partial x}dx + \frac{\partial \xi}{\partial y}dy, \quad d\eta = \frac{\partial \eta}{\partial x}dx + \frac{\partial \eta}{\partial y}dy,$$

or in matrix form

$$\begin{bmatrix} d\xi \\ d\eta \end{bmatrix} = \begin{bmatrix} \frac{\partial \xi}{\partial x} & \frac{\partial \xi}{\partial y} \\ \frac{\partial \eta}{\partial x} & \frac{\partial \eta}{\partial y} \end{bmatrix} \begin{bmatrix} dx \\ dy \end{bmatrix}. \tag{4.15}$$

Similarly, for the inverse transformation from (ξ, η) to (x, y), we have that

$$dx = \frac{\partial x}{\partial \xi}d\xi + \frac{\partial x}{\partial \eta}d\eta, \quad dy = \frac{\partial y}{\partial \xi}d\xi + \frac{\partial y}{\partial \eta}d\eta,$$

or

$$\begin{bmatrix} dx \\ dy \end{bmatrix} = \begin{bmatrix} \frac{\partial x}{\partial \xi} & \frac{\partial x}{\partial \eta} \\ \frac{\partial y}{\partial \xi} & \frac{\partial y}{\partial \eta} \end{bmatrix} \begin{bmatrix} d\xi \\ d\eta \end{bmatrix}.$$

From the latter expression, we can solve for $\begin{bmatrix} d\xi & d\eta \end{bmatrix}^T$ by multiplying by the inverse of the 2×2 matrix as follows:

$$\begin{aligned}
\begin{bmatrix} d\xi \\ d\eta \end{bmatrix} &= \begin{bmatrix} \frac{\partial x}{\partial \xi} & \frac{\partial x}{\partial \eta} \\ \frac{\partial y}{\partial \xi} & \frac{\partial y}{\partial \eta} \end{bmatrix}^{-1} \begin{bmatrix} dx \\ dy \end{bmatrix} \\
&= \frac{\begin{bmatrix} \frac{\partial y}{\partial \eta} & -\frac{\partial x}{\partial \eta} \\ -\frac{\partial y}{\partial \xi} & \frac{\partial x}{\partial \xi} \end{bmatrix}}{\begin{vmatrix} \frac{\partial x}{\partial \xi} & \frac{\partial x}{\partial \eta} \\ \frac{\partial y}{\partial \xi} & \frac{\partial y}{\partial \eta} \end{vmatrix}} \begin{bmatrix} dx \\ dy \end{bmatrix} \\
\begin{bmatrix} d\xi \\ d\eta \end{bmatrix} &= \frac{1}{\mathcal{J}} \begin{bmatrix} \frac{\partial y}{\partial \eta} & -\frac{\partial x}{\partial \eta} \\ -\frac{\partial y}{\partial \xi} & \frac{\partial x}{\partial \xi} \end{bmatrix} \begin{bmatrix} dx \\ dy \end{bmatrix},
\end{aligned}$$

where the *Jacobian* of the transformation is the determinant in the denominator

$$\mathcal{J} = \begin{vmatrix} \dfrac{\partial x}{\partial \xi} & \dfrac{\partial x}{\partial \eta} \\ \dfrac{\partial y}{\partial \xi} & \dfrac{\partial y}{\partial \eta} \end{vmatrix} = \frac{\partial x}{\partial \xi}\frac{\partial y}{\partial \eta} - \frac{\partial x}{\partial \eta}\frac{\partial y}{\partial \xi}. \tag{4.16}$$

Therefore, for the transformation to exist and be one-to-one, it must be such that $\mathcal{J} \neq 0$.

Now let us consider how to transform derivatives from one coordinate system to the other. The transformation laws, which arise from the chain rule, are

$$\frac{\partial}{\partial x} = \frac{\partial \xi}{\partial x}\frac{\partial}{\partial \xi} + \frac{\partial \eta}{\partial x}\frac{\partial}{\partial \eta}, \tag{4.17}$$

$$\frac{\partial}{\partial y} = \frac{\partial \xi}{\partial y}\frac{\partial}{\partial \xi} + \frac{\partial \eta}{\partial y}\frac{\partial}{\partial \eta}, \tag{4.18}$$

where $\partial q_j/\partial u_i$ are called the *metrics* of the transformation. Comparing our two expressions for $\begin{bmatrix} d\xi & d\eta \end{bmatrix}^T$, that is, (4.15) and the preceding expression, we see that

$$\begin{bmatrix} \dfrac{\partial \xi}{\partial x} & \dfrac{\partial \xi}{\partial y} \\ \dfrac{\partial \eta}{\partial x} & \dfrac{\partial \eta}{\partial y} \end{bmatrix} = \frac{1}{\mathcal{J}} \begin{bmatrix} \dfrac{\partial y}{\partial \eta} & -\dfrac{\partial x}{\partial \eta} \\ -\dfrac{\partial y}{\partial \xi} & \dfrac{\partial x}{\partial \xi} \end{bmatrix}.$$

Consequently, the transformation metrics are

$$\frac{\partial \xi}{\partial x} = \frac{1}{\mathcal{J}}\frac{\partial y}{\partial \eta}, \qquad \frac{\partial \xi}{\partial y} = -\frac{1}{\mathcal{J}}\frac{\partial x}{\partial \eta},$$

$$\frac{\partial \eta}{\partial x} = -\frac{1}{\mathcal{J}}\frac{\partial y}{\partial \xi}, \qquad \frac{\partial \eta}{\partial y} = \frac{1}{\mathcal{J}}\frac{\partial x}{\partial \xi}.$$

Substituting into the transformation laws (4.17) and (4.18) gives

$$\frac{\partial}{\partial x} = \frac{1}{\mathcal{J}}\left(\frac{\partial y}{\partial \eta}\frac{\partial}{\partial \xi} - \frac{\partial y}{\partial \xi}\frac{\partial}{\partial \eta}\right),$$

$$\frac{\partial}{\partial y} = \frac{1}{\mathcal{J}}\left(\frac{\partial x}{\partial \xi}\frac{\partial}{\partial \eta} - \frac{\partial x}{\partial \eta}\frac{\partial}{\partial \xi}\right).$$

These transformation laws can be used to obtain the various versions of the gradient, divergence, curl, and Laplacian given by (4.6)–(4.9) in cylindrical and spherical coordinates from those in Cartesian coordinates. From the transformation laws for the first derivatives, we obtain the second derivative transformation laws in Cartesian coordinates as follows:

$$\begin{aligned}\frac{\partial^2}{\partial x^2} &= \frac{\partial^2 \xi}{\partial x^2}\frac{\partial}{\partial \xi} + \frac{\partial^2 \eta}{\partial x^2}\frac{\partial}{\partial \eta} + \left(\frac{\partial \xi}{\partial x}\right)^2 \frac{\partial^2}{\partial \xi^2} + \left(\frac{\partial \eta}{\partial x}\right)^2 \frac{\partial^2}{\partial \eta^2} + 2\frac{\partial \eta}{\partial x}\frac{\partial \xi}{\partial x}\frac{\partial^2}{\partial \eta \partial \xi} \\ &= \frac{1}{\mathcal{J}}\left[\frac{\partial y}{\partial \eta}\frac{\partial}{\partial \xi}\left(\frac{1}{\mathcal{J}}\frac{\partial y}{\partial \eta}\frac{\partial}{\partial \xi}\right) - \frac{\partial y}{\partial \eta}\frac{\partial}{\partial \xi}\left(\frac{1}{\mathcal{J}}\frac{\partial y}{\partial \xi}\frac{\partial}{\partial \eta}\right)\right. \\ &\quad \left. - \frac{\partial y}{\partial \xi}\frac{\partial}{\partial \eta}\left(\frac{1}{\mathcal{J}}\frac{\partial y}{\partial \eta}\frac{\partial}{\partial \xi}\right) + \frac{\partial y}{\partial \xi}\frac{\partial}{\partial \eta}\left(\frac{1}{\mathcal{J}}\frac{\partial y}{\partial \xi}\frac{\partial}{\partial \eta}\right)\right],\end{aligned} \tag{4.19}$$

and

$$\begin{aligned}\frac{\partial^2}{\partial y^2} &= \frac{\partial^2 \xi}{\partial y^2}\frac{\partial}{\partial \xi} + \frac{\partial^2 \eta}{\partial y^2}\frac{\partial}{\partial \eta} + \left(\frac{\partial \xi}{\partial y}\right)^2 \frac{\partial^2}{\partial \xi^2} + \left(\frac{\partial \eta}{\partial y}\right)^2 \frac{\partial^2}{\partial \eta^2} + 2\frac{\partial \eta}{\partial y}\frac{\partial \xi}{\partial y}\frac{\partial^2}{\partial \eta \partial \xi} \\ &= \frac{1}{\mathcal{J}}\left[\frac{\partial x}{\partial \eta}\frac{\partial}{\partial \xi}\left(\frac{1}{\mathcal{J}}\frac{\partial x}{\partial \eta}\frac{\partial}{\partial \xi}\right) - \frac{\partial x}{\partial \eta}\frac{\partial}{\partial \xi}\left(\frac{1}{\mathcal{J}}\frac{\partial x}{\partial \xi}\frac{\partial}{\partial \eta}\right)\right. \\ &\quad \left. - \frac{\partial x}{\partial \xi}\frac{\partial}{\partial \eta}\left(\frac{1}{\mathcal{J}}\frac{\partial x}{\partial \eta}\frac{\partial}{\partial \xi}\right) + \frac{\partial x}{\partial \xi}\frac{\partial}{\partial \eta}\left(\frac{1}{\mathcal{J}}\frac{\partial x}{\partial \xi}\frac{\partial}{\partial \eta}\right)\right].\end{aligned} \tag{4.20}$$

As you can see, such transformations can get rather messy, and one needs to take great care in being sure that they are carried out correctly.

Similar to the Hessian matrix defined in the next section, which contains the second derivatives of $f(x_1, x_2, \ldots, x_N)$, the *Jacobian matrix* is a matrix of the first derivatives of $f_1(x_1, x_2, \ldots, x_N)$, $f_2(x_1, x_2, \ldots, x_N)$, $\ldots$, $f_N(x_1, x_2, \ldots, x_N)$ according to

$$\mathcal{J} = \begin{bmatrix} \frac{\partial f_1}{\partial x_1} & \frac{\partial f_1}{\partial x_2} & \cdots & \frac{\partial f_1}{\partial x_N} \\ \frac{\partial f_2}{\partial x_1} & \frac{\partial f_2}{\partial x_2} & \cdots & \frac{\partial f_2}{\partial x_N} \\ \vdots & \vdots & \ddots & \vdots \\ \frac{\partial f_N}{\partial x_1} & \frac{\partial f_N}{\partial x_2} & \cdots & \frac{\partial f_N}{\partial x_N} \end{bmatrix}.$$

In order to further exercise your skills in vector calculus and marinate in some additional applications, the reader is encouraged to work through chapters 8 and 9 in Cassel (2013) on electromagnetics and fluid mechanics, respectively.

4.1.5 Derivatives in Higher Dimensions

Our focus thus far has been on derivative and integral operations in two- or three-dimensional coordinate systems corresponding to physical problems in continuum mechanics, electromagnetics, and physics. As you will see shortly, there will be occasion to extend these derivative and gradient operators to vectors of arbitrary dimension. Suppose we have an N-dimensional vector

$$\mathbf{x} = \begin{bmatrix} x_1 & x_2 & \cdots & x_n & \cdots & x_N \end{bmatrix}^T.$$

How can we differentiate a scalar function $\phi(x_1, x_2, \ldots, x_N)$ with respect to this vector? It turns out to be precisely what you would expect, namely

$$\frac{d\phi}{d\mathbf{x}} = \nabla\phi = \begin{bmatrix} \frac{\partial\phi}{\partial x_1} & \frac{\partial\phi}{\partial x_2} & \cdots & \frac{\partial\phi}{\partial x_n} & \cdots & \frac{\partial\phi}{\partial x_N} \end{bmatrix}^T,$$

which is the gradient of the scalar function ϕ.

Let us consider two examples to illustrate how differentiation with respect to a vector works. First, we differentiate the inner product of two N-dimensional vectors $\mathbf{x}$ and $\mathbf{y}$. Recalling that the inner product of two vectors is a scalar, so here $\phi = \langle \mathbf{x}, \mathbf{y} \rangle$, then

$$\frac{d}{d\mathbf{x}} \langle \mathbf{x}, \mathbf{y} \rangle = \frac{d}{d\mathbf{x}} (x_1 y_1 + x_2 y_2 + \cdots + x_N y_N) = \begin{bmatrix} y_1 & y_2 & \cdots & y_N \end{bmatrix}^T = \mathbf{y},$$

which is what you might suspect intuitively. Next, consider the special case when $\mathbf{y} = \mathbf{x}$. In this case, $\phi = \langle \mathbf{x}, \mathbf{x} \rangle = \|\mathbf{x}\|^2$, and

$$\frac{d}{d\mathbf{x}} \langle \mathbf{x}, \mathbf{x} \rangle = \frac{d}{d\mathbf{x}} \left(x_1^2 + x_2^2 + \cdots + x_N^2 \right) = \begin{bmatrix} 2x_1 & 2x_2 & \cdots & 2x_N \end{bmatrix}^T = 2\mathbf{x},$$

which is again what you might suspect intuitively.

4.2 Tensors

4.2.1 Scalars, Vectors, and Tensors

The primary applications in which vector calculus plays a central role are in the physical sciences, such as mechanics and electromagnetics. As touched on periodically throughout the book thus far, mechanics draws on a special case of vector and matrix algebra in which the vectors are three-dimensional at most, corresponding to the three spatial coordinates, and represent specific constructs in mechanics. These include vectors representing position, velocity, acceleration, force, and moment, for example.

Recall that a *scalar* is a quantity that requires only one component, that is, magnitude, to specify its state. Physical quantities, such as pressure, temperature, and density, are scalars; note that they are independent of the choice of coordinate system. A three-dimensional *vector* is a quantity that requires three components to specify its state, that is, it requires both a magnitude and direction in three-dimensional space. Physical quantities, such as position, force, and velocity, are given by vectors in three-dimensional space. A *tensor* in three dimensions is a quantity that requires nine components to specify its state. Physical quantities, such as stress, strain, and strain rate, are defined by tensors.

Although subtle, it is important to understand how vectors and tensors are used differently in the physical sciences than in mathematics more generally. Mathematically, a vector or matrix is simply a collection of scalar values; for example, a three-dimensional vector could consist of your age, height, and weight. In mechanics, however, we utilize vectors and tensors to represent a single physical quantity. The three components of a three-dimensional position vector, for example, are the distances along the three directions of a coordinate system; the three components of a velocity vector are the velocities in each of the coordinate directions, and so forth. Therefore, there is a prescribed geometric and physical relationship between each of the components; they are not simply a collection of three unrelated scalar values. Similarly, mathematically, a 3×3 matrix is an ordered collection of nine quantities that may or may not be related to one another, but a tensor, while also being a 3×3 matrix, implies that the nine quantities are related in a particular manner.

In order to clearly distinguish between scalars, vectors, and matrices as used in mathematics and mechanics, they are referred to by their *rank* – or *order* – in mechanics based on the number of basis vectors required to specify them. Scalars have only a magnitude, but no direction, so they are *rank-zero tensors*. Vectors have magnitude and a single direction; therefore, they are *rank-one tensors*, and a tensor as described earlier requires two directions to define; therefore, it is a *rank-two tensor*. A rank-two tensor specifies a magnitude and two directions – one specifying the normal direction of the face on which the component of the tensor acts and one specifying that component's direction. Normally, we can easily tell from the context whether a quantity is truly a rank-zero, -one, or -two tensor, and it is not necessary to distinguish them from standard mathematical scalars, vectors, and matrices. In particular, it is rarely necessary to distinguish a rank-one tensor from a run-of-the-mill vector, but it is quite common in mechanics to refer to rank-two tensors simply as *tensors* in order to imply their additional properties.

Let us be more specific. As we saw in Section 1.6.2, vectors in mechanics have additional properties and restrictions then their purely mathematical counterparts. Mathematically, for example, a vector can be moved about within a coordinate system as long as it maintains its same magnitude and direction. In mechanics, however, where the vector may represent a force, it matters where that force acts. Consequently, it cannot be simply moved about the coordinate system; it must maintain its line of action – for nondeformable bodies – or point of action – for deformable bodies. Because our force vector represents a physical quantity that is independent of the frame of reference within which it is expressed, that is, it appears the same to all observers, it must maintain its magnitude and absolute direction when expressed in terms of an alternative coordinate system. The values of each component would be different when the force vector is expressed relative to the alternative coordinate system, but the vector maintains its absolute direction; a one Newton force acting east remains a one Newton force acting east regardless of whether the coordinate system is such that the x-direction coincides with the north, east, or any other direction.[2]

Similarly, tensors are a special case of a matrix. They inherit all the mathematical properties of matrices, but with additional requirements owing to their physical

[2] It is important to note that a position vector is by definition measured relative to the origin of its coordinate system and has no absolute definition; therefore, position vectors are not rank-one tensors.

meaning. As with force vectors, the physical state of stress at a point remains the same in an absolute sense as the coordinate system is altered, while the specific values of each of the components of stress change accordingly relative to the coordinate system used. In this way, every (second-order) tensor is a 3×3 matrix, but every such matrix is not a tensor. Summarizing, then, a tensor is a quantity that is invariant to a coordinate transformation; it does not depend on the vantage point of the observer.

The stress tensor in solid mechanics has been derived from static equilibrium considerations in Section 2.1.2 and is defined by

$$\tau = \begin{bmatrix} \tau_{xx} & \tau_{xy} & \tau_{xz} \\ \tau_{yx} & \tau_{yy} & \tau_{yz} \\ \tau_{zx} & \tau_{zy} & \tau_{zz} \end{bmatrix}.$$

Recall that the first subscript indicates the face on which the stress acts – as defined by its outward facing normal – and the second subscript indicates the direction that the quantity acts. Thus, the τ_{ii} components on the diagonal are the normal stresses that act perpendicular to the faces, and the τ_{ij} $(i \neq j)$ components are the shear stresses that act tangent to the faces of an infinitesimally small tetrahedral element of the substance as illustrated in Figure 4.6. In general, such tensors have nine components; however, static equilibrium of the stress tensor in solid mechanics requires that it be symmetric, with $\tau_{ij} = \tau_{ji}$ in order for the net moment on the infinitesimal element to be zero, in which case there are only six unique components. Consequently, shear stresses must occur in pairs with the magnitude of shear stresses on perpendicular faces being equal.

In Chapter 2, we highlighted the fact that any stress state can be geometrically transformed using a rotation such that only normal stresses – and no shear stresses – act on the infinitesimally small element; these are called the *principal stresses*. In such a case, the stress tensor becomes a diagonal matrix. Because the stress tensor is symmetric, it is always possible to diagonalize it in order to identify the orientation of the element that produces such a state of stress via the eigenvalues and eigenvectors of the stress tensor.

As with stress, the state of strain of an infinitesimally small element of a solid requires nine quantities forming the *strain tensor*. The quantities along the diagonal of the strain tensor are *linear strains* that act to elongate the element in the coordinate directions, while the off-diagonal strains are *shear strains* that act to change the orientation of the surfaces of the element such that they are no longer perpendicular. The strain tensor is

$$\epsilon = \begin{bmatrix} \epsilon_{xx} & \epsilon_{xy} & \epsilon_{xz} \\ \epsilon_{yx} & \epsilon_{yy} & \epsilon_{yz} \\ \epsilon_{zx} & \epsilon_{zy} & \epsilon_{zz} \end{bmatrix},$$

and, like the stress tensor, is symmetric such that $\epsilon_{ij} = \epsilon_{ji}$. As with stresses, the state of strain at a point can be diagonalized using the eigenvalues and eigenvectors of the strain tensor so that only linear strains act on the faces.

While many tensors are symmetric based on physical reasoning, not all are. However, any tensor can be expressed as the sum of a symmetric and a skew-symmetric

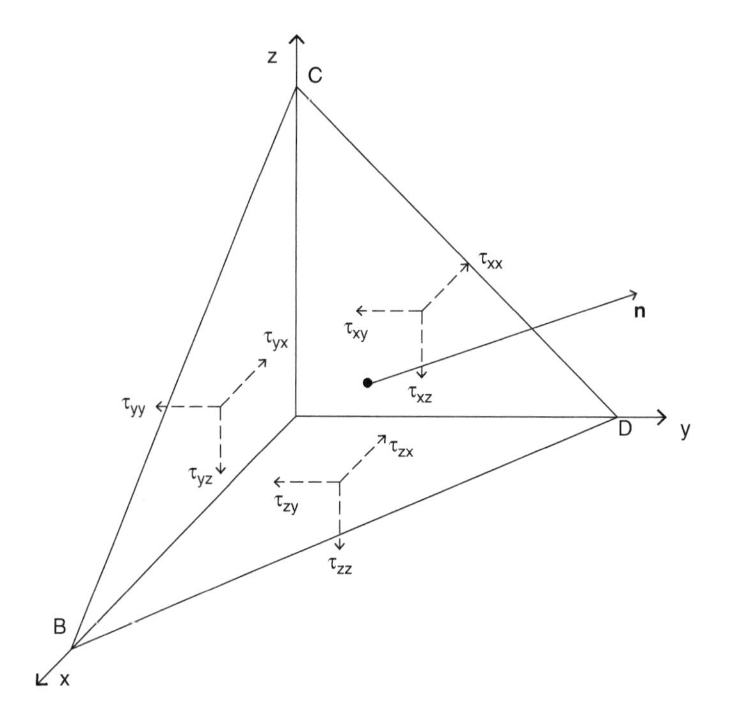

Figure 4.6 Tetrahedral element of a solid or fluid with the stresses on each of the three orthogonal faces to counterbalance the force on the inclined face.

matrix. Not only does such a decomposition allow one to take advantage of their properties in any further analysis, in many cases, each of these matrices has physical meaning. Recall that a symmetric matrix is such that

$$\mathbf{T} = \mathbf{T}^T,$$

while for a skew-symmetric matrix

$$\mathbf{T} = -\mathbf{T}^T.$$

Note that the diagonal elements of a skew-symmetric matrix must be zero in order to satisfy $T_{ij} = -T_{ji}$. Therefore, a symmetric tensor has six independent components, while a skew-symmetric tensor has three. Every square matrix can be expressed as the sum of two matrices according to

$$\mathbf{T} = \frac{1}{2}\left(\mathbf{T} - \mathbf{T}^T\right) + \frac{1}{2}\left(\mathbf{T} + \mathbf{T}^T\right),$$

where the first term results in a symmetric matrix, while the second term results in a skew-symmetric matrix. For example, if tensor $\mathbf{T}$ is the velocity-gradient tensor of a moving fluid, then the symmetric part is called the strain-rate tensor, and the skew-symmetric part is the vorticity tensor, which quantifies the rotation of points within a fluid flow. Because a skew-symmetric matrix only has three unique magnitudes,

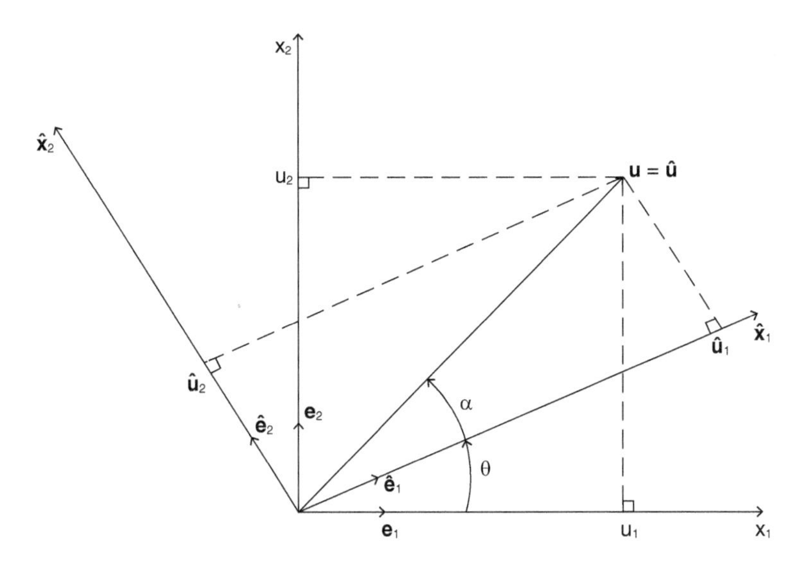

Figure 4.7 Change of coordinate system used to represent a two-dimensional fixed vector.

its components can be expressed as a vector, called the *vorticity*, which gives the magnitude and direction of the rotation of each fluid particle.

In order to illustrate the special properties of tensors under a change of coordinate system, let us consider a fixed vector (rank-one tensor) $\mathbf{u}$ relative to a two-dimensional coordinate system (x_1, x_2), which is rotated counterclockwise an angle θ to an alternative coordinate system $(\hat{x}_1, \hat{x}_2)$ as shown in Figure 4.7. In terms of the (x_1, x_2) coordinate system, the vector is given by

$$\mathbf{u} = u_1\mathbf{e}_1 + u_2\mathbf{e}_2, \tag{4.21}$$

and in terms of the $(\hat{x}_1, \hat{x}_2)$ coordinate system, it is written as

$$\hat{\mathbf{u}} = \hat{u}_1\hat{\mathbf{e}}_1 + \hat{u}_2\hat{\mathbf{e}}_2. \tag{4.22}$$

But $\mathbf{u} = \hat{\mathbf{u}}$ is the same vector in an absolute sense; therefore, (4.21) and (4.22) require that

$$\hat{u}_1\hat{\mathbf{e}}_1 + \hat{u}_2\hat{\mathbf{e}}_2 = u_1\mathbf{e}_1 + u_2\mathbf{e}_2. \tag{4.23}$$

Because the unit vectors in the two coordinate directions are orthogonal, $\langle \hat{\mathbf{e}}_1, \hat{\mathbf{e}}_2 \rangle = 0$, and taking the inner product of $\hat{\mathbf{e}}_1$ with (4.23) yields

$$\hat{u}_1 = u_1 \langle \hat{\mathbf{e}}_1, \mathbf{e}_1 \rangle + u_2 \langle \hat{\mathbf{e}}_1, \mathbf{e}_2 \rangle. \tag{4.24}$$

Similarly, taking the inner product of $\hat{\mathbf{e}}_2$ with (4.23) yields

$$\hat{u}_2 = u_1 \langle \hat{\mathbf{e}}_2, \mathbf{e}_1 \rangle + u_2 \langle \hat{\mathbf{e}}_2, \mathbf{e}_2 \rangle. \tag{4.25}$$

Equations (4.24) and (4.25) are a system of two equations as follows

$$\begin{bmatrix} \hat{u}_1 \\ \hat{u}_2 \end{bmatrix} = \begin{bmatrix} Q_{11} & Q_{12} \\ Q_{21} & Q_{22} \end{bmatrix} \begin{bmatrix} u_1 \\ u_2 \end{bmatrix}, \tag{4.26}$$

where from the geometry of the transformation

$$\begin{aligned} Q_{11} &= \langle \hat{\mathbf{e}}_1, \mathbf{e}_1 \rangle = \cos\theta, \\ Q_{12} &= \langle \hat{\mathbf{e}}_1, \mathbf{e}_2 \rangle = \sin\theta, \\ Q_{21} &= \langle \hat{\mathbf{e}}_2, \mathbf{e}_1 \rangle = -\sin\theta, \\ Q_{22} &= \langle \hat{\mathbf{e}}_2, \mathbf{e}_2 \rangle = \cos\theta. \end{aligned}$$

Consequently, the components of the vector $\mathbf{u} = \hat{\mathbf{u}}$ in the two coordinate systems are related by

$$\begin{bmatrix} \hat{u}_1 \\ \hat{u}_2 \end{bmatrix} = \mathbf{Q} \begin{bmatrix} u_1 \\ u_2 \end{bmatrix}, \tag{4.27}$$

where

$$\mathbf{Q} = \begin{bmatrix} \cos\theta & \sin\theta \\ -\sin\theta & \cos\theta \end{bmatrix}.$$

Noting the similarity between $\mathbf{Q}$ and the rotation matrix in Example 1.13, we see that $\mathbf{Q}$ is a rotation matrix.[3]

REMARKS:

1. *The gradient of a vector is a tensor, and the divergence of a tensor is a vector.*
2. *The integral theorems previously expressed for vectors carry over to second-order tensors as well.*
3. *Just as we have the scalar – or dot or inner – product and the cross product of vectors, there is a* tensor product *of vectors as well. It results in a rank-two tensor and is called a* dyad*; it corresponds to the outer product of two three-dimensional vectors. Recall from Section 1.6.3 that the outer product of vectors* $\mathbf{u}$ *and* $\mathbf{v}$ *is given by* $\mathbf{u}\mathbf{v}^T$*, which produces a* 3×3 *symmetric matrix.*

4.2.2 Continuum Mechanics

Continuum mechanics is the bridge between the mathematics of vector and matrix methods – linear algebra – and the physics of continuous substances, such as solids, fluids, and heat transfer. Continuum mechanics largely consists of developing the mathematical relationships between the stress and strain – or stress and strain-rate – tensors via *constitutive laws* for various types of materials and substances. These constitutive relations formalize how various materials and substances respond to applied internal and external loads. For example, in Hooke's law for elastic solids, stress is

[3] The difference between the two is because here we are rotating the coordinate system, while in Example 1.13, the vector is being rotated.

directly proportional to strain, and in Newtonian fluids, stress is directly proportional to strain rate.

Proficiency with the algebra and calculus of vectors and matrices (tensors) is necessary for the study of continuum mechanics. Continuum mechanics involves a beautiful interplay where the mathematics of linear algebra greatly informs our physical understanding of the behavior of continuous substances and unifies solid mechanics, fluid mechanics, heat transfer, electromagnetics, and general relativity. For much more on tensors and their relationship to mechanics, see books on continuum mechanics, such as Reddy (2013) or Nair (2009).

4.3 Extrema of Functions and Optimization Preview

One of the common uses of calculus is in determining minimums, maximums, or stationary points of functions. Essential in its own right, this also provides the basis for optimization of systems governed by algebraic equations, which will be discussed in Chapter 12. We must address cases with and without constraints; we consider unconstrained functions of one, two, then N independent variable(s) first.

4.3.1 General Considerations

First, let us consider a one-dimensional function $f(x)$ as shown in Figure 4.8. The point x_0 is a *stationary*, or *critical*, *point* of $f(x)$ if

$$\frac{df}{dx} = 0 \quad \text{at} \quad x = x_0,$$

or equivalently

$$df = \frac{df}{dx}dx = 0 \quad \text{at} \quad x = x_0,$$

where df is the *total differential* of $f(x)$. That is, the slope of $f(x)$ is zero at a stationary point $x = x_0$. The following possibilities exist at a stationary point:

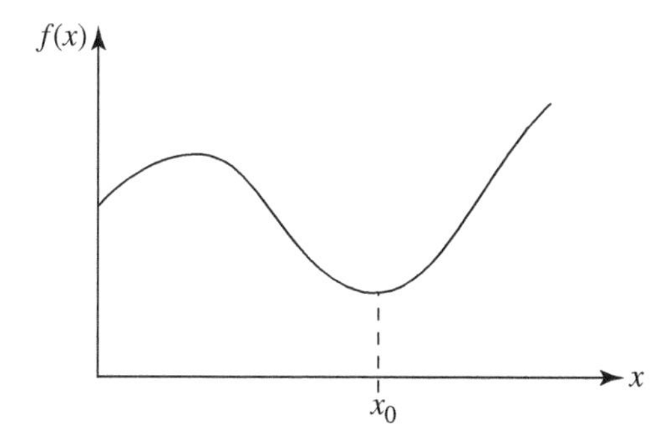

Figure 4.8 One-dimensional function $f(x)$ with a local minimum at $x = x_0$.

1. If $\dfrac{d^2 f}{dx^2} < 0$ at x_0, the function $f(x)$ has a *local maximum* at $x = x_0$.
2. If $\dfrac{d^2 f}{dx^2} > 0$ at x_0, the function $f(x)$ has a *local minimum* at $x = x_0$.
3. If $\dfrac{d^2 f}{dx^2} = 0$, then $f(x)$ may still have a local minimum (for example, $f(x) = x^4$ at $x = 0$) or a local maximum (for example, $f(x) = -x^4$ at $x = 0$), or it may have neither (for example, $f(x) = x^3$ at $x = 0$).

This is known as the *second-derivative test*. The important point here is that the requirement that $f'(x_0) = 0$, in which case x_0 is a stationary point, provides a *necessary* condition for a local extremum,[4] while possibilities (1) and (2) provide additional *sufficient* conditions for a local extremum at x_0.

Now consider the two-dimensional function $f(x, y)$. For an extremum to occur at (x_0, y_0), it is necessary – but not sufficient – that

$$\frac{\partial f}{\partial x} = \frac{\partial f}{\partial y} = 0 \quad \text{at} \quad x = x_0,\ y = y_0,$$

or equivalently

$$df = \frac{\partial f}{\partial x} dx + \frac{\partial f}{\partial y} dy = 0 \quad \text{at} \quad x = x_0, y = y_0,$$

where df is the *total differential* of $f(x, y)$. The point (x_0, y_0) is a *stationary point* of $f(x, y)$ if $df = 0$ at (x_0, y_0), for which the rate of change of $f(x, y)$ at (x_0, y_0) in all directions is zero.

It is convenient to express the second-derivative test for stationary points of multivariable functions in terms of the *Hessian* defined by the following determinant:

$$|\mathbf{H}| = \begin{vmatrix} \dfrac{\partial^2 f}{\partial x^2} & \dfrac{\partial^2 f}{\partial x \partial y} \\ \dfrac{\partial^2 f}{\partial y \partial x} & \dfrac{\partial^2 f}{\partial y^2} \end{vmatrix} = \frac{\partial^2 f}{\partial x^2} \frac{\partial^2 f}{\partial y^2} - \left(\frac{\partial^2 f}{\partial x \partial y} \right)^2.$$

The Hessian is the determinant of the *Hessian matrix* $\mathbf{H}$, which includes all possible second-order derivatives of the function.

At a stationary point (x_0, y_0), the following possibilities exist:

1. If $|\mathbf{H}| > 0$ and $\dfrac{\partial^2 f}{\partial x^2} < 0$ at (x_0, y_0), then (x_0, y_0) is a *local maximum*.
2. If $|\mathbf{H}| > 0$ and $\dfrac{\partial^2 f}{\partial x^2} > 0$ at (x_0, y_0), then (x_0, y_0) is a *local minimum*.
3. If $|\mathbf{H}| < 0$ at (x_0, y_0), then (x_0, y_0) is a *saddle point*.
4. If $|\mathbf{H}| = 0$ at (x_0, y_0), then (1), (2), or (3) are all possible.

Possibilities (1) and (2) provide sufficient conditions, along with the necessary condition that $df = 0$ at (x_0, y_0), for the existence of an extremum. See Hildebrand (1976) for a derivation of these conditions.

[4] An extremum is a minimum or maximum of a function.

The preceding criteria can be generalized to N-dimensions, in which case we have a *stationary point* of $f(x_1, x_2, \ldots, x_n)$ if

$$df = \frac{\partial f}{\partial x_1}dx_1 + \frac{\partial f}{\partial x_2}dx_2 + \cdots + \frac{\partial f}{\partial x_N}dx_N = 0.$$

The Hessian matrix is the gradient of the gradient of the scalar function f and is now $N \times N$ and given by[5]

$$\mathbf{H} = \boldsymbol{\nabla}(\boldsymbol{\nabla} f) = \begin{bmatrix} \frac{\partial^2 f}{\partial x_1^2} & \frac{\partial^2 f}{\partial x_1 \partial x_2} & \cdots & \frac{\partial^2 f}{\partial x_1 \partial x_N} \\ \frac{\partial^2 f}{\partial x_2 \partial x_1} & \frac{\partial^2 f}{\partial x_2^2} & \cdots & \frac{\partial^2 f}{\partial x_2 \partial x_N} \\ \vdots & \vdots & \ddots & \vdots \\ \frac{\partial^2 f}{\partial x_N \partial x_1} & \frac{\partial^2 f}{\partial x_N \partial x_2} & \cdots & \frac{\partial^2 f}{\partial x_N^2} \end{bmatrix}.$$

If all of the derivatives of f are continuous, then the Hessian matrix is symmetric. Using this Hessian matrix, the *second-derivative test* is used to determine what type of extrema exists at each stationary point as follows:

1. If the Hessian matrix is negative definite (all eigenvalues are negative), then the function $f(x_1, \ldots, x_N)$ has a local maximum.
2. If the Hessian matrix is positive definite (all eigenvalues are positive), then the function $f(x_1, \ldots, x_N)$ has a local minimum.
3. If the Hessian matrix has both positive and negative eigenvalues, then the function $f(x_1, \ldots, x_N)$ has a saddle point.
4. If the Hessian matrix is semidefinite (at least one eigenvalue is zero), then the second-derivative test is inconclusive.

Note that the cases with $f(x)$ and $f(x, y)$ considered earlier are special cases of this more general result.

Example 4.2 Obtain the location (x_1, x_2, x_3) at which a minimum value of the function

$$f(x_1, x_2, x_3) = \frac{f_1(x_1, x_2, x_3)}{f_2(x_1, x_2, x_3)} = \frac{x_1^2 + 6x_1x_2 + 4x_2^2 + x_3^2}{x_1^2 + x_2^2 + x_3^2}$$

occurs.

Solution

At a point (x_1, x_2, x_3) where f has zero slope, we have

$$\frac{\partial f}{\partial x_i} = 0, \quad i = 1, 2, 3.$$

[5] Some authors denote the Hessian by $\boldsymbol{\nabla}^2 f$.

With $f = f_1/f_2$, this requires that (using the quotient rule)

$$\frac{1}{f_2}\left(\frac{\partial f_1}{\partial x_i} - \frac{f_1}{f_2}\frac{\partial f_2}{\partial x_i}\right) = 0, \quad i = 1,2,3,$$

or letting $\lambda = f = f_1/f_2$ (and canceling $1/f_2$)

$$\frac{\partial f_1}{\partial x_i} - \lambda\frac{\partial f_2}{\partial x_i} = 0, \quad i = 1,2,3. \tag{4.28}$$

Recall that

$$f_1 = x_1^2 + 6x_1x_2 + 4x_2^2 + x_3^2, \quad f_2 = x_1^2 + x_2^2 + x_3^2.$$

Therefore, the partial derivatives in (4.28) are given by

$$\begin{aligned}
\frac{\partial f_1}{\partial x_1} &= 2x_1 + 6x_2, & \frac{\partial f_2}{\partial x_1} &= 2x_1,\\
\frac{\partial f_1}{\partial x_2} &= 6x_1 + 8x_2, & \frac{\partial f_2}{\partial x_2} &= 2x_2,\\
\frac{\partial f_1}{\partial x_3} &= 2x_3, & \frac{\partial f_2}{\partial x_3} &= 2x_3.
\end{aligned}$$

Substituting into (4.28) gives the following three equations for the variables at a stationary point

$$\begin{aligned}
2x_1 + 6x_2 - \lambda 2x_1 &= 0,\\
6x_1 + 8x_2 - \lambda 2x_2 &= 0,\\
2x_3 - \lambda 2x_3 &= 0,
\end{aligned}$$

or canceling the common factor 2 from each of the equations and rearranging gives the eigenproblem

$$\begin{bmatrix} 1 & 3 & 0 \\ 3 & 4 & 0 \\ 0 & 0 & 1 \end{bmatrix}\begin{bmatrix} x_1 \\ x_2 \\ x_3 \end{bmatrix} = \lambda\begin{bmatrix} x_1 \\ x_2 \\ x_3 \end{bmatrix}.$$

For a nontrivial solution, the required eigenvalues are

$$\lambda_1 = 1, \quad \lambda_2 = \frac{1}{2}(5 + 3\sqrt{5}), \quad \lambda_3 = \frac{1}{2}(5 - 3\sqrt{5}),$$

and the corresponding eigenvectors are

$$\mathbf{u}_1 = \begin{bmatrix} 0 \\ 0 \\ 1 \end{bmatrix}, \quad \mathbf{u}_2 = \begin{bmatrix} -\frac{4}{3} + \frac{1}{6}(5 + 3\sqrt{5}) \\ 1 \\ 0 \end{bmatrix}, \quad \mathbf{u}_3 = \begin{bmatrix} -\frac{4}{3} + \frac{1}{6}(5 - 3\sqrt{5}) \\ 1 \\ 0 \end{bmatrix}.$$

Of the three eigenvalues, $f = \lambda_3 = \frac{1}{2}(5 - 3\sqrt{5})$ has the minimum value, which occurs at

$$\begin{bmatrix} x_1 \\ x_2 \\ x_3 \end{bmatrix} = \mathbf{u}_3 = \begin{bmatrix} -\frac{4}{3} + \frac{1}{6}(5 - 3\sqrt{5}) \\ 1 \\ 0 \end{bmatrix}.$$

Alternatively, the nature of each stationary point could be determined using the Hessian matrix and the second-derivative test.

Interestingly, we were able to convert this nonlinear optimization problem into an eigenproblem, which is a linear system of algebraic equations, where the eigenvalues are possible values of the objective function and the eigenvectors are the corresponding values of the design variables at stationary points, thereby identifying possible extremum points. See Section 4.3.3 for another, more general, situation for which a reduction from a nonlinear optimization problem to an eigenproblem occurs.

4.3.2 Constrained Extrema and Lagrange Multipliers

It is often necessary to determine an extremum of a function subject to some constraint(s). As in the previous section, let us first consider the case with a single function before extending to systems of equations.

Recall from the previous section that a necessary condition for an extremum of a function $f(x, y, z)$ at (x_0, y_0, z_0) without constraints is

$$df = \frac{\partial f}{\partial x}dx + \frac{\partial f}{\partial y}dy + \frac{\partial f}{\partial z}dz = 0. \tag{4.29}$$

Because dx, dy, and dz are arbitrary (x, y, and z are independent variables), this requires that

$$\frac{\partial f}{\partial x} = 0, \quad \frac{\partial f}{\partial y} = 0, \quad \frac{\partial f}{\partial z} = 0,$$

which is solved to find x_0, y_0, and z_0.

Now let us determine the stationary point(s) of a function $f(x, y, z)$ subject to an algebraic constraint, say

$$g(x, y, z) = c,$$

where c is a specified constant. The constraint provides an algebraic relationship between the independent variables x, y, and z that must be satisfied. In addition to (4.29), we have that the total differential of $g(x, y, z)$ is

$$dg = \frac{\partial g}{\partial x}dx + \frac{\partial g}{\partial y}dy + \frac{\partial g}{\partial z}dz = 0, \tag{4.30}$$

which is zero because g is equal to a constant. Because both (4.29) and (4.30) equal zero at a stationary point (x_0, y_0, z_0), it follows that we can add them according to

$$df + \Lambda dg = \left(\frac{\partial f}{\partial x} + \Lambda \frac{\partial g}{\partial x}\right) dx + \left(\frac{\partial f}{\partial y} + \Lambda \frac{\partial g}{\partial y}\right) dy + \left(\frac{\partial f}{\partial z} + \Lambda \frac{\partial g}{\partial z}\right) dz = 0, \tag{4.31}$$

where Λ is an arbitrary constant, which we call the *Lagrange multiplier.*[6]

Note that because of the constraint $g = c$, the variables x, y, and z are no longer independent. With one constraint, for example, we can only have two of the three variables being independent. As a result, we cannot regard dx, dy, and dz as all being arbitrary and set the three coefficients equal to zero as in the unconstrained case. Instead, suppose that $\partial g/\partial z \neq 0$ at (x_0, y_0, z_0). Then the last term in (4.31) may be eliminated by specifying the arbitrary Lagrange multiplier to be $\Lambda = -(\partial f/\partial z)/(\partial g/\partial z)$, giving

$$\left(\frac{\partial f}{\partial x} + \Lambda \frac{\partial g}{\partial x}\right) dx + \left(\frac{\partial f}{\partial y} + \Lambda \frac{\partial g}{\partial y}\right) dy = 0.$$

The remaining variables, x and y, may now be regarded as independent, and the coefficients of dx and dy must each vanish. This results in four equations for the four unknowns x_0, y_0, z_0, and Λ as follows:

$$\frac{\partial f}{\partial x} + \Lambda \frac{\partial g}{\partial x} = 0, \quad \frac{\partial f}{\partial y} + \Lambda \frac{\partial g}{\partial y} = 0, \quad \frac{\partial f}{\partial z} + \Lambda \frac{\partial g}{\partial z} = 0, \quad g = c.$$

Consequently, determining the stationary point of the function $f(x, y, z)$ subject to the constraint $g(x, y, z) = c$ is equivalent to determining the stationary point of the *augmented function*

$$\tilde{f}(x, y, z) = f(x, y, z) + \Lambda \left[g(x, y, z) - c\right]$$

subject to no constraint. Thus, we seek the point (x_0, y_0, z_0), where

$$\frac{\partial \tilde{f}}{\partial x} = \frac{\partial \tilde{f}}{\partial y} = \frac{\partial \tilde{f}}{\partial z} = 0.$$

These provide the necessary conditions for an extremum of $f(x, y, z)$ subject to the constraint that $g(x, y, z) = c$ is satisfied.

REMARKS:

1. *Observe that because c is a constant, we may write the augmented function as $\tilde{f} = f + \Lambda(g - c)$, as before, or as $\tilde{f} = f + \Lambda g$.*
2. *Additional constraints may be imposed in the same manner, with each constraint having its own Lagrange multiplier.*

Example 4.3 Find the semimajor and semiminor axes of the ellipse defined by

$$(x_1 + x_2)^2 + 2(x_1 - x_2)^2 = 8,$$

[6] In some applications, Lagrange multipliers turn out to also be eigenvalues; therefore, many authors use λ to denote Lagrange multipliers. Throughout this text, however, we use λ to denote eigenvalues (as is common) and Λ for Lagrange multipliers in order to clearly distinquish between the two uses.

which may be written

$$3x_1^2 - 2x_1x_2 + 3x_2^2 = 8.$$

Solution

To determine the semimajor (minor) axis, calculate the farthest (nearest) point on the ellipse from the origin. Therefore, we maximize (minimize) $d = x_1^2 + x_2^2$, the square of the distance from the origin, subject to the constraint that the coordinates (x_1, x_2) be on the ellipse. To accomplish this, we define an augmented function as follows

$$\tilde{d} = d + \Lambda(3x_1^2 - 2x_1x_2 + 3x_2^2 - 8),$$

where Λ is the Lagrange multiplier and is multiplied by the constraint that the extrema be on the ellipse. To determine the extrema of the algebraic function $\tilde{d}$, we evaluate

$$\frac{\partial \tilde{d}}{\partial x_1} = 0, \quad \frac{\partial \tilde{d}}{\partial x_2} = 0,$$

with

$$\tilde{d} = x_1^2 + x_2^2 + \Lambda\left(3x_1^2 - 2x_1x_2 + 3x_2^2 - 8\right).$$

Evaluating the partial derivatives and setting equal to zero gives

$$\frac{\partial \tilde{d}}{\partial x_1} = 2x_1 + \Lambda\,(6x_1 - 2x_2) = 0,$$

$$\frac{\partial \tilde{d}}{\partial x_2} = 2x_2 + \Lambda\,(-2x_1 + 6x_2) = 0.$$

Thus, we have two equations for x_1 and x_2 given by

$$3x_1 - x_2 = \lambda_n x_1,$$
$$-x_1 + 3x_2 = \lambda_n x_2,$$

where $\lambda_n = -1/\Lambda$. This is an eigenproblem of the form $\mathbf{A}\mathbf{x}_n = \lambda_n \mathbf{x}_n$, where

$$\mathbf{A} = \begin{bmatrix} 3 & -1 \\ -1 & 3 \end{bmatrix}.$$

Observe that the matrix $\mathbf{A}$ is the same as that defining the quadratic as considered in Example 2.4. The eigenvalues of the symmetric matrix $\mathbf{A}$ are $\lambda_1 = 2$ and $\lambda_2 = 4$, with the corresponding eigenvectors being

$$\mathbf{x}_1 = \begin{bmatrix} 1 \\ 1 \end{bmatrix}, \quad \mathbf{x}_2 = \begin{bmatrix} -1 \\ 1 \end{bmatrix}.$$

The two eigenvectors $\mathbf{x}_1$ and $\mathbf{x}_2$ (along with $-\mathbf{x}_1$ and $-\mathbf{x}_2$) give the directions of the semimajor and semiminor axes, which are along lines that bisect the first and third quadrants and second and fourth quadrants. Note that because $\mathbf{A}$ is real and symmetric with distinct eigenvalues, the eigenvectors are mutually orthogonal as expected for a quadratic.

In order to determine which eigenvectors correspond to the semimajor and semiminor axes, we recognize that a point on the ellipse must satisfy

$$(x_1 + x_2)^2 + 2(x_1 - x_2)^2 = 8.$$

Considering $\mathbf{x}_1^T = \begin{bmatrix} 1 & 1 \end{bmatrix}$, let us set $x_1 = c_1$ and $x_2 = c_1$. Substituting into the equation for the ellipse yields

$$4c_1^2 + 0 = 8,$$

in which case $c_1 = \pm\sqrt{2}$. Therefore, $x_1 = \sqrt{2}$ and $x_2 = \sqrt{2}$ or ($x_1 = -\sqrt{2}$ and $x_2 = -\sqrt{2}$), and the length of the corresponding axis is $\sqrt{x_1^2 + x_2^2} = 2$. Similarly, considering $\mathbf{x}_2^T = \begin{bmatrix} -1 & 1 \end{bmatrix}$, let us set $x_1 = -c_2$ and $x_2 = c_2$. Substituting into the equation for the ellipse yields

$$0 + 8c_2^2 = 8,$$

in which case $c_2 = \pm 1$. Therefore, $x_1 = -1$ and $x_2 = 1$ (or $x_1 = 1$ and $x_2 = -1$), and the length of the corresponding axis is $\sqrt{x_1^2 + x_2^2} = \sqrt{2}$. As a result, the eigenvector $\mathbf{x}_1$ corresponds to the semimajor axis, and $\mathbf{x}_2$ corresponds to the semiminor axis.

In this example, both the objective function and constraint are quadratic functions. In such *quadratic programming* problems, the optimal solution for the augmented function always produces a "generalized" eigenproblem of the form

$$\mathbf{A}\mathbf{x}_n = \lambda_n \mathbf{B}\mathbf{x}_n$$

to be solved for the Lagrange multipliers, which are the eigenvalues, and the corresponding optimal design variables, which are the eigenvectors. In this example, $\mathbf{B} = \mathbf{I}$, which corresponds to a regular eigenproblem.

4.3.3 Quadratic Programming

Quadratic programming has been introduced in Example 4.3. We further consider this important class of optimization problems in a more general manner. The reader should remind themselves of the general results for quadratics in Section 2.3.3.

Suppose that the objective is to minimize the quadratic function[7]

$$\begin{aligned} J(x_1, x_2, \ldots, x_N) &= A_{11}x_1^2 + A_{22}x_2^2 + \cdots + A_{NN}x_N^2 \\ &+ 2A_{12}x_1x_2 + 2A_{13}x_1x_3 + \cdots + 2A_{N-1,N}x_{N-1}x_N, \end{aligned}$$

which is referred to as the *objective function*. The constraint is also in the form of a quadratic

$$\begin{aligned} &B_{11}x_1^2 + B_{22}x_2^2 + \cdots + B_{NN}x_N^2 + 2B_{12}x_1x_2 \\ &+ 2B_{13}x_1x_3 + \cdots + 2B_{N-1,N}x_{N-1}x_N = 1. \end{aligned}$$

[7] Note that a minimum can be changed to a maximum or vice versa simply by changing the sign of the objective function.

Note that both the objective function and the constraint consist of a single quadratic function. Alternatively, in matrix form we have the objective function

$$J(\mathbf{x}) = \mathbf{x}^T \mathbf{A} \mathbf{x} \tag{4.32}$$

subject to the quadratic constraint that

$$\mathbf{x}^T \mathbf{B} \mathbf{x} = 1. \tag{4.33}$$

Here, $\mathbf{A}$ and $\mathbf{B}$ are $N \times N$ symmetric matrices containing constants, and $\mathbf{x}$ is an $N \times 1$ vector containing the variables. As before, the constraint is incorporated using a Lagrange multiplier λ.[8] Thus, we now want to minimize the *augmented function* expressed in the matrix form

$$\tilde{J} = \mathbf{x}^T \mathbf{A} \mathbf{x} + \lambda \left(1 - \mathbf{x}^T \mathbf{B} \mathbf{x} \right). \tag{4.34}$$

The necessary conditions for an extremum of the augmented function are that its derivative with respect to each of the variables must be zero as follows:

$$\frac{\partial \tilde{J}}{\partial x_i} = 0, \quad i = 1, 2, \ldots, N. \tag{4.35}$$

In order to evaluate (4.35) for (4.34), recall that $\mathbf{A}$ is symmetric and

$$\mathbf{x}^T \mathbf{A} \mathbf{x} = \begin{bmatrix} x_1 & x_2 & \cdots & x_N \end{bmatrix} \begin{bmatrix} A_{11} & A_{12} & \cdots & A_{1N} \\ A_{12} & A_{22} & \cdots & A_{2N} \\ \vdots & \vdots & \ddots & \vdots \\ A_{1N} & A_{2N} & \cdots & A_{NN} \end{bmatrix} \begin{bmatrix} x_1 \\ x_2 \\ \vdots \\ x_N \end{bmatrix}$$

$$\begin{aligned} \mathbf{x}^T \mathbf{A} \mathbf{x} = {} & A_{11} x_1^2 + A_{22} x_2^2 + \cdots + A_{NN} x_N^2 \\ & + 2A_{12} x_1 x_2 + 2A_{13} x_1 x_3 + \cdots + 2A_{N-1,N} x_{N-1} x_N, \end{aligned}$$

where each term is quadratic in x_i. Differentiating with respect to x_i, $i = 1, 2, \ldots, N$ yields

$$\begin{aligned} \frac{\partial}{\partial x_1}\left(\mathbf{x}^T \mathbf{A} \mathbf{x}\right) &= 2A_{11} x_1 + 2A_{12} x_2 + 2A_{13} x_3 + \cdots + 2A_{1N} x_N, \\ \frac{\partial}{\partial x_2}\left(\mathbf{x}^T \mathbf{A} \mathbf{x}\right) &= 2A_{21} x_1 + 2A_{22} x_2 + 2A_{23} x_3 + \cdots + 2A_{2N} x_N, \\ &\vdots \\ \frac{\partial}{\partial x_i}\left(\mathbf{x}^T \mathbf{A} \mathbf{x}\right) &= 2A_{i1} x_1 + 2A_{i2} x_2 + \cdots + 2A_{ii} x_i + \cdots + 2A_{iN} x_N, \\ &\vdots \\ \frac{\partial}{\partial x_N}\left(\mathbf{x}^T \mathbf{A} \mathbf{x}\right) &= 2A_{N1} x_1 + 2A_{N2} x_2 + \cdots + 2A_{N,N-1} x_{N-1} + 2A_{NN} x_N, \end{aligned}$$

which is simply $2\mathbf{A}\mathbf{x}$. For the augmented objective function (4.34), therefore, (4.35) leads to

[8] We use λ, rather than the usual Λ, for the Lagrange multiplier here because it turns out to be the eigenvalues of an eigenproblem.

$$2(\mathbf{Ax} - \lambda \mathbf{Bx}) = 0.$$

Because of the constraint (4.33), we are seeking nontrivial solutions for $\mathbf{x}$. Such solutions correspond to solving the *generalized eigenproblem*

$$\mathbf{Ax}_n = \lambda_n \mathbf{Bx}_n, \tag{4.36}$$

where the eigenvalues λ_n are the values of the Lagrange multiplier, and the eigenvectors $\mathbf{x}_n$ are the candidate stationary points for which (4.32) is an extrema. The resulting eigenvectors must be checked against the constraint (4.33) to be sure that it is satisfied and are thus stationary points. To determine whether the stationary points are a minimum or maximum would require checking the behavior of the second derivatives via the Hessian.

Example 4.4 Determine the stationary points of the quadratic function

$$f(x_1, x_2) = x_1^2 + x_2^2 \tag{4.37}$$

subject to the quadratic constraint

$$x_1 x_2 = 1. \tag{4.38}$$

Solution

In this case, we have (4.36) with

$$\mathbf{A} = \begin{bmatrix} 1 & 0 \\ 0 & 1 \end{bmatrix}, \quad \mathbf{B} = \begin{bmatrix} 0 & \frac{1}{2} \\ \frac{1}{2} & 0 \end{bmatrix}, \quad \mathbf{x} = \begin{bmatrix} x_1 \\ x_2 \end{bmatrix}.$$

The generalized eigenproblem (4.36) may be rewritten as a regular eigenproblem of the form

$$\mathbf{B}^{-1}\mathbf{Ax}_n = \lambda_n \mathbf{x}_n,$$

which in this case is

$$\begin{bmatrix} 0 & 2 \\ 2 & 0 \end{bmatrix} \mathbf{x}_n = \lambda_n \mathbf{x}_n.$$

The corresponding eigenvalues and eigenvectors are

$$\lambda_1 = -2, \quad \mathbf{x}_1 = c_1 \begin{bmatrix} -1 \\ 1 \end{bmatrix},$$

$$\lambda_2 = 2, \quad \mathbf{x}_2 = c_2 \begin{bmatrix} 1 \\ 1 \end{bmatrix}.$$

Observe that $\mathbf{x}_1$ does not satisfy the constraint for any c_1, and that $c_2 = \pm 1$ in order for $\mathbf{x}_2$ to satisfy the constraint. Therefore, the stationary points are $(x_1, x_2) = (1, 1)$ and $(x_1, x_2) = (-1, -1)$ corresponding to $\lambda = 2$. It can be confirmed that both of these points are local minimums.

Let us reconsider Example 4.3 involving determining the major and minor axes of an ellipse.

Example 4.5 Determine the stationary points of the quadratic function

$$f(x_1, x_2) = x_1^2 + x_2^2 \tag{4.39}$$

subject to the quadratic constraint

$$3x_1^2 - 2x_1x_2 + 3x_2^2 = 8. \tag{4.40}$$

Solution

In this case, we have (4.36) with

$$\mathbf{A} = \begin{bmatrix} 1 & 0 \\ 0 & 1 \end{bmatrix}, \quad \mathbf{B} = \begin{bmatrix} 3 & -1 \\ -1 & 3 \end{bmatrix}, \quad \mathbf{x} = \begin{bmatrix} x_1 \\ x_2 \end{bmatrix}.$$

The generalized eigenproblem (4.36) may be rewritten as a regular eigenproblem of the form

$$\mathbf{B}^{-1}\mathbf{A}\mathbf{x}_n = \lambda_n \mathbf{x}_n,$$

which in this case is

$$\frac{1}{8}\begin{bmatrix} 3 & 1 \\ 1 & 3 \end{bmatrix}\mathbf{x}_n = \lambda_n \mathbf{x}_n.$$

The corresponding eigenvalues and eigenvectors are

$$\lambda_1 = \frac{1}{4}, \quad \mathbf{x}_1 = c_1 \begin{bmatrix} -1 \\ 1 \end{bmatrix},$$

$$\lambda_2 = \frac{1}{2}, \quad \mathbf{x}_2 = c_2 \begin{bmatrix} 1 \\ 1 \end{bmatrix}.$$

Observe that $c_1 = \pm 1$ and $c_2 = \pm\sqrt{2}$ to satisfy the constraint. Therefore, the stationary points are $(x_1, x_2) = \left\{(-1,1), (1,-1), (\sqrt{2}, \sqrt{2}), (-\sqrt{2}, -\sqrt{2})\right\}$; the first two are minimums corresponding to the semiminor axes, and the last two are maximums corresponding to the semimajor axes.

4.3.4 Linear Programming

The quadratic programming problem considered in the last section reduces to solving a generalized eigenproblem for the extrema that minimize the objective function and satisfy the constraints. Let us consider application of this same approach to determining extrema of a similar problem, except with a linear algebraic objective function and linear algebraic constraints. This is known as *linear programming*, which is an important topic in optimization.

Suppose that now the objective function having N independent variables to be minimized is given by the linear function

$$J(x_1, x_2, \ldots, x_N) = c_1x_1 + c_2x_2 + \cdots + c_Nx_N.$$

The M linear constraints, where $M < N$, are

$$\begin{aligned} A_{11}x_1 + A_{12}x_2 + \cdots + A_{1N}x_N &= b_1, \\ A_{21}x_1 + A_{22}x_2 + \cdots + A_{2N}x_N &= b_2, \\ &\vdots \\ A_{M1}x_1 + A_{M2}x_2 + \cdots + A_{MN}x_N &= b_M. \end{aligned}$$

In vector form, we have the objective function

$$J(\mathbf{x}) = \mathbf{c}^T\mathbf{x} = \langle \mathbf{c}, \mathbf{x} \rangle \tag{4.41}$$

subject to the constraints given by $\mathbf{Ax} = \mathbf{b}$, where $\mathbf{A}$ is $M \times N$ with $M < N$. That is, there are fewer constraint equations than independent variables. These are equality constraints; inequality constraints can be accommodated within this framework using *slack variables*, which are discussed in Chapter 12.

Let us first pursue the approach outlined in the previous sections for optimization of algebraic equations with constraints. The constraint equations are incorporated using Lagrange multipliers $\mathbf{\Lambda} = \begin{bmatrix} \Lambda_1 & \Lambda_2 & \cdots & \Lambda_M \end{bmatrix}^T$. Thus, we now want to minimize the *augmented function* expressed in matrix form

$$\tilde{J} = \langle \mathbf{c}, \mathbf{x} \rangle + \mathbf{\Lambda}^T (\mathbf{b} - \mathbf{Ax}). \tag{4.42}$$

Again, the necessary condition for an extremum of a function is that its derivative with respect to each of the independent variables must be zero. This requires that

$$\frac{\partial \tilde{J}}{\partial x_i} = 0, \quad i = 1, 2, \ldots, N. \tag{4.43}$$

Applying (4.43) to the augmented function (4.42) yields the result that

$$\mathbf{A}^T\mathbf{\Lambda} = \mathbf{c}$$

must be satisfied.[9] Given the vector $\mathbf{c}$ from the objective function and matrix $\mathbf{A}$ from the constraint equations, this is a system of linear algebraic equations for the Lagrange multipliers $\mathbf{\Lambda}$. However, we do not seek the Lagrange multipliers; we seek the stationary solution(s) $\mathbf{x}$ that minimizes the objective function and satisfies the constraints.

The reason why the usual approach does not work for linear programming is because the objective function is a straight line (in two dimensions), plane (in three dimensions), or linear function (in N dimensions) that does not have finite extrema. In order to have a finite optimal solution, therefore, requires additional inequality constraints that bound the so-called *feasible space*, which is the region of x_1, x_2, ..., x_N space that is consistent with the constraints where an optimal solution could

[9] We do not include the details here because the method does not work. The details for similar problems – that do work – will be given in Chapters 10 and 12.

reside. Linear programming requires introduction of additional techniques that do not simply follow from using Lagrange multipliers and differentiating the algebraic functions and setting equal to zero. These will be covered in Chapter 12.

4.4 Summary of Vector and Matrix Derivatives

The previous section gives us a small taste of the power and flexibility – and limitations – of the methods now in our arsenal. Although that for linear programming was unsuccessful and that for quadratic programming appears quite narrow in scope, this lays the foundation for the least-squares, optimization, and control methods to be considered in Part III. Not only will we develop methods for solving linear programming problems, the least-squares method will provide a general framework with which to tackle a wide variety of applications in optimization, as well as curve fitting and interpolation. Optimization and root-finding algorithms to be discussed in Chapter 12 are a direct extension of the topics covered in this chapter. In fact, the reader is fully equipped to progress directly to Part III if desired.

Because algebra and calculus are so integral[10] to nearly all areas of mathematics, science, and engineering, their extensions to matrices are central to so many of the applications of interest to scientists and engineers. Part I provides the necessary background for the numerical methods in Part II and the least-squares and optimization techniques in Part III. Perhaps the most vital and far reaching applications are numerical methods as they extend theoretical methods to be able to treat systems of any size and complexity – within the limits of our computational resources.

Finally, let us summarize the vector derivatives that have been evaluated throughout this chapter for future reference; one will be derived in Chapter 10 but is included here for reference:

1. Gradient: Vector derivative of a scalar function $f(x_1, x_2, \ldots, x_n, \ldots, x_N)$:

$$\frac{df}{d\mathbf{x}} = \boldsymbol{\nabla} f = \begin{bmatrix} \dfrac{\partial f}{\partial x_1} & \dfrac{\partial f}{\partial x_2} & \cdots & \dfrac{\partial f}{\partial x_n} & \cdots & \dfrac{\partial f}{\partial x_N} \end{bmatrix}^T. \tag{4.44}$$

2. Hessian matrix: Gradient of the gradient of a scalar function $f(x_1, x_2, \ldots, x_n, \ldots, x_N)$:

$$\mathbf{H} = \boldsymbol{\nabla}\left(\boldsymbol{\nabla} f\right) = \begin{bmatrix} \dfrac{\partial^2 f}{\partial x_1^2} & \dfrac{\partial^2 f}{\partial x_1 \partial x_2} & \cdots & \dfrac{\partial^2 f}{\partial x_1 \partial x_N} \\ \dfrac{\partial^2 f}{\partial x_2 \partial x_1} & \dfrac{\partial^2 f}{\partial x_2^2} & \cdots & \dfrac{\partial^2 f}{\partial x_2 \partial x_N} \\ \vdots & \vdots & \ddots & \vdots \\ \dfrac{\partial^2 f}{\partial x_N \partial x_1} & \dfrac{\partial^2 f}{\partial x_N \partial x_2} & \cdots & \dfrac{\partial^2 f}{\partial x_N^2} \end{bmatrix}. \tag{4.45}$$

[10] Pun intended.

3. Vector derivative of the inner product of two vectors:

$$\frac{d}{d\mathbf{x}}\langle \mathbf{x}, \mathbf{b}\rangle = \frac{d}{d\mathbf{x}}\left(\mathbf{x}^T\mathbf{b}\right) = \mathbf{b}. \tag{4.46}$$

4. Vector derivative of the norm of a vector:

$$\frac{d}{d\mathbf{x}}\|\mathbf{x}\|^2 = \frac{d}{d\mathbf{x}}\langle \mathbf{x}, \mathbf{x}\rangle = \frac{d}{d\mathbf{x}}\left(\mathbf{x}^T\mathbf{x}\right) = 2\mathbf{x}. \tag{4.47}$$

5. Vector derivative of a vector-matrix product:

$$\frac{d}{d\mathbf{x}}\left(\mathbf{b}^T\mathbf{A}\mathbf{x}\right) = \mathbf{A}^T\mathbf{b}. \tag{4.48}$$

6. Vector derivative of a quadratic, for which $\mathbf{A}$ is symmetric:

$$\frac{d}{d\mathbf{x}}\left(\mathbf{x}^T\mathbf{A}\mathbf{x}\right) = 2\mathbf{A}\mathbf{x}. \tag{4.49}$$

4.5 Briefly on Bases

In this chapter, we return to the idea of determining the best basis vectors – or coordinate system – for representing a physical system as discussed in Chapter 1 and how to translate differential operators between such coordinate systems. This is essential so that we can write the governing differential equations for a system in one coordinate system with respect to any other coordinate system that may be convenient for that particular problem. Such conversions are very tedious but greatly expand one's repertoire of bases within which to solve physical problems governed by differential equations.

In Chapter 10, we will build on the calculus-based groundwork established here in order to formulate the least-squares method. This is then used in Chapter 11 for polynomial regression curve fitting and interpolation of data. It will be shown how the selection of various basis functions is at the core of least-squares regression, curve fitting, and interpolation. In particular, polynomial regression and interpolation, of which linear regression is a special case, Fourier series, and spline interpolation are distinguished by the number and type of basis functions included in the trial function.

Exercises

4.1 Let $\mathbf{V}(x, y, z)$ be a vector function in Cartesian coordinates. Prove the following vector identities (note the use of the dot, rather than inner, product notation):
(a) $\nabla \cdot (\nabla \times \mathbf{V}) = 0$
(b) $(\mathbf{V} \cdot \nabla)\mathbf{V} = (\nabla \times \mathbf{V}) \times \mathbf{V} + \nabla\left(\frac{\mathbf{V}^2}{2}\right)$
(c) $(\nabla \times \mathbf{V}) \times \mathbf{V}$ is normal to $\mathbf{V}$

4.2 In cylindrical coordinates, the gradient operator is given by

$$\nabla = \frac{\partial}{\partial r}\mathbf{e}_r + \frac{1}{r}\frac{\partial}{\partial \theta}\mathbf{e}_\theta + \frac{\partial}{\partial z}\mathbf{e}_z,$$

and a scalar and vector function are

$$\phi = \phi(r,\theta,z), \quad \text{and} \quad \mathbf{V} = v_r(r,\theta,z)\mathbf{e}_r + v_\theta(r,\theta,z)\mathbf{e}_\theta + v_z(r,\theta,z)\mathbf{e}_z.$$

Noting the use of the dot, rather than inner, product notation, perform the following:
(a) Determine $\nabla \cdot \nabla\phi$ and $\nabla \cdot \mathbf{V}$ and check using the results given in Section 4.1.2.
(b) Show that $\nabla \cdot (\nabla \times \mathbf{V}) = 0$ in cylindrical coordinates.

4.3 Let $\mathbf{V}(x,y) = y^2\mathbf{i}+x\mathbf{j}$. Show that Stokes' theorem holds over the square with corners at $(x,y) = (2,2), (2,3), (3,3)$, and $(3,2)$.

4.4 Consider the one-dimensional, unsteady scalar-transport equation

$$\frac{\partial \phi}{\partial t} = \alpha\frac{\partial^2 \phi}{\partial x^2} - u\frac{\partial \phi}{\partial x},$$

where $u = u(x)$. Transform the equation from the x coordinate to the ξ coordinate using the transformations $\xi = \ln(x+1)$.

4.5 Consider the two-dimensional, unsteady diffusion equation

$$\frac{\partial u}{\partial t} = \alpha\left(\frac{\partial^2 u}{\partial x^2} + \frac{\partial^2 u}{\partial y^2}\right).$$

Transform the equation from (x,y) coordinates to (ξ,η) coordinates using the transformations $\xi = x^2$, $\eta = y^2$.

4.6 Consider the two-dimensional scalar-transport equation

$$u\frac{\partial \phi}{\partial x} + v\frac{\partial \phi}{\partial y} = \alpha\left(\frac{\partial^2 \phi}{\partial x^2} + \frac{\partial^2 \phi}{\partial y^2}\right).$$

Transform the equation from (x,y) coordinates to (ξ,η) coordinates using the transformations $\xi = x^2$, $\eta = y^2$.

4.7 Consider the Poisson equation

$$\frac{\partial^2 u}{\partial x^2} + \frac{\partial^2 u}{\partial y^2} = f(x,y).$$

Transform the equation from (x,y) coordinates to (ξ,η) coordinates using the transformations $\xi = \ln(x+1)$, $\eta = y^2$.

4.8 Consider the Poisson equation

$$\frac{\partial^2 u}{\partial x^2} + \frac{\partial^2 u}{\partial y^2} = f(x,y).$$

Transform the equation from (x,y) coordinates to (ξ,η) coordinates using the transformations $\xi = xy$, $\eta = y$.

4.9 Determine the point on the plane

$$c_1x_1 + c_2x_2 + c_3x_3 = c_4$$

that is closest to the origin. What is this distance?

4.10 Determine the points on the sphere

$$x_1^2 + x_2^2 + x_3^2 = 4$$

that are closest and farthest from the point $(x_1, x_2, x_3) = (1, -1, 1)$.

4.11 Determine the maximum value of the function $f(x_1, x_2, x_3) = x_1^2 + 2x_2 - x_3^2$ on the line along which the planes $2x_1 - x_2 = 0$ and $x_2 + x_3 = 0$ intersect.

4.12 Evaluate the gradient vector and Hessian matrix for the following multidimensional functions:

(a) $f(x_1, x_2) = 2x_1 x_2^2 + 3e^{x_1 x_2}$

(b) $f(x_1, x_2, x_3) = x_1^2 + x_2^2 + 2x_3^2$

(c) $f(x_1, x_2) = \ln\left(x_1^2 + 2x_1 x_2 + 3x_2^2\right)$

5 Analysis of Discrete Dynamical Systems

> It may happen that slight differences in the initial conditions produce very great differences in the final phenomena; a slight error in the former would make an enormous error in the latter. Prediction becomes impossible and we have the fortuitous phenomena. (Jules Henri Poincaré)

In Section 2.6.3, we saw that matrix methods can be used to both solve dynamical systems problems as well as determine their natural frequencies, with the general solution being a linear combination of the natural frequencies. Determination of the natural frequencies of a system without the necessity to fully solve the governing equations points to one of the central features of dynamical systems theory. It is a loosely defined and broad collection of mathematical methods and techniques for characterizing the motion of linear and nonlinear mechanical, electrical, chemical, and biological systems that evolve with time. In particular, these methods consist of a set of tools that provide understanding, interpretation, and insight into complex dynamical systems and their solutions, even when the governing equations are not known or are too complex to fully solve.

In some cases, this additional insight is sought from raw data produced by the system itself, whether experimentally or numerically. In other cases, we seek understanding of a system for which such solutions or results are not available, in which case the analysis of the dynamical system is in lieu of obtaining the solution of the governing equations. For example, it is often possible to determine the number, type, and stability of solutions as certain physical parameters are varied within the system without actually obtaining any solutions.

What distinguishes dynamical systems theory from most other mathematical approaches is that it retains a significant emphasis on analysis of time-evolving systems using qualitative methods of description and classification. While it is becoming increasingly quantitative, the emphasis remains on what *types* of solutions are possible and their stability, not necessarily on the details of the solutions themselves. Dynamical systems theory is the modern extension of "classical mechanics" and includes recent developments in deterministic chaos (to be discussed later in this chapter), control theory (see chapter 10 of Cassel, 2013), and reduced-order modeling (to be covered in Chapter 13) applied to both discrete and continuous systems.

We use the terms *discrete* and *continuous* in the manner typical of classical mechanics, in which it is the spatial distribution of the mass that is the distinguishing feature. A discrete system is one in which the mass is concentrated in discrete, rigid (nondeformable) masses. A continuous system has its mass distributed continuously throughout the deformable system. Discrete dynamical systems are comprised of point masses, pendulums, springs, et cetera, and are generally governed by ordinary differential equations in time, and their stability is governed by algebraic eigenproblems. Continuous dynamical systems are comprised of deformable strings, membranes, beams, et cetera, and are generally governed by partial differential equations in time and space, and their stability is governed by differential eigenproblems. Thus, we talk in terms of the *system* itself as being discrete or continuous.

This is not to be confused with the way in which these terms are used in mathematics generally and differential equations specifically. A mathematician speaks of the mathematical *model* as being discrete or continuous, where the distinction suggests the nature of the independent variable(s) as being discrete or continuous. In our treatment, all of the mathematical models for our discrete and continuous systems are *continuous* in this sense. Analysis techniques for discrete systems build on Chapters 1 and 2, while those for continuous systems build on Chapter 3. We will discuss in Part II how numerical methods can be used to discretize the continuous mathematical models – governing equations – of our discrete and continuous systems in order to facilitate obtaining approximate solutions in cases when an analytical solution is not available. Although this may seem confusing at first, simply be clear on whether it is the *physical system* or the *mathematical model* of the system that is being described as being *discrete* or *continuous*. Our focus in this chapter is on methods applied to continuous models of discrete systems.

The development and application of dynamical systems theory to continuous systems governed by partial differential equations is an area of active research. Because such systems exhibit a broad range of behaviors, the techniques often must be developed for specific topical domains, such as fluid mechanics, and there are few general-purpose methods for analyzing stability and bifurcation, for example. Two promising approaches that have broad applicability are *Lagrangian coherent structures*, which are based on finite-time Lyapunov exponents (see Pikovsky and Politi, 2016), and *Lagrangian descriptors* (see Mancho et al., 2013). Both of these methods are capable of demarcating dynamically distinct regions of a moving substance.

This chapter on dynamical systems will serve as a bridge between the matrix methods discussed in Part I and the numerical and optimization methods to be covered in Parts II and III. The examples to be encountered in this chapter provide ideal illustrations and applications of methods emphasized in Chapters 1 and 2, such as solving systems of linear algebraic equations, eigenproblems, and solving systems of first-order ordinary differential equations, for instance. They will motivate the need for numerical methods to solve linear and nonlinear ordinary differential equations in Part II and provide the tools necessary for analyzing and categorizing common behaviors in such solutions. Finally, it will serve to illustrate the range of behaviors that are possible in linear and nonlinear discrete systems in order to give an idea of what we would like to optimize or control in Part III.

5.1 Introduction

In this chapter, we consider linear and nonlinear, autonomous and nonautonomous discrete systems. Let us begin by defining some common terms and concepts:

- **Dynamical system**: A mechanical, electrical, chemical, or biological system evolving with respect to time or a timelike variable.
- **Degrees of freedom**: Minimum number of dependent variables required to fix the state of a system.
- **Linear system**: The governing equation(s) of motion along with initial and boundary conditions are all linear. Linear systems are such that as a parameter is changed slowly, the solution changes slowly and smoothly as well; the "output" is proportional to its "input." Superposition of solutions applies.
- **Nonlinear system**: The governing equation(s) of motion and/or initial and boundary conditions are nonlinear. When a parameter is changed slowly, the solution may change slowly and continuously or abruptly and discontinuously, and small changes to the input can lead to large changes in the system behavior; the "output" is not proportional to its "input." Superposition of solutions does not apply.
- **Autonomous system**: The equations do not involve the independent variable, typically time, explicitly. For example, the equation(s) of motion is of the form

$$\dot{\mathbf{u}} = \mathbf{f}(\mathbf{u}).$$

- **Nonautonomous system**: The equations include time explicitly, typically through time-dependent forcing and/or coefficients in the equations. For example, the equation(s) of motion are of the form

$$\dot{\mathbf{u}} = \mathbf{f}(\mathbf{u}, t).$$

- **Symmetry**: The system is invariant with respect to a particular geometric transformation.
- **Phase-plane portrait**: A graphical representation of the solution trajectory of a dynamical system with time. For a second-order, linear, or nonlinear system governed by two coupled first-order differential equations

$$\begin{aligned} \dot{u}_1 &= f_1(u_1, u_2), \\ \dot{u}_2 &= f_2(u_1, u_2), \end{aligned}$$

  the phase portrait is the plot of the trajectory of $u_2(t)$ versus $u_1(t)$ with time. The dependent variables $u_1(t)$ and $u_2(t)$ typically represent position and velocity, respectively, of second-order systems.
- **Equilibrium point**: A steady, that is, time-invariant, solution represented by a single point on the phase plane. For autonomous dynamical systems governed by first-order systems of differential equations of the form $\dot{\mathbf{u}} = \mathbf{f}(\mathbf{u})$, the equilibrium points are the zeros of $\mathbf{f}(\mathbf{u})$, where $\dot{\mathbf{u}} = \mathbf{0}$ such that $\mathbf{u}$ does not change with time. For two-dimensional autonomous systems, two orbits can only cross at an equilibrium point.

- **Orbit**: Solution trajectory in the phase plane.
- **Periodic orbit**: A periodic solution represented by a closed curve in the phase plane with $u(t + P) = u(t)$ for all t, where P is the period.
- **Limit cycle**: Isolated periodic orbit to which nearby orbits spiral toward or away.
- **Floquet theory**: The study of periodic solutions.
- **Separatrix**: A closed curve in the phase plane that separates closed periodic solutions – limit cycles – from unbounded trajectories.
- **Attractor**: A region in the phase plane to which the solution is attracted as $t \to \infty$. Such regions comprise the *basin*, or *domain*, *of attraction*. A *strange attractor* may arise in nonlinear systems as a result of their sensitivity to initial conditions.
- **Bifurcation**: A qualitative change in the behavior of the solution as a system parameter changes. The topology of the solution trajectory in the phase plane changes. The qualitative change could be marked by a change in the number and type of equilibrium points or limit-cycle solutions, or it could be a change in their stability properties, for example.
- **Bifurcation diagram**: A plot of equilibrium points with respect to changes in a system parameter.
- **Poincaré section or map**: Whereas the phase plane shows the solution trajectory for all time, the Poincaré section only shows the solution trajectory as a point in the phase plane where it is located at integer multiples of a specified period. A limit-cycle solution is represented by a single point on a Poincaré section.
- **Deterministic chaos**: Irregular motion that may be generated by nonlinear systems, which may be characterized by sensitivity to initial conditions, fractal behavior, and strange attractors.

The most important distinction with regard to the types of solution behavior that may be observed is whether the system is linear or nonlinear. Linear systems theory is so powerful, and the theory so well developed, that when dealing with a nonlinear system, it is common to first analyze the system by linearizing the model in the vicinity of a characteristic – often equilibrium – state or solution trajectory, thereby bringing to bear the full range of tools available for analyzing linear systems. Although such an analysis only applies when the system state is close to that about which it has been linearized, this approach is often very fruitful and provides a wealth of knowledge about the system's behaviors.

5.2 Phase-Plane Analysis – Linear Systems

Recall that when solving systems of first-order, linear ordinary differential equations in Chapter 2, we simply plotted the dependent variable(s) as a function of time in order to display the solution behavior. This is typically sufficient for autonomous linear systems of equations as the possible solution behaviors are relatively limited and easy to identify from such plots. For more complex systems, such as nonautonomous and/or nonlinear systems, however, the range of possible solution behaviors is far more diverse, and additional diagnostic tools are necessary to discern the characteristic behaviors displayed in the solution of a particular system. The *phase plane* provides a helpful means of analyzing solutions of dynamical systems that often leads to patterns and structures that can be identified within the phase plane leading to greater understanding and insight into the underlying system. In order to introduce some terminology and basic approaches, we first consider autonomous linear systems in the next two sections before undertaking a more comprehensive treatment that includes nonautonomous and nonlinear behavior.

5.2.1 Linear Oscillator with Damping

Consider the oscillation of a *linear spring with damping* as illustrated in Figure 5.1. As derived in section 5.1.1 of Cassel (2013), a linear oscillator with damping is governed by the ordinary differential equation

$$m\frac{d^2\hat{u}}{d\hat{t}^2} + c\frac{d\hat{u}}{d\hat{t}} + k\hat{u} = 0, \quad \hat{u}(0) = 0, \quad \dot{\hat{u}}(0) = A, \tag{5.1}$$

where m is the mass, c is the damping coefficient, k is the linear spring constant, A is the initial velocity, and the dependent variable $\hat{u}(\hat{t})$ gives the location of the mass relative to its neutral point at which there is no force in the spring. The initial conditions indicate that the mass is initially given a specified velocity while held at its neutral position at $t = 0$.

For convenience, we nondimensionalize time based on the oscillatory frequency according to

$$t = \frac{\hat{t}}{\sqrt{m/k}}.$$

The position is nondimensionalized based on the initial velocity A and oscillatory time scale as follows:

$$u(t) = \frac{\hat{u}(\hat{t})}{A\sqrt{m/k}}.$$

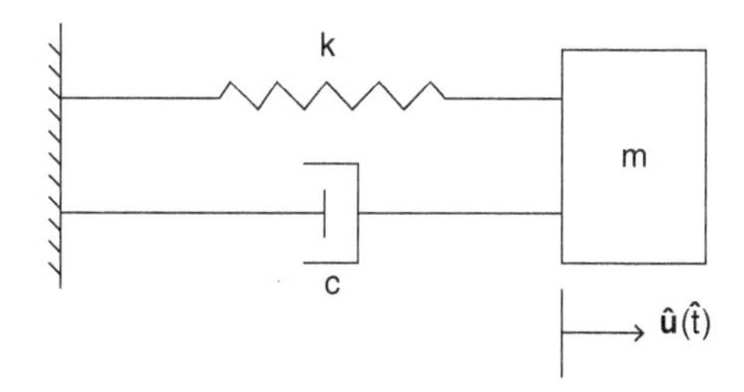

Figure 5.1 Schematic of a linear spring with damping.

Substituting into (5.1) leads to the nondimensional form of the equation of motion

$$\frac{d^2u}{dt^2} + 2d\frac{du}{dt} + u = 0, \tag{5.2}$$

with the initial conditions

$$u(0) = 0, \quad \dot{u}(0) = 1, \tag{5.3}$$

and the scaled damping parameter is given by

$$d = \frac{c}{2\sqrt{km}}.$$

The exact solution of (5.2) with initial conditions (5.3) for all d and time is

$$u(t) = \frac{e^{-dt}}{\sqrt{1-d^2}} \sin\left(\sqrt{1-d^2}t\right).$$

In order to plot the solution trajectory in the phase plane, note that the original equation (5.2) is equivalent to the system of first-order equations (see Section 2.6.2)

$$\begin{aligned} \frac{du_1}{dt} &= u_2, \\ \frac{du_2}{dt} &= -u_1 - 2du_2. \end{aligned} \tag{5.4}$$

In this formulation, $u_1(t) = u(t)$ gives the position, and $u_2(t) = \dot{u}(t)$ gives the velocity of the mass. The initial conditions are

$$u_1(0) = 0, \quad u_2(0) = 1.$$

The first-order formulation (5.4) is called the *state-space form.*

5.2.2 Equilibrium Points

Equilibrium points of a system are points on the phase plane – corresponding to fixed values of $u_1(t)$ and $u_2(t)$ – for which the system will stay permanently for all time if the system is placed carefully in that state, for example, via the initial conditions. Such equilibrium points are also referred to as *steady* or *time-independent* solutions. Of course, all real systems are subject to small disturbances from their environment; for now, we will imagine that no such disturbances exist and consider such stability questions in Section 5.3.

For a dynamical system written in the state-space form

$$\begin{aligned} \dot{u}_1 &= f_1(u_1, u_2), \\ \dot{u}_2 &= f_2(u_1, u_2), \end{aligned}$$

the equilibrium point(s) can be found by determining the values of $u_1(t)$ and $u_2(t)$ for which $f_1(u_1, u_2) = 0$ and $f_2(u_1, u_2) = 0$. Because the solutions at equilibrium points do not change with time, their time derivatives on the left-hand side of the equations vanish, and we are left with an algebraic, rather than differential, system of equations to solve for the equilibrium points.

If the system is linear, then the system of first-order differential equations can be written in the matrix form

$$\dot{\mathbf{u}} = \mathbf{A}\mathbf{u},$$

where for the linear oscillator with damping

$$\mathbf{u} = \begin{bmatrix} u_1(t) \\ u_2(t) \end{bmatrix}, \quad \mathbf{A} = \begin{bmatrix} 0 & 1 \\ -1 & -2d \end{bmatrix}.$$

The equilibrium point for the linear oscillator with damping – where $\mathbf{Au} = \mathbf{0}$ – is simply the origin of the phase plane

$$u_1(t) = 0, \quad u_2(t) = 0.$$

Observe that from the point of view of the second-order equation (5.2), equilibrium points are those that have no velocity $\dot{u}(t)$ or acceleration $\ddot{u}(t)$. This leads to the same conclusion that the equilibrium point for the damped oscillator occurs when $u(t) = 0$, which corresponds to $u_1(t) = 0$ and $u_2(t) = 0$.

5.2.3 Limit-Cycle Solutions

A *limit-cycle* solution is a closed curve in the phase plane and represents a periodic solution that repeats indefinitely, such as the motion of an undamped harmonic oscillator or simple pendulum. For example, with no damping ($d = 0$), (5.2) is simply the equation governing a linear, harmonic oscillator with solutions as shown in Figure 5.2. The plot on the right is the phase-plane plot with velocity $\dot{u}(t) = u_2(t)$ versus position $u(t) = u_1(t)$. Here, the harmonic oscillation forms a closed limit-cycle solution in the phase plane. By adding a small amount of damping, with $d = 0.02$, we get the typical behavior shown in Figure 5.3. Thus, the solution is a slowly damped oscillation that spirals inward on the phase plane owing to the damping.

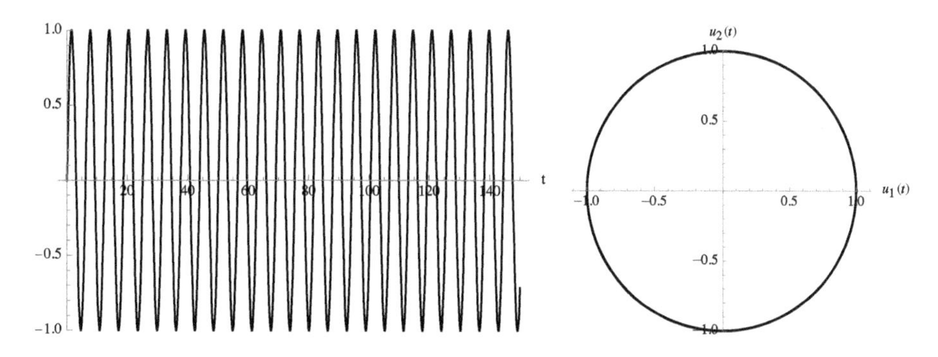

Figure 5.2 Solution of the linear-spring equation with no damping ($d = 0$). On the left is $u(t)$ versus t; on the right is the phase-plane plot.

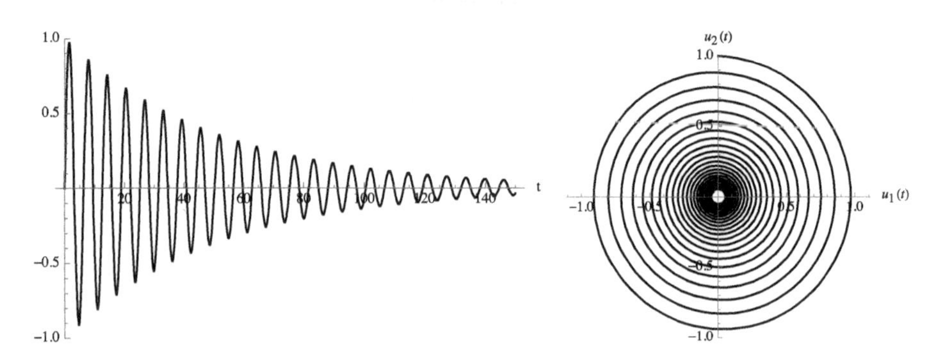

Figure 5.3 Solution of the linear-spring equation with damping ($d = 0.02$).

5.3 Bifurcation and Stability Theory – Linear Systems

As discussed in the last section, equilibrium points are states of a system at which the system will remain unchanged for all time if not disturbed. Real systems are subject to disturbances from their environment, and it is of interest to determine the behavior of the system when acted upon by such disturbances. Here, we focus on *linear stability*, in which we assume that the disturbances are infinitesimally small. Our focus here is on what is known as *asymptotic stability*, which in this context simply means stability behavior as $t \to \infty$.

For a given dynamical system, the procedure to determine its stability is as follows: (1) determine the equilibrium state(s), and (2) determine the stability characteristics of the system about each equilibrium position. When subject to a small disturbance, if the motion of the system remains bounded[1] as $t \to \infty$, the system is said to be *linearly stable* at that equilibrium point. If the motion grows and becomes unbounded as $t \to \infty$, then the system is linearly unstable at that equilibrium point.

Let us consider a real-life example to illustrate why we should be concerned with the fate of a system subject to small disturbances. On January 21, 2014, a landslide in northern Italy released several large boulders that proceeded to roll down a hill. One of these boulders rolled through the barn and courtyard of a family's vineyard, narrowly missing the house, as shown in Figure 5.4. Even more traumatic is the other large boulder that came to rest mere feet behind the house itself, as shown in Figure 5.5. Thankfully, no one was hurt during this incident; however, imagine the influence that small disturbances may have had on the paths that these two boulders followed and the even more severe destruction that may have ensued. Lest you feel too bad for the family that occupies this house, observe the large warning sign that has existed in their vineyard for many years in Figure 5.6.

[1] For a solution to remain bounded, it either decays with time and returns to the equilibrium point, such as a damped oscillator or pendulum, or at least the motion does not grow in amplitude with time, such as an undamped oscillator or simple pendulum.

Figure 5.4 Boulder from northern Italy landslide (image: Tareom).

Figure 5.5 Boulder from northern Italy landslide (image: Tareom).

Figure 5.6 Boulder from northern Italy landslide (image: Tareom).

5.3.1 Illustrative Example

We illustrate the procedure for determining a system's asymptotic stability in the following example.

Example 5.1 Let us consider stability of the three-spring and two-mass system in Example 2.12 and shown in Figure 2.7 (also see section 5.1.1 of Cassel, 2013).

Solution

Recall that the governing equations are ($K = k/m$)

$$\frac{d^2x_1}{dt^2} = -3Kx_1 + Kx_2,$$

$$\frac{d^2x_2}{dt^2} = \frac{1}{2}Kx_1 - Kx_2,$$

or when converted to a system of first-order equations $\dot{\mathbf{u}} = \mathbf{A}\mathbf{u}$ ($u_1 = x_1, u_2 = x_2, u_3 = \dot{x}_1, u_4 = \dot{x}_2$) as in Section 2.6.3, they are

$$\begin{bmatrix} \dot{u}_1 \\ \dot{u}_2 \\ \dot{u}_3 \\ \dot{u}_4 \end{bmatrix} = \begin{bmatrix} 0 & 0 & 1 & 0 \\ 0 & 0 & 0 & 1 \\ -3K & K & 0 & 0 \\ \frac{1}{2}K & -K & 0 & 0 \end{bmatrix} \begin{bmatrix} u_1 \\ u_2 \\ u_3 \\ u_4 \end{bmatrix}. \tag{5.5}$$

First, we obtain the equilibrium positions for which the velocities and accelerations of the masses are both zero. Thus,

$$\{\dot{x}_1, \dot{x}_2, \ddot{x}_1, \ddot{x}_2\} = \{\dot{u}_1, \dot{u}_2, \dot{u}_3, \dot{u}_4\} = \{0, 0, 0, 0\}.$$

This is equivalent to setting $\dot{\mathbf{u}} = \mathbf{0}$ in (5.5) and solving $\mathbf{A}\mathbf{u} = \mathbf{0}$. Because $\mathbf{A}$ is invertible, the only solution of this homogeneous system is the trivial solution

$$\mathbf{s} = \mathbf{u} = \mathbf{0},$$

or

$$\mathbf{s} = \begin{bmatrix} u_1 & u_2 & u_3 & u_4 \end{bmatrix}^T = \begin{bmatrix} 0 & 0 & 0 & 0 \end{bmatrix}^T, \tag{5.6}$$

where $\mathbf{s} = \begin{bmatrix} s_1 & s_2 & s_3 & s_4 \end{bmatrix}^T$ denotes an equilibrium point, where the solution remains stationary in the phase plane, at which the masses could remain indefinitely. Note that $\mathbf{s} = \mathbf{0}$ is the neutral position of the masses for which the forces in the three springs are zero.

The second step is to consider the stability behavior of the system about this equilibrium point subject to small disturbances

$$\mathbf{u} = \mathbf{s} + \epsilon\hat{\mathbf{u}},$$

where $\epsilon \ll 1$ in order to evaluate linear stability. Given that $\dot{\mathbf{u}} = \epsilon\dot{\hat{\mathbf{u}}}$ and $\mathbf{s} = \mathbf{0}$ for this case, the linear system $\dot{\mathbf{u}} = \mathbf{A}\mathbf{u}$ becomes

$$\dot{\hat{\mathbf{u}}} = \mathbf{A}\hat{\mathbf{u}}. \tag{5.7}$$

Observe that the equations for $\mathbf{u}$ and the disturbances $\hat{\mathbf{u}}$ are the same in the linear case; therefore, introduction of the disturbances is often implied, but not done explicitly, for linear systems.

The solution of (5.7) determines stability of the system about the equilibrium point at the origin. As in Example 2.12, the solution is determined by the eigenvalues of the coefficient matrix $\mathbf{A}$ through diagonalization. Recall that the eigenvalues of $\mathbf{A}$ are the complex conjugate pairs

$$\lambda_{1,2} = \pm 1.7958\mathrm{i}, \quad \lambda_{3,4} = \pm 0.8805\mathrm{i}.$$

Recalling that the eigenvalues are the natural frequencies of the system, the solution in the vicinity of $\mathbf{s}$ is a linear combination of $\sin(\lambda_i t)$ and $\cos(\lambda_i t)$, $i = 1,2,3,4$ corresponding to harmonic motion. Harmonic motion remains bounded for all time and is stable. We say that the equilibrium point $\mathbf{s}$ of the system is *linearly stable* in the form of a *stable center* as discussed in the next section. Note that in this case, the motion does not decay toward $\mathbf{s}$ owing to the lack of damping in the system.

REMARKS:

1. *Asymptotic stability as $t \to \infty$ of linear, autonomous dynamical systems is determined solely by the nature of its eigenvalues as outlined in the next section.*
2. *More degrees of freedom leads to larger systems of equations, additional natural frequencies, and more stability modes corresponding to each equilibrium point.*

5.3.2 Stability of Linear, Second-Order, Autonomous Systems

As we have seen in the previous section, the stability of linear, autonomous dynamical systems is determined from the nature of the eigenvalues of the coefficient matrix in the equations of motion expressed as a system of first-order linear equations $\dot{\mathbf{u}} = \mathbf{A}\mathbf{u}$. This is the case when the coefficient matrix $\mathbf{A}$ is comprised of constants that do not change with time; such systems are called *autonomous*. Stability analysis that is based on the behavior of the eigenvalues, or modes, of the system is called *modal stability analysis*.

In order to illustrate the possible types of stable and unstable equilibrium points, we evaluate stability of the general second-order system expressed as a system of first-order equations in the form

$$\begin{bmatrix} \dot{u}_1(t) \\ \dot{u}_2(t) \end{bmatrix} = \begin{bmatrix} A_{11} & A_{12} \\ A_{21} & A_{22} \end{bmatrix} \begin{bmatrix} u_1(t) \\ u_2(t) \end{bmatrix}.$$

The original system may be nonlinear; however, it has been linearized about the equilibrium point(s) as illustrated for the simple pendulum in Section 5.4.1. For autonomous systems, the equilibrium points are steady and do not change with time. Therefore, $\dot{\mathbf{u}} = 0$, and the steady equilibrium points are given by $\mathbf{Au} = \mathbf{0}$.

In order to characterize the stability of the system, we obtain the eigenvalues of $\mathbf{A}$ as follows:

$$|\mathbf{A} - \lambda\mathbf{I}| = 0,$$

$$\begin{vmatrix} A_{11} - \lambda & A_{12} \\ A_{21} & A_{22} - \lambda \end{vmatrix} = 0,$$

$$(A_{11} - \lambda)(A_{22} - \lambda) - A_{12}A_{21} = 0,$$

$$\lambda^2 - (A_{11} + A_{22})\lambda + A_{11}A_{22} - A_{12}A_{21} = 0,$$

$$\lambda^2 - \text{tr}(\mathbf{A})\lambda + |\mathbf{A}| = 0,$$

where from the quadratic formula

$$\lambda_{1,2} = \frac{1}{2}\left[\text{tr}(\mathbf{A}) \pm \sqrt{\text{tr}(\mathbf{A})^2 - 4|\mathbf{A}|}\right].$$

Many references summarize the various possibilities graphically by demarcating the various stable and unstable regions on a plot of $|\mathbf{A}|$ versus $\text{tr}(\mathbf{A})$. It is also informative to plot solution trajectories in the *phase plane*, which is a plot of $u_2(t)$ versus $u_1(t)$.

Each type of stable or unstable point is considered in turn by plotting several solution trajectories on the phase plane. The dots on each trajectory indicate the initial condition for that trajectory. In each case, the origin of the phase plane is the equilibrium point around which the system has been linearized for consideration of stability. In general, complicated systems may have numerous equilibrium points, each of which exhibits one of the stable or unstable behaviors illustrated as follows:

1. **Unstable Saddle Point**: $\text{tr}(\mathbf{A})^2 - 4|\mathbf{A}| > 0, |\mathbf{A}| < 0$
   If the eigenvalues of $\mathbf{A}$ are positive and negative real values, the equilibrium point is an *unstable saddle* as shown in Figure 5.7. Observe that while some trajectories initially move toward the equilibrium point, all trajectories eventually move away from the origin and become unbounded. Therefore, it is called an *unstable saddle point*. This is the behavior exhibited by the simple pendulum with n odd ($\theta = \pi$) as shown in Section 5.4.1.
2. **Unstable Node or Source**: $\text{tr}(\mathbf{A})^2 - 4|\mathbf{A}| > 0, \text{tr}(\mathbf{A}) > 0, |\mathbf{A}| > 0$
   If the eigenvalues of $\mathbf{A}$ are both positive real values, the equilibrium point is an *unstable node* as shown in Figure 5.8. In the case of an unstable node, all trajectories move away from the origin to become unbounded.
3. **Stable Node or Sink**: $\text{tr}(\mathbf{A})^2 - 4|\mathbf{A}| > 0, \text{tr}(\mathbf{A}) < 0, |\mathbf{A}| > 0$
   If the eigenvalues of $\mathbf{A}$ are both negative real values, the equilibrium point is a *stable node* as shown in Figure 5.9. In the case of a stable node, all trajectories move toward the equilibrium point and remain there.

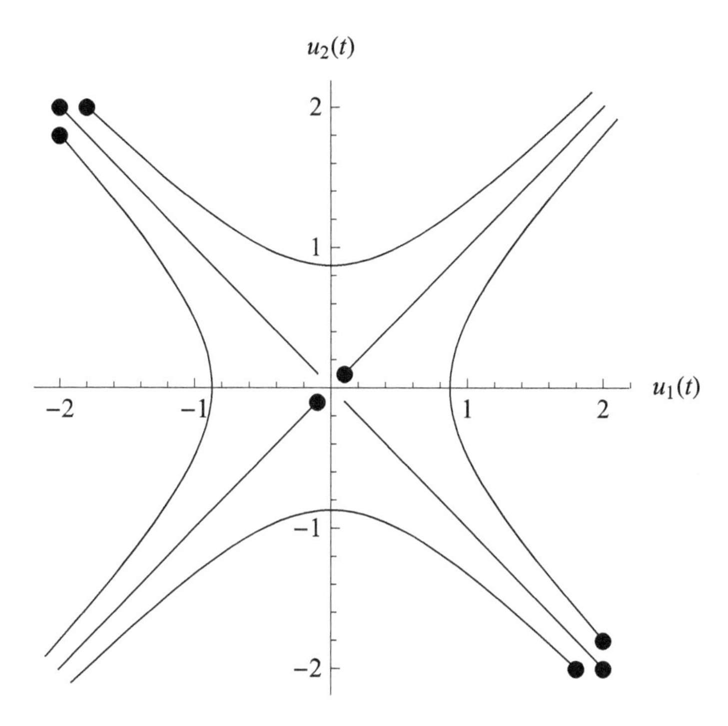

Figure 5.7 Phase-plane plot of an unstable saddle point (dots denote initial condition for each trajectory).

4. **Unstable Spiral or Focus**: $\text{tr}(\mathbf{A})^2 - 4|\mathbf{A}| < 0, \text{Re}[\lambda_{1,2}] > 0$ ($\text{tr}(\mathbf{A}) \neq 0$)
   If the eigenvalues of **A** are complex, but with positive real parts, the equilibrium point is an *unstable spiral.* A plot of position versus time is shown in Figure 5.10, and the phase-plane plot is shown in Figure 5.11. As can be seen from both plots, the trajectory begins at the origin, which is the equilibrium point, and spirals outward, becoming unbounded.
5. **Stable Spiral or Focus**: $\text{tr}(\mathbf{A})^2 - 4|\mathbf{A}| < 0, \text{Re}[\lambda_{1,2}] < 0$ ($\text{tr}(\mathbf{A}) \neq 0$)
   If the eigenvalues of **A** are complex, but with negative real parts, the equilibrium point is a *stable spiral.* A plot of position versus time is shown in Figure 5.12, and the phase-plane plot is shown in Figure 5.13. In the case of a stable focus, the trajectory spirals in toward the equilibrium point from any initial condition.
6. **Stable Center**: $\text{tr}(\mathbf{A})^2 - 4|\mathbf{A}| < 0, \text{tr}(\mathbf{A}) = 0$
   If the eigenvalues of **A** are purely imaginary, the equilibrium point is a *stable center.* A plot of position versus time is shown in Figure 5.14, and the phase-plane plot is shown in Figure 5.15. A stable center is comprised of a periodic limit-cycle solution centered at the equilibrium point. Because the trajectory remains bounded, the solution is stable. The undamped spring–mass system considered in Example 5.1 and the simple pendulum with n even ($\theta = 0$) considered in Section 5.4.1 has a stable center.

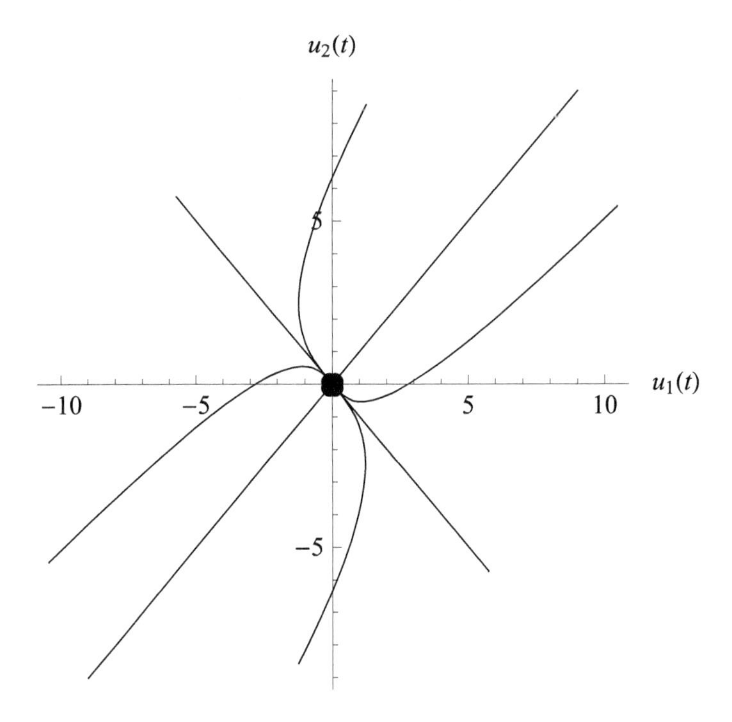

Figure 5.8 Phase-plane plot of an unstable node or source (dots denote initial condition for each trajectory).

There are situations for which such a modal analysis is incomplete, as for non-normal systems illustrated in the next section, or not possible, as for nonautonomous systems.

5.3.3 Nonnormal Systems – Transient Growth

The conventional modal stability analysis discussed thus far considers each individual stability mode, as characterized by its eigenvalue, separately. This is why it is called a *modal stability analysis*, and the asymptotic stability characteristics as $t \rightarrow \infty$ are determined from each corresponding eigenvalue, which determines whether the associated system behavior grows or decays in time in response to a small disturbance. A modal analysis is sufficient for systems governed by *normal matrices* as discussed in Section 2.4, for which the eigenvectors are mutually orthogonal and, therefore, do not interact with one another, thereby allowing for each mode to be considered individually. A normal matrix is one for which the coefficient matrix commutes with its conjugate transpose according to

$$\mathbf{A}\overline{\mathbf{A}}^T = \overline{\mathbf{A}}^T\mathbf{A}.$$

Recall that a real, symmetric matrix, for example, is one type of normal matrix.

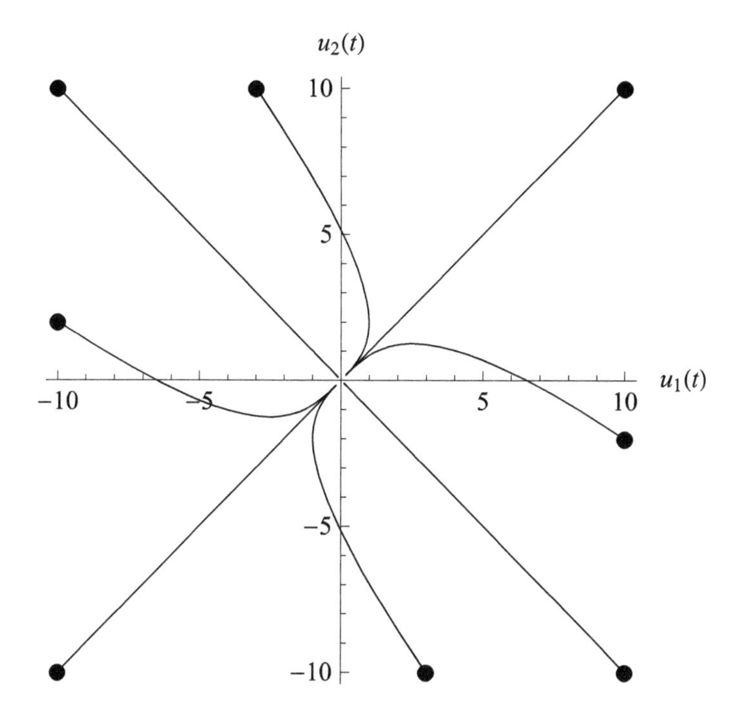

Figure 5.9 Phase-plane plot of a stable node or sink (dots denote initial condition for each trajectory).

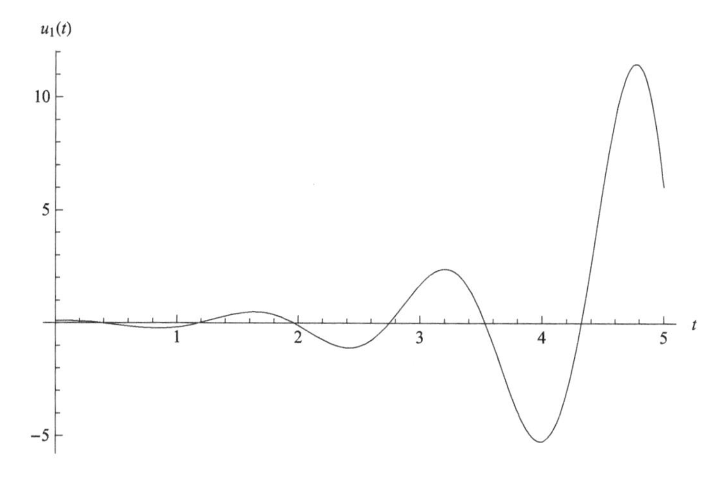

Figure 5.10 Plot of position versus time for an unstable spiral or focus.

For systems governed by nonnormal matrices, the asymptotic stability behavior for large times is still determined by the eigenvalues of the system's coefficient matrix. However, owing to exchange of energy between the nonorthogonal modes, it is possible that the system will exhibit a different behavior for $O(1)$ times as compared to the asymptotic behavior observed for large times. This is known as *transient growth.*

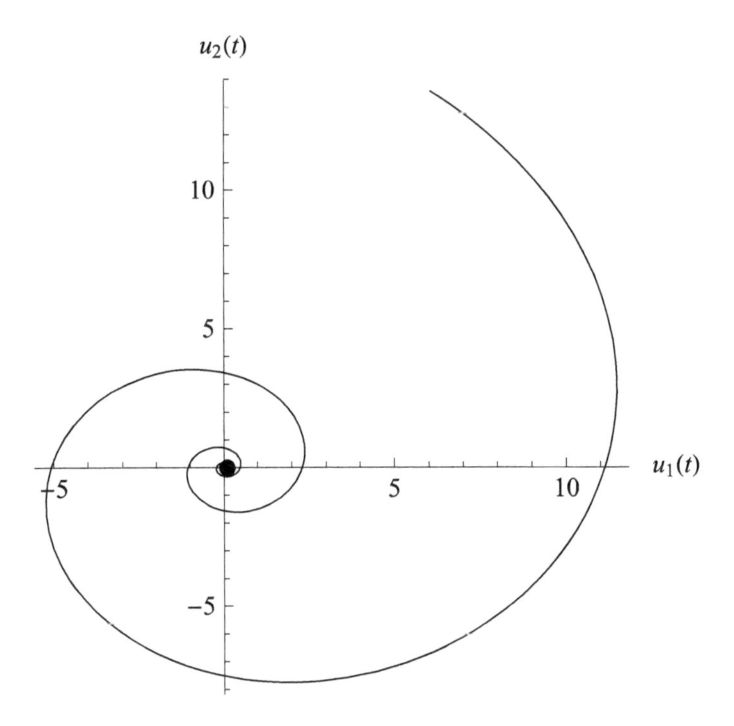

Figure 5.11 Phase-plane plot of an unstable spiral or focus (dot denotes initial condition for trajectory).

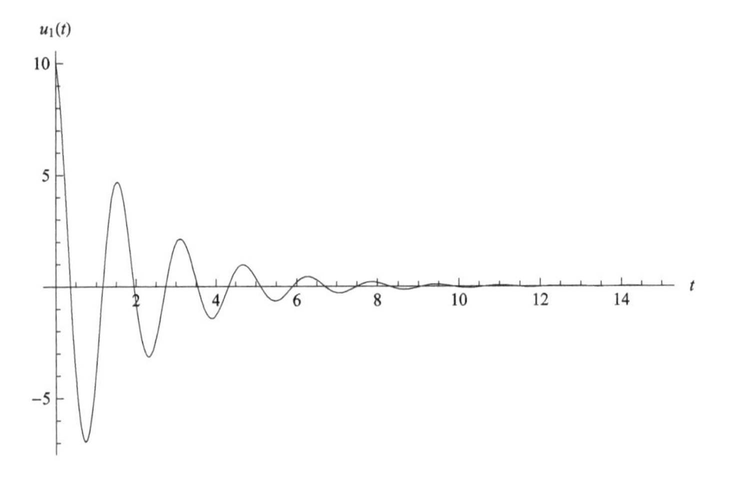

Figure 5.12 Plot of position versus time for a stable spiral or focus.

For example, a system may be asymptotically stable, meaning that none of the eigenvalues lead to a growth of their corresponding modes (eigenvectors) as $t \to \infty$, but it displays a transient growth for finite times in which the amplitude of a perturbation grows for some period of time before yielding to the large-time asymptotic behavior.

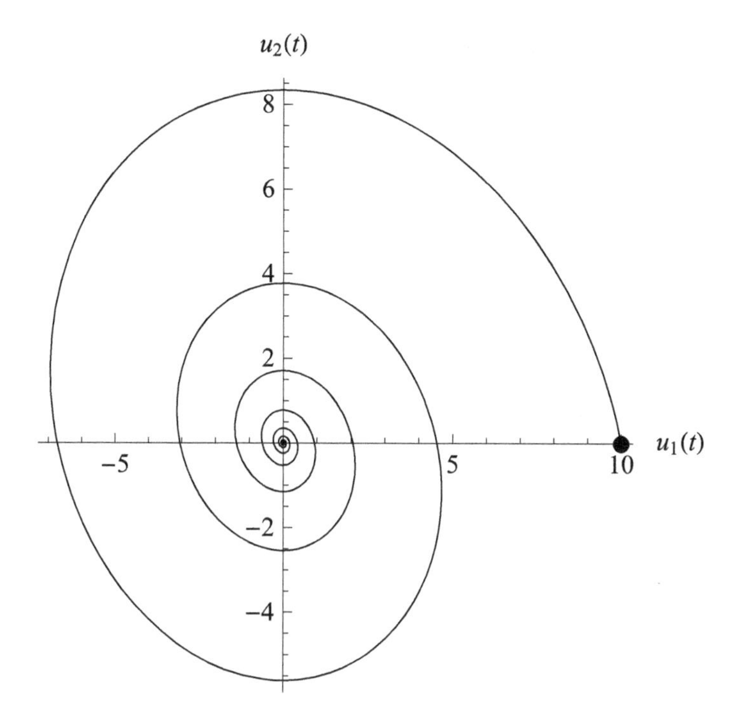

Figure 5.13 Phase-plane plot of a stable spiral or focus (dot denotes initial condition for trajectory).

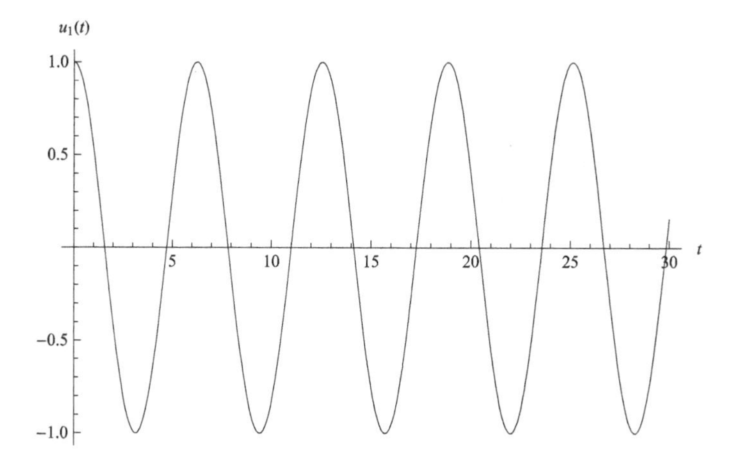

Figure 5.14 Plot of position versus time for a stable center.

Consider a second-order example given in Farrell and Ioannou (1996) in order to illustrate the effect of nonnormality of the system on the transient growth behavior for finite times before the solution approaches its asymptotically stable equilibrium point for large time. The system is governed by the following first-order ordinary differential equations:

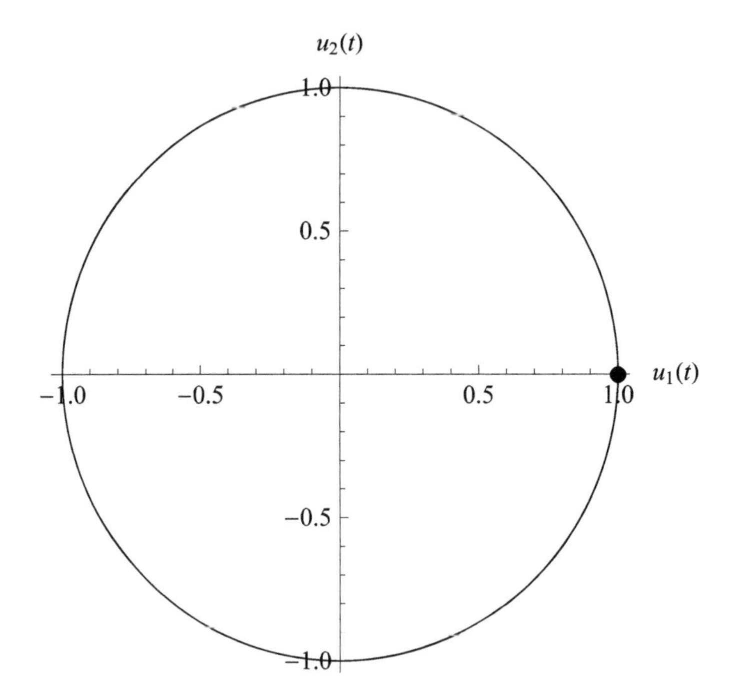

Figure 5.15 Phase-plane plot of a stable center (dot denotes initial condition for trajectory).

$$\begin{bmatrix} \dot{u}_1 \\ \dot{u}_2 \end{bmatrix} = \begin{bmatrix} -1 & -\cot\theta \\ 0 & -2 \end{bmatrix} \begin{bmatrix} u_1 \\ u_2 \end{bmatrix}. \tag{5.8}$$

The coefficient matrix $\mathbf{A}$ is normal if $\theta = \pi/2$ and nonnormal for all other θ. The system has an equilibrium point at $(u_1, u_2) = (0, 0)$.

Let us first consider the behavior of the normal system with $\theta = \pi/2$. In this case, the eigenvalues are $\lambda_1 = -2$ and $\lambda_2 = -1$. Because the eigenvalues are negative and real, the system is asymptotically stable in the form of a stable node. The corresponding eigenvectors are given by

$$\mathbf{v}_1 = \begin{bmatrix} 0 \\ 1 \end{bmatrix}, \quad \mathbf{v}_2 = \begin{bmatrix} 1 \\ 0 \end{bmatrix},$$

which are mutually orthogonal as expected for a normal system. The solution for the system (5.8) with $\theta = \pi/2$ is shown in Figures 5.16 and 5.17 for the initial conditions $u_1(0) = 1$ and $u_2(0) = 1$. For such a stable normal system, the perturbed solution decays exponentially toward the stable equilibrium point $(u_1, u_2) = (0, 0)$ according to the rates given by the eigenvalues. In phase space, the trajectory starts at the initial condition $(u_1(0), u_2(0)) = (1, 1)$ and moves progressively toward the equilibrium point $(0, 0)$.

Now consider the system (5.8) with $\theta = \pi/100$, for which the system is no longer normal. While the eigenvalues $\lambda_1 = -2$ and $\lambda_2 = -1$ remain the same, the corresponding eigenvectors are now

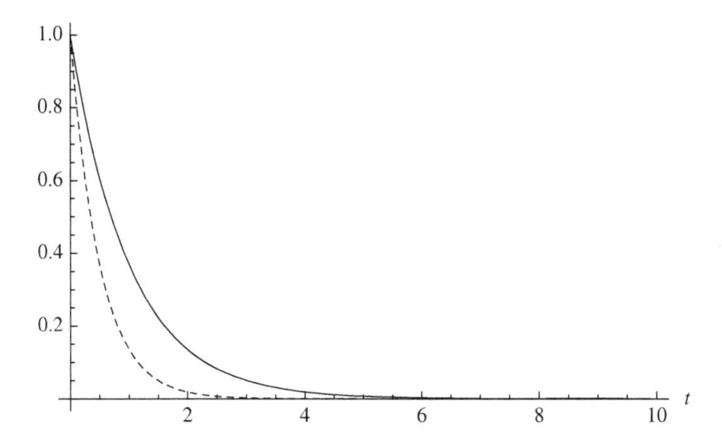

Figure 5.16 Solution of the normal system (5.8) with $\theta = \pi/2$ for $u_1(t)$ (solid line) and $u_2(t)$ (dashed line).

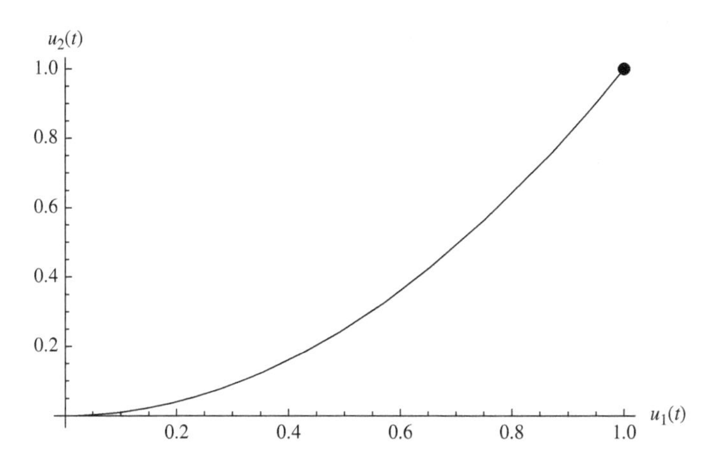

Figure 5.17 Solution of the normal system (5.8) with $\theta = \pi/2$ for $u_1(t)$ and $u_2(t)$ in phase space (the dot indicates the initial condition).

$$\mathbf{v}_1 = \begin{bmatrix} \cot\left(\frac{\pi}{100}\right) \\ 1 \end{bmatrix}, \quad \mathbf{v}_2 = \begin{bmatrix} 1 \\ 0 \end{bmatrix},$$

which are no longer orthogonal, giving rise to the potential for transient growth behavior. The solution for the system (5.8) with $\theta = \pi/100$ is shown in Figures 5.18 and 5.19 for the same initial conditions $u_1(0) = 1$ and $u_2(0) = 1$ as before. Recall that the system is asymptotically stable as predicted by its eigenvalues, and indeed the perturbations do decay toward zero for large times. Whereas $u_2(t)$ again decays monotonically[2] toward the equilibrium point, $u_1(t)$ first grows to become quite large before decaying toward the stable equilibrium point. In phase space, the trajectory starts at the initial condition $(u_1(0), u_2(0)) = (1, 1)$ but moves farther away from the

[2] If a function decays monotonically, it never increases over any interval.

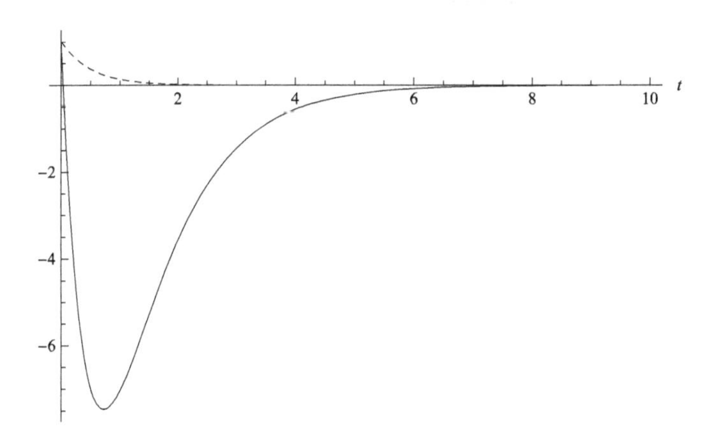

Figure 5.18 Solution of the nonnormal system (5.8) with $\theta = \pi/100$ for $u_1(t)$ (solid line) and $u_2(t)$ (dashed line).

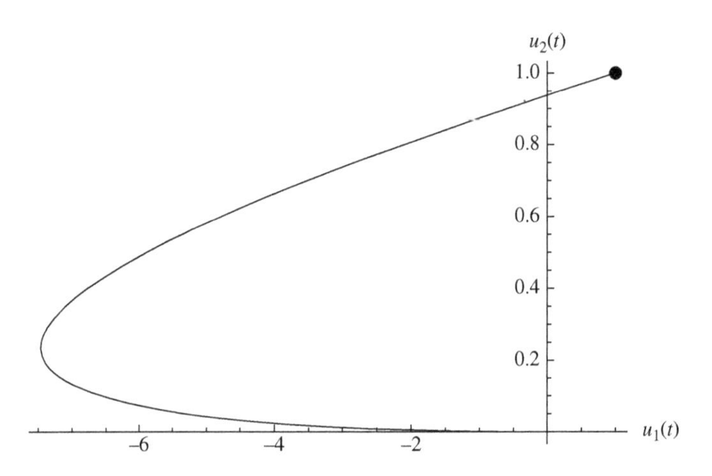

Figure 5.19 Solution of the nonnormal system (5.8) with $\theta = \pi/100$ for $u_1(t)$ and $u_2(t)$ in phase space (the dot indicates the initial condition).

equilibrium point at the origin before being attracted to it, essentially succumbing to the asymptotically stable large-time behavior.

Transient growth behavior can occur for nonnormal systems because the nonorthogonal eigenvectors lead to the possibility that the individual modes can exchange energy leading to transient growth. In other words, even though each eigenvalue leads to exponential decay of their respective modes, the corresponding nonorthogonal eigenvectors may interact to induce a temporary growth of the solution owing to the different rates of decay of each mode.

Not all nonnormal systems exhibit such a transient growth behavior. For example, try the system (5.8) with $\theta = \pi/4$. Also, observe that the coefficient matrices are nonnormal for both the simple and forced pendulum examples considered later in the chapter; however, neither exhibits transient growth behavior. Therefore, nonnormality is a necessary, but not sufficient, condition for transient growth behavior.

For nonnormal systems that do exhibit transient growth, the question becomes, "What is the initial perturbation that leads to the maximum possible transient growth of the system within finite time before succumbing to the asymptotic stability behavior as $t \to \infty$?" The initial perturbation that experiences maximum growth over a specified time range is called the *optimal perturbation* and is found using variational methods as shown in section 6.5 of Cassel (2013).

5.4 Phase-Plane and Stability Analysis – Nonlinear Systems

Autonomous linear systems can only exhibit a limited range of behaviors comprising various stable and unstable equilibrium points and limit cycles, where each has a distinct signature in the phase plane as discussed in Section 5.3.2. Nonlinear systems, however, display a number of far more complex – and more interesting – behaviors. In addition to the phase plane used for linear systems, we will make use of Poincaré sections, attractors, and bifurcation diagrams to help us better understand the behavior of the solutions of nonlinear systems of equations. We will use a series of increasingly complex systems to illustrate these behaviors in the following sections. Our goal here is to illustrate some important methods, terminology, and phenomena. For a more formal and comprehensive treatment of these topics, including heteroclinic and homoclinic orbits, manifolds, Lyapunov exponents and stability, global bifurcations, and Hamiltonian dynamics, see Drazen (1992), Meiss (2007), and Pikovsky and Politi (2016).

5.4.1 Stability of the Simple Pendulum

Let us begin by analyzing the simple pendulum. We will illustrate how the nonlinear system is linearized and analyzed for stability. As illustrated in Figure 5.20, a simple pendulum – with no damping – is a one-degree of freedom system with the angle $\theta(t)$ being the dependent variable. As derived in section 6.2 of Cassel (2013), its equation of motion is given by

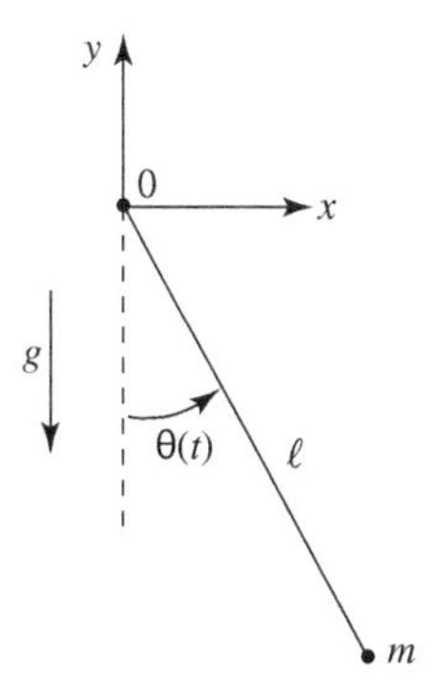

Figure 5.20 Schematic of the simple pendulum.

$$\ddot{\theta} + \frac{g}{\ell} \sin \theta = 0. \tag{5.9}$$

To transform the nonlinear governing equation to a system of first-order equations, let

$$x_1(t) = \theta(t), \quad x_2(t) = \dot{\theta}(t),$$

such that $x_1(t)$ and $x_2(t)$ give the angular position and velocity, respectively. Differentiating and substituting the governing equation gives

$$\begin{aligned} \dot{x}_1 &= \dot{\theta} = x_2, \\ \dot{x}_2 &= \ddot{\theta} = -\frac{g}{\ell} \sin \theta = -\frac{g}{\ell} \sin x_1. \end{aligned} \tag{5.10}$$

Therefore, we have a system of first-order *nonlinear* equations owing to the sine function.

Equilibrium positions occur where the angular velocity $\dot{\theta}$ and angular acceleration $\ddot{\theta}$ are zero; that is,

$$\{\dot{\theta}, \ddot{\theta}\} = \{\dot{x}_1, \dot{x}_2\} = \{0, 0\}.$$

Thus, from the system (5.10)

$$x_2 = 0, \qquad -\frac{g}{\ell} \sin x_1 = 0,$$

and the stationary points are given by

$$s_1 = x_1 = n\pi, \quad s_2 = x_2 = 0, \quad n = 0, 1, 2, \ldots.$$

The equilibrium states of the system $\mathbf{s} = \begin{bmatrix} s_1 & s_2 \end{bmatrix}^T$ are then

$$\mathbf{s} = \begin{bmatrix} s_1 & s_2 \end{bmatrix}^T = \begin{bmatrix} n\pi & 0 \end{bmatrix}^T, \quad n = 0, 1, 2, \ldots.$$

Therefore, there are two equilibrium points corresponding to n even and n odd. The equilibrium point with n even corresponds to when the pendulum is hanging vertically downward ($\theta = 0, 2\pi, \ldots$), and the equilibrium point with n odd corresponds to when the pendulum is located vertically above the pivot point ($\theta = \pi, 3\pi, \ldots$).

In order to evaluate stability of the equilibrium states, let us impose small disturbances about the equilibrium points according to

$$\begin{aligned} x_1(t) &= s_1 + \epsilon u_1(t) = n\pi + \epsilon u_1(t), \\ x_2(t) &= s_2 + \epsilon u_2(t) = \epsilon u_2(t), \end{aligned}$$

where $\epsilon \ll 1$. Substituting into (5.10) leads to

$$\begin{aligned} \epsilon \dot{u}_1 &= \epsilon u_2, \\ \epsilon \dot{u}_2 &= -\frac{g}{\ell} \sin(n\pi + \epsilon u_1). \end{aligned} \tag{5.11}$$

Note that

$$\sin(n\pi + \epsilon u_1) = \sin(n\pi)\cos(\epsilon u_1) + \cos(n\pi)\sin(\epsilon u_1) = \begin{cases} \sin(\epsilon u_1), & n \text{ even} \\ -\sin(\epsilon u_1), & n \text{ odd} \end{cases}.$$

Because ϵ is small, we may expand the sine function in terms of a Taylor series

$$\sin(\epsilon u_1) = \epsilon u_1 - \frac{(\epsilon u_1)^3}{3!} + \cdots = \epsilon u_1 + O(\epsilon^3),$$

where we have neglected higher-order-terms in ϵ. Essentially, we have linearized the system about the equilibrium points by considering an infinitesimally small perturbation to the equilibrium solutions of the system. Substituting into the system of equations (5.11) and canceling ϵ leads to the linear system of equations

$$\begin{aligned} \dot{u}_1 &= u_2, \\ \dot{u}_2 &= \mp \frac{g}{\ell} u_1, \end{aligned} \tag{5.12}$$

where the minus sign corresponds to the equilibrium point with n even, and the plus sign to n odd. Let us consider each case in turn.

For n even, the system (5.12) in matrix form $\dot{\mathbf{u}} = \mathbf{A}\mathbf{u}$ is

$$\begin{bmatrix} \dot{u}_1 \\ \dot{u}_2 \end{bmatrix} = \begin{bmatrix} 0 & 1 \\ -\frac{g}{\ell} & 0 \end{bmatrix} \begin{bmatrix} u_1 \\ u_2 \end{bmatrix}.$$

In order to diagonalize the system, determine the eigenvalues of $\mathbf{A}$. We write

$$(\mathbf{A} - \lambda_n \mathbf{I}) = \mathbf{0}.$$

For a nontrivial solution, the determinant must be zero, such that

$$\begin{vmatrix} -\lambda_n & 1 \\ -\frac{g}{\ell} & -\lambda_n \end{vmatrix} = 0,$$

or

$$\lambda_n^2 + \frac{g}{\ell} = 0.$$

Factoring the characteristic equation gives the eigenvalues

$$\lambda_{1,2} = \pm\sqrt{-\frac{g}{\ell}} = \pm\sqrt{\frac{g}{\ell}}\,\mathrm{i},$$

which are plus and minus an imaginary number. Then in uncoupled variables $\mathbf{v}$ ($\mathbf{u} = \mathbf{U}\mathbf{v}$), where

$$\dot{\mathbf{v}} = \mathbf{U}^{-1}\mathbf{A}\mathbf{U}\mathbf{v}, \quad \mathbf{U}^{-1}\mathbf{A}\mathbf{U} = \begin{bmatrix} \lambda_1 & 0 \\ 0 & \lambda_2 \end{bmatrix},$$

the solution near the equilibrium point corresponding to n even is of the form

$$u(t) = \hat{c}_1 e^{\sqrt{\frac{g}{\ell}}\mathrm{i}t} + \hat{c}_2 e^{-\sqrt{\frac{g}{\ell}}\mathrm{i}t},$$

or

$$u(t) = c_1 \sin\left(\sqrt{\frac{g}{\ell}}t\right) + c_2 \cos\left(\sqrt{\frac{g}{\ell}}t\right).$$

This oscillatory solution is *linearly stable* in the form of a *stable center* as the solution remains bounded for all time.

In the case of $\underline{n \text{ odd}}$, the system (5.12) is

$$\begin{bmatrix} \dot{u}_1 \\ \dot{u}_2 \end{bmatrix} = \begin{bmatrix} 0 & 1 \\ \frac{g}{\ell} & 0 \end{bmatrix} \begin{bmatrix} u_1 \\ u_2 \end{bmatrix}.$$

Determining the eigenvalues produces

$$\begin{vmatrix} -\lambda_n & 1 \\ \frac{g}{\ell} & -\lambda_n \end{vmatrix} = 0,$$

$$\lambda_n^2 - \frac{g}{\ell} = 0,$$

$$\lambda_{1,2} = \pm\sqrt{\frac{g}{\ell}},$$

which are plus and minus a real number. Thus, in uncoupled variables

$$v_1 = c_1 e^{\sqrt{\frac{g}{\ell}}t}, \quad v_2 = c_2 e^{-\sqrt{\frac{g}{\ell}}t}.$$

Then the solution near the equilibrium point corresponding to n odd is of the form

$$u(t) = c_1 e^{\sqrt{\frac{g}{\ell}}t} + c_2 e^{-\sqrt{\frac{g}{\ell}}t}.$$

Observe that while the second term decays exponentially as $t \to \infty$, the first term grows exponentially, eventually becoming unbounded. Therefore, this equilibrium point is *linearly unstable* in the form of an *unstable saddle*.

A very interesting problem results if a small change to the simple pendulum is made by adding vertical forcing of the pivot point. Because the forcing leads to coefficients in the differential equation of motion that depend explicitly on time, it becomes a nonautonomous system. While the forcing does not alter the equilibrium points at $\theta = 0$ and $\theta = \pi$, it has a dramatic effect on the stability of these two equilibrium points. The linear equation that governs stability of the forced pendulum near the equilibrium points is called *Mathieu's equation*. Interestingly, forcing at just the right frequencies leads to the normally stable $\theta = 0$ equilibrium point becoming unstable for narrow ranges of the forcing frequency. Even more surprisingly, forcing can stabilize the normally unstable $\theta = \pi$ equilibrium point if performed at just the right frequency. See section 6.4 of Cassel (2013) for more discussion and results for this fascinating case.

5.4.2 Types of Bifurcations

Now consider a series of nonlinear examples that illustrate additional features that can arise in nonlinear systems, in particular various bifurcation behaviors.

Saddle-Node Bifurcation

Consider the second-order system given by the nonlinear ordinary differential equations

$$\begin{aligned} \dot{u}_1 &= \alpha - u_1^2, \\ \dot{u}_2 &= -u_2. \end{aligned} \tag{5.13}$$

Equilibrium points occur when $\dot{u}_1 = \dot{u}_2 = 0$, which from (5.13) occur when

$$u_1 = \pm\sqrt{\alpha}, \quad u_2 = 0.$$

Whereas for $\alpha < 0$, there are no equilibrium points (u_1 and u_2 cannot be imaginary), for $\alpha > 0$, there are two equilibrium points given by

$$(u_1, u_2) = \left\{ \left(\sqrt{\alpha}, 0\right), \left(-\sqrt{\alpha}, 0\right) \right\}. \tag{5.14}$$

First, let us linearize about the equilibrium point $(u_1, u_2) = \left(\sqrt{\alpha}, 0\right)$. To do so, we define the new variables

$$u_1 = \sqrt{\alpha} + \epsilon\hat{u}_1, \quad u_2 = 0 + \epsilon\hat{u}_2, \tag{5.15}$$

where $\epsilon \ll 1$ is small. Substituting into the original equations (5.13) yields

$$\epsilon\dot{\hat{u}}_1 = \alpha - \left(\sqrt{\alpha} + \epsilon\hat{u}_1\right)^2 = \alpha - \alpha - 2\epsilon\sqrt{\alpha}\hat{u}_1 - \epsilon^2\hat{u}_1^2, \quad \epsilon\dot{\hat{u}}_2 = -\epsilon\hat{u}_2,$$

which after neglecting small terms of $O(\epsilon^2)$ simplifies to

$$\dot{\hat{u}}_1 = -2\sqrt{\alpha}\hat{u}_1, \quad \dot{\hat{u}}_2 = -\hat{u}_2.$$

In matrix form, this is the system of first-order ordinary differential equations

$$\dot{\hat{\mathbf{u}}} = \mathbf{A}\hat{\mathbf{u}}, \quad \mathbf{A} = \begin{bmatrix} -2\sqrt{\alpha} & 0 \\ 0 & -1 \end{bmatrix}. \tag{5.16}$$

The linearized equations (5.16) govern the behavior of the system (5.13) in the vicinity of the equilibrium point $(u_1, u_2) = \left(\sqrt{\alpha}, 0\right)$, and the stability of these solutions is determined by the eigenvalues of $\mathbf{A}$ according to Section 5.3.2. Recalling that $\alpha > 0$, the eigenvalues of the diagonal matrix $\mathbf{A}$ are given by

$$\lambda_1 = -2\sqrt{\alpha} < 0, \quad \lambda_2 = -1. \tag{5.17}$$

Because both of the eigenvalues are real and negative, the equilibrium point $(u_1, u_2) = \left(\sqrt{\alpha}, 0\right)$ is a *stable node*.

Now, let us linearize about the equilibrium point $(u_1, u_2) = \left(-\sqrt{\alpha}, 0\right)$. To do so, we define the new variables

$$u_1 = -\sqrt{\alpha} + \epsilon\hat{u}_1, \quad u_2 = 0 + \epsilon\hat{u}_2. \tag{5.18}$$

Substituting into the original equations (5.13) yields

$$\epsilon\dot{\hat{u}}_1 = \alpha - \left(-\sqrt{\alpha} + \epsilon\hat{u}_1\right)^2 = \alpha - \alpha + 2\epsilon\sqrt{\alpha}\hat{u}_1 - \epsilon^2\hat{u}_1^2, \quad \epsilon\dot{\hat{u}}_2 = -\epsilon\hat{u}_2,$$

which after neglecting small terms of $O(\epsilon^2)$ simplifies to

$$\dot{\hat{u}}_1 = 2\sqrt{\alpha}\hat{u}_1, \quad \dot{\hat{u}}_2 = -\hat{u}_2.$$

In matrix form, this is the system of first-order ordinary differential equations

$$\dot{\hat{\mathbf{u}}} = \mathbf{A}\hat{\mathbf{u}}, \quad \mathbf{A} = \begin{bmatrix} 2\sqrt{\alpha} & 0 \\ 0 & -1 \end{bmatrix}. \tag{5.19}$$

The linearized equations (5.19) govern the behavior of the system (5.13) in the vicinity of the equilibrium point $(u_1, u_2) = \left(-\sqrt{\alpha}, 0\right)$. Recalling that $\alpha > 0$, the eigenvalues of the diagonal matrix $\mathbf{A}$ are given by

$$\lambda_1 = 2\sqrt{\alpha} > 0, \quad \lambda_2 = -1. \tag{5.20}$$

Because both of the eigenvalues are real and of opposite sign, the equilibrium point $(u_1, u_2) = \left(-\sqrt{\alpha}, 0\right)$ is an *unstable saddle.*

Summarizing, there are no equilibrium points for $\alpha < 0$, and as $\alpha \to 0$, the unstable saddle and stable node approach each other and merge when $\alpha = 0$ to form the *bifurcation point*. This is illustrated in the $\hat{u}_1$ versus α bifurcation diagram as shown in Figure 5.21 and is known as a *saddle-node bifurcation*.

For stability of nonlinear systems, therefore, we (1) determine the equilibrium points of the system, (2) obtain equations near the equilibrium points that govern evolution of the disturbances, (3) linearize the system about the equilibrium points, and (4) evaluate the eigenvalues for asymptotic stability near each equilibrium point.

Transcritical Bifurcation

Consider the second-order system given by the nonlinear ordinary differential equations

$$\begin{aligned} \dot{u}_1 &= \alpha u_1 - u_1^2, \\ \dot{u}_2 &= -u_2, \end{aligned} \tag{5.21}$$

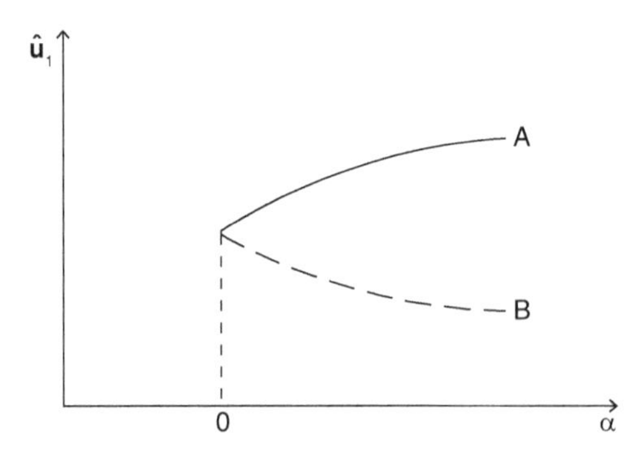

Figure 5.21 Bifurcation diagram for a saddle-node bifurcation (solid curves indicate stable solutions; dashed curves indicate unstable solutions).

which is similar to the previous system except with u_1 multiplying the α parameter. Equilibrium points occur when $\dot{u}_1 = \dot{u}_2 = 0$, which from (5.21) occur when

$$u_1 = 0, \alpha, \quad u_2 = 0.$$

Therefore, there are two equilibrium points given by

$$(u_1, u_2) = \{(0,0), (\alpha, 0)\}. \tag{5.22}$$

First, let us linearize about the equilibrium point $(u_1, u_2) = (0,0)$. To do so, we define the new variables

$$u_1 = 0 + \epsilon \hat{u}_1, \quad u_2 = 0 + \epsilon \hat{u}_2, \tag{5.23}$$

where $\epsilon \ll 1$ is small. Substituting into the original equations (5.21) yields

$$\dot{\hat{u}}_1 = \alpha \hat{u}_1, \quad \dot{\hat{u}}_2 = -\hat{u}_2.$$

In matrix form, this is the system of first-order ordinary differential equations

$$\dot{\hat{\mathbf{u}}} = \mathbf{A}\hat{\mathbf{u}}, \quad \mathbf{A} = \begin{bmatrix} \alpha & 0 \\ 0 & -1 \end{bmatrix}. \tag{5.24}$$

The linearized equations (5.24) govern the behavior of the system (5.21) in the vicinity of the equilibrium point $(u_1, u_2) = (0,0)$. The eigenvalues of the diagonal matrix $\mathbf{A}$ are given by

$$\lambda_1 = \alpha, \quad \lambda_2 = -1. \tag{5.25}$$

When $\alpha < 0$, both of the eigenvalues are real and negative, and the equilibrium point $(u_1, u_2) = (0,0)$ is a *stable node*. When $\alpha > 0$, both of the eigenvalues are real and of opposite sign, and the equilibrium point is an *unstable saddle*.

Now, let us linearize about the equilibrium point $(u_1, u_2) = (\alpha, 0)$. To do so, we define the new variables

$$u_1 = \alpha + \epsilon \hat{u}_1, \quad u_2 = 0 + \epsilon \hat{u}_2. \tag{5.26}$$

Substituting into the original equations (5.21) yields

$$\epsilon \dot{\hat{u}}_1 = \alpha\left(\alpha + \epsilon \hat{u}_1\right) - \left(\alpha + \epsilon \hat{u}_1\right)^2 = \alpha^2 + \epsilon \alpha \hat{u}_1 - \alpha^2 - 2\epsilon \alpha \hat{u}_1 - \epsilon^2 \hat{u}_1^2, \; \epsilon \dot{\hat{u}}_2 = -\epsilon \hat{u}_2,$$

which after neglecting small terms of $O(\epsilon^2)$ simplifies to

$$\dot{\hat{u}}_1 = -\alpha \hat{u}_1, \quad \dot{\hat{u}}_2 = -\hat{u}_2.$$

In matrix form, this is the system of first-order ordinary differential equations

$$\dot{\hat{\mathbf{u}}} = \mathbf{A}\hat{\mathbf{u}}, \quad \mathbf{A} = \begin{bmatrix} -\alpha & 0 \\ 0 & -1 \end{bmatrix}. \tag{5.27}$$

The linearized equations (5.27) govern the behavior of the system (5.21) in the vicinity of the equilibrium point $(u_1, u_2) = (\alpha, 0)$. The eigenvalues of the diagonal

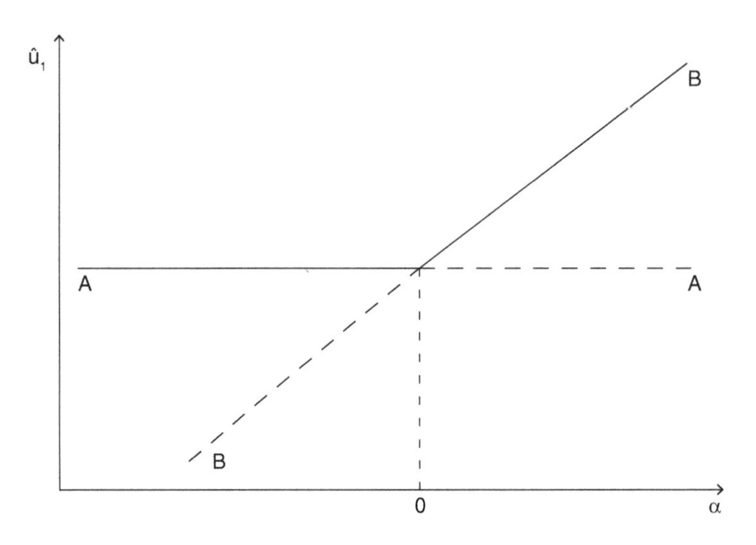

Figure 5.22 Bifurcation diagram for a transcritical bifurcation (solid curves indicate stable solutions; dashed curves indicate unstable solutions).

matrix **A** are given by

$$\lambda_1 = -\alpha, \quad \lambda_2 = -1. \tag{5.28}$$

When $\alpha < 0$, both of the eigenvalues are real and of opposite sign, and the equilibrium point $(u_1, u_2) = (\alpha, 0)$ is an *unstable saddle*. When $\alpha > 0$, both of the eigenvalues are real and of the same sign, and the equilibrium point is a *stable node*.

Summarizing, there is an unstable saddle and stable node for $\alpha < 0$ as well as when $\alpha > 0$. At the bifurcation point $\alpha = 0$, the two equilibrium points switch respectively from stable to unstable and unstable to stable. This is illustrated in the $\hat{u}_1$ versus α bifurcation diagram as shown in Figure 5.22 and is known as a *transcritical bifurcation*.

Pitchfork Bifurcation

Consider the second-order system given by the nonlinear ordinary differential equations

$$\begin{aligned} \dot{u}_1 &= \alpha u_1 - u_1^3, \\ \dot{u}_2 &= -u_2, \end{aligned} \tag{5.29}$$

which is similar to the previous system except that the exponent on u_1 in the first equation has been changed to 3. Equilibrium points occur when $\dot{u}_1 = \dot{u}_2 = 0$, which from (5.29) occur when

$$u_1 = 0, \pm\sqrt{\alpha}, \quad u_2 = 0.$$

There are now three equilibrium points given by

$$(u_1, u_2) = \left\{(0,0), \left(\sqrt{\alpha}, 0\right), \left(-\sqrt{\alpha}, 0\right)\right\}. \tag{5.30}$$

First, let us linearize about the equilibrium point $(u_1, u_2) = (0,0)$. To do so, we define the new variables

$$u_1 = 0 + \epsilon \hat{u}_1, \quad u_2 = 0 + \epsilon \hat{u}_2, \tag{5.31}$$

where $\epsilon \ll 1$ is small. Substituting into the original equations (5.29) yields

$$\dot{\hat{u}}_1 = \alpha \hat{u}_1, \quad \dot{\hat{u}}_2 = -\hat{u}_2.$$

In matrix form, this is the system of first-order ordinary differential equations

$$\dot{\hat{\mathbf{u}}} = \mathbf{A}\hat{\mathbf{u}}, \quad \mathbf{A} = \begin{bmatrix} \alpha & 0 \\ 0 & -1 \end{bmatrix}. \tag{5.32}$$

The linearized equations (5.32) govern the behavior of the system (5.29) in the vicinity of the point $(u_1, u_2) = (0,0)$. The eigenvalues of the diagonal matrix $\mathbf{A}$ are given by

$$\lambda_1 = \alpha, \quad \lambda_2 = -1. \tag{5.33}$$

When $\alpha < 0$, both of the eigenvalues are real and negative, and the equilibrium point $(u_1, u_2) = (0,0)$ is a *stable node*. When $\alpha > 0$, both of the eigenvalues are real and of opposite sign, and the equilibrium point is an *unstable saddle*.

Next, let us linearize about the equilibrium point $(u_1, u_2) = \left(\sqrt{\alpha}, 0\right)$, where $\alpha > 0$. To do so, we define the new variables

$$u_1 = \sqrt{\alpha} + \epsilon \hat{u}_1, \quad u_2 = 0 + \epsilon \hat{u}_2. \tag{5.34}$$

Substituting into the original equations (5.29) yields

$$\begin{aligned} \epsilon \dot{\hat{u}}_1 &= \alpha\left(\sqrt{\alpha} + \epsilon \hat{u}_1\right) - \left(\sqrt{\alpha} + \epsilon \hat{u}_1\right)^3 = \alpha^{3/2} + \epsilon \alpha \hat{u}_1 - \alpha^{3/2} - 3\epsilon \alpha \hat{u}_1 + H.O.T., \\ \epsilon \dot{\hat{u}}_2 &= -\epsilon \hat{u}_2, \end{aligned}$$

which after neglecting small terms of $O(\epsilon^2)$ and higher simplifies to

$$\dot{\hat{u}}_1 = -2\alpha \hat{u}_1, \quad \dot{\hat{u}}_2 = -\hat{u}_2.$$

In matrix form, this is the system of first-order ordinary differential equations

$$\dot{\hat{\mathbf{u}}} = \mathbf{A}\hat{\mathbf{u}}, \quad \mathbf{A} = \begin{bmatrix} -2\alpha & 0 \\ 0 & -1 \end{bmatrix}. \tag{5.35}$$

The linearized equations (5.35) govern the behavior of the system (5.29) in the vicinity of the point $(u_1, u_2) = \left(\sqrt{\alpha}, 0\right)$. Recalling that $\alpha > 0$, the eigenvalues of the diagonal matrix $\mathbf{A}$ are given by

$$\lambda_1 = -2\alpha < 0, \quad \lambda_2 = -1. \tag{5.36}$$

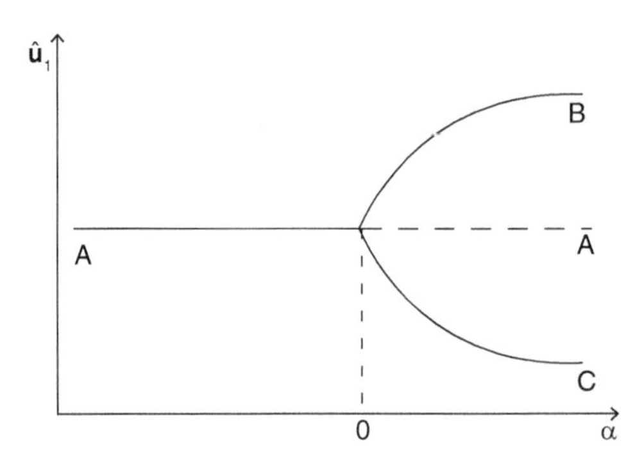

Figure 5.23 Bifurcation diagram for a pitchfork bifurcation (solid curves indicate stable solutions; dashed curves indicate unstable solutions).

Because both of the eigenvalues are real and negative, the equilibrium point $(u_1, u_2) = \left(\sqrt{\alpha}, 0\right)$ is a *stable node*.

Finally, let us linearize about the equilibrium point $(u_1, u_2) = \left(-\sqrt{\alpha}, 0\right)$. To do so, we define the new variables

$$u_1 = -\sqrt{\alpha} + \epsilon \hat{u}_1, \quad u_2 = 0 + \epsilon \hat{u}_2. \tag{5.37}$$

Substituting into the original equations (5.29) yields

$$\begin{aligned} \epsilon \dot{\hat{u}}_1 &= \alpha\left(-\sqrt{\alpha} + \epsilon \hat{u}_1\right) - \left(-\sqrt{\alpha} + \epsilon \hat{u}_1\right)^3 \\ &= -\alpha^{3/2} + \epsilon \alpha \hat{u}_1 + \alpha^{3/2} - 3\epsilon \alpha \hat{u}_1 + H.O.T., \\ \epsilon \dot{\hat{u}}_2 &= -\epsilon \hat{u}_2, \end{aligned}$$

which after neglecting small terms of $O(\epsilon^2)$ and higher simplifies to

$$\dot{\hat{u}}_1 = -2\alpha \hat{u}_1, \quad \dot{\hat{u}}_2 = -\hat{u}_2.$$

In matrix form, this is the system of first-order ordinary differential equations

$$\dot{\hat{\mathbf{u}}} = \mathbf{A}\hat{\mathbf{u}}, \quad \mathbf{A} = \begin{bmatrix} 2\sqrt{\alpha} & 0 \\ 0 & -1 \end{bmatrix}. \tag{5.38}$$

The system of equations (5.38) is the same as that for the previous equilibrium point; therefore, we have a *stable node* for $\alpha > 0$.

Summarizing, there is one stable equilibrium point when $\alpha < 0$ that becomes unstable for $\alpha > 0$. Also for $\alpha > 0$, there are two additional stable equilibrium points. This is illustrated in the $\hat{u}_1$ versus α bifurcation diagram as shown in Figure 5.23 and is known as a *pitchfork bifurcation*.

Hopf Bifurcation

Observe from the preceding bifurcations that the stable or unstable equilibrium points bifurcate into other stable or unstable equilibrium points as the parameter α changes across the bifurcation point. The *Hopf bifurcation* considered next involves a limit-cycle solution around an equilibrium point. When the limit-cycle solution is stable, but the equilibrium point is unstable, it is known as a *supercritical Hopf bifurcation.* When the limit-cycle solution is unstable, but the equilibrium point is stable, it is known as a *subcritical Hopf bifurcation.*

To illustrate the Hopf bifurcation, let us consider the van der Pol oscillator, which is governed by the second-order, nonlinear equation

$$\ddot{u} - \left(\alpha - u^2\right)\dot{u} + u = 0. \tag{5.39}$$

In order to convert to a system of first-order differential equations, let $u_1 = u$ and $u_2 = \dot{u}$, then

$$\begin{aligned} \dot{u}_1 &= u_2, \\ \dot{u}_2 &= -u_1 + \left(\alpha - u_1^2\right)u_2. \end{aligned} \tag{5.40}$$

Equilibrium points occur when $\dot{u}_1 = \dot{u}_2 = 0$, which from (5.40) occur when

$$u_1 = 0, \quad u_2 = 0.$$

Thus, the equilibrium point is given by

$$(u_1, u_2) = \{(0,0)\}. \tag{5.41}$$

To linearize about the equilibrium point $(u_1, u_2) = (0,0)$, let us define the new variables

$$u_1 = 0 + \epsilon\hat{u}_1, \quad u_2 = 0 + \epsilon\hat{u}_2, \tag{5.42}$$

where $\epsilon \ll 1$ is small. Substituting into (5.40) yields

$$\epsilon\dot{\hat{u}}_1 = \epsilon\hat{u}_2, \quad \epsilon\dot{\hat{u}}_2 = -\epsilon\hat{u}_1 + \left(\alpha - \epsilon\hat{u}_1^2\right)\epsilon\hat{u}_2,$$

which after neglecting small terms of $O(\epsilon^2)$ simplifies to

$$\dot{\hat{u}}_1 = \hat{u}_2, \quad \dot{\hat{u}}_2 = -\hat{u}_1 + \alpha\hat{u}_2.$$

In matrix form, this is the system of first-order ordinary differential equations

$$\dot{\hat{\mathbf{u}}} = \mathbf{A}\hat{\mathbf{u}}, \quad \mathbf{A} = \begin{bmatrix} 0 & 1 \\ -1 & \alpha \end{bmatrix}. \tag{5.43}$$

The eigenvalues of matrix $\mathbf{A}$ that determine the stability of the system near the equilibrium point at the origin are given by

$$\lambda_{1,2} = \frac{\alpha}{2} \pm \frac{1}{2}\sqrt{\alpha^2 - 4}. \tag{5.44}$$

There are four possible cases, as follows:

1. **Stable Node:** $\alpha \leq -2$
   For the case with $\alpha \leq -2$, the origin is a *stable node*, and the solution trajectories proceed toward the origin as illustrated by the four trajectories in Figure 5.24.

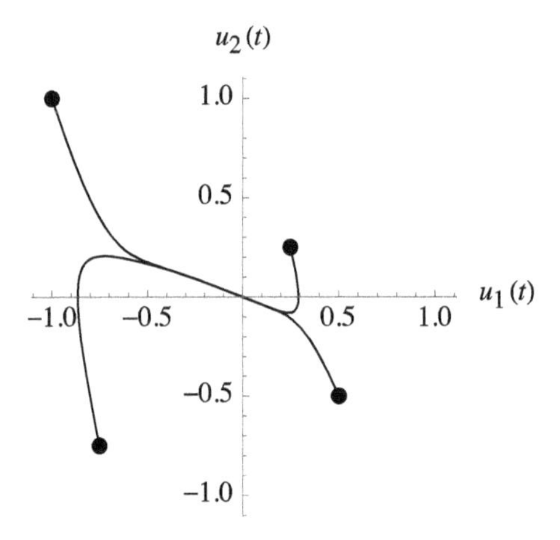

Figure 5.24 Phase-plane plot of trajectories for *stable node* corresponding to $\alpha = -3.0$ in a Hopf bifurcation (dots denote initial condition for each trajectory).

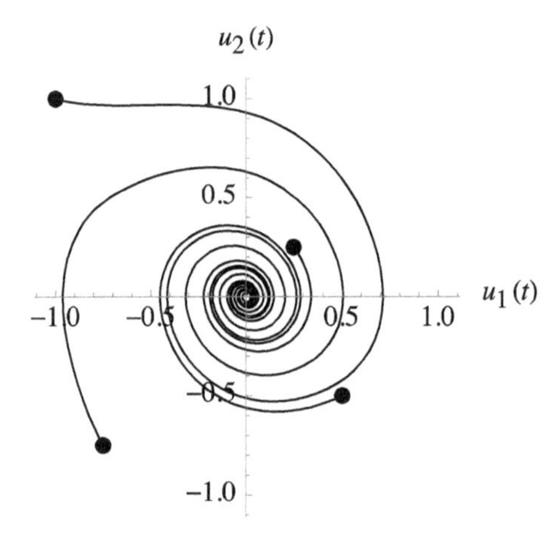

Figure 5.25 Phase-plane plot of trajectories for *stable spiral* corresponding to $\alpha = -0.25$ in a Hopf bifurcation (dots denote initial condition for each trajectory).

2. **Stable Spiral**: $-2 < \alpha < 0$
   For the case with $-2 < \alpha < 0$, the origin is a *stable spiral*, and the solution trajectories spiral toward the origin as illustrated by the four trajectories in Figure 5.25.
3. **Unstable Spiral**: $0 < \alpha < 2$
   For the case with $0 < \alpha < 2$, the origin is an *unstable spiral*, and the solution trajectories spiral outward away from the origin as illustrated by the four trajectories in Figure 5.26. Although the trajectories spiral outward away from the origin for initial conditions close to the equilibrium point at the origin consistent with an unstable spiral, it appears that they all converge on a limit-cycle solution. Let us choose four additional trajectories that begin outside this limit-cycle solution as

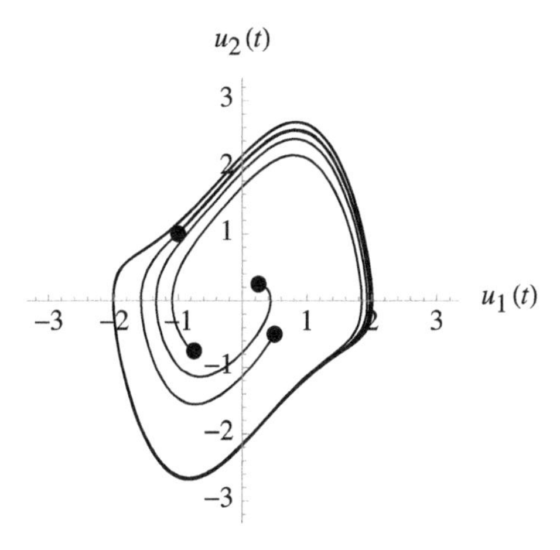

Figure 5.26 Phase-plane plot of trajectories for *unstable spiral* corresponding to $\alpha = 1.0$ in a Hopf bifurcation (dots denote initial condition for each trajectory).

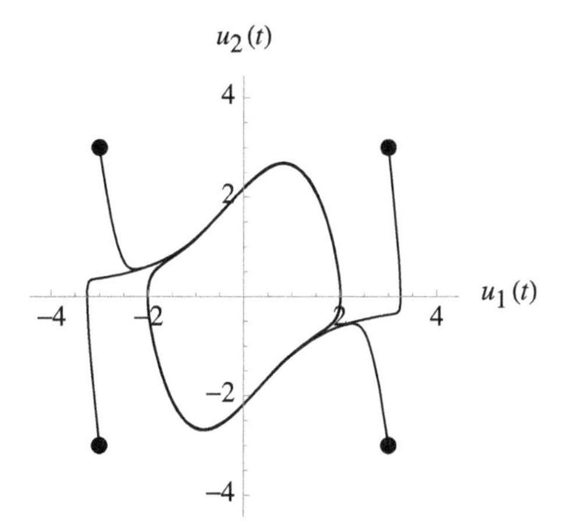

Figure 5.27 Phase-plane plot of trajectories for $\alpha = 1.0$ in a Hopf bifurcation (dots denote initial condition for each trajectory).

shown in Figure 5.27. As you can see, the trajectories all spiral inward and converge on the same limit cycle.

4. **Unstable Node**: $\alpha \geq 2$

   For the case with $\alpha \geq 2$, the origin is an *unstable node*, and the solution trajectories proceed outward away from the origin as illustrated by the four trajectories in Figure 5.28. Although the trajectories proceed outward away from the origin for initial conditions close to the equilibrium point at the origin consistent with an unstable node, it appears that they all converge on a limit-cycle solution. Let us choose four additional trajectories that begin outside this limit-cycle solution as shown in Figure 5.29. As you can see, the trajectories all proceed inward and converge on the same limit cycle.

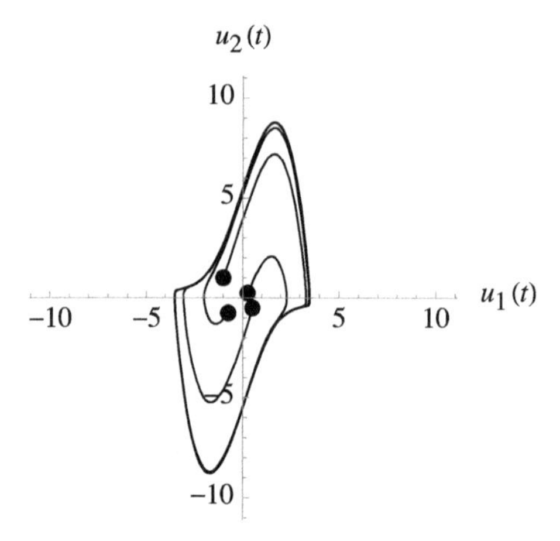

Figure 5.28 Phase-plane plot of trajectories for *unstable node* corresponding to $\alpha = 3.0$ in a Hopf bifurcation (dots denote initial condition for each trajectory).

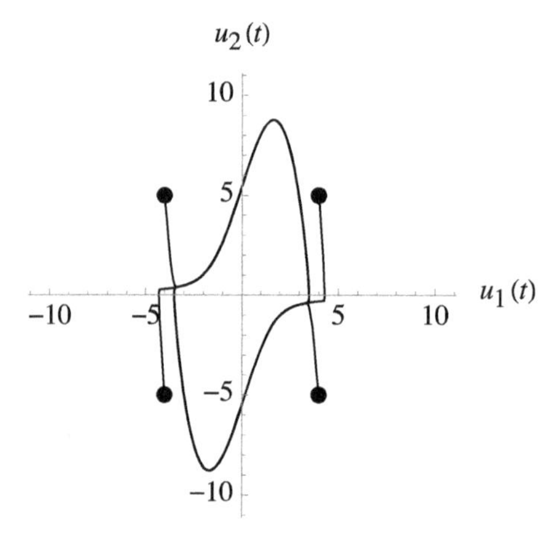

Figure 5.29 Phase-plane plot of trajectories for $\alpha = 3.0$ in a Hopf bifurcation (dots denote initial condition for each trajectory).

Summarizing, the equilibrium point at the origin is stable for $\alpha < 0$ and unstable for $\alpha > 0$, and for the latter case, a stable limit-cycle solution exists around the equilibrium point. Therefore, there is a *supercritical Hopf bifurcation* at $\alpha = 0$. Note that the presence of the periodic limit-cycle solution is not revealed from the stability analysis about the equilibrium point alone; it is necessary to explore solutions of the full nonlinear system throughout the phase plane to expose its presence.

5.5 Poincaré and Bifurcation Diagrams – Duffing Equation

Next, we analyze solutions of the nonlinear *Duffing equation*, which exhibits an intriguing array of behaviors that are common in nonlinear systems. In doing so, the Poincaré section will be introduced and illustrated.

5.5.1 Derivation

We begin by deriving the nonlinear Duffing equation. The nonlinearity can arise in two alternative applications: (1) a spring with a nonlinear restoring force, or (2) a simple pendulum for which a small amount of nonlinearity is included. For example, consider an oscillator with a nonlinear restoring force owing to the spring given by

$$f = -[k\hat{u}(\hat{t}) + \hat{\alpha}\hat{u}^3(\hat{t})],$$

where k is the usual linear spring constant, and $\hat{\alpha}$ represents the degree of nonlinearity in the spring. Recalling Newton's second law

$$ma = f,$$

we then have the second-order, nonlinear differential equation

$$m\frac{d^2\hat{u}}{d\hat{t}^2} = -k\hat{u} - \hat{\alpha}\hat{u}^3 \tag{5.45}$$

for the position $\hat{u}(t)$. We nondimensionalize time $\hat{t}$ and displacement $\hat{u}$ as follows:

$$t = \frac{\hat{t}}{\sqrt{m/k}}, \quad u(t) = \frac{\hat{u}(\hat{t})}{A},$$

where $\sqrt{k/m}$ is the natural frequency of the linear oscillator (with $\hat{\alpha} = 0$), and A is the initial displacement, such that $\hat{u}(0) = A$. Substituting into (5.45) gives

$$\frac{d^2u}{dt^2} + u + \alpha u^3 = 0, \tag{5.46}$$

where $\alpha = A^2\hat{\alpha}/k > 0$. This is known as the *Duffing equation*. The initial conditions are the initial position and velocity of the mass, for example,

$$u(0) = 1, \quad \dot{u}(0) = 0.$$

Observe that the forced Duffing equation is the nonlinear version of (7.1) derived in Section 7.1.1 for the forced spring–mass problem.

Alternatively, recall from Section 5.4.1 that the governing equation for the *simple pendulum* is

$$\frac{d^2\theta}{d\hat{t}^2} + \frac{g}{\ell}\sin\theta = 0. \tag{5.47}$$

To approximate the nonlinear term for small angles, let us expand $\sin\theta$ as a Taylor series

$$\sin\theta = \theta - \frac{\theta^3}{3!} + \frac{\theta^5}{5!} + \cdots.$$

Substituting into (5.47) gives

$$\frac{d^2\theta}{d\hat{t}^2} + \frac{g}{\ell}\left(\theta - \frac{\theta^3}{3!} + \frac{\theta^5}{5!} + \cdots\right) = 0.$$

Let us nondimensionalize time and define the angle according to

$$t = \frac{\hat{t}}{\sqrt{\ell/g}}, \quad u(t) = \theta(\hat{t}).$$

Substituting into the pendulum equation and retaining the first two terms in the Taylor series yields

$$\frac{d^2u}{dt^2} + u - \frac{1}{6}u^3 = 0, \tag{5.48}$$

which is also a Duffing equation (5.46) with $\alpha = -1/6$.

We may write the general form of the Duffing equation (5.46) as a system of first-order ordinary differential equations using the transformations

$$u_1(t) = u(t), \quad u_2(t) = \dot{u}(t).$$

Differentiating and substituting the Duffing equation (5.46) gives

$$\begin{aligned} \dot{u}_1(t) &= \dot{u} = u_2, \\ \dot{u}_2(t) &= \ddot{u} = -u - \alpha u^3 = -u_1 - \alpha u_1^3. \end{aligned}$$

Thus, the equivalent system of first-order ordinary differential equations is

$$\begin{aligned} \frac{du_1}{dt} &= u_2, \\ \frac{du_2}{dt} &= -u_1 - \alpha u_1^3. \end{aligned} \tag{5.49}$$

Owing to the nonlinearity, solutions are obtained numerically using the methods of Part II. With no nonlinearity ($\alpha = 0$), this is simply the equation governing a linear harmonic oscillator with a limit-cycle solution as shown in Figure 5.30.

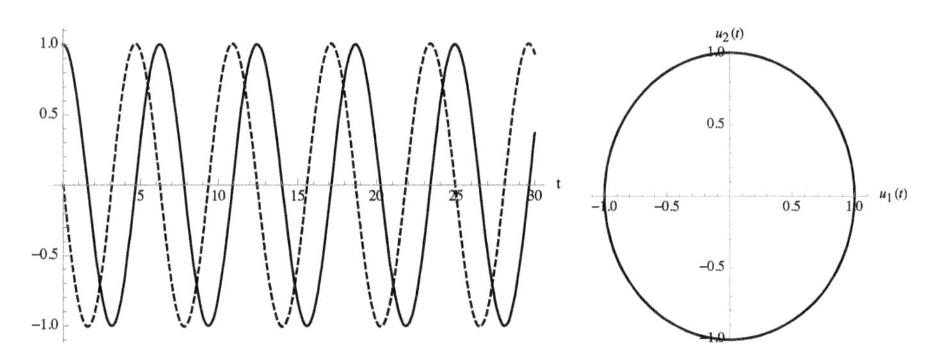

Figure 5.30 Limit-cycle solution of the Duffing equation with $\alpha = 0.02$. On the left is a plot of $u_1(t) = u(t)$ (solid line) and $u_2(t) = \dot{u}(t)$ (dashed line); on the right is the same solution plotted on the phase plane.

5.5.2 Equilibrium Points

Before considering more interesting solutions of the Duffing equation, let us first determine the equilibrium points of the system at which the system does not change with time, that is, where the velocity and acceleration are zero. As before, this occurs when $\dot{u}_1 = \dot{u}_2 = 0$ in (5.49), which requires that $u_2 = 0$, and

$$u_1 \left(u_1^2 + \frac{1}{\alpha} \right) = 0,$$

or

$$u_1 \left(u_1 + \frac{i}{\sqrt{\alpha}} \right) \left(u_1 - \frac{i}{\sqrt{\alpha}} \right) = 0.$$

Thus, the three equilibrium points are

$$(u_1, u_2) = \left\{ (0,0), \left(\frac{1}{\sqrt{-\alpha}}, 0 \right), \left(-\frac{1}{\sqrt{-\alpha}}, 0 \right) \right\},$$

where the origin is an equilibrium point for all α, and the other two are equilibrium points for $\alpha < 0$.

REMARKS:

1. *For the nonlinear oscillator ($\alpha > 0$):*
   - *Equilibrium point $(u_1, u_2) = (0,0)$ corresponds to the neutral position of the spring, for which $f = 0$.*
2. *For the simple pendulum with nonlinearity ($\alpha < 0$):*
   - *Equilibrium point $(u_1, u_2) = (0,0)$ corresponds to the stable position $\theta = 0$.*
   - *Equilibrium points $(u_1, u_2) = \left(\pm 1/\sqrt{-\alpha}, 0\right)$ correspond to unstable positions $\theta = \pm\pi$.*
3. *As with other nonlinear vibrations models, the period of oscillation for periodic solutions depends on the trajectory in the phase plane. That is, there is a dependence of the period on amplitude; compare this to a linear pendulum having a constant period for any amplitude oscillation.*

5.5.3 Solutions with $\alpha > 0$

For $\alpha > 0$, the solution is periodic in time represented by closed curves on the phase plane, and the equilibrium point is at the origin. Which limit cycle occurs is determined by the initial conditions $[u_1(0), u_2(0)]$. Let us consider the case when $\alpha = 0.02 > 0$ shown in Figure 5.30. Observe that the solution is periodic in time as expected.

5.5.4 Solutions with $\alpha < 0$

For $\alpha < 0$, the solution is also periodic in time for trajectories that begin near the origin; however, additional equilibrium points are located at

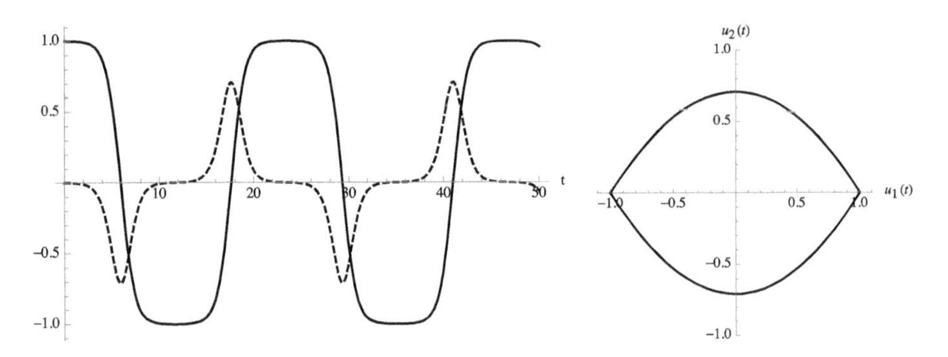

Figure 5.31 Solution of the Duffing equation with $\alpha = -0.9998$. On the left is a plot of $u(t)$ (solid line) and $\dot{u}(t)$ (dashed line); on the right is the same solution plotted on the phase plane.

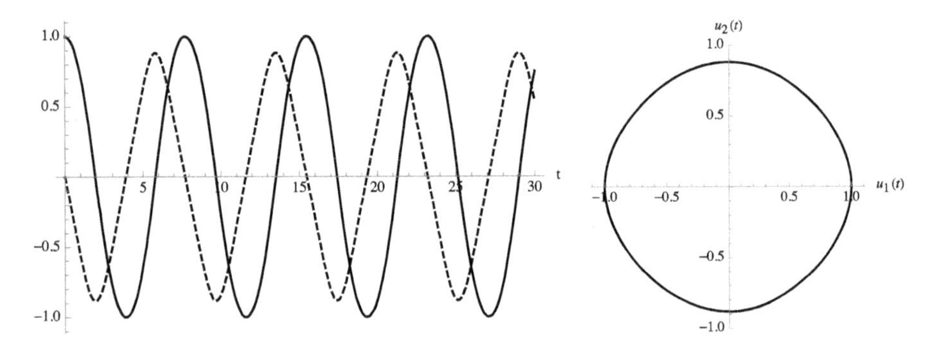

Figure 5.32 Solution of the Duffing equation with $\alpha = -0.4444$. On the left is a plot of $u(t)$ (solid line) and $\dot{u}(t)$ (dashed line); on the right is the same solution plotted on the phase plane.

$$u_1 = \pm\frac{1}{\sqrt{-\alpha}}, \quad u_2 = 0.$$

The trajectory through these points is called the *separatrix* and separates periodic solutions, which are stable, from those that grow to become unbounded, which are unstable. Whether the trajectory is inside or outside the separatrix is determined by the initial conditions $[u_1(0), u_2(0)]$. Motions that begin outside the separatrix that passes through these points grow to become unbounded.

Because we have fixed the initial position at $u(0) = 1$, we will choose a value of α that is just inside the corresponding separatrix, such that the equilibrium points are at $u_1 = \pm 1$. The trajectory for $\alpha = -0.9998 < 0$ is shown in Figure 5.31. This trajectory in the phase plane approximates the separatrix. Again, trajectories that begin inside the separatrix, through specification of α, will generate limit-cycle solutions, and trajectories that begin outside the separatrix will become unbounded. Let us choose a value of α that gives a trajectory that is well inside the separatrix, such as $\alpha = -0.4444 < 0$ shown in Figure 5.32. As expected, we have a limit-cycle solution similar to those for $\alpha > 0$. Now we choose α such that the trajectory is just

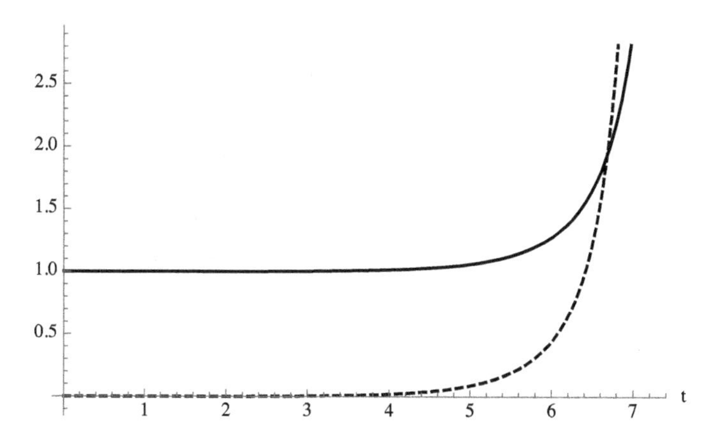

Figure 5.33 Solution of the Duffing equation with $\alpha = -1.0002$; plot of $u(t)$ (solid line) and $\dot{u}(t)$ (dashed line).

outside the separatrix. For $\alpha = -1.0002 < 0$, for example, the solution is shown in Figure 5.33. As expected, the solution eventually becomes unbounded.

There is an important result in dynamical systems theory known as the Poincaré–Bendixson theorem. It states that *second-order autonomous systems do not have chaotic solutions*. It applies for linear and nonlinear systems and restricts the only attractors to being equilibrium points and limit cycles. Note that the Duffing equation exhibits equilibrium points, limit cycles, and a separatrix as allowed for by this theorem. In the next section, we will add nonautonomous forcing, such that the Poincaré–Bendixson theorem no longer applies, allowing for fractals and chaotic behavior.

5.5.5 Duffing Equation with Damping and Forcing

In order to add in some additional physics, let us augment the Duffing equation (5.46) with a damping term and periodic forcing as follows:

$$\frac{d^2u}{dt^2} + d\frac{du}{dt} + u + \alpha u^3 = F\cos(\Omega t).$$

The first-order derivative term corresponds to the effect of damping, with damping coefficient d (see (5.2) for the linear damped oscillator), and the right-hand side accounts for the periodic forcing with amplitude F and frequency Ω. The presence of the nonautonomous forcing term renders the Poincaré–Bendixson theorem invalid and allows for deterministic chaos.

With the limit-cycle solution highlighted in the previous section as a baseline, consider the case with no damping ($d = 0$) or nonlinearity ($\alpha = 0$) and a small amount of forcing ($F = 0.1$) at the natural frequency of the system ($\Omega = 1$) as shown in Figure 5.34. Owing to resonance, the solution grows in time and eventually becomes unbounded. Interestingly, just a small amount of nonlinearity can break this

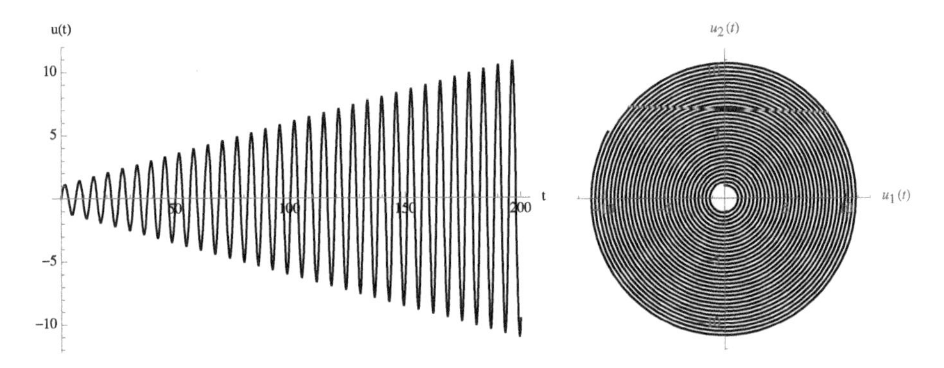

Figure 5.34 Solution of the forced Duffing equation with $\alpha = 0$, $d = 0$, $F = 0.1$, and $\Omega = 1$.

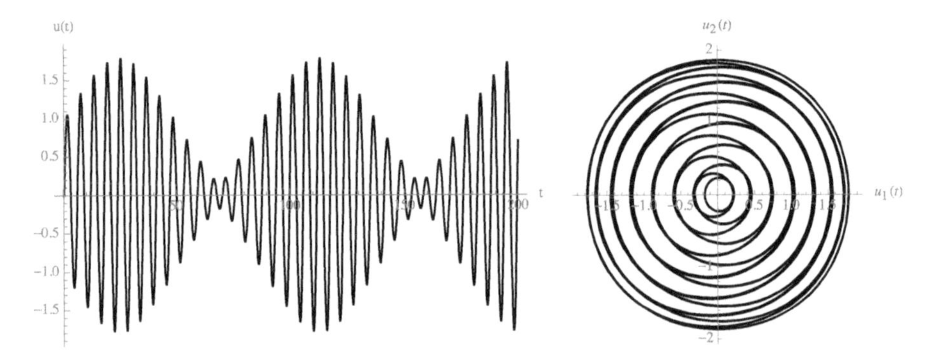

Figure 5.35 Solution of the forced Duffing equation with $\alpha = 0.1$, $d = 0$, $F = 0.1$, and $\Omega = 1$.

resonance. For example, the solution with $\alpha = 0.1$ is shown in Figure 5.35. The nonlinearity "breaks" the resonance and introduces a modulation in the amplitude of the oscillation.

Previously, we considered stability of equilibrium points of a linear or nonlinear system for which the solution remains unchanged for all time. Although limit-cycle solutions do change with time, they do so in a manner that is consistent for all time. Therefore, we may ask if such a limit-cycle solution is subject to a small disturbance, will the oscillation remain bounded (stable) or become unbounded (unstable)?

Recall that the Duffing equation with no damping or forcing results in a limit-cycle solution for any set of initial conditions when the nonlinear parameter $\alpha > 0$. When small amplitude forcing is included, a limit-cycle solution still exists, but only for a particular set of initial conditions. The initial conditions for which this is the case can be determined using a *multiple-scales analysis* in *perturbation methods*,[3] which can facilitate solutions of systems governed by differential equations having a small parameter – here the forcing amplitude and nonlinearity. For our limit-cycle solution, the multiple-scales analysis indicates how the curves bounding the oscillation

[3] Perturbation methods are sometimes called *asymptotic methods* in contexts, such as fluid dynamics, where the term "perturbation" implies a stability analysis.

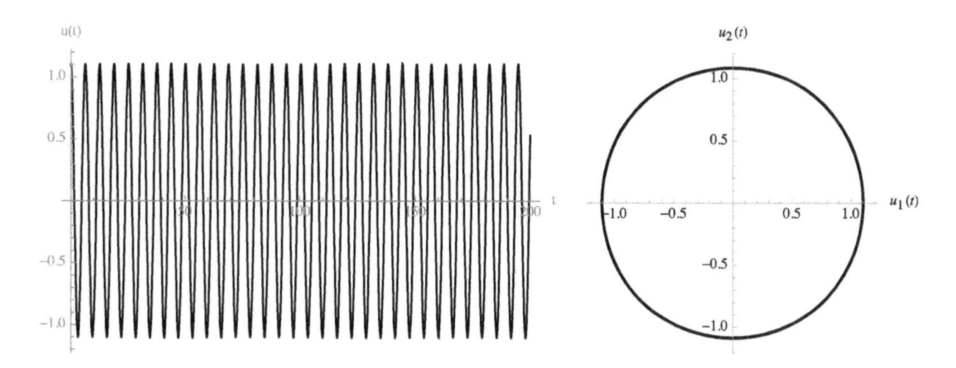

Figure 5.36 Solution of the forced Duffing equation with $\alpha = 0.1$, $d = 0$, $F = 0.1$, and $\Omega = 1$.

amplitude behave for time scales that are much longer than the oscillatory time scale. Because such bounding curves do not change with time for a limit-cycle solution, we can consider them to be equilibrium points and evaluate their long-time stability. In essence, we evaluate stability of the bounding curves of the oscillatory solution.

A multiple-scales analysis of the forced Duffing equation with small forcing amplitude and nonlinearity, but without damping, indicates that the initial conditions

$$u_1(0) = c_{1E} = \pm \left| \frac{4F}{3\alpha} \right|^{1/3}, \quad u_2(0) = 0,$$

result in a limit-cycle solution. A plot of the numerical solution for these initial conditions is given in Figure 5.36 for the case with $d = 0$ (no damping), $\alpha = 0.1$, $F = 0.1$, and $\Omega = 1$, which displays a periodic limit-cycle solution as expected.

To investigate stability of the limit-cycle solution that would result, we perturb the system and determine the behavior about the equilibrium points for the bounding curves. A stability analysis with the equilibrium values of c_{1E} gives the eigenvalues

$$\lambda_{1,2} = \pm \frac{3\sqrt{3}}{8} \alpha \left| \frac{4F}{3\alpha} \right|^{2/3} \mathrm{i} = \pm \Lambda \mathrm{i},$$

where Λ is real. Therefore, we have two purely imaginary eigenvalues, which corresponds to a *stable center* in the behavior of the bounding curves of the oscillation near the equilibrium points. Let us perturb our limit-cycle solution slightly by adding a small disturbance to the initial conditions corresponding to the limit-cycle solution according to

$$u_1(0) = c_{1E} + 0.05, \quad u_2(0) = 0.05.$$

The numerical solution for this case is shown in Figure 5.37. As expected from a perturbation analysis, the oscillatory solution exhibits a small amplitude modulation corresponding to the slow frequency. This solution remains bounded and is, therefore, stable.

Let us observe the influence of small damping with $d = 0.2$ on the previous perturbed limit-cycle solution as shown in Figure 5.38. Note that damping causes the

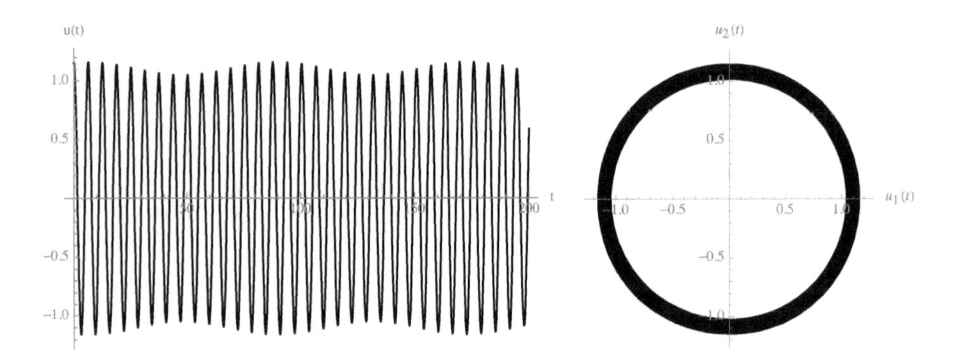

Figure 5.37 Solution of the forced Duffing equation with $\alpha = 0.1$, $d = 0$, $F = 0.1$, and $\Omega = 1$ and perturbed initial condition.

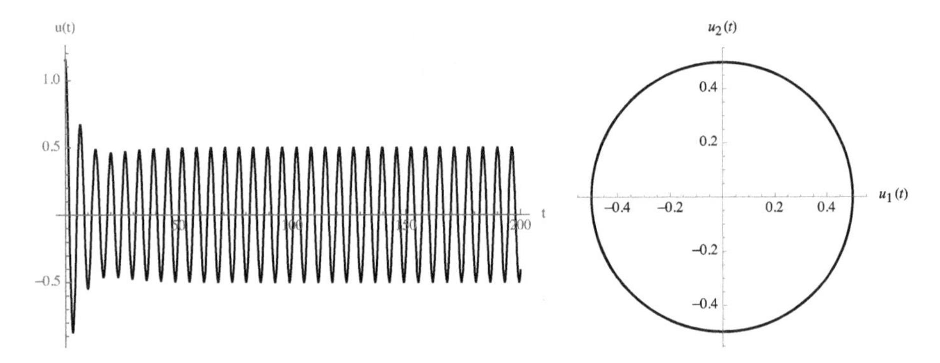

Figure 5.38 Solution of the forced Duffing equation with $\alpha = 0.1$, $d = 0.2$, $F = 0.1$, and $\Omega = 1$ and perturbed initial condition.

overall amplitude of the oscillation to diminish initially, but a new limit-cycle solution is eventually obtained when the forcing and damping are in balance. We can see the new limit cycle in the phase plane if we begin plotting the trajectory after the initial transients are removed as shown on the right.

One of the most remarkable, and least "linear," behaviors that nonlinear systems may exhibit is *deterministic chaos*. This will be explored in more detail in the next section using the Saltzman–Lorenz model, but we can get a taste of its richness via solutions of the forced Duffing equation. Let us increase the magnitude of the coefficients on the nonlinear and forcing terms such that they are $O(1)$, while keeping the damping small. The following results are for $d = 0.1$, $\alpha = 5$, $\Omega = 1.4$, and increasing forcing amplitudes F. The initial conditions are $u_1(0) = 0, u_2(0) = 0$.

The solution with $F = 0.1$ given in Figure 5.39 shows that after an initial transient, the solution settles down to a limit-cycle solution – the initial transient behavior has been eliminated from the phase-plane plot.

As we increase the forcing amplitude F, we expect the nonlinearity of the system to produce increasingly complicated trajectories in the phase plane. Because it is difficult to characterize these complicated solution trajectories in the phase plane alone, it is

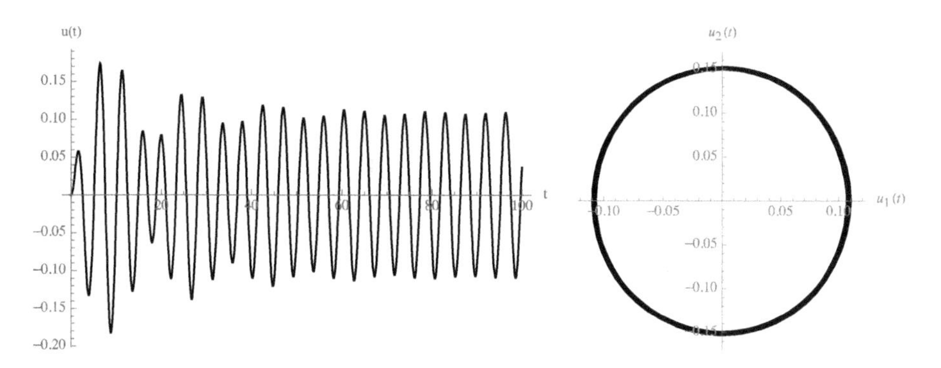

Figure 5.39 Solution of the forced Duffing equation with $\alpha = 5, d = 0.1, F = 0.1$, and $\Omega = 1.4$.

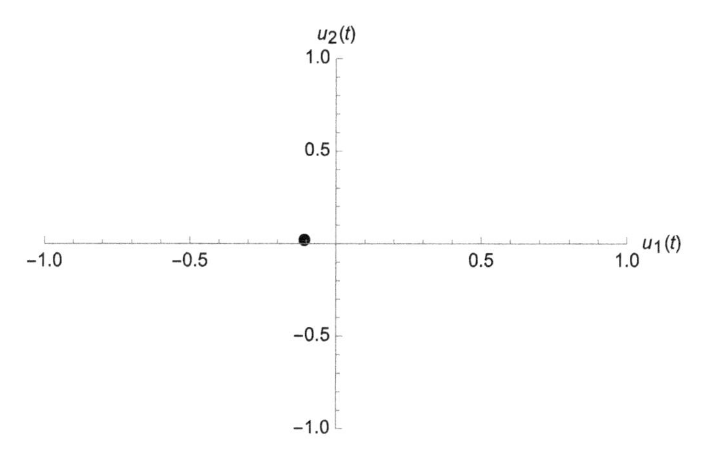

Figure 5.40 Poincaré section for the forced Duffing equation with $\alpha = 5, d = 0.1, F = 0.1$, and $\Omega = 1.4$.

useful to introduce the *Poincaré section*, or *map*, as an alternative means of showing the solution behavior. Whereas the phase plane shows the solution trajectory for all time, the Poincaré section only shows the solution trajectory as a point in the phase plane where it is located at integer multiples of the period corresponding to the forcing frequency, that is, at $t = 2\pi/\Omega, 4\pi/\Omega, \ldots$.[4] Note that for large forcing amplitude, the period of the solution is dominated by the period that is related to the forcing frequency, not the natural frequency of the oscillator. For $F = 0.1$ shown in Figure 5.39, the solution is a periodic limit cycle consisting of a single loop in the phase plane; therefore, the Poincaré section is simply a single point, as shown in Figure 5.40.

Increasing the forcing amplitude to $F = 10$ yields the solution in Figure 5.41. Upon removal of the initial transients in the phase plane, we see that the solution is a limit cycle consisting of two distinct loops (periods) that the trajectory traverses. The two periods show up as two points on the Poincaré section in Figure 5.42. We say that the system has undergone a "period doubling" in going from $F = 0.1$ to $F = 10$.

[4] This is similar to the effect of a strobe light.

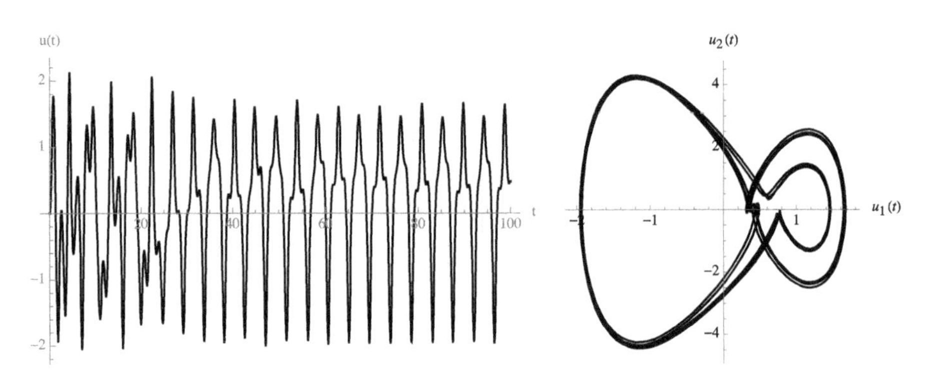

Figure 5.41 Solution of the forced Duffing equation with $\alpha = 5, d = 0.1, F = 10$, and $\Omega = 1.4$.

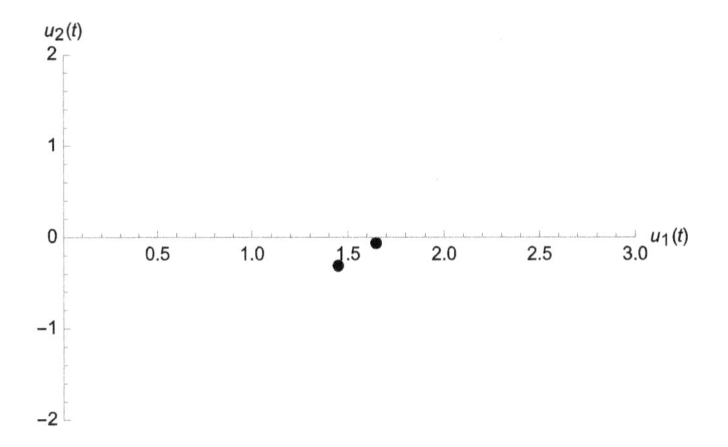

Figure 5.42 Poincaré section for the forced Duffing equation with $\alpha = 5, d = 0.1, F = 10$, and $\Omega = 1.4$.

Increasing the amplitude eventually leads to chaotic behavior, for which the phase-plane plot becomes very complex, even after removing the initial transient behavior. For example, the solution for $F = 100$ is shown in Figure 5.43. Although the dynamics of the system have become very complicated, and appear quite random, the Poincaré section reveals an underlying order in the now chaotic solution that is remarkable, not to mention rather beautiful, as shown in Figure 5.44. Notice that the trajectory is bounded within a finite region of the phase plane, which is referred to as an *attractor*. The concept of an attractor is that no matter where the initial conditions place us in the phase plane, the solution will eventually be "attracted" toward the region of the attractor as long as it is within the *basin of attraction*. Let us zoom in on the Poincaré section in the region near $(u, \dot{u}) = (3, -10)$. As we see in Figure 5.45, the local Poincaré section shows that the rich structure of the overall solution appears in a very similar manner on a smaller scale. Zooming still more in the region bounded by $2.95 < u < 3.0$ and $-10.4 < \dot{u} < -9.6$ again reveals similar structures on very small scales as shown in Figure 5.46. Such behavior that appears in a similar manner

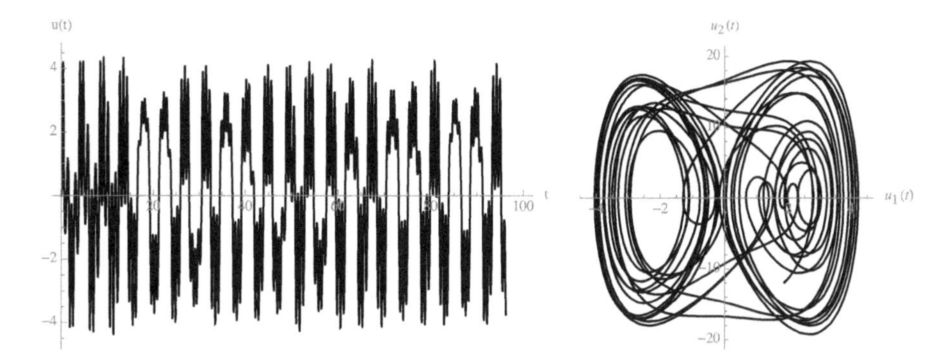

Figure 5.43 Solution of the forced Duffing equation with $\alpha = 5, d = 0.1, F = 100$, and $\Omega = 1.4$.

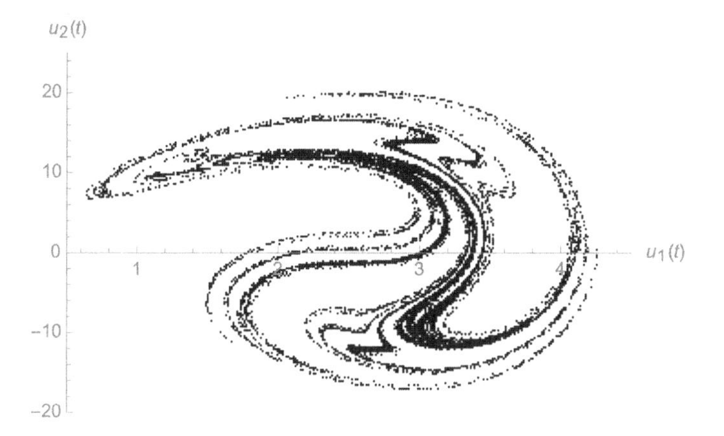

Figure 5.44 Poincaré section for the forced Duffing equation with $\alpha = 5, d = 0.1, F = 100$, and $\Omega = 1.4$.

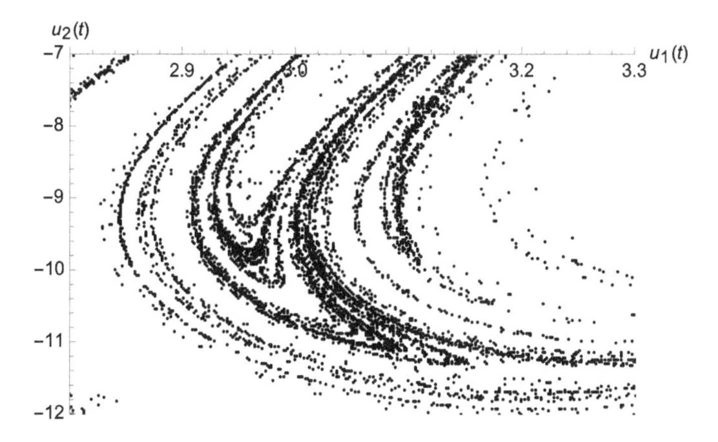

Figure 5.45 Poincaré section for the forced Duffing equation with $\alpha = 5, d = 0.1, F = 100$, and $\Omega = 1.4$.

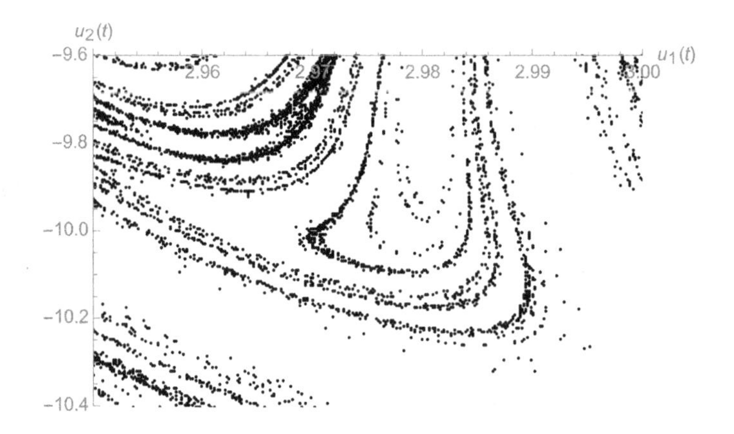

Figure 5.46 Poincaré section for the forced Duffing equation with $\alpha = 5, d = 0.1, F = 100$, and $\Omega = 1.4$.

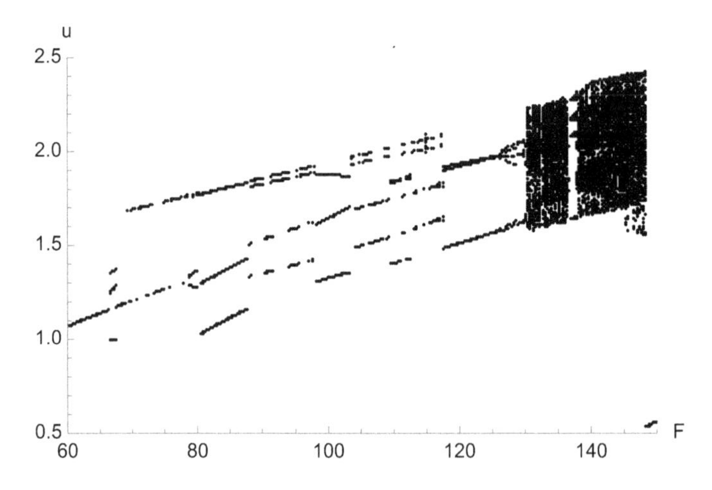

Figure 5.47 Bifurcation diagram for the forced Duffing equation with $\alpha = 5, d = 0.1$, $60 < F < 150$, and $\Omega = 1.4$.

on many scales is referred to as a *fractal* and is a characteristic of certain chaotic systems.

To understand how the system behaves as the forcing amplitude F is increased and the system becomes chaotic, it is helpful to generate a *bifurcation diagram*. A bifurcation point for a system is a point in parameter space where the solution changes its qualitative behavior, for example going from a solution trajectory with one period to one with two periods or where a stable solution becomes unstable. A bifurcation diagram indicates such points in parameter space where the solution bifurcates.

Let us first generate the bifurcation diagram for a large range of forcing amplitudes F as shown in Figure 5.47. Observe that there are ranges of F for which a limit cycle with one period exists, a limit cycle with several periods exists, and where the system exhibits chaos. Let us zoom in on the region between $125 < F < 133$ and the "upper

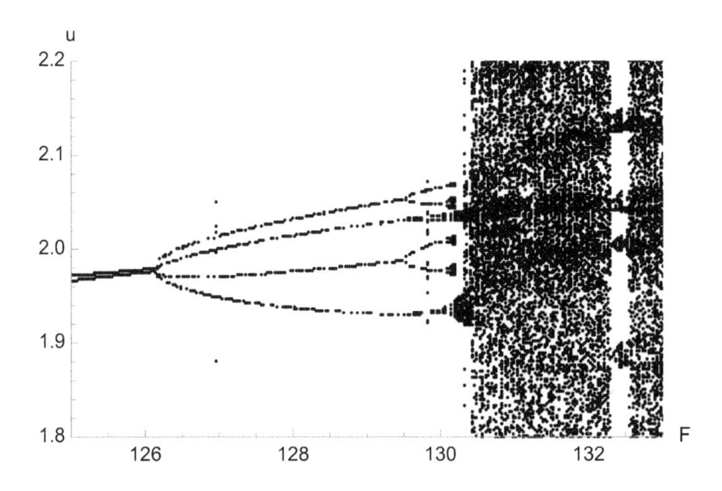

Figure 5.48 Bifurcation diagram for the forced Duffing equation with $\alpha = 5$, $d = 0.1$, $125 < F < 133$, and $\Omega = 1.4$.

branch" of the bifurcation diagram to more clearly see how the system approaches chaos. This is shown in Figure 5.48. We see that the solution goes through a series of "period doublings" after which it becomes chaotic. Interestingly, as we continue to increase the forcing amplitude F beyond the range for which chaos ensues, the solution returns to a periodic limit cycle before again becoming chaotic. In the overall bifurcation diagram shown in Figure 5.47, we can see that the solution bifurcates from a chaotic solution to one consisting of a single limit cycle with a single period near $F = 148$ (see the lower right portion of the figure). The forced Duffing equation exhibits a number of such bifurcations alternating between limit cycle and chaotic behavior.

REMARKS:

1. *In addition to the traditional plot of the dependent variable(s) with respect to time t, observe the usefulness of plotting the solution trajectory in the phase plane, Poincaré sections, and bifurcation diagrams.*
2. *Period doubling is just one of several routes to chaos. Each one has a unique signature in the phase plane, Poincaré section, and bifurcation diagram.*

5.6 Attractors and Periodic Orbits – Saltzman–Lorenz Model

Much of our modern understanding of deterministic chaos[5] in nonlinear systems arises from the seminal work of the meteorologist Edward Lorenz at MIT in the early 1960s. As is so often the case, there is an element of serendipity coupled with astute

[5] Many simply refer to these phenomena as "chaos" for short. However, it is important to emphasize that they are the result of solutions of deterministic differential equations, not simply some intrusion of random effects. Therefore, we use the more accurate "deterministic chaos" moniker.

observation and creative investigation in his discoveries. He was willing to allow his results to challenge prevailing wisdom at the time, and the result was a sea change in our thinking about nonlinear systems.

The Saltzman–Lorenz model[6] was investigated by Lorenz (1963). The equations were developed as a simple model for the so-called Rayleigh–Bénard convection problem, which governs the formation of convection rolls of a fluid between two parallel, horizontal surfaces at different temperatures. His interest was in predicting atmospheric dynamics with the ultimate goal of developing long-term weather prediction tools. What he found through obtaining numerical solutions of these seemingly simple equations has had a far-reaching influence on our understanding of the possible behaviors in nonlinear equations.

The Saltzman–Lorenz model was derived from the Navier–Stokes equations governing fluid dynamics and heat transfer of a fluid by making several simplifications, and it results in a system of three coupled, nonlinear, first-order ordinary differential equations:

$$\begin{aligned} \frac{dx}{dt} &= \sigma\left(y - x\right), \\ \frac{dy}{dt} &= rx - y - xz, \\ \frac{dz}{dt} &= xy - bz, \end{aligned} \tag{5.50}$$

where σ is the Prandtl number of the fluid, b is related to the aspect ratio (shape) of the convection rolls, and r is related to the Rayleigh number, which is the nondimensional temperature difference between the surfaces. For now, all solutions shown are for

$$\sigma = 3, \quad b = 1.$$

That is, we will keep the same fluid, σ, and maintain the shape of the convection rolls, b, to be the same and only adjust the temperature difference between the plates via the Rayleigh number r. The dependent variables $x(t)$, $y(t)$, and $z(t)$ essentially contain the time dependence of the streamfunction, which encapsulates the fluid motion, and temperature distributions expressed as truncated Fourier expansions. Note that the equations are nonlinear owing to the xz and xy terms in the second and third equations, respectively. Therefore, the solutions shown throughout the remainder of this section are obtained numerically using methods from Part II.

5.6.1 Equilibrium Points and Stability

In order to evaluate stability of the Saltzman–Lorenz system, let us determine the equilibrium points for which the system is independent of time (steady). Setting $\dot{x} = \dot{y} = \dot{z} = 0$ in the Saltzman–Lorenz model (5.50) gives a system of nonlinear algebraic

[6] Most refer to this model simply as the "Lorenz model" owing to his seminal contributions to our understanding of deterministic chaos. However, the mathematical model was first derived by Saltzman (1962).

equations. From the first equation, it is clear that any equilibrium point must be such that

$$x = y.$$

Substituting this into the third equation requires that

$$z = \frac{1}{b}x^2.$$

Substituting both of these results into the second equation yields

$$rx - x - \frac{1}{b}x^3 = 0,$$

which when factored gives

$$x\left[x^2 - b(r-1)\right] = 0.$$

Therefore, x has the three possible values

$$x = 0 \quad \text{or} \quad x = \pm\sqrt{b(r-1)}$$

at equilibrium points. As a result, the equilibrium points of the Saltzman–Lorenz system are

$$x = y = z = 0, \tag{5.51}$$

for all r, and

$$x = y = \pm\sqrt{b(r-1)}, \quad z = r - 1, \tag{5.52}$$

for $r > 1$.

Equation (5.51) corresponds to a stationary solution, in which there is no fluid motion and only conduction of heat through the fluid. We will refer to it as point

$$\mathcal{S} = (0, 0, 0) \tag{5.53}$$

in phase space. The equilibrium points indicated by (5.52) correspond to steady clockwise ($\mathcal{C}_1$) and counterclockwise ($\mathcal{C}_2$) rotating convection rolls and are given by

$$\mathcal{C}_1 = \left(-\sqrt{b(r-1)}, -\sqrt{b(r-1)}, r-1\right), \quad \mathcal{C}_2 = \left(\sqrt{b(r-1)}, \sqrt{b(r-1)}, r-1\right). \tag{5.54}$$

Alvarez-Ramirez et al. (2005) summarize the stability of the equilibrium points $\mathcal{S}$ and $\mathcal{C}_{1,2}$ as follows:

- $0 < r < 1$: The conduction-only solution $\mathcal{S}$ is the only equilibrium point and is globally linearly stable. The convection solutions $\mathcal{C}_1$ and $\mathcal{C}_2$ do not exist.
- $r \geq 1$: $\mathcal{S}$ is linearly unstable.
- $1 < r < r_H$: $\mathcal{C}_1$ and $\mathcal{C}_2$ are linearly stable.
- $r \geq r_H$: $\mathcal{C}_1$ and $\mathcal{C}_2$ become linearly unstable, and this is the range of Rayleigh number r for which the Saltzman–Lorenz model exhibits deterministic chaos.

Recall that a bifurcation is a point in parameter space where the qualitative character of a solution changes, such as the onset of an instability. In this case, there is a pitchfork bifurcation at $r = 1$, and a Hopf bifurcation at $r = r_H$, where

$$r_H = \sigma \frac{\sigma + b + 3}{\sigma - b - 1}.$$

A pitchfork bifurcation is such that an equilibrium point loses stability and two new equilibrium points are formed. It gets its name from its bifurcation diagram, in which one equilibrium point bifurcates into three (see Figure 5.23). A Hopf bifurcation occurs when the eigenvalue of an equilibrium point crosses the imaginary axis, which signals the loss (subcritical) or generation (supercritical) of a limit-cycle solution as the parameter increases.

Because $\mathcal{S}$, $\mathcal{C}_1$, and $\mathcal{C}_2$ are equilibrium solutions, the solution of the Saltzman–Lorenz equations should stay at these points if specified as the initial condition. If they are stable equilibrium points, then the solution should stay close to the equilibrium point even when a small perturbation is introduced. Let us first consider the stationary solution $\mathcal{S}$, and apply a small perturbation to the initial condition for $\mathcal{S}$ as follows:

$$x(0) = 0.000000001, \quad y(0) = 0, \quad z(0) = 0.$$

Because the $\mathcal{S}$ equilibrium point is stable for $r < 1$, let us consider a solution with $r = 0.9$ as shown in Figure 5.49. Note that indeed the solution remains very near to the stable equilibrium point at the origin.

Using the same initial condition as before, let us change the Rayleigh number such that $r = 10 > 1$, for which the stationary point $\mathcal{S}$ is unstable. Observe from Figure 5.50 that the solution remains very close to the equilibrium point $\mathcal{S}$ at the origin until approximately $t = 5$, after which the instability causes the solution to diverge substantially from $\mathcal{S}$.

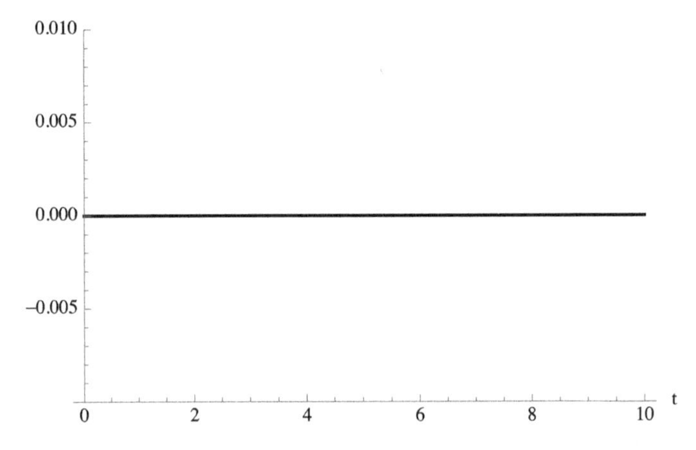

Figure 5.49 Solution of the Saltzman–Lorenz model with $\sigma = 3$, $b = 1$, and $r = 0.9$ and provided initial condition.

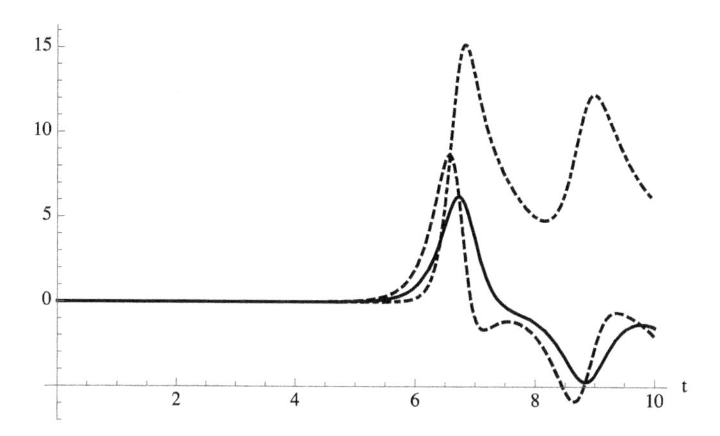

Figure 5.50 Solution of the Saltzman–Lorenz model with $\sigma = 3$, $b = 1$, and $r = 10$ and provided initial condition.

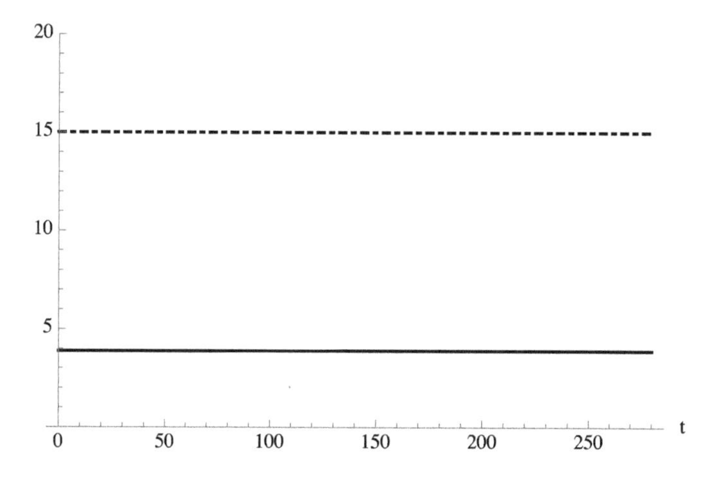

Figure 5.51 Solution of the Saltzman–Lorenz model with $\sigma = 3$, $b = 1$, and $r = 16$ and provided initial condition.

For $1 < r < r_H$, the convection equilibrium points C_1 and C_2 are stable. For the parameters specified, the threshold for a Hopf bifurcation is given by $r_H = 21$. Let us set the value of the Rayleigh number to be $r = 16 < r_H$. The initial conditions corresponding to the equilibrium point C_2, where a small perturbation $x(0)$, are given by

$$x(0) = \sqrt{b(r-1)} + 0.00001, \quad y(0) = \sqrt{b(r-1)}, \quad z(0) = r - 1.$$

The solution is shown in Figure 5.51. Because the equilibrium point C_2 is stable for the specified value of r, the solution remains close to C_2 for a long period of time.

For $r > r_H$, the convection equilibrium points C_1 and C_2 are unstable. Let us set $r = 26 > r_H$, such that it is above this critical value. Keeping the initial conditions the same as for the previous case, the solution is given in Figure 5.52. Observe once again that the solution remains very close to the equilibrium point C_2 until approximately

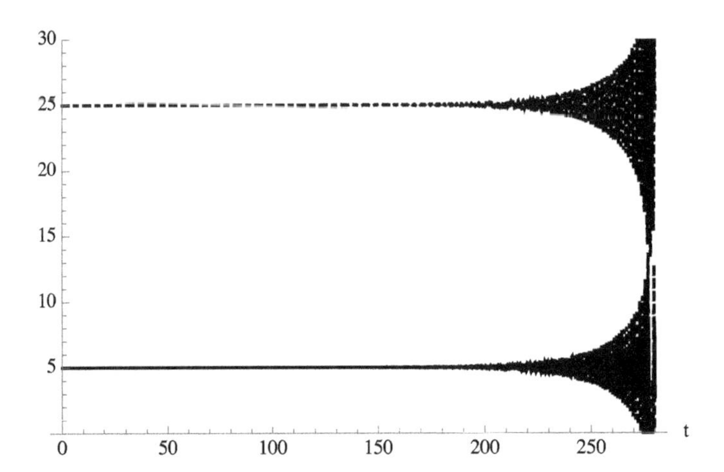

Figure 5.52 Solution of the Saltzman–Lorenz model with $\sigma = 3$, $b = 1$, and $r = 26$ and provided initial condition.

$t = 200$, at which time the instability causes the solution to diverge dramatically from $\mathcal{C}_2$.

Recall that the forced Duffing equation considered in Section 5.5 experiences a period-doubling route to chaos, whereas the Saltzman–Lorenz model considered here experiences a sudden bifurcation from a steady, stable solution to a chaotic solution via a Hopf bifurcation at $r = r_H$. The Saltzman–Lorenz model is a so-called reduced-order model of a more general coupled fluid dynamics and heat transfer problem. To see how to treat stability of the more general case of continuous systems governed by partial differential equations, see chapter 9 of Cassel (2013) for hydrodynamic stability of moving fluids.

5.6.2 Steady Convection Solution

Identifying the equilibrium and bifurcation points of the Saltzman–Lorenz model and the stability characteristics of its various solutions provides a road map for predicting the types of solutions that will occur for various choices of the system parameters and initial conditions. For example, let us consider the case with $r = 10 < r_H = 21$, for which the convection rolls $\mathcal{C}_1$ and $\mathcal{C}_2$ are both steady and stable. The initial conditions are chosen to be[7]

$$x(0) = 6, \quad y(0) = 6, \quad z(0) = 6.$$

In order to see how the solution behaves, let us first plot each of the dependent variables with time in Figure 5.53. Observe that each variable is oscillating toward a fixed value $(x, y, z) = (3, 3, 9)$ for which the solution does not change for large times. In fact, this is the steady equilibrium point corresponding to the convection roll $\mathcal{C}_2$ as

[7] Despite the apparently "devilish" nature of these initial conditions, there is nothing particularly special about this choice.

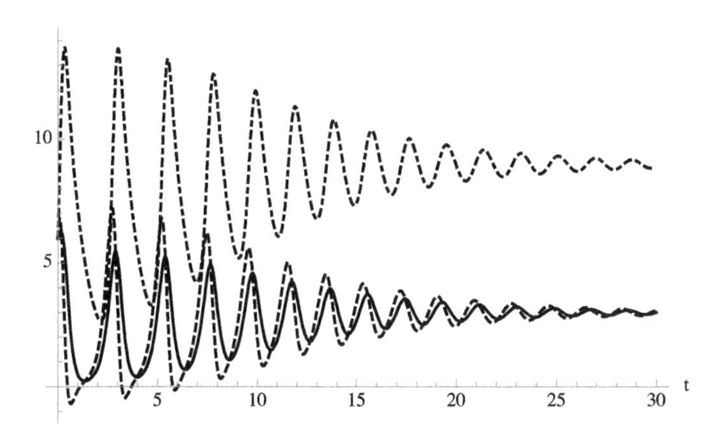

Figure 5.53 Solution of the Saltzman–Lorenz model with $\sigma = 3$, $b = 1$, and $r = 10$.

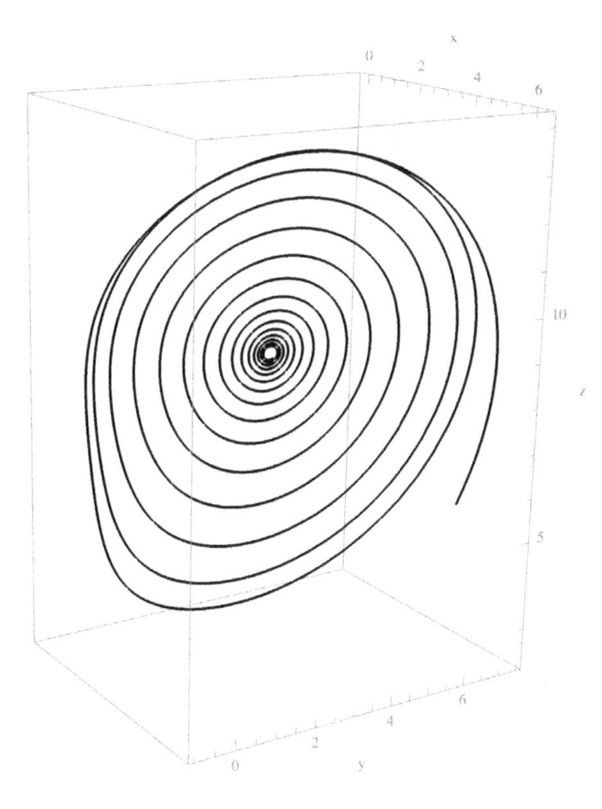

Figure 5.54 Solution of the Saltzman–Lorenz model with $\sigma = 3$, $b = 1$, and $r = 10$: x-y-z plane.

given by (5.54) with $\sigma = 3$ and $b = 1$. Such damped oscillations are reminiscent of linear behavior.

Because we have three, rather than two, dependent variables, we can also view the solution as a three-dimensional plot in phase space. This is provided in Figure 5.54.

The solution trajectory simply spirals inward from $(x, y, z) = (6, 6, 6)$ toward the C_2 point in a manner very similar to a damped oscillation in linear systems. Because the convection solutions are stable for $r = 10$, the solution remains steady and unchanged at the point $(x, y, z) = (3, 3, 9)$ for all time – even when perturbed.

5.6.3 Chaotic Solution

Now let us observe what happens when we increase the temperature difference between the surfaces – the Rayleigh number – such that

$$r = 27 > r_H = 21.$$

The other parameters and initial conditions remain the same as before. A plot of the three dependent variables is shown in Figure 5.55. Note that the trajectory for this case oscillates in a much less regular fashion and is not approaching a steady solution as for the case with $r = 10$. We can more easily see how the solution is behaving by observing the three-dimensional parametric plot in Figure 5.56. Note that again the solution starts at $(x, y, z) = (6, 6, 6)$, but then it switches back and forth between two lobes and no longer approaches a steady solution at large times.

The three-dimensional phase-plane plot in Figure 5.56 likely looks familiar. It is called the *Lorenz attractor*, and is one of many so-called *strange attractors*[8] that are observed in the solutions of various nonlinear systems of equations. It is called an "attractor" because, given any initial condition within the *basin of attraction*, the solution trajectory will be captured and remain in the attractor. It is called "strange" because the attractor has a geometrically complicated shape. The two lobes of the Lorenz attractor correspond to opposite directions of rotation of the convection rolls – clockwise and counterclockwise – with the centers of the lobes corresponding to the unstable, steady convection equilibrium points C_1 and C_2 given by (5.54). Therefore,

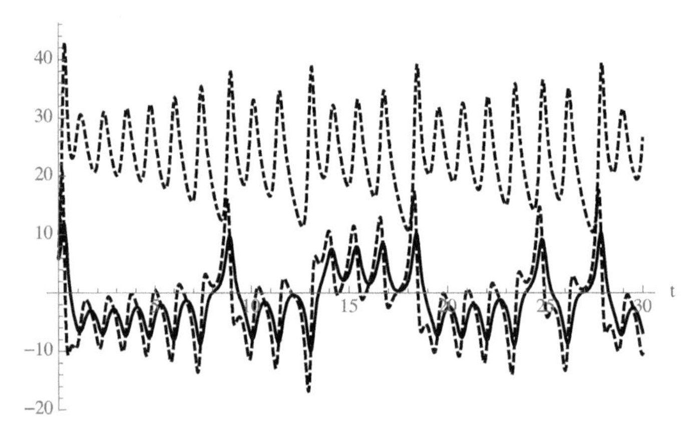

Figure 5.55 Solution of the Saltzman–Lorenz model with $\sigma = 3$, $b = 1$, and $r = 27$.

[8] Although chaotic attractors are typically "strange," they are not synonymous. There are chaotic attractors that are not strange, and strange attractors that are not chaotic.

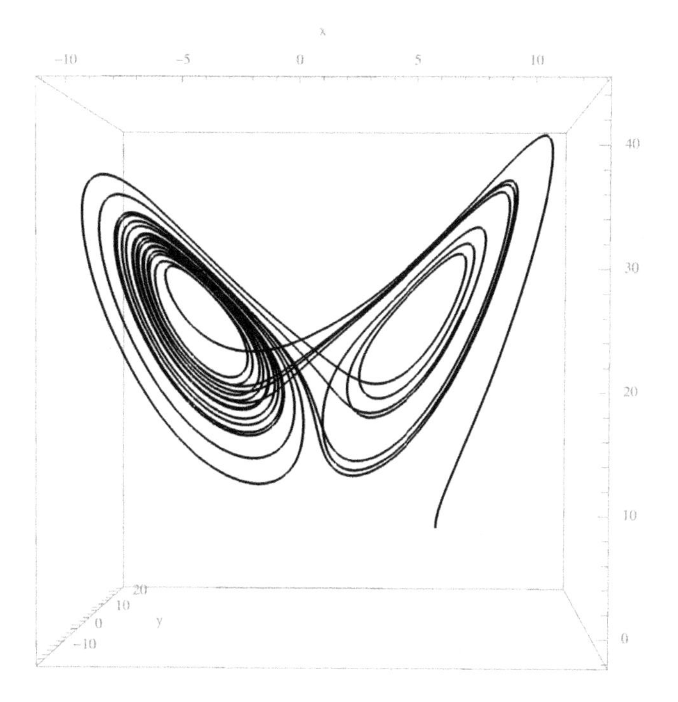

Figure 5.56 Solution of the Saltzman–Lorenz model with $\sigma = 3$, $b = 1$, and $r = 27$: x-y-z plane.

the solution indicates that the direction of rotation switches in a seemingly random manner between clockwise and counterclockwise.

5.6.4 Sensitivity to Initial Conditions

One of the unusual properties of chaotic solutions is their sensitivity to initial conditions. This was discovered accidentally by Lorenz when he wanted to recreate a previously completed numerical solution. To do so, he took the values of x, y, and z at a particular time from a printout of the original solution that showed the values to three decimal places. Thinking that this level of accuracy would be sufficient to recreate the original solution, he instead found that the solution eventually diverged significantly from the original solution as time progressed. The problem was that the computer stored numbers to the sixth decimal place. Therefore, if the value of one of the variables was 0.753441, for example, at the time he entered the values to use as initial conditions in the subsequent run, the printout from which he obtained the initial conditions would have read 0.753. This slight difference in initial conditions was enough to dramatically alter the solution of the nonlinear Lorenz model after some period of time.

To observe this sensitivity to initial conditions, let us evaluate the final solution of the previous case at $t = 30$, which is given by

$$(x, y, z) = (-6.8155, \ -10.5432, 26.4584);$$

therefore, the trajectory is in the left lobe of the attractor corresponding to $\mathcal{C}_1$ at $t = 30$. Now we will solve the equations with the same parameters and initial conditions except that for z, which we will alter slightly to $z(0) = 5.999$ instead of $z(0) = 6.000$ as before. While the solution trajectory is indistinguishable from the strange attractor shown in Figure 5.56, observe the values of the solution at $t = 30$, which are now

$$(x, y, z) = (7.45987, 12.4114, 26.9402).$$

As you can see, the solution is very different and in fact is in the right lobe of the attractor, which corresponds to the $\mathcal{C}_2$ convection solution having the opposite direction of rotation, even though only a small change has been made in the initial condition.

The sensitivity to initial conditions illustrated here may be interpreted physically as two points that are initially very close to each other in the phase plane eventually become separated. This exponential separation can be quantified using Lyapunov exponents as discussed in Pikovsky and Politi (2016). Such behavior gives rise to the so-called *butterfly effect*, in which a butterfly flapping its wings in India is imagined to eventually completely alter the weather patterns in Chicago. While this may be an overstatement, it illustrates the point that small changes in nonlinear systems can lead to large changes in system behavior, which is a hallmark of deterministic chaos. In the context of weather prediction, Lorenz concluded that this sensitivity to initial conditions means that the accuracy of atmospheric models is severely limited, restricting our predictive capabilities to relatively short periods of time. In fact, despite dramatic improvements in the mathematical models used for weather prediction and the computers on which such models are simulated since the early 1960s, accurate forecasting still only exists for a matter of days, as predicted by Lorenz (1963).[9]

In summary, as illustrated using the Saltzman–Lorenz model, three of the hallmarks of nonlinear systems as compared to linear systems are (1) very different qualitative behavior that may be observed depending upon the input parameters; (2) seemingly random behavior that exhibits some underlying order – deterministic chaos; and (3) sensitivity to initial conditions.

5.6.5 Periodic Orbits

A universal theme within deterministic chaos is the underlying order that is exhibited by "chaotic" solutions. Nothing exemplifies this more than the presence of periodic orbits within chaotic solutions. Viswanath (2003) showed that chaotic solutions of the Saltzman–Lorenz model are comprised of an infinity of *periodic orbits*. Periodic orbits are trajectories that repeat themselves over and over again and come back on themselves repeatedly. More specifically, a periodic orbit with period P is such that $u(t + P) = u(t)$ for all t. Periodic orbits can be stable or unstable as determined

[9] Although Lorenz was the first to illustrate this behavior using numerical results in 1963, it was anticipated much earlier by Henri Poincaré as highlighted in the quote at the beginning of the chapter from 1914.

using Floquet theory. For the Saltzman–Lorenz model, each periodic orbit is denoted by a sequence of letters A and B, where A corresponds to one of the lobes of the Lorenz attractor (left in the following figures), and B the other. The desired periodic orbits are selected by imposing a fixed initial condition for z, say $z(0) = 27$, and then choosing x and y initial conditions that fall on the desired periodic orbit as determined by Viswanath (2003).

Let us adjust the parameters a bit from what was used previously. Now we use

$$\sigma = 10, \quad b = \frac{8}{3}, \quad r = 28.$$

We illustrate by showing a few of the simplest periodic orbits.

AB Orbit: First, we consider the AB orbit that alternately traverses each of the A and B lobes of the Lorenz attractor. For the given parameters, the initial conditions for this orbit are

$$x(0) = -13.763610682134, \quad y(0) = -19.578751942452, \quad z(0) = 27.0.$$

The AB orbit is illustrated in Figure 5.57, and the period of the AB orbit is $P = 1.559$.

AAB Orbit: Next, we consider the AAB orbit that traverses the A lobe twice in sequence and then the B lobe once during each period. For the given parameters, the

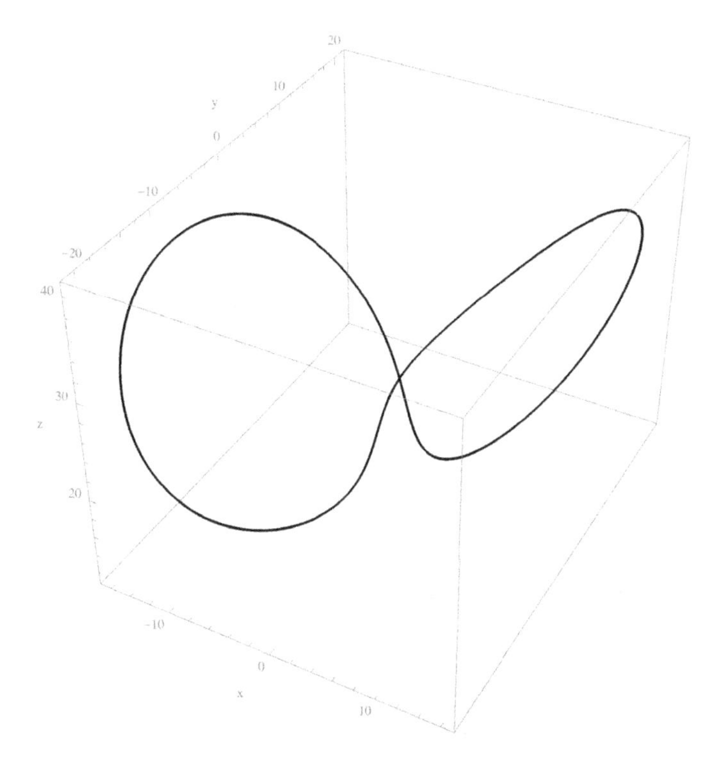

Figure 5.57 Solution of the Saltzman–Lorenz model with $\sigma = 10, b = 8/3$, and $r = 28$ showing the AB periodic orbit.

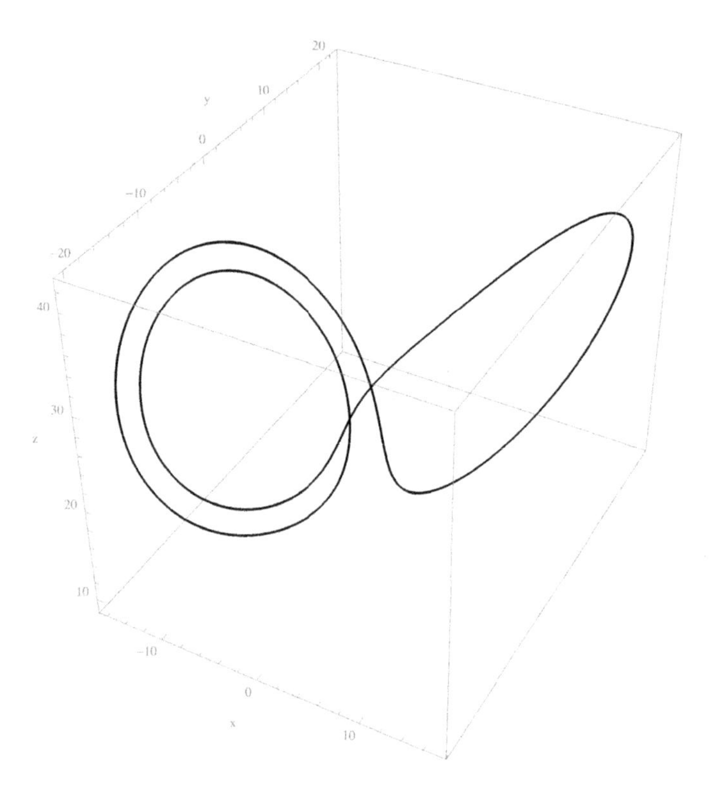

Figure 5.58 Solution of the Saltzman–Lorenz model with $\sigma = 10$, $b = 8/3$, and $r = 28$ showing the AAB periodic orbit.

initial conditions for this orbit are

$$x(0) = -12.595115397689, \quad y(0) = -16.970525307084, \quad z(0) = 27.0.$$

The AAB orbit is illustrated in Figure 5.58, and the period of the AAB orbit is $P = 2.306$.

AABBB Orbit: Finally, we consider the AABBB orbit that traverses the A lobe twice in sequence followed by the B lobe three times during each period. For the given parameters, the initial conditions for this orbit are

$$x(0) = -13.056930146345, \quad y(0) = -17.987214049281, \quad z(0) = 27.0.$$

The AABBB orbit is illustrated in Figure 5.59, and the period of the AABBB orbit is $P = 3.802$.

There are an infinity of such periodic orbits, the first 111,011 of which have been identified by Viswanath (2003) involving symbol sequences of length 20 or less. He used an iterative numerical technique based on Linstedt's method of strained coordinates, which is another perturbation method that is capable of identifying periodic limit-cycle solutions. Again, although the nonlinear Saltzman–Lorenz model exhibits

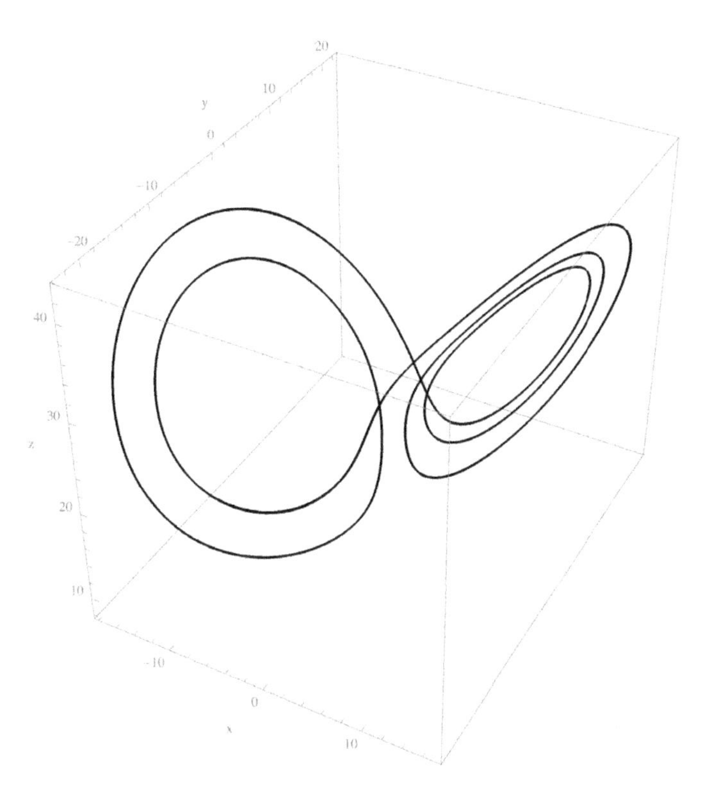

Figure 5.59 Solution of the Saltzman–Lorenz model with $\sigma = 10$, $b = 8/3$, and $r = 28$ showing the AABBB periodic orbit.

chaotic solutions, there is more order than one might expect in its solutions. It turns out that a property of all chaotic attractors is that they are comprised of an *infinite* number of *unstable* periodic orbits. In fact, periodic orbits have been identified in turbulent Couette flow, which is governed by the nonlinear Navier–Stokes partial differential equations, by Viswanath (2007). The fact that similar dynamics is occurring in nonlinear systems governed by both ordinary differential equations, having a finite number of degrees of freedom, as well as partial differential equations, characterized by an infinite number of degrees of freedom, is rather remarkable and highlights the potential for unification of methods developed for both types of mathematical models spanning a vast array of physical applications. This is the hope and promise provided by dynamical systems theory!

Part II

Numerical Methods

6 Computational Linear Algebra

> I often say that when you can measure what you are speaking about, and express it in numbers, you know something about it; but when you cannot express it in numbers, your knowledge is of a meager and unsatisfactory kind; it may be the beginning of knowledge, but you have scarcely, in your thoughts, advanced to the stage of science, whatever the matter may be. (Sir William Thomson – Lord Kelvin)

Congratulations! Having mastered the material in Part I, you have the fundamental knowledge and techniques for solving the linear systems of algebraic equations, algebraic eigenproblems, and linear differential equations that are necessary to address a broad range of applications. In Chapters 1 and 2, for example, we developed methods for determining exact solutions to small matrix and eigenproblems. While these methods are essential knowledge for everything that is to come, the methods themselves are only appropriate or efficient for small systems if solved by hand or moderate numbers of unknowns if solved exactly using mathematical software, such as MATLAB or Mathematica. In scientific and engineering applications, however, we will frequently have occasion to perform similar operations on systems involving hundreds, thousands, or even millions of unknowns. Therefore, we must consider how these methods can be extended – scaled – efficiently, and typically approximately, to such large systems. This comprises the subjects of *computational*, or *numerical, linear algebra* and *numerical methods* and provides the methods and algorithms used by general-purpose numerical software codes and libraries. Specifically, computational linear algebra primarily focuses on the ubiquitous tasks of solving large linear systems of algebraic equations and the algebraic eigenproblem. More often than not, these are the methods being used by the internal functions and procedures contained in mathematical software, such as MATLAB and Mathematica, and numerical libraries, such as BLAS and LAPACK.[1] Part II is devoted to the development of numerical methods for solving large-scale systems of linear algebraic equations, algebraic eigenproblems, and ordinary and partial differential equations.

In Chapter 1, one method sufficed for solving small systems of linear algebraic equations – Gaussian elimination. When solving very large systems of equations on a computer, however, we need an assortment of methods for dealing with all kinds of

[1] BLAS stands for Basic Linear Algebra Subprograms, and LAPACK stands for Linear Algebra PACKage. BLAS is a set of low-level functions that are used within mathematical libraries, such as LAPACK, to perform all common linear algebra operations within Fortran, C, and Python codes. The underlying routines are written very efficiently in Fortran.

Table 6.1 Summary of methods for solving systems of linear algebraic equations: $\mathbf{Au} = \mathbf{b}$ ($\rho(\mathbf{M})$ is the spectral radius of the iteration matrix).

Method	Limitations	Best for	Section(s)
Gaussian elimination	Any **A**	Small/moderate **A**	1.5.1
Matrix inverse	Nonsingular **A**	Small/moderate **A**	1.5.2
Cramer's rule	Nonsingular **A**	Small **A**	1.5.3
Linear combination of eigenvectors	Real, symmetric **A**	Small/moderate **A**	2.3.2, 2.3.4
Singular-value decomposition	Any **A**	Large, ill-conditioned **A**	2.8, 6.5.4, 10.2
QR decomposition	Nonsingular **A**	Large **A**	2.10, 10.2
LU decomposition	Nonsingular **A**	Large **A**	6.3.1
Cholesky decomposition	Symmetric, positive definite **A**	Large **A**	6.3.2
Thomas algorithm	Tridiagonal **A**	Large **A**	6.3.3
Jacobi iteration	$\rho(\mathbf{M}) < 1$	Large, diagonally dominant **A**	6.4.4
Gauss–Seidel iteration	$\rho(\mathbf{M}) < 1$	Large, diagonally dominant **A**	6.4.5
SOR	$\rho(\mathbf{M}) < 1$	Large, diagonally dominant **A**	6.4.6
Least-squares method	Any **A**	Over/underdetermined systems	10.2
Conjugate-gradient method	Symmetric, positive-definite **A**	Large, sparse **A**	6.4.7, 10.6
GMRES method	Nonsingular **A**	Large, sparse **A**	6.4.7, 10.7
Galerkin projection	Nonsingular **A**	Reduced-order modeling	13.4.1

eventualities and produce computationally efficient algorithms for problems that may be dense or sparse, structured or unstructured, well conditioned or ill conditioned, et cetera. Our choice of what method is best in a given situation will be informed by the application, how it is discretized into a system of equations, the computer architecture on which it will be solved (serial or parallel), along with numerous other practical considerations. There is increased emphasis on iterative methods as they provide additional flexibility as compared to direct methods for developing efficient algorithms for "solving" very large problems by being content with an approximate solution. Techniques for solving very large systems of equations computationally are discussed throughout Parts II and III, and all of the methods discussed in this text for solving systems of linear algebraic equations are summarized in Table 6.1 with reference to their respective sections. Similarly, computer algorithms for calculating the eigenvalues and eigenvectors of large matrices numerically are typically based on QR decomposition, which was introduced in Section 2.10. Such algorithms, and their variants, are described in Section 6.5 and summarized in Table 6.2.

Table 6.2 Summary of methods for solving algebraic eigenproblems: $\mathbf{Au}_n = \lambda_n \mathbf{u}_n$.

Method	Limitations	Best for	Section
Factor characteristic equation	Square **A**	Small **A**	2.2
Iterative QR algorithm	Square **A**	Large **A**	6.5
Arnoldi method	Square **A**	Large, sparse **A**	6.5.3
Lanczos method	Symmetric **A**	Large, sparse **A**	6.5.3

The third major type of problem that we face using matrix methods is solving differential equations. We encountered systems of linear ordinary differential equations in Chapter 2, which can be solved using diagonalization, and more general ordinary and partial differential equations that can be solved using eigenfunction expansions as discussed in Chapter 3. In general, the techniques developed in Chapters 2 and 3 result in exact solutions, but they are restricted to certain classes of linear equations. Therefore, these methods are powerful, but limited in scope. For more complicated linear, as well as nonlinear, equations, numerical methods must be employed to obtain approximate solutions. Solution of the full range of ordinary and partial differential equations that arise in science and engineering applications requires a comprehensive arsenal of numerical methods. Numerical methods for boundary-value problems are covered in Chapter 8, and numerical methods for initial-value problems are covered in Chapter 9. All of the methods discussed in this text for exactly or numerically solving ordinary and partial differential equations are summarized in Table 6.3 with reference to their respective sections.

Because most of the methods discussed in this chapter have been made available in a wide variety of programming environments through mathematical libraries and software, our emphasis here is on the underlying linear algebra and the methods themselves. For the most part, we will not be concerned with implementation details as these have been taken care of for us in the various tools available, and it is not likely that as scientists or engineers we will be called upon to write our own computational linear algebra codes. Being discerning users of the available libraries and software does require us to be knowledgable about these methods, however, so that we can make wise choices about which methods are most appropriate in a given application and so that we can use these methods in their most effective and efficient manners.

6.1 Introduction to Numerical Methods

Our usual mathematics sequence taught to future engineers and scientists typically progresses from arithmetic, to algebra and trigonometry, to calculus, and differential equations.[2] In arithmetic, the focus is on various types of numbers (integer, real, and complex) and their basic operations (addition, subtraction, multiplication, and

[2] This list is supplemented for some fields by statistics and linear algebra as well.

Table 6.3 Summary of methods for solving ordinary and partial differential equations: $\mathcal{L}u = f$.

Method	Limitations	Best for	Section(s)
Diagonalization	Linear, ordinary $\mathcal{L}$	Small/moderate systems: $\dot{\mathbf{u}} = \mathbf{Au} + \mathbf{f}$	2.6
Linear combination of eigenfunctions	Linear, ordinary $\mathcal{L}$	Self-adjoint $\mathcal{L}$	3.2, 3.3
Separation of variables	Linear, separable, partial $\mathcal{L}$	Simple domains	3.4
Finite-difference methods	Any $\mathcal{L}$	Simple domains	7, 8, 9
Spectral methods	Any $\mathcal{L}$	Smooth solutions in simple domains	7.3, 13.2
Finite-element methods	Any $\mathcal{L}$	Complex domains	7.4
Fourier transform methods	Linear, partial $\mathcal{L}$	Moderate systems from BVP	8.4.2
Cyclic reduction	Linear, partial $\mathcal{L}$	Moderate systems from BVP	8.4.2
Jacobi iteration	Any $\mathcal{L}$	Large systems from BVP	8.5.1
Gauss–Seidel iteration	Any $\mathcal{L}$	Large systems from BVP	8.5.2
SOR	Any $\mathcal{L}$	Large systems from BVP	8.5.3
ADI	Any $\mathcal{L}$	Large systems from BVP or IVP	8.7, 9.10.3, 9.10.4
Multigrid methods	Any $\mathcal{L}$	Large systems from BVP or IVP	8.8
First-order explicit method	Any $\mathcal{L}$	Large systems from IVP	9.2.1, 9.2.4, 9.5.1, 9.9.1, 9.10.1
First-order implicit method	Any $\mathcal{L}$	Large systems from IVP	9.2.2, 9.2.4, 9.7.1, 9.10.2
Crank–Nicolson method	Any $\mathcal{L}$	Large systems from IVP	9.2.3, 9.2.4, 9.7.2, 9.9.2, 9.9.3
Predictor-corrector methods	Ordinary differential $\mathcal{L}$	Moderate/large systems of IVP	9.3
Multistep methods	Ordinary differential $\mathcal{L}$	Moderate/large systems of IVP	9.3
Galerkin projection	Any $\mathcal{L}$	Reduced-order modeling	13.3, 13.4.2, 13.4.3, 13.5

division). In algebra and trigonometry, the emphasis is on functions and how they are manipulated and represented. In calculus, the attention is on differentiation and integration of those functions. Finally, differential equations provide our primary connection between calculus and the physical world. All of these topics are primarily taught from the point of view of obtaining exact closed-form solutions. In calculus, for example, we perform differentiation and integration on smooth and continuous functions; we seek exact solutions of algebraic, trigonometric, and differential equations.

As we quickly learn as budding engineers and scientists, however, these exact analytical methods only take us so far when considering practical applications in our chosen fields. It is not long before it is necessary to employ numerical techniques in order to obtain useful information about a set of data or a system. *Numerical*, or *computational, methods* is the general term given to approaches for obtaining approximate solutions to algebraic, differential, and integral equations. Because modern numerical methods lead to algorithms that are typically designed to be performed by digital computers, there is also an intimate connection between numerical methods and computer architectures.

A prominent theme in numerical methods is development of techniques for turning complex problems, such as nonlinear and/or very large problems, into ones that are amenable to solution using standard matrix methods. The trade-off in doing so is nearly always that the resulting solution is only an approximation of the exact solution. This takes us from the *exact* to the *approximate*, and from the *continuous* to the *discrete*. For example, a continuous function is represented as a smooth curve for the dependent variable with respect to the independent variable, say $u(x)$. This function may represent the continuous temperature as a function of time or space, for example. More often than not, however, such measurements are obtained at discrete times or locations. Rather than continuous functions, therefore, such data would be most naturally represented as a vector, with each element of the vector being the temperature measurement at each discrete time or location. Thus, one-dimensional continuous functions become discrete vectors. Similarly, two- and three-dimensional functions would become matrices. As a result, linear algebra plays a central role in approximate/discrete mathematics, which is why the whole of Part II will focus on this connection between approximate numerical methods and matrix methods.

Taking these ideas and concepts into the discrete and approximate world of numerical methods requires an appreciation for how functions can be represented as a series of numbers, how those numbers are represented on computers, and how they can be manipulated in the form of vectors and matrices. More than simply developing approximate analogs to familiar continuous operations, however, numerical methods are an essential set of tools that extend what is possible, or even relevant, to continuous functions and systems. For example, how do we represent our discrete, and error-prone, temperature versus time or location data measurements such that useful information can be extracted about the behavior of the system or to facilitate subsequent analysis? The story starts with how we represent approximations of numbers and functions, which will be discussed in Section 6.2.

In fact, one could argue that it is the widespread applicability and remarkable efficiency of matrix methods that led to the development of numerical methods in order to extend their scope to additional classes of problems. For example, numerical methods allow us to leverage matrix methods to approximately solve all manner of ordinary and partial differential equations that cannot be solved using diagonalization (Chapter 2) or eigenfunction expansions (Chapter 3). Therefore, one of the most common and important "applications" of matrix methods in modern science and engineering contexts is in the development and execution of numerical methods. This association is quite natural both in terms of applications and the methods themselves. Therefore, the reader will benefit from learning these numerical methods in association with the mathematics of vectors, matrices, and differential operators as articulated in Part I.

6.1.1 Impact on Research and Practice

Traditionally, scientists and engineers have approached problems using *analytical methods* for solving the governing differential equations and *experimental modeling* and prototyping of the system itself. The first is based on mathematical modeling and analysis, and the second is based on careful observation. Unfortunately, analytical solutions are limited to simple problems in terms of the physics as well as the geometry. Therefore, experimental methods have traditionally formed the basis for engineering design and research of complex problems in science and engineering. Modern digital computers, however, are giving rise to an increasingly viable alternative using computational methods. One can view the computational approach as either an extension of the analytical and theoretical approach, as it involves obtaining approximate numerical solutions of the appropriate governing equations, or as a separate approach altogether. Some even speak in terms of conducting "numerical experiments" as an analog to physical experiments. Whatever one's point of view, there is no question that numerical methods are continuing to dramatically advance all areas of research and practice in science and engineering.

The computational[3] approach to research and practice in science and engineering is both very old and very new. It has its roots in approximate methods developed many years ago for solving differential equations, such as Euler's method (see Section 9.2.1), and even obtaining estimates for π. However, the advent of the digital computer in the middle of the twentieth century ushered in a true revolution in scientific and engineering research and practice. These approximate methods moved from hand-executed devices – such as pencil and paper, abacus, or slide rule – to electronic devices.[4] Unshackled from the speed constraints and accuracy limitations – not to mention shear tedium[5] – of hand calculations, this revolution led to an unprecedented

[3] We will use the terms "numerical" and "computational" interchangeably.

[4] I have neglected here the relatively short-lived reign of mechanical computing devices owing to their limited applications and lack of widespread use.

[5] The most popular method for calculating digits of π for several hundred years before the availability of computing devices was literally called the "method of exhaustion." It involved dividing a circle into smaller and smaller triangles and summing their areas to approximate that of a circle.

rate of development in both computing hardware and computational algorithms that have combined to dramatically extend the theoretical pursuits of yesteryear as well as lead to whole new areas of scientific endeavor and engineering practice. Progressively more sophisticated algorithms and software paired with ever more powerful hardware continues to provide for increasingly realistic simulations of mechanical, electrical, chemical, and biological systems. This revolution depends upon the remarkable efficiency of the computer algorithms that have been developed to solve larger and larger systems of equations, thereby building on the fundamental mathematics of vectors and matrices covered in Chapters 1 and 2. Much of what we know scientifically and do practically would not be possible apart from this revolution.

To show how computational methods complement analytical and experimental ones, let us consider some advantages (+) and disadvantages (−) of each approach to engineering problems:

Analytical:

+ Provides *exact* solutions to the governing equations.
+ Gives physical insight, such as the relative importance of different effects.
+ Can consider hypothetical problems by neglecting friction, gravity, et cetera.
− Exact solutions are only available for simple problems and geometries.
− Of limited value in design.

Computational:

+ Addresses more complex problems, including physics and geometries.
+ Provides detailed solutions from which a good understanding of the physics can be discerned.
+ Can easily try different configurations, such as geometries or boundary conditions, which is important in design.
+ Computers are becoming faster and cheaper; therefore, the range of applicability of computational methods continues to expand.
+ Generally more cost effective and faster than experimental prototyping.
− Requires accurate governing equations and models, which are not always available.
− Boundary conditions are sometimes difficult to implement.
− Difficult to do in certain parameter regimes, for example, when highly nonlinear physics is present.

Experimental:

+ Easier to get overall quantities for a problem, such as lift and drag on an airfoil.
+ No "modeling" or assumptions necessary.
− Often requires intrusive measurement probes.
− Limited measurement accuracy.
− Some quantities are difficult to measure, for example, the stress in the interior of a beam.
− Experimental equipment is often expensive.
− Difficult and costly to test full-scale models.

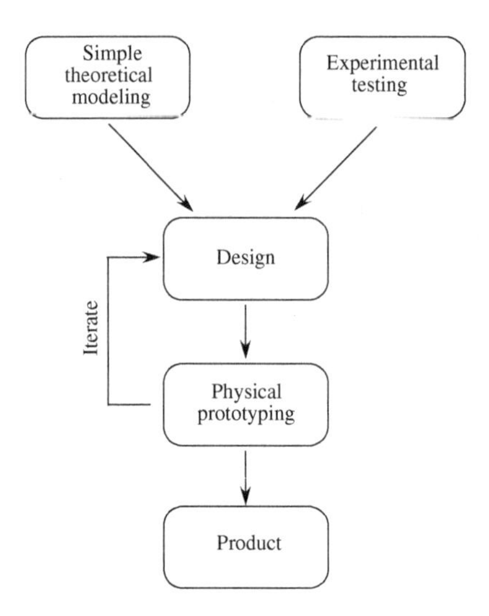

Figure 6.1 The traditional design approach.

As an example of how the rise of computational methods is altering engineering practice, consider the typical design process. Figure 6.1 shows the traditional approach used before computers became ubiquitous, where the heart of the design process consists of iteration between the design and physical prototyping stages until a satisfactory product or process is obtained. This approach can be both time consuming and costly as it involves repeated prototype building and testing. The modern approach to design is illustrated in Figure 6.2, where computational modeling is inserted in two steps of the design process. Along with theoretical modeling and experimental testing, computational modeling can be used in the early stages of the design process in order to better understand the underlying physics of the process or product before the initial design phase. Computational prototyping can then be used to "test" various designs, thereby narrowing down the possible designs before performing physical prototyping.

Using computational modeling, the modern design approach provides for more knowledge and understanding being incorporated into the initial design. In addition, it allows the engineer to explore various ideas and options before the first physical prototype is built. For example, whereas the traditional approach may have required on the order of ten to 15 wind tunnel tests to develop a wing design, the modern approach incorporating computational modeling may only require on the order of two to four wind tunnel tests. As a bonus, computational modeling and "prototyping" are generally faster and cheaper than experimental testing and physical prototyping. This reduces time-to-market and design costs as fewer physical prototypes and design iterations are required, but at the same time, it holds the potential to result in a better final product.

6.1.2 Scientific Computing

Before getting started, let us clearly define some frequently used terminology in computational linear algebra and numerical methods that will better aid us in putting these

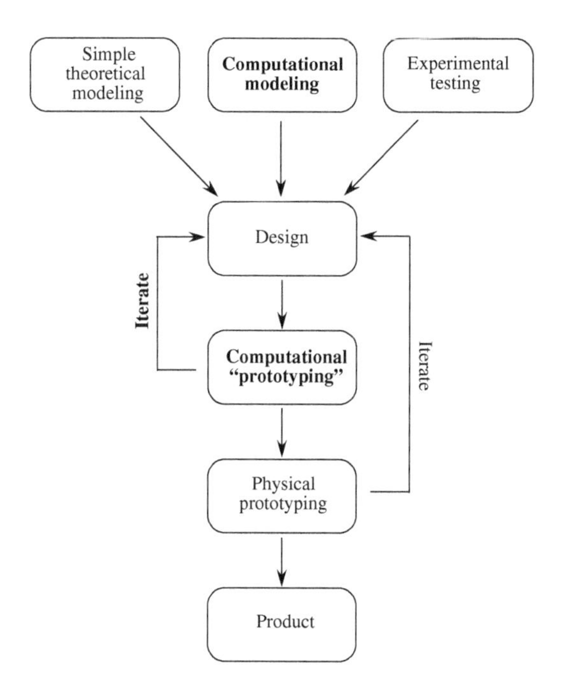

Figure 6.2 The modern design approach incorporating computational modeling.

topics into their overall context. The progression from mathematics to applications is as follows: *mathematical operations* to *mathematical methods* to *numerical algorithms* to *computer codes* to *applications*.

A mathematical method is the sequence of mathematical operations required to accomplish some task. These operations and methods are expressed in symbolic form using our mathematical vocabulary. Numerical algorithms are more specific and detail how all aspects of a method, including special cases and extensions, would be implemented. Whereas numerical methods are computer architecture and programming language agnostic, computer codes involve *software engineering* in order to address all of the essential aspects of developing and testing large-scale application codes. As scientists and engineers, we are primarily interested in the applications at the end of the sequence. However, we must fully appreciate and understand each of these ingredients that underly our applications in order to best utilize them – not viewing them simply as a black box with no knowledge of their inner workings.

Numerical analysis spans the mathematics of linear algebra, including operations and methods, through to algorithms, while *numerical methods* would typically be thought of as including mathematical methods and numerical algorithms. The term *scientific computing* normally encompasses the sequence from mathematical methods through to applications; it is the broadest of terms, including all but the purest of mathematics. It is equal elements numerical analysis and software engineering and it is built on the mathematical foundations of linear algebra. As you can see, there is a great deal of overlap in how these terms are used; this is a reflection of the inherently multidisciplinary nature of scientific computing and the strong interconnectedness of the entire enterprise.

The sequence from operations to methods to algorithms to codes to applications is not a one-way street. One of the aspects that makes computational linear algebra such an interesting topic in its own right is that as new programming languages, programming paradigms, applications, and hardware architectures are developed, we are often forced to rethink which algorithms are "best" in particular applications. Such advancements often upset the prevailing balance in the interplay between accuracy and efficiency considerations. Parallel computing, which comes in numerous different forms and is rapidly evolving, is an important example of this. One is likely to choose very different algorithms for a problem that will be solved on thousands or millions of processors simultaneously as compared to only one, and changes in parallel architectures can sometimes lead to dramatic shifts in the programming paradigms required to maximize such resources. In a similar manner, the rise of new science and engineering applications stretches the need for even more advanced algorithms and codes, sometime drawing on completely different areas of mathematics, methods, or algorithms.

Let us use the example of the Gram–Schmidt orthogonalization procedure to illustrate this terminology. The procedure itself is a *mathematical method* that involves several *mathematical operations*, including inner products, vector additions, and evaluating vector norms. The method, which is given in Section 1.7.3, articulates the sequence of operations that comprises Gram–Schmidt orthogonalization. A *numerical algorithm* designed to perform the procedure on a computer would be devised to carry out all operations accurately and efficiently for large numbers of large dimensional vectors. Mathematical methods typically do not account for the approximate nature of numbers as stored and used in digital computers; algorithms do, and algorithms are the connection between mathematical methods and computer codes. The primary difference between a *numerical algorithm* and *computer code* is that algorithms are written to be independent of the hardware and software; they do not account for the specifics of implementation in a particular programming language or mathematical software or the hardware architecture on which it will be run. Whereas the mathematician is not concerned with operation counts and round-off errors, the numerical analyst most certainly is as these are important determiners in choosing the most appropriate algorithm for a given application – along with numerical stability, convergence rates, computer storage requirements, et cetera.[6] The Gram–Schmidt orthogonalization procedure would then be a single component of a much larger code designed to address a particular application.

So is Gaussian elimination a method or an algorithm? The answer is a resounding yes! As applied to small systems of equations in Chapter 1, the *method* can be outlined in a single page of text (see Section 1.5.1). Recall that we did not need to worry ourselves with issues of stability and convergence, conditioning and round-off errors,

[6] Because of the recursive nature of Gram–Schmidt orthogonalization, it turns out to be a terrible method for obtaining large sets of orthonormal basis vectors. The most robust method to accumulated round-off errors is based on singular-value decomposition (Section 2.8).

or even computational efficiency and storage requirements. We carried out the steps of the method and got the answer we sought. As it is extended to large systems of equations, we need to consider all of these issues and more. A good *algorithm* will work equally well when the system is well or ill conditioned, it will be able to adapt to situations when preconditioning or row interchanges are required, and it will give us the solution as quickly as possible and with as little error as practical.

Note that mathematical operations rarely change, methods sometimes change, algorithms often change, and codes frequently change. This is reflected in the progression of the present text. The underlying matrix methods covered in Part I have remained largely unchanged for well over a century and predate development of the digital computer in the middle of the nineteenth century, while the numerical methods that are the focus of Part II have been in place for mere decades and some are still active areas of research, and the applications of matrix and numerical methods covered in Part III to assist in solving, understanding, optimizing, and controlling systems are very much active areas of research.

Once again, do not think of this sequence as being a one-way progression; there is a great deal of feedback between each of these aspects of scientific computing. In fact, the development of the digital computer has played an immeasurable part in spurring on developments in numerical methods and algorithms, and the rise of new applications drives progress in all of these areas. Thankfully, the mathematics of linear algebra has proven extremely resilient to new advances and developments in computer architectures, programming languages, and scientific and engineering applications that often serve to bring to the fore mathematical operations that have existed for many years. As we move from continuous to discrete, and from exact to approximate, therefore, we are progressing through this sequence from mathematical operators, to methods, to algorithms, to the computer codes used in applications, in a natural, but complex, manner.

The objective of Part II of this text is to address the issues that arise when solving large-scale computational problems that arise in science and engineering. Because of the broad interests of scientists and engineers, such a treatment of numerical methods must be comprehensive and span the solution of systems of linear algebraic equations and the eigenproblem to solution of ordinary and partial differential equations, which are so central to scientific and engineering research and practice. In addition, a given science or engineering application typically involves several algorithms, whether to solve different equations in the mathematical model or different aspects of a larger problem. Therefore, it is often the case that a certain algorithm is ideal in one setting but not another given the other aspects of the overall application that must be addressed.

Because we have mentioned computer programming languages, let us briefly summarize those that are most popular in scientific computing (listed in chronological order of their introduction):

- **Fortran ("FORmula TRANslation")**: Developed and optimized for numerical programming, produces the most efficient code, and is more modern than you think.

- **C++**: General-purpose programming language used for almost all large-scale software development in the computer science community.
- **MATLAB ("MATrix LABoratory")**: High-level mathematical programming language with syntax similar to C. Very popular in engineering.
- **Mathematica**: High-level mathematical programming language that can accommodate several different programming styles, for example, procedural, functional, et cetera. Includes exact symbolic mathematics.
- **Python**: Started as a scripting language. Is now a general-purpose, high-level programming language that is increasingly popular in many settings, including scientific computing and data science. Because of its roots as a scripting language, Python is particularly adept at combining Fortran and/or C++ codes, which are commonly used to write mathematical libraries, into a single coding framework.

6.2 Approximation and Its Effects

When solving small systems of equations and eigenproblems by hand, we typically emphasize ease of execution over algorithmic efficiency. That is, we prefer methods that are straightforward to carry out even if they require more steps than an alternative method that may be more efficient in an automated algorithm. When solving large systems, however, the opposite is true. The primary emphasis will be placed on efficiency so as to be able to obtain solutions as quickly as possible via computers, even at the expense of simplicity of the actual algorithm. Because a computer will be carrying out the steps, we would favor the more complex algorithm if it produces a solution more quickly as we only have to program the method once – the computer does the work to solve each problem.

Moreover, this drive for efficiency will often suggest that we favor *iterative*, rather than *direct*, *methods*. Although iterative methods only provide approximate solutions, they typically scale better to large systems than direct methods. The methods considered in Chapters 1 and 2 are all direct methods that produce an exact solution of the system of equations or eigenproblem in a finite number of steps carried out in a predefined sequence. Think, for example, about Gaussian elimination. In contrast, iterative methods are step-by-step algorithms that produce a sequence of approximations to the sought-after solution, which when repeated many times converges toward the exact solution of the problem – at least we hope so.

Before discussing methods and algorithms for large problems of interest, it is necessary to first consider some of the unique issues that arise owing to the fact that the mathematics will be carried out approximately by a digital computer. This includes operation counts, round-off errors, and a brief mention of numerical stability. As suggested by the preceding, the tone of our discussion in describing numerical methods often will seem quite different than in previous chapters; however, keep in mind that we have the same goals, that is, solving systems of algebraic equations, eigenproblems, systems of first-order linear ordinary differential equations, et cetera. It is just that the systems will be much larger and solved by a computer.

6.2.1 Operation Counts

In order to compare various methods and algorithms, both direct and iterative, against one another, we will often appeal to operation counts that show how the methods scale with the size of the problem. Such operation counts allow us to compare different methods for executing the same matrix operation as well as determining which matrix method is most computationally efficient for a given application. For example, we have observed several uses of the matrix inverse in Chapter 1; however, evaluating the matrix inverse is notoriously inefficient for large matrices. As a result, it is actually quite rare for computational methods to be based on the inverse; instead, we use alternatives whenever possible.

Operation counts provide an exact, or more often order-of-magnitude, indication of the number of additions, subtractions, multiplications, and divisions required by a particular algorithm or method. Take, for example, the inner product of two N-dimensional vectors

$$\langle \mathbf{u}, \mathbf{v} \rangle = \begin{bmatrix} u_1 & u_2 & \cdots & u_N \end{bmatrix} \begin{bmatrix} v_1 & v_2 & \cdots & v_N \end{bmatrix}^T = u_1 v_1 + u_2 v_2 + \cdots + u_N v_N.$$

Evaluating the inner product requires N multiplications and $N-1$ additions; therefore, the total operation count is $2N - 1$, which we say is $2N$ or $O(N)$ for large N.[7] By extension, matrix multiplication of two $N \times N$ matrices requires N^2 inner products, one for each element in the resulting $N \times N$ matrix. Consequently, the total operation count for matrix multiplication is $N^2(2N-1)$, that is, $2N^3$ or $O(N^3)$ for large N. The operation count for Gram–Schmidt orthogonalization of K vectors that are of size $N \times 1$ is

$$\frac{1}{2}(4N-1)K(K-1) + 3NK,$$

which for large N and K is approximately $2NK^2$ (the $3NK$ term is smaller than the first term when K is large).

Gaussian elimination of an $N \times N$ matrix has an operation count that is $\frac{2}{3}N^3$ or $O(N^3)$ for large N. If, for example, we could come up with a method that accomplishes the same thing as Gaussian elimination but is $O(N^2)$, then this would represent a dramatic improvement in efficiency and scalability as N^2 grows much more slowly than N^3 with increasing N. We say that $O(N^2) \ll O(N^3)$ for large N, where $\ll$ means "much smaller (or less) than." For future reference, note that $N^M \ll N^M \log N \ll N^{M+1}$ for large N.

Note that many of the operations covered in Chapters 1 and 2 can be specialized for cases when the matrices are large but sparse. For example, if two large, but sparse, matrices are multiplied, there is no need to perform the multiplications when one of the corresponding elements in either of the matrices is zero. Specialized algorithms for such cases can be developed that dramatically reduce the storage requirements (no need to store the many zeros) and operation counts (no need to operate with the

[7] $O(N)$ is read "is of the same order as N," or simply, "order N," and indicates that the operation count is directly proportional to N for large N.

many zeros) that significantly reduce the operation counts given previously for the general cases.

Operation counts for numerical algorithms can be used along with computer hardware specifications to estimate how long it would take for a particular algorithm to be completed on a particular computer's *central processing unit* (CPU). For numerical calculations, the important figure of merit is the FLOPS[8] rating of the CPU or its cores. In addition, because of the large number of operations required when dealing with large systems, we must be concerned with the consequences of round-off errors inherent to calculations on digital computers.

6.2.2 Ill Conditioning and Round-Off Errors

Here we discuss some issues unique to large-scale calculations that are common to both direct and iterative methods. In particular, there are two issues that arise when solving large systems of equations with digital computers.[9] First, real, that is floating point, numbers are not represented exactly on digital computers; in other words, all numbers, including rational numbers, are represented as decimals. This leads to *round-off errors*. More specifically, the continuous number line, with an infinity of real numbers within any interval, has as its smallest intervals $2^{-51} = 4.4 \times 10^{-16}$ when using double precision.[10] Consequently, any number smaller than this is effectively, but not exactly, zero. Every calculation, for example, addition or multiplication, introduces a round-off error of roughly this magnitude. Second, because of the large number of operations required to solve large systems of algebraic equations, we must be concerned with whether our numerical method causes these small round-off errors to grow and pollute the final solution that is obtained; this is the subject of *numerical stability*, which will be considered in detail in Chapter 9.

When solving small to moderate-sized systems of equations, the primary diagnostic required is the determinant, which tells us whether the square matrix $\mathbf{A}$ is singular and, thus, not invertible. This is the case if $|\mathbf{A}| = 0$. When dealing with round-off errors in solving large systems of equations, however, we also must be concerned when the

[8] FLOPS stands for FLoating point Operations Per Second. A *floating point operation* is an addition, subtraction, multiplication, or division involving two floating point numbers. Real numbers as represented by computers are called *floating point numbers* because they contain a fixed number of significant figures, and the decimal point "floats." In contrast, a *fixed point number* is an integer as the decimal point is fixed and does not float.

[9] Nearly all computers today are *digital computers*. However, the fact that they are called "digital computers" suggests that there are also *analog computers*. Before digital computers were developed in the 1940s and 1950s, analog computers had become rather sophisticated. For example, an analog computer called a Norden bombsight was used by US bomber planes in World War II to calculate bomb trajectories based on the plane's current flight conditions, and it actually flew planes during the bombing process in order to more accurately target the bombs. These devices were considered so advanced that, if a bomber was shot down or to land in enemy territory, the bombardier was instructed to first destroy the bombsight before seeking to save himself.

[10] By default, most general-purpose software uses single precision for calculations; however, double precision is standard for numerical calculations of the sort we are interested in throughout this text.

matrix is *nearly* singular, that is, when the determinant is "close" to zero. How close? We need a way to quantify how "wrong" our solution is.

Let us define the *error vector* **e** as follows:

$$\mathbf{e} = \hat{\mathbf{u}} - \mathbf{u},$$

where $\hat{\mathbf{u}}$ is the exact solution of the system of algebraic equations, and **u** is the approximate solution obtained numerically.[11] In general, of course, we do not know the exact solution, so we need an alternative measure of the error. A common alternative is the *residual*. For the system of equations $\mathbf{A}\hat{\mathbf{u}} = \mathbf{b}$, the residual vector **r** is given by

$$\mathbf{r} = \mathbf{b} - \mathbf{A}\mathbf{u}.$$

Note that if $\mathbf{u} = \hat{\mathbf{u}}$, that is, the numerical solution equals the exact solution, then the residual is zero.

The residual has two deficiencies: (1) for some **A**, the residual **r** may be small even though the error **e** is large and (2) the residual here is a vector; we would like a single scalar measure of the accuracy. To address the second deficiency, we can obtain a single scalar number that represents an entire vector using the *norm* as discussed in Section 1.9. For a vector, we use the length of the vector, which is called the L_2-norm.

To address the first deficiency of the residual, we can use norms to quantify the loss of accuracy expected owing to round-off errors when performing matrix operations, such as inverting a matrix, by defining the *condition number*. Using any norm, the condition number cond(**A**), or $\kappa(\mathbf{A})$, of an invertible matrix **A** is defined by

$$\text{cond}(\mathbf{A}) = \kappa(\mathbf{A}) = \|\mathbf{A}\|\,\|\mathbf{A}^{-1}\|, \tag{6.1}$$

where $1 \leq \kappa(\mathbf{A}) \leq \infty$. Using the L_2-norm, the condition number is given by

$$\text{cond}_2(\mathbf{A}) = \kappa_2(\mathbf{A}) = \|\mathbf{A}\|_2\|\mathbf{A}^{-1}\|_2 = \frac{\sigma_1}{\sigma_N}, \tag{6.2}$$

where σ_i are the singular values of **A** (see Section 2.8). That is, the condition number is the ratio of the largest to the smallest singular values of **A**. Recall that for normal, such as symmetric, matrices, the singular values of **A** are equal to the absolute values of its eigenvalues. In this case, then,

$$\text{cond}_2(\mathbf{A}) = \kappa_2(\mathbf{A}) = \frac{|\lambda_i|_{\text{max}}}{|\lambda_i|_{\text{min}}},$$

where λ_i are the eigenvalues of **A**.

A matrix with a condition number close to one is said to be *well conditioned*, and a matrix with a large condition number is called *ill conditioned*. For example, a singular matrix with $|\lambda_i|_{\text{min}} = 0$, in which case $|\mathbf{A}| = 0$, has $\kappa(\mathbf{A}) = \infty$. For condition numbers close to one, the residual and error are of the same order of magnitude, and only small errors are expected in inverting **A**. The larger the condition number, the

[11] We will use **u** generically for the dependent variable when the meaning is clear from the context, and $\hat{\mathbf{u}}$ for the exact solution and **u** for the approximate solution when it is necessary to distinguish between them throughout Parts II and III.

larger the magnitude of the error in the solution $\mathbf{u}$ as compared to that of an error in the coefficient matrix $\mathbf{A}$ as shown in the following.

What is considered to be a "small" or "large" condition number depends upon the problem being considered. Using the condition number, we can estimate the number of decimal places of accuracy that may be lost owing to round-off errors in the calculation. The number of decimal places d that we expect to lose is

$$d = \log_{10} \kappa(\mathbf{A}).$$

Generally, we use *double precision* in numerical calculations, which includes 16 decimal places of accuracy, and it is reasonable to lose no more than say five to six decimal places of accuracy owing to round-off errors – obviously, this depends upon the application under consideration and what will be done with the result. This is why we generally use double precision for numerical computations, because it gives us more digits of accuracy to work with.

One may wonder why we do not simply use the determinant of a matrix to quantify how close to being singular it is. Recall that the determinant of a matrix is the product of its eigenvalues, but these depend upon the scale of the matrix. In contrast, the condition number is scale invariant. For example, consider the $N \times N$ matrix $\mathbf{A} = a\mathbf{I}$ with constant a. The determinant of this diagonal matrix is $|\mathbf{A}| = a^N$, which is small if $|a| < 1$ and large N, whereas $\kappa_2(\mathbf{A}) = a/a = 1$ for any a and N. As we develop various numerical methods for solving systems of algebraic equations, we must be cognizant of the influence that each decision has on the solution obtained such that it leads to well-conditioned matrices for the equations that we are interested in solving.

In order to observe the influence of the condition number, consider how a system of algebraic equations responds to the introduction of small perturbations (errors) in the coefficient matrix $\mathbf{A}$. A system is considered to be well conditioned if small perturbations to the problem only lead to small changes in the solution vector $\mathbf{u}$. When this is not the case, the system is said to be ill conditioned. Let us perturb the coefficient matrix $\mathbf{A}$ by the small amount $\epsilon\mathbf{A}_1$ and evaluate the resulting perturbation $\epsilon\mathbf{u}_1$ in the solution vector $\mathbf{u}$, where $\epsilon \ll 1$. From the system of equations $\mathbf{Au} = \mathbf{b}$, we have

$$(\mathbf{A} + \epsilon\mathbf{A}_1)(\mathbf{u} + \epsilon\mathbf{u}_1) = \mathbf{b}.$$

Carrying out the multiplication produces

$$\mathbf{Au} + \mathbf{A}\epsilon\mathbf{u}_1 + \epsilon\mathbf{A}_1\mathbf{u} + \epsilon\mathbf{A}_1\epsilon\mathbf{u}_1 = \mathbf{b}.$$

Because the first and last terms cancel from our system of equations and the $\epsilon\mathbf{A}_1\epsilon\mathbf{u}_1 = O(\epsilon^2)$ term is much smaller than the others, as it is the product of two small perturbations, solving for $\epsilon\mathbf{u}_1$ yields

$$\epsilon\mathbf{u}_1 = -\mathbf{A}^{-1}\epsilon\mathbf{A}_1\mathbf{u}. \tag{6.3}$$

It is a property of vector and matrix norms that

$$\|\mathbf{Au}\| \leq \|\mathbf{A}\|\,\|\mathbf{u}\|;$$

therefore, applying to (6.3) requires that

$$\|\epsilon \mathbf{u}_1\| \leq \|\mathbf{A}^{-1}\| \, \|\epsilon \mathbf{A}_1\| \, \|\mathbf{u}\|,$$

or

$$\frac{\|\epsilon \mathbf{u}_1\|}{\|\mathbf{u}\|} \leq \|\mathbf{A}^{-1}\| \, \|\epsilon \mathbf{A}_1\|.$$

From the definition of the condition number (6.1), this becomes

$$\frac{\|\epsilon \mathbf{u}_1\|}{\|\mathbf{u}\|} \leq \kappa(\mathbf{A}) \frac{\|\epsilon \mathbf{A}_1\|}{\|\mathbf{A}\|}. \tag{6.4}$$

Therefore, the magnitude of the resulting inaccuracy in the solution $\epsilon \mathbf{u}_1$ is related to the magnitude of the perturbation of the coefficient matrix $\epsilon \mathbf{A}_1$ through the condition number, where each is normalized by the norm of its parent vector or matrix. The larger the condition number, the larger the magnitude of the error in the solution as compared to that of the disturbance to the coefficient matrix. Alternatively, one could perturb the right-hand-side vector $\mathbf{b}$ by the small amount $\epsilon \mathbf{b}_1$ and evaluate the resulting perturbation in the solution vector $\epsilon \mathbf{u}_1$. The result (6.4) is the same but with $\mathbf{A}$ replaced by $\mathbf{b}$.

Although the condition number provides a definitive measure of the "conditioning" of a matrix, its determination requires calculation of the full spectrum of eigenvalues (or singular values) of a very large matrix in order to evaluate (6.2). As an alternative, it is much easier to check if a matrix is *diagonally dominant*. Diagonal dominance simply requires comparing the sums of the magnitudes of the off-diagonal elements with that of the main diagonal element of each row of the matrix. The matrix $\mathbf{A}$ is diagonally dominant if

$$|A_{ii}| \geq \sum_{j=1,\, j\neq i}^{N} |A_{ij}| \tag{6.5}$$

holds for each row of the matrix. If the greater than sign applies, then the matrix is said to be *strictly*, or *strongly, diagonally dominant*. If the equal sign applies, it is *weakly diagonally dominant*. It can be shown that diagonally dominant matrices are well conditioned. More specifically, it can be proven that for any strictly diagonally dominant matrix $\mathbf{A}$:

1. $\mathbf{A}$ is nonsingular; therefore, it is invertible.
2. Gaussian elimination does not require row interchanges, that is, pivoting, for conditioning.
3. Computations are stable with respect to round-off errors.

For weakly diagonally dominant matrices, the preceding properties are *probably* true in practice, but it cannot be proven formally. As various methods are discussed, we will check our algorithms for diagonal dominance in order to ensure that we have well-conditioned matrices that lead to stable numerical solutions.

The condition number and diagonal dominance addresses how *mathematically* amenable a particular system is to producing an accurate solution. In addition, we also must be concerned with whether the numerical algorithm that is used to actually obtain this mathematical solution on a computer is *numerically* amenable to producing an accurate solution. This is determined by the algorithm's *numerical stability*, which will be discussed in Chapter 9. That is, conditioning is a mathematical property of the algebraic system itself, and numerical stability is a property of the numerical algorithm used to obtain its solution. See Trefethen and Bau (1997) and Higham (2002) for much more on numerical approximation and stability of common operations and algorithms in numerical linear algebra, including many of those treated in the remainder of this chapter.

Example 6.1 Let us illustrate the influence of the condition number of a matrix on the accumulation of round-off errors when carrying out calculations with matrices. The basic principles can be dramatically illustrated by taking the inverse of the notoriously ill-conditioned Hilbert matrix. The Hilbert matrix is defined by

$$H[i,j] = \frac{1}{i+j-1}.$$

For example, the 10×10 Hilbert matrix is given by

$$\mathbf{H}_{10} = \begin{bmatrix} 1 & \frac{1}{2} & \frac{1}{3} & \frac{1}{4} & \frac{1}{5} & \frac{1}{6} & \frac{1}{7} & \frac{1}{8} & \frac{1}{9} & \frac{1}{10} \\ \frac{1}{2} & \frac{1}{3} & \frac{1}{4} & \frac{1}{5} & \frac{1}{6} & \frac{1}{7} & \frac{1}{8} & \frac{1}{9} & \frac{1}{10} & \frac{1}{11} \\ \frac{1}{3} & \frac{1}{4} & \frac{1}{5} & \frac{1}{6} & \frac{1}{7} & \frac{1}{8} & \frac{1}{9} & \frac{1}{10} & \frac{1}{11} & \frac{1}{12} \\ \frac{1}{4} & \frac{1}{5} & \frac{1}{6} & \frac{1}{7} & \frac{1}{8} & \frac{1}{9} & \frac{1}{10} & \frac{1}{11} & \frac{1}{12} & \frac{1}{13} \\ \frac{1}{5} & \frac{1}{6} & \frac{1}{7} & \frac{1}{8} & \frac{1}{9} & \frac{1}{10} & \frac{1}{11} & \frac{1}{12} & \frac{1}{13} & \frac{1}{14} \\ \frac{1}{6} & \frac{1}{7} & \frac{1}{8} & \frac{1}{9} & \frac{1}{10} & \frac{1}{11} & \frac{1}{12} & \frac{1}{13} & \frac{1}{14} & \frac{1}{15} \\ \frac{1}{7} & \frac{1}{8} & \frac{1}{9} & \frac{1}{10} & \frac{1}{11} & \frac{1}{12} & \frac{1}{13} & \frac{1}{14} & \frac{1}{15} & \frac{1}{16} \\ \frac{1}{8} & \frac{1}{9} & \frac{1}{10} & \frac{1}{11} & \frac{1}{12} & \frac{1}{13} & \frac{1}{14} & \frac{1}{15} & \frac{1}{16} & \frac{1}{17} \\ \frac{1}{9} & \frac{1}{10} & \frac{1}{11} & \frac{1}{12} & \frac{1}{13} & \frac{1}{14} & \frac{1}{15} & \frac{1}{16} & \frac{1}{17} & \frac{1}{18} \\ \frac{1}{10} & \frac{1}{11} & \frac{1}{12} & \frac{1}{13} & \frac{1}{14} & \frac{1}{15} & \frac{1}{16} & \frac{1}{17} & \frac{1}{18} & \frac{1}{19} \end{bmatrix}.$$

Observe that the Hilbert matrix is not diagonally dominant. First, let us evaluate the determinant[12] of the 10×10 Hilbert matrix, which is[13]

$$|\mathbf{H}_{10}| = 2.16418 \times 10^{-53}.$$

As you can see, the value of the determinant is very close to zero. That is, the Hilbert matrix is nearly singular. Because the matrix is not singular, however, we are able to obtain its inverse, but because it is nearly singular, we expect that there may be some difficulties owing to round-off errors.

[12] We have used Mathematica for all calculations. Because there are no other approximations other than those owing to round-off error, other software, such as MATLAB, would give very similar results.

[13] This is the numerical value from an exact calculation of the determinant, which is why it is smaller than 10^{-16}.

For the reasons discussed previously, the condition number is used to quantify the level of ill conditioning. In other words, the fact that the determinant is very small serves as a warning, but it does not allow us to quantify the effect of performing operations with this nearly singular matrix. First, we obtain the condition number using its definition based on the eigenvalues of the matrix.[14] Recall that the condition number of a symmetric matrix is the ratio of the magnitudes of the largest to the smallest eigenvalues by magnitude. These are

$$|\lambda_i|_{max} = 1.75192, \quad |\lambda_i|_{min} = 1.09315 \times 10^{-13}.$$

Consequently, the condition number is

$$\kappa(\mathbf{H}_{10}) = \frac{|\lambda_i|_{max}}{|\lambda_i|_{min}} = \frac{1.75192}{1.09315 \times 10^{-13}} = 1.60263 \times 10^{13}.$$

Observe that the very small determinant corresponds to a very large condition number; we say that the Hilbert matrix is ill conditioned. The significance of such matrices is that performing calculations with an ill-conditioned matrix can lead to growth of round-off errors that can contaminate the solution.

Recall that using the condition number, we can estimate the number of significant digits that will be lost in performing calculations with the 10×10 Hilbert matrix by taking the base 10 logarithm of the condition number, which gives $d = 13.2048$, so we expect to lose approximately 13 digits of accuracy when we invert the 10×10 Hilbert matrix, for example. In order to illustrate the loss of accuracy, we first invert the Hilbert matrix numerically using the internal Mathematica procedure. This inverse is then multiplied by the original matrix, which if done exactly would produce the 10×10 identity matrix. Doing the inverse numerically, however, produces a result in which none of the main diagonal elements are exactly unity, and none of the off-diagonal elements are exactly zero. To illustrate these inaccuracies, observe the main diagonal elements of the resulting matrix[15]

$$\begin{bmatrix} 1.00002 \\ 1.00001 \\ 0.999912 \\ 0.999979 \\ 0.999985 \\ 1.00019 \\ 0.999641 \\ 1.00039 \\ 0.999847 \\ 1.00004 \end{bmatrix}.$$

Thus, using numbers that are only accurate to machine precision, we do not get back the identity matrix exactly. In fact, we only get the identity matrix accurate to three

[14] Mathematical software has built-in functions to compute the condition number directly without having to first determine the eigenvalues.

[15] Note that not all digits are shown for the sake of space, and the off-diagonal elements, which should be exactly zero, display similar inaccuracies.

digits to the right of the decimal point even though all of the calculations were done using double precision, that is, 16 digits. This is owing to the round-off errors that build up during the calculation. The three digits of accuracy precisely corresponds to the predicted loss of 13 of the 16 digits of accuracy in performing the calculations.

6.3 Systems of Linear Algebraic Equations – Direct Methods

With a better understanding of how computer approximations may affect numerical calculations and an appreciation for operation counts, we are ready to consider the primary task of computational linear algebra, which is solving large systems of linear algebraic equations. We begin with a series of direct methods that are faster – having smaller operation counts – than Gaussian elimination, which is $\sim\frac{2}{3}N^3$ for large N. Two are decomposition methods similar to polar and singular-value decomposition discussed in Chapter 2. These methods are appropriate for implementation in computer algorithms that determine the solution for large systems. We then discuss iterative, or relaxation, methods in the next section for obtaining approximate solutions of systems of linear algebraic equations. The essential issue of whether these iterative methods converge toward the exact solution is also addressed.

Finally, we discuss methods for systems of equations having sparse coefficient matrices. A *sparse* matrix is one in which relatively few of the elements of the matrix are nonzero. This is often the case for systems of equations that arise from numerical methods, for example. In some cases, these sparse matrices also have a particular structure. For example, we will encounter tridiagonal, block tridiagonal, and other types of banded matrices.

6.3.1 LU Decomposition

LU decomposition provides a general method for solving systems governed by any nonsingular matrix $\mathbf{A}$ that has several advantages over Gaussian elimination. We decompose (factor) $\mathbf{A}$ into the product of two matrices[16]

$$\mathbf{A} = \mathbf{LR},$$

where $\mathbf{L}$ is *left triangular*, and $\mathbf{R}$ is *right triangular*. For example, if $\mathbf{A}$ is 4×4, then $\mathbf{LR} = \mathbf{A}$ is

$$\begin{bmatrix} 1 & 0 & 0 & 0 \\ L_{21} & 1 & 0 & 0 \\ L_{31} & L_{32} & 1 & 0 \\ L_{41} & L_{42} & L_{43} & 1 \end{bmatrix} \begin{bmatrix} R_{11} & R_{12} & R_{13} & R_{14} \\ 0 & R_{22} & R_{23} & R_{24} \\ 0 & 0 & R_{33} & R_{34} \\ 0 & 0 & 0 & R_{44} \end{bmatrix} = \begin{bmatrix} A_{11} & A_{12} & A_{13} & A_{14} \\ A_{21} & A_{22} & A_{23} & A_{24} \\ A_{31} & A_{32} & A_{33} & A_{34} \\ A_{41} & A_{42} & A_{43} & A_{44} \end{bmatrix}.$$

[16] The "L" and "U" in LU decomposition refer to the resulting "lower" and "upper" triangular matrices. In order to maintain consistent notation and terminology with the balance of the text, we will use "L" and "R" for "left" and "right" triangular matrices. However, the method will be referred to as LU decomposition as is universally the case.

It will be shown that if such an LU decomposition exists for the matrix **A**, then the solution of the system of equation $\mathbf{Au} = \mathbf{b}$ can be obtained simply by performing a forward and then backward substitution on two successive systems of equations having triangular coefficient matrices.

In order to obtain the elements in the **L** and **R** triangular matrices, note that from matrix multiplication for the first row of **L** that

$$R_{11} = A_{11}, \quad R_{12} = A_{12}, \quad R_{13} = A_{13}, \quad R_{14} = A_{14},$$

which gives the first row of **R** for a known matrix **A**. Continuing the matrix multiplication, the second row of **L** leads to the elements

$$\begin{aligned}
L_{21}R_{11} &= A_{21} & &\Rightarrow & L_{21} &= \tfrac{A_{21}}{R_{11}},\\
L_{21}R_{12} + R_{22} &= A_{22} & &\Rightarrow & R_{22} &= A_{22} - L_{21}R_{12},\\
L_{21}R_{13} + R_{23} &= A_{23} & &\Rightarrow & R_{23} &= A_{23} - L_{21}R_{13},\\
L_{21}R_{14} + R_{24} &= A_{24} & &\Rightarrow & R_{24} &= A_{24} - L_{21}R_{14}.
\end{aligned}$$

Note that R_{11}, R_{12}, R_{13}, and R_{14} are known from the first row; therefore, these expressions give the second rows of **L** and **R**. This procedure is continued to obtain all L_{ij} and R_{ij}.

Once the LU decomposition of **A** is completed, the system $\mathbf{Au} = \mathbf{b}$ may be solved in the two-step procedure as follows. Given the LU decomposition, we have

$$\mathbf{LRu} = \mathbf{b}.$$

Let us introduce the vector $\mathbf{v} = \mathbf{Ru}$. Then

$$\mathbf{Lv} = \mathbf{b},$$

or

$$\begin{bmatrix} 1 & 0 & 0 & \cdots & 0 \\ L_{21} & 1 & 0 & \cdots & 0 \\ L_{31} & L_{32} & 1 & \cdots & 0 \\ \vdots & \vdots & \vdots & \ddots & \\ L_{N1} & L_{N2} & L_{N3} & \cdots & 1 \end{bmatrix} \begin{bmatrix} v_1 \\ v_2 \\ v_3 \\ \vdots \\ v_N \end{bmatrix} = \begin{bmatrix} b_1 \\ b_2 \\ b_3 \\ \vdots \\ b_N \end{bmatrix}.$$

This system of equations may be solved by straightforward *forward substitution*, for example,

$$v_1 = b_1, \quad v_2 = b_2 - L_{21}v_1, \ldots,$$

or more generally

$$v_i = b_i - \sum_{j=1}^{i-1} L_{ij}v_j, \quad i = 2, 3, \ldots, N,$$

where N is the size of **A**. Given the solution **v**, we may now solve

$$\mathbf{Ru} = \mathbf{v},$$

or

$$\begin{bmatrix} R_{11} & R_{12} & R_{13} & \cdots & R_{1N} \\ 0 & R_{22} & R_{23} & \cdots & R_{2N} \\ 0 & 0 & R_{33} & \cdots & R_{3N} \\ \vdots & \vdots & \vdots & \ddots & \\ 0 & 0 & 0 & \cdots & R_{NN} \end{bmatrix} \begin{bmatrix} u_1 \\ u_2 \\ u_3 \\ \vdots \\ u_N \end{bmatrix} = \begin{bmatrix} v_1 \\ v_2 \\ v_3 \\ \vdots \\ v_N \end{bmatrix},$$

for the desired solution **u** using *backward substitution* as follows:

$$u_N = \frac{v_N}{R_{NN}}, \quad u_i = \frac{1}{R_{ii}} \left(v_i - \sum_{j=i+1}^{N} R_{ij} u_j \right), \quad i = N-1, N-2, \dots, 1.$$

Given the LU decomposition of matrix **A**, the inverse $\mathbf{A}^{-1}$ also may be obtained as follows. Recall that

$$\mathbf{A}\mathbf{A}^{-1} = \mathbf{I}.$$

Let us consider the 3×3 case for illustration, for which we have

$$\begin{bmatrix} A_{11} & A_{12} & A_{13} \\ A_{21} & A_{22} & A_{23} \\ A_{31} & A_{32} & A_{33} \end{bmatrix} \begin{bmatrix} B_{11} & B_{12} & B_{13} \\ B_{21} & B_{22} & B_{23} \\ B_{31} & B_{32} & B_{33} \end{bmatrix} = \begin{bmatrix} 1 & 0 & 0 \\ 0 & 1 & 0 \\ 0 & 0 & 1 \end{bmatrix},$$

where B_{ij} are the elements of $\mathbf{A}^{-1}$. We may then solve for the columns of $\mathbf{B} = \mathbf{A}^{-1}$ separately by solving the three systems:

$$\mathbf{A} \begin{bmatrix} B_{11} \\ B_{21} \\ B_{31} \end{bmatrix} = \begin{bmatrix} 1 \\ 0 \\ 0 \end{bmatrix}, \quad \mathbf{A} \begin{bmatrix} B_{12} \\ B_{22} \\ B_{32} \end{bmatrix} = \begin{bmatrix} 0 \\ 1 \\ 0 \end{bmatrix}, \quad \mathbf{A} \begin{bmatrix} B_{13} \\ B_{23} \\ B_{33} \end{bmatrix} = \begin{bmatrix} 0 \\ 0 \\ 1 \end{bmatrix}.$$

Because **A** does not change, these three systems of equations may be solved efficiently using the LU decomposition of **A** as described earlier.

Example 6.2 Let us reconsider the system of linear algebraic equations $\mathbf{Au} = \mathbf{b}$ in Examples 1.7 and 1.8 given in matrix form by

$$\begin{bmatrix} 1 & 2 & 3 \\ 2 & 3 & 4 \\ 3 & 4 & 6 \end{bmatrix} \begin{bmatrix} u_1 \\ u_2 \\ u_3 \end{bmatrix} = \begin{bmatrix} 1 \\ 1 \\ 1 \end{bmatrix}.$$

Obtain the LU decomposition of the coefficient matrix **A** and use it to solve the system of algebraic equations for **u**.

Solution

For the 3×3 matrix **A**, the LU decomposition is given by

$$\begin{bmatrix} 1 & 0 & 0 \\ L_{21} & 1 & 0 \\ L_{31} & L_{32} & 1 \end{bmatrix} \begin{bmatrix} R_{11} & R_{12} & R_{13} \\ 0 & R_{22} & R_{23} \\ 0 & 0 & R_{33} \end{bmatrix} = \begin{bmatrix} 1 & 2 & 3 \\ 2 & 3 & 4 \\ 3 & 4 & 6 \end{bmatrix}.$$

Applying matrix multiplication for the first row of **A** requires that

$$R_{11} = 1, \quad R_{12} = 2, \quad R_{13} = 3.$$

For the second row of **A**, matrix multiplication leads to

$$\begin{aligned}
L_{21}R_{11} &= 2 && \Rightarrow \quad L_{21} = \frac{2}{1} = 2,\\
L_{21}R_{12} + R_{22} &= 3 && \Rightarrow \quad R_{22} = 3 - 2(2) = -1,\\
L_{21}R_{13} + R_{23} &= 4 && \Rightarrow \quad R_{23} = 4 - 2(3) = -2.
\end{aligned}$$

Similarly, for the third row of **A**, we obtain

$$\begin{aligned}
L_{31}R_{11} &= 3 && \Rightarrow \quad L_{31} = \frac{3}{1} = 3,\\
L_{31}R_{12} + L_{32}R_{22} &= 4 && \Rightarrow \quad L_{32} = \frac{1}{-1}\left[4 - 3(2)\right] = 2,\\
L_{31}R_{13} + L_{32}R_{23} + R_{33} &= 6 && \Rightarrow \quad R_{33} = 6 - 3(3) - 2(-2) = 1.
\end{aligned}$$

Consequently, the LU decomposition of matrix **A** is given by

$$\mathbf{L} = \begin{bmatrix} 1 & 0 & 0 \\ 2 & 1 & 0 \\ 3 & 2 & 1 \end{bmatrix}, \quad \mathbf{R} = \begin{bmatrix} 1 & 2 & 3 \\ 0 & -1 & -2 \\ 0 & 0 & 1 \end{bmatrix}.$$

Given the LU decomposition of the coefficient matrix **A**, we seek the solution of the system of linear algebraic equations $\mathbf{Au} = \mathbf{b}$ as follows. Because $\mathbf{A} = \mathbf{LR}$, the system of equations can be written in the form

$$\mathbf{LRu} = \mathbf{b}.$$

Letting $\mathbf{v} = \mathbf{Ru}$, this can be written in the form

$$\mathbf{Lv} = \mathbf{b}.$$

From the LU decomposition, this is

$$\begin{bmatrix} 1 & 0 & 0 \\ 2 & 1 & 0 \\ 3 & 2 & 1 \end{bmatrix} \begin{bmatrix} v_1 \\ v_2 \\ v_3 \end{bmatrix} = \begin{bmatrix} 1 \\ 1 \\ 1 \end{bmatrix}.$$

Forward substitution leads to the solution for **v** as follows:

$$\begin{aligned}
v_1 &= 1,\\
2v_1 + v_2 &= 1 && \Rightarrow \quad v_2 = 1 - 2(1) = -1,\\
3v_1 + 2v_2 + v_3 &= 1 && \Rightarrow \quad v_3 = 1 - 3(1) - 2(-1) = 0.
\end{aligned}$$

Then the system $\mathbf{Ru} = \mathbf{v}$ is given by

$$\begin{bmatrix} 1 & 2 & 3 \\ 0 & -1 & -2 \\ 0 & 0 & 1 \end{bmatrix} \begin{bmatrix} u_1 \\ u_2 \\ u_3 \end{bmatrix} = \begin{bmatrix} 1 \\ -1 \\ 0 \end{bmatrix}.$$

Backward substitution leads to the final solution:

$$
\begin{aligned}
u_3 &= 0, \\
-u_2 - 2u_3 &= -1 \quad \Rightarrow \quad u_2 = 1 - 2(0) = 1, \\
u_1 + 2u_2 + 3u_3 &= 1 \quad \Rightarrow \quad u_1 = 1 - 2(1) - 3(0) = -1.
\end{aligned}
$$

Thus, the solution to the original system of linear algebraic equations is

$$
\mathbf{u} = \begin{bmatrix} -1 \\ 1 \\ 0 \end{bmatrix}.
$$

REMARKS:

1. *The approach used in this example is called* Doolittle's method *and results in ones on the main diagonal of* **L**. *Alternatively,* Crout's method *leads to ones on the main diagonal of* **R**.
2. *LU decomposition is a direct method as it requires a prescribed number of known steps, but it is approximately twice as fast as Gaussian elimination ($N^3/3$ operations compared to $2N^3/3$ for large N). It exists for any nonsingular matrix* **A**.
3. *Both matrices* **L** *and* **R** *may be stored as a single matrix, for example,*

$$
\begin{bmatrix} R_{11} & R_{12} & R_{13} & R_{14} \\ L_{21} & R_{22} & R_{23} & R_{24} \\ L_{31} & L_{32} & R_{33} & R_{34} \\ L_{41} & L_{42} & L_{43} & R_{44} \end{bmatrix},
$$

   where we know that **L** *has ones on the main diagonal. In fact, the procedure is such that the elements of* **A** *can be replaced by the corresponding elements of* **L** *and* **R** *as they are determined, thereby minimizing computer memory requirements.*
4. *This procedure is particularly efficient if* **A** *remains the same, while* **b** *is changed (for example, recall the truss and electrical circuits examples in Section 1.4.1). In this case, the* **L** *and* **R** *matrices only must be determined once.*
5. *Sometimes it is necessary to use pivoting, in which rows of* **A** *are interchanged, in order to avoid division by zero (or small numbers) as discussed in Trefethen and Bau (1997). This is analogous to exchanging rows (equations) in Gaussian elimination.*
6. *Recall that the determinant of a triangular matrix is the product of the elements along the main diagonal, which are its eigenvalues. Because the determinant of* **L** *is unity, the determinant of* **A** *is simply*

$$
|\mathbf{A}| = |\mathbf{R}| = \prod_{i=1}^{N} R_{ii},
$$

   which is the product of the main diagonal elements of the triangular matrix **R**.
7. *For more details on the implementation of LU decomposition, see Press et al. (2007).*

6.3.2 Cholesky Decomposition

LU decomposition applies for any nonsingular matrix. If $\mathbf{A}$ is also symmetric (or Hermitian more generally) and positive definite, whereby it has all positive eigenvalues, a special case of LU decomposition can be devised that is even more efficient to compute. This is known as *Cholesky decomposition* (factorization) and is given by

$$\mathbf{A} = \mathbf{R}^T\mathbf{R},$$

or alternatively $\mathbf{A} = \mathbf{L}\mathbf{L}^T$. Whereas LU decomposition requires determination of two triangular matrices, Cholesky decomposition only requires determination of one; the other triangular matrix is simply its transpose, in which case $\mathbf{L} = \mathbf{R}^T$.

Because $\mathbf{R}$ is right triangular of the form

$$\mathbf{R} = \begin{bmatrix} R_{11} & R_{12} & \cdots & R_{1N} \\ 0 & R_{22} & \cdots & R_{2N} \\ \vdots & \vdots & \ddots & \vdots \\ 0 & 0 & \cdots & R_{NN} \end{bmatrix},$$

then $\mathbf{A} = \mathbf{R}^T\mathbf{R}$ is

$$\begin{bmatrix} A_{11} & A_{12} & \cdots & A_{1N} \\ A_{21} & A_{22} & \cdots & A_{2N} \\ \vdots & \vdots & \ddots & \vdots \\ A_{N1} & A_{N2} & \cdots & A_{NN} \end{bmatrix} = \begin{bmatrix} R_{11} & 0 & \cdots & 0 \\ R_{12} & R_{22} & \cdots & 0 \\ \vdots & \vdots & \ddots & \vdots \\ R_{1N} & R_{2N} & \cdots & R_{NN} \end{bmatrix} \begin{bmatrix} R_{11} & R_{12} & \cdots & R_{1N} \\ 0 & R_{22} & \cdots & R_{2N} \\ \vdots & \vdots & \ddots & \vdots \\ 0 & 0 & \cdots & R_{NN} \end{bmatrix}.$$

Multiplying the matrices on the right-hand side, we obtain

$$\begin{aligned} A_{11} &= R_{11}^2 & \Rightarrow \quad R_{11} &= \sqrt{A_{11}}, \\ A_{12} &= R_{11}R_{12} & \Rightarrow \quad R_{12} &= \frac{A_{12}}{R_{11}}, \\ &\vdots \\ A_{1N} &= R_{11}R_{1N} & \Rightarrow \quad R_{1N} &= \frac{A_{1N}}{R_{11}}, \end{aligned}$$

which is the first row of $\mathbf{R}$. This procedure is continued to calculate all $\mathbf{R}_{ij}$ in a manner similar to LU decomposition. Once $\mathbf{R}$ is determined, the solution of $\mathbf{Au} = \mathbf{b}$ simply requires a backward substitution as in LU decomposition.

If the inverse of $\mathbf{A}$ is desired, observe that

$$\mathbf{A}^{-1} = \left(\mathbf{R}^T\mathbf{R}\right)^{-1} = \mathbf{R}^{-1}\left(\mathbf{R}^T\right)^{-1},$$

where

$$\mathbf{R}^{-1} = \mathbf{B} = \begin{bmatrix} B_{11} & B_{12} & \cdots & B_{1N} \\ 0 & B_{22} & \cdots & B_{2N} \\ \vdots & \vdots & \ddots & \vdots \\ 0 & 0 & \cdots & B_{NN} \end{bmatrix},$$

which is right triangular. We determine the $\mathbf{B}_{ij}$ by evaluating

$$\mathbf{RB} = \mathbf{I},$$

where both $\mathbf{R}$ and $\mathbf{B}$ are right triangular, and $\mathbf{R}$ is already known.

REMARKS:

1. *Cholesky decomposition can be thought of as taking the "square root" of the matrix* $\mathbf{A}$.
2. *When it applies, solving systems of equations using Cholesky decomposition is another two times faster than using LU decomposition, which is approximately twice as fast as Gaussian elimination. That is, for large N, the operation counts are:*
   - *Gaussian elimination:* $\frac{2}{3}N^3$
   - *LU decomposition:* $\frac{1}{3}N^3$
   - *Cholesky decomposition:* $\frac{1}{6}N^3$

   Because of symmetry, it is also only necessary to store one half of the matrix as compared to LU decomposition. In addition, Cholesky decomposition has improved numerical stability properties as compared to alternative algorithms. Because of these properties, it is the preferred algorithm for obtaining the solution of systems with symmetric, positive-definite coefficient matrices.
3. *The Cholesky decomposition will fail if the matrix* $\mathbf{A}$ *is not positive definite. Because of its efficiency, therefore, it is often the best technique for determining if a matrix is positive definite unless the eigenvalues are desired for some other reason.*

6.3.3 Tridiagonal Systems of Equations

We will commonly encounter tridiagonal systems of equations when solving differential equations using finite-difference methods as treated in Chapters 8 and 9. For example, recall Couette flow in Example 1.3, which resulted in the need to solve a tridiagonal system of algebraic equations. Other numerical techniques, such as cubic-spline interpolation in Section 11.6, also result in tridiagonal systems of equations. As such, it is worthwhile considering their properties and an efficient technique for their solution. In particular, we develop the *Thomas algorithm* that only requires $O(N)$ operations, which is as good a scaling as one could hope for.

Properties of Tridiagonal Matrices

Consider an $N \times N$ tridiagonal matrix with constants a, b, and c along the lower, main, and upper diagonals, respectively, of the form

$$\mathbf{A} = \begin{bmatrix} b & c & 0 & \cdots & 0 & 0 \\ a & b & c & \cdots & 0 & 0 \\ 0 & a & b & \cdots & 0 & 0 \\ \vdots & \vdots & \vdots & \ddots & \vdots & \vdots \\ 0 & 0 & 0 & \cdots & b & c \\ 0 & 0 & 0 & \cdots & a & b \end{bmatrix}.$$

This is a tridiagonal Toeplitz matrix. It can be shown that the eigenvalues of such a tridiagonal matrix with constants along each diagonal are

$$\lambda_n = b + 2\sqrt{ac}\cos\left(\frac{n\pi}{N+1}\right), \quad n = 1, \dots, N. \tag{6.6}$$

The eigenvalues with the largest and smallest magnitudes are (which is largest or smallest depends upon the values of a, b, and c)

$$|\lambda_1| = \left| b + 2\sqrt{ac}\cos\left(\frac{\pi}{N+1}\right)\right|,$$

$$|\lambda_N| = \left| b + 2\sqrt{ac}\cos\left(\frac{N\pi}{N+1}\right)\right|.$$

Let us consider N large. In this case, expanding cosine in a Taylor series gives (the first is expanded about $\pi/(N+1) \to 0$ and the second about $N\pi/(N+1) \to \pi$)

$$\cos\left(\frac{\pi}{N+1}\right) = 1 - \frac{1}{2!}\left(\frac{\pi}{N+1}\right)^2 + \frac{1}{4!}\left(\frac{\pi}{N+1}\right)^4 + \dots,$$

$$\cos\left(\frac{N\pi}{N+1}\right) = -1 + \frac{1}{2!}\left(\frac{N\pi}{N+1} - \pi\right)^2 - \dots = -1 + \frac{1}{2}\left(\frac{\pi}{N+1}\right)^2 - \dots .$$

Consider the common case that may result from the use of central differences for a second-order derivative as in Example 1.3:

$$a = 1, \quad b = -2, \quad c = 1,$$

which is *weakly diagonally dominant*. Then with the large N expansions, we have

$$|\lambda_1| = \left| -2 + 2\sqrt{(1)(1)}\left[1 - \frac{1}{2}\left(\frac{\pi}{N+1}\right)^2 + \cdots\right]\right| = \left(\frac{\pi}{N+1}\right)^2 + \dots,$$

$$|\lambda_N| = \left| -2 + 2\sqrt{(1)(1)}\left[-1 + \frac{1}{2}\left(\frac{\pi}{N+1}\right)^2 - \cdots\right]\right| = 4 + \cdots .$$

Thus, the condition number of this symmetric matrix for large N is approximately

$$\kappa_2(\mathbf{A}) \approx \frac{4}{(\pi/(N+1))^2} = \frac{4(N+1)^2}{\pi^2}, \quad \text{for large } N.$$

Therefore, the condition number increases proportional to N^2 with increasing N.

Now consider the case with

$$a = 1, \quad b = -4, \quad c = 1,$$

which is *strictly diagonally dominant*. Following the same procedure, the condition number for large N is approximately

$$\kappa_2(\mathbf{A}) \approx \frac{6}{2} = 3, \quad \text{for large } N,$$

in which case it is constant with increasing N.

In our previous example, we illustrated the influence of round-off errors on inversion of the notoriously ill-conditioned Hilbert matrix. Lest we be discouraged that matrix inversion is an inherently inaccurate numerical procedure, let us consider a matrix that behaves more typically and is more representative of what we will encounter in numerical methods.

Example 6.3 Consider the 10×10 tridiagonal matrix with -4's on the main diagonal and 1's on the upper and lower diagonals, such that it is strongly diagonally dominant. The condition number is $\kappa = 2.84428$. This is a much more reasonable condition number than what we had for the Hilbert matrix (recall that the closer to one the better). Note that this is surprisingly close to the large-N estimate for this case, which is 3, with only $N = 10$. Estimating the number of significant digits that will be lost in inverting our tridiagonal matrix, we get $d = \log_{10}(2.84428) = 0.453972$, so we expect to lose less than one digit of accuracy. Indeed, performing the same procedure as in Example 6.1 produces ones along the main diagonal and zeros everywhere else to machine precision.

For a matrix of the same form but with $N = 1{,}000$, the condition number is $\kappa = 2.99998$, which is very close to our large-N estimate. Observe that for this strongly diagonally dominant case, the size of the matrix has very little influence on the condition number and loss of accuracy owing to round-off errors.

Example 6.4 Now consider the 10×10 tridiagonal matrix with -2's on the main diagonal and 1's on the upper and lower diagonals, such that it is weakly diagonally dominant. The condition number is now $\kappa = 48.3742$, which is much larger than for the strongly diagonally dominant case considered in the previous example. Using our large-N estimate gives 49.04, which is fairly close to the actual condition number even for only $N = 10$. Estimating the number of significant digits that will be lost in inverting our tridiagonal matrix, we get $d = \log_{10}(48.3742) = 1.68461$, so we expect to lose one to two digits of accuracy. Indeed, performing the same procedure as in Example 6.1 produces ones along the main diagonal and zeros everywhere else within 14 digits of accuracy (recall that 1×10^{-16} is effectively zero in double precision calculations).

For a matrix of the same form but with $N = 1{,}000$, the condition number is $\kappa = 406{,}095$; the large-N estimate predicts $\kappa = 406{,}095.7$. Observe that for this weakly diagonally dominant case, the number of digits expected to be lost is $d = \log_{10}(406{,}095) = 5.60863$. While losing five to six digits of accuracy is not nearly as bad as for the Hilbert matrix, such loss in accuracy may be important depending upon what will be done with the resulting values of the inverse. Moreover, note that the size of the matrix has a significant influence on the condition number and loss of accuracy for the weakly diagonally dominant case.

Thomas Algorithm

Now we generalize the tridiagonal system of algebraic equations $\mathbf{Au} = \mathbf{d}$ to encompass scenarios common in various numerical methods by allowing for different values along each diagonal and on the right-hand side.[17] In other words, it is no longer a Toeplitz matrix. As for Couette flow in Example 1.3, let us imagine that we have a case in which the end values u_1 and u_{I+1} are known at $x = 0$ and $x = \ell$, respectively. That is, we have so called Dirichlet boundary conditions. In this situation, the tridiagonal system of equations is given by

$$\begin{bmatrix} b_2 & c_2 & 0 & 0 & \cdots & 0 & 0 & 0 \\ a_3 & b_3 & c_3 & 0 & \cdots & 0 & 0 & 0 \\ 0 & a_4 & b_4 & c_4 & \cdots & 0 & 0 & 0 \\ \vdots & \vdots & \vdots & \vdots & \ddots & \vdots & \vdots & \vdots \\ 0 & 0 & 0 & 0 & \cdots & a_{I-1} & b_{I-1} & c_{I-1} \\ 0 & 0 & 0 & 0 & \cdots & 0 & a_I & b_I \end{bmatrix} \begin{bmatrix} u_2 \\ u_3 \\ u_4 \\ \vdots \\ u_{I-1} \\ u_I \end{bmatrix} = \begin{bmatrix} d_2 - a_2 u_1 \\ d_3 \\ d_4 \\ \vdots \\ d_{I-1} \\ d_I - c_I u_{I+1} \end{bmatrix}.$$

This tridiagonal form is typical of other finite-difference methods having compact stencils and Dirichlet boundary conditions as we will encounter in Chapters 8 and 9. Because of its central importance in several numerical algorithms in Part II, we will dissect the Thomas algorithm in detail.

Tridiagonal systems may be solved directly, and efficiently, using the *Thomas algorithm*, which is based on Gaussian elimination. Recall that Gaussian elimination consists of forward elimination and backward substitution steps. First, consider the *forward elimination*, which eliminates the a_i coefficients along the lower diagonal. Let us begin by dividing the first equation through by b_2 to give

$$\begin{bmatrix} 1 & F_2 & 0 & 0 & \cdots & 0 & 0 & 0 \\ a_3 & b_3 & c_3 & 0 & \cdots & 0 & 0 & 0 \\ 0 & a_4 & b_4 & c_4 & \cdots & 0 & 0 & 0 \\ \vdots & \vdots & \vdots & \vdots & \ddots & \vdots & \vdots & \vdots \\ 0 & 0 & 0 & 0 & \cdots & a_{I-1} & b_{I-1} & c_{I-1} \\ 0 & 0 & 0 & 0 & \cdots & 0 & a_I & b_I \end{bmatrix} \begin{bmatrix} u_2 \\ u_3 \\ u_4 \\ \vdots \\ u_{I-1} \\ u_I \end{bmatrix} = \begin{bmatrix} \delta_2 \\ d_3 \\ d_4 \\ \vdots \\ d_{I-1} \\ d_I - c_I u_{I+1} \end{bmatrix},$$

[17] We use $\mathbf{d}$ for the right-hand-side vector, and I for the index for consistency with how the Thomas algorithm is applied throughout Part II.

where

$$F_2 = \frac{c_2}{b_2}, \quad \delta_2 = \frac{d_2 - a_2 u_1}{b_2}. \tag{6.7}$$

To eliminate a_3 in the second equation, subtract a_3 times the first equation from the second equation to produce

$$\begin{bmatrix} 1 & F_2 & 0 & 0 & \cdots & 0 & 0 & 0 \\ 0 & b_3 - a_3 F_2 & c_3 & 0 & \cdots & 0 & 0 & 0 \\ 0 & a_4 & b_4 & c_4 & \cdots & 0 & 0 & 0 \\ \vdots & \vdots & \vdots & \vdots & \ddots & \vdots & \vdots & \vdots \\ 0 & 0 & 0 & 0 & \cdots & a_{I-1} & b_{I-1} & c_{I-1} \\ 0 & 0 & 0 & 0 & \cdots & 0 & a_I & b_I \end{bmatrix} \begin{bmatrix} u_2 \\ u_3 \\ u_4 \\ \vdots \\ u_{I-1} \\ u_I \end{bmatrix} = \begin{bmatrix} \delta_2 \\ d_3 - a_3 \delta_2 \\ d_4 \\ \vdots \\ d_{I-1} \\ d_I - c_I u_{I+1} \end{bmatrix}.$$

Dividing the second equation through by $b_3 - a_3 F_2$ then leads to

$$\begin{bmatrix} 1 & F_2 & 0 & 0 & \cdots & 0 & 0 & 0 \\ 0 & 1 & F_3 & 0 & \cdots & 0 & 0 & 0 \\ 0 & a_4 & b_4 & c_4 & \cdots & 0 & 0 & 0 \\ \vdots & \vdots & \vdots & \vdots & \ddots & \vdots & \vdots & \vdots \\ 0 & 0 & 0 & 0 & \cdots & a_{I-1} & b_{I-1} & c_{I-1} \\ 0 & 0 & 0 & 0 & \cdots & 0 & a_I & b_I \end{bmatrix} \begin{bmatrix} u_2 \\ u_3 \\ u_4 \\ \vdots \\ u_{I-1} \\ u_I \end{bmatrix} = \begin{bmatrix} \delta_2 \\ \delta_3 \\ d_4 \\ \vdots \\ d_{I-1} \\ d_I - c_I u_{I+1} \end{bmatrix},$$

where

$$F_3 = \frac{c_3}{b_3 - a_3 F_2}, \quad \delta_3 = \frac{d_3 - a_3 \delta_2}{b_3 - a_3 F_2}. \tag{6.8}$$

Similarly, to eliminate a_4 in the third equation, subtract a_4 times the second equation from the third equation to give

$$\begin{bmatrix} 1 & F_2 & 0 & 0 & \cdots & 0 & 0 & 0 \\ 0 & 1 & F_3 & 0 & \cdots & 0 & 0 & 0 \\ 0 & 0 & b_4 - a_4 F_3 & c_4 & \cdots & 0 & 0 & 0 \\ \vdots & \vdots & \vdots & \vdots & \ddots & \vdots & \vdots & \vdots \\ 0 & 0 & 0 & 0 & \cdots & a_{I-1} & b_{I-1} & c_{I-1} \\ 0 & 0 & 0 & 0 & \cdots & 0 & a_I & b_I \end{bmatrix} \begin{bmatrix} u_2 \\ u_3 \\ u_4 \\ \vdots \\ u_{I-1} \\ u_I \end{bmatrix} = \begin{bmatrix} \delta_2 \\ \delta_3 \\ d_4 - a_4 \delta_3 \\ \vdots \\ d_{I-1} \\ d_I - c_I u_{I+1} \end{bmatrix}.$$

Dividing the third equation through by $b_4 - a_4 F_3$ then leads to

$$\begin{bmatrix} 1 & F_2 & 0 & 0 & \cdots & 0 & 0 & 0 \\ 0 & 1 & F_3 & 0 & \cdots & 0 & 0 & 0 \\ 0 & 0 & 1 & F_4 & \cdots & 0 & 0 & 0 \\ \vdots & \vdots & \vdots & \vdots & \ddots & \vdots & \vdots & \vdots \\ 0 & 0 & 0 & 0 & \cdots & a_{I-1} & b_{I-1} & c_{I-1} \\ 0 & 0 & 0 & 0 & \cdots & 0 & a_I & b_I \end{bmatrix} \begin{bmatrix} u_2 \\ u_3 \\ u_4 \\ \vdots \\ u_{I-1} \\ u_I \end{bmatrix} = \begin{bmatrix} \delta_2 \\ \delta_3 \\ \delta_4 \\ \vdots \\ d_{I-1} \\ d_I - c_I u_{I+1} \end{bmatrix},$$

where

$$F_4 = \frac{c_4}{b_4 - a_4 F_3}, \qquad \delta_4 = \frac{d_4 - a_4 \delta_3}{b_4 - a_4 F_3}. \tag{6.9}$$

Comparing (6.7) through (6.9), observe that we can define the following recursive coefficients to perform the forward elimination:

$$F_1 = 0, \quad \delta_1 = u_1 = u_b,$$

$$F_i = \frac{c_i}{b_i - a_i F_{i-1}}, \qquad \delta_i = \frac{d_i - a_i \delta_{i-1}}{b_i - a_i F_{i-1}}, \qquad i = 2, \ldots, I.$$

Upon completion of the forward elimination step, we have the row-echelon form of the system of equations given by

$$\begin{bmatrix} 1 & F_2 & 0 & 0 & \cdots & 0 & 0 & 0 \\ 0 & 1 & F_3 & 0 & \cdots & 0 & 0 & 0 \\ 0 & 0 & 1 & F_4 & \cdots & 0 & 0 & 0 \\ \vdots & \vdots & \vdots & \vdots & \ddots & \vdots & \vdots & \vdots \\ 0 & 0 & 0 & 0 & \cdots & 0 & 1 & F_{I-1} \\ 0 & 0 & 0 & 0 & \cdots & 0 & 0 & 1 \end{bmatrix} \begin{bmatrix} u_2 \\ u_3 \\ u_4 \\ \vdots \\ u_{I-1} \\ u_I \end{bmatrix} = \begin{bmatrix} \delta_2 \\ \delta_3 \\ \delta_4 \\ \vdots \\ \delta_{I-1} \\ \delta_I - F_I u_{I+1} \end{bmatrix}.$$

We then apply *backward substitution* to obtain the solutions for u_i. Starting with the last equation, we have

$$u_I = \delta_I - F_I u_{I+1},$$

where $u_{I+1} = u_\ell$ is known from the specified Dirichlet boundary condition at $x = \ell$. Using this result, the second to last equation then gives

$$u_{I-1} = \delta_{I-1} - F_{I-1} u_I.$$

Generalizing yields

$$u_i = \delta_i - F_i u_{i+1}, \quad i = I, \ldots, 2,$$

where we note the order starting at $x = \ell$ and ending at $x = 0$.

Summarizing, the Thomas algorithm consists of the two recursive stages that must be performed in order:

1. Forward elimination:

$$F_1 = 0, \quad \delta_1 = u_1 = u_b = \text{boundary condition}$$
$$F_i = \frac{c_i}{b_i - a_i F_{i-1}}, \qquad \delta_i = \frac{d_i - a_i \delta_{i-1}}{b_i - a_i F_{i-1}}, \qquad i = 2, \ldots, I.$$

2. Backward substitution:

$$u_{I+1} = u_\ell = \text{boundary condition}$$
$$u_i = \delta_i - F_i u_{i+1}, \quad i = I, \ldots, 2.$$

Again, note the order of evaluation for the backward substitution. Whereas the arrays are built up from left to right in the forward elimination step, the solution is built up from right to left in the backward elimination step.

The Thomas algorithm only requires $O(I)$ operations, which is as good a scaling as one could hope for as it scales linearly with the number of points – doubling the number of points only doubles the number of operations. In contrast, recall that Gaussian elimination of a dense matrix requires $O(I^3)$ operations. The Thomas algorithm is so efficient that numerical methods for more complex situations, such as two- and three-dimensional partial differential equations, are often designed specifically to convert ordinary and partial differential equations into tridiagonal matrices in order to take advantage of the Thomas algorithm. See, for example, finite-difference methods for ordinary differential equations in Sections 8.1 and 8.2, the alternating-direction-implicit (ADI) method for elliptic partial differential equations in Section 8.7, and the first-order implicit and Crank–Nicolson methods for parabolic partial differential equations in Sections 9.7.2, 9.9.2, and 9.10.

REMARKS:

1. *Observe that it is only necessary to store each of the three diagonals in a vector (one-dimensional array) and not the entire matrix owing to the tridiagonal structure of the matrix.*
2. *The Thomas algorithm successfully produces the solution when the coefficient matrix is diagonally dominant, that is, when*

$$|b_i| \geq |a_i| + |c_i| .$$

3. *Similar algorithms are available for other* banded, *such as pentadiagonal, matrices.*
4. *Notice how round-off errors could accumulate in the F_i and δ_i coefficients during the forward elimination step, and u_i in the backward substitution step.*

6.4 Systems of Linear Algebraic Equations – Iterative Methods

Other than the Thomas algorithm for tridiagonal systems of equations, the direct methods for solving linear systems of algebraic equations all scale according to $O(N^3)$. Therefore, the operation counts increase dramatically as the size of the matrix problem grows. Consequently, it is often advantageous to utilize iterative techniques when solving very large systems on a computer. Although we cannot formally prove that iterative methods will be faster than direct methods, this is generally found to be the case in practice as they often scale well with the size of the problem. They also form the basis for popular techniques for numerically solving ordinary and partial differential equations as discussed in Chapters 8 and 9.

In addition to being faster than direct methods, iterative methods are also typically much simpler conceptually and easier to implement. Such advantages, however, must come at a cost; this cost has to do with the additional question as to whether our iterative process will converge and how fast it will do so. That is, how many iterations must be executed in order to obtain a solution of sufficient accuracy.

When we execute the direct methods covered in Chapter 1 for solving systems of linear algebraic equations by hand, we obtain the exact solution if we do so without making any approximations.[18] When these direct methods, as well as those in the previous section, are performed using floating point arithmetic on a computer, however, one encounters round-off errors as discussed in Section 6.2.2.

In addition to round-off errors, iterative methods are also subject to *iterative convergence errors* that result from the iterative process itself. Insofar as iterative methods produce a sequence of successive approximations to the solution of the original system of equations, one now also must be concerned with whether this sequence of iterative approximations will converge toward the true exact solution. Here, we seek to obtain a criterion for determining if an iterative algorithm will converge toward the exact solution or not and then articulate a series of popular iterative methods for solving systems of linear algebraic equations.

6.4.1 General Framework

Let us say we have a system of linear algebraic equations of the form

$$\begin{aligned} A_{11}u_1 + A_{12}u_2 + A_{13}u_3 + \cdots + A_{1N}u_N &= \hat{b}_1, \\ A_{21}u_1 + A_{22}u_2 + A_{23}u_3 + \cdots + A_{2N}u_N &= \hat{b}_2, \\ &\vdots \\ A_{N1}u_1 + A_{N2}u_2 + A_{N3}u_3 + \cdots + A_{NN}u_N &= \hat{b}_N, \end{aligned}$$

or $\mathbf{Au} = \hat{\mathbf{b}}$ in matrix form. A simple iterative technique, known as the *Jacobi method*, can be obtained by simply solving the first equation for u_1, the second equation for u_2, and so forth, as follows:

$$\begin{aligned} u_1 &= -\frac{A_{12}}{A_{11}}u_2 - \frac{A_{13}}{A_{11}}u_3 - \cdots - \frac{A_{1N}}{A_{11}}u_N + \frac{\hat{b}_1}{A_{11}}, \\ u_2 &= -\frac{A_{21}}{A_{22}}u_1 - \frac{A_{23}}{A_{22}}u_3 - \cdots - \frac{A_{2N}}{A_{22}}u_N + \frac{\hat{b}_2}{A_{22}}, \\ &\vdots \\ u_N &= -\frac{A_{N1}}{A_{NN}}u_1 - \frac{A_{N2}}{A_{NN}}u_2 - \frac{A_{N3}}{A_{NN}}u_3 - \cdots + \frac{\hat{b}_N}{A_{NN}}. \end{aligned}$$

The $u_i, i = 1, \ldots, N$, on the right-hand sides are taken from the previous iteration in order to update the solution vector on the left-hand side at the current iteration. In matrix form, this and other iterative numerical schemes may be expressed as

$$\mathbf{u}^{(n+1)} = \mathbf{M}\mathbf{u}^{(n)} + \mathbf{b}, \tag{6.10}$$

[18] Note that Mathematica and MATLAB are also capable of obtaining exact solutions using symbolic arithmetic, which is the computer analog to hand calculations in that no round-off errors are incurred.

where superscript (n) is the iteration number; $\mathbf{u}^{(n)}, n = 0, 1, 2, \ldots$, is the sequence of iterative approximations to the exact solution; $\mathbf{u}^{(0)}$ is the initial guess; and $\mathbf{M}$ is the $N \times N$ iteration matrix, which is determined by the iterative method. Note that the iteration number (n) is placed in parentheses so as not to be confused with powers of the vectors or matrix.

6.4.2 Requirement for Iterative Convergence

We want the preceding iterative procedure to *converge* to the exact solution $\hat{\mathbf{u}}$ of

$$\hat{\mathbf{u}} = \mathbf{M}\hat{\mathbf{u}} + \mathbf{b}, \tag{6.11}$$

as $n \to \infty$ for any initial guess $\mathbf{u}^{(0)}$. Consequently, exact iterative convergence, which would take an infinity of iterations, occurs when the exact solution $\hat{\mathbf{u}}$ input on the right-hand side is returned precisely upon execution of an iteration. In practice, we terminate the iterative process when the approximation returned has changed very little by an iteration as discussed in the next section.

In order to determine a convergence criteria, let $\lambda_1, \ldots, \lambda_N$ be the eigenvalues of the iteration matrix $\mathbf{M}$ and $\mathbf{u}_1, \ldots, \mathbf{u}_N$ the corresponding eigenvectors, which we assume are linearly independent. We define the error at the nth iteration by

$$\mathbf{e}^{(n)} = \hat{\mathbf{u}} - \mathbf{u}^{(n)}.$$

Substituting into (6.10) gives

$$\hat{\mathbf{u}} - \mathbf{e}^{(n+1)} = \mathbf{M}\left(\hat{\mathbf{u}} - \mathbf{e}^{(n)}\right) + \mathbf{b},$$

or

$$\mathbf{e}^{(n+1)} = \hat{\mathbf{u}} - \mathbf{M}\hat{\mathbf{u}} + \mathbf{M}\mathbf{e}^{(n)} - \mathbf{b},$$

but from (6.11)

$$\mathbf{e}^{(n+1)} = \mathbf{M}\mathbf{e}^{(n)}, \quad n = 0, 1, 2, \ldots,$$

which provides an alternative way to write (6.10) in terms of the error. It follows that

$$\mathbf{e}^{(n)} = \mathbf{M}\mathbf{e}^{(n-1)} = \mathbf{M}\left(\mathbf{M}\mathbf{e}^{(n-2)}\right) = \cdots = \mathbf{M}^n\mathbf{e}^{(0)}. \tag{6.12}$$

We may write the error of the initial guess $\mathbf{e}^{(0)}$ as a linear combination of the linearly independent eigenvectors as follows

$$\mathbf{e}^{(0)} = \alpha_1\mathbf{u}_1 + \alpha_2\mathbf{u}_2 + \cdots + \alpha_N\mathbf{u}_N.$$

Substituting for $\mathbf{e}^{(0)}$ in (6.12) gives

$$\mathbf{e}^{(n)} = \mathbf{M}^n\left(\alpha_1\mathbf{u}_1 + \alpha_2\mathbf{u}_2 + \cdots + \alpha_N\mathbf{u}_N\right),$$

or

$$\mathbf{e}^{(n)} = \alpha_1\mathbf{M}^n\mathbf{u}_1 + \alpha_2\mathbf{M}^n\mathbf{u}_2 + \cdots + \alpha_N\mathbf{M}^n\mathbf{u}_N. \tag{6.13}$$

But $\mathbf{M}\mathbf{u}_i = \lambda_i \mathbf{u}_i$, and multiplying by $\mathbf{M}$ gives

$$\mathbf{M}\left(\mathbf{M}\mathbf{u}_i\right) = \lambda_i \mathbf{M}\mathbf{u}_i = \lambda_i^2 \mathbf{u}_i.$$

Generalizing to any integer power of $\mathbf{M}$, we have the general result

$$\mathbf{M}^n \mathbf{u}_i = \lambda_i^n \mathbf{u}_i,$$

where now n does represent the nth power. Consequently, if λ_i are the eigenvalues of $\mathbf{M}$, the eigenvalues of $\mathbf{M}^n$ are λ_i^n, and the eigenvectors of $\mathbf{M}$ and $\mathbf{M}^n$ are the same (also see Section 2.2). Substituting into (6.13) leads to

$$\mathbf{e}^{(n)} = \alpha_1 \lambda_1^n \mathbf{u}_1 + \alpha_2 \lambda_2^n \mathbf{u}_2 + \cdots + \alpha_N \lambda_N^n \mathbf{u}_N.$$

For convergence, $\left|\mathbf{e}^{(n)}\right|$ must go to zero as $n \to \infty$. Because $|\alpha_i \mathbf{u}_i| \neq 0$, $|\lambda_i^n|$ must go to zero as $n \to \infty$. Thus, we have the general result that for convergence of the iterative numerical method

$$\rho = |\lambda_i|_{\max} < 1, \quad i = 1, \ldots, N,$$

where ρ is the *spectral radius* of the iteration matrix $\mathbf{M}$. Therefore, iterative methods must be devised such that the magnitude of the eigenvalues of the iteration matrix are *all* less than one.

A more general iterative scheme may be obtained by decomposing the coefficient matrix $\mathbf{A}$ as follows:

$$\mathbf{A} = \mathbf{M}_1 - \mathbf{M}_2.$$

Then we construct an iterative technique of the form

$$\mathbf{M}_1 \mathbf{u}^{(n+1)} = \mathbf{M}_2 \mathbf{u}^{(n)} + \mathbf{b}.$$

To be efficient numerically, $\mathbf{M}_1$ should be easily invertible. If we write the preceding equation as

$$\mathbf{u}^{(n+1)} = \mathbf{M}_1^{-1} \mathbf{M}_2 \mathbf{u}^{(n)} + \mathbf{M}_1^{-1} \mathbf{b},$$

we can follow the same analysis as before, with $\mathbf{M} = \mathbf{M}_1^{-1}\mathbf{M}_2$, leading to the *necessary and sufficient condition* that

$$\rho = |\lambda_i|_{\max} < 1.$$

REMARK:

1. *The λ_i are the eigenvalues of the iterative matrix $\mathbf{M} = \mathbf{M}_1^{-1}\mathbf{M}_2$, and ρ is its spectral radius.*
2. *While the spectral radius provides a necessary and sufficient condition, strict diagonal dominance provides a sufficient condition. Because it is not a necessary condition, an iterative method may converge even if it is not strictly diagonally dominant; however, we cannot guarantee this to be the case – unless we check the spectral radius.*

3. *In addition to determining whether an iterative method will converge, the spectral radius also determines the rate at which they converge, in which case smaller is better.*
4. *For the Jacobi method,* $\mathbf{M}_1$ *is a diagonal matrix, and for the commonly used Gauss–Seidel method,* $\mathbf{M}_1$ *is a triangular matrix (see Sections 6.4.4 and 6.4.5).*
5. *This requirement for convergence applies for all iterative methods for which the eigenvectors of the iteration matrix* $\mathbf{M}$ *are linearly independent.*

6.4.3 Practical Test for Iterative Convergence

Having a criterion that ensures that an iterative process will converge toward the exact solution is critical; however, we still need a convergence criterion in order to know when to terminate the iterative process. In the preceding analysis, we imagine that the number of iterations $n \to \infty$. In practice, however, we must devise a criterion for deciding when the iterative process has reached a sufficient level of convergence in order to balance our desire for an accurate solution obtained within the minimum number of iterations and, thus, computational time.

In practice, we will terminate the iterative process once each successive iteration $(n+1)$ produces a result that is very close to the previous one (n). This is determined using a convergence test. We could simply test the absolute difference in the solution between two successive iterations according to

$$\left| u_i^{(n+1)} - u_i^{(n)} \right| < \epsilon \quad \text{for} \quad i = 1, \ldots, N,$$

where ϵ is a small tolerance value that we choose, for example $\epsilon = 10^{-5}$. However, this convergence criterion does not take into account the magnitude of the solution u_i. Therefore, a relative difference is more suitable, in which we divide by the value at each point in order to normalize the difference. For example, we could test

$$\left| \frac{u_i^{(n+1)} - u_i^{(n)}}{u_i^{(n+1)}} \right| < \epsilon \quad \text{for} \quad i = 1, \ldots, N.$$

Now, however, we must be concerned with the possibility of a divide by zero. To eliminate this possibility, we use

$$\frac{\max\left(|u_i^{(n+1)} - u_i^{(n)}| \right)}{\max\left(|u_i^{(n+1)}| \right)} < \epsilon \quad \text{for} \quad i = 1, \ldots, N.$$

By utilizing the maximum magnitude over all of the points in the denominator, we avoid dividing by zero, producing a robust iterative convergence test.

6.4.4 Jacobi Method

Recall from Section 6.4.1 that in the Jacobi method, we solve each linear algebraic equation as an explicit equation for one of $u_i, i = 1, \ldots, N$. As indicated in (6.10), the values of u_i on the right-hand side are then taken from the previous iteration, denoted by superscript (n), while the single value on the left-hand side is taken at the current iteration, denoted by superscript $(n+1)$.

Although we do not actually implement iterative methods in matrix form, it is instructive to view them in the form $\mathbf{Au} = \mathbf{b}$ in order to analyze their iterative convergence properties. As in Section 6.4.2, we write the coefficient matrix $\mathbf{A} = \mathbf{M}_1 - \mathbf{M}_2$, such that an iterative scheme may be devised by writing $\mathbf{Au} = \mathbf{b}$ in the form

$$\mathbf{M}_1\mathbf{u}^{(n+1)} = \mathbf{M}_2\mathbf{u}^{(n)} + \mathbf{b}.$$

Multiplying by $\mathbf{M}_1^{-1}$ gives

$$\mathbf{u}^{(n+1)} = \mathbf{M}_1^{-1}\mathbf{M}_2\mathbf{u}^{(n)} + \mathbf{M}_1^{-1}\mathbf{b},$$

or

$$\mathbf{u}^{(n+1)} = \mathbf{M}\mathbf{u}^{(n)} + \mathbf{M}_1^{-1}\mathbf{b},$$

where $\mathbf{M} = \mathbf{M}_1^{-1}\mathbf{M}_2$ is the iteration matrix.

Let

$$\begin{aligned} \mathbf{D} &= \text{main diagonal elements of } \mathbf{A}, \\ -\mathbf{L} &= \text{left triangular elements of } \mathbf{A} \text{ less main diagonal}, \\ -\mathbf{R} &= \text{right triangular elements of } \mathbf{A} \text{ less main diagonal}. \end{aligned}$$

Therefore, $\mathbf{Au} = \mathbf{b}$ becomes

$$(\mathbf{D} - \mathbf{L} - \mathbf{R})\mathbf{u} = \mathbf{b}.$$

Using this notation, the Jacobi iteration, with $\mathbf{M}_1 = \mathbf{D}$ and $\mathbf{M}_2 = \mathbf{L} + \mathbf{R}$, is of the form

$$\mathbf{D}\mathbf{u}^{(n+1)} = (\mathbf{L} + \mathbf{R})\,\mathbf{u}^{(n)} + \mathbf{b},$$

or

$$\mathbf{u}^{(n+1)} = \mathbf{D}^{-1}\,(\mathbf{L} + \mathbf{R})\,\mathbf{u}^{(n)} + \mathbf{D}^{-1}\mathbf{b},$$

such that the iteration matrix is

$$\mathbf{M} = \mathbf{M}_1^{-1}\mathbf{M}_2 = \mathbf{D}^{-1}\,(\mathbf{L} + \mathbf{R}).$$

We then check to be sure that the spectral radius of the iteration matrix $\mathbf{M}$ satisfies the requirement that $\rho < 1$ to ensure convergence of the iterative scheme according to the general result in Section 6.4.2. In addition to checking for iterative convergence, a smaller spectral radius results in more rapid convergence, that is, in fewer iterations. For the Jacobi method, as $N \to \infty$, $\rho_{Jac}(N) \to 1$. As a result, we experience slower iterative convergence as N is increased. In other words, there is a disproportionate increase in computational time as the size of the system is increased. This is why the Jacobi method is not used in practice.

6.4.5 Gauss–Seidel Method

In addition to the slow iterative convergence properties of the Jacobi method, it also requires one to store both the current and previous iterations of the solution. The Gauss–Seidel method addresses both issues by using the most recently updated information during the iterative process. In this way, it is no longer necessary to store $u_i^{(n)}$ at the previous iteration as with Jacobi iteration. In fact, the values of u_i are all stored in the same array, and it is not necessary to distinguish between the (n)th or $(n+1)$st iterates. We simply use the most recently updated information as each new value of $u_i^{(n+1)}$ overwrites its corresponding value $u_i^{(n)}$ from the previous iteration.

In matrix form, the Gauss–Seidel method consists of $\mathbf{M}_1 = \mathbf{D} - \mathbf{L}$ and $\mathbf{M}_2 = \mathbf{R}$, in which case

$$(\mathbf{D} - \mathbf{L})\,\mathbf{u}^{(n+1)} = \mathbf{R}\mathbf{u}^{(n)} + \mathbf{b},$$

or

$$\mathbf{u}^{(n+1)} = (\mathbf{D} - \mathbf{L})^{-1}\,\mathbf{R}\mathbf{u}^{(n)} + (\mathbf{D} - \mathbf{L})^{-1}\,\mathbf{b}. \tag{6.14}$$

Therefore, the iteration matrix is

$$\mathbf{M} = \mathbf{M}_1^{-1}\mathbf{M}_2 = (\mathbf{D} - \mathbf{L})^{-1}\,\mathbf{R}.$$

It can be shown that

$$\rho_{GS}(N) = \rho_{Jac}^2(N).$$

Because $\rho_{Jac} < 1$ for an iterative problem that converges, the spectral radius for Gauss–Seidel will be smaller and converge in fewer iterations. For example, it will be shown in Section 8.5.2 that Gauss–Seidel converges twice as fast, that is, requires half as many iterations, as the Jacobi method for a particular model problem.

6.4.6 Successive Over-Relaxation (SOR)

In Gauss–Seidel iteration, the sign of the error typically does not change from iteration to iteration. That is, the approximate iterative solution tends to approach the exact solution from the positive or negative side exclusively. In addition, it normally does so relatively slowly. Iterative convergence can often be accelerated by "over-relaxing,"[19] or magnifying, the change at each iteration. This over-relaxation is accomplished by taking a weighted average (linear combination) of the previous iterate $\mathbf{u}_i^{(n)}$ and the Gauss–Seidel iterate $\mathbf{u}_i^{(n+1)}$ to form the new approximation at each iteration.

If we denote the Gauss–Seidel iterate (6.14) by $\mathbf{u}_i^*$, the new successive over-relaxation (SOR) iterate is given by

$$\mathbf{u}_i^{(n+1)} = (1 - \omega)\mathbf{u}_i^{(n)} + \omega\mathbf{u}_i^*, \tag{6.15}$$

where ω is the acceleration, or relaxation, parameter, and $1 < \omega < 2$ for convergence of linear systems. Note that $\omega = 1$ corresponds to the Gauss–Seidel method.

[19] The terms "iteration" and "relaxation" are used interchangeably.

In matrix form, the SOR method is given by $\mathbf{M}_1 = \mathbf{D} - \omega\mathbf{L}$ and $\mathbf{M}_2 = (1-\omega)\mathbf{D} + \omega\mathbf{R}$. Then

$$(\mathbf{D} - \omega\mathbf{L})\,\mathbf{u}^{(n+1)} = [(1-\omega)\mathbf{D} + \omega\mathbf{R}]\,\mathbf{u}^{(n)} + \omega\mathbf{b},$$

or

$$\mathbf{u}^{(n+1)} = (\mathbf{D} - \omega\mathbf{L})^{-1}\,[(1-\omega)\mathbf{D} + \omega\mathbf{R}]\,\mathbf{u}^{(n)} + \omega\,(\mathbf{D} - \omega\mathbf{L})^{-1}\,\mathbf{b}.$$

Therefore, the iteration matrix is

$$\mathbf{M} = \mathbf{M}_1^{-1}\mathbf{M}_2 = (\mathbf{D} - \omega\mathbf{L})^{-1}\,[(1-\omega)\mathbf{D} + \omega\mathbf{R}]\,.$$

It can be shown that the optimal value of ω that minimizes the spectral radius, and consequently the number of iterations, is (see, for example, Morton & Mayers, 1994 and Moin, 2010)

$$\omega_{opt}(N) = \frac{2}{1 + \sqrt{1 - \rho_{Jac}^2(N)}}, \tag{6.16}$$

and for this ω_{opt}, the spectral radius for SOR is

$$\rho_{SOR}(N) = \omega_{opt}(N) - 1. \tag{6.17}$$

Example 6.5 Given the following system of linear algebraic equations

$$\begin{aligned} 2x_1 + x_3 &= 9, \\ 2x_2 + x_3 &= 3, \\ x_1 + 2x_3 &= 3, \end{aligned}$$

perform the first four iterations of the Gauss–Seidel method with the initial guess $(x_1, x_2, x_3) = (0, 0, 0)$.

Solution

In the Gauss–Seidel method, we solve the first equation for x_1, the second equation for x_2, and the third equation for x_3 as follows:

$$x_1 = \frac{1}{2}\,(9 - x_3)\,,$$

$$x_2 = \frac{1}{2}\,(3 - x_3)\,,$$

$$x_3 = \frac{1}{2}\,(3 - x_1)\,.$$

For the initial guess

$$x_1^{(0)} = 0, \quad x_2^{(0)} = 0, \quad x_3^{(0)} = 0,$$

Table 6.4 Sequence of iterative approximations to the solution for Example 6.5 using the Gauss–Seidel method ($x_i^{(0)}$ is the initial guess).

Iteration	$x_1^{(n)}$	$x_2^{(n)}$	$x_3^{(n)}$
$n = 0$	0	0	0
$n = 1$	4.5	4.5	−0.75
$n = 2$	4.875	1.875	−0.9375
$n = 3$	4.96875	1.96875	−0.984375
$n = 4$	4.9921875	1.9921875	−0.99609375

the first iteration yields

$$\begin{aligned} x_1^{(1)} &= \frac{1}{2}\left(9 - x_3^{(0)}\right) &= \frac{1}{2}(9-0) &= \frac{9}{2}, \\ x_2^{(1)} &= \frac{1}{2}\left(3 - x_3^{(0)}\right) &= \frac{1}{2}(3-0) &= \frac{9}{2}, \\ x_3^{(1)} &= \frac{1}{2}\left(3 - x_1^{(1)}\right) &= \frac{1}{2}\left(3 - \frac{9}{2}\right) &= -\frac{3}{4}. \end{aligned}$$

Note that in the Gauss–Seidel method, updated values of x_1, x_2, and x_3 are used as soon as they are available within the current iteration. For example, in the last equation for $x_3^{(1)}$, the value of $x_1^{(1)} = 9/2$ is used. In the same way, the second, third, and fourth iterations of Gauss–Seidel produce the sequence of approximations shown in Table 6.4. These values are converging toward the exact solution of the system of equations, which is $(x_1, x_2, x_3) = (5, 2, -1)$.

REMARKS:

1. *For direct methods, we only need to be concerned with the influence of round-off errors as determined by the condition number or diagonal dominance. For iterative methods, we are also concerned with iterative convergence rate as determined by the spectral radius and the criterion for terminating the iteration process.*
2. *We will discuss iterative methods, including Jacobi, Gauss–Seidel, SOR, and ADI methods, in more detail in the context of obtaining numerical solutions of elliptic partial differential equations in Chapter 8.*

6.4.7 Conjugate-Gradient and Generalized Minimum Residual Methods

The iterative techniques outlined in the previous sections perform their work directly on the original system of algebraic equations and are very simple to implement. They are simple because the operations performed during each iteration are both straightforward and computationally efficient. However, they require a large number of iterations in order to converge to a solution that is close to the exact solution. A very different class of techniques can be obtained by formulating the system of linear algebraic equations as an optimization problem. While the computations required during each iteration are significantly increased as compared to Jacobi, Gauss–Seidel, and SOR, the number of iterations until convergence is dramatically reduced.

The conjugate-gradient and generalized minimum residual (GMRES) methods are based on solving minimization problems as an indirect means to solve the system of algebraic equations. We have two choices – minimize a norm of the error or the residual. The former applies to symmetric, positive-definite matrices, while the latter applies to any nonsingular matrix. Because they are based on least-squares methods, we must put off considering them in more detail until Section 10.6. Your patience will be rewarded as these methods are typically the most computationally efficient algorithms for solving large, sparse systems of linear algebraic equations.

6.5 Numerical Solution of the Algebraic Eigenproblem

Along with solving systems of linear algebraic equations, the other workhorse of computational linear algebra is algorithms for obtaining the eigenvalues and eigenvectors of large matrices. For large matrices, the procedure outlined in Section 2.2 is too inefficient and memory intensive for implementation on a computer. For example, it is not practical or efficient to try to factor the high-order characteristic polynomials numerically. Therefore, we require approximate numerical solution techniques. The standard, and nearly universal, method for numerically approximating the eigenvalues and eigenvectors of a matrix is based on *QR decomposition*,[20] which entails performing a series of *similarity transformations* that leave the eigenvalues unchanged. QR decomposition is also the basis for numerical algorithms designed to solve least-squares problems, perform orthogonalization as discussed in the next section, and obtain the singular-value decomposition of a matrix.

The algorithm for approximating the eigenvalues of a matrix consists of two stages. The first stage is to perform a series of similarity transformations that convert the original matrix to a Hessenberg matrix, or tridiagonal matrix if the matrix is symmetric, that has the same eigenvalues as the original matrix. The second stage is to determine the eigenvalues of the resulting Hessenberg (or tridiagonal) matrix. This second stage is an iterative algorithm that consists of performing a QR decomposition during each iteration. The perceptive reader will wonder why we need the first stage; why not simply execute the algorithm in stage two to the original matrix? We will address this along the way. An excellent resource for further reading on many of the subtleties of the methods presented here is Trefethen and Bau (1997).

As we navigate through the various components of obtaining approximations to the eigenvalues of a matrix, take note of the distinction between the QR *decomposition* and the QR *algorithm*.

[20] This is the approach used by the built-in Mathematica and MATLAB functions **Eigenvalues[]/ Eigenvectors[]** and **eig()**, respectively.

6.5.1 Similarity Transformations

Recall that the diagonalization procedure outlined in Section 2.5 relies on the fact that the diagonal matrix is *similar* to the matrix from which it was obtained. Two matrices are similar if they have the same eigenvalues. Similarity transformations form the basis for the iterative QR algorithm of obtaining eigenvalues and eigenvectors as shown in the next section.

Consider the eigenproblem

$$\mathbf{A}\mathbf{u}_n = \lambda_n \mathbf{u}_n, \tag{6.18}$$

where $\mathbf{A}$ is a real, square $N \times N$ matrix. Suppose that $\mathbf{Q}$ is an $N \times N$ orthogonal matrix such that $\mathbf{Q}^{-1} = \mathbf{Q}^T$, but not necessarily a modal matrix for matrix $\mathbf{A}$. Let us consider the transformation

$$\mathbf{B} = \mathbf{Q}^T\mathbf{A}\mathbf{Q}. \tag{6.19}$$

Postmultiplying both sides by $\mathbf{Q}^T\mathbf{u}_n$ leads to

$$\begin{aligned}
\mathbf{B}\mathbf{Q}^T\mathbf{u}_n &= \mathbf{Q}^T\mathbf{A}\mathbf{Q}\mathbf{Q}^T\mathbf{u}_n \\
&= \mathbf{Q}^T\mathbf{A}\mathbf{u}_n \\
&= \mathbf{Q}^T\lambda_n\mathbf{u}_n \\
\mathbf{B}\mathbf{Q}^T\mathbf{u}_n &= \lambda_n\mathbf{Q}^T\mathbf{u}_n.
\end{aligned}$$

Defining $\mathbf{v}_n = \mathbf{Q}^T\mathbf{u}_n$, this can be written as

$$\mathbf{B}\mathbf{v}_n = \lambda_n\mathbf{v}_n, \tag{6.20}$$

which is an eigenproblem for the matrix $\mathbf{B}$ defined by the transformation (6.19). Note that the eigenproblems (6.18) and (6.20) have the same eigenvalues λ_n; therefore, we call (6.19) a *similarity transformation* because $\mathbf{A}$ and $\mathbf{B} = \mathbf{Q}^T\mathbf{A}\mathbf{Q}$ have the same eigenvalues (even though Q is not necessarily a modal matrix for $\mathbf{A}$). This is the case because $\mathbf{Q}$ is orthogonal. The eigenvectors $\mathbf{u}_n$ of $\mathbf{A}$ and $\mathbf{v}_n$ of $\mathbf{B}$ are related by

$$\mathbf{v}_n = \mathbf{Q}^T\mathbf{u}_n \quad \text{or} \quad \mathbf{u}_n = \mathbf{Q}\mathbf{v}_n. \tag{6.21}$$

If in addition to being real and square, $\mathbf{A}$ is symmetric, such that $\mathbf{A} = \mathbf{A}^T$, observe from (6.19) that

$$\mathbf{B}^T = \left(\mathbf{Q}^T\mathbf{A}\mathbf{Q}\right)^T = \mathbf{Q}^T\mathbf{A}^T\left(\mathbf{Q}^T\right)^T = \mathbf{Q}^T\mathbf{A}\mathbf{Q} = \mathbf{B}.$$

Therefore, if $\mathbf{A}$ is symmetric, then $\mathbf{B} = \mathbf{Q}^T\mathbf{A}\mathbf{Q}$ is symmetric as well when $\mathbf{Q}$ is orthogonal. In summary, for $\mathbf{A}$ real and symmetric, the similarity transformation (6.19) preserves the eigenvalues and symmetry of $\mathbf{A}$, and the eigenvectors are related by (6.21).

6.5.2 Iterative QR Algorithm to Obtain Eigenvalues

Having confirmed the properties of similarity transformations, we now turn our attention to the iterative QR algorithm for obtaining the eigenvalues – and eigenvectors if desired – of a matrix $\mathbf{A}$, which requires such transformations.

Basic Approach

According to Section 2.10, a QR decomposition of a matrix $\mathbf{A}$ exists such that

$$\mathbf{A} = \mathbf{QR},$$

where $\mathbf{Q}$ is an orthogonal matrix, and $\mathbf{R}$ is a right (upper) triangular matrix. Letting

$$\mathbf{A}^{(0)} = \mathbf{A}, \quad \mathbf{Q}^{(0)} = \mathbf{Q}, \quad \mathbf{R}^{(0)} = \mathbf{R},$$

the QR decomposition of the given matrix is

$$\mathbf{A}^{(0)} = \mathbf{Q}^{(0)}\mathbf{R}^{(0)}. \tag{6.22}$$

Let us form the product

$$\mathbf{A}^{(1)} = \mathbf{R}^{(0)}\mathbf{Q}^{(0)}, \tag{6.23}$$

where we note the reverse order. Because $\mathbf{Q}^{(0)}$ is orthogonal, premultiplying (6.22) by $\left(\mathbf{Q}^{(0)}\right)^{-1} = \left(\mathbf{Q}^{(0)}\right)^T$ gives

$$\left(\mathbf{Q}^{(0)}\right)^T \mathbf{A}^{(0)} = \left(\mathbf{Q}^{(0)}\right)^T \mathbf{Q}^{(0)}\mathbf{R}^{(0)} = \mathbf{R}^{(0)}.$$

Substituting this expression for $\mathbf{R}^{(0)}$ into (6.23), therefore, we may determine $\mathbf{A}^{(1)}$ from

$$\mathbf{A}^{(1)} = \left(\mathbf{Q}^{(0)}\right)^T \mathbf{A}^{(0)}\mathbf{Q}^{(0)}, \tag{6.24}$$

which is a similarity transformation according to the previous section. That is, taking the product $\mathbf{R}^{(0)}\mathbf{Q}^{(0)}$ is equivalent to the similarity transformation (6.24), and $\mathbf{A}^{(1)}$ has the same eigenvalues as $\mathbf{A}^{(0)} = \mathbf{A}$. Thus, generalizing (6.23), we have

$$\mathbf{A}^{(k+1)} = \mathbf{R}^{(k)}\mathbf{Q}^{(k)}, \quad k = 0, 1, 2, \ldots, \tag{6.25}$$

where the sequence of matrices $\mathbf{A}^{(1)}, \mathbf{A}^{(2)}, \ldots, \mathbf{A}^{(k)}, \ldots$ are all similar to $\mathbf{A}^{(0)} = \mathbf{A}$. Not only do they all have the same eigenvalues, the similarity transformations maintain the same structure as $\mathbf{A}$; for example, they remain symmetric or tridiagonal.

It can be shown (not easily) that the sequence of similar matrices

$$\mathbf{A}^{(0)}, \mathbf{A}^{(1)}, \mathbf{A}^{(2)}, \ldots$$

gets progressively closer, that is, converges, to a diagonal matrix if $\mathbf{A}$ is symmetric or a right triangular matrix if $\mathbf{A}$ is nonsymmetric. In either case, the eigenvalues of $\mathbf{A}$ (and $\mathbf{A}^{(1)}, \mathbf{A}^{(2)}, \ldots$) are on the main diagonal in decreasing order by magnitude.

Example 6.6 Let us illustrate the iterative QR algorithm to approximate the eigenvalues of the symmetric 4×4 matrix

$$\mathbf{A} = \begin{bmatrix} 4 & 2 & 2 & 1 \\ 2 & -3 & 1 & 1 \\ 2 & 1 & 3 & 1 \\ 1 & 1 & 1 & 2 \end{bmatrix}.$$

For symmetric $\mathbf{A}$, the final converged result is expected to be a diagonal matrix with the eigenvalues on the diagonal in order of decreasing magnitude. For comparison, the actual eigenvalues of matrix $\mathbf{A}$ are

$$\lambda_1 = 6.64575, \quad \lambda_2 = -3.64575, \quad \lambda_3 = 1.64575, \quad \lambda_4 = 1.35425.$$

The iterative algorithm summarized in (6.25) simply consists of obtaining the QR decomposition of each successive similar matrix $\mathbf{A}^{(k)}, k = 0, 1, 2, \ldots$ followed by taking the product $\mathbf{R}^{(k)}\mathbf{Q}^{(k)}$ to obtain the next estimate $\mathbf{A}^{(k+1)}$, whose diagonal elements converge toward the eigenvalues of $\mathbf{A}$. After five such iterations, we have the following matrix, which is similar to $\mathbf{A}$:

$$\mathbf{A}^{(5)} = \mathbf{R}^{(4)}\mathbf{Q}^{(4)} = \begin{bmatrix} 6.64339 & -0.15562 & 0.00330 & -0.00073 \\ -0.15562 & -3.64334 & -0.01525 & 0.00690 \\ 0.00330 & -0.01525 & 1.59603 & -0.10959 \\ -0.00073 & 0.00690 & -0.10959 & 1.40392 \end{bmatrix}.$$

Observe that after only five iterations, the diagonal elements are converging toward the actual eigenvalues, and the off-diagonal elements are converging toward zero. After five additional iterations, the algorithm yields

$$\mathbf{A}^{(10)} = \mathbf{R}^{(9)}\mathbf{Q}^{(9)} = \begin{bmatrix} 6.64575 & 0.00773 & 0 & 0 \\ 0.00773 & -3.64575 & 0.00030 & 0.00005 \\ 0 & 0.00030 & 1.63747 & 0.04843 \\ 0 & 0.00005 & 0.04843 & 1.36253 \end{bmatrix}.$$

The diagonal elements now capture at least the first two significant figures of the actual eigenvalues, and the off-diagonal elements continue to converge toward zero. A total of 25 iterations results in

$$\mathbf{A}^{(25)} = \mathbf{R}^{(24)}\mathbf{Q}^{(24)} = \begin{bmatrix} 6.64575 & 0 & 0 & 0 \\ 0 & -3.64575 & 0 & 0 \\ 0 & 0 & 1.64573 & -0.00267 \\ 0 & 0 & -0.00267 & 1.35427 \end{bmatrix}.$$

We now have at least the first five significant figures of the actual eigenvalues. After 60 total iterations, the eigenvalues have converged to the actual values within six significant figures as shown here:

$$\mathbf{A}^{(60)} = \mathbf{R}^{(59)}\mathbf{Q}^{(59)} = \begin{bmatrix} 6.64575 & 0 & 0 & 0 \\ 0 & -3.64575 & 0 & 0 \\ 0 & 0 & 1.64575 & 0 \\ 0 & 0 & 0 & 1.35425 \end{bmatrix}.$$

As expected, the iterative QR algorithm produces a diagonal matrix for symmetric $\mathbf{A}$ with the eigenvalues along the diagonal in decreasing order by magnitude.

But how do we determine the $\mathbf{Q}$ and $\mathbf{R}$ matrices in the QR decomposition for each iteration? This requires *plane rotations*. In addition, the QR algorithm presented earlier is relatively inefficient computationally; therefore, it would be advantageous to first transform the matrix into another similar form for which the number of iterations required by the QR algorithm is reduced sufficiently to offset the additional overhead of the transformation. It turns out that the application of the QR algorithm to a Hessenberg matrix[21] or tridiagonal matrix is the sweet spot in terms of minimizing computational cost of the overall algorithm. Therefore, matrices are typically first transformed using a series of similarity transformations to a tridiagonal or Hessenberg form if the original matrix A is symmetric or nonsymmetric, respectively. The similarity transformations are based on Householder rotations for dense matrices and Givens rotations for sparse matrices. This process is discussed in the next section.

Plane Rotations

Consider the $N \times N$ transformation matrix $\mathbf{P}$ comprised of the identity matrix with only four elements changed in the pth and qth rows and columns according to

$$P_{pp} = P_{qq} = c, \quad P_{pq} = -s, \quad P_{qp} = s,$$

where $c = \cos\phi$ and $s = \sin\phi$. That is,

$$\mathbf{P} = \begin{bmatrix} 1 & & & & & & & & & \\ & \ddots & & & & & & & & \\ & & 1 & & & & & & & \\ & & & c & 0 & \cdots & 0 & -s & & \\ & & & 0 & 1 & & & 0 & & \\ & & & \vdots & & \ddots & & \vdots & & \\ & & & 0 & & & 1 & 0 & & \\ & & & s & 0 & \cdots & 0 & c & & \\ & & & & & & & & 1 & \\ & & & & & & & & & \ddots & \\ & & & & & & & & & & 1 \end{bmatrix}.$$

[21] Recall from Section 1.2 that a Hessenberg matrix is one in which all of the elements below the first lower diagonal are zero.

Observe the effect of transforming an N-dimensional vector $\mathbf{x}$ according to the transformation

$$\mathbf{y} = \mathbf{P}\mathbf{x}, \tag{6.26}$$

where $\mathbf{x} = \begin{bmatrix} x_1 & x_2 & \cdots & x_p & \cdots & x_q & \cdots & x_N \end{bmatrix}^T$. Then

$$\mathbf{y} = \mathbf{P}\mathbf{x} = \begin{bmatrix} x_1 & x_2 & \cdots & y_p & \cdots & y_q & \cdots & x_N \end{bmatrix}^T,$$

where the only two elements that are altered are

$$y_p = cx_p - sx_q, \tag{6.27}$$

$$y_q = sx_p + cx_q. \tag{6.28}$$

For example, consider the case with $N = 2$, for which $p = 1, q = 2$, and

$$y_1 = cx_1 - sx_2,$$

$$y_2 = sx_1 + cx_2,$$

or

$$\begin{bmatrix} y_1 \\ y_2 \end{bmatrix} = \begin{bmatrix} \cos\phi & -\sin\phi \\ \sin\phi & \cos\phi \end{bmatrix} \begin{bmatrix} x_1 \\ x_2 \end{bmatrix}.$$

This transformation rotates the vector $\mathbf{x}$ through an angle ϕ to obtain $\mathbf{y}$ as illustrated in Example 1.10. Note that $\mathbf{y} = \mathbf{P}^T\mathbf{x}$ rotates the vector $\mathbf{x}$ through an angle $-\phi$.[22] In addition, recall from Section 2.4 that the rotation matrix $\mathbf{P}$ is orthogonal, such that $\mathbf{P}^T = \mathbf{P}^{-1}$.

Thus, in the general N-dimensional case (6.26), $\mathbf{P}$ rotates the vector $\mathbf{x}$ through an angle ϕ in the $x_p x_q$-plane. The angle ϕ may be chosen with one of several objectives in mind. For example, it could be used to zero all elements below (or to the right of) a specified element, for example $\mathbf{y} = \begin{bmatrix} y_1 & y_2 & \cdots & y_j & 0 & \cdots & 0 \end{bmatrix}^T$. This could be accomplished using the *Householder transformation*, or *reflection*, which is efficient for dense matrices. Alternatively, the transformation could be used to zero a single element, for example y_p or y_q (see (6.27) and (6.28)). This could be accomplished using the *Givens transformation*, or *rotation*, which is efficient for sparse, structured (for example, banded) matrices. Alternatively, we can generalize to rotate a set of vectors, in the form of a matrix, by taking $\mathbf{Y} = \mathbf{P}\mathbf{X}$.

We can imagine a series of Givens or Householder transformations that reduce the matrix $\mathbf{A}$ to a matrix that is right (upper) triangular, which is the $\mathbf{R}$ matrix in its QR decomposition. Thus, if m projections are required to produce an upper triangular matrix, $\mathbf{R}$ is given by

$$\mathbf{R} = \mathbf{P}_m \cdots \mathbf{P}_2\mathbf{P}_1\mathbf{A}. \tag{6.29}$$

[22] Recall that a positive rotation is counterclockwise.

Because $\mathbf{A} = \mathbf{QR}$, in which case $\mathbf{R} = \mathbf{Q}^T\mathbf{A}$, the orthogonal matrix $\mathbf{Q}$ is then obtained from

$$\mathbf{Q}^T = \mathbf{P}_m \cdots \mathbf{P}_2\mathbf{P}_1.$$

Taking the transpose leads to

$$\mathbf{Q} = \mathbf{P}_1^T\mathbf{P}_2^T \cdots \mathbf{P}_m^T. \tag{6.30}$$

In this manner, the QR decomposition (6.29) and (6.30) is obtained from a series of plane Givens or Householder rotations. Givens transformations are most efficient for large, sparse, structured matrices, which can be configured to only zero the elements that are not already zero. There is a "fast Givens transformation," for which the $\mathbf{P}$ matrices are not orthogonal, but the QR decompositions can be obtained two times faster than in the standard Givens transformation illustrated here. Convergence of the iterative QR algorithm may be accelerated using shifting (see, for examplé, section 11.3 of Press et al., 2007, Trefethen and Bau, 1997).

The order of operations for the QR algorithm per iteration are as follows: $O(N^3)$ for a dense matrix, $O(N^2)$ for a Hessenberg matrix, and $O(N)$ for a tridiagonal matrix. Thus, the most efficient procedure is as follows:

1. Transform $\mathbf{A}$ to a similar tridiagonal or Hessenberg form if $\mathbf{A}$ is symmetric or nonsymmetric, respectively. This is done using a series of similarity transformations based on Householder rotations for dense matrices or Givens rotations for sparse matrices.
2. Use the iterative QR algorithm to obtain the eigenvalues of the tridiagonal or Hessenberg matrix.

The iterative QR algorithm is the workhorse of the vast majority of eigenproblem solvers. Carried to completion, the iterative QR algorithm provides approximations for the full spectrum consisting of all eigenvalues of a matrix.

Example 6.7 Let us obtain the QR decomposition for the same symmetric 4×4 matrix as considered in Example 6.6:

$$\mathbf{A} = \begin{bmatrix} 4 & 2 & 2 & 1 \\ 2 & -3 & 1 & 1 \\ 2 & 1 & 3 & 1 \\ 1 & 1 & 1 & 2 \end{bmatrix}.$$

The orthogonal matrix $\mathbf{Q}$ will be built up as each Givens rotation is applied, and the original matrix $\mathbf{A}$ will become the right triangular matrix $\mathbf{R}$ when the Givens rotations have all been applied. Obviously, we would normally write the QR decomposition algorithm as a standalone function. However, to illustrate the step-by-step process and show the intermediate results along the way, we will carry out the steps "manually" for the small matrix being considered.

Recall that a Givens rotation zeros a single element of the original matrix via a rotation through some angle.[23] We first determine the necessary Givens matrix **P** that zeros the desired element of matrix **A** and then multiply **PA** to obtain the modified matrix according to (6.29). If we start in the lower-left corner of the matrix and work our way up the columns until we reach the diagonal element, we get the following sequence of Givens rotations and modified matrices:

$$\mathbf{P}_1 = \begin{bmatrix} 0.970 & 0 & 0 & 0.242 \\ 0 & 1 & 0 & 0 \\ 0 & 0 & 1 & 0 \\ -0.242 & 0 & 0 & 0.970 \end{bmatrix}, \quad \hat{\mathbf{A}}_1 = \mathbf{P}_1\mathbf{A} = \begin{bmatrix} 4.123 & 2.182 & 2.182 & 1.455 \\ 2 & -3 & 1 & 1 \\ 2 & 1 & 3 & 1 \\ 0 & 0.485 & 0.485 & 1.697 \end{bmatrix}.$$

Application of the Givens rotation matrix transforms the matrix **A** such that the $(4, 1)$ element is now zero. Note that the transformed matrix is not similar to the original matrix; that is, it does not have the same eigenvalues. In order, the remaining sequence of transformation and modified matrices are:

$$\mathbf{P}_2 = \begin{bmatrix} 0.899 & 0 & 0.436 & 0 \\ 0 & 1 & 0 & 0 \\ -0.436 & 0 & 0.899 & 0 \\ 0 & 0 & 0 & 1 \end{bmatrix}, \quad \hat{\mathbf{A}}_2 = \mathbf{P}_2\hat{\mathbf{A}}_1 = \begin{bmatrix} 4.582 & 2.400 & 3.273 & 1.745 \\ 2 & -3 & 1 & 1 \\ 0 & -0.052 & 1.746 & 0.264 \\ 0 & 0.485 & 0.485 & 1.697 \end{bmatrix}.$$

$$\mathbf{P}_3 = \begin{bmatrix} 0.916 & 0.4 & 0 & 0 \\ -0.4 & 0.916 & 0 & 0 \\ 0 & 0 & 1 & 0 \\ 0 & 0 & 0 & 1 \end{bmatrix}, \quad \hat{\mathbf{A}}_3 = \mathbf{P}_3\hat{\mathbf{A}}_2 = \begin{bmatrix} 5 & 1 & 3.4 & 2 \\ 0 & -3.709 & -0.392 & 0.218 \\ 0 & -0.052 & 1.746 & 0.264 \\ 0 & 0.485 & 0.485 & 1.697 \end{bmatrix}.$$

$$\mathbf{P}_4 = \begin{bmatrix} 1 & 0 & 0 & 0 \\ 0 & 0.991 & 0 & -0.129 \\ 0 & 0 & 1 & 0 \\ 0 & 0.129 & 0 & 0.991 \end{bmatrix}, \quad \hat{\mathbf{A}}_4 = \mathbf{P}_4\hat{\mathbf{A}}_3 = \begin{bmatrix} 5 & 1 & 3.4 & 2 \\ 0 & -3.741 & -0.452 & -0.003 \\ 0 & -0.052 & 1.746 & 0.264 \\ 0 & 0 & 0.430 & 1.711 \end{bmatrix}.$$

$$\mathbf{P}_5 = \begin{bmatrix} 1 & 0 & 0 & 0 \\ 0 & 0.999 & 0.014 & 0 \\ 0 & -0.014 & 0.999 & 0 \\ 0 & 0 & 0 & 1 \end{bmatrix}, \quad \hat{\mathbf{A}}_5 = \mathbf{P}_5\hat{\mathbf{A}}_4 = \begin{bmatrix} 5 & 1 & 3.4 & 2 \\ 0 & -3.741 & -0.427 & 0 \\ 0 & 0 & 1.752 & 0.264 \\ 0 & 0 & 0.430 & 1.711 \end{bmatrix}.$$

$$\mathbf{P}_6 = \begin{bmatrix} 1 & 0 & 0 & 0 \\ 0 & 1 & 0 & 0 \\ 0 & 0 & 0.971 & 0.238 \\ 0 & 0 & -0.238 & 0.971 \end{bmatrix}, \quad \mathbf{R} = \mathbf{P}_6\hat{\mathbf{A}}_5 = \begin{bmatrix} 5 & 1 & 3.4 & 2 \\ 0 & -3.741 & -0.427 & 0 \\ 0 & 0 & 1.804 & 0.664 \\ 0 & 0 & 0 & 1.599 \end{bmatrix}.$$

Note that the final modified matrix is the right triangular matrix **R** in the QR decomposition. The orthogonal matrix **Q** is then produced by multiplying the transposes of

[23] It is not so important for our purposes to know how this angle is obtained, just that such an angle exists.

the Givens projection matrices in reverse order according to (6.30) yielding

$$\mathbf{Q} = \mathbf{P}_1^T \mathbf{P}_2^T \mathbf{P}_3^T \mathbf{P}_4^T \mathbf{P}_5^T \mathbf{P}_6^T = \begin{bmatrix} 0.8 & -0.320 & -0.474 & -0.177 \\ 0.4 & 0.908 & 0.015 & 0.118 \\ 0.4 & -0.160 & 0.870 & -0.236 \\ 0.2 & -0.213 & 0.126 & 0.947 \end{bmatrix}.$$

One can confirm that $\mathbf{Q}$ is indeed orthogonal by multiplying it by its transpose to produce the identity matrix. We can also confirm that these matrices form the QR decomposition by taking their product to restore the original matrix $\mathbf{A}$.

REMARKS:

1. *Once again, it is important to distinguish between QR decomposition and the QR algorithm. The QR decomposition itself is a direct method involving a prescribed sequence and number of steps to produce the* $\mathbf{Q}$ *and* $\mathbf{R}$ *matrices, while the QR algorithm is an iterative method to obtain the eigenvalues, where each iteration requires a QR decomposition.*
2. *The right triangular matrix* $\mathbf{R}$ *is equivalent to a Schur decomposition of the matrix* $\mathbf{A}$ *as discussed in Section 2.7.*
3. *Although only square matrices have eigenvalues and eigenvectors, the QR decomposition exists for any* $M \times N$ *matrix. Obtaining the QR decomposition of an* $M \times N$ *matrix requires* $\sim 2MN^2$ *operations and results in* $M \times N$ *matrix* $\mathbf{Q}$ *and* $N \times N$ *matrix* $\mathbf{R}$.
4. *Once we have the eigenvalues* $\lambda_n, n = 1, 2, \ldots, N$ *using the QR algorithm, the corresponding eigenvectors* $\mathbf{u}_n, n = 1, 2, \ldots, N$ *can be obtained from solutions of the singular systems of equations*

$$(\mathbf{A} - \lambda_n \mathbf{I})\, \mathbf{u}_n = \mathbf{0}.$$

In practice, this is accomplishing using an algorithm called inverse iteration *as described in Trefethen and Bau (1997) and Press et al. (2007).*

6.5.3 Arnoldi Method

Recall that when testing for convergence of an iterative method, we only need the spectral radius – the largest eigenvalue by magnitude – of the iteration matrix. Similarly, when evaluating stability of a dynamical system, we are only concerned with the "least stable mode." In these and other applications, therefore, it is often only necessary to obtain a small number of eigenvalues of a large sparse matrix, and it is not necessary to obtain the full spectrum. This is done efficiently using the *Arnoldi method*.

Suppose we seek the K eigenvalues that are largest by magnitude of the large sparse $N \times N$ matrix $\mathbf{A}$, where $K \ll N$. Given an arbitrary N-dimensional vector $\mathbf{q}_0$, we define the *Krylov subspace* by

$$\mathcal{K}_K(\mathbf{A}, \mathbf{q}_0) = \text{span}\left\{\mathbf{q}_0, \mathbf{A}\mathbf{q}_0, \mathbf{A}^2\mathbf{q}_0, \ldots, \mathbf{A}^{K-1}\mathbf{q}_0\right\},$$

which has dimension K and is a subspace of $\mathcal{R}^N$. The Arnoldi method is based on constructing an orthonormal basis, for example, using Gram–Schmidt orthogonalization, of the Krylov subspace $\mathcal{K}_K$ that can be used to project a general $N \times N$ matrix $\mathbf{A}$ onto the K-dimensional Krylov subspace $\mathcal{K}_K(\mathbf{A}, \mathbf{q}_0)$.

We form the orthogonal projection matrix $\mathbf{Q}$ using the following step-by-step direct method that produces a Hessenberg matrix $\mathbf{H}$ whose eigenvalues approximate the largest K eigenvalues of $\mathbf{A}$:

1. Specify the starting Arnoldi vector $\mathbf{q}_0$.
2. Normalize: $\mathbf{q}_1 = \mathbf{q}_0/\|\mathbf{q}_0\|$.
3. Set $\mathbf{Q}_1 = \mathbf{q}_1$.
4. Do for $k = 2, \ldots, K$:
   (i) Multiply $\mathbf{q}_k = \mathbf{A}\mathbf{q}_{k-1}$.
   (ii) Orthonormalize $\mathbf{q}_k$ against $\mathbf{q}_1, \mathbf{q}_2, \ldots, \mathbf{q}_{k-1}$.
   (iii) Append $\mathbf{q}_k$ to $\mathbf{Q}_{k-1}$ to form $\mathbf{Q}_k$.
   (iv) Form the Hessenberg matrix $\mathbf{H}_k = \mathbf{Q}_k^T\mathbf{A}\mathbf{Q}_k$.
   (v) Determine the eigenvalues of $\mathbf{H}_k$ using the iterative QR algorithm given in Section 6.5.2.
5. End Do

At each step $k = 2, \ldots, K$, an $N \times k$ orthogonal matrix $\mathbf{Q}_k$ is produced, the columns of which form an orthonormal basis for the Krylov subspace $\mathcal{K}_k(\mathbf{A}, \mathbf{q}_0)$. Using the projection matrix $\mathbf{Q}_k$, we transform $\mathbf{A}$ to produce a $k \times k$ Hessenberg matrix $\mathbf{H}_k$ (or tridiagonal for symmetric $\mathbf{A}$), which is an orthogonal projection of $\mathbf{A}$ onto the Krylov subspace $\mathcal{K}_k$. The eigenvalues of $\mathbf{H}_k$, sometimes called the Ritz eigenvalues, approximate the largest k eigenvalues of $\mathbf{A}$. The approximations of the eigenvalues improve as each step is incorporated, and we obtain the approximation of one additional eigenvalue.

Because $K \ll N$, we only require the determination of eigenvalues of Hessenberg (or tridiagonal) matrices that are no larger than $K \times K$ as opposed to the original $N \times N$ matrix $\mathbf{A}$. In addition, the more sparse the matrix $\mathbf{A}$ is, the smaller K can be and still obtain a good approximation of the largest K eigenvalues of $\mathbf{A}$. In practice, however, it may be necessary to perform more than K steps in order to obtain the desired accuracy for the first K eigenvalues.

Although the outcome of each step depends upon the starting Arnoldi vector $\mathbf{q}_0$ used, the procedure converges to the correct eigenvalues of matrix $\mathbf{A}$ for any $\mathbf{q}_0$. Recall that the Arnoldi method approximates the K largest eigenvalues by magnitude. Alternatively, one can designate a particular portion of the complex plane where the eigenvalues are sought. For example, it can be designed to determine the K eigenvalues with the largest real or imaginary part. This can be accomplished using a *shift and invert* approach. When seeking a set of eigenvalues in a particular portion of the full spectrum, it is desirable that the starting Arnoldi vector $\mathbf{q}_0$ be in – or "nearly" in – the subspace spanned by the eigenvectors corresponding to the sought after eigenvalues.

As the Arnoldi method progresses, we get better approximations of the desired eigenvectors that can then be used to form a more desirable starting vector. This is known as the *implicitly restarted Arnoldi method* and is based on the implicitly shifted QR decomposition method. Restarting also reduces storage requirements by keeping K small. This approach has been implemented in ARPACK, which was developed at Rice University in the mid-1990s. It was first developed as a Fortran 77 library of subroutines, and subsequently it has been implemented in C++ as ARPACK++. It also has been implemented in MATLAB via the **eigs()** function, where the "s" denotes "sparse." In addition, it has been implemented in Mathematica, where one includes the option "Method → Arnoldi" in the **Eigenvalues[]** function. ARPACK is described in Radke (1996).

REMARKS:

1. *When applied to symmetric matrices, the Arnoldi method reduces to the* Lanczos method, *in which the Hessenberg matrix is tridiagonal.*
2. *The Arnoldi method can be designed to apply to the* generalized eigenproblem

$$\mathbf{A}\mathbf{u}_n = \lambda_n \mathbf{B}\mathbf{u}_n,$$

where it is required that **B** *be positive definite, that is, have all positive eigenvalues. For example, the generalized eigenproblem is encountered in structural design problems, in which* **A** *is called the* stiffness matrix, *and* **B** *is called the* mass matrix. *It also arises in hydrodynamic stability (see chapter 9 of Cassel, 2013).*

3. *QR decomposition and the Arnoldi method may also be adapted to solve linear systems of equations; this is called the* generalized minimum residual (GMRES) method *and is described in Section 10.7.*
4. *Interestingly, the method was introduced by Arnoldi (1951), but it was not originally applied to the eigenproblem! For additional references on Arnoldi's method, see Nayar and Ortega (1993) and Saad (2003).*

Example 6.8 We illustrate the step-by-step approach by applying the Arnoldi method to the 4×4 symmetric matrix considered in the previous two examples. Although this is not a large sparse matrix, it serves to illustrate the approach. The matrix is given by

$$\mathbf{A} = \begin{bmatrix} 4 & 2 & 2 & 1 \\ 2 & -3 & 1 & 1 \\ 2 & 1 & 3 & 1 \\ 1 & 1 & 1 & 2 \end{bmatrix}.$$

For comparison, the eigenvalues of matrix **A** once again are

$$\lambda_1 = 6.64575, \quad \lambda_2 = -3.64575, \quad \lambda_3 = 1.64575, \quad \lambda_4 = 1.35425.$$

In the Arnoldi method, we begin with an arbitrary N-dimensional starting Arnoldi vector. Here, $N = 4$, so let us start with the vector

$$\mathbf{q}_0 = \begin{bmatrix} 1 \\ 1 \\ 1 \\ 1 \end{bmatrix}.$$

In its most compact form, the Arnoldi method consists of the following. The first step is to form a basis for the Krylov subspace of matrix $\mathbf{A}$ and vector $\mathbf{q}_0$, which are the column vectors of $\hat{\mathbf{Q}}$ given by

$$\hat{\mathbf{Q}} = \begin{bmatrix} \vdots & \vdots & \vdots & \vdots \\ \mathbf{q}_0 & \mathbf{A}\mathbf{q}_0 & \mathbf{A}^2\mathbf{q}_0 & \mathbf{A}^3\mathbf{q}_0 \\ \vdots & \vdots & \vdots & \vdots \end{bmatrix} = \begin{bmatrix} 1 & 9 & 57 & 399 \\ 1 & 1 & 27 & 105 \\ 1 & 7 & 45 & 303 \\ 1 & 5 & 27 & 183 \end{bmatrix}.$$

We then orthonormalize these basis vectors, for example using Gram–Schmidt, to produce

$$\mathbf{Q} = \begin{bmatrix} 0.5 & 0.591608 & 0.391695 & 0.496564 \\ 0.5 & -0.760639 & 0.405684 & 0.0827606 \\ 0.5 & 0.253546 & 0.0279782 & -0.827606 \\ 0.5 & -0.0845154 & -0.825356 & 0.248282 \end{bmatrix},$$

where the column vectors are the orthonormal basis vectors. Because matrix $\mathbf{A}$ is symmetric, the $k \times k$ Hessenberg matrix is also tridiagonal and is determined as follows:

$$\mathbf{H} = \mathbf{Q}^T\mathbf{A}\mathbf{Q} = \begin{bmatrix} 5.5 & 2.95804 & 0 & 0 \\ 2.95804 & -1.72857 & 2.07138 & 0 \\ 0 & 2.07138 & 0.824462 & 0.121563 \\ 0 & 0 & 0.121563 & 1.40411 \end{bmatrix}.$$

Finally, the eigenvalues of the tridiagonal Hessenberg matrix are determined using the iterative QR algorithm from Section 6.5.2. This produces the eigenvalues

$$\lambda_1 = 6.64575, \quad \lambda_2 = -3.64575, \quad \lambda_3 = 1.64575, \quad \lambda_4 = 1.35425,$$

which are the same as the actual eigenvalues given earlier.

Now let us illustrate the step-by-step procedure outlined earlier to see how the eigenvalue approximations develop. The first step is to specify the starting vector

$$\mathbf{q}_0 = \begin{bmatrix} 1 \\ 1 \\ 1 \\ 1 \end{bmatrix},$$

which is the same as before. The second step is to normalize the starting vector

$$\mathbf{q}_1 = \frac{\mathbf{q}_0}{\|\mathbf{q}_0\|} = \begin{bmatrix} 0.5 \\ 0.5 \\ 0.5 \\ 0.5 \end{bmatrix}.$$

Third, we place this normalized vector $\mathbf{q}_1$ as the first column of a matrix $\mathbf{Q}$.

$$\mathbf{Q} = \mathbf{q}_1.$$

This is the first in a sequence of orthonormal vectors built up through the step-by-step process. The Arnoldi method begins by applying a transformation to the original matrix $\mathbf{A}$ that produces a Hessenberg (tridiagonal) matrix that approximates the largest magnitude eigenvalue of $\mathbf{A}$. Note that because $\mathbf{Q}$ is an $N \times 1$ matrix, the Hessenberg matrix $\mathbf{H}$ is 1×1, and is given by

$$\mathbf{H} = \mathbf{Q}^T \mathbf{A} \mathbf{Q} = \begin{bmatrix} 5.5 \end{bmatrix}.$$

The single eigenvalue of $\mathbf{H}$ approximates the largest eigenvalue of $\mathbf{A}$.

We next carry out the series of steps with $k = 2, \dots, K$ that leads to increasingly accurate estimates for the eigenvalues. Because our matrix is small, we will set $K = N = 4$ in order to see the process all the way through to determine all four eigenvalues. Each subsequent step follows the following procedure to produce an additional vector $\mathbf{q}$, which is orthogonal to the previous $\mathbf{q}$ vectors, to append to the orthonormal matrix $\mathbf{Q}$. Because the number of columns of $\mathbf{Q}$ increases by one during each step, the resulting Hessenberg (tridiagonal) matrix $\mathbf{H}$ increases in size by one row and column during each step.

In the first step with $k = 2$, we form a new Arnoldi vector $\mathbf{q}_2$ by multiplying the matrix $\mathbf{A}$ by the normalized vector $\mathbf{q}_1$ according to $\mathbf{q}_2 = \mathbf{A}\mathbf{q}_1$. We then orthonormalize this $\mathbf{q}_2$ relative to $\mathbf{q}_1$, for example using the Gram–Schmidt method. This produces an updated orthogonal matrix $\mathbf{Q}$ that is 4×2, and the similarity transformation produces a Hessenberg (tridiagonal) matrix that is 2×2 of the form

$$\mathbf{H} = \mathbf{Q}^T \mathbf{A} \mathbf{Q} = \begin{bmatrix} 5.5 & 2.95804 \\ 2.95804 & -1.72857 \end{bmatrix}.$$

The eigenvalues of $\mathbf{H}$, which are

$$\lambda_1 = 6.55616, \quad \lambda_2 = -2.78473,$$

approximate the first two eigenvalues of $\mathbf{A}$. Observe that after only one step, the approximation to the first eigenvalue is surprisingly close to the actual eigenvalue determined earlier, and we now have an approximation to the second eigenvalue.

Following the same procedure, but with $k = 3$, produces a 4×3 orthogonal matrix $\mathbf{Q}$ after appending an additional column with $\mathbf{q}_3 = \mathbf{A}\mathbf{q}_2$ and a 3×3 tridiagonal Hessenberg matrix $\mathbf{H}$ given by

Table 6.5 Sequence of approximations to the first two eigenvalues of a 4×4 matrix using Arnoldi's method in Example 6.8 with the starting vector $\mathbf{q}_0 = [1\ 1\ 1\ 1]^T$.

Step k	λ_1	λ_2
2	6.556159791712	−2.784731220283
3	6.645705479657	−3.645274757472
4	6.645751311065	−3.645751311065

$$\mathbf{H} = \mathbf{Q}^T\mathbf{A}\mathbf{Q} = \begin{bmatrix} 5.5 & 2.95804 & 0 \\ 2.95804 & -1.72857 & 2.07138 \\ 0 & 2.07138 & 0.824462 \end{bmatrix}.$$

The eigenvalues of $\mathbf{H}$, which are

$$\lambda_1 = 6.64571, \quad \lambda_2 = -3.64527, \quad \lambda_3 = 1.59546,$$

provide estimates for the first two eigenvalues that are quite accurate, and the third is now being approximated.

Now with $k = 4$, the next step produces a 4×4 orthogonal matrix $\mathbf{Q}$ and a 4×4 tridiagonal Hessenberg matrix $\mathbf{H}$

$$\mathbf{H} = \mathbf{Q}^T\mathbf{A}\mathbf{Q} = \begin{bmatrix} 5.5 & 2.95804 & 0 & 0 \\ 2.95804 & -1.72857 & 2.07138 & 0 \\ 0 & 2.07138 & 0.824462 & 0.121563 \\ 0 & 0 & 0.121563 & 1.40411 \end{bmatrix}.$$

The eigenvalues of $\mathbf{H}$ are

$$\lambda_1 = 6.64575, \quad \lambda_2 = -3.64575, \quad \lambda_3 = 1.64575, \quad \lambda_4 = 1.35425.$$

Because $\mathbf{H}$ is similar to $\mathbf{A}$, their eigenvalues are the same.

Carrying out all four steps for this 4×4 example produces all four eigenvalues of the original matrix $\mathbf{A}$. To summarize, the sequence of approximations for the first two eigenvalues are given in Table 6.5. Recall that the starting Arnoldi vector $\mathbf{q}_0$ is arbitrary; therefore, let us observe the impact of using a different choice:

$$\mathbf{q}_0 = \begin{bmatrix} 1 \\ 2 \\ 3 \\ 4 \end{bmatrix}.$$

The sequence of approximations for the first two eigenvalues with this starting vector is given in Table 6.6. Although the approximations in the second and third steps are different as compared to the preceding example with different $\mathbf{q}_0$, they converge quickly to the correct eigenvalues.

Table 6.6 Sequence of approximations to the first two eigenvalues of a 4×4 matrix using Arnoldi's method in Example 6.8 with the starting vector $\mathbf{q}_0 = [1\ 2\ 3\ 4]^T$.

Step k	λ_1	λ_2
2	6.276667800135	0.074057230357
3	6.645665516800	−3.644445293403
4	6.645751311065	−3.645751311065

The power of Arnoldi's method is that one can obtain a good approximation of the largest eigenvalue(s) of $\mathbf{A}$ after only a small number of steps, where each step only requires multiplication of a matrix times a vector, orthonormalization of one vector with respect to the previous vectors, and determination of the eigenvalues of a small matrix. For large sparse matrices, this procedure is much more efficient than the general procedures illustrated previously for determining the full spectrum of eigenvalues of dense matrices. This is illustrated in the next example.

Example 6.9 The power of the Arnoldi method is in approximating small numbers of eigenvalues for large sparse matrices. To illustrate this, let us begin with a 20×20 sparse matrix, which happens to be symmetric. We form the matrix by modifying an identity matrix as follows:

$$
\mathbf{A} = \begin{bmatrix}
1 & 0 & 0 & 0 & 0 & 0 & 0 & 0 & 0 & 0 & 0 & 0 & 0 & 0 & 0 & 0 & 0 & 0 & 0 & 0 \\
0 & 1 & 0 & 0 & 0 & 0 & 0 & 0 & 2 & 0 & 0 & 0 & 0 & 0 & 0 & 0 & 0 & 0 & 0 & 0 \\
0 & 0 & 1 & 0 & 2 & 0 & 0 & 0 & 0 & 0 & 0 & 0 & 0 & 0 & 3 & 0 & 0 & 0 & 0 & 0 \\
0 & 0 & 0 & 1 & 0 & 0 & 4 & 0 & 0 & 0 & 0 & 0 & 0 & 0 & 0 & 0 & 0 & 0 & 0 & 0 \\
0 & 0 & 2 & 0 & 1 & 0 & 0 & 0 & 0 & 0 & 0 & 0 & 0 & 0 & 0 & 0 & 0 & 0 & 0 & 0 \\
0 & 0 & 0 & 0 & 0 & 1 & 0 & 0 & 0 & 0 & 0 & 0 & 0 & 0 & 0 & 0 & 0 & 0 & 0 & 0 \\
0 & 0 & 0 & 4 & 0 & 0 & 1 & 4 & 0 & 0 & 0 & 0 & 0 & 0 & 0 & 0 & 0 & 0 & 0 & 0 \\
0 & 0 & 0 & 0 & 0 & 0 & 4 & 1 & 0 & 0 & 0 & 1 & 0 & 0 & 0 & 0 & 0 & 0 & 0 & 0 \\
0 & 2 & 0 & 0 & 0 & 0 & 0 & 0 & 1 & 0 & 0 & 0 & 0 & 0 & 0 & 0 & 0 & 0 & 0 & 0 \\
0 & 0 & 0 & 0 & 0 & 0 & 0 & 0 & 0 & 1 & 0 & 0 & 0 & 0 & 0 & 0 & 2 & 0 & 0 & 0 \\
0 & 0 & 0 & 0 & 0 & 0 & 0 & 0 & 0 & 0 & 1 & 0 & 0 & 0 & 0 & 0 & 0 & 0 & 0 & 0 \\
0 & 0 & 0 & 0 & 0 & 0 & 0 & 1 & 0 & 0 & 0 & 1 & 0 & 0 & 0 & 0 & 0 & 0 & 0 & 0 \\
0 & 0 & 0 & 0 & 0 & 0 & 0 & 0 & 0 & 0 & 0 & 0 & 1 & 0 & 0 & 0 & 0 & 0 & 0 & 0 \\
0 & 0 & 0 & 0 & 0 & 0 & 0 & 0 & 0 & 0 & 0 & 0 & 0 & 1 & 0 & 0 & 0 & 0 & 0 & 0 \\
0 & 0 & 3 & 0 & 0 & 0 & 0 & 0 & 0 & 0 & 0 & 0 & 0 & 0 & 1 & 0 & 0 & 0 & 0 & 0 \\
0 & 0 & 0 & 0 & 0 & 0 & 0 & 0 & 0 & 0 & 0 & 0 & 0 & 0 & 0 & 1 & 0 & 0 & 0 & 0 \\
0 & 0 & 0 & 0 & 0 & 0 & 0 & 0 & 0 & 2 & 0 & 0 & 0 & 0 & 0 & 0 & 1 & 0 & 0 & 0 \\
0 & 0 & 0 & 0 & 0 & 0 & 0 & 0 & 0 & 0 & 0 & 0 & 0 & 0 & 0 & 0 & 0 & 1 & 0 & 0 \\
0 & 0 & 0 & 0 & 0 & 0 & 0 & 0 & 0 & 0 & 0 & 0 & 0 & 0 & 0 & 0 & 0 & 0 & 1 & 0 \\
0 & 0 & 0 & 0 & 0 & 0 & 0 & 0 & 0 & 0 & 0 & 0 & 0 & 0 & 0 & 0 & 0 & 0 & 0 & 1
\end{bmatrix}.
$$

Table 6.7 Sequence of approximations to the first two eigenvalues of a 20×20 sparse matrix using Arnoldi's method in Example 6.9 with the starting vector $\mathbf{q}_0 = [1\ 1\ 1\ 1]^T$.

Step i	λ_1	λ_2
2	5.682994085848	1.148938687261
3	6.281974563001	1.943318072524
4	6.631368463146	−4.129075192591
5	6.682386603130	−4.483111939425
6	6.701023345082	−4.690975899377
7	6.701556864037	−4.701458332998
8	6.701562118716	−4.701562118716
9	6.701562118716	−4.701562118716
10	6.701562118716	−4.701562118716

We use the Arnoldi algorithm to approximate the largest two eigenvalues. Recall that at each step k, we are determining the eigenvalues of a $k \times k$ Hessenberg (or tridiagonal) matrix. Table 6.7 shows the results for ten steps in the procedure. Observe that for the precision shown, we have converged approximations of the largest two eigenvalues after only eight steps of the Arnoldi method, in which case it is only necessary to obtain the eigenvalues of an 8×8 Hessenberg (or tridiagonal) matrix, rather than the original 20×20 matrix. The less sparse – more dense – the matrix $\mathbf{A}$, the more steps are required to produce a good approximation of the largest eigenvalues.

6.5.4 Singular-Value Decomposition Revisited

As we discovered in Section 2.8, it is useful to extend the ideas of eigenvalues and eigenvectors to rectangular matrices. This is accomplished via singular-value decomposition (SVD) of any $M \times N$ matrix $\mathbf{A}$ according to

$$\mathbf{A} = \mathbf{U}\mathbf{\Sigma}\mathbf{V}^T,$$

where $\mathbf{\Sigma}$ is an $M \times N$ diagonal matrix containing the singular values of $\mathbf{A}$, $\mathbf{U}$ is an $M \times M$ orthogonal matrix, and $\mathbf{V}$ is an $N \times N$ orthogonal matrix. Recall that the singular values σ_n and the columns $\mathbf{u}_n$ and $\mathbf{v}_n$ of the matrices $\mathbf{U}$ and $\mathbf{V}$, respectively, can be obtained from the two eigenproblems

$$\mathbf{A}\mathbf{A}^T\mathbf{u}_n = \sigma_n^2\mathbf{u}_n, \quad \mathbf{A}^T\mathbf{A}\mathbf{v}_n = \sigma_n^2\mathbf{v}_n.$$

Consequently, it would appear that obtaining the SVD of a matrix $\mathbf{A}$ would be as simple as determining the eigenvalues and eigenvectors of $\mathbf{A}\mathbf{A}^T$ or $\mathbf{A}^T\mathbf{A}$. While this approach is suitable for small to moderate-sized matrices, as illustrated in Example 2.13, this is often not the case for large matrices. This is because the condition number for $\mathbf{A}\mathbf{A}^T$ and $\mathbf{A}^T\mathbf{A}$ is the square of that for the original matrix $\mathbf{A}$, in which case numerical errors are more likely to be amplified sufficiently to pollute the solution.

An alternative method, which is broadly similar to the two-stage process in the iterative QR algorithm, is based on forming a symmetric matrix from the original matrix $\mathbf{A}$ and then determining its eigenvalues and eigenvectors. This matrix is given by

$$\hat{\mathbf{A}} = \begin{bmatrix} \mathbf{0} & \mathbf{A}^T \\ \mathbf{A} & \mathbf{0} \end{bmatrix},$$

which is an $(M+N)\times(M+N)$ square, symmetric matrix for any $\mathbf{A}$. Then the absolute values of the eigenvalues of $\hat{\mathbf{A}}$ are the singular values of $\mathbf{A}$, and the eigenvectors of $\hat{\mathbf{A}}$ can be used to obtain the singular vectors of $\mathbf{A}$. Note that although the size of $\hat{\mathbf{A}}$ is larger than that of $\mathbf{A}$, it is more sparse given the zero blocks. See Trefethen and Bau (1997), Demmel (1997), and Golub and Van Loan (2013) for additional details.

Calculating the SVD of a matrix $\mathbf{A}$ has numerous applications in numerical linear algebra and often provides the most efficient method for the following:

- Determining the rank of a matrix, which is equal to the number of nonzero singular values.
- Orthogonalization of vectors without the round-off errors inherent in the Gram–Schmidt orthogonalization technique.
- The L_2-norm of a matrix is equal to its largest singular value.
- Calculating the coefficients in least-squares curve fitting.

6.6 Epilogue

As indicated at the beginning of the chapter, it is not likely that the scientist or engineer will be called upon to write computer codes that perform orthogonalization, SVD, QR decomposition, the QR iterative algorithm, the Arnoldi method, or any of the other methods covered in this chapter. However, given that our application codes use computational linear algebra methods and algorithms as their building blocks, it is essential to have a good understanding of the methods themselves and some of the issues that can arise when executing such algorithms. For example, when do we need to be concerned with adverse round-off effects, and how can we mitigate them? Should direct or iterative methods be used to solve a particular system of algebraic equations? What is an appropriate convergence test and criterion? How does one choose the relaxation parameter in SOR? What should we choose for K in Arnoldi's method in order to obtain good approximations for a prescribed number of eigenvalues? Insofar as several of the applications throughout the remainder of Parts II and III utilize QR decomposition or SVD, how do we know which will be more effective in a given situation?

We now turn our attention to numerical methods for approximating solutions of ordinary and partial differential equations throughout the remainder of Part II. These methods are central to many applications in science and engineering.

Exercises

Unless stated otherwise, perform the exercises using hand calculations. Exercises to be completed using "built-in functions" should be completed using the built-in functions within Python, MATLAB, or Mathematica for the vector and matrix operation(s) or algorithm. A "user-defined function" is a standalone computer program to accomplish a task written in a programming language, such as Python, MATLAB, Mathematica, Fortran, or C++.

6.1 The eigenvalues of a symmetric matrix $\mathbf{A}$ are given by

$$\lambda_1 = -1.08,\ \lambda_2 = -0.00045,\ \lambda_3 = 1.72,\ \lambda_4 = 4.09,\ \lambda_5 = 5.70,\ \lambda_6 = 6.01.$$

What is the condition number of this matrix? Would you say it is well conditioned or ill conditioned? If you were to invert the matrix $\mathbf{A}$, how many digits of accuracy would you expect to lose owing to round-off errors?

6.2 Determine the condition number for the tridiagonal system

$$2u_{i-1} - 5u_i + 2u_{i+1} = 1, \quad i = 1, \ldots, N,$$

with N large. How many digits of accuracy would you expect to lose owing to round-off errors when inverting this tridiagonal matrix?

6.3 Consider the matrix

$$\mathbf{A} = \begin{bmatrix} 1 & 3 & 2 \\ 2 & 1 & 2 \\ 1 & 0 & 2 \end{bmatrix}.$$

Obtain the LU decomposition of the matrix $\mathbf{A}$.

6.4 Consider the matrix and vector

$$\mathbf{A} = \begin{bmatrix} 5 & -1 & 0 \\ -1 & 5 & -1 \\ 0 & -1 & 5 \end{bmatrix}, \quad \mathbf{b} = \begin{bmatrix} 9 \\ 4 \\ -6 \end{bmatrix}.$$

Obtain the LU decomposition of the matrix $\mathbf{A}$, and use the decomposition to solve the system of linear algebraic equations $\mathbf{Au} = \mathbf{b}$.

6.5 Determine the Cholesky decomposition of the matrix

$$\mathbf{A} = \begin{bmatrix} 4 & 2 \\ 2 & 4 \end{bmatrix}.$$

Determine the inverse of $\mathbf{A}$ using this decomposition.

6.6 A system of algebraic equations $\mathbf{Au} = \mathbf{b}$ is to be solved using two iterative algorithms. By writing $\mathbf{A} = \mathbf{M}_1 - \mathbf{M}_2$, the iterative schemes may be expressed as

$$\mathbf{M}_1\mathbf{u}^{(n+1)} = \mathbf{M}_2\mathbf{u}^{(n)} + \mathbf{b},$$

where the n's are successive iterations. For the matrix

$$\mathbf{A} = \begin{bmatrix} a & b \\ c & d \end{bmatrix},$$

where a and d are the same sign, and b and c are the same sign, the two schemes are such that[24]

$$\text{(i) } \mathbf{M}_1 = \begin{bmatrix} a & 0 \\ 0 & d \end{bmatrix}, \quad \text{(ii) } \mathbf{M}_1 = \begin{bmatrix} a & b \\ 0 & d \end{bmatrix}.$$

(a) Determine the criterion in terms of the elements of **A** required for each iterative algorithm to converge.
(b) Compare the iterative convergence rates of the two algorithms.

6.7 Suppose that the system of linear algebraic equations

$$\begin{array}{cccc} u_1 & -au_2 & & = b_1, \\ -au_1 & +u_2 & -au_3 & = b_2, \\ & -au_2 & +u_3 & = b_3, \end{array}$$

is solved using the following iterative scheme:

$$\begin{aligned} u_1^{(n+1)} &= au_2^{(n)} + b_1 \\ u_2^{(n+1)} &= a\left(u_1^{(n)} + u_3^{(n)}\right) + b_2 \\ u_3^{(n+1)} &= au_2^{(n)} + b_3 \end{aligned} .$$

Show that iterative convergence is guaranteed if and only if $|a| < \frac{1}{\sqrt{2}}$.

6.8 The system of linear algebraic equations $\mathbf{Au} = \mathbf{b}$ is to be solved using the following iterative method:[25]

$$\mathbf{u}^{(n+1)} = (\mathbf{I} + a\mathbf{A})\,\mathbf{u}^{(n)} - a\mathbf{b},$$

where a is a real constant, $\mathbf{I}$ is the identity matrix, and $\mathbf{A}$ is an $N \times N$ tridiagonal matrix of the form

$$\mathbf{A} = \begin{bmatrix} -2 & 1 & & \\ 1 & -2 & 1 & \\ & \ddots & \ddots & \ddots \\ & & 1 & -2 \end{bmatrix}.$$

What is the criterion for a such that this iterative algorithm will converge for large N?

6.9 An iterative solution of the system of algebraic equations $\mathbf{Au} = \mathbf{b}$ is sought using the Jacobi method, where

$$\mathbf{A} = \begin{bmatrix} 2 & 1 \\ a & 2 \end{bmatrix}, \quad \mathbf{b} = \begin{bmatrix} 1 \\ 2 \end{bmatrix}.$$

Obtain the necessary and sufficient condition on a for iterative convergence.

6.10 An iterative solution of the system of algebraic equations $\mathbf{Au} = \mathbf{b}$ is sought using the Gauss–Seidel method, where

$$\mathbf{A} = \begin{bmatrix} 2 & 1 \\ a & 2 \end{bmatrix}, \quad \mathbf{b} = \begin{bmatrix} 1 \\ 1 \end{bmatrix}.$$

Obtain the necessary and sufficient condition on a for iterative convergence.

[24] Adapted from Moin (2010).
[25] Adapted from Moin (2010).

6.11 Given the following system of equations

$$u_1 + 2u_2 + 3u_3 = 1,$$
$$2u_1 + 3u_2 + 4u_3 = 1,$$
$$3u_1 + 4u_2 + 6u_3 = 1,$$

perform the first iteration of the Gauss–Seidel method with the initial guess

$$u_1^{(0)} = -1, \quad u_2^{(0)} = 1, \quad u_3^{(0)} = 0.$$

Compare your answer to the initial guess; what does this mean?

6.12 Consider the system of linear algebraic equations $\mathbf{Au} = \mathbf{b}$ given by

$$8u_1 + 2u_2 + 3u_3 = 51,$$
$$2u_1 + 5u_2 + u_3 = 23,$$
$$-3u_1 + u_2 + 6u_3 = 20.$$

For the following, use the initial guess $(u_1, u_2, u_3) = (0, 0, 0)$:

(a) Perform the first three iterations of the Gauss–Seidel method.
(b) Write a user-defined function to perform the Gauss–Seidel iteration. Perform the first three iterations and compare with part (a).
(c) Obtain the converged solution using your user-defined function from part (b). Compare your converged solution to the exact solution using built-in functions. Observe and discuss the influence of the convergence tolerance on the Gauss–Seidel iterative solution.

6.13 Consider the following matrix:

$$\mathbf{A} = \begin{bmatrix} 2 & 4 & 1 & 3 & 2 \\ 5 & 3 & 4 & 1 & 2 \\ 2 & 2 & 1 & 2 & 3 \\ 4 & 1 & 3 & 2 & 5 \\ 4 & 3 & 1 & 2 & 3 \end{bmatrix}.$$

Indicate whether the following statements are true or false (if false, explain why):

(a) Determination of the exact eigenvalues of $\mathbf{A}$ using the QR algorithm requires five QR decompositions to be performed.
(b) Performing the QR algorithm on the matrix $\mathbf{A}$ results in a diagonal matrix.
(c) The most efficient way to obtain the eigenvalues of $\mathbf{A}$ is to first transform $\mathbf{A}$ into a similar triangular matrix to which the QR algorithm is then applied.
(d) The Arnoldi method cannot be used to determine the eigenvalues of $\mathbf{A}$.

6.14 Applying the QR algorithm to the matrix

$$\mathbf{A} = \begin{bmatrix} -2 & 0 & 0 & 5 \\ 1 & 3 & 0 & 0 \\ 0 & 4 & 4 & 0 \\ 2 & 0 & 0 & -3 \end{bmatrix}$$

produces

$$\hat{\mathbf{A}} = \begin{bmatrix} -5.70 & 0.37 & 0.65 & 2.93 \\ 0 & 4 & -4.01 & -0.11 \\ 0 & 0 & 3 & -0.87 \\ 0 & 0 & 0 & 0.70 \end{bmatrix}.$$

(a) How many QR decompositions would have been required to produce the result $\hat{\mathbf{A}}$ from $\mathbf{A}$?

(b) What are the eigenvalues of $\mathbf{A}$?

(c) Rather than using the QR algorithm to directly produce $\hat{\mathbf{A}}$ from $\mathbf{A}$, describe a more efficient procedure.

(d) Is the Arnoldi method a good alternative to the QR algorithm for determining the full spectrum of eigenvalues of $\mathbf{A}$? Why or why not?

6.15 Use a built-in function to determine the QR decomposition of matrix $\mathbf{A}$ in Example 6.7.

7 Numerical Methods for Differential Equations

> Perfection is achieved, not when there is nothing more to add, but when there is nothing left to take away. (Antoine de Saint-Exupéry)

One of the central goals of applied mathematics is to develop methods for solving differential equations as they form the governing equations for many topics in the sciences and engineering. In Chapter 2, we addressed the solution of systems of linear first-order and higher-order ordinary differential equations via diagonalization. In Chapter 3, we used eigenfunction expansions to develop methods for solving self-adjoint ordinary differential equations, with extension to certain linear partial differential equations through the method of separation of variables. While these methods represent fundamental techniques in applied mathematics, the scope of their application is very limited, primarily owing to their restriction to linear differential equations. As a complement to these analytical techniques, therefore, numerical methods open up the full spectrum of ordinary and partial differential equations for solution. Although these solutions are necessarily approximate, the techniques are adaptable to ordinary differential equations in the form of initial- and boundary-value problems as well as large classes of partial differential equations.

Insofar as many topics in science and engineering involve solving differential equations, and insofar as very few practical problems are amenable to exact closed form solution, numerical methods form an essential tool in the researcher's and practitioner's arsenal. In this chapter, we introduce finite-difference methods and focus on aspects common to solving both ordinary and partial differential equations. In addition, brief introductions to spectral and finite-element methods are given for comparison. A simple initial-value problem and a boundary-value problem are used to develop many of the ideas and provide a framework for thinking about numerical methods as applied to differential equations. The following chapters then address methods for boundary-value and initial-value differential equations in turn.

There is nothing fundamentally different about numerical methods for ordinary and partial differential equations – we simply have more than one independent variable to account for in partial differential equations. The primary distinction, and the one that matters most when selecting appropriate numerical methods, is whether the differential equation is an initial-value problem or a boundary-value problem. The mathematical structure of each is fundamentally different, and this influences the development of finite-difference methods for each type of problem.

7.1 General Considerations

7.1.1 Numerical Solution Procedure

Before getting into a detailed discussion of numerical methods for solving differential equations, it is helpful to discuss the *numerical solution procedure* in the form of a general framework as illustrated in Figure 7.1. We begin with the physical system of interest to be modeled. The first step is to apply appropriate physical laws and models in order to derive the mathematical model, or governing equations, of the system. These are typically in the form of ordinary or partial differential equations. The physical laws include, for example, conservation of mass, momentum, and energy, and models include any assumptions or idealizations applied in order to simplify the governing equations. When possible, analytical solutions of the mathematical model are sought. If this is not possible, we turn to numerical methods, which is the focus of Part II. The second step is to discretize the mathematical model, which involves approximation of the continuous differential equation(s) by a system of algebraic equations for the dependent variables at discrete locations in the independent variables (space and time). In this way, the exact and continuous mathematical model is converted into a system of linear algebraic equations that can be solved using the methods outlined in the previous chapter. For example, see Figure 7.2. The discretization step leads to a system of linear algebraic equations, whose numerical solution comprises step 3 of the numerical solution procedure. The method of discretization often produces a large, sparse matrix problem with a particular structure. For example, we will

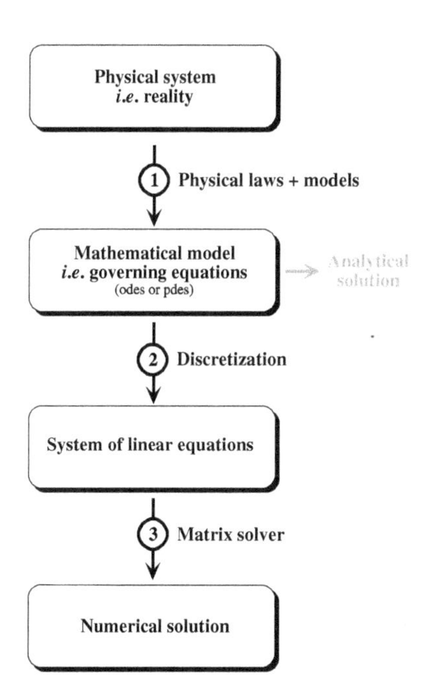

Figure 7.1 The general numerical solution procedure.

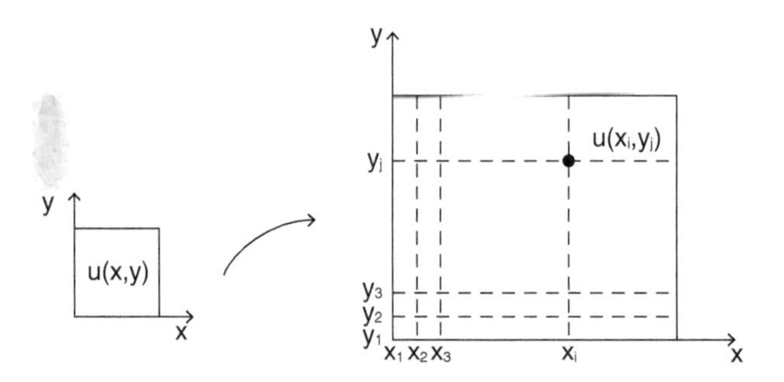

Figure 7.2 Schematic of a discretized two-dimensional domain.

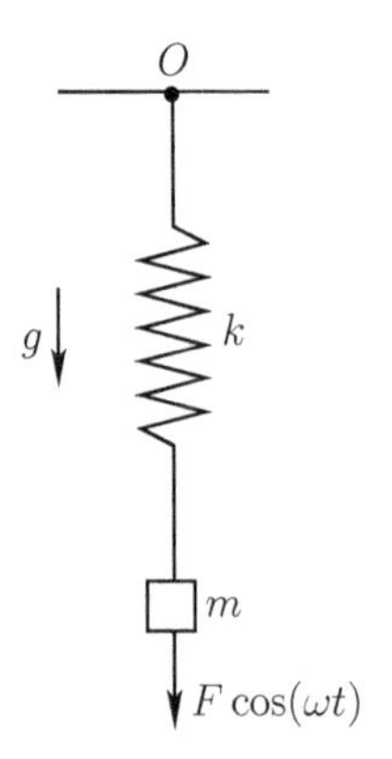

Figure 7.3 Schematic of the forced spring–mass system.

see that second-order accurate, central differences lead to a tridiagonal system(s) of equations. This large matrix problem is then solved using direct or iterative methods.

Introducing a simple example will allow us to discuss the steps in the numerical solution procedure in a more concrete fashion. For this purpose, let us consider the forced spring–mass system shown in Figure 7.3 to be our physical system. The height of the mass $u(t)$ is measured vertically upward from its neutral position at which no spring force f_s acts on the mass; that is, $f_s = 0$ at $u = 0$. We assume that the spring is linear, such that the spring force is directly proportional to its elongation according to

$$f_s = -ku,$$

where k is the linear spring constant. We neglect the mass of the spring but account for the drag on the moving mass. The drag force is assumed to behave according to the Stokes' model for low speeds given by

$$f_d = -cv,$$

where c is the drag coefficient, and $v(t)$ is the velocity of the mass. The minus sign arises because the drag force is in the opposite direction of the motion of the mass.

In order to obtain the equation of motion, let us consider Newton's second law

$$ma(t) = \sum f_i(t),$$

where $a(t)$ is the acceleration of the mass, and $f_i(t)$ are the forces acting on it. Newton's second law enforces conservation of momentum. Recall that the velocity $v(t)$ and position $u(t)$ are related to the acceleration via

$$v = \frac{du}{dt}, \qquad a = \frac{dv}{dt} = \frac{d^2u}{dt^2}.$$

From a free-body diagram of the forces acting on the mass, with $W = mg$ being the weight, and $f_f(t) = F\cos(\omega t)$ being the force owing to the prescribed forcing, Newton's second law leads to

$$ma = f_d + f_s - W - f_f,$$

or

$$ma = -cv - ku - mg - F\cos(\omega t).$$

Writing in terms of the mass position $u(t)$ only, the governing equation is

$$\frac{d^2u}{dt^2} + \frac{c}{m}\frac{du}{dt} + \frac{k}{m}u = -g - \frac{F}{m}\cos(\omega t), \tag{7.1}$$

which is a second-order, linear, nonhomogeneous ordinary differential equation in the form of an initial-value problem. As such, we require initial conditions on the position and velocity at $t = 0$. For example,

$$u(0) = u_0, \qquad \left(\frac{du}{dt}\right)_{t=0} = v_0,$$

where u_0 and v_0 are the specified initial position and velocity, respectively.

Now let us consider the forced spring–mass system in the context of the general numerical solution procedure. Step 1 is application of the physical law and models, namely conservation of momentum in the form of Newton's second law. In this case, we assume that the spring is linear elastic and that the moving mass is subject to low-speed Stokes' drag. This results in the mathematical model given by (7.1) in the form of a second-order, linear ordinary differential equation. Although it is possible to obtain an exact solution analytically in this case, let us continue with the numerical solution procedure.

The second step of the numerical solution procedure involves discretizing the continuous governing equation and time domain. We approximate the *continuous* time domain and differential equation at *discrete* locations in time t_i, which are separated by the small time step Δt such that $t_i = t_{i-1} + \Delta t$. In order to see how the derivatives in the governing equation (7.1) are discretized, recall the definition of the derivative with Δt small, but finite:

$$\frac{du}{dt} = \dot{u}(t) = \lim_{\Delta t \to 0} \frac{u(t + \Delta t) - u(t)}{\Delta t} \approx \frac{u(t + \Delta t) - u(t)}{\Delta t}.$$

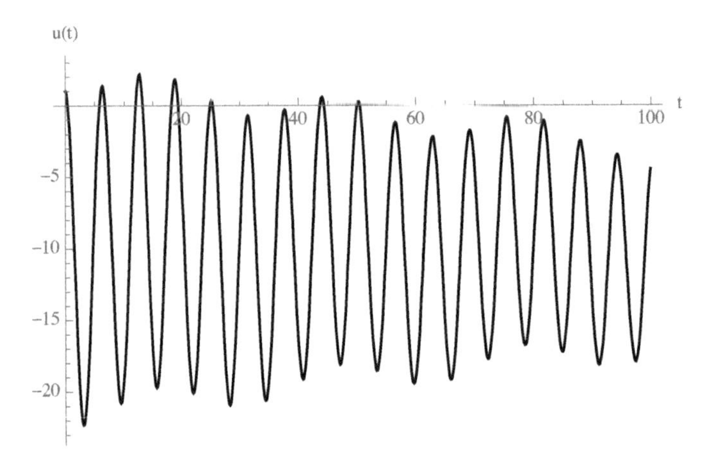

Figure 7.4 Solution of the forced spring–mass system with $g = 9.81$, $m = 1$, $k = 1$, $c = 0.01$, $F = 1$, and $\omega = 0.2$.

As suggested by this definition, the derivative of $u(t)$ can be approximated by linear combinations of the values of u at adjacent time steps – more on this later. Such *finite differences*, that is, differences of the dependent variable between adjacent finite time steps, allow for calculation of the position u at the current time step in terms of values at previous time steps. In this way, the value of the position is calculated at successive time steps in turn. See Figure 7.4 for a sample solution.

7.1.2 Properties of a Numerical Solution

Each step of the numerical solution procedure leads to its own source of error. Step 1 produces *modeling errors*, which are the differences between the actual physical system and the exact solution of the mathematical model. The difference between the exact solution of the governing equations and the exact solution of the system of algebraic equations is the *discretization error*. This error, which is produced by step 2, is comprised of two contributing factors: (1) the inherent error of the method of discretization, namely the *truncation error*, and (2) the error owing to the resolution of the computational grid used in the discretization. Finally, unless a direct method is used, there is the *iterative convergence error*, which is the difference between the iterative numerical solution and the exact solution of the algebraic equations. In both direct and iterative methods, *round-off errors* arise as discussed in Section 6.2.2.

As we discuss various numerical methods applied to a variety of types of problems and equations, there are several properties of successful numerical solution methods that must be considered. The first is *consistency*, which requires that the discretized equations formally become the continuous governing equations as the grid size goes to zero; this is a property of the discretization method. Specifically, the truncation error, which is the difference between the solution of the discretized equations and the exact solution of the governing equations, must go to zero as the grid size goes to zero. For example, we will see that one possible finite-difference approximation to the first-order derivative of $u(x)$ with respect to x is given by

$$\frac{du}{dx} = \frac{u_{i+1} - u_{i-1}}{2\Delta x} + O(\Delta x^2),$$

where $u_{i-1} = u(x_i - \Delta x)$ and $u_{i+1} = u(x_i + \Delta x)$. The last term indicates that the truncation error is $O(\Delta x^2)$, which is read "on the order of $(\Delta x)^2$" or simply "order $(\Delta x)^2$." Therefore, from the definition of the derivative, as $\Delta x \to 0$, we see that $(u_{i+1} - u_{i-1}) / (2\Delta x) \to du/dx$ for consistency.

The second property is *numerical stability*, which requires that the numerical procedure must not magnify the small round-off errors that are inevitably produced by the numerical solution such that the numerical solution diverges from the exact solution. This will be discussed in more detail in Chapter 9 in the context of initial-value problems. Note the similarity between stability and conditioning discussed in Section 6.2.2 as they both involve examining the effect of disturbances. The distinction is that conditioning quantifies the effect of disturbances in the system of equations itself regardless of the numerical solution used, whereas stability determines the effect of disturbances in the algorithm that is used to solve the system of equations.

The third property is *convergence*, whereby the numerical solution of the discretized equations approaches the exact solution of the governing equations as the grid size goes to zero; this is not the same as iterative convergence. Of course, we generally do not have the exact solution; therefore, we refine the grid until a *grid-independent* solution is obtained. The truncation error gives the *rate of convergence* to the exact solution as the grid size is reduced. For example, for an $O(\Delta x^2)$ truncation error, halving the grid size reduces the error by a factor of four.

Finally, we desire *correctness*, whereby the numerical solution should compare favorably with available analytical solutions, experimental results, or other computational solutions within the limitations of each. This is addressed through validation.

Given the numerous sources of errors and the potential for numerical instability, how do we determine whether a numerical solution is accurate and reliable? This is accomplished through *validation* and *verification*. Validation is quantification of the modeling errors in the computational model, that is, step 1 of the numerical solution procedure. It seeks to answer the questions: "Am I solving the correct equations?", "Am I capturing the physics correctly?", and "Is the solution qualitatively correct?" Validation is accomplished through comparison with highly accurate experiments carried out on the actual system. Verification is quantification of the discretization and iterative convergence errors in the computational model and its solution, that is, steps 2 and 3 of the numerical solution procedure. It seeks to answer the questions: "Am I solving the equations correctly?", "Am I capturing the mathematics correctly?", and "Is the solution quantitatively correct?" Verification is accomplished through comparison with existing benchmark solutions in the form of exact or highly accurate numerical solutions to which the numerical solutions can be compared.

7.1.3 Numerical Solution Approaches

The discretization step – step 2 in the numerical solution procedure – may be accomplished using various methods. The most common approaches used in science and engineering applications are finite-difference methods, finite-element methods, and spectral methods. Our focus in this text is primarily on finite-difference methods owing to their widespread use, flexibility, and fundamental nature relative to other numerical methods. We summarize the basic approach, advantages, and disadvantages of each of these methods in the remainder of this section, and then discuss their mathematical basis in the following sections.

Finite-Difference Methods

Basic approach:

- Discretize the governing equations in differential form using Taylor series–based finite-difference approximations at each grid point.
- Produces linear algebraic equations involving each grid point and surrounding points.
- *Local* approximation method.
- Provides a general and flexible framework for discretizing ordinary and partial differential equations of any form in any field.
- Popular in research.

Advantages:

- Relatively straightforward to understand and implement (based on Taylor series).
- Utilizes the familiar differential form of the governing equations.
- Very general as it applies to a wide variety of problems, including complex physics.
- Typically implemented on straightforward structured grids.
- Can be extended to higher-order approximations.
- Typically produce sparse, highly structured matrices that can be solved efficiently.

Disadvantages:

- Difficult to implement for complex geometries.
- Increasing accuracy of finite-difference approximations requires including additional grid points, thereby increasing the overall complexity of the code and adversely impacting computational efficiency.
- Algebraic convergence rate.

Finite-Element Methods

Basic approach:

- Apply conservation equations in variational form with shape functions to a set of *finite elements*.
- Produces a set of linear or nonlinear algebraic equations for solution of each element.

- *Local* approximation method.
- Popular in commercial codes – particularly for solid mechanics and heat transfer.

Advantages:

- Easy to treat complex geometries using inherently unstructured grids.
- Elements are naturally unstructured and nonuniform.
- Can increase order of accuracy by using higher-order shape functions and/or increasing the number of elements.

Disadvantages:

- Unstructured grids significantly increase computational complexity as compared to structured grids.
- Solution methods are comparatively inefficient for the dense and unstructured matrices typically resulting from finite-element discretizations.
- Algebraic convergence rate.

Spectral Numerical Methods

Basic approach:

- Solution of the governing equations in differential form are approximated using truncated – usually orthogonal – eigenfunction expansions.
- Produces system of algebraic equations (steady) or system of ordinary differential equations (unsteady) involving the coefficients in the eigenfunction expansion.
- *Global* approximation method, that is, we solve for the solution in the entire domain simultaneously.
- Popular for situations in which a highly accurate solution is required.

Advantages:

- Obtains highly accurate solutions when the underlying solution is smooth.
- Can achieve rapid spectral (exponential) convergence.
- Being a global method, no grid is required. Increased accuracy is achieved by simply increasing the number of terms in the expansions.
- Can be combined with finite-element methods to produce the *spectral-element method* that combines many of the advantages of each.

Disadvantages:

- Less straightforward to implement than finite-difference methods.
- More difficult to treat complicated geometries and boundary conditions.
- Small changes in the problem, for example, boundary conditions, can cause large changes in the algorithm.
- Not well suited for solutions having large gradients.

A perusal of the relative advantages and disadvantages of each method reveals that they are almost completely complementary in the sense that a weakness of one is a strength of another. Consequently, each has a prominent place in scientific computing, and the scientist or engineer is likely to utilize some combination of these methods if they perform simulations. Although these are the most common general-purpose numerical methods, a wide variety of more specialized methods are in use in various fields in the sciences and engineering. For example, in computational fluid dynamics (CFD) and heat transfer, finite-volume, boundary-element, vortex-based, and immersed boundary methods are used in various applications.

Our primary focus in this text is on finite-difference methods. However, we will provide background on finite-element and spectral methods in order to get a feel for how each is implemented. Finite-element methods are discussed further in Section 7.4, and spectral numerical methods will be discussed further in Sections 7.3 as well as 13.2.5 in the context of Galerkin projection and reduced-order modeling.

7.2 Formal Basis for Finite-Difference Methods

In Section 1.4.1, we motivated the finite-difference approximation using the definition of the derivative with small, but finite, step size. Here, we will formalize this approach using Taylor series expansions, but first reconsider the definition of the derivative for the function $u(x)$ at a point x_i as follows:[1]

$$u'(x_i) = \left(\frac{du}{dx}\right)_{x=x_i} = \left(\frac{du}{dx}\right)_i = \lim_{\Delta x \to 0} \frac{u(x_i + \Delta x) - u(x_i)}{\Delta x}.$$

This may be interpreted as a *forward difference* if Δx is small, but not going all the way to zero, thereby resulting in a *finite difference*. If, for example, $x_{i+1} = x_i + \Delta x$ and $u_{i+1} = u(x_{i+1})$, then we can interpret this graphically as shown in Figure 7.5. Thus, we could intuit three approximations to the first derivative:

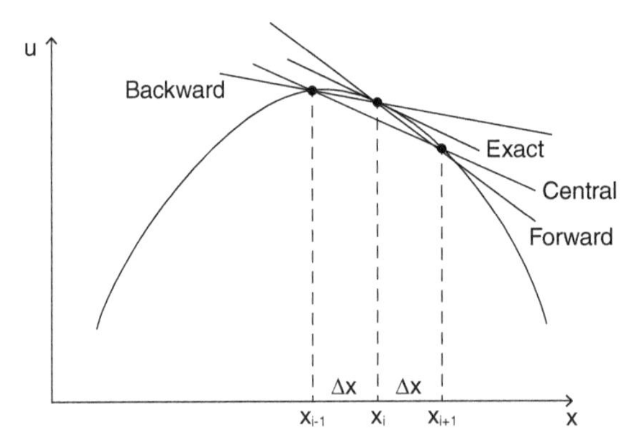

Figure 7.5 Graphical representation of the forward, backward, and central difference approximations to the first derivative.

[1] In an initial-value problem, the independent variable is time t rather than position x, and we will denote the derivative by a dot rather than a prime.

$$\text{Forward difference:} \qquad \left(\frac{du}{dx}\right)_i \approx \frac{u_{i+1} - u_i}{\Delta x},$$

$$\text{Backward difference:} \qquad \left(\frac{du}{dx}\right)_i \approx \frac{u_i - u_{i-1}}{\Delta x},$$

$$\text{Central difference:} \qquad \left(\frac{du}{dx}\right)_i \approx \frac{u_{i+1} - u_{i-1}}{2\Delta x}.$$

Intuitively, we might expect that the central difference will provide a more accurate approximation than the forward and backward differences. Indeed, this is the case as shown formally using Taylor series expansions.

7.2.1 Taylor Series

Finite-difference approximations are based on truncated Taylor series expansions, which allow us to express the local behavior of a function in the vicinity of some point in terms of the value of the function and its derivatives at that point.[2] Consider the Taylor series expansion of $u(x)$ in the vicinity of the point x_i

$$\begin{aligned} u(x) = u(x_i) &+ (x - x_i)\left(\frac{du}{dx}\right)_i + \frac{(x-x_i)^2}{2!}\left(\frac{d^2u}{dx^2}\right)_i \\ &+ \frac{(x-x_i)^3}{3!}\left(\frac{d^3u}{dx^3}\right)_i + \cdots + \frac{(x-x_i)^n}{n!}\left(\frac{d^nu}{dx^n}\right)_i + \cdots. \end{aligned} \tag{7.2}$$

Let us apply the Taylor series at $x = x_{i+1}$, with $x_{i+1} - x_i = \Delta x$, as follows:

$$\begin{aligned} u_{i+1} = u_i &+ \Delta x\left(\frac{du}{dx}\right)_i + \frac{\Delta x^2}{2}\left(\frac{d^2u}{dx^2}\right)_i + \frac{\Delta x^3}{6}\left(\frac{d^3u}{dx^3}\right)_i \\ &+ \cdots + \frac{\Delta x^n}{n!}\left(\frac{d^nu}{dx^n}\right)_i + \cdots. \end{aligned} \tag{7.3}$$

Solving for $(du/dx)_i$ gives

$$\left(\frac{du}{dx}\right)_i = \frac{u_{i+1} - u_i}{\Delta x} - \frac{\Delta x}{2}\left(\frac{d^2u}{dx^2}\right)_i - \cdots - \frac{\Delta x^{n-1}}{n!}\left(\frac{d^nu}{dx^n}\right)_i + \cdots. \tag{7.4}$$

Similarly, apply the Taylor series at $x = x_{i-1}$, with $x_{i-1} - x_i = -\Delta x$, to give

$$\begin{aligned} u_{i-1} = u_i &- \Delta x\left(\frac{du}{dx}\right)_i + \frac{\Delta x^2}{2}\left(\frac{d^2u}{dx^2}\right)_i - \frac{\Delta x^3}{6}\left(\frac{d^3u}{dx^3}\right)_i \\ &+ \cdots + \frac{(-1)^n\Delta x^n}{n!}\left(\frac{d^nu}{dx^n}\right)_i + \cdots. \end{aligned} \tag{7.5}$$

[2] Have you ever stopped to realize how powerful the Taylor series is? It says that you can know all there is to know about an entire function everywhere simply by knowing all of the derivatives of the function at a single point and the distance from that point!

Solving again for $(du/dx)_i$ yields

$$\left(\frac{du}{dx}\right)_i = \frac{u_i - u_{i-1}}{\Delta x} + \frac{\Delta x}{2}\left(\frac{d^2u}{dx^2}\right)_i - \frac{\Delta x^2}{6}\left(\frac{d^3u}{dx^3}\right)_i + \cdots + \frac{(-1)^n \Delta x^{n-1}}{n!}\left(\frac{d^nu}{dx^n}\right)_i + \cdots. \tag{7.6}$$

Alternatively, subtract the Taylor series (7.5) at x_{i-1} from the Taylor series (7.3) at x_{i+1} to obtain

$$u_{i+1} - u_{i-1} = 2\Delta x\left(\frac{du}{dx}\right)_i + \frac{\Delta x^3}{3}\left(\frac{d^3u}{dx^3}\right)_i + \cdots + \frac{2\Delta x^{2n+1}}{(2n+1)!}\left(\frac{d^{2n+1}u}{dx^{2n+1}}\right)_i + \cdots, \tag{7.7}$$

and solve for $(du/dx)_i$ to obtain

$$\left(\frac{du}{dx}\right)_i = \frac{u_{i+1} - u_{i-1}}{2\Delta x} - \frac{\Delta x^2}{6}\left(\frac{d^3u}{dx^3}\right)_i - \cdots - \frac{\Delta x^{2n}}{(2n+1)!}\left(\frac{d^{2n+1}u}{dx^{2n+1}}\right)_i - \cdots. \tag{7.8}$$

If all of the terms are retained in the expansions, (7.4), (7.6), and (7.8) are *exact* expressions for the first derivative $(du/dx)_i$. Approximate *finite-difference* expressions for the first derivative may then be obtained by truncating the series after the first term:

Forward difference: $$\left(\frac{du}{dx}\right)_i = \frac{u_{i+1} - u_i}{\Delta x} + O(\Delta x),$$

Backward difference: $$\left(\frac{du}{dx}\right)_i = \frac{u_i - u_{i-1}}{\Delta x} + O(\Delta x),$$

Central difference: $$\left(\frac{du}{dx}\right)_i = \frac{u_{i+1} - u_{i-1}}{2\Delta x} + O(\Delta x^2).$$

The $O(\Delta x)$ and $O(\Delta x^2)$ terms represent the *truncation error* of the corresponding approximation. For small Δx, successive terms in the Taylor series get smaller, and the order of the truncation error is given by the first truncated term. We say that the forward- and backward-difference approximations are "first-order accurate," and the central-difference approximation is "second-order accurate." Note that the central-difference approximation is indeed more accurate than the forward and backward differences as expected. Observe that the truncation error arises because of our choice of algorithm, whereas the round-off error arises because of the way calculations are carried out on a computer. Therefore, truncation error would result even on a "perfect" computer using exact arithmetic.

Higher-order approximations and/or higher-order derivatives may be obtained by various manipulations of the Taylor series at additional points. For example, to obtain a *second-order accurate forward-difference approximation to the first derivative*, apply the Taylor series at x_{i+2} as follows ($x_{i+2} - x_i = 2\Delta x$):

$$u_{i+2} = u_i + 2\Delta x \left(\frac{du}{dx}\right)_i + \frac{(2\Delta x)^2}{2!}\left(\frac{d^2u}{dx^2}\right)_i + \frac{(2\Delta x)^3}{3!}\left(\frac{d^3u}{dx^3}\right)_i + \cdots. \tag{7.9}$$

The $(d^2u/dx^2)_i$ term can be eliminated by taking $4\times(7.3) - (7.9)$ to obtain

$$4u_{i+1} - u_{i+2} = 3u_i + 2\Delta x \left(\frac{du}{dx}\right)_i - \frac{2\Delta x^3}{3}\left(\frac{d^3u}{dx^3}\right)_i + \cdots.$$

Solving for $(du/dx)_i$ yields

$$\left(\frac{du}{dx}\right)_i = \frac{-3u_i + 4u_{i+1} - u_{i+2}}{2\Delta x} + \frac{\Delta x^2}{3}\left(\frac{d^3u}{dx^3}\right)_i + \cdots, \tag{7.10}$$

which is second-order accurate and involves the point of interest and the next two points to the right.

For a *second-order accurate central-difference approximation to the second derivative*, add (7.3) and (7.5) for u_{i+1} and u_{i-1}, respectively, to eliminate the $(du/dx)_i$ term. This gives

$$u_{i+1} + u_{i-1} = 2u_i + \Delta x^2 \left(\frac{d^2u}{dx^2}\right)_i + \frac{\Delta x^4}{12}\left(\frac{d^4u}{dx^4}\right)_i + \cdots.$$

Solving for $(d^2u/dx^2)_i$ leads to

$$\left(\frac{d^2u}{dx^2}\right)_i = \frac{u_{i+1} - 2u_i + u_{i-1}}{\Delta x^2} - \frac{\Delta x^2}{12}\left(\frac{d^4u}{dx^4}\right)_i + \cdots, \tag{7.11}$$

which is second-order accurate and involves the point of interest and its nearest neighbors to the left and right.

We call finite-difference approximations that only involve the point of interest and its two nearest neighbors on either side *compact*. Therefore, the preceding second-order accurate central-difference approximations for both the first and second derivatives are compact, whereas the second-order accurate forward-difference approximation to the first derivative is not compact.

7.2.2 Finite-Difference Stencil

In comparing the first-order and second-order accurate forward-difference approximations, observe that increasing the order of accuracy of a finite-difference approximation requires including additional grid points. Thus, as more complex situations are encountered, for example, involving higher-order derivatives and/or approximations, determining how to combine the linear combination of Taylor series to produce such finite-difference formulae can become very difficult. Alternatively, it is sometimes easier to frame the question as follows: For a given set of adjacent grid points, called the *finite-difference stencil*, what is the highest-order finite-difference approximation possible? Or stated slightly differently: What is the "best" finite-difference approximation using a given pattern of grid points, in other words, the one with the smallest truncation error? For a similar, but alternative, approach using Padé approximations, see Lele (1992).

We illustrate the procedure using an example. Equation (7.10) provides a second-order accurate, forward-difference approximation to the first derivative, and it involves three adjacent points at x_i, x_{i+1}, and x_{i+2}. Let us instead determine the most accurate approximation to the first derivative that involves the four points x_i, x_{i+1}, x_{i+2}, and x_{i+3}. This will be of the form

$$u_i' + c_0 u_i + c_1 u_{i+1} + c_2 u_{i+2} + c_3 u_{i+3} = T.E., \tag{7.12}$$

where primes denote derivatives of u with respect to x, and $T.E.$ is the truncation error. The objective is to determine the constants c_0, c_1, c_2, and c_3 that produce the highest-order truncation error. The Taylor series approximations for $u(x)$ at x_{i+1}, x_{i+2}, and x_{i+3} about x_i are

$$u_{i+1} = u_i + \Delta x u_i' + \frac{\Delta x^2}{2!} u_i'' + \frac{\Delta x^3}{3!} u_i''' + \frac{\Delta x^4}{4!} u_i'''' + \cdots,$$

$$u_{i+2} = u_i + 2\Delta x u_i' + \frac{(2\Delta x)^2}{2!} u_i'' + \frac{(2\Delta x)^3}{3!} u_i''' + \frac{(2\Delta x)^4}{4!} u_i'''' + \cdots,$$

$$u_{i+3} = u_i + 3\Delta x u_i' + \frac{(3\Delta x)^2}{2!} u_i'' + \frac{(3\Delta x)^3}{3!} u_i''' + \frac{(3\Delta x)^4}{4!} u_i'''' + \cdots.$$

Substituting these expansions into (7.12) and collecting terms leads to

$$\begin{aligned}
&(c_0 + c_1 + c_2 + c_3)\, u_i + [1 + \Delta x\,(c_1 + 2c_2 + 3c_3)]\, u_i' \\
&+ \Delta x^2 \left(\frac{1}{2}c_1 + 2c_2 + \frac{9}{2}c_3\right) u_i'' + \Delta x^3 \left(\frac{1}{6}c_1 + \frac{4}{3}c_2 + \frac{9}{2}c_3\right) u_i''' \\
&+ \Delta x^4 \left(\frac{1}{24}c_1 + \frac{2}{3}c_2 + \frac{27}{8}c_3\right) u_i'''' + \cdots = T.E.
\end{aligned} \tag{7.13}$$

The highest-order truncation error will occur when the maximum number of lower-order derivative terms are eliminated in this equation. Because we have four constants to determine in this case, we can eliminate the first four terms in the expansion. This requires the following four simultaneous linear algebraic equations to be solved for the coefficients:

$$\begin{aligned}
c_0 + c_1 + c_2 + c_3 &= 0, \\
c_1 + 2c_2 + 3c_3 &= -\frac{1}{\Delta x}, \\
\frac{1}{2}c_1 + 2c_2 + \frac{9}{2}c_3 &= 0, \\
\frac{1}{6}c_1 + \frac{4}{3}c_2 + \frac{9}{2}c_3 &= 0.
\end{aligned}$$

The solution to this system of equations (with the least common denominator) is

$$c_0 = \frac{11}{6\Delta x}, \quad c_1 = -\frac{18}{6\Delta x}, \quad c_2 = \frac{9}{6\Delta x}, \quad c_3 = -\frac{2}{6\Delta x}.$$

The remaining term that has not been zeroed provides the leading-order truncation error. Substituting the solutions for the coefficients just obtained, this term becomes

$$T.E. = -\frac{1}{4}\Delta x^3 u_i'''',$$

which indicates that the approximation is third-order accurate. Therefore, the approximation (7.12) that has the highest-order truncation error is

$$\left(\frac{du}{dx}\right)_i = \frac{-11u_i + 18u_{i+1} - 9u_{i+2} + 2u_{i+3}}{6\Delta x} + O(\Delta x^3). \tag{7.14}$$

Using this procedure with a five-point stencil that involves the points x_{i-2}, x_{i-1}, x_i, x_{i+1}, and x_{i+2}, the following central-difference formulae can be derived for the first four derivatives:

$$\left(\frac{du}{dx}\right)_i = \frac{u_{i-2} - 8u_{i-1} + 8u_{i+1} - u_{i+2}}{12\Delta x} + O(\Delta x^4).$$

$$\left(\frac{d^2u}{dx^2}\right)_i = \frac{-u_{i-2} + 16u_{i-1} - 30u_i + 16u_{i+1} - u_{i+2}}{12\Delta x^2} + O(\Delta x^4).$$

$$\left(\frac{d^3u}{dx^3}\right)_i = \frac{-u_{i-2} + 2u_{i-1} - 2u_{i+1} + u_{i+2}}{2\Delta x^3} + O(\Delta x^2).$$

$$\left(\frac{d^4u}{dx^4}\right)_i = \frac{u_{i-2} - 4u_{i-1} + 6u_i - 4u_{i+1} + u_{i+2}}{\Delta x^4} + O(\Delta x^2).$$

Observe that the first two are fourth-order accurate, while the latter two are second-order accurate.

7.2.3 Nonuniform and Unstructured Grids

Thus far, we have used what are called "uniform, collocated, structured grids." *Uniform* means that the grid spacings Δx in each direction are uniform. *Collocated* means that all dependent variables are approximated at the same grid points. Finally, *structured* means that there is an ordered relationship between neighboring grid cells. Let us consider alternatives to uniform and structured grids.

In many physical problems, the solution has *local* regions of intense gradients. This is particularly common in fluid dynamics, where thin boundary layers near solid surfaces exhibit large gradients in velocity, for example. A uniform grid with sufficient resolution in such regions would waste computational resources elsewhere where they are not needed. Therefore, it would be beneficial to be able to impose a finer grid in regions with large gradients in order to resolve them, but without having to increase the resolution elsewhere in the domain.

A nonuniform grid would allow us to refine the grid only where it is needed. Let us obtain the finite-difference approximations using Taylor series as before, but without

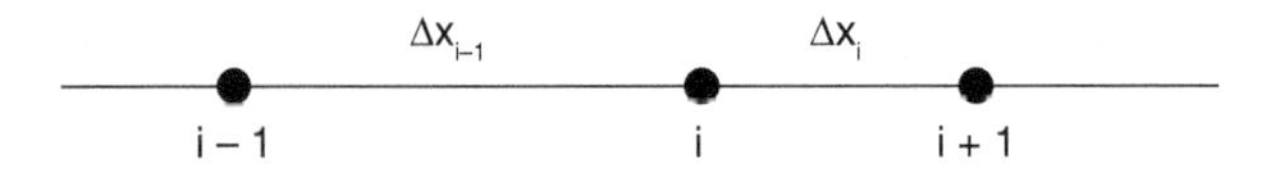

Figure 7.6 Schematic of a nonuniform grid.

assuming that all of the grid sizes Δx are equal. For example, consider the first-derivative term $(du/dx)_i$ as illustrated in Figure 7.6. The Taylor series (7.3) applied at $x = x_{i+1}$, with $x_{i+1} - x_i = \Delta x_i$, is now

$$u_{i+1} = u_i + \Delta x_i \left(\frac{du}{dx}\right)_i + \frac{\Delta x_i^2}{2}\left(\frac{d^2u}{dx^2}\right)_i + \frac{\Delta x_i^3}{6}\left(\frac{d^3u}{dx^3}\right)_i + \cdots + \frac{\Delta x_i^n}{n!}\left(\frac{d^nu}{dx^n}\right)_i + \cdots . \tag{7.15}$$

Similarly, the Taylor series (7.5) at $x = x_{i-1}$, with $x_{i-1} - x_i = -\Delta x_{i-1}$, is then

$$u_{i-1} = u_i - \Delta x_{i-1} \left(\frac{du}{dx}\right)_i + \frac{\Delta x_{i-1}^2}{2}\left(\frac{d^2u}{dx^2}\right)_i - \frac{\Delta x_{i-1}^3}{6}\left(\frac{d^3u}{dx^3}\right)_i + \cdots + \frac{(-1)^n\Delta x_{i-1}^n}{n!}\left(\frac{d^nu}{dx^n}\right)_i + \cdots . \tag{7.16}$$

Subtracting (7.15) from (7.16) and solving for $(du/dx)_i$ leads to

$$\frac{du}{dx} = \frac{u_{i+1} - u_{i-1}}{\Delta x_{i-1} + \Delta x_i} - \frac{\Delta x_i^2 - \Delta x_{i-1}^2}{2(\Delta x_{i-1} + \Delta x_i)}\left(\frac{d^2u}{dx^2}\right)_i - \frac{\Delta x_i^3 + \Delta x_{i-1}^3}{6(\Delta x_{i-1} + \Delta x_i)}\left(\frac{d^3u}{dx^3}\right)_i + \cdots .$$

If the grid is uniform, that is, $\Delta x_{i-1} = \Delta x_i$, then the second term vanishes, and the approximation reduces to the usual $O(\Delta x^2)$-accurate central difference approximation for the first derivative. However, for a nonuniform grid, the truncation error is only $O(\Delta x)$. We could restore second-order accuracy by using an appropriate approximation to $\left(d^2u/dx^2\right)_i$ in the second term, which results in

$$\frac{du}{dx} = \frac{u_{i+1}\Delta x_{i-1}^2 - u_{i-1}\Delta x_i^2 + u_i(\Delta x_i^2 - \Delta x_{i-1}^2)}{\Delta x_i \Delta x_{i-1}(\Delta x_{i-1} + \Delta x_i)} + O(\Delta x^2).$$

As one can imagine, this gets very complicated, and it is difficult to ensure consistent accuracy for all approximations in an equation.

An alternative approach for concentrating grid points in certain regions of a domain employs grid transformations that map the physical domain to a computational domain that may have a simple overall shape and/or cluster grid points in regions of the physical domain where the solution varies rapidly – where large gradients occur. The mapping is such that a uniform grid in the computational domain corresponds to a nonuniform grid in the physical domain, thereby allowing for utilization of all of the finite-difference approximations throughout Part II for the numerical solution carried out in the computational domain. The solution is then transformed back to the physical domain.

There are three general approaches to grid generation: algebraic, elliptic, and variational. In algebraic grid generation, an algebraic function is chosen a priori that accomplishes the desired transformation. In elliptic grid generation, an elliptic partial differential equation(s) is solved for the mapping. In variational grid generation, an optimization problem is formulated that when solved produces the mapping; this is called solution-adaptive grid generation. See chapter 12 of Cassel (2013) for more on algebraic, elliptic, and variational grid generation.

Thus far, we have been considering structured grids, which are comprised of an ordered arrangement of cells or elements – typically rectangles in two dimensions or quadrilaterals in three dimensions. They are commonly used for problems in which the geometry of the domain has a regular shape, and they can be extended to more complex domains through the use of grid transformations from a complex to simple domain. Grid transformations maintain the original structure of the computational grid; therefore, they still limit the flexibility to accommodate very complex geometries. In such cases, unstructured grids provide a powerful alternative to treat complex domain shapes, including those that have irregular shapes and/or cutouts. The cells of an unstructured grid are not constrained to be arranged in an ordered fashion, thereby providing additional flexibility in arranging the cells to accommodate complex shapes; the only requirement is that faces of adjacent cells must coincide. In this way, one can locally refine the grid as desired simply by defining smaller grid cells. For two-dimensional domains, the elements are typically quadrilateral, defined by four nodes, or triangular, defined by three nodes. In three-dimensional domains, the elements would be generalized to become hexahedral or tetrahedral. Unstructured grids present a significant challenge – and additional computational overhead – in accounting for the irregular orientation and arrangement of cells; however, it obviates the need for complicated grid generation strategies. Unstructured grids are more easily accommodated using finite-element methods, which is one of its fundamental advantages over finite-difference methods. We will utilize structured grids throughout Part II as they are commonly used for finite-difference methods.

7.3 Formal Basis for Spectral Numerical Methods

In finite-difference methods, the continuous governing differential equation is approximated *locally* in such a way that a large system of algebraic equations is produced for the solution at each point in a grid defined by the discretization process. Spectral methods are a class of numerical techniques based on the principles of eigenfunction expansions introduced in Chapter 3, in which the unknown solution is approximated *globally* in the form of a truncated series expansion. As discussed in Section 7.1.3, the primary virtue of spectral methods is that they converge very rapidly toward the exact solution as the number of terms in the expansion is increased. At its best, this convergence is exponential, which is referred to as *spectral convergence*.

Along with finite-element methods, spectral methods utilize the *method of weighted residuals*. We begin by approximating the unknown solution of a differential equation $\mathcal{L}u = f$ using the expansion

$$u(x,y,z,t) = \phi_0(x,y,z) + \sum_{n=1}^{N} a_n(t)\phi_n(x,y,z), \tag{7.17}$$

where $u(x,y,z,t)$ is the approximate solution of the differential equation; $\phi_n(x,y,z)$ are the chosen spatial *basis*, or *trial, functions*; $a_n(t)$ are the time-dependent "coefficients"; and $\phi_0(x,y,z)$ is chosen to satisfy the boundary conditions, such that $\phi_n = 0$, $n = 1,\ldots,N$ at the boundaries. Take note that the spatial dependence is taken care of in the basis functions, and the time dependence, if present, is handled by the coefficients.

Similar to the solution, we expand the forcing function $f(x,y,z)$ in terms of the same basis functions according to

$$f(x,y,z) = \sum_{n=1}^{N} \hat{a}_n \phi_n(x,y,z), \tag{7.18}$$

where the $\hat{a}_n$ coefficients are determined as in Section 3.2.2 according to the inner product

$$\hat{a}_n = \big\langle f(x,y,z), \phi_n(x,y,z) \big\rangle$$

if the basis functions $\phi_n(x,y,z)$ are orthonormal. If the basis functions are not orthogonal, a coupled system of equations results that must be solved for the coefficients in the expansion (7.18). Using orthonormal basis functions for linear problems leads to an uncoupled system of equations that is particularly straightforward to solve for the coefficients $a_n(t)$. Moreover, adding additional terms in the expansion does not alter the previous terms already computed as would be the case if a coupled system of equations was to be solved for the coefficients.

Consider the exact solution $\hat{u}(x,y,z,t)$ of the differential equation

$$\mathcal{L}\hat{u} = f. \tag{7.19}$$

Because the numerical solution $u(x,y,z,t)$ is only approximate, it does not exactly satisfy the original differential equation. Substituting its expansion (7.17) into the differential equation (7.19), we define the *residual* as

$$r = f - \mathcal{L}u.$$

Note that as $N \to \infty$, $u \to \hat{u}$, and $r \to 0$.

In order to determine the coefficients $a_n(t)$ for the given set of basis functions in the expansion (7.17), we require that the integral of the weighted residual be zero over the spatial domain. That is, the inner product of r and w_i are zero (they are orthogonal to one another), where $w_i(x)$ are the *weight*, or *test, functions*. In the one-dimensional case, for example, this requires that

$$\langle r(x,t), w_i(x)\rangle = \int_{x_0}^{x_1} r(x,t)w_i(x)dx = 0, \quad i = 1,\ldots,N. \tag{7.20}$$

Substituting the trial functions (7.17) into (7.20) requires that

$$\int_{x_0}^{x_1} \left\{ \sum_{n=1}^{N} \hat{a}_n \phi_n(x) - \mathcal{L}\left[\phi_0(x) + \sum_{n=1}^{N} a_n(t)\phi_n(x) \right] \right\} w_i(x)dx = 0, \quad i = 1,\ldots,N. \tag{7.21}$$

Different *weighted residual methods* correspond to different choices for the weight (test) functions:

- Galerkin:

$$w_i(x) = \phi_i(x);$$

  that is, the weight (test) functions are the same as the basis (trial) functions.
- Least squares:

$$w_i(x) = \frac{\partial r}{\partial a_i},$$

  which results in the square of the norm of the residual $\int r^2 dx$ being a minimum.
- Collocation:

$$w_i(x) = \delta(x - x_i),$$

  where δ is the Dirac delta function centered at the collocation points x_i.

For more on weighted residual methods and their connections to Galerkin projection and least squares; see Section 13.2.4. In spectral numerical methods, the trial functions are chosen such that they are mutually orthogonal for reasons that will become apparent in the following one-dimensional example.

Example 7.1 Let us consider the ordinary differential equation

$$\mathcal{L}u = \frac{d^2u}{dx^2} + u = 0, \quad 0 \le x \le 1, \tag{7.22}$$

with the boundary conditions

$$u(0) = 0, \quad u(1) = 1.$$

For the trial functions, use sines, which are mutually orthogonal over the specified domain, of the form

$$\phi_n(x) = \sin(n\pi x), \quad n = 1,\ldots,N.$$

Note that $\phi_n = 0$ at $x = 0, 1$. To satisfy the boundary conditions, we choose $\phi_0(x) = x$. Thus, the spectral solution (7.17) becomes

$$u(x) = x + \sum_{n=1}^{N} a_n \sin(n\pi x), \tag{7.23}$$

where the a_n coefficients are constants as there is no time dependence. Differentiating gives

$$u'(x) = 1 + \sum_{n=1}^{N} a_n n\pi \cos(n\pi x),$$

and again

$$u''(x) = -\sum_{n=1}^{N} a_n (n\pi)^2 \sin(n\pi x).$$

Substituting into the differential equation (7.22) to obtain the residual yields

$$r = f - \mathcal{L}u = -u'' - u = \sum_{n=1}^{N} a_n (n\pi)^2 \sin(n\pi x) - x - \sum_{n=1}^{N} a_n \sin(n\pi x),$$

or

$$r(x) = -x - \sum_{n=1}^{N} a_n \left[1 - (n\pi)^2\right] \sin(n\pi x).$$

Applying the Galerkin method, in which case $w_i(x) = \phi_i(x)$, equation (7.20) becomes

$$\int_0^1 r(x)\phi_i(x)dx = 0, \quad i = 1, \ldots, N,$$

which results in N equations for the N coefficients a_n as follows. Substituting the weight functions and residual gives

$$\int_0^1 \left\{ x + \sum_{n=1}^{N} a_n \left[1 - (n\pi)^2\right] \sin(n\pi x) \right\} \{\sin(i\pi x)\}\, dx = 0,$$

or

$$\int_0^1 x \sin(i\pi x)dx + \sum_{n=1}^{N} a_n \left[1 - (n\pi)^2\right] \int_0^1 \sin(i\pi x)\sin(n\pi x)dx = 0 \qquad (7.24)$$

for $i = 1, \ldots, N$. Owing to orthogonality of sines, for which $\int_0^1 \sin(i\pi x)\sin(n\pi x)dx = 0$ for $n \neq i$, the only contribution to the summation in (7.24) arises when $n = i$. Let us evaluate the resulting integrals:

$$\int_0^1 x \sin(i\pi x)dx = \left[\frac{\sin(i\pi x)}{(i\pi)^2} - \frac{x\cos(i\pi x)}{i\pi}\right]_0^1$$

$$= \begin{cases} \dfrac{1}{i\pi}, & i = 1,3,5,\ldots \\ -\dfrac{1}{i\pi}, & i = 2,4,6,\ldots \end{cases} = -\frac{(-1)^i}{i\pi},$$

and

$$\int_0^1 \sin^2(i\pi x)dx = \left[\frac{x}{2} - \frac{\sin(2i\pi x)}{4i\pi}\right]_0^1 = \frac{1}{2}.$$

Thus, equation (7.24) becomes

$$-\frac{(-1)^i}{i\pi} + \frac{1}{2}a_i\left[1 - (i\pi)^2\right] = 0, \quad i = 1, \ldots, N.$$

Solving for the coefficients yields

$$a_i = \frac{2(-1)^i}{i\pi\left[1 - (i\pi)^2\right]}, \quad i = 1, \ldots, N, \tag{7.25}$$

which completes the expansion (7.23)

$$u(x) = x + \sum_{n=1}^{N} \frac{2(-1)^n}{n\pi\left[1 - (n\pi)^2\right]} \sin(n\pi x). \tag{7.26}$$

For comparison, the exact solution of (7.22) is

$$\hat{u}(x) = \frac{\sin x}{\sin 1}.$$

A plot of (7.26), even with only one term in the spectral solution, is indistinguishable from the exact solution. Only small N is required in this example because the underlying solution is a sine function. This, of course, is not normally the case.

As illustrated in the example, spectral numerical methods convert an ordinary or partial differential equation into a system of linear algebraic equations or a system of simplified ordinary differential equations for the time-dependent coefficients in the trial function. For steady problems, the a_n coefficients are in general determined by solving a system of N algebraic equations. When orthogonal basis functions are used, however, the equations for the coefficients are uncoupled. Therefore, inclusion of additional terms does not change the coefficients on the previous terms as each basis function is unaffected by the others. For unsteady problems, on the other hand, the time-dependent $a_n(t)$ "coefficients" are in general determined by solving a system of N ordinary differential equations in time. These can be solved using finite-difference methods for time derivatives as discussed in Chapter 9. The spatial aspects of the problem are accounted for in the basis functions.

The two critical decisions when applying spectral numerical methods are (1) which method of weighted residuals to use, and (2) what eigenfunctions to use in the basis (trial) function. Generally, the Galerkin method is used for linear problems, and collocation is used for nonlinear problems, which is referred to as a *pseudospectral method* (see Fletcher 1991a). With regard to the basis functions, trigonometric functions are used as the basis functions for periodic problems. For nonperiodic problems, *Chebyshev* or *Legendre* polynomials are typically used. Recall from Section 3.3.3 that

Chebyshev polynomials are orthogonal over the domain $-1 \leq x \leq 1$ with respect to the weight function $w = \left(1 - x^2\right)^{-1/2}$. When using Chebyshev polynomials as the basis functions, the boundary conditions must be homogeneous. In the Galerkin method, the trial functions must satisfy the boundary conditions, whereas in the *tau method*, which is often used for nonperiodic problems, they do not.

Spectral numerical methods give highly accurate approximate solutions when the underlying solution is *smooth*, and it is when the solution is smooth that exponential (spectral) convergence is achieved with increasing N. The number of terms required becomes very large when the solution contains locally large gradients, such as shockwaves. Finite-difference and finite-element methods, on the other hand, only experience geometric convergence. For example, in a second-order accurate method, the error scales according to $1/N^2$, which is much slower than exponential decay.

REMARKS:

1. *Because each basis function spans the entire domain, spectral methods provide* global *approximations.*
2. *The method of weighted residuals and spectral numerical methods are very closely aligned with Galerkin projection, which is the basis for reduced-order models of discrete and continuous systems as discussed in Chapter 13.*
3. *Note that the finite-volume method, which is popular in computational fluid dynamics and heat transfer, may be framed in terms of the method of weighted residuals (see Fletcher 1991a and Ferziger et al. 2020). Similarly, finite-difference methods can be interpreted as a method of weighted residuals approach using the collocation method.*
4. *As will be illustrated in Section 13.2.5, when the eigenfunctions of the differential operator in the equation are used with the Galerkin method, a spectral numerical method will produce the same coefficients as obtained in the separation of variables solutions of Chapter 3. Therefore, a solution obtained in such a case using spectral methods is just a truncated eigenfunction expansion.*
5. *For more on spectral numerical methods, see Aref and Balachandar (2018), Canuto et al. (1988), and Fletcher (1984).*

7.4 Formal Basis for Finite-Element Methods

As with spectral methods, finite-element methods are based on weighted-residual methods. After choosing the basis (trial) and weight (test) functions, the weighted-residual (7.21) is integrated by parts to produce the so-called *weak form* of the problem; the original differential equation is known as the *strong form.*

In the commonly used *Galerkin method*, the weight functions are the same as the trial functions; therefore, (7.21) for the one-dimensional case becomes

$$\int_{x_0}^{x_1} \left\{ f(x) - \mathcal{L}\left[\phi_0(x) + \sum_{n=1}^{N} a_n(t)\phi_n(x) \right] \right\} \phi_i(x)dx = 0, \quad i = 0, \ldots, N.$$

The primary difference between finite-element methods as compared to spectral methods is that N is small – typically $N = 1$ (linear) or 2 (quadratic) – for finite-element methods. However, these *shape functions* are applied individually across many small "elements" that make up the entire domain. In other words, just like finite-difference methods, finite-element methods are *local* approximation methods. Recall that in spectral methods, each basis function is applied across the entire domain, which is why N must in general be larger than in finite-element methods.

Finite-element methods can be combined with spectral numerical methods to obtain the *spectral-element method*, for which the shape functions across each element in the finite-element method are replaced by spectral expansion approximations. This typically allows for larger elements as compared to finite-element methods alone that have only linear or quadratic shape functions. One can then take advantage of the flexibility of finite elements in representing complex domains with the spectral accuracy and convergence rate of spectral methods.

Because finite-element methods are based on variational methods, see chapter 3 of Cassel (2013) for more details on the variational underpinnings of finite-element methods. For additional details of finite-element methods and the method of weighted residuals, see Moin (2010) and Chung (2010).

Informally, there is a *principle of conservation of effort* in numerical methods. The total effort required to solve an ordinary or partial differential equation problem consists of that for us to formulate the problem and discretize it using one of the preceding methods (steps 1 and 2 in the numerical solution procedure) and then that required for the computer to solve this problem (step 3). As you can probably sense from our earlier discussion, finite-difference methods are typically the easiest to set up on paper and to program, but they require the greatest amount of computational effort owing to the very large number of iterations required. Finite-element methods tend to require additional effort on our part to formulate and code the method but are somewhat less demanding computationally. For example, solving the same problem generally requires more grid points and iterations using a finite-difference method as compared to the number of elements and iterations required using a finite-element method. Spectral numerical methods, on the other hand, build in a great deal of information about the problem in the formulation itself – being a global method – and can typically be solved very quickly owing to its spectral convergence properties. The reduced-order models to be discussed in Chapter 13 hold even greater potential than spectral methods to reduce the computational time as the basis functions are determined for the specific problem under consideration. However, additional effort is required to determine these basis functions ahead of time. In addition, each discretization approach has advantages and disadvantages when solved on various computer architectures; this is why each of these methods continues to be used and further developed, and it is not likely that one single method will ever prove to be the "best" for all problems when solved on all computers.

7.5 Classification of Second-Order Partial Differential Equations

Throughout the remainder of Part II, our focus will be on developing numerical methods for solving ordinary and partial differential equations governing initial- and boundary-value problems. Recall that in *initial-value problems* (IVP), we start with initial conditions at $t = 0$ and "march" the solution forward in time. In contrast, a *boundary-value problem* (BVP) is such that the solution throughout the entire domain is determined simultaneously subject to boundary conditions specified on all parts of the domain boundary. The distinction between initial- and boundary-value problems applies to both ordinary and partial differential equations. An additional classification, that applies to second-order partial differential equations, is whether the equation is hyperbolic, parabolic, or elliptic. The first two are initial-value problems and the latter is a boundary-value problem. This classification is considered next.

7.5.1 Mathematical Classification

As we develop numerical methods for solving differential equations, it is essential that they be faithful to the physical behavior inherent within the various types of equations. This behavior is determined by its classification, which depends upon the nature of its *characteristics*. These are the curves within the domain along which information propagates in space and/or time. The nature of the characteristics depend upon the coefficients of the highest-order derivatives. Owing to their prominence in applications, we focus on second-order partial differential equations.

Consider the general second-order partial differential equation for $u(x, y)$ given by

$$au_{xx} + bu_{xy} + cu_{yy} + du_x + eu_y + fu = g, \tag{7.27}$$

where subscripts denote partial differentiation with respect to the indicated variable. The equation is *linear* if the coefficients a, b, c, d, e, and f are only functions of (x, y). If they are functions of (x, y, u, u_x, u_y), then the equation is said to be *quasilinear*, in which case the equation is linear in the highest derivatives, and any nonlinearity is confined to the lower-order derivatives.

Let us determine the criteria necessary for the existence of a smooth (differentiable) and unique (single-valued) solution along a characteristic curve C as illustrated in Figure 7.7. Along C, we define the parametric functions

$$\begin{gathered} \phi_1(\tau) = u_{xx}, \quad \phi_2(\tau) = u_{xy}, \quad \phi_3(\tau) = u_{yy}, \\ \psi_1(\tau) = u_x, \quad \psi_2(\tau) = u_y, \end{gathered} \tag{7.28}$$

where τ is a variable along a characteristic. Substituting into (7.27) gives (keeping the second-order terms on the left-hand side)

$$a\phi_1 + b\phi_2 + c\phi_3 = H, \tag{7.29}$$

where

$$H = g - d\psi_1 - e\psi_2 - fu.$$

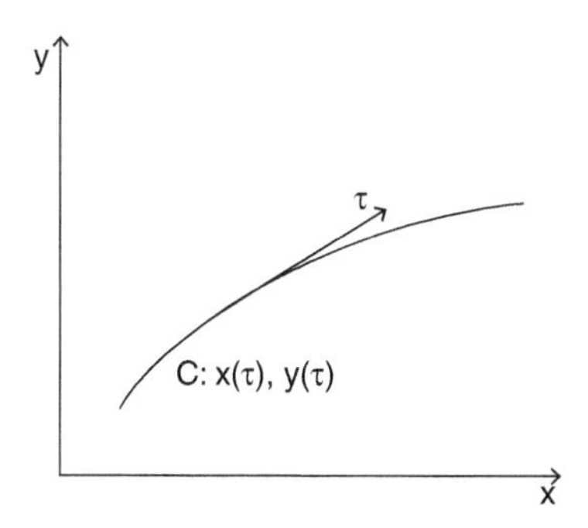

Figure 7.7 The characteristic C in a two-dimensional domain along which the solution propagates.

Transforming the independent variables from (x, y) to τ, we have

$$\frac{d}{d\tau} = \frac{dx}{d\tau}\frac{\partial}{\partial x} + \frac{dy}{d\tau}\frac{\partial}{\partial y}.$$

Thus, from (7.28) we have

$$\frac{d\psi_1}{d\tau} = \frac{d}{d\tau}u_x = \frac{dx}{d\tau}u_{xx} + \frac{dy}{d\tau}u_{xy} = \frac{dx}{d\tau}\phi_1 + \frac{dy}{d\tau}\phi_2, \tag{7.30}$$

and

$$\frac{d\psi_2}{d\tau} = \frac{d}{d\tau}u_y = \frac{dx}{d\tau}u_{xy} + \frac{dy}{d\tau}u_{yy} = \frac{dx}{d\tau}\phi_2 + \frac{dy}{d\tau}\phi_3. \tag{7.31}$$

Equations (7.29) through (7.31) are three equations for three unknowns, namely the second-order derivatives ϕ_1, ϕ_2, and ϕ_3. Written in matrix form, they are

$$\begin{bmatrix} a & b & c \\ \dfrac{dx}{d\tau} & \dfrac{dy}{d\tau} & 0 \\ 0 & \dfrac{dx}{d\tau} & \dfrac{dy}{d\tau} \end{bmatrix} \begin{bmatrix} \phi_1 \\ \phi_2 \\ \phi_3 \end{bmatrix} = \begin{bmatrix} H \\ \dfrac{d\psi_1}{d\tau} \\ \dfrac{d\psi_2}{d\tau} \end{bmatrix}.$$

Because the system is nonhomogeneous, if the determinant of the coefficient matrix is *not* equal to zero, a unique solution exists for the second derivatives along the characteristic curve C. It also can be shown that if the second-order derivatives exist, then derivatives of all orders exist along C as well, in which case they are smooth.

On the other hand, if the determinant of the coefficient matrix *is* equal to zero, then the solution is not unique, and the second derivatives are discontinuous along C. Setting the determinant equal to zero requires that

$$a\left(\frac{dy}{d\tau}\right)^2 - b\left(\frac{dx}{d\tau}\right)\left(\frac{dy}{d\tau}\right) + c\left(\frac{dx}{d\tau}\right)^2 = 0,$$

Table 7.1 Types of partial differential equations based on number of characteristics.

$b^2 - 4ac$	Real roots	Characteristics	Type of PDE
$b^2 - 4ac > 0$	2	2	Hyperbolic
$b^2 - 4ac = 0$	1	1	Parabolic
$b^2 - 4ac < 0$	None	None	Elliptic

or multiplying by $(d\tau/dx)^2$ yields

$$a\left(\frac{dy}{dx}\right)^2 - b\left(\frac{dy}{dx}\right) + c = 0.$$

This is a quadratic equation for dy/dx, which is the slope of the characteristic curve C in the (x, y) plane. Consequently, from the quadratic formula, the slope is

$$\frac{dy}{dx} = \frac{b \pm \sqrt{b^2 - 4ac}}{2a}. \tag{7.32}$$

The characteristic curves C of (7.27), for which $y(x)$ satisfies (7.32), are curves along which the second-order derivatives are discontinuous. Because the characteristics must be real, their behavior is determined by the sign of $b^2 - 4ac$ as summarized in Table 7.1.

Alternatively, note that if from (7.27), we let $x = x_1$, $y = x_2$, then $au_{11} + \frac{b}{2}u_{12} + \frac{b}{2}u_{21} + cu_{22} = H$, and we can define the matrix

$$\mathbf{A} = \begin{bmatrix} a & \frac{b}{2} \\ \frac{b}{2} & c \end{bmatrix}, \tag{7.33}$$

which is analogous to *quadratic forms* in Section 2.3.3. Taking the determinant then gives

$$|\mathbf{A}| = ac - \frac{b^2}{4},$$

or upon rearranging we have

$$-4|\mathbf{A}| = b^2 - 4ac \begin{cases} > 0, & \text{hyperbolic} \\ = 0, & \text{parabolic} \\ < 0, & \text{elliptic} \end{cases}.$$

Thus, the classification of a second-order partial differential equation can be determined from the determinant of $\mathbf{A}$ as defined by (7.33).

REMARKS:

1. *The terminology arises from classification of the second-degree algebraic equation $ax^2 + bxy + cy^2 + dx + ey + f = 0$, which are conic sections.*
2. *The classification only depends on a, b, and c, which are the coefficients of the highest-order derivatives.*

3. *It can be shown that the classification of a partial differential equation is independent of the coordinate system (see, for example, Tannehill et al. 1997). For example, if it is elliptic in Cartesian coordinates, then it is elliptic in all curvilinear coordinate systems.*

Let us consider each type of second-order partial differential equation to see how their respective solutions behave. In particular, take note of the nature of initial and boundary conditions along with the *domain of influence* (DoI) and *domain of dependence* (DoD) for each type of equation. Each type of equation will be represented by a canonical partial differential equation representative of its classification. In general, the coefficients a, b, and c in (7.32) can be functions of the coordinates x and y, in which case the characteristics C are curves in the two-dimensional domain. For the sake of simplicity, however, let us consider the case when they are constants, such that the characteristics are straight lines. The conclusions drawn for each type of equation are then naturally generalizable.

7.5.2 Hyperbolic Equations (Initial-Value Problems)

When $b^2 - 4ac > 0$ in (7.32), the partial differential equation is hyperbolic, and there are two real roots corresponding to the two characteristics. For a, b, and c constant, let these two real roots be given by

$$\frac{dy}{dx} = \lambda_1, \quad \frac{dy}{dx} = \lambda_2, \tag{7.34}$$

where λ_1 and λ_2 are real constants. Then we may integrate to obtain the straight lines

$$y = \lambda_1 x + \gamma_1, \quad y = \lambda_2 x + \gamma_2.$$

Therefore, the solution propagates along two linear characteristic curves.

For example, consider the wave equation

$$\frac{\partial^2 u}{\partial t^2} = \sigma^2 \frac{\partial^2 u}{\partial x^2},$$

where $u(x,t)$ is the amplitude of the wave, and σ is the wave speed within the medium. In this case, the independent variable y becomes time t, and $a = \sigma^2$, $b = 0$, and $c = -1$. Comparing with (7.32) and (7.34), we have

$$\lambda_1 = \frac{1}{\sigma}, \quad \lambda_2 = -\frac{1}{\sigma}.$$

Therefore, the characteristics of the wave equation with a, b, and c constant are straight lines with slopes $1/\sigma$ and $-1/\sigma$ as shown in Figure 7.8. Take particular notice of the domains of influence and dependence. The domain of influence of point P indicates the region within (x,t) space whose solution can influence that at P. The domain of dependence of point P indicates the region within (x,t) space whose solution depends on that at P.

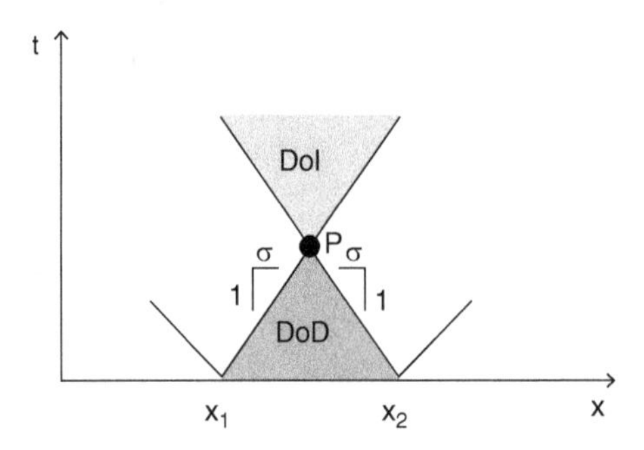

Figure 7.8 The characteristics of the wave equation indicating the DoI and DoD from disturbances at x_1 and x_2.

The solution of the wave equation with constant wave speed σ is of the form

$$u(x,t) = F_1(x + \sigma t) + F_2(x - \sigma t),$$

with F_1 being a right-moving traveling wave, and F_2 being a left-moving traveling wave. Initial conditions are required at say $t = 0$, such as

$$u(x,0) = f(x), \quad u_t(x,0) = g(x).$$

The first is an initial condition on the amplitude, and the second is on its velocity. Note that no boundary conditions are necessary unless there are boundaries at finite x.

7.5.3 Parabolic Equations (Initial-Value Problems)

When $b^2 - 4ac = 0$ in (7.32), the partial differential equation is parabolic, and there is one real root corresponding to a single characteristic. For a, b, and c constant, the single root is given by

$$\frac{dy}{dx} = \frac{b}{2a}. \tag{7.35}$$

Integrating yields

$$y = \frac{b}{2a}x + \gamma_1,$$

which is a straight line. Therefore, the solution propagates along one linear characteristic direction (usually time).

For example, consider the one-dimensional, unsteady diffusion equation

$$\frac{\partial u}{\partial t} = \alpha \frac{\partial^2 u}{\partial x^2},$$

which governs unsteady heat conduction, for example. Here, $u(x,t)$ is the quantity undergoing diffusion (for example, temperature), and α is the diffusivity of the medium. Again, the independent variable y becomes time t, and $a = \alpha$ and $b = c = 0$. Because $b = 0$ in (7.35), the characteristics are lines of constant t, corresponding to a solution that marches forward in time as illustrated in Figure 7.9. Observe that the DoI is every position and time prior to the current time, and the DoD is every position

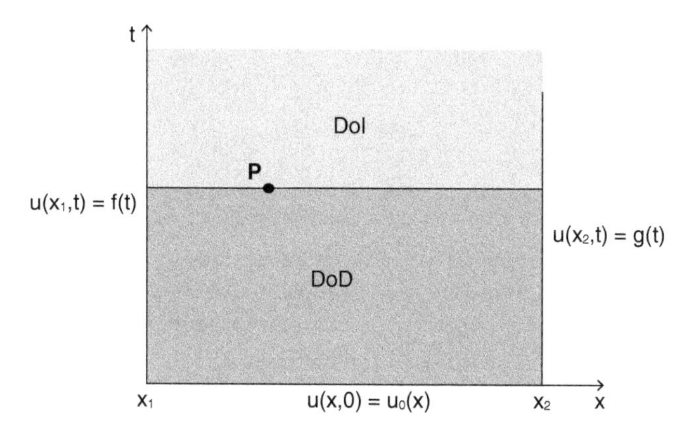

Figure 7.9 The characteristic of the unsteady diffusion equation indicating the DoI and DoD.

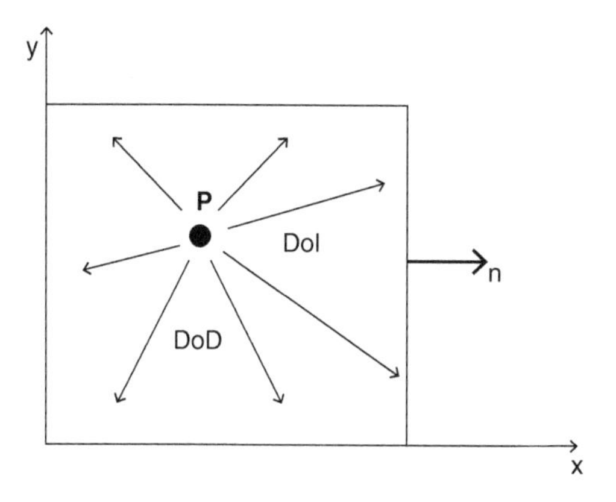

Figure 7.10 The DoI and DoD for an elliptic partial differential equation.

and time subsequent to the current time. Initial and boundary conditions are required, such as

$$u(x,0) = u_0(x), \quad u(x_1,t) = f(t), \quad u(x_2,t) = g(t),$$

which can be interpreted as the initial temperature throughout the domain and the temperatures at the boundaries in the context of unsteady heat conduction.

7.5.4 Elliptic Equations (Boundary-Value Problems)

When $b^2 - 4ac < 0$ in (7.32), the partial differential equation is elliptic, and there are no real roots or characteristics. Consequently, disturbances have infinite speed of propagation in all directions. In other words, a disturbance *anywhere* affects the solution *everywhere* instantaneously. As illustrated in Figure 7.10, the DoI and DoD are the entire domain.

For example, consider the Laplace equation

$$\frac{\partial^2 u}{\partial x^2} + \frac{\partial^2 u}{\partial y^2} = 0,$$

which governs steady heat conduction, potential fluid flow, electrostatic fields, et cetera. In this case, $a = 1, b = 0$, and $c = 1$. Because of the instantaneous propagation of the solution in all directions, elliptic equations require a global solution strategy, and the boundary conditions, which typically specify u or its normal derivative at the boundary, must be specified on a closed contour bounding the entire domain.

7.5.5 Mixed Equations

If $a(x,y)$, $b(x,y)$, and $c(x,y)$ are variable coefficients, then $b^2 - 4ac$ may change sign with space and/or time, in which case the character of the partial differential equation may be different in certain regions. For example, consider transonic fluid flow that occurs near the speed of sound of the medium. The governing equation for two-dimensional, steady, compressible, potential flow about a slender body is

$$\left(1 - M^2\right)\frac{\partial^2 \phi}{\partial x^2} + \frac{\partial^2 \phi}{\partial y^2} = 0,$$

where $\phi(x,y)$ is the velocity potential, and M is the local Mach number. Here, $a = 1 - M^2$, $b = 0$, and $c = 1$. To determine the nature of the equation, observe that

$$b^2 - 4ac = 0 - 4\left(1 - M^2\right)(1) = -4\left(1 - M^2\right);$$

therefore, the type of partial differential equation is determined by the Mach number in each region of the transonic flow according to Table 7.2. For example, transonic flow past an airfoil is illustrated in Figure 7.11. Observe that in this case, all three types of equations are present, each one exhibiting very different physical behavior. Thus, we have the same equation, but different behavior in various regions. In some cases, we have different equations in different regions of the domain owing to the local governing physics.

Table 7.2 Types of partial differential equations based on Mach number in transonic flow.

Mach number	$b^2 - 4ac$	Type of PDE
$M < 1$ (subsonic)	$b^2 - 4ac < 0$	Elliptic
$M = 1$ (sonic)	$b^2 - 4ac = 0$	Parabolic
$M > 1$ (supersonic)	$b^2 - 4ac > 0$	Hyperbolic

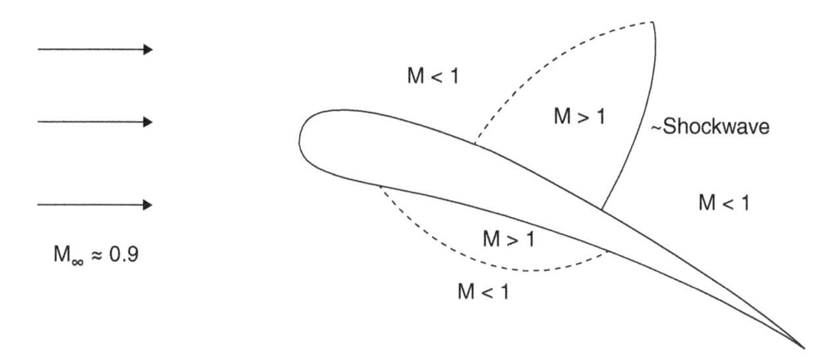

Figure 7.11 Transonic fluid flow past an airfoil with distinct regions where the equation is elliptic, parabolic, and hyperbolic.

Exercises

Unless stated otherwise, perform the exercises using hand calculations. A "user-defined function" is a standalone computer program to accomplish a task written in a programming language, such as Python, MATLAB, Mathematica, Fortran, or C++.

7.1 Write a user-defined function that approximates the second derivative of a function $u(x)$ at each point on a uniform grid for which discrete data are given. Use second-order accurate finite differences for all interior and end points. Test the user-defined function on the following data:

x	−1.0	−0.5	0.0	0.5	1.0	1.5	2.0
$u(x)$	−3.632	−0.3935	1.000	0.6487	−1.282	−4.518	−8.611

x	2.5	3.0	3.5	4.0	4.5
$u(x)$	−12.82	−15.91	−15.88	−9.402	9.017

7.2 Suppose the temperature data along the left portion of a 10 *cm* extended fin (the extended fin problem will be considered in Section 8.1) are as follows:

x (cm)	0.0	1.0	2.0	3.0	4.0	5.0
T (K)	473.0	446.3	422.6	401.2	382.0	364.3

The fin has constant cross-sectional area $A_c = 1.6 \times 10^{-5}\,\text{m}^2$ and thermal conductivity $k = 240\,\text{W/m}\cdot\text{K}$. Write a user-defined function to approximate the heat flux from the base of the fin at $x = 0$, which is given by

$$q_b = -kA_c \left(\frac{dT}{dx} \right)_{x=0},$$

using first-order, second-order, and third-order accurate forward difference approximations.

7.3 Obtain a second-order accurate backward-difference approximation for du/dx at x_i discretized on a uniform grid.

7.4 Obtain a third-order accurate forward-difference approximation for du/dx at x_i discretized on a uniform grid.

7.5 Using central differences, obtain a finite-difference approximation for

$$\nabla^4 u = \nabla^2\left(\nabla^2 u\right), \quad u = u(x)$$

at a typical grid point x_i, which only involves values of $u(x_i)$ at the grid points $x_{i-2}, x_{i-1}, x_i, x_{i+1}$ and x_{i+2}.

7.6 Use central differences to obtain a second-order accurate finite-difference approximation for

$$\frac{d}{dx}\left[\sigma(x)\frac{du}{dx}\right]$$

at a typical grid point x_i, which only involves values of $\sigma(x_i)$ and $u(x_i)$ at the grid points x_{i-1}, x_i, and x_{i+1} ($\sigma(x)$ is only known at the grid points).

7.7 Obtain a fourth-order accurate central-difference approximation for d^2u/dx^2 at a typical grid point x_i, which involves values of $u(x_i)$ at the grid points $x_{i-2}, x_{i-1}, x_i, x_{i+1}$, and x_{i+2}.

7.8 Obtain a central-difference approximation for d^3u/dx^3 at a typical grid point x_i, which involves values of $u(x_i)$ at the grid points $x_{i-2}, x_{i-1}, x_i, x_{i+1}$, and x_{i+2}. What is the truncation error of the approximation?

7.9 Consider the following second-order, linear partial differential equation:

$$C_1(x,y)\frac{\partial^2\phi}{\partial x^2} + C_2(x,y)\frac{\partial^2\phi}{\partial x\partial y} + C_3(x,y)\frac{\partial\phi}{\partial x} + C_4(x,y)\frac{\partial\phi}{\partial y} + C_5(x,y)\phi = F(x,y).$$

What condition(s) must be met in order for this equation to be hyperbolic?

7.10 Consider the following second-order, linear partial differential equation:

$$C_1(x,y)\frac{\partial^2\phi}{\partial x^2} + C_2(x,y)\frac{\partial\phi}{\partial x} + C_3(x,y)\frac{\partial^2\phi}{\partial y^2} + C_4(x,y)\frac{\partial\phi}{\partial y} + C_5(x,y)\phi = F(x,y).$$

What condition(s) must be met in order for this equation to be elliptic?

7.11 Consider the following second-order, linear partial differential equation:

$$C_1(x,y)\frac{\partial^2\phi}{\partial x^2} + C_2(x,y)\frac{\partial\phi}{\partial x} + C_3(x,y)\frac{\partial^2\phi}{\partial y^2} + C_4(x,y)\frac{\partial\phi}{\partial y} + C_5(x,y)\phi = F(x,y).$$

What condition(s) must be met in order for this equation to be parabolic? What is the direction(s) of propagation of the solution?

7.12 Consider the following second-order, linear partial differential equation:

$$\left[1 - C_1^2(x,y)\right]\left(\frac{\partial^2 u}{\partial x^2} + \frac{\partial u}{\partial x}\right) + C_2(x,y)\frac{\partial^2 u}{\partial y^2} + C_3(x,y)\frac{\partial u}{\partial y} + C_4(x,y)u = F(x,y),$$

where $C_1(x,y) > 0$, $C_2(x,y) > 0$, $C_3(x,y) > 0$, and $C_4(x,y) > 0$ are positive for all (x,y). Under what conditions is this equation elliptic, parabolic, or hyperbolic?

8 Finite-Difference Methods for Boundary-Value Problems

> I recommend this method to you for imitation. You will hardly ever again eliminate directly, at least not when you have more than two unknowns. The indirect procedure can be done while half asleep, or while thinking about other things. (Carl Friedrich Gauss[1])

In this chapter, attention is focused on obtaining numerical solutions to ordinary and partial differential equations in the form of boundary-value problems. In the case of partial differential equations, this corresponds to elliptic equations. We begin with ordinary boundary-value problems to illustrate the basic approaches to discretizing the differential equations and applying the boundary conditions. These approaches are then extended naturally to elliptic partial differential equations.

8.1 Illustrative Example from Heat Transfer

Let us begin by considering a one-dimensional model of the heat conduction in an extended fin, an array of which may be used to cool an electronic device, for example. This example will assist in further solidifying our understanding of the general numerical solution procedure as well as introduce a number of the issues in finite-difference methods, such as handling different types of boundary conditions. Figure 8.1 is a schematic of the extended fin with heat conduction in the fin and convection between the fin and ambient air.

8.1.1 Governing Equation

Step 1 of the numerical solution procedure consists of applying conservation of energy within the fin along with any simplifying assumptions. In this case, we assume that the heat transfer is one-dimensional – varying axially along the length of the fin. This is a good assumption for a fin with small cross-sectional area relative to its length and for which the cross-sectional area changes gradually along the length of the fin. It is further assumed that the convective heat transfer coefficient is a constant at every

[1] In a letter to his student, Christian Ludwig Gerling, dated December 26, 1823, describing the "indirect procedure," what we would now call an iterative relaxation – and apparently relaxing – method that would become known as the Gauss–Seidel method. By "eliminate directly," he is referring to his Gaussian elimination procedure.

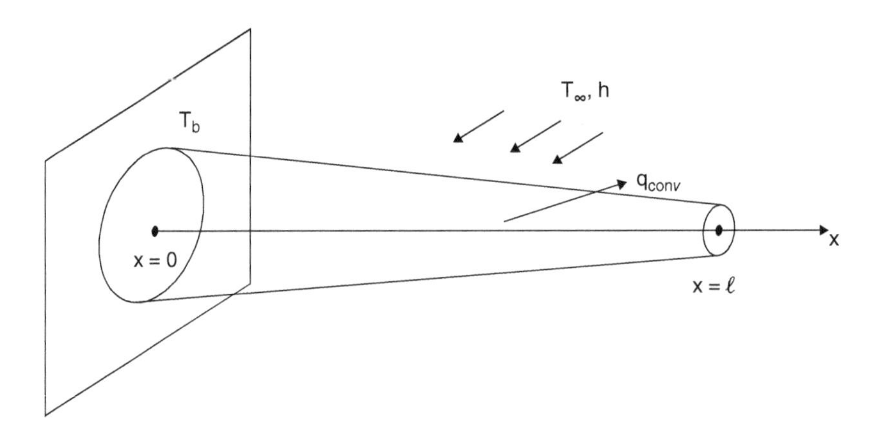

Figure 8.1 Schematic of the extended fin.

point on the surface of the fin. Based on these assumptions, the heat transfer within the extended fin is governed by the one-dimensional ordinary-differential equation (see, for example, Incropera et al. 2007)

$$\frac{d^2T}{dx^2} + \left(\frac{1}{A_c}\frac{dA_c}{dx}\right)\frac{dT}{dx} - \left(\frac{1}{A_c}\frac{h}{k}\frac{dA_s}{dx}\right)(T - T_\infty) = 0, \tag{8.1}$$

where $T(x)$ is the temperature distribution along the length of the fin, T_∞ is the ambient air temperature far away from the fin, $A_c(x)$ is the cross-sectional area, $A_s(x)$ is the cumulative surface area from the base, h is the convective heat transfer coefficient at the surface of the fin, and k is the thermal conductivity of the fin material. Observe that (8.1) is a second-order, linear ordinary differential equation with variable coefficients, and it is a *boundary-value problem* requiring boundary conditions at both ends of the domain.

Letting $u(x) = T(x) - T_\infty$, rewrite (8.1) as

$$\frac{d^2u}{dx^2} + f(x)\frac{du}{dx} + g(x)u = h(x), \tag{8.2}$$

where

$$f(x) = \frac{1}{A_c}\frac{dA_c}{dx}, \quad g(x) = -\frac{1}{A_c}\frac{h}{k}\frac{dA_s}{dx}, \quad h(x) = 0.$$

Because $h(x) = 0$, the differential equation is said to be *homogeneous*. For now, consider the case where we have a specified temperature at both the base and tip of the fin, such that

$$\begin{aligned} u &= u_b = T_b - T_\infty \quad \text{at} \quad x = 0, \\ u &= u_\ell = T_\ell - T_\infty \quad \text{at} \quad x = \ell. \end{aligned} \tag{8.3}$$

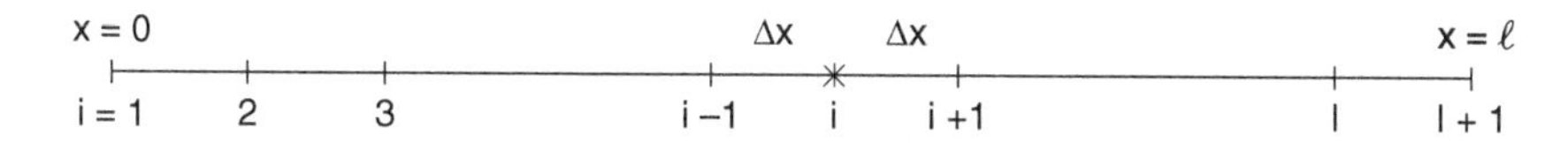

Figure 8.2 Schematic of the uniform, one-dimensional grid for the extended fin example.

These are called *Dirichlet* boundary conditions. Equation (8.2) with boundary conditions (8.3) represents the mathematical model – step 1 in the numerical solution procedure.

Step 2 consists of discretizing the domain and governing differential equation. Dividing the interval $0 \leq x \leq \ell$ into I equal subintervals of length $\Delta x = \ell/I$ gives the uniform grid illustrated in Figure 8.2. Here, $f_i = f(x_i)$, $g_i = g(x_i)$, and $h_i = h(x_i)$ are known at each grid point, and the solution $u_i = u(x_i)$ is to be determined for all interior points $i = 2, \ldots, I$.[2] Approximating the derivatives in (8.2) using second-order accurate central differences gives

$$\left(\frac{du}{dx}\right)_i = \frac{u_{i+1} - u_{i-1}}{2\Delta x} + O\left(\Delta x^2\right),$$

$$\left(\frac{d^2u}{dx^2}\right)_i = \frac{u_{i+1} - 2u_i + u_{i-1}}{(\Delta x)^2} + O\left(\Delta x^2\right).$$

Substituting into (8.2) yields

$$\frac{u_{i+1} - 2u_i + u_{i-1}}{\Delta x^2} + f_i \frac{u_{i+1} - u_{i-1}}{2\Delta x} + g_i u_i = h_i,$$

for the point x_i. Multiplying by $(\Delta x)^2$ and collecting terms leads to

$$a_i u_{i-1} + b_i u_i + c_i u_{i+1} = d_i, \quad i = 2, \ldots, I, \tag{8.4}$$

where this *difference equation* is applied at each interior point of the domain, and

$$a_i = 1 - \frac{1}{2}\Delta x f_i, \quad b_i = -2 + (\Delta x)^2 g_i, \quad c_i = 1 + \frac{1}{2}\Delta x f_i, \quad d_i = (\Delta x)^2 h_i. \tag{8.5}$$

Recall that the Dirichlet boundary conditions (8.3) are

$$u_1 = u_b, \quad u_{I+1} = u_\ell. \tag{8.6}$$

Observe that for $i = 2$, (8.4) becomes

$$a_2 \overset{u_b}{\cancel{u_1}} + b_2 u_2 + c_2 u_3 = d_2,$$

or

$$b_2 u_2 + c_2 u_3 = d_2 - a_2 u_b. \tag{8.7}$$

[2] We will begin our grid indices at one in agreement with the typical notation used mathematically for vectors and matrices as in Chapters 1 and 2; that is, $i = 1$ corresponds to $x = 0$. This is despite the fact that some programming languages begin array indices at zero by default. This is discussed further at the end of Section 8.4.1.

Similarly, for $i = I$, (8.4) becomes

$$a_I u_{I-1} + b_I u_I + c_I u_{I+1} = d_I,$$

or

$$a_I u_{I-1} + b_I u_I = d_I - c_I u_\ell. \tag{8.8}$$

Note that because we have discretized the differential equation at each interior grid point, we obtain a set of $(I-1)$ algebraic equations for the $(I-1)$ unknown values of the temperature $u_i, i = 2, \ldots, I$ at each interior grid point. In matrix form, this $(I-1) \times (I-1)$ system of algebraic equations $\mathbf{Au} = \mathbf{d}$ is given by

$$\begin{bmatrix} b_2 & c_2 & 0 & 0 & \cdots & 0 & 0 & 0 \\ a_3 & b_3 & c_3 & 0 & \cdots & 0 & 0 & 0 \\ 0 & a_4 & b_4 & c_4 & \cdots & 0 & 0 & 0 \\ \vdots & \vdots & \vdots & \vdots & \ddots & \vdots & \vdots & \vdots \\ 0 & 0 & 0 & 0 & \cdots & a_{I-1} & b_{I-1} & c_{I-1} \\ 0 & 0 & 0 & 0 & \cdots & 0 & a_I & b_I \end{bmatrix} \begin{bmatrix} u_2 \\ u_3 \\ u_4 \\ \vdots \\ u_{I-1} \\ u_I \end{bmatrix} = \begin{bmatrix} d_2 - a_2 u_b \\ d_3 \\ d_4 \\ \vdots \\ d_{I-1} \\ d_I - c_I u_\ell \end{bmatrix},$$

where we note that the right-hand-side coefficients have been adjusted to account for the known values of u at the boundaries. The coefficient matrix for the difference equation (8.4) is tridiagonal; therefore, the system of equations can be solved efficiently using the Thomas algorithm given in Section 6.3.3, which is step 3 in the numerical solution procedure.

REMARKS:

1. *As in* (8.4), *it is customary to write difference equations with the unknowns on the left-hand side and knowns on the right-hand side.*
2. *We multiply through by* $(\Delta x)^2$ *in* (8.4) *such that the resulting coefficients in the difference equation are* $O(1)$.
3. *To prevent ill conditioning, the tridiagonal system of equations must be diagonally dominant, such that*

$$|b_i| \geq |a_i| + |c_i|.$$

8.1.2 Convection Boundary Condition

In order to illustrate treatment of derivative boundary conditions, let us consider a more realistic convection boundary condition at the tip of the fin in place of the Dirichlet boundary condition, which assumes that we know the temperature there. The convection condition at the tip is given by

$$-k\frac{dT}{dx} = h(T - T_\infty) \quad \text{at} \quad x = \ell,$$

which is a balance between conduction (left side) and convection (right side). That is, the heat conducted out the tip of the fin must be convected away. Substituting $u(x) = T(x) - T_\infty$ gives

$$-k\left(\frac{du}{dx}\right)_{x=\ell}=hu(\ell).$$

Note that the convection condition results in specifying a linear combination of the temperature and its derivative at the tip. This is known as a *Robin*, or *mixed, boundary condition* and is of the general form

$$pu+q\frac{du}{dx}=r \quad \text{at} \quad x=\ell, \tag{8.9}$$

where for the convection condition $p=h$, $q=k$, and $r=0$.

To apply the boundary condition (8.9) at the tip of the fin, first write the general difference equation (8.4) at the right boundary, where $i=I+1$, as follows:

$$a_{I+1}u_I+b_{I+1}u_{I+1}+c_{I+1}u_{I+2}=d_{I+1}. \tag{8.10}$$

Observe that u_{I+2} is outside the domain as x_{I+1} corresponds to the boundary $x=\ell$; therefore, it must be eliminated from the preceding equation. To do so, let us approximate the boundary condition (8.9) at the tip of the fin as well using a second-order accurate, central difference for the derivative. This yields

$$pu_{I+1}+q\frac{u_{I+2}-u_I}{2\Delta x}=r,$$

which also contains the value u_{I+2} outside the domain. Hence, solving for this gives

$$u_{I+2}=u_I+\frac{2\Delta x}{q}\left(r-pu_{I+1}\right).$$

Substituting into (8.10) in order to eliminate the point outside the domain and collecting terms yields

$$q\left(a_{I+1}+c_{I+1}\right)u_I+\left(qb_{I+1}-2\Delta xpc_{I+1}\right)u_{I+1}=qd_{I+1}-2\Delta xrc_{I+1}.$$

This equation is appended to the end of the tridiagonal system of equations to allow for determination of the additional unknown u_{I+1}. Doing so adjusts the a_{I+1}, b_{I+1}, and d_{I+1} coefficients from those defined in (8.5).

REMARKS:

1. *The solution procedure is the same as that for Dirichlet boundary conditions except that there is one additional equation corresponding to the right boundary ($i=I+1$) in the tridiagonal system of equations.*
2. *A Neumann boundary condition is just a special case of the Robin (mixed) boundary condition considered here, in which $p=0$ and $q=1$.*

8.1.3 Heat Flux

Finally, once the solution for the temperature along the length of the fin has been obtained by solving the tridiagonal system of equation, the heat flux can be evaluated at the base of the fin. This is given by Fourier's law, which gives the rate of heat transfer into the base of the fin owing to conduction according to

$$q_b = -kA_c(0)\left(\frac{dT}{dx}\right)_{x=0} = -kA_c(0)\left(\frac{du}{dx}\right)_{x=0}.$$

In order to evaluate du/dx at the base $x = 0$, we must use a forward difference. From (7.4) applied at $i = 1$, we have the first-order accurate forward-difference approximation

$$\left(\frac{du}{dx}\right)_{x=0} = \frac{u_2 - u_1}{\Delta x} + O(\Delta x).$$

For a more accurate estimate, we may use the second-order accurate approximation from (7.10) applied at $i = 1$ as follows

$$\left(\frac{du}{dx}\right)_{x=0} = \frac{-3u_1 + 4u_2 - u_3}{2\Delta x} + O(\Delta x^2).$$

Even higher-order finite-difference estimates may be formed. For example, the third-order, forward-difference approximation from (7.14) is

$$\left(\frac{du}{dx}\right)_{x=0} = \frac{-11u_1 + 18u_2 - 9u_3 + 2u_4}{6\Delta x} + O(\Delta x^3).$$

Observe that each successive approximation requires one additional point in the interior of the domain.

8.2 General Second-Order Ordinary Differential Equation

Consider the general second-order boundary-value problem

$$\frac{d^2u}{dx^2} = F\left(x, u, \frac{du}{dx}\right), \quad a \le x \le b, \tag{8.11}$$

where $F\left(x, u, u'\right)$ may consist of a linear or nonlinear relationship involving the independent variable x, the dependent variable $u(x)$, and its first derivative. In the extended-fin equation, for example, F is a linear function of the dependent variable $u(x)$ and its first derivative $u'(x)$. A second-order boundary-value problem requires two boundary conditions, which may be of the form:

1. Dirichlet:

$$u(a) = A, \quad u(b) = B, \tag{8.12}$$

2. Neumann:

$$\left.\frac{du}{dx}\right|_{x=a} = A, \quad \left.\frac{du}{dx}\right|_{x=b} = B, \tag{8.13}$$

3. Robin (mixed):

$$p_a u(a) + q_a \left.\frac{du}{dx}\right|_{x=a} = r_a, \quad p_b u(b) + q_b \left.\frac{du}{dx}\right|_{x=b} = r_b, \tag{8.14}$$

where A, B, p_a, ... are specified constants.

Note that if the boundary-value problem (8.11) is nonlinear, then a system of nonlinear algebraic equations will result. Such a nonlinear system of algebraic equations can be solved using Newton's method in Section 12.2.4.

Example 8.1 Consider the linear, second-order boundary-value problem

$$\frac{d^2u}{dx^2} + 2\frac{du}{dx} + u = 0, \quad 0 \leq x \leq 1, \tag{8.15}$$

with the Dirichlet boundary conditions

$$u(0) = 1, \quad u(1) = 0. \tag{8.16}$$

For comparison with the numerical solution, the exact solution of this linear ordinary differential equation with constant coefficients is given by

$$u(x) = e^{-x}(1 - x). \tag{8.17}$$

Solution

In the general linear form (8.2), the coefficients are

$$f = 2, \quad g = 1, \quad h = 0.$$

Therefore, the tridiagonal system of equations $\mathbf{Au} = \mathbf{d}$ is given by

$$au_{i-1} + bu_i + cu_{i+1} = d, \quad i = 2, \ldots, I,$$

where

$$\begin{aligned} a &= 1 - \frac{1}{2}\Delta x f = 1 - \Delta x, \\ b &= -2 + (\Delta x)^2 g = -2 + (\Delta x)^2, \\ c &= 1 + \frac{1}{2}\Delta x f = 1 + \Delta x, \\ d &= (\Delta x)^2 h = 0. \end{aligned}$$

From the boundary conditions (8.16), the endpoints (8.6) are

$$u_1 = A = 1, \quad u_{I+1} = B = 0.$$

Thus, the d_i coefficients are adjusted as follows

$$d_2 = d - aA = d - a, \quad d_I = d - cB = d.$$

The numerical solution with $I = 20$ and the exact solution are shown Table 8.1 and Figure 8.3. Observe that the maximum error for problems with two Dirichlet boundary

Table 8.1 Comparison of exact and numerical solutions with $I = 20$ for Example 8.1.

i	x	Numerical solution	Exact solution	Absolute error
1	0.00	1.00000000	1.00000000	0.00000000
2	0.05	0.90362104	0.90366795	0.00004690
3	0.10	0.81426956	0.81435368	0.00008411
4	0.15	0.73148900	0.73160178	0.00011277
5	0.20	0.65485067	0.65498460	0.00013393
6	0.25	0.58395205	0.58410059	0.00014853
7	0.30	0.51841532	0.51857275	0.00015743
8	0.35	0.45788587	0.45804726	0.00016139
9	0.40	0.40203092	0.40219203	0.00016111
10	0.45	0.35053827	0.35069548	0.00015721
11	0.50	0.30311507	0.30326533	0.00015026
12	0.55	0.25948666	0.25962741	0.00014075
13	0.60	0.21939551	0.21952465	0.00012914
14	0.65	0.18260020	0.18271602	0.00011582
15	0.70	0.14887444	0.14897559	0.00010115
16	0.75	0.11800619	0.11809164	0.00008544
17	0.80	0.08979681	0.08986579	0.00006898
18	0.85	0.06406024	0.06411224	0.00005200
19	0.90	0.04062224	0.04065697	0.00003472
20	0.95	0.01931971	0.01933705	0.00001733
21	1.00	0.00000000	0.00000000	0.00000000

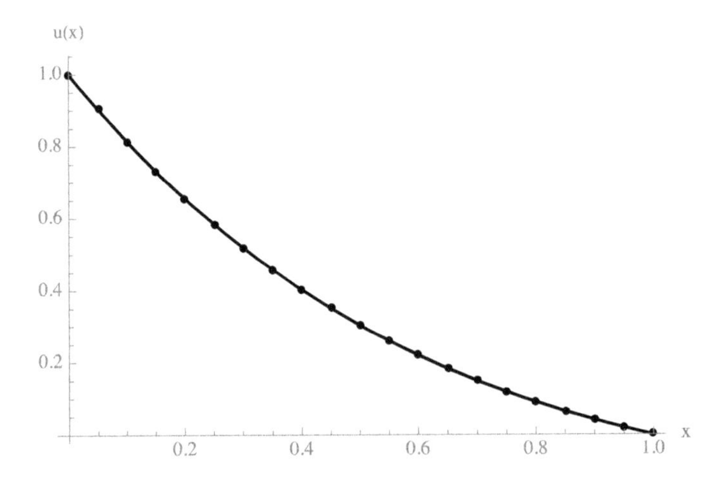

Figure 8.3 Comparison of exact (solid) and numerical (dots) solutions with $I = 20$ for Example 8.1.

Table 8.2 Influence of grid size on maximum error for Example 8.1.

I	$(\Delta x)^2$	Maximum error
5	0.040000	0.0026359542
10	0.010000	0.0006472965
20	0.002500	0.0001613921
40	0.000625	0.0000403906

conditions occurs in the interior of the domain. Table 8.2 illustrates the effect of the number of grid intervals I on the error. In particular, recall that the finite-difference approximations are all second-order accurate, that is, $O(\Delta x)^2$. Observe that each time the number of grid intervals is doubled, the maximum error is reduced by a factor of four. In other words, halving the grid size Δx by doubling the number of grid intervals I reduces the error by a factor of four owing to the second-order accuracy of the derivative approximations.

8.3 Partial Differential Equations

The ordinary differential equations considered thus far correspond to one-dimensional boundary-value problems in space. Let us now turn our attention to elliptic partial differential equations, which correspond to two- or three-dimensional boundary-value problems. The distinguishing feature of elliptic problems is that they have no preferred direction of propagation; therefore, they require a global solution strategy and boundary conditions on a closed contour surrounding the entire domain as illustrated in Figure 8.4.

We focus primarily on the Poisson equation throughout the remainder of the chapter as it is important in its own right in many applications, as well as being representative of elliptic partial differential equations. Therefore, numerical methods developed for the Poisson equation can be applied and extended to other elliptic partial differential equations as well.

8.3.1 Model Problem – Poisson Equation

Recall from Section 7.5 that the canonical second-order, elliptic partial differential equation is the *Laplace equation*, which in two-dimensional Cartesian coordinates is

$$\frac{\partial^2 u}{\partial x^2} + \frac{\partial^2 u}{\partial y^2} = 0. \tag{8.18}$$

The nonhomogeneous version of the Laplace equation,

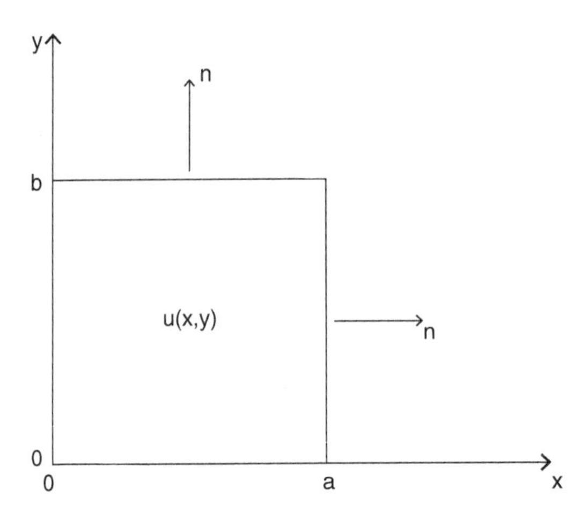

Figure 8.4 Schematic of a two-dimensional elliptic problem with boundary conditions around the entire closed domain.

$$\frac{\partial^2 u}{\partial x^2} + \frac{\partial^2 u}{\partial y^2} = f(x, y), \tag{8.19}$$

is called the *Poisson equation.*

The types of boundary conditions include *Dirichlet*, in which the values of u on the boundary are specified, *Neumann*, in which the normal derivative of u is specified on the boundary, and *Robin*, or *mixed*, in which a linear combination of u and its normal derivative is specified along the boundary. In the context of heat transfer, for example, a Dirichlet condition corresponds to an isothermal boundary condition, a Neumann condition corresponds to a specified heat flux, and a Robin condition arises from a convection condition at the boundary as exemplified in Section 8.1.2. Combinations of the preceding boundary conditions may be applied on different portions of the boundary as long as some boundary condition is applied at every point along the boundary contour.

An elliptic problem is linear if it consists of a linear equation and boundary conditions, whereas a nonlinear problem is such that either the equation and/or boundary conditions are nonlinear. For example, the Laplace or Poisson equation with Dirichlet, Neumann, or Robin boundary conditions is linear. An example of a nonlinear boundary condition that would render a problem nonlinear is a radiation boundary condition, such as

$$\frac{\partial u}{\partial n} = D(u^4 - u_{sur}^4),$$

where n represents the outward facing normal to the boundary. The nonlinearity arises owing to the fourth power on the temperature – even through the governing Laplace equation remains linear.

In order to define the computational grid, the x domain $0 \leq x \leq a$ is divided into I equal subintervals of length Δx, with $x_i = \Delta x(i - 1)$. Likewise, the y domain

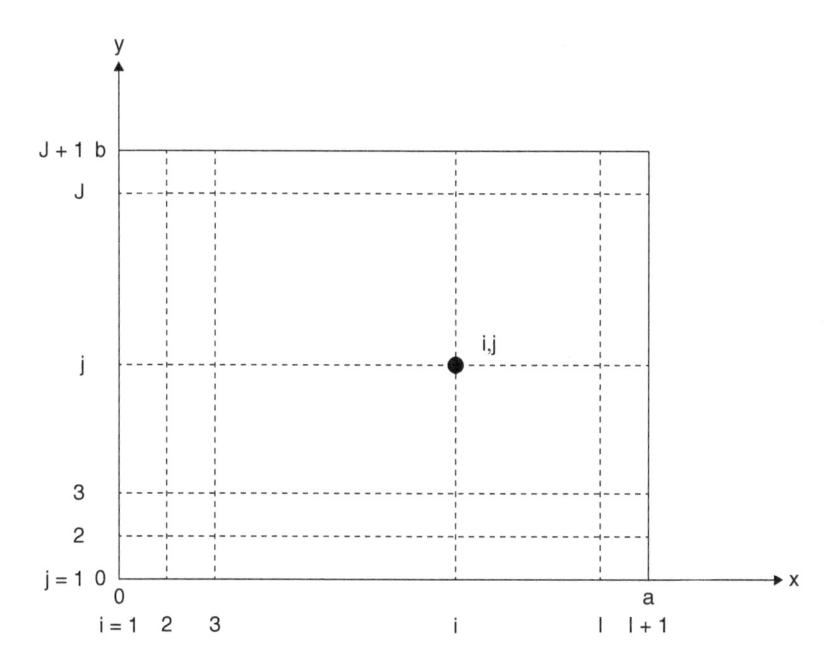

Figure 8.5 Schematic of the finite-difference grid for the Poisson equation in a two-dimensional rectangular domain.

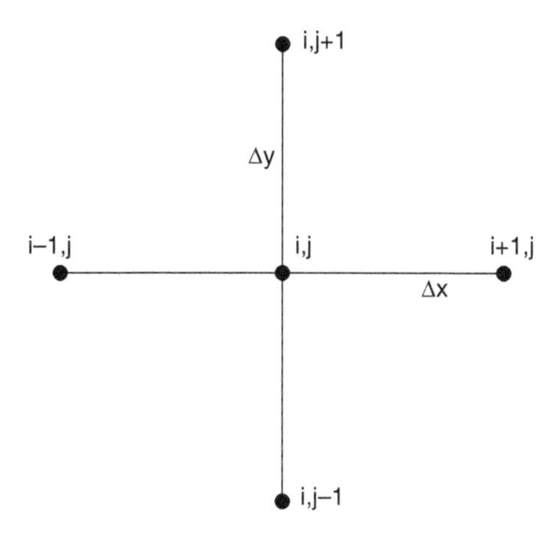

Figure 8.6 Finite-difference stencil for second-order accurate approximations.

$0 \leq y \leq b$ is divided into J equal subintervals of length Δy, with $y_j = \Delta y(j-1)$. See Figure 8.5 and note that the domain has been discretized into a two-dimensional grid intersecting at $(I+1) \times (J+1)$ points, including along the boundaries.

Consider second-order accurate, central difference approximations to the derivatives in the Poisson equation (8.19) at a typical point (i, j); the five-point *finite-difference stencil* is shown in Figure 8.6. With $u_{i,j} = u(x_i, y_j)$, the second-order accurate, central difference approximations are given by

$$\frac{\partial^2 u}{\partial x^2} = \frac{u_{i+1,j} - 2u_{i,j} + u_{i-1,j}}{\Delta x^2} + O(\Delta x^2),$$

$$\frac{\partial^2 u}{\partial y^2} = \frac{u_{i,j+1} - 2u_{i,j} + u_{i,j-1}}{\Delta y^2} + O(\Delta y^2).$$

Substituting into the Poisson equation (8.19), multiplying by $(\Delta x)^2$, and collecting terms gives the final form of the finite-difference equation

$$u_{i+1,j} - 2\left[1 + \left(\frac{\Delta x}{\Delta y}\right)^2\right] u_{i,j} + u_{i-1,j} + \left(\frac{\Delta x}{\Delta y}\right)^2 \left(u_{i,j+1} + u_{i,j-1}\right) = \Delta x^2 f_{i,j}. \tag{8.20}$$

As is typical, we collect the unknowns on the left-hand side and the knowns on the right-hand side of the finite-difference equation. Multiplying through by $(\Delta x)^2$ ensures that the coefficients of each term on the left-hand side are $O(1)$.

When applied for $i = 1, \ldots, I+1$ and $j = 1, \ldots, J+1$, the difference equation (8.20) results in a system of $(I+1) \times (J+1)$ linear algebraic equations for the $(I+1) \times (J+1)$ unknowns at each grid point. To see this system of equations in the usual matrix form $\mathbf{Au} = \mathbf{d}$, it is necessary to "stack" the two-dimensional variables $u_{i,j}$ to form a single vector; this will be illustrated in Section 8.4.1. Once in this form, the full arsenal of direct and iterative methods from Chapter 6, such as LU decomposition, Gauss–Seidel, and SOR, are available to solve the typically large system of linear algebraic equations. Alternatively, it is often more straightforward to devise direct and iterative methods that can be applied directly to the difference equation (8.20) in two- and three-dimensional systems, without first converting it to a traditional matrix problem. This latter approach will be our primary focus throughout the remainder of the chapter.

REMARK:

1. *Solutions to the Laplace and Poisson equations with Neumann boundary conditions on the entire boundary can only be determined relative to an unknown constant, that is, $u(x, y) + c$ also is a solution. We will need to take this into account when devising numerical algorithms.*

8.3.2 Overview of Direct and Iterative Methods

As outlined in Chapter 6, there are two classes of techniques for solving systems of linear algebraic equations:

1. Direct methods
   - A set procedure with a predefined number of steps that leads to the solution of the system of algebraic equations, such as Gaussian elimination and LU decomposition.
   - Incurs discretization errors, but no iterative convergence errors.

  - Efficient for linear systems with certain structures, such as tridiagonal and block tridiagonal.
  - Become less efficient for large, dense systems of equations, which typically require $O(N^3)$ operations.
  - Typically cannot adapt to nonlinear problems.
  - Algorithms are often difficult to implement on parallel computer architectures with large numbers of CPUs and/or cores (see Section 9.13).
2. Iterative (relaxation) methods
  - Beginning with an initial guess, an iterative process is repeated that – hopefully – results in successively closer approximations to the exact solution of the system of algebraic equations.
  - Introduces iterative convergence errors in addition to discretization errors.
  - Generally more efficient for large, sparse systems of equations. In particular, they allow us to get closer to $O(N^2)$ operations, such that it scales with the total number of points as for the Thomas algorithm.
  - Can apply to linear and nonlinear problems.
  - Algorithms are generally more amenable to implementation on parallel computer architectures.

The "best" direct method for solving the system $\mathbf{Au} = \mathbf{d}$ depends upon the form of the coefficient matrix **A**. If the matrix is full and dense, then LU decomposition is likely the best approach (or Cholesky decomposition for symmetric positive definite matrices); however, this is very expensive computationally. If the matrix is sparse and structured, for example, banded with only a small number of nonzero diagonals, then it can often be solved using more efficient algorithms. For example, tridiagonal systems may be solved very effectively using the Thomas algorithm. There are two additional direct methods in widespread use. Fast Fourier transform (FFT) and cyclic reduction are generally the fastest methods for problems in which they apply, such as block tridiagonal matrices. As such, these methods are often referred to as *fast Poisson solvers*.

Although these direct methods remain in general use in many applications, very large systems involving thousands or millions of unknowns demand methods that display improved scaling properties and have more modest storage requirements. This brings us to iterative methods as introduced in Section 6.4. In order to gain something – improved scalability in this case – there is almost always a price to pay. In the case of iterative methods, the price is that we will no longer obtain the exact solution of the algebraic systems of equations that results from our numerical discretization. Instead, we will carry out the iteration process until an approximate solution is obtained having an acceptably small amount of iterative convergence error. As in Section 6.4, we will begin with a discussion of the Jacobi, Gauss–Seidel, and SOR methods. This will be followed by the alternating-direction-implicit (ADI) method, which improves even more on scalability properties. Finally, the multigrid framework will be described, which accelerates the underlying iteration technique, such as Gauss–Seidel or ADI, and results in the fastest and most flexible algorithms currently available. As the number of unknowns increases in our numerical problems, and as the computer architectures on which we solve them become increasingly parallel, iterative methods are gaining in prominence over direct methods.

8.4 Direct Methods for Linear Systems

Before describing the iterative methods that are our primary focus owing to their improved scalability, let us see how we could solve the system of linear algebraic equations resulting from the discretized Poisson equation using standard direct methods and briefly describe two additional direct methods that are particularly efficient – when they apply. As mentioned previously, there are two basic strategies for solving two- and three-dimensional discretized problems. The system of equations can be converted into the usual matrix form $\mathbf{Au} = \mathbf{d}$ and solved using standard methods; alternatively, direct and iterative methods can be devised to directly treat the two- or three-dimensional form. Let us consider the first option in this section, and the second option will be pursued throughout the remainder of the chapter.

8.4.1 Conversion to Standard Matrix Form

Repeating the difference equation (8.20) for the Poisson equation using second-order accurate central differences, we have

$$u_{i+1,j} - 2(1+\hat{\Delta})u_{i,j} + u_{i-1,j} + \hat{\Delta}\left(u_{i,j+1} + u_{i,j-1}\right) = \Delta x^2 f_{i,j}, \qquad (8.21)$$

where $\hat{\Delta} = (\Delta x/\Delta y)^2$, and $i = 1, \dots, I+1; j = 1, \dots, J+1$. In order to write the system of difference equations in a convenient matrix form $\mathbf{Au} = \mathbf{d}$, where the solution is in vector form, let us renumber the two-dimensional mesh[3] (i, j) into a one-dimensional array (n) so that $\mathbf{u}$ is a vector rather than a matrix as illustrated in Figure 8.7. Thus, the relationship between the one-dimensional and two-dimensional indices is

$$n = (i-1)(J+1) + j,$$

where $i = 1, \dots, I+1$, $j = 1, \dots, J+1$, and $n = 1, \dots, (I+1)(J+1)$. Therefore, our five-point finite-difference stencil becomes that shown in Figure 8.8, and the finite-difference equation (8.21) for the Poisson equation becomes (with $\Delta = \Delta x = \Delta y$; therefore, $\hat{\Delta} = 1$)

$$u_{n+J+1} + u_{n-(J+1)} + u_{n+1} + u_{n-1} - 4u_n = \Delta^2 f_n, \qquad (8.22)$$

where $n = 1, \dots, (I+1)(J+1)$.

[3] The terms *grid* and *mesh* are used interchangeably.

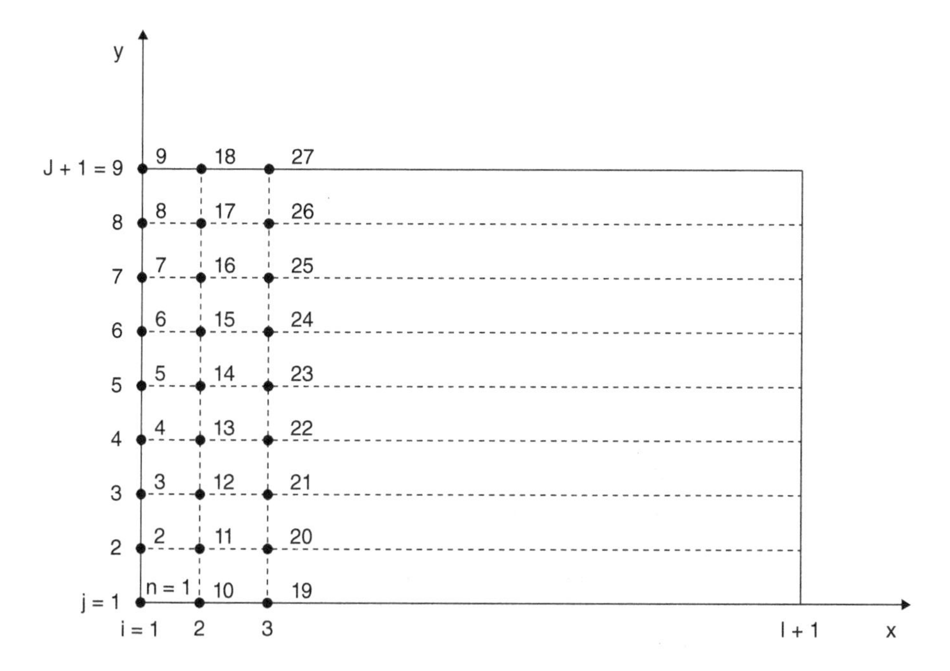

Figure 8.7 Conversion of the two-dimensional indices (i, j) into the one-dimensional index n with $J = 8$.

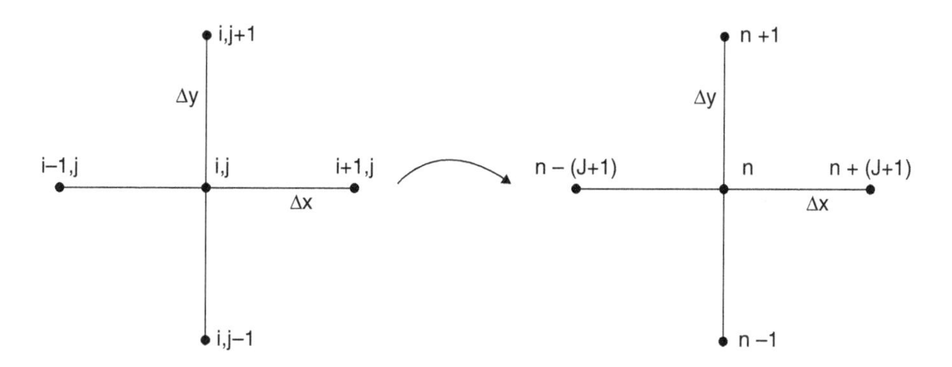

Figure 8.8 Finite-difference stencil with two- and one-dimensional indices.

Thus, the system of equations $\mathbf{Au} = \mathbf{d}$ is of the form

$$\begin{bmatrix} \mathbf{D} & \mathbf{I} & \mathbf{0} & \cdots & \mathbf{0} & \mathbf{0} \\ \mathbf{I} & \mathbf{D} & \mathbf{I} & \cdots & \mathbf{0} & \mathbf{0} \\ \mathbf{0} & \mathbf{I} & \mathbf{D} & \cdots & \mathbf{0} & \mathbf{0} \\ \vdots & \vdots & \vdots & \ddots & \vdots & \vdots \\ \mathbf{0} & \mathbf{0} & \mathbf{0} & \cdots & \mathbf{D} & \mathbf{I} \\ \mathbf{0} & \mathbf{0} & \mathbf{0} & \cdots & \mathbf{I} & \mathbf{D} \end{bmatrix} \begin{bmatrix} u_1 \\ u_2 \\ u_3 \\ \vdots \\ u_{(I+1)(J+1)-1} \\ u_{(I+1)(J+1)} \end{bmatrix} = \begin{bmatrix} d_1 \\ d_2 \\ d_3 \\ \vdots \\ d_{(I+1)(J+1)-1} \\ d_{(I+1)(J+1)} \end{bmatrix}. \tag{8.23}$$

The coefficient matrix $\mathbf{A}$ is a block tridiagonal matrix with $(I+1)$ blocks in both directions, where each block is a $(J+1)\times(J+1)$ matrix, and $\mathbf{d}$ contains all known information from the boundary conditions. Recall that $\mathbf{0}$ is a zero matrix block, $\mathbf{I}$ is an identity matrix block, and the tridiagonal blocks are

$$\mathbf{D} = \begin{bmatrix} -4 & 1 & 0 & \cdots & 0 & 0 \\ 1 & -4 & 1 & \cdots & 0 & 0 \\ 0 & 1 & -4 & \cdots & 0 & 0 \\ \vdots & \vdots & \vdots & \ddots & \vdots & \vdots \\ 0 & 0 & 0 & \cdots & -4 & 1 \\ 0 & 0 & 0 & \cdots & 1 & -4 \end{bmatrix}.$$

The linear system of algebraic equations (8.23) is sparse and structured. That is, most of the elements in the coefficient matrix $\mathbf{A}$ are zero, but the nonzero elements have a banded structure. Any of the general direct methods, such as LU decomposition, could be used to solve the block tridiagonal system of equations, and they can be adapted to the banded structure of the system similar to the way that Gaussian elimination is adapted to tridiagonal systems in the Thomas algorithm.

In the preceding approach, a finite-difference equation on a two- or three- dimensional computational grid is recast as a single system of linear algebraic equations, where the solution at each grid point is stacked to form a solution vector in our usual matrix form $\mathbf{Au} = \mathbf{d}$. Alternatively, more efficient and conceptually simpler algorithms can be devised by treating the two- or three-dimensional difference equation directly. Rather than stacking the solution $u_{i,j}$ into a single solution vector u_n, the solution is maintained as a matrix, with each element in the matrix being the approximate solution at a single grid point in the domain. In this way, the matrix of solution values directly represents the solution on the structured grid. This is the basis for the methods to be investigated throughout the remainder of the chapter. This includes direct methods, such as Fourier transform methods and cyclic reduction, as well as iterative methods, such as Gauss–Seidel, alternating-direction-implicit, and multigrid methods.

At this point, it is helpful to consider briefly how these methods would be implemented in a typical programming language. In programming languages, such as Python, C++, and Fortran, vectors and matrices are stored as *arrays*. A vector is equivalent to a one-dimensional array, and a matrix is a two-dimensional array. For the most part, this amounts to simple semantics. However, there are two issues that arise when translating between the mathematical constructs of vectors and matrices and the programming construct of arrays. First, while vectors and matrices are always indexed starting with one, some programming languages start indexing arrays at zero by default. Fortran, MATLAB, and Mathematica, for example, index arrays from one by default, while Python and C++ start with zero.[4] The second issue has to do with

[4] You will notice that those programming languages that index arrays starting at one were developed for mathematicians, scientists, and engineers, while those that do not were developed for a broader computer science audience.

how the array is superimposed on the domain. Recall that a matrix has its "origin," that is, the $(1,1)$ location, in the upper left-hand corner of the matrix. In all of our schematics, however, we imagine the "origin" of the domain and corresponding two-dimensional grid array being in the lower left-hand corner in order to correspond to the physical geometry of the domain. This does not have any bearing on performing the calculations, but it is necessary to keep in mind when plotting and interpreting results. Finally, storing the solution in arrays has significant advantages when extending the methods to three dimensions. Whereas a two-dimensional solution can be imagined to directly correspond to a matrix, there is no analogous construct for three-dimensional solutions. However, arrays can have any number of dimensions. Therefore, a three-dimensional solution $u_{i,j,k}$ can be stored in a three-dimensional array as a direct extension of the two-dimensional case, and numerical methods extend naturally to three dimensions. Moreover, expressing the discretized solution in a two- or three-dimensional array allows for it to directly correspond to the computational grid, such that there is no confusion with regard to the orientation of the results with respect to the domain.

8.4.2 Cyclic Reduction

The two additional direct methods are based on discrete Fourier transforms and reducing the problem to the solution of a recursive series of tridiagonal solutions. Although very efficient, Fourier transforms can only be used for partial differential equations with constant coefficients in the direction(s) for which the Fourier transform is applied. Because Fourier transform methods do not draw on matrix methods, they are not discussed further here; the reader is referred to Press et al. (2007) for details on Fourier transform methods and the FFT at its heart. Such methods are very common in spectral, or frequency, analysis in image and signal processing. It is important to understand that Fourier transforms are applied to solve the *difference* equation, not the *differential* equation; therefore, they are *not* a spectral numerical method (see Section 7.3).

Let us consider cyclic reduction, which is somewhat more general than Fourier transform methods. It is based on converting the multidimensional problem into a series of simple tridiagonal solves. To see how this is done, consider the Poisson equation

$$\frac{\partial^2 u}{\partial x^2} + \frac{\partial^2 u}{\partial y^2} = f(x,y)$$

discretized on a two-dimensional grid with $\Delta = \Delta x = \Delta y$ and $i = 0, \dots, I; j = 0, \dots, J$, where $I = 2^m$ with integer m. This results in $(I+1) \times (J+1)$ points as shown in Figure 8.9.

Applying central differences to the Poisson equation, the difference equation for constant x-lines becomes

$$\mathbf{u}_{i-1} - 2\mathbf{u}_i + \mathbf{u}_{i+1} + \hat{\mathbf{B}}\mathbf{u}_i = \Delta^2 \mathbf{f}_i, \quad i = 0, \dots, I, \tag{8.24}$$

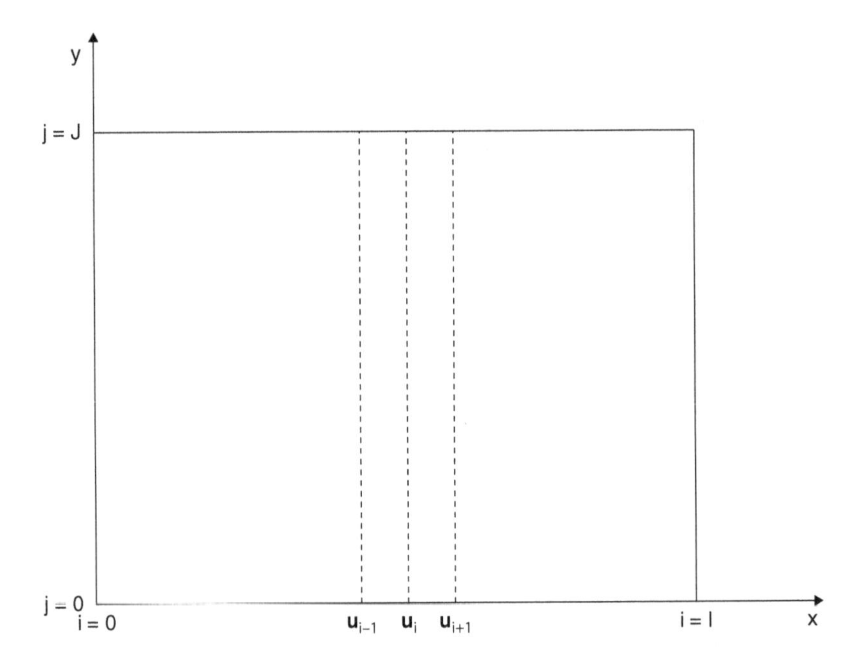

Figure 8.9 Two-dimensional grid for cyclic reduction.

where

$$\mathbf{u}_i = \begin{bmatrix} u_{i,0} \\ u_{i,1} \\ u_{i,2} \\ \vdots \\ u_{i,J-1} \\ u_{i,J} \end{bmatrix}, \quad \mathbf{f}_i = \begin{bmatrix} f_{i,0} \\ f_{i,1} \\ f_{i,2} \\ \vdots \\ f_{i,J-1} \\ f_{i,J} \end{bmatrix}, \quad \hat{\mathbf{B}} = \begin{bmatrix} -2 & 1 & 0 & \cdots & 0 & 0 \\ 1 & -2 & 1 & \cdots & 0 & 0 \\ 0 & 1 & -2 & \cdots & 0 & 0 \\ \vdots & \vdots & \vdots & \ddots & \vdots & \vdots \\ 0 & 0 & 0 & \cdots & -2 & 1 \\ 0 & 0 & 0 & \cdots & 1 & -2 \end{bmatrix}.$$

The first three terms in (8.24) correspond to the central difference approximation to the second derivative in the x-direction, and the fourth term corresponds to the central difference in the y-direction (see $\hat{\mathbf{B}}$). Taking $\mathbf{B} = -2\mathbf{I} + \hat{\mathbf{B}}$, where $\mathbf{I}$ is the identity matrix, (8.24) becomes

$$\mathbf{u}_{i-1} + \mathbf{B}\mathbf{u}_i + \mathbf{u}_{i+1} = \Delta^2 \mathbf{f}_i, \quad i = 0, \dots, I, \tag{8.25}$$

where the $(J+1) \times (J+1)$ matrix $\mathbf{B}$ is

$$\mathbf{B} = \begin{bmatrix} -4 & 1 & 0 & \cdots & 0 & 0 \\ 1 & -4 & 1 & \cdots & 0 & 0 \\ 0 & 1 & -4 & \cdots & 0 & 0 \\ \vdots & \vdots & \vdots & \ddots & \vdots & \vdots \\ 0 & 0 & 0 & \cdots & -4 & 1 \\ 0 & 0 & 0 & \cdots & 1 & -4 \end{bmatrix}.$$

Note that (8.25) corresponds to the block tridiagonal form (8.23) for (8.22), where $\mathbf{B}$ is the tridiagonal portion and the coefficients of the $\mathbf{u}_{i-1}$ and $\mathbf{u}_{u+1}$ terms are the "fringes." Writing three successive equations of (8.25) for $i-1$, i, and $i+1$ gives

$$\begin{aligned}\mathbf{u}_{i-2} + \mathbf{B}\mathbf{u}_{i-1} + \mathbf{u}_i &= \Delta^2 \mathbf{f}_{i-1},\\ \mathbf{u}_{i-1} + \mathbf{B}\mathbf{u}_i + \mathbf{u}_{i+1} &= \Delta^2 \mathbf{f}_i,\\ \mathbf{u}_i + \mathbf{B}\mathbf{u}_{i+1} + \mathbf{u}_{i+2} &= \Delta^2 \mathbf{f}_{i+1},\end{aligned}$$

respectively. Multiplying $-\mathbf{B}$ times the middle equation and adding all three gives

$$\mathbf{u}_{i-2} + \mathbf{B}^*\mathbf{u}_i + \mathbf{u}_{i+2} = \Delta^2 \mathbf{f}_i^*, \tag{8.26}$$

where

$$\begin{aligned}\mathbf{B}^* &= 2\mathbf{I} - \mathbf{B}^2,\\ \mathbf{f}_i^* &= \mathbf{f}_{i-1} - \mathbf{B}\mathbf{f}_i + \mathbf{f}_{i+1}.\end{aligned}$$

This is an equation of the same form as (8.25); therefore, applying this procedure to all even-numbered i equations in (8.25) reduces the number of equations by a factor of two. This *cyclic reduction* procedure can be repeated recursively until a single equation remains for the middle line of variables $\mathbf{u}_{I/2}$, which is tridiagonal. Using the solution for $\mathbf{u}_{I/2}$, solutions for all other i are obtained by successively solving the tridiagonal problems at each level in reverse as illustrated in Figure 8.10. This results in a total of I tridiagonal problems to obtain $\mathbf{u}_i$, $i = 0, \ldots, I$. The recursive nature of the cyclic reduction algorithm in forming the tridiagonal problems is why the number of grid points in the x-direction must be such that $I = 2^m$, with integer m. That is, the number of grid points in the direction for which cyclic reduction is applied must be an integer power of two in order to accommodate the recursive nature of the algorithm.

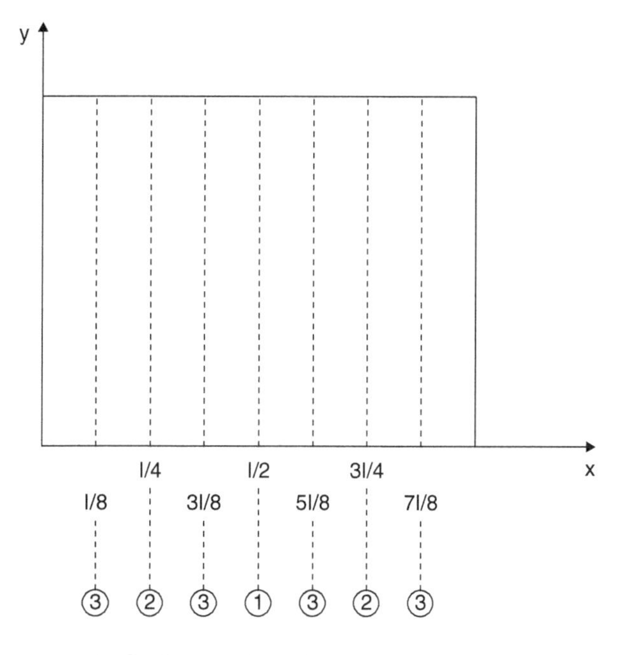

Figure 8.10 Cyclic reduction pattern.

Whereas the Fourier transform method can only be applied to linear differential equations with constant coefficients, cyclic reduction may be applied to linear equations with variable coefficients. With Fourier transform methods being slightly faster than cyclic reduction, one can use Fourier transform methods in directions with constant coefficients and cyclic reduction in directions with variable coefficients to produce the fastest direct methods for elliptic partial differential equations. These so-called *fast Poisson solvers* have $O(N \log N)$ operations, where $N = I \times J$ in the two-dimensional case, and $O(N) \ll O(N \log N) \ll O(N^2)$.

8.5 Iterative (Relaxation) Methods

Once again, direct methods have the advantage that no iterative convergence errors enter the numerical solution; however, they generally do not scale well to a system having very large numbers of unknowns, and they are difficult to adapt to nonlinear partial differential equations. These weaknesses are addressed using iterative methods but at the cost of additional errors owing to the iterative process. Here, we proceed through the Jacobi, Gauss–Seidel, and SOR methods introduced in Chapter 6, but now adapted to two-dimensional problems, with extension to three-dimensional problems being straightforward. These methods are further improved upon using techniques that are specific to multidimensional problems.

Returning to the difference equation (8.20) for the Poisson equation (8.19), recall that we have

$$u_{i+1,j} - 2\left(1 + \hat{\Delta}\right) u_{i,j} + u_{i-1,j} + \hat{\Delta}\left(u_{i,j+1} + u_{i,j-1}\right) = \Delta x^2 f_{i,j}, \tag{8.27}$$

where $\hat{\Delta} = (\Delta x/\Delta y)^2$. Iterative methods consist of beginning with an initial guess and iteratively "relaxing" equation (8.27) at each grid point until convergence. Note that in the subsequent discussion of each method, reference to the "model problem" consists of solving the Poisson equation (8.27) with Dirichlet boundary conditions using second-order accurate central differences with $\Delta x = \Delta y$ ($\hat{\Delta} = 1$) and $I = J$.

8.5.1 Jacobi Method

Recall from Section 6.4.4 that the Jacobi method simply consists of solving the difference equation as an explicit equation for $u_{i,j}$ at each grid point in the domain. The values of $u_{i,j}$ on the right-hand side are then taken from the previous iteration, denoted by superscript (n), while the single value on the left-hand side is taken at the current iteration, denoted by superscript $(n+1)$. Solving (8.27) for $u_{i,j}$ yields

$$u_{i,j}^{(n+1)} = \frac{1}{2\left(1 + \hat{\Delta}\right)} \left[u_{i+1,j}^{(n)} + u_{i-1,j}^{(n)} + \hat{\Delta}\left(u_{i,j+1}^{(n)} + u_{i,j-1}^{(n)}\right) - \Delta x^2 f_{i,j}\right]. \tag{8.28}$$

The procedure is as follows:

1. Provide an initial guess $u_{i,j}^{(0)}$ for $u_{i,j}$ at each point $i = 1, \ldots, I+1, j = 1, \ldots, J+1$.
2. Iterate (relax) by applying (8.28) at each grid point to produce successive approximations:
$$u_{i,j}^{(1)}, u_{i,j}^{(2)}, \ldots, u_{i,j}^{(n)}, \ldots.$$
3. Continue until the iteration process converges, which is determined, for example, by the requirement that
$$\frac{\max\left(\left|u_{i,j}^{(n+1)} - u_{i,j}^{(n)}\right|\right)}{\max\left(\left|u_{i,j}^{(n+1)}\right|\right)} < \epsilon,$$
where ϵ is a small tolerance value set by the user (see Section 6.4.3 for a description of this convergence test).

Recall from Section 6.4.2 that for an iterative method to converge, the spectral radius ρ of the iteration matrix $\mathbf{M}$ must be less than one. In addition, a smaller spectral radius results in more rapid iterative convergence, that is, in fewer iterations. It can be shown that for (8.28) with $\hat{\Delta} = 1$ $(\Delta x = \Delta y)$ and Dirichlet boundary conditions that
$$\rho_{Jac}(I, J) = \frac{1}{2}\left[\cos\left(\frac{\pi}{I+1}\right) + \cos\left(\frac{\pi}{J+1}\right)\right].$$
If $J = I$ and I is large, then from the Taylor series for cosine
$$\rho_{Jac}(I) = \cos\left(\frac{\pi}{I+1}\right) = 1 - \frac{1}{2}\left(\frac{\pi}{I+1}\right)^2 + \cdots < 1; \tag{8.29}$$
therefore, as $I \to \infty$, $\rho_{Jac}(I) \to 1$. As a result, we observe slower iterative convergence as the number of grid points I is increased. In other words, there is a disproportionate increase in computational time as I is increased. Because it is too slow, the Jacobi method is not used in practice; however, it is commonly used as a basis for comparison with other methods to follow.

In addition to being very slow to converge, the Jacobi method requires that both $u_{i,j}^{(n+1)}$ and $u_{i,j}^{(n)}$ be stored for all $i = 1, \ldots, I+1, j = 1, \ldots, J+1$. Consequently, it is necessary to store two values for each grid point, thereby doubling the storage requirement for a given problem as compared to direct methods, for example.

8.5.2 Gauss–Seidel Method

Recall that the Gauss–Seidel method improves on the Jacobi method by using the most recently updated information. This not only eliminates the need to store the full solution array at both the previous and current iterations, it also speeds up the iteration process substantially. For example, consider sweeping through the grid with increasing i and j as illustrated in Figure 8.11. As one can see, when updating $u_{i,j}$, the points

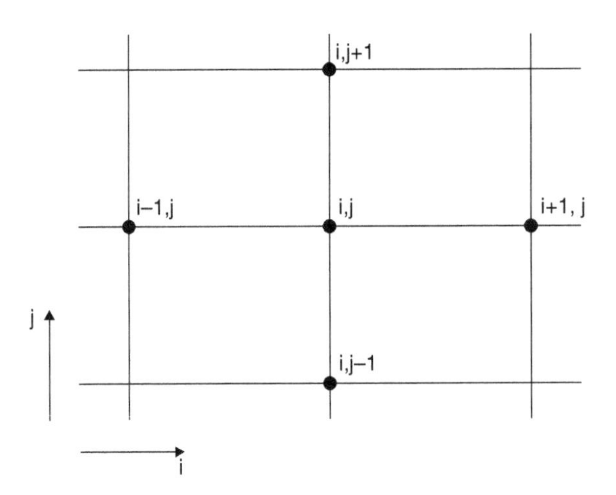

Figure 8.11 Sweeping directions for Gauss-Seidel iteration.

$u_{i,j-1}$ and $u_{i-1,j}$ have already been updated during the current iteration, and using these updated values, (8.28) is changed to

$$u_{i,j}^{(n+1)} = \frac{1}{2\left(1+\hat{\Delta}\right)}\left[u_{i+1,j}^{(n)} + u_{i-1,j}^{(n+1)} + \hat{\Delta}\left(u_{i,j+1}^{(n)} + u_{i,j-1}^{(n+1)}\right) - \Delta x^2 f_{i,j}\right]. \quad (8.30)$$

The values of $u_{i,j}$ are all stored in the same array, and it is not necessary to distinguish between the (n)th or $(n+1)$st iterates. We simply use the most recently updated information as it becomes available.

It can be shown that for our model problem, the spectral radius of the Gauss–Seidel method is such that

$$\rho_{GS}(I) = \rho_{Jac}^2(I);$$

therefore, from (8.29), the spectral radius for large I is

$$\rho_{GS}(I) = \rho_{Jac}^2(I) = \left[1 - \frac{1}{2}\left(\frac{\pi}{I+1}\right)^2 + \cdots\right]^2 = 1 - \left(\frac{\pi}{I+1}\right)^2 + \cdots. \quad (8.31)$$

Consequently, the rate of iterative convergence is *twice* as fast as that for the Jacobi method for large I, that is, the Gauss–Seidel method requires one half the iterations for the same level of accuracy for our model problem. Moreover, it only requires one half the storage of the Jacobi method. Finally, recall from Section 6.4.5 that it can be shown that strong diagonal dominance of the coefficient matrix **A** is a *sufficient*, but not necessary, condition for iterative convergence of the Jacobi and Gauss–Seidel methods.

8.5.3 Successive Over-Relaxation (SOR)

Recall from Section 6.4.6 that SOR accelerates Gauss–Seidel iteration by magnifying the change in the solution accomplished by each iteration. By taking a weighted average of the previous iterate $u_{i,j}^{(n)}$ and the Gauss–Seidel iterate $u_{i,j}^{(n+1)}$, the iteration process may be accelerated toward the exact solution.

If we denote the Gauss–Seidel iterate (8.30) by $u_{i,j}^*$, the new SOR iterate is given by

$$u_{i,j}^{(n+1)} = (1-\omega)u_{i,j}^{(n)} + \omega u_{i,j}^*, \tag{8.32}$$

where ω is the relaxation parameter and $0 < \omega < 2$ for convergence (Morton and Mayers 1994). We then have the following three possibilities:

$$\begin{aligned} 0 < \omega < 1 &\quad\Rightarrow\quad \text{Under-relaxation,} \\ \omega = 1 &\quad\Rightarrow\quad \text{Gauss–Seidel,} \\ 1 < \omega < 2 &\quad\Rightarrow\quad \text{Over-relaxation.} \end{aligned}$$

Over-relaxation typically – and often dramatically – accelerates iterative convergence of linear problems, while under-relaxation is often required for convergence when dealing with nonlinear problems.

Recall from Section 6.4.6 that the optimal value of ω that minimizes the spectral radius for SOR is given by

$$\omega_{opt}(I) = \frac{2}{1+\sqrt{1-\rho_{Jac}^2(I)}}, \tag{8.33}$$

and for this ω_{opt}, the spectral radius for SOR is

$$\rho_{SOR}(I) = \omega_{opt}(I) - 1. \tag{8.34}$$

Recall that $\rho_{Jac} = 1 - \frac{1}{2}\left(\frac{\pi}{I+1}\right)^2 + \cdots$ for large I for the model problem; thus,

$$\begin{aligned} \omega_{opt} &= \frac{2}{1+\sqrt{1-\rho_{Jac}^2}} \\ &= \frac{2}{1+\sqrt{1-\left[1-\frac{1}{2}\left(\frac{\pi}{I+1}\right)^2+\cdots\right]^2}} \\ &= \frac{2}{1+\sqrt{1-\left[1-\left(\frac{\pi}{I+1}\right)^2+\cdots\right]}} \\ &= \frac{2}{1+\frac{\pi}{I+1}} + \cdots . \\ \omega_{opt} &= 2\left(1-\frac{\pi}{I+1}+\cdots\right) \quad \text{for large } I. \end{aligned} \tag{8.35}$$

Then for large I, the spectral radius for SOR is

$$\rho_{SOR} = \omega_{opt} - 1 = 2\left(1-\frac{\pi}{I+1}+\cdots\right) - 1 = 1 - \frac{2\pi}{I+1} + \cdots . \tag{8.36}$$

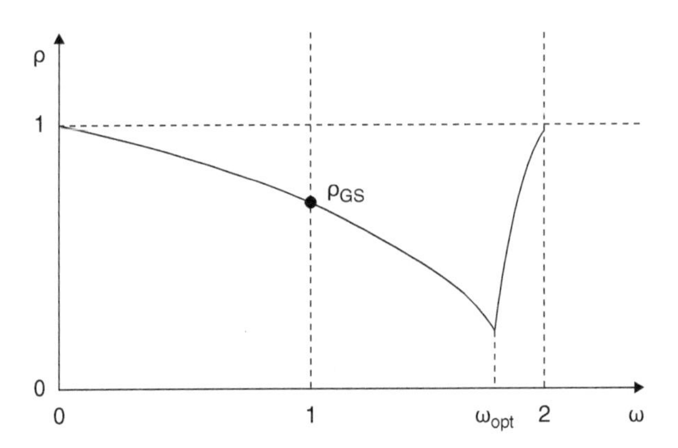

Figure 8.12 Spectral radius versus acceleration parameter for SOR.

Thus, as $I \to \infty, \omega_{opt} \to 2$, and $\rho_{SOR} \to 1$. However, from a comparison of (8.36) and (8.31), we see that

$$\rho_{SOR}(I) < \rho_{GS}(I),$$

such that SOR converges at a rate $2(I + 1)/\pi$ times faster than Gauss–Seidel if the optimal value of the relaxation parameter is used for the model problem under consideration. Therefore, the convergence rate improves linearly with increasing I relative to Gauss–Seidel.

The preceding analysis assumes that we know ω_{opt}. Typically when solving partial differential equations we do not, and the rate of convergence depends significantly on the choice of ω. For example, the typical behavior for linear problems with given I is shown in Figure 8.12. In addition, because ρ_{Jac} depends on the number of grid intervals I, the optimal acceleration parameter will change with the number of intervals. Moreover, although ω_{opt} does not depend on the right-hand side $f(x, y)$ of the Poisson equation, it does depend upon the shape of the domain, the differential equation, the boundary conditions, and the method of discretization. For a given problem, therefore, ω_{opt} often must be estimated from a similar problem and/or trial and error. If one is willing to spend the time to find the ω_{opt} for a given problem, then SOR can result in dramatic reductions in the number of iterations required as compared to Gauss–Seidel. If such an exercise is not warranted, for example, because only one solution is sought for the system of equations, then simply use Gauss–Seidel.

8.6 Boundary Conditions

Let us consider how Dirichlet and Neumann boundary conditions are implemented within the various iterative methods discussed. These are essentially two-dimensional extensions of the methods used for the one-dimensional boundary-value problems considered in Sections 8.1 and 8.2.

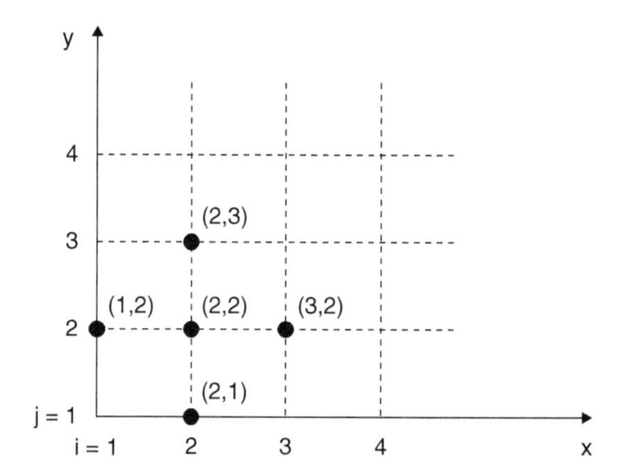

Figure 8.13 Schematic of Jacobi or Gauss–Seidel iterate at the point $(i, j) = (2, 2)$.

8.6.1 Dirichlet Boundary Conditions

Implementation of specified Dirichlet boundary conditions is very straightforward as no calculation is required at the boundary. Simply apply the specified values at the boundaries to $u_{1,j}$, $u_{I+1,j}$, $u_{i,1}$, and $u_{i,J+1}$, and iterate on the Jacobi (8.28), Gauss–Seidel (8.30), or SOR (8.32) equation in the interior, $i = 2, \ldots, I, j = 2, \ldots, J$. For example, the Jacobi or Gauss–Seidel iterate at $i = 2, j = 2$ is illustrated in Figure 8.13 and given by

$$u_{2,2}^{(n+1)} = \frac{1}{2\left(1+\hat{\Delta}\right)}\left[u_{3,2}^{(n)} + u_{1,2} + \hat{\Delta}\left(u_{2,3}^{(n)} + u_{2,1}\right) - \Delta x^2 f_{2,2}\right].$$

Hence, we simply apply the general finite-difference equation for Jacobi or Gauss–Seidel in the interior as usual, and the values on the boundary are picked up as necessary.

8.6.2 Neumann Boundary Conditions

As an example of a Neumann boundary condition, consider

$$\frac{\partial u}{\partial x} = c \quad \text{at} \quad x = 0 \tag{8.37}$$

as shown in Figure 8.14. The simplest treatment would be to use the Jacobi (8.28), Gauss–Seidel (8.30), or SOR (8.32) equation to update $u_{i,j}$ in the interior for $i = 2, \ldots, I$, and then to approximate the boundary condition (8.37) by a forward difference applied at $i = 1$ according to

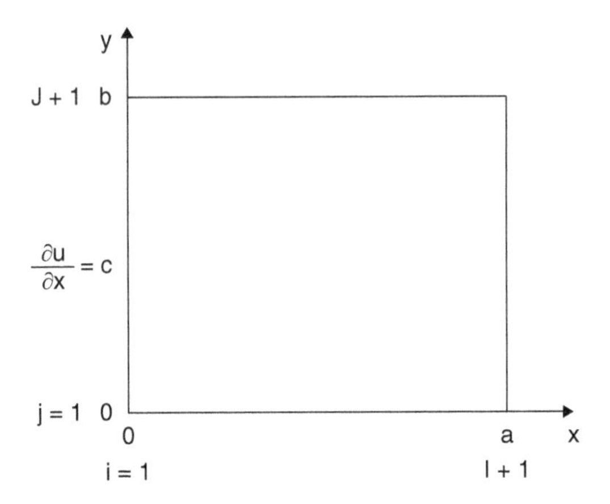

Figure 8.14 Neumann boundary condition at $x = 0$.

$$\frac{u_{2,j} - u_{1,j}}{\Delta x} + O(\Delta x) = c. \tag{8.38}$$

This could then be used to update $u_{1,j}, j = 2, \dots, J$ using

$$u_{1,j} = u_{2,j} - c\Delta x.$$

This is ill advised, however, because (8.38) is only first-order accurate. If a lower-order approximation, such as (8.38), is used at a boundary, its truncation error generally dominates the convergence rate of the entire scheme.

A better alternative is to use the same method as in Section 8.1.2 for a Robin boundary condition. In this approach, the interior points are updated as before, but we now apply the difference equation at the boundary. For example, we could apply Jacobi (8.28) at $i = 1$ as follows:

$$u_{1,j}^{(n+1)} = \frac{1}{2\left(1+\hat{\Delta}\right)}\left[u_{2,j}^{(n)} + u_{0,j}^{(n)} + \hat{\Delta}\left(u_{1,j+1}^{(n)} + u_{1,j-1}^{(n)}\right) - \Delta x^2 f_{1,j}\right]. \tag{8.39}$$

However, this involves a value $u_{0,j}^{(n)}$ that is outside the domain. A second-order accurate, central-difference approximation for the boundary condition (8.37) is

$$\frac{u_{2,j}^{(n)} - u_{0,j}^{(n)}}{2\Delta x} + O(\Delta x^2) = c, \tag{8.40}$$

which also involves the value $u_{0,j}^{n}$. Therefore, solving (8.40) for $u_{0,j}^{n}$ gives

$$u_{0,j}^{(n)} = u_{2,j}^{(n)} - 2c\Delta x,$$

and substituting into the difference equation (8.39) to eliminate $u_{0,j}^{(n)}$ leads to

$$u_{1,j}^{(n+1)} = \frac{1}{2\left(1+\hat{\Delta}\right)}\left[2\left(u_{2,j}^{(n)} - c\Delta x\right) + \hat{\Delta}\left(u_{1,j+1}^{(n)} + u_{1,j-1}^{(n)}\right) - \Delta x^2 f_{1,j}\right]. \tag{8.41}$$

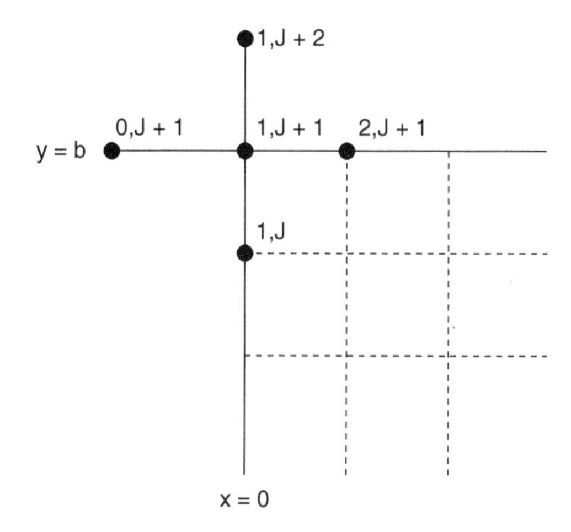

Figure 8.15 Boundary conditions in the top-left corner of the domain.

Thus, we use (8.41) to update $u_{1,j}^{(n+1)}, j = 2, \ldots, J$. This is the same procedure used for a Dirichlet condition but with an additional sweep along the left boundary using (8.41) for $u_{1,j}^{(n+1)}, j = 2, \ldots, J$.

This approach requires special treatment at corners depending upon the boundary condition along the adjacent boundary. For example, consider if in addition to the boundary condition (8.37) that we have

$$\frac{\partial u}{\partial y} = d \quad \text{at} \quad y = b. \tag{8.42}$$

This is illustrated in Figure 8.15. Applying (8.41), which already includes the boundary condition along $x = 0$, at the corner $i = 1, j = J + 1$ yields

$$u_{1,J+1}^{(n+1)} = \frac{1}{2\left(1+\hat{\Delta}\right)}\left[2\left(u_{2,J+1}^{(n)} - c\Delta x\right) + \hat{\Delta}\left(u_{1,J+2}^{(n)} + u_{1,J}^{(n)}\right) - \Delta x^2 f_{1,J+1}\right], \tag{8.43}$$

where $u_{1,J+2}^{(n)}$ is outside the domain. Approximating (8.42) using a central difference in the same manner as (8.40) gives

$$\frac{u_{1,J+2}^{(n)} - u_{1,J}^{(n)}}{2\Delta y} = d,$$

which leads to

$$u_{1,J+2}^{(n)} = u_{1,J}^{(n)} + 2d\Delta y.$$

Substituting into (8.43) to eliminate the point outside the domain $u_{1,J+2}^{n}$ gives

$$u_{1,J+1}^{(n+1)} = \frac{1}{2\left(1+\hat{\Delta}\right)}\left[2\left(u_{2,J+1}^{(n)} - c\Delta x\right) + 2\hat{\Delta}\left(u_{1,J}^{(n)} + d\Delta y\right) - \Delta x^2 f_{1,J+1}\right],$$

which is used to update the corner value $i = 1$, $j = J + 1$ for each iteration. Note that solving the Poisson equation subject to Neumann boundary conditions on all boundaries would require nine unique difference equations, one for the interior, four for the sides, and four for the corners.

8.7 Alternating-Direction-Implicit (ADI) Method

Recall that in elliptic problems, the solution anywhere depends on the solution everywhere – it has an infinite speed of propagation in all directions. However, in Jacobi, Gauss–Seidel, and SOR, information only propagates through the mesh one point at a time. For example, if we are sweeping from left to right along lines of constant y with $0 \leq x \leq a$, it takes I iterations before the boundary condition at $x = a$ is "felt" at $x = 0$. Consequently, these point-by-point relaxation techniques are not very "elliptic-like." Nor are they very matrix-like. While this makes them particularly straightforward to program, their simplicity comes at the expense of numerical efficiency. A method that is more faithful to the elliptic character of the equation could be obtained by solving entire lines in the grid in an implicit manner. For example, sweeping along lines of constant y and solving each constant y-line implicitly – all at once – would allow for the boundary condition at $x = a$ to influence the solution in the entire domain after only one sweep through the grid; rather than sweeping point by point, we sweep line by line.

In order to see how such a scheme might be devised, let us return to the difference equation (8.27) for the Poisson equation

$$u_{i+1,j} - 2\left(1 + \hat{\Delta}\right) u_{i,j} + u_{i-1,j} + \hat{\Delta}\left(u_{i,j+1} + u_{i,j-1}\right) = \Delta x^2 f_{i,j}. \tag{8.44}$$

Consider the jth constant y-line and assume that values along the $(j+1)$st and $(j-1)$st lines are taken from the previous iterate. Rewriting (8.44) as an implicit equation for the values of $u_{i,j}$ along the jth line gives

$$u_{i+1,j}^{(n+1)} - 2\left(1 + \hat{\Delta}\right) u_{i,j}^{(n+1)} + u_{i-1,j}^{(n+1)} = \Delta x^2 f_{i,j} - \hat{\Delta}\left(u_{i,j+1}^{(n)} + u_{i,j-1}^{(n)}\right), \quad i = 2, \ldots, I. \tag{8.45}$$

Therefore, we have a tridiagonal problem for $u_{i,j}$ along the jth line, which can be solved using the Thomas algorithm.

REMARKS:

1. *If sweeping through j-lines, $j = 2, \ldots, J$, then $u_{i,j-1}^{(n)}$ becomes $u_{i,j-1}^{(n+1)}$ in (8.45), as it has already been updated. In other words, we might as well use the most recently updated values as in Gauss–Seidel.*
2. *SOR can also be incorporated after each tridiagonal solve to accelerate iterative convergence for linear problems.*
3. *This approach is more efficient at spreading information throughout the domain; therefore, it reduces the number of iterations required for convergence, but there is more computation per iteration.*
4. *ADI also takes advantage of the efficient Thomas algorithm.*

The preceding illustration provides the motivation for the *alternating-direction-implicit (ADI) method*. In the ADI method, we sweep along lines but in alternating directions. Although the order is arbitrary, let us sweep along lines of constant y first followed by lines of constant x.

In the first half of the iteration from (n) to $(n + 1/2)$, we perform a sweep along constant y-lines by solving the series of tridiagonal problems for each of $j = 2, \ldots, J$ given by

$$\begin{aligned} u_{i+1,j}^{(n+1/2)} - (2+\sigma)\, u_{i,j}^{(n+1/2)} + u_{i-1,j}^{(n+1/2)} &= \Delta x^2 f_{i,j} \\ -\hat{\Delta}\left[u_{i,j+1}^{(n)} - \left(2 - \frac{\sigma}{\hat{\Delta}}\right) u_{i,j}^{(n)} + u_{i,j-1}^{(n+1/2)}\right], &\quad i = 2, \ldots, I. \end{aligned} \tag{8.46}$$

This results in a tridiagonal system of equations to solve for each constant y line, that is, for each j and all i. The tridiagonal system corresponding to each value of j is solved in succession as we sweep along each constant y line. This is why the $u_{i,j-1}^{(n+1/2)}$ term on the right-hand side has been updated from the previous line solved.

Unlike in (8.45), differencing in the x- and y-directions are kept separate to mimic diffusion in each direction. This is why $u_{i,j}$ appears on both sides of the equation – one from the derivative in the x-direction and one from the derivative in the y-direction. This approach is called a *splitting method*, in which all of the terms associated with the derivatives in each direction are kept together on one side of the difference equation. Observe that we have subtracted a term $\sigma u_{i,j}$ on each side of the equation. The numerical parameter σ is an acceleration parameter to enhance diagonal dominance $(\sigma \geq 0)$ on the left-hand side of the equation by increasing the magnitude of the coefficient on the main diagonal; $\sigma = 0$ corresponds to no acceleration. Note that the σ terms on each side of the equation cancel so as not to alter the solution for $u_{i,j}$. Observe that while splitting is physically motivated, acceleration is numerically motivated.

In the second half of the iteration from $(n+1/2)$ to $(n+1)$, we sweep along constant x-lines by solving the series of tridiagonal problems for $i = 2, \ldots, I$ given by

$$\begin{aligned} \hat{\Delta} u_{i,j+1}^{(n+1)} - \left(2\hat{\Delta} + \sigma\right) u_{i,j}^{(n+1)} + \hat{\Delta} u_{i,j-1}^{(n+1)} &= \Delta x^2 f_{i,j} \\ -\left[u_{i+1,j}^{(n+1/2)} - (2-\sigma)\, u_{i,j}^{(n+1/2)} + u_{i-1,j}^{(n+1)}\right], &\quad j = 2, \ldots, J, \end{aligned} \tag{8.47}$$

where $u_{i-1,j}^{(n+1)}$ has been updated from the previous line. Once again, the acceleration parameter σ is incorporated in order to enhance diagonal dominance of the tridiagonal system that is solved for each constant x line, and splitting has been used.

For our model problem, it can be shown that the acceleration parameter that gives the best speedup is

$$\sigma = 2\cos(\pi/R),$$

where $R = \max(I + 1, J + 1)$. This result can be used as a starting point for selecting σ in similar problems. While the acceleration parameter is introduced to enhance diagonal dominance, there is no basis to expect that being "more" diagonally dominant – with larger σ – is better.

Each ADI iteration involves $(I-1)+(J-1)$ tridiagonal solves (for Dirichlet boundary conditions in which it is not necessary to solve along the boundaries), and each point in the grid is updated twice per iteration – once during the constant y-line sweeps and once during the constant x-line sweeps. The ADI method with splitting is typically significantly faster than Gauss–Seidel or SOR for equations in which the terms are easily separated into x- and y-directions. Although there is more computational work per iteration, the number of iterations is reduced significantly. Note that some of the efficiency gains are lost if there are mixed derivative terms, such as $\partial^2 u/\partial x \partial y$.

In addition to faster convergence, the ADI method more naturally accommodates the various types of boundary conditions. In particular, when using the Thomas algorithm as described in Section 6.3.3 to solve the tridiagonal system along each constant x- or y-line, the vector coefficients **a**, **b**, **c**, and **d** for the constant y-lines are defined for $i = 1, \ldots, I+1$, and the necessary adjustments for the boundary conditions at $x = 0$ and $x = a$ are made within the Thomas algorithm, with the same holding along lines of constant x. If we have Neumann or Robin boundary conditions, then it is necessary to solve a tridiagonal system along such boundaries, with the points outside the domain on the right-hand side of the difference equation being eliminated as for Gauss–Seidel, such that the **d** values are modified. Unlike Gauss–Seidel, no special treatment is required at the corners; they are handled automatically by the Thomas algorithm.

In order to emphasize the splitting of derivative directions, and introduce notation to be used later, let us write the finite-difference equation for the Poisson equation

$$\frac{\partial^2 u}{\partial x^2} + \frac{\partial^2 u}{\partial y^2} = f$$

in the abbreviated form

$$\delta_x^2 u_{i,j} + \delta_y^2 u_{i,j} = f_{i,j},$$

where the second-order accurate central-difference operators are given by

$$\delta_x^2 u_{i,j} = \frac{u_{i+1,j} - 2u_{i,j} + u_{i-1,j}}{(\Delta x)^2},$$

$$\delta_y^2 u_{i,j} = \frac{u_{i,j+1} - 2u_{i,j} + u_{i,j-1}}{(\Delta y)^2}.$$

Then in the ADI method with splitting, along constant y lines, we solve

$$\Delta x^2 \delta_x^2 u_{i,j} = \Delta x^2 \left(f_{i,j} - \delta_y^2 u_{i,j} \right),$$

and along constant x lines we solve

$$\Delta x^2 \delta_y^2 u_{i,j} = \Delta x^2 \left(f_{i,j} - \delta_x^2 u_{i,j} \right).$$

These are (8.46) and (8.47), respectively, with $\sigma = 0$ and written in our abbreviated notation.

8.8 Multigrid Methods

As we have emphasized, iterative methods have several advantages over direct methods; they scale better to very large systems of equations, and they can be adapted to most any differential equation, including nonlinear ones. As a result, they are by far the most commonly used in scientific computing applications requiring solutions of partial differential equations. However, the standard iterative methods considered thus far have speeds that trail those of the best fast Poisson solvers, such as fast Fourier transforms and cyclic reduction as discussed in Section 8.4.2. Normally, we would be content with a general-purpose method that is somewhat slower than more efficient methods that only work for linear problems. That did not stop Achi Brandt[5] from developing a strategy in the 1970s for accelerating iterative methods that makes them comparable to fast Poisson solvers but apply to nonlinear equations with minimal loss of efficiency – the best of both worlds.

Multigrid methods recognize and take advantage of an interesting property in the way that iterative methods progress toward the exact solution. If one were to carefully examine the iterative convergence history of the typical iterative techniques, such as Gauss–Seidel or ADI, the following property would be observed: high-frequency modes of the solution (or error) experience fast iterative convergence; that is, their errors are reduced quickly with each iteration, whereas low-frequency modes of the solution (or error) exhibit relatively slower iterative convergence. Thus, there is rapid iterative convergence of the overall solution until the high-frequency modes are smoothed out followed by slow iterative convergence when only lower-frequency modes are present. This is reflected in the propensity for iterative methods to reduce the error of the iterative solution dramatically during the early stages of the iteration process followed by a long period involving many iterations during which the error reduction is stubbornly slow. Although this behavior can be very frustrating, by recognizing its source, a very efficient framework can be devised that accelerates the underlying iterative technique used.

Multigrid methods take advantage of this property of relaxation techniques by recognizing that smooth, low-frequency components of the error become more oscillatory and high frequency with respect to the grid size on a coarser grid. That is, there are fewer grid points per wavelength with respect to a coarser grid as compared to a finer grid, and the error appears more oscillatory on the coarser grid. Thus, relaxation would be expected to be more effective on a coarse-grid representation of the error. Note that it is also faster as there are fewer points to compute on the coarser grid. This property of iterative methods is clearly illustrated in Briggs et al. (2000).

It is important to understand at the outset that multigrid methods are not so much a specific method as they are a *framework* for accelerating existing relaxation (iterative) techniques, such as the Gauss–Seidel or ADI method. As such, they take advantage of the scalability and flexibility of iterative methods but allow for further efficiency gains with respect to iterative convergence behavior. Therefore, they are currently

[5] See original references on "multilevel adaptive techniques."

among the best methods for solving partial differential equations numerically, both in terms of efficiency and flexibility. Multigrid methods are comparable in speed with fast direct methods, such as Fourier methods and cyclic reduction, but they can be adapted to solve general elliptic equations with variable coefficients and even nonlinear equations.

8.8.1 Coarse-Grid Correction

In order to illustrate the multigrid framework and how it takes advantage of the property of iterative methods stated earlier, let us consider a general difference equation that has been obtained from a linear – for now – partial differential equation

$$\mathcal{L}\hat{u}(x,y) = f(x,y), \tag{8.48}$$

where $\mathcal{L}$ represents the *finite-difference operator*, and $\hat{u}(x,y)$ is the exact solution of the difference equation. The *error* of a numerical approximation $u(x,y)$ is defined by

$$e(x,y) = \hat{u}(x,y) - u(x,y). \tag{8.49}$$

Given the numerical solution $u(x,y)$, the *residual* is defined by[6]

$$r(x,y) = f(x,y) - \mathcal{L}u(x,y). \tag{8.50}$$

Observe from (8.48) that if $u(x,y) = \hat{u}(x,y)$, then the residual is zero; therefore, the residual is a measure of how "wrong" the approximate solution is. Substituting (8.49) into (8.48) gives

$$\mathcal{L}e(x,y) + \mathcal{L}u(x,y) = f(x,y),$$

or

$$\mathcal{L}e(x,y) = f(x,y) - \mathcal{L}u(x,y),$$

which is the error equation

$$\mathcal{L}e(x,y) = r(x,y). \tag{8.51}$$

This error equation is an alternative, but equivalent, expression of the original difference equation (8.48). Observe that the dependent variable is now the error in the approximate solution, and the right-hand side is the residual of the approximate numerical solution.

For the multigrid method, it is necessary to have a means of clearly identifying operators and quantities on successive grids. As illustrated in Figure 8.16, the fine grid is denoted by Ω^h and the course grid by Ω^{2h}.[7] Note that grids are typically reduced by a factor of two in each direction, such that the coarser grid is comprised

[6] One will note that the residual is variously defined as $r = f - \mathcal{L}u$ and $r = \mathcal{L}u - f$ by other authors. Either definition works fine in most contexts. Here, however, this definition leads to a more straightforward formulation of the multigrid methodology.

[7] In the numerical methods literature, h is often used to indicate the grid size, here given by Δx and Δy. Thus, a grid with $2h$ has grid sizes that are twice that of h.

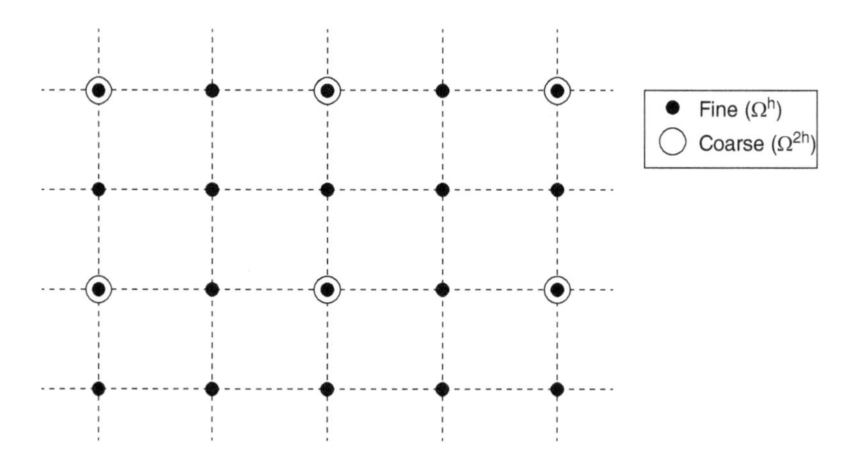

Figure 8.16 The coarse and fine grids, with the coarse grid consisting of every other point in the fine grid.

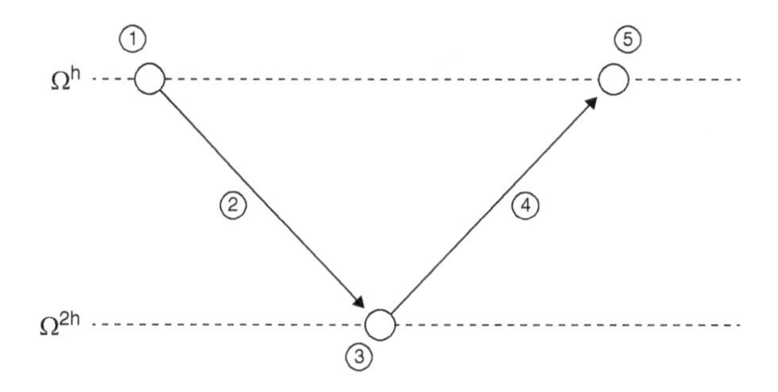

Figure 8.17 Coarse-grid correction sequence between a fine grid Ω^h and a coarse grid Ω^{2h}.

of every other point in the finer grid. In order to shift information between the fine and coarse grids, we use the restriction and interpolation operators. The *restriction operator*, given by I_h^{2h}, moves information from the fine grid Ω^h to the coarse grid Ω^{2h}. The *interpolation*, or *prolongation, operator*, given by I_{2h}^h, moves information from the coarse grid Ω^{2h} to the fine grid Ω^h. The most commonly used interpolation operator is based on *bilinear interpolation*. In the operators, observe that the subscript indicates the grid from which information is moved, and the superscript indicates the grid to which the information is moved.

From these definitions, we can devise a scheme with which to *correct* the solution on a fine grid by solving for the error on a coarse grid. This is known as *coarse-grid correction* (CGC) and consists of the following steps as illustrated in Figure 8.17:

1. Relax the original difference equation $\mathcal{L}^h u^h = f^h$ on the fine grid Ω^h using Gauss–Seidel, ADI, et cetera, ν_1 times with an initial guess for u^h.
2. Compute the residual on the fine grid Ω^h and restrict it to the coarse grid Ω^{2h}:

$$r^{2h} = I_h^{2h} r^h = I_h^{2h}(f^h - \mathcal{L}^h u^h).$$

3. Using the residual as the right-hand side, "solve" the error equation $\mathcal{L}^{2h}e^{2h} = r^{2h}$ on the coarse grid Ω^{2h}.
4. Interpolate the error to the fine grid and correct the fine-grid approximation according to

$$u^h \leftarrow u^h + I_{2h}^h e^{2h}.$$

5. Relax $\mathcal{L}^h u^h = f^h$ on the fine grid Ω^h ν_2 times with the corrected approximation u^h as the initial guess.

This CGC scheme is the primary component of all the many multigrid algorithms. In practice, ν_1 and ν_2 are small, typically 1, 2, or 3, and in the CGC scheme, it is the residual $r(x, y)$ that is being restricted and the error $e(x, y)$ that is interpolated. However, we will present these as operating on the dependent variable $u(x, y)$ for generality.

An obvious question at this point is, "How do we obtain the coarse-grid solution for $e^{2h}(x, y)$ in step 3?" Actually, we already know the answer to this – perform additional CGCs. That is, if the CGC works between two grid levels, it should be even more effective between three, four, or as many grid levels as we can accommodate. To implement this, then, we recursively replace step 3 by additional CGCs on progressively coarser grids until it is no longer possible to further reduce the grid. This leads to the so-called *V-cycle* as illustrated in Figure 8.18. The V-cycles are then repeated until iterative convergence; each V of the V-cycle is essentially a multigrid iteration.

Although we simply denote the error at each grid level as $e(x, y)$, it is important to realize that on each successively coarser grid, the quantity that is being relaxed is the error of the error on the next finer grid. Thus, on the finest grid, say Ω^h, relaxation is carried out on the original equation for $u(x, y)$; on the next coarser grid

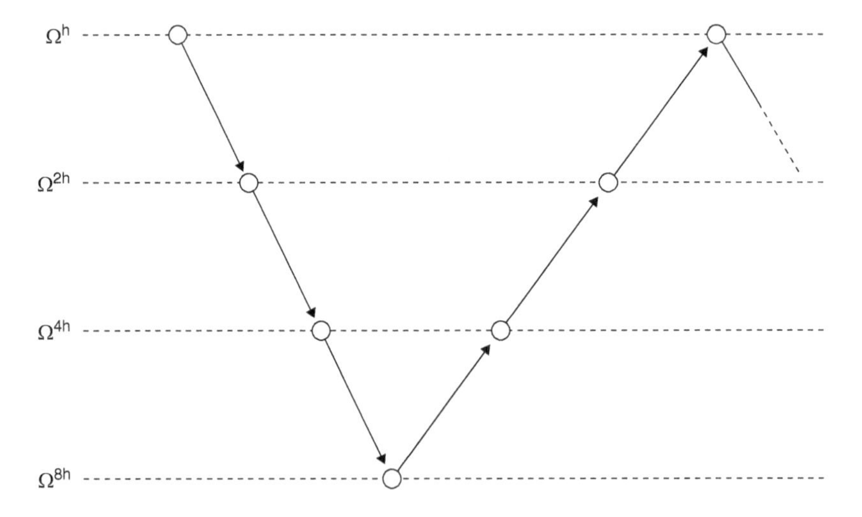

Figure 8.18 Multigrid V-cycle with four grid levels; down arrows represent restriction, up arrows represent interpolation, and circles denote relaxation.

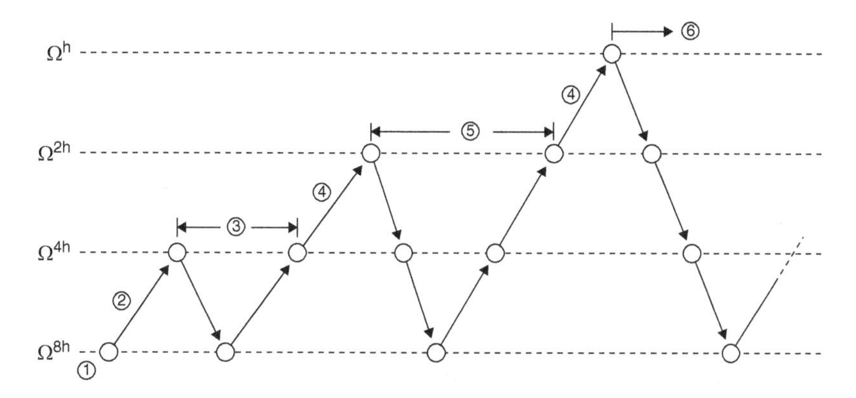

Figure 8.19 Schematic of full multigrid V-cycle to obtain a good initial guess before commencing V-cycles.

Ω^{2h}, relaxation is on the equation for the error on Ω^h; on the next coarser grid Ω^{4h}, relaxation is on the equation for the error on Ω^{2h}; and so forth. As with the restriction and prolongation operators, however, the relaxation is presented as being applied to the dependent variable $u(x, y)$.

This simple V-cycle scheme is appropriate when a good initial guess is available to start the V-cycles. For example, when considering a solution of (8.48) in the context of an unsteady calculation, in which case the solution for $u^h(x, y)$ from the previous time step is a good initial guess for the current time step. If no good initial guess is available, then *full multigrid* (FMG)[8] *V-cycle* may be applied according to the following procedure, which utilizes the same components as in the coarse-grid correction sequence:

1. Solve $\mathcal{L}u = f$ on the coarsest grid (note: u, not e).
2. Interpolate u to next finer grid.
3. Perform V-cycle to correct u.
4. Interpolate u to next finer grid.
5. Repeat (3) and (4) until the finest grid is reached.
6. Perform V-cycles until iterative convergence.

Note that $f(x, y)$ is required on all grid levels for FMG. This procedure is illustrated in Figure 8.19.

The coarse-grid correction sequence represents the core of the multigrid method; however, there are numerous additional details that one must account for when implementing multigrid methods. The reader is referred to the following excellent references for these details: Briggs et al. (2000) and Thomas and Brandt (2003). Press et al. (2007) also has a good introduction to multigrid methods. Let us say a few words about grid definitions and relaxation methods.

Because each successive grid differs by a factor of two in each direction, the finest grid size is often taken as $2^n + 1$ points, where n is an integer. This is overly restrictive,

[8] Also called *nested iteration*.

particularly when we get to large grid sizes. Somewhat more general grids may be obtained using the following grid definitions. The differential equation is discretized on a uniform grid having $N_x \times N_y$ points, which are defined by

$$N_x = m_x 2^{(n_x-1)} + 1, \quad N_y = m_y 2^{(n_y-1)} + 1, \tag{8.52}$$

where n_x and n_y determine the number of grid levels, and m_x and m_y determine the size of the coarsest grid, which is $(m_x + 1) \times (m_y + 1)$. In order to maximize the benefits of the multigrid methodology, we want to maximize the number of grid levels between which the algorithm will move. Therefore, for a given grid, n_x and n_y should be as large as possible, and m_x and m_y should be as small as possible for maximum efficiency. Typically, m_x and m_y are 2, 3, or 5.

At the heart of the multigrid method is an iterative (relaxation) scheme that is used to update the numerical approximation to the solution. Typically, *red-black* Gauss–Seidel iteration is used to relax the difference equation; every other grid point is labeled red or black as on a checker game board. By performing the relaxation on all of the alternating red and black grid points separately, it eliminates data dependencies such that it is easily implemented on parallel computers (see Section 9.13). Note that when Gauss–Seidel is used, SOR should not be implemented because it destroys the high-frequency smoothing of the multigrid approach. Although Gauss–Seidel is most commonly used owing to its ease of implementation, particularly in parallel, it is better to use ADI relaxation for the same reason that ADI is better than Gauss–Seidel as a standalone iterative technique. The multigrid method scales to larger grid sizes more effectively than the underlying ADI relaxation alone; this is manifest by the fact that only small increases in the number of V-cycles are necessary with increasing N.

Improvements in a numerical method or algorithm usually come at a cost. Aside from the additional programming complexity, which is considerable, the only "cost" for the dramatic speed and scalability improvements of multigrid techniques is a doubling of memory requirements in order to store the solution $u(x, y)$ or error $e(x, y)$ at all grid levels. Hence, the multigrid method is essentially a trade-off between computational time and memory requirements. This is generally a worthwhile trade-off on modern computers, for which memory is cheap and abundant while increases in processor speed come with considerable cost, both financially and in terms of programming complexity for parallel-processing architectures.

8.8.2 Full-Approximation-Storage (FAS) Method for Nonlinear Equations

Because of their speed and generality, multigrid methods are currently the preferred framework for solving elliptic partial differential equations, including nonlinear equations. The coarse-grid correction sequence developed previously for linear equations must be adjusted for nonlinear equations. To see how, let us go back and consider the difference equation

$$\mathcal{L}\hat{u}(x, y) = f(x, y), \tag{8.53}$$

where $\mathcal{L}$ represents the *difference operator*, which is now nonlinear, and $\hat{u}(x, y)$ is the exact solution of the difference equation. The *error* of a numerical approximation $u(x, y)$ is defined as before by

$$e(x, y) = \hat{u}(x, y) - u(x, y). \tag{8.54}$$

Given the numerical solution $u(x, y)$, the *residual* is also defined as before by

$$r(x, y) = f(x, y) - \mathcal{L}u(x, y). \tag{8.55}$$

Because the difference operator is now nonlinear, the definition of the error (8.54) cannot be substituted into the difference equation (8.53) to obtain the error equation as before. Instead, the residual (8.55) is substituted into the original difference equation (8.53) as follows:

$$\mathcal{L}\hat{u}(x, y) = \mathcal{L}u(x, y) + r(x, y). \tag{8.56}$$

Use of (8.56) requires that we have both the residual $r(x, y)$ as well as the approximate solution $u(x, y)$. Therefore, the CGC procedure is modified as follows (in particular, take note of the modifications in steps 2 and 3):

1. Relax $\mathcal{L}^h u^h = f^h$ for ν_1 times on the fine grid Ω^h.
2. Compute the residual on the fine grid Ω^h and restrict the residual *and* the approximate solution to the coarse grid Ω^{2h}:
$$r^{2h} = I_h^{2h} r^h = I_h^{2h}(f^h - \mathcal{L}^h u^h),$$
$$u^{2h} = I_h^{2h} u^h.$$
3. Consider (8.56) on the coarse grid, that is,
$$\mathcal{L}^{2h}\hat{u}^{2h} = \mathcal{L}^{2h}u^{2h} + r^{2h}.$$
This is used to "solve" for $\hat{u}^{2h}$ on the coarse grid. The error on the coarse grid is then given by
$$e^{2h} = \hat{u}^{2h} - u^{2h}.$$
4. Interpolate the error to the fine grid and correct the fine-grid approximation according to
$$u^h \leftarrow u^h + I_{2h}^h e^{2h}.$$
5. Relax $\mathcal{L}^h u^h = f^h$ ν_2 times on the fine grid Ω^h with the corrected approximation u^h as the initial guess.

The CGC sequence is implemented recursively in step 3 as before using V- or W-cycles to maximize the multigrid method's ability to accelerate the iterative process. This is called the *full-approximation-storage* (*FAS*) *method* because the solution and its residual must both be stored and translated between grid levels. Observe that the difference equation for the dependent variable $\hat{u}(x, y)$ is relaxed, rather than for the error $e(x, y)$, on each grid level. Because the difference equation is nonlinear, Newton linearization is used to linearize the nonlinear terms where necessary as discussed in Section 8.10.2. Finally, FMG is often incorporated into the FAS method to obtain a good initial gucss.

8.9 Compact Higher-Order Methods

Thus far, we have used second-order accurate finite differences throughout, which results in a five-point stencil for two-dimensional problems. In some cases, however, it is desirable to incorporate higher-order approximations. For example, this may allow for numerical solution of a problem with fewer grid points while maintaining the same level of accuracy. In other words, this would offset a decrease in the truncation error with an increase in discretization error to produce the same level of accuracy. This is particularly important in simulations within three-dimensional domains, for which any reduction in the number of grid points required is welcome, and is commonly used in simulations of acoustics and turbulence in fluid mechanics, for example.

Recall that in order to obtain higher-order finite-difference approximations, it is necessary to incorporate additional points in the finite-difference stencil. However, this produces matrices with additional nonzero "bands" and requires special treatment near boundaries. If possible, it is advantageous to maintain the *compact* nature of second-order accurate central differences because they produce tridiagonal systems of equations that may be solved efficiently and do not require special treatment near boundaries.

Here, we derive a *compact, fourth-order accurate, central-difference approximation* for the second-order derivatives in the Poisson equation using a method based on that in E and Liu (1996). As in Section 8.7, we define the following second-order accurate, central-difference operators

$$\delta_x^2 u_{i,j} = \frac{u_{i-1,j} - 2u_{i,j} + u_{i+1,j}}{\Delta x^2},$$

$$\delta_y^2 u_{i,j} = \frac{u_{i,j-1} - 2u_{i,j} + u_{i,j+1}}{\Delta y^2}.$$

Recall from (7.11) in Section 7.2.1 that from the Taylor series, we have

$$\frac{\partial^2 u}{\partial x^2} = \delta_x^2 u - \frac{\Delta x^2}{12}\frac{\partial^4 u}{\partial x^4} + O(\Delta x^4), \tag{8.57}$$

where the second term is the truncation error for the second-order accurate approximation, which we will now include in our approximation. Therefore,

$$\delta_x^2 u = \frac{\partial^2 u}{\partial x^2} + \frac{\Delta x^2}{12}\frac{\partial^4 u}{\partial x^4} + O(\Delta x^4) = \left(1 + \frac{\Delta x^2}{12}\frac{\partial^2}{\partial x^2}\right)\frac{\partial^2 u}{\partial x^2} + O(\Delta x^4). \tag{8.58}$$

But from (8.57), observe that

$$\frac{\partial^2}{\partial x^2} = \delta_x^2 + O(\Delta x^2).$$

Substituting into (8.58) gives

$$\delta_x^2 u = \left\{1 + \frac{\Delta x^2}{12}\left[\delta_x^2 + O(\Delta x^2)\right]\right\}\frac{\partial^2 u}{\partial x^2} + O(\Delta x^4) = \left(1 + \frac{\Delta x^2}{12}\delta_x^2\right)\frac{\partial^2 u}{\partial x^2} + O(\Delta x^4).$$

Solving for $\partial^2 u/\partial x^2$ yields

$$\frac{\partial^2 u}{\partial x^2} = \left(1 + \frac{\Delta x^2}{12}\delta_x^2\right)^{-1}\delta_x^2 u + \left(1 + \frac{\Delta x^2}{12}\delta_x^2\right)^{-1} O(\Delta x^4). \tag{8.59}$$

From a binomial expansion (with Δx sufficiently small) observe that

$$\left(1 + \frac{\Delta x^2}{12}\delta_x^2\right)^{-1} = 1 - \frac{\Delta x^2}{12}\delta_x^2 + O(\Delta x^4). \tag{8.60}$$

Because the last term in (8.59) is still $O(\Delta x^4)$ after substitution of (8.60), we can write (8.59) as

$$\frac{\partial^2 u}{\partial x^2} = \left(1 + \frac{\Delta x^2}{12}\delta_x^2\right)^{-1}\delta_x^2 u + O(\Delta x^4). \tag{8.61}$$

Substituting the expression (8.60) into (8.61) leads to an $O(\Delta x^4)$ accurate central-difference approximation for the second derivative given by

$$\frac{\partial^2 u}{\partial x^2} = \left(1 - \frac{\Delta x^2}{12}\delta_x^2\right)\delta_x^2 u + O(\Delta x^4).$$

Expanding the finite-difference operators leads to the corresponding finite-difference approximation given at the end of Section 7.2.2. Owing to the $\delta_x^2(\delta_x^2 u)$ operator, however, this approximation involves the five points $u_{i-2}, u_{i-1}, u_i, u_{i+1}$, and u_{i+2}; therefore, it is not compact. In order to obtain a compact scheme, we also consider the second derivative in the y-direction. Similar to (8.61), we have in the y-direction

$$\frac{\partial^2 u}{\partial y^2} = \left(1 + \frac{\Delta y^2}{12}\delta_y^2\right)^{-1}\delta_y^2 u + O(\Delta y^4). \tag{8.62}$$

Now consider the Poisson equation

$$\frac{\partial^2 u}{\partial x^2} + \frac{\partial^2 u}{\partial y^2} = f(x, y).$$

Substituting (8.61) and (8.62) into the Poisson equation leads to

$$\left(1 + \frac{\Delta x^2}{12}\delta_x^2\right)^{-1}\delta_x^2 u + \left(1 + \frac{\Delta y^2}{12}\delta_y^2\right)^{-1}\delta_y^2 u + O(\Delta^4) = f(x, y),$$

where $\Delta = \max(\Delta x, \Delta y)$. Multiplying by $\left(1 + \frac{\Delta x^2}{12}\delta_x^2\right)\left(1 + \frac{\Delta y^2}{12}\delta_y^2\right)$ gives

$$\begin{aligned} &\left(1 + \frac{\Delta y^2}{12}\delta_y^2\right)\delta_x^2 u + \left(1 + \frac{\Delta x^2}{12}\delta_x^2\right)\delta_y^2 u + O(\Delta^4) \\ &= \left[1 + \frac{\Delta x^2}{12}\delta_x^2 + \frac{\Delta y^2}{12}\delta_y^2 + O(\Delta^4)\right] f(x,y), \end{aligned} \tag{8.63}$$

which is a fourth-order accurate finite-difference approximation to the Poisson equation. Expanding the first term in (8.63) yields

$$\begin{aligned} \left(1 + \frac{\Delta y^2}{12}\delta_y^2\right)\delta_x^2 u &= \left(1 + \frac{\Delta y^2}{12}\delta_y^2\right)\frac{u_{i-1,j} - 2u_{i,j} + u_{i+1,j}}{\Delta x^2} \\ &= \frac{1}{\Delta x^2}\left(u_{i-1,j} - 2u_{i,j} + u_{i+1,j}\right) \\ &\quad + \frac{1}{12\Delta x^2}\left[\left(u_{i-1,j-1} - 2u_{i-1,j} + u_{i-1,j+1}\right)\right. \\ &\quad - 2\left(u_{i,j-1} - 2u_{i,j} + u_{i,j+1}\right) \\ &\quad \left. + \left(u_{i+1,j-1} - 2u_{i+1,j} + u_{i+1,j+1}\right)\right] \\ &= \frac{1}{12\Delta x^2}\left[-20u_{i,j} + 10\left(u_{i-1,j} + u_{i+1,j}\right)\right. \\ &\quad - 2\left(u_{i,j-1} + u_{i,j+1}\right) \\ &\quad \left. + u_{i-1,j-1} + u_{i-1,j+1} + u_{i+1,j-1} + u_{i+1,j+1}\right]. \end{aligned}$$

Therefore, we have a nine-point stencil, but the approximation only requires three points in each direction; thus, it is *compact*.

Similarly, expanding the second term in (8.63) yields

$$\begin{aligned} \left(1 + \frac{\Delta x^2}{12}\delta_x^2\right)\delta_y^2 u = \frac{1}{12\Delta y^2}&\left[-20u_{i,j} + 10\left(u_{i,j-1} + u_{i,j+1}\right) - 2\left(u_{i-1,j} + u_{i+1,j}\right)\right. \\ &\left. + u_{i-1,j-1} + u_{i-1,j+1} + u_{i+1,j-1} + u_{i+1,j+1}\right], \end{aligned}$$

and the right-hand side of (8.63) is

$$\begin{aligned} \left[1 + \frac{\Delta x^2}{12}\delta_x^2 + \frac{\Delta y^2}{12}\delta_y^2\right] f(x,y) &= f_{i,j} + \frac{1}{12}\left[f_{i-1,j} - 2f_{i,j} + f_{i+1,j}\right. \\ &\quad \left. + f_{i,j-1} - 2f_{i,j} + f_{i,j+1}\right] \\ &= \frac{1}{12}\left[8f_{i,j} + f_{i-1,j} + f_{i+1,j} + f_{i,j-1} + f_{i,j+1}\right]. \end{aligned}$$

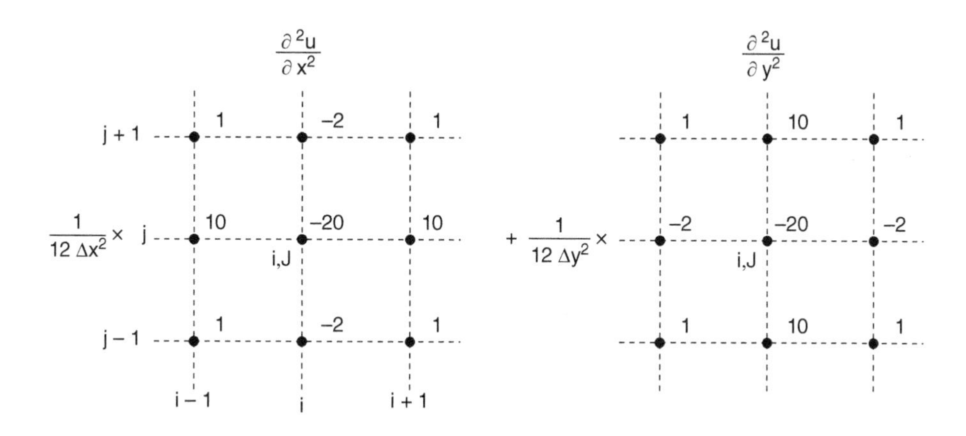

Figure 8.20 Graphical representation of the coefficients in the nine-point stencil for $u_{i,j}$.

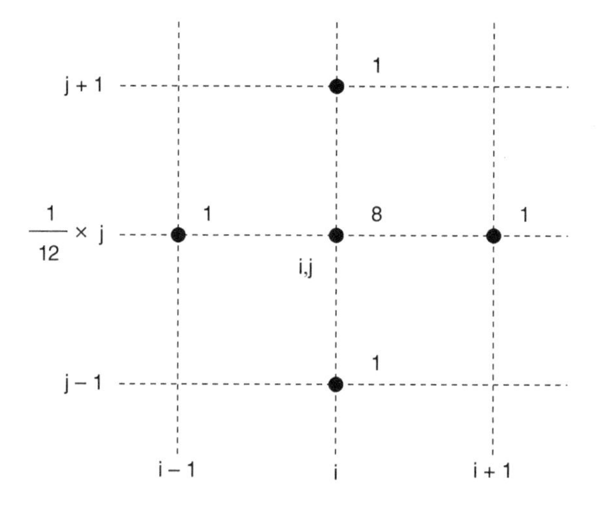

Figure 8.21 Graphical representation of the coefficients in the nine-point stencil for $f_{i,j}$.

The coefficients in the nine-point finite-difference stencils for $u(x, y)$ and $f(x, y)$ are illustrated graphically in Figures 8.20 and 8.21, respectively.

Because the finite-difference stencil is *compact* – only involving three points in each direction – application of the ADI method as in Section 8.7 results in a set of tridiagonal problems to solve. Thus, we can still use the Thomas algorithm for the tridiagonal solves. In this manner, the fourth-order, compact finite-difference approach is no less efficient than that for the second-order ADI scheme, except that there are additional terms on the right-hand side of the equation that must be evaluated.

REMARKS:

1. *Observe that in* (8.63), *the two-dimensionality of the equation has been exploited to obtain the compact finite-difference stencil. That is, the* $\delta_x^2\left(\delta_x^2 u\right)$ *and* $\delta_y^2\left(\delta_y^2 u\right)$ *operators have been converted to* $\delta_x^2\left(\delta_y^2 u\right)$ *and* $\delta_y^2\left(\delta_x^2 u\right)$ *difference operators.*

2. *The primary disadvantage of using higher-order schemes of this type is that it is generally necessary to use lower-order approximations for derivative boundary conditions, which may effect the overall convergence rate. This is not a problem, however, for Dirichlet boundary conditions.*
3. *This is just one example of a higher-order finite-difference method. For more on this important topic, see Tolstykh (1994) and Gustafsson (2008).*

8.10 Treatment of Nonlinear Terms

The methods discussed thus far apply to general *linear* elliptic partial differential equations. In certain applications, the elliptic partial differential equations include nonlinear terms. For example, in fluid mechanics and heat transfer, the Laplacian operator $\nabla^2 u = \nabla \cdot \nabla u$ accounts for the diffusion of momentum and/or heat throughout the domain, respectively. In applications that also involve convection, additional terms are required that are nonlinear. These are of a particular form that can be treated using *Picard linearization* and *upwind-downwind differencing* as discussed in the next section. More general forms of nonlinear terms are treated using Newton linearization as discussed in Section 8.10.2.

8.10.1 Nonlinear Convection Terms

Picard Linearization

Consider the two-dimensional, steady Burgers equations

$$Re\left(u\frac{\partial u}{\partial x} + v\frac{\partial u}{\partial y}\right) = \frac{\partial^2 u}{\partial x^2} + \frac{\partial^2 u}{\partial y^2}, \tag{8.64}$$

$$Re\left(u\frac{\partial v}{\partial x} + v\frac{\partial v}{\partial y}\right) = \frac{\partial^2 v}{\partial x^2} + \frac{\partial^2 v}{\partial y^2}, \tag{8.65}$$

which represent a simplified prototype of the Navier–Stokes equations governing fluid flow as there are no pressure terms. The velocity components are $u(x, y)$ and $v(x, y)$ in the x- and y-directions, respectively. The Reynolds number Re is a nondimensional parameter representing the ratio of convective to viscous forces in the flow; larger Reynolds numbers result in increased nonlinearity of the equations. The terms on the left-hand side are the *convection*, *inertial*, or *acceleration* terms, and those on the right-hand side are the *viscous*, or *diffusion*, terms. The Burgers equations are elliptic owing to the nature of the second-order viscous terms, but the convection terms render the equations nonlinear – actually quasilinear.

A simple approach to dealing with such nonlinear convective terms is known as *Picard linearization*, in which we take the coefficients of the nonlinear first-derivative terms to be known from the previous iteration denoted by $u^*_{i,j}$ and $v^*_{i,j}$. Let us begin by approximating (8.64) using second-order accurate central differences for all derivatives as follows:

$$Re\left(u^*_{i,j}\frac{u_{i+1,j}-u_{i-1,j}}{2\Delta x}+v^*_{i,j}\frac{u_{i,j+1}-u_{i,j-1}}{2\Delta y}\right)$$
$$=\frac{u_{i+1,j}-2u_{i,j}+u_{i-1,j}}{\Delta x^2}+\frac{u_{i,j+1}-2u_{i,j}+u_{i,j-1}}{\Delta y^2}.$$

Multiplying by Δx^2 and collecting terms leads to the difference equation

$$\begin{aligned}&\left(1-\frac{1}{2}Re\,\Delta x\,u^*_{i,j}\right)u_{i+1,j}+\left(1+\frac{1}{2}Re\,\Delta x\,u^*_{i,j}\right)u_{i-1,j}\\&+\left(\hat{\Delta}-\frac{1}{2}Re\,\Delta x\,\hat{\Delta}^{1/2}v^*_{i,j}\right)u_{i,j+1}+\left(\hat{\Delta}+\frac{1}{2}Re\,\Delta x\,\hat{\Delta}^{1/2}v^*_{i,j}\right)u_{i,j-1}\\&-2(1+\hat{\Delta})u_{i,j}=0,\end{aligned}\tag{8.66}$$

where $\hat{\Delta}=(\Delta x/\Delta y)^2$. Note that we have linearized the difference equation, not the differential equation, in order to obtain a linear system of algebraic equations. Consequently, the physical nonlinearity is still being accounted for in the numerical solution. We can solve (8.66) using any of the iterative methods discussed thus far.[9]

For the iteration to converge, we must check to be sure that (8.66) is diagonally dominant. To be diagonally dominant, we must have

$$|1-p|+|1+p|+\left|\hat{\Delta}-q\right|+\left|\hat{\Delta}+q\right|\le\left|2(1+\hat{\Delta})\right|,$$

where

$$p=\frac{1}{2}Re\,\Delta x\,u^*_{i,j},\quad q=\frac{1}{2}Re\,\Delta x\,\hat{\Delta}^{1/2}\,v^*_{i,j}.$$

Suppose, for example, that $p>1$ and $q>\hat{\Delta}$; then this requires that

$$(p-1)+(1+p)+(q-\hat{\Delta})+(\hat{\Delta}+q)\le 2(1+\hat{\Delta}),$$

or

$$2(p+q)\le 2(1+\hat{\Delta}),$$

but with $p>1$ and $q>\hat{\Delta}$ this condition cannot be satisfied, and (8.66) is not diagonally dominant. The same result holds for $p<-1$ and $q<-\hat{\Delta}$. Therefore, we must have $|p|\le 1$ and $|q|\le\hat{\Delta}$ or

$$\left|\frac{1}{2}Re\,\Delta x\,u^*_{i,j}\right|\le 1,\quad\text{and}\quad\left|\frac{1}{2}Re\,\Delta x\,\hat{\Delta}^{1/2}\,v^*_{i,j}\right|\le\hat{\Delta}^{1/2},$$

which is a restriction on the mesh size for a given Reynolds number and velocity field.

With respect to maintaining diagonal dominance, there are two difficulties with this approach:

1. As the Reynolds number Re increases, the grid sizes Δx and Δy must correspondingly decrease.

[9] In SOR, we generally need under-relaxation for nonlinear problems.

2. The velocities $u^*_{i,j}$ and $v^*_{i,j}$ vary throughout the domain and are unknown when choosing the grid sizes.

The underlying cause of these numerical convergence problems is the use of central-difference approximations for the first-order derivatives as they contribute to the off-diagonal terms $u_{i-1,j}$, $u_{i+1,j}$, $u_{i,j-1}$, and $u_{i,j+1}$ but not the main diagonal term $u_{i,j}$, thereby adversely affecting diagonal dominance as compared to the case without convection.

Upwind-Downwind Differencing

In order to restore diagonal dominance free of mesh-size restrictions, we use forward or backward differences for the first-derivative terms depending upon the signs of their coefficients, which are determined by the velocities. For example, consider the $u^* \partial u/\partial x$ term:

1. If $u^*_{i,j} > 0$, then using a backward difference gives

$$u^* \frac{\partial u}{\partial x} = u^*_{i,j} \frac{u_{i,j} - u_{i-1,j}}{\Delta x} + O(\Delta x),$$

which gives a positive addition to the $u_{i,j}$ term to promote diagonal dominance (note the sign of $u_{i,j}$ from the viscous terms on the right-hand side of the difference equation).

2. If $u^*_{i,j} < 0$, then using a forward difference gives

$$u^* \frac{\partial u}{\partial x} = u^*_{i,j} \frac{u_{i+1,j} - u_{i,j}}{\Delta x} + O(\Delta x),$$

which again gives a positive addition to the $u_{i,j}$ term to promote diagonal dominance.

Similarly, for the $v^* \partial u/\partial y$ term:

1. If $v^*_{i,j} > 0$, then use a backward difference according to

$$v^* \frac{\partial u}{\partial y} = v^*_{i,j} \frac{u_{i,j} - u_{i,j-1}}{\Delta y} + O(\Delta y).$$

2. If $v^*_{i,j} < 0$, then use a forward difference according to

$$v^* \frac{\partial u}{\partial y} = v^*_{i,j} \frac{u_{i,j+1} - u_{i,j}}{\Delta y} + O(\Delta y).$$

Let us consider diagonal dominance of the approximations to the x-derivative terms in the Burgers equation (8.64) with upwind-downwind differencing of the first-derivative terms according to

$$T_x = \frac{\partial^2 u}{\partial x^2} - Re\, u \frac{\partial u}{\partial x} = \frac{u_{i+1,j} - 2u_{i,j} + u_{i-1,j}}{\Delta x^2} - \frac{Re\, u^*_{i,j}}{\Delta x} \begin{cases} u_{i,j} - u_{i-1,j}, & u^*_{i,j} > 0 \\ u_{i+1,j} - u_{i,j}, & u^*_{i,j} < 0 \end{cases};$$

therefore, multiplying by $(\Delta x)^2$ to keep the coefficients $O(1)$ and collecting terms gives

$$\Delta x^2 T_x = \begin{cases} u_{i+1,j} + \left(1 + Re\,\Delta x\, u^*_{i,j}\right) u_{i-1,j} - \left(2 + Re\,\Delta x\, u^*_{i,j}\right) u_{i,j}, & u^*_{i,j} > 0 \\ \left(1 - Re\,\Delta x\, u^*_{i,j}\right) u_{i+1,j} + u_{i-1,j} - \left(2 - Re\,\Delta x\, u^*_{i,j}\right) u_{i,j}, & u^*_{i,j} < 0 \end{cases}.$$

As a result, if we are using ADI with splitting of x- and y-derivative terms, diagonal dominance of the tridiagonal problems along lines of constant y requires the following:

1. For $u^*_{i,j} > 0$, in which case $Re\,\Delta x\, u^*_{i,j} > 0$, we must have

$$|1| + |1 + Re\,\Delta x\, u^*_{i,j}| \leq |-(2 + Re\,\Delta x\, u^*_{i,j})|,$$

in which case

$$1 + 1 + Re\,\Delta x\, u^*_{i,j} = 2 + Re\,\Delta x\, u^*_{i,j},$$

which is weakly diagonally dominant.

2. For $u^*_{i,j} < 0$, in which case $Re\,\Delta x\, u^*_{i,j} < 0$, we must have

$$|1| + |1 - Re\,\Delta x\, u^*_{i,j}| \leq |-(2 - Re\,\Delta x\, u^*_{i,j})|,$$

in which case

$$1 + 1 - Re\,\Delta x\, u^*_{i,j} = 2 - Re\,\Delta x\, u^*_{i,j},$$

which is also weakly diagonally dominant.

The same is true for the y-derivative terms when sweeping along lines of constant x. Because the tridiagonal problems solved along each constant x and y line of the grid are weakly diagonally dominant, the iteration is expected to converge with no mesh restrictions. Therefore, the grid size is chosen solely based on accuracy considerations, without regard for iterative convergence.

It is important to appreciate that upwind-downwind differencing is motivated by numerical, not physical, considerations. It is simply a scheme to increase the magnitude of the diagonal coefficients $u_{i,j}$ relative to the off-diagonal terms. In order to maintain consistency of approximations, use upwind-downwind differencing for both first-order derivative terms whether sweeping along lines of constant x or y. Consequently, there are four possible cases at each point in the grid depending on the signs of $u^*_{i,j}$ and $v^*_{i,j}$.

Although upwind-downwind differencing lifts the mesh-size restrictions required for diagonal dominance, the forward and backward differences are only first-order accurate. Therefore, the method is only $O(\Delta x, \Delta y)$ accurate. To see the potential consequences of the first-order accuracy in the convective terms, consider the one-dimensional Burgers equation

$$Re\, u \frac{du}{dx} = \frac{d^2u}{dx^2}. \tag{8.67}$$

Recall from Section 7.2.1 that, for example, the first-order, backward-difference approximation to the first-order derivative is

$$\left(\frac{du}{dx}\right)_i = \frac{u_i - u_{i-1}}{\Delta x} + \frac{\Delta x}{2}\left(\frac{d^2u}{dx^2}\right)_i + \dots,$$

where we have included the truncation error. Substituting into (8.67) gives

$$Re\, u_i \left[\frac{u_i - u_{i-1}}{\Delta x} + \frac{\Delta x}{2}\left(\frac{d^2u}{dx^2}\right)_i + \cdots\right] = \left(\frac{d^2u}{dx^2}\right)_i,$$

or

$$Re\, u_i \frac{u_i - u_{i-1}}{\Delta x} = \left[1 - \frac{Re}{2}\Delta x\, u_i\right]\left(\frac{d^2u}{dx^2}\right)_i.$$

Therefore, depending upon the values of $Re, \Delta x$, and $u(x)$, the truncation error from the first-derivative term, which is not included in the numerical solution, may be of the same order, or even larger than, the physical diffusion term. This is often referred to as *artificial*, or *numerical, diffusion*, the effects of which increase with increasing Reynolds number.

There are two potential remedies to the first-order approximations inherent in the upwind-downwind differencing approach as usually implemented:

1. Second-order, that is, $O(\Delta x^2, \Delta y^2)$, accuracy can be restored using *deferred correction*, in which we use the approximate solution for $u(x, y)$ to evaluate the leading term of the truncation error, which is then added to the original discretized equation as a source term.
2. Second-order accurate forward and backward differences could be used, but the resulting system of equations would no longer be compact and tridiagonal.

8.10.2 Newton Linearization

Picard linearization is applicable to differential equations that are nonlinear owing to convection-type terms, in which case we simply lag the velocity multiplying the derivative by taking its value at the previous iteration. In more general cases involving nonlinear terms in the governing ordinary or partial differential equation, *Newton linearization* is required.

To illustrate the method, reconsider the $u\partial u/\partial x$ term in the Burgers equation in order to see how it is different from Picard linearization. Let us use a second-order accurate, central-difference approximation for the first derivative as follows:

$$u_{i,j}\frac{u_{i+1,j} - u_{i-1,j}}{2\Delta x} + O(\Delta x^2). \tag{8.68}$$

In Newton linearization, a truncated Taylor series for $u_{i,j}$ about its value at the previous iteration is applied as follows. The value of $u_{i,j}$ at the current iteration is given by

$$u_{i,j} = u^*_{i,j} + \Delta u_{i,j}, \tag{8.69}$$

where again $u^*_{i,j}$ is the velocity from the previous iteration, and $\Delta u_{i,j}$ is assumed to be small. Substituting into (8.68) yields (neglecting $2\Delta x$ for now)

$$\left(u^*_{i,j} + \Delta u_{i,j}\right)\left(u^*_{i+1,j} + \Delta u_{i+1,j} - u^*_{i-1,j} - \Delta u_{i-1,j}\right).$$

Multiplying and neglecting terms that are quadratic in Δu gives

$$u^*_{i,j}u^*_{i+1,j} - u^*_{i,j}u^*_{i-1,j} + u^*_{i,j}\Delta u_{i+1,j} - u^*_{i,j}\Delta u_{i-1,j} + u^*_{i+1,j}\Delta u_{i,j} - u^*_{i-1,j}\Delta u_{i,j}. \tag{8.70}$$

Substituting (8.69) for the Δu terms produces

$$u^*_{i,j}u^*_{i+1,j} - u^*_{i,j}u^*_{i-1,j} + u^*_{i,j}\left(u_{i+1,j} - u^*_{i+1,j}\right) - u^*_{i,j}\left(u_{i-1,j} - u^*_{i-1,j}\right)$$
$$+ u^*_{i+1,j}\left(u_{i,j} - u^*_{i,j}\right) - u^*_{i-1,j}\left(u_{i,j} - u^*_{i,j}\right),$$

which simplifies to

$$u^*_{i,j}u_{i+1,j} - u^*_{i,j}u_{i-1,j} + u^*_{i+1,j}\left(u_{i,j} - u^*_{i,j}\right) - u^*_{i-1,j}\left(u_{i,j} - u^*_{i,j}\right). \tag{8.71}$$

Observe that because all of the starred terms are known from the previous iteration, there is now no more than one unknown in each term, and they appear linearly. Thus, we have linearized the nonlinear algebraic term (8.68). The term (8.68) is then replaced with (8.71) divided by $(2\Delta x)$ in the full finite-difference equation and iterated until convergence as usual.

REMARKS:

1. *Newton's method exhibits a quadratic convergence rate if a good initial guess is used, such that Δu is small.*
2. *If Δu is too large, then Newton's method may diverge from the true solution.*

Exercises

Unless stated otherwise, perform the exercises using hand calculations. Exercises to be completed using "built-in functions" should be completed using the built-in functions within Python, MATLAB, or Mathematica for the vector and matrix operation(s) or algorithm. A "user-defined function" is a standalone computer program to accomplish a task written in a programming language, such as Python, MATLAB, Mathematica, Fortran, or C++.

8.1 Consider the ordinary differential equation

$$\frac{d^2u}{dx^2} - \frac{du}{dx} = 0, \quad u(0) = 1, \quad u(1) = 2.$$

(a) Using second-order accurate, central-difference approximations for all derivatives, determine the coefficients a, b, c, and d in the tridiagonal matrix to be solved:

$$au_{i-1} + bu_i + cu_{i+1} = d.$$

Is the system strongly diagonally dominant, weakly diagonally dominant, or not diagonally dominant?

(b) When solving the tridiagonal problem given in part (a) using the Thomas algorithm with the domain discretized using 100 intervals (101 points), how many digits of accuracy would you expect to lose?

8.2 Suppose that the movement of rush-hour traffic on a typical expressway can be modeled using the ordinary differential equation

$$\frac{d^2u}{dx^2} - x\frac{du}{dx} + 2x^2 = 0, \quad 0 \leq x \leq 5,$$

where $u(x)$ is the density of cars (vehicles per mile), and x is distance (in miles) in the direction of traffic flow. We want to solve this equation subject to the boundary conditions

$$u(0) = 300, \quad u(5) = 400.$$

(a) Use second-order accurate, central-difference approximations to discretize the differential equation and write down the finite-difference equation in the form that it would be solved for a typical point x_i in the domain.

(b) If the domain is divided into five equal subintervals, write down the matrix problem to be solved to determine $u(x_i), i = 1, \ldots, 6$.

(c) If the system of equations is to be solved iteratively using the Gauss–Seidel method, write down the equation to be iterated upon.

8.3 Jacobi iteration is to be used to solve the following ordinary differential equation:

$$\frac{d^2u}{dx^2} + a\frac{du}{dx} + bu = 0.$$

(a) Using second-order accurate, central differences, determine the difference equation to be used to update $u(x_i)$ within the interior of the computational domain.

(b) If $b = 3/(\Delta x)^2$, what is the necessary and sufficient condition on a that is required for iterative convergence?

8.4 The Thomas algorithm is to be used to solve the tridiagonal system of equations

$$a_i u_{i-1} + b_i u_i + c_i u_{i+1} = d_i, \quad i = 1, \ldots, I+1.$$

For a boundary condition

$$pu + q\frac{du}{dx} = r, \quad (p, q, r \text{ constant})$$

applied at the left boundary $i = 1$, determine the adjusted coefficients for the first equation in the tridiagonal system to maintain second-order accuracy.

8.5 The Thomas algorithm is to be used to solve the tridiagonal system of equations

$$a_i u_{i-1} + b_i u_i + c_i u_{i+1} = d_i, \quad i = 1, \ldots, I+1.$$

For a boundary condition

$$pu + q\frac{du}{dx} = r, \quad (p, q, r \text{ constant})$$

applied at the right boundary $i = I + 1$, determine the adjusted coefficients for the last equation in the tridiagonal system to maintain second-order accuracy.

8.6 Recall the general second-order linear ordinary differential equation

$$\frac{d^2u}{dx^2} + f(x)\frac{du}{dx} + g(x)u = h(x),$$

with the Dirichlet boundary conditions

$$u(a) = A, \quad u(b) = B.$$

In Section 8.1, we showed how to discretize the domain $a \leq x \leq b$ into I equal subintervals and the differential equation using second-order accurate, central-difference approximations to convert the differential equation into a tridiagonal system of linear algebraic equations.

(a) Write a user-defined function to obtain a numerical solution of the general differential equation. The inputs are the values of $f(x_i)$, $g(x_i)$, and $h(x_i)$ at each grid point in the domain along with the values of a, b, A, B, and the number of intervals I. The function should form the tridiagonal system of algebraic equations, solve it using any built-in matrix solver, and plot the results.

(b) Test your function using Example 8.1.

8.7 Use your user-defined function from Exercise 8.6 to solve the extended-fin problem in Section 8.1. Note that if the fin has constant cross-sectional area, then $A_s(x) = Px$, where P is the perimeter of the cross-section of the fin. For this case, the fin equation simplifies to

$$\frac{d^2u}{dx^2} - m^2u = 0,$$

where

$$m^2 = \frac{hP}{kA_c},$$

which is constant. Consider a cylindrically shaped fin with the following fin configuration (note the units):

$$T(0) = T_b = 200°\text{C}, \quad T(\ell) = T_\ell = 60°\text{C}, \quad T_\infty = 20°\text{C}, \tag{8.72}$$

$$\ell = 10 \text{ cm}, \quad P = 9 \text{ cm}, \quad k = 80 \text{ W/m °C}, \quad h = 15 \text{ W/m}^2 \text{ °C}. \tag{8.73}$$

Obtain and plot the numerical solution with $I = 40$.

8.8 Reconsider the extended-fin problem as defined in Exercise 8.7 and perform the following using your user-defined function from Exercise 8.6:

(a) Obtain a grid-independent numerical solution for $u(x), 0 \leq x \leq \ell$.

(b) Compare your grid-independent numerical solution from part (a) with the analytical solution given by

$$u(x) = \frac{u_\ell \sinh(mx) + u_b \sinh[m(\ell - x)]}{\sinh(m\ell)}, \tag{8.74}$$

by computing the error.

(c) Illustrate how the numerical solution converges to the exact solution at a rate proportional to the truncation error of the finite-difference approximations, that is, $O(\Delta x^2)$, as the grid size is reduced.

(d) From the numerical solution obtained in part (a), determine the heat flux from the base of the fin, which is defined by

$$q_f = q_b = -kA_c(0)\left(\frac{dT}{dx}\right)_{x=0} = -kA_c(0)\left(\frac{du}{dx}\right)_{x=0}, \tag{8.75}$$

using first-order, second-order, and third-order accurate finite-difference approximations. Compare these results with the exact solution given by

$$q_f = kA_c(0)m\frac{u_b \cosh(m\ell) - u_\ell}{\sinh(m\ell)}. \tag{8.76}$$

Discuss your results.

(e) Replace the fixed temperature condition (Dirichlet) at the tip of the fin ($x = \ell$) with a convection condition (Robin) of the form

$$-k\left(\frac{dT}{dx}\right)_{x=\ell} = h[T(\ell) - T_\infty], \tag{8.77}$$

or in terms of u

$$-k\left(\frac{du}{dx}\right)_{x=\ell} = hu(\ell). \tag{8.78}$$

Obtain the grid-independent temperature distribution in the fin.

(f) Compare your numerical solution from part (e) with the analytical solution given by

$$u(x) = u_b\frac{\cosh[(m(\ell - x)] + \bar{h}\sinh[m(\ell - x)]}{\cosh(m\ell) + \bar{h}\sinh(m\ell)}, \tag{8.79}$$

where $\bar{h} = \frac{h}{mk}$.

(g) Another possible fin condition is to make the fin long enough so that the tip is at the same temperature as the ambient air [$u(\ell) = 0$], in which case all of the heat has been transferred out of the fin by convection before it reaches the tip. Theoretically, this would require a fin of infinite length ($\ell \to \infty$). In practice, what is the approximate minimum length of the fin necessary for this condition to be a good approximation? Is this practical?

8.9 In Section 8.6.2, we applied the Gauss–Seidel method to the Poisson partial differential equation with Dirichlet boundary conditions and a Neumann boundary condition along $x = 0$. Applying the same procedure, determine the equation to be solved using the Gauss–Seidel method along the $y = 0$ boundary with the Neumann boundary condition (excluding the corner points)

$$\frac{\partial u}{\partial y} = A.$$

8.10 Jacobi or Gauss–Seidel iteration is to be used to solve the steady heat conduction equation

$$\frac{\partial^2 T}{\partial x^2} + \frac{\partial^2 T}{\partial y^2} = 0,$$

with a convection condition at the upper boundary of the domain:

$$-k\frac{\partial T}{\partial y} = h(T - T_\infty) \quad \text{at} \quad y = y_{max},$$

where k, h, and T_∞ are constants. Using central differences with $\Delta = \Delta x = \Delta y$, determine the equation to be used to update the temperature T along the upper boundary at $j = J + 1$ (excluding the corners).

8.11 The following equation is to be solved using the ADI method:

$$\frac{\partial^2 u}{\partial x^2} + \frac{\partial u}{\partial x} + \frac{\partial^2 u}{\partial y^2} = 0.$$

Using second-order accurate, central differences and $\Delta = \Delta x = \Delta y$, write down the tridiagonal problems to be solved along the constant x- and y-lines (use the splitting method). Evaluate each of the tridiagonal problems to determine if they are diagonally dominant.

8.12 Jacobi iteration is to be used to solve the following differential equation

$$\frac{\partial^2 u}{\partial x^2} + \frac{\partial^2 u}{\partial x \partial y} + \frac{\partial^2 u}{\partial y^2} = 0.$$

(a) Using second-order accurate, central differences, determine the difference equation to be used to update $u(x, y)$ within the interior of the computational domain.

(b) Determine the difference equation to be used to update $u(x, y)$ on the upper boundary of the computational domain at $j = J + 1$ (excluding the corners), where the following Neumann boundary condition is applied:

$$\frac{\partial u}{\partial y} = 0.$$

(c) Determine the difference equation to be used to update $u(x, y)$ on the lower boundary of the computational domain at $j = 1$ (excluding the corners), where the following Robin boundary condition is applied:

$$u + 2\frac{\partial u}{\partial y} = 0.$$

8.13 We seek to solve the Helmholtz equation

$$\frac{\partial^2 u}{\partial x^2} + \frac{\partial^2 u}{\partial y^2} + u = 0, \quad 0 \le x \le a, \quad 0 \le y \le b,$$

iteratively using the Gauss–Seidel method. Using second-order accurate, central-difference approximations for all derivatives, determine the equation to be used to update the corner point at $(x, y) = (0, 0)$ if the boundary conditions are

$$\frac{\partial u}{\partial x} = 1 \quad \text{at} \quad x = 0,$$

$$\frac{\partial u}{\partial y} = -1 \quad \text{at} \quad y = 0.$$

8.14 A common test problem for elliptic solvers is the Poisson equation

$$\frac{\partial^2 u}{\partial x^2} + \frac{\partial^2 u}{\partial y^2} = \cos(m\pi x)\cos(n\pi y), \tag{8.80}$$

over the interval $0 \leq x \leq 1, 0 \leq y \leq 1$. The constants m and n are odd integers. For the Neumann boundary conditions

$$\frac{\partial u}{\partial x} = 0 \quad \text{at} \quad x = 0, x = 1, \tag{8.81}$$

and

$$\frac{\partial u}{\partial y} = 0 \quad \text{at} \quad y = 0, y = 1, \tag{8.82}$$

the exact solution is

$$u(x, y) = -\frac{1}{(m\pi)^2 + (n\pi)^2}\cos(m\pi x)\cos(n\pi y). \tag{8.83}$$

Write a user-defined function that solves this problem using Gauss–Seidel iteration with SOR. Provide the numerical solution for $m = n = 1$ on a 51×51 mesh, and compare your results with the exact solution. Use a convergence criterion of $\epsilon = 1 \times 10^{-5}$.

Note that this is a Poisson equation with Neumann boundary conditions; therefore, the solution can only be determined relative to an arbitrary constant. In order to prevent divergence (and to make comparison with the exact solution easier), normalize the solution after completion of each iteration by taking $u_{i,j} = u_{i,j} - u(x = 1/2, y = 1/2)$. This sets the value of u at the center of the domain to zero (as in the exact solution).

8.15 Repeat Exercise 8.14 using an ADI method that uses the Thomas algorithm to solve the tridiagonal system of equations along each line. Use splitting and acceleration as in Section 8.7.

8.16 Modify the ADI algorithm from Exercise 8.15 to solve for the potential fluid flow in the vicinity of a stagnation point on a wall. In potential (incompressible, inviscid, irrotational) flow, the streamfunction $\psi(x, y)$ is governed by Laplace's equation

$$\frac{\partial^2 \psi}{\partial x^2} + \frac{\partial^2 \psi}{\partial y^2} = 0.$$

Consider the domain shown in Figure 8.22, where the stagnation point is at the origin, the left boundary is the wall, and one half of the stagnation point flow is considtered with the x-axis being an axis of symmetry. The computational domain is $0 \leq x \leq 1, 0 \leq y \leq 1$, and the Dirichlet boundary conditions are

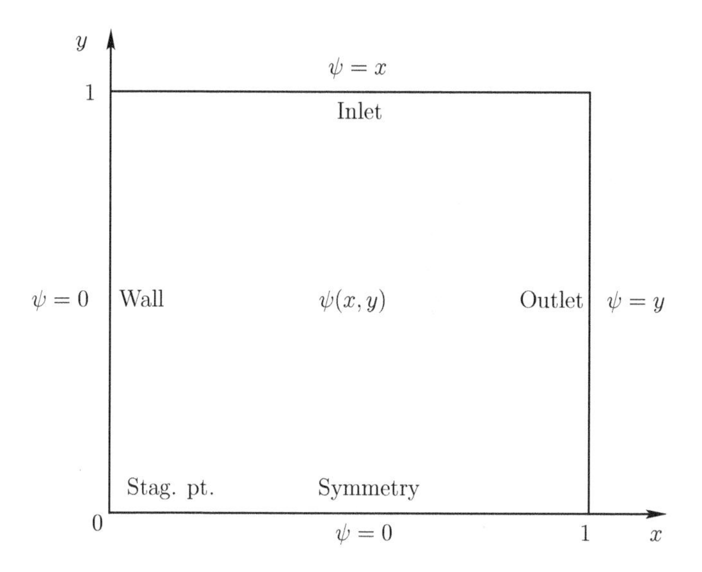

Figure 8.22 Schematic for Exercise 8.16.

$$\psi = 0, \quad \text{at} \quad x = 0,$$

$$\psi = y, \quad \text{at} \quad x = 1,$$

$$\psi = 0, \quad \text{at} \quad y = 0,$$

$$\psi = x, \quad \text{at} \quad y = 1.$$

For comparison with your numerical solution, note that the exact solution of the Laplace equation subject to these boundary conditions is $\psi(x, y) = xy$, corresponding to the velocity components $u = x$ and $v = -y$ ($u = \partial\psi/\partial y, v = -\partial\psi/\partial x$).

8.17 Consider steady heat conduction, which is governed by the Laplace equation

$$\frac{\partial^2 T}{\partial x^2} + \frac{\partial^2 T}{\partial y^2} = 0, \tag{8.84}$$

in the domain $0 \leq x \leq 1, 0 \leq y \leq 1$. The boundary conditions are

$$T = 0, \quad \text{at} \quad x = 0, \tag{8.85}$$

$$T = 0, \quad \text{at} \quad x = 1, \tag{8.86}$$

$$T = T_0(x), \quad \text{at} \quad y = 0, \tag{8.87}$$

$$T = 0, \quad \text{at} \quad y = 1. \tag{8.88}$$

(a) Using the ADI method, obtain the finite-difference equation(s) necessary to solve this problem.

(b) Write a user-defined function to solve the finite-difference equation(s) from part (a) for the case when

$$T_0(x) = \exp[-\sigma(x - 1/2)^2]. \tag{8.89}$$

This corresponds to a Gaussian distribution centered at $x = 1/2$ with its width determined by setting σ; take $\sigma = 25$.

(c) Obtain a grid-independent solution for the case given in part (b). Plot the temperature distributions along the lines $y = 0, 0.25, 0.5$ and $x = 0.25, 0.5$. Plot contours of constant temperature (isotherms) throughout the domain. Discuss your results.

8.18 The following equation is to be solved using the ADI method with splitting and upwind-downwind differencing:

$$\frac{\partial^2 u}{\partial x^2} + P(x,y)\frac{\partial u}{\partial x} + \frac{\partial^2 u}{\partial y^2} = 0.$$

Write down the tridiagonal problems to be solved along constant x and y lines. Check each of these sets of problems for diagonal dominance. What is the effect of using upwind-downwind differencing on the truncation error of an iterative scheme as compared to using central differences for all terms?

8.19 The convection-diffusion equation

$$u\frac{\partial \phi}{\partial x} + v\frac{\partial \phi}{\partial y} = \alpha\left(\frac{\partial^2 \phi}{\partial x^2} + \frac{\partial^2 \phi}{\partial y^2}\right),$$

is to be solved on the computational domain $0 \le x \le 1, 0 \le y \le 1$ using a Gauss–Seidel method with first-order accurate upwind-downwind differencing (take $\Delta = \Delta x = \Delta y$).

(a) Determine the equation to be used to update points in the interior of the domain.

(b) Determine the equation to be used to update the points along the bottom boundary if a Neumann condition is applied:

$$\frac{\partial \phi}{\partial y} = 0 \quad \text{at} \quad y = 0.$$

8.20 The convection-diffusion equation

$$u\frac{\partial T}{\partial x} + v\frac{\partial T}{\partial y} = \alpha\left(\frac{\partial^2 T}{\partial x^2} + \frac{\partial^2 T}{\partial y^2}\right),$$

is to be solved on the computational domain $0 \le x \le 1, 0 \le y \le 1$, using an ADI method with first-order accurate upwind-downwind differencing (use splitting of the x and y derivatives and take $\Delta = \Delta x = \Delta y$). Determine the tridiagonal equation to be solved on the bottom boundary if a convection condition is applied as follows:

$$-k\frac{\partial T}{\partial y} = h(T - T_\infty) \quad \text{at} \quad y = 0,$$

where k, h, and T_∞ are constants. Is this system of equations diagonally dominant?

8.21 Consider the problem of the transport (via both convection and diffusion) of a scalar quantity, such as temperature, in the velocity field near a stagnation point flow as shown in Figure 8.23. Recall from Exercise 8.16 that the velocities in the x- and y-directions in one-half of the symmetric stagnation point flow from a potential flow solution are given by $u = x$ and $v = -y$, respectively, that is, the streamlines

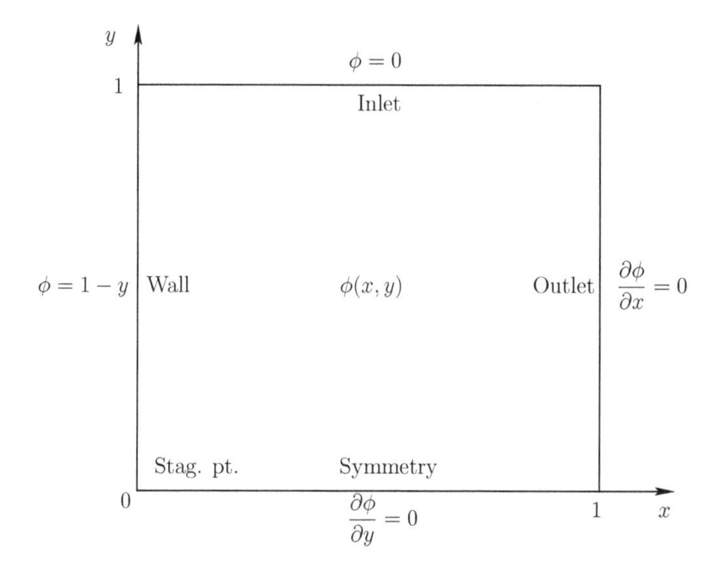

Figure 8.23 Schematic for Exercise 8.21.

are lines of $\psi(x, y) = xy =$ constant. Note that this is somewhat artificial in that the velocities are taken from a potential flow solution, which is inviscid, while the convection-diffusion equation for $\phi(x, y)$ is viscous. The steady scalar-transport equation is given by

$$u \frac{\partial \phi}{\partial x} + v \frac{\partial \phi}{\partial y} = \alpha \left(\frac{\partial^2 \phi}{\partial x^2} + \frac{\partial^2 \phi}{\partial y^2} \right), \tag{8.90}$$

where α is the diffusion parameter. The computational domain is $0 \leq x \leq 1, 0 \leq y \leq 1$. The boundary conditions are

$$\phi = 1 - y, \quad \text{at} \quad x = 0, \tag{8.91}$$

$$\frac{\partial \phi}{\partial x} = 0, \quad \text{at} \quad x = 1, \tag{8.92}$$

$$\frac{\partial \phi}{\partial y} = 0, \quad \text{at} \quad y = 0, \tag{8.93}$$

$$\phi = 0, \quad \text{at} \quad y = 1. \tag{8.94}$$

Modify your ADI algorithm from Exercise 8.15 to include upwind-downwind differencing of the convective terms on the left-hand side of the convection-diffusion equation. Do not assume that $u > 0$ and $v < 0$; that is, include all four possible velocity scenarios at each point. Obtain grid-independent solutions for $\phi(x, y)$ with values of α equal to 0.1 and 0.01. Note that you may need to use a different convergence criterion and/or acceleration parameters as compared to those in Exercise 8.15. In addition, different values of α may require different acceleration parameters.

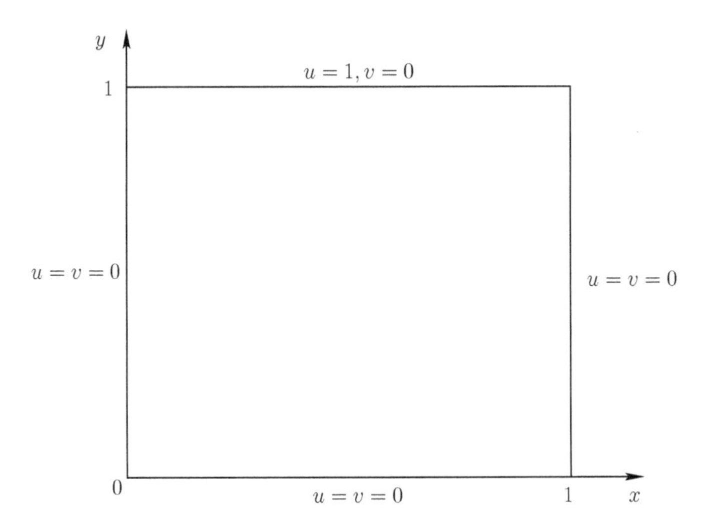

Figure 8.24 Schematic for Exercise 8.22.

8.22 A popular test problem for Navier–Stokes solvers is the flow in a driven cavity shown in Figure 8.24. It is common to test new numerical algorithms on this problem because it avoids the treatment of inlet, outlet, and symmetry boundary conditions, and many numerical solutions are available in the literature for various Reynolds numbers. Note, however, that there are singularities at the upper corners owing to the abrupt change in boundary conditions for u between the side walls and the moving surface at the top of the cavity.

In the vorticity-streamfunction formulation, the steady, incompressible flow in the cavity is governed by the vorticity-transport equation

$$u\frac{\partial \omega}{\partial x} + v\frac{\partial \omega}{\partial y} = \frac{1}{Re}\left(\frac{\partial^2 \omega}{\partial x^2} + \frac{\partial^2 \omega}{\partial y^2}\right), \tag{8.95}$$

and a Poisson equation for the streamfunction

$$\frac{\partial^2 \psi}{\partial x^2} + \frac{\partial^2 \psi}{\partial y^2} = -\omega. \tag{8.96}$$

The boundary condition for this Poisson equation is that $\psi = 0$ on all boundaries.

Using the ADI Poisson solver you developed in Exercise 8.15 and your ADI algorithm for the scalar-transport equation developed in Exercise 8.21, develop an iterative algorithm to solve the coupled equations (8.95) and (8.96). Research *Thom's method* and use it to obtain Dirichlet boundary conditions for the vorticity-transport equation, and use central difference approximations to evaluate the definitions of the streamfunction, $u = \partial\psi/\partial y, v = -\partial\psi/\partial x$, to obtain updated values of the velocities $u(x, y)$ and $v(x, y)$ for use in this equation. Obtain solutions for $Re = 0$ (no convection) and $Re = 100$ on a 101×101 grid. Plot the streamlines and the isovorticity lines. Note that it may be necessary to use different acceleration parameters for each equation. Obtain what you regard to be grid-independent solutions for higher Reynolds numbers and compare with results in the literature.

8.23 A difference equation is to be solved using a multigrid algorithm. The grid is defined by

$$N_x = m_x 2^{(n_x-1)} + 1, \quad N_y = m_y 2^{(n_y-1)} + 1,$$

where $m_x = 3, n_x = 6, m_y = 2$, and $n_y = 4$. Give the number of grid levels to be used in the multigrid algorithm and the grid sizes for each grid.

8.24 A difference equation is to be solved on a 97×33 grid using a multigrid algorithm, where the grid is defined by

$$N_x = m_x 2^{(n_x-1)} + 1, \quad N_y = m_y 2^{(n_y-1)} + 1.$$

Determine the m_x, n_x, m_y, and n_y that will result in the most efficient multigrid solution. Give the number of grid levels to be used in the multigrid algorithm and the grid sizes for each grid.

8.25 A multigrid method is to be developed for the following differential equation:

$$\frac{d^2u}{dx^2} + \alpha \frac{du}{dx} = f(x), \quad 0 \le x \le 1,$$

where α is a constant, $f(x)$ is a known function, and the boundary conditions are

$$u = a \quad \text{at} \quad x = 0, \qquad \frac{du}{dx} = b \quad \text{at} \quad x = 1,$$

where a and b are constants.

(a) The equation is to be discretized with the finest grid having 97 points (96 intervals). Give the number of grid levels to be used in the multigrid algorithm and the number of grid points for each grid level.

(b) Using second-order accurate central differences, write down the equations to be evaluated using Gauss–Seidel relaxation on the finest (Ω^h) and the next coarser (Ω^{2h}) grid levels, including the expressions to be evaluated at the endpoints. Also write down the expression for the residual on the finest grid.

8.26 A multigrid method is to be developed for the following Poisson equation:

$$\frac{\partial^2 u}{\partial x^2} + \frac{\partial^2 u}{\partial y^2} = 4.$$

Assuming that an approximate solution $u(x, y)$ has been obtained on the finest grid Ω^h, write down the error equation on the next coarser grid Ω^{2h} that is to be relaxed using Gauss–Seidel iteration. Use central differences and take $\Delta = \Delta x = \Delta y$.

8.27 A multigrid method is to be developed for the following differential equation:

$$\frac{\partial^2 u}{\partial x^2} + \frac{\partial u}{\partial y} + \frac{\partial^2 u}{\partial y^2} = 4.$$

(a) Assuming that an approximate solution $u(x, y)$ has been obtained on the finest grid Ω^h, write down the error equation on the next coarser grid Ω^{2h} that is to be relaxed using Gauss–Seidel iteration. Use central differences and take $\Delta = \Delta x = \Delta y$.

(b) If the boundary condition on u is

$$\frac{\partial u}{\partial x} = y \quad \text{at} \quad x = 0,$$

what is the equation to be used to update the error along this boundary on the coarser grid?

8.28 A multigrid method is to be developed for the following Helmholtz equation:

$$\frac{\partial^2 u}{\partial x^2} + \frac{\partial^2 u}{\partial y^2} - 16u = 0.$$

Assuming that an approximate solution $u(x, y)$ has been obtained on the finest grid Ω^h, write down the error equation on the next coarser grid Ω^{2h} that is to be relaxed using Gauss–Seidel iteration. Use central differences and take $\Delta = \Delta x = \Delta y$.

8.29 A multigrid method is to be developed for the following Poisson equation:

$$\frac{\partial^2 u}{\partial x^2} + \frac{\partial^2 u}{\partial y^2} = 4.$$

Assuming that an approximate solution $u(x, y)$ has been obtained on the finest grid Ω^h, write down the error equation on the next coarser grid Ω^{2h} that is to be relaxed along lines of constant y using ADI iteration. Use central differences with splitting and take $\Delta = \Delta x = \Delta y$.

8.30 Recall from Chapter 5 that analyzing the physical stability of discrete systems, which are governed by ordinary differential equations, results in the need to solve a generalized algebraic eigenproblem. Stability of continuous systems, which are governed by partial differential equations, requires solution of differential eigenproblems. For example, the *Orr–Sommerfeld equation* results from a local, linear, normal-mode stability analysis of viscous fluid flows, which is derived in section 9.5.3 of Cassel (2013) and is given by

$$(u_0 - c)(v_1'' - \alpha^2 v_1) - u_0'' v_1 = \frac{1}{\alpha i}\frac{1}{Re}(v_1'''' - 2\alpha^2 v_1'' + \alpha^4 v_1),$$

for the disturbance velocity $v_1(y)$. The known base flow is given by $u_0(y)$, the Reynolds number of the flow is denoted by Re, the wavenumber of the disturbance is given by α, and the wave speeds c are the complex eigenvalues that are sought. Wave speeds with a positive imaginary part are unstable to small disturbances and grow to become unbounded. Note that this is a continuous differential eigenproblem of the generalized form

$$\mathcal{L}_1 v_1 = c\mathcal{L}_2 v_1.$$

Because the Orr–Sommerfeld equation is fourth-order, two boundary conditions are needed on the disturbance velocity at each boundary. For solid surfaces at $y = a, b$, we set

$$v_1 = v_1' = 0, \quad \text{at} \quad y = a, b.$$

(a) Using second-order accurate central differences, approximate the Orr–Sommerfeld equation to produce the corresponding generalized algebraic eigenproblem

$$\mathbf{M}(\alpha, Re)\mathbf{v} = c\mathbf{N}(\alpha)\mathbf{v}.$$

Show that matrix $\mathbf{M}(\alpha, Re)$ is symmetric and pentadiagonal, while matrix $\mathbf{N}(\alpha)$ is symmetric and tridiagonal. This (large) generalized eigenproblem must be solved to obtain the complex wave speeds $c = c_r + \mathrm{i}c_\mathrm{i}$, which are the eigenvalues, and the discretized eigenfunctions $v_1(y)$ for a given α, Re, and $u_0(y)$. These discrete eigenvalues and eigenvectors of the algebraic eigenproblem approximate the eigenvalues and eigenfunctions of the original differential eigenproblem.

(b) Consider stability of Couette flow (see Example 1.3), for which the base-flow solution for the velocity profile is given by

$$u_0(y) = y(2 - y), \quad 0 \leq y \leq 2. \tag{8.97}$$

Using 200 intervals (201 points), calculate the complex wave speeds for a given wavenumber $\alpha = 1$ and Reynolds number $Re = 10{,}000$ in order to evaluate stability. A flow is unstable if the imaginary part of one of the discrete eigenvalues is positive, such that $\max(c_\mathrm{i}) > 0$. The growth rate of the instability is then $\alpha \max(c_\mathrm{i})$.

In addition to the fact that the matrices are sparse, in stability contexts such as this, we only need the *most unstable* (or least stable) *mode*, not the entire spectrum of eigenvalues. Recall that this requires the mode with the largest imaginary part. Currently, the state of the art in such situations is the Arnoldi method discussed in Section 6.5.3. Note that **N** is symmetric; it also must be positive definite for use in the Arnoldi method. That is, it must have all positive eigenvalues. In our case, this requires us to take the negatives of the matrices **M** and **N** as determined here.

9 Finite-Difference Methods for Initial-Value Problems

> The scientist does not study nature because it is useful; he studies it because he delights in it, and he delights in it because it is beautiful. (Jules Henri Poincaré)

In contrast to boundary-value problems, which have no preferred direction of propagation of solution and require a global solution strategy, initial-value problems evolve in time. They have a particular direction of propagation of solution; therefore, they require algorithms that "march" forward in time from an initial condition. Because such problems evolve in time, they are referred to as *unsteady* or *transient* differential equations. As in the previous chapter, we develop numerical methods for obtaining solutions of ordinary and partial differential equations. In the case of partial differential equations, this corresponds to parabolic or hyperbolic equations. We begin with ordinary initial-value problems to illustrate the basic approaches to time-marching algorithms. These approaches are then extended naturally to parabolic and hyperbolic partial differential equations. While there are a whole host of methods that can be applied to ordinary differential initial-value problems, we focus primarily on methods for ordinary differential equations that can be extended to partial differential equations. Methods that are only applicable to ordinary differential equations are mentioned briefly.

9.1 Introduction

Recall the initial-value problem obtained for the forced spring–mass system with drag in Section 7.1.1. The equation of motion governing the location $u(t)$ of the mass is given by the second-order, linear differential equation

$$\frac{d^2u}{dt^2} + \frac{c}{m}\frac{du}{dt} + \frac{k}{m}u = -g - \frac{F}{m}\cos(\omega t), \tag{9.1}$$

with the initial conditions

$$u(0) = u_0, \qquad \left.\frac{du}{dt}\right|_{t=0} = v_0, \tag{9.2}$$

where u_0 is the initial position of the mass, and v_0 is its initial velocity. Numerical algorithms are sought that allow us to "march" such initial-value problems forward in time from the initial condition as illustrated in Figure 9.1.

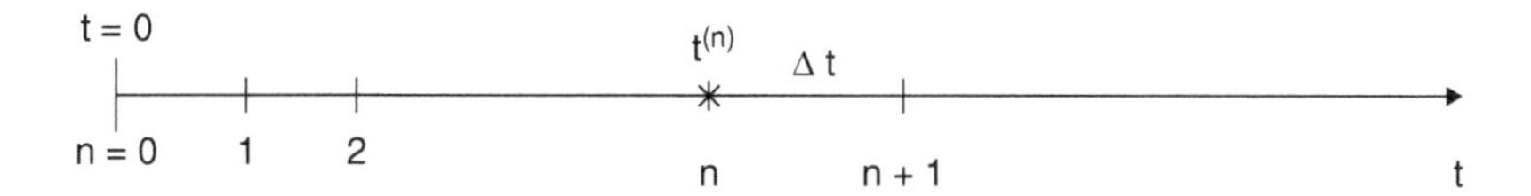

Figure 9.1 Discretization of the time domain.

Such methods can be classified as *single-step* or *multistep* and *explicit* or *implicit*. Single-step methods lead to the solution at the current time step using the known solution only at the previous time step. Multistep methods lead to the solution at the current time step using known solutions at multiple previous time steps. The explicit and implicit distinction describes the actual form of the equation that is solved at each time step in order to obtain the solution. In explicit methods, the solution at the current time step can be written in terms of an explicit expression of the known solution(s) at the previous time step(s); there is only one unknown, which is the solution at the current time step. In implicit methods, the solution at the current time step must be solved for using an equation that contains the unknown solution implicitly, that is, there are multiple unknowns to be solved for at each time step.

In contrast to boundary-value problems, initial-value problems have a different set of numerical properties that must be accounted for. Recall from our treatment of boundary-value problems that we are primarily concerned with accuracy, which is dictated by the truncation error used in the finite-difference approximations, and iterative convergence. Accuracy is concerned with how close the numerical solution is to the exact solution, for example, is the truncation error $O(\Delta x)$ or $O(\Delta x^2)$. In the case of initial-value problems, we are concerned with both accuracy and numerical stability. Numerical stability addresses the question, "Does the numerical solution become unbounded owing to small disturbances – from round-off errors – in the numerical solution as time progresses?" If so, it is numerically unstable. If not, it is numerically stable.

We focus our attention on developing methods for solving first-order initial-value problems and then extend to more general scenarios. We begin by considering single, first-order initial-value problems for a scalar function $u(t)$. We then proceed to extend such methods to solving systems of first-order and higher-order initial-value problems, for which $\mathbf{u}(t)$ is a vector function.

9.2 Single-Step Methods for Ordinary Differential Equations

Single-step methods only require the solution at the previous time step in order to calculate the solution at the current time step. Three such single-step methods are discussed here that can be extended to systems of ordinary differential equations and partial differential equations.

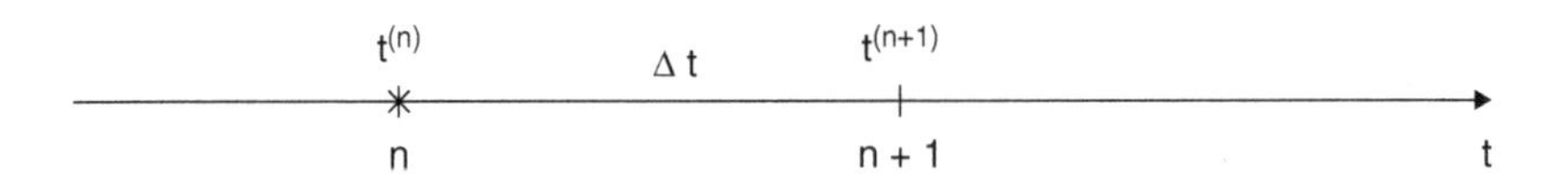

Figure 9.2 Finite-difference stencil for the first-order explicit method. The "×" marks the location where the differential equation is approximated.

9.2.1 First-Order Explicit Method

Consider the first-order initial-value problem

$$\frac{du}{dt} = f(t,u), \tag{9.3}$$

with the initial condition

$$u(0) = u_0. \tag{9.4}$$

Note that $f(t,u)$ may be a linear or nonlinear function of $u(t)$.

We begin by discretizing the time domain, with $\Delta t = t^{(n+1)} - t^{(n)}$, as illustrated in Figure 9.2.[1] Let us approximate (9.3) at the previous time step $t^{(n)}$ using a first-order accurate forward-difference approximation according to

$$\frac{u^{(n+1)} - u^{(n)}}{\Delta t} + O(\Delta t) = f\left(t^{(n)}, u^{(n)}\right),$$

where $u^{(n)} = u\left(t^{(n)}\right)$ is known from the previous time step. Solving for the only unknown in terms of the knowns gives

$$u^{(n+1)} = u^{(n)} + \Delta t\, f\left(t^{(n)}, u^{(n)}\right) + O(\Delta t^2), \tag{9.5}$$

which is an explicit expression for $u^{(n+1)}$ at the current time step in terms of the known solution $u^{(n)}$ and function evaluations $f\left(t^{(n)}, u^{(n)}\right)$ at the previous time step.

Example 9.1 Apply the first-order explicit method to the first-order, linear initial-value problem

$$\frac{du}{dt} = -au, \quad u(0) = u_0, \tag{9.6}$$

where $a > 0$ to prevent exponential growth of the solution. Thus, $f(t,u) = -au$. The exact solution is

$$u(t) = u_0 e^{-at}, \tag{9.7}$$

which exhibits exponential decay with increasing time, in which case the solution is bounded for all time.

[1] Whereas superscript (n) indicates the iteration number in boundary-value problems, it denotes the time step in initial-value problems.

Solution

Using the first-order explicit method, (9.5) dictates that the time steps advance according to

$$\begin{aligned} u^{(n+1)} &= u^{(n)} + \Delta t\, f\left(t^{(n)}, u^{(n)}\right) \\ &= u^{(n)} - a\Delta t u^{(n)} \\ u^{(n+1)} &= (1 - a\Delta t)u^{(n)}. \end{aligned} \tag{9.8}$$

Example 9.2 Obtain the numerical solution for the first-order initial-value problem

$$\frac{du}{dt} = -2u, \quad u(0) = 1,$$

using the first-order explicit method. Thus, $a = 2$, $f(t,u) = -2u$, and $u_0 = 1$ in (9.6). Compare the numerical solution to the exact solution

$$u(t) = e^{-2t},$$

and illustrate the influence of the time step on the numerical solution.

Solution

From (9.8), the solution is advanced at each time step according to

$$u^{(0)} = 1, \quad u^{(n+1)} = (1 - 2\Delta t)\, u^{(n)}, \quad n = 0, 1, 2, 3, \ldots.$$

The numerical solution with time step $\Delta t = 0.1$ is shown in Figure 9.3. The absolute error at $t_{max} = 2$ is 0.006786. The numerical solution with time step $\Delta t = 0.01$ is shown in Figure 9.4. Observe that the solution is in significantly better agreement with

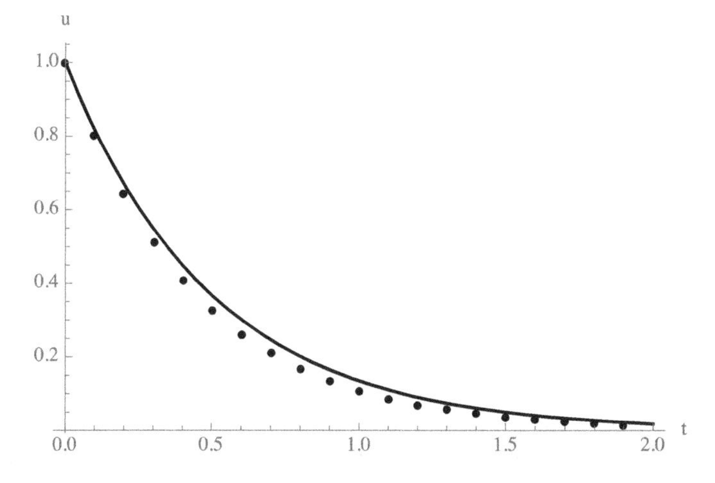

Figure 9.3 Comparison of exact (solid) and numerical solutions (dots) of (9.6) with $a = 2$ using the first-order explicit method with $\Delta t = 0.1$.

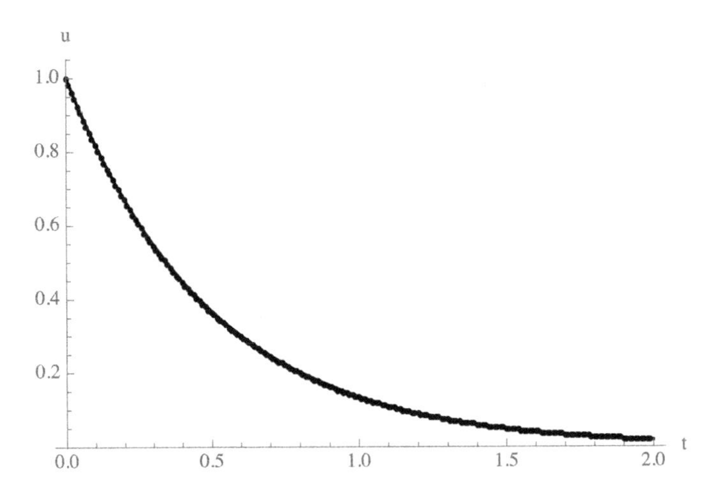

Figure 9.4 Comparison of exact (solid) and numerical solutions (dots) of (9.6) with $a = 2$ using the first-order explicit method with $\Delta t = 0.01$.

the exact solution than the solution with $\Delta t = 0.1$. The absolute error at $t_{max} = 2$ is 0.000728. This is approximately a factor of 10 smaller than that for $\Delta t = 0.1$ reflecting the $O(\Delta t)$ accuracy of the first-order explicit method.

Let us revisit the two properties relevant to initial-value problems, namely accuracy and numerical stability. Regarding accuracy, there are two types of truncation error: (1) local truncation error, and (2) global (propagated) truncation error. Local truncation error is the error incurred during each time step owing to the truncation error of the derivative approximation in the finite-difference equation. Recall that a first-order accurate forward-difference approximation was used for the time derivative. In the resulting explicit expression (9.5) used to advance each time step, it is observed that we incur an $O(\Delta t^2)$ error during each time step; this is the local truncation error. The global error is the truncation error accumulated over a series of time steps. For example, if the local truncation error per time step is $O(\Delta t^2)$, as in the first-order explicit method, then the global truncation error after N time steps is

$$O(\Delta t^2) \times N = O(\Delta t^2)\frac{t^{(N)} - t^{(0)}}{\Delta t} = O(\Delta t)$$

for $t^{(N)} - t^{(0)} = O(1)$. Thus, the first-order explicit method has a "first-order accurate," that is, $O(\Delta t)$, global truncation error.

Regarding numerical stability, a numerical solution is unstable if it becomes unbounded and goes to infinity for large times, while the exact solution remains bounded and finite for all times. To check for numerical stability of the linear initial-value problem (9.6), for example, apply (9.8) for N time steps (u_0 is the initial condition)

$$\begin{aligned}
n = 0: \quad & u^{(1)} &= (1 - a\Delta t)u^{(0)} = (1 - a\Delta t)u_0, \\
n = 1: \quad & u^{(2)} &= (1 - a\Delta t)u^{(1)} = (1 - a\Delta t)^2 u_0, \\
\vdots \\
n = N - 1: \quad & u^{(N)} &= (1 - a\Delta t)^N u_0.
\end{aligned}$$

The factor $G = (1 - a\Delta t)$ is called the *gain* G and represents the increase or decrease in the magnitude of the solution $u^{(n)}$ during each time step. Therefore, to be stable and prevent unbounded growth as N increases requires that the gain be such that

$$|G| = |1 - a\Delta t| \leq 1.$$

Recalling that $a > 0$, this is the case if

$$\begin{aligned}
-1 \leq 1 - a\Delta t \leq 1 \\
-2 \leq -a\Delta t \leq 0 \\
2 \geq a\Delta t \geq 0.
\end{aligned}$$

Observe that $a\Delta t \geq 0$ for all $a > 0$ and Δt, but for the left inequality to be satisfied requires that

$$\Delta t \leq \frac{2}{a}. \tag{9.9}$$

This provides a limitation on the time step in order to maintain numerical stability of the first-order explicit method. Therefore, we say that the first-order explicit method is *conditionally stable*, as numerical stability requires a limitation on the time step to be satisfied.

Example 9.3 Illustrate numerical stability of the first-order initial-value problem

$$\frac{du}{dt} = -2u, \quad u(0) = 1,$$

solved using the first-order explicit method. Again, $a = 2$, $f(t,u) = -2u$, and $u_0 = 1$ in (9.6).

Solution

Recall that the first-order explicit method is numerically stable for this problem when

$$\Delta t \leq \frac{2}{a} = 1.$$

Figure 9.5 illustrates the unstable, oscillatory numerical solution that results for a time step greater than this threshold. The amplitude of oscillation increases with time and becomes unbounded. Recall that the exact solution decays exponentially with time.

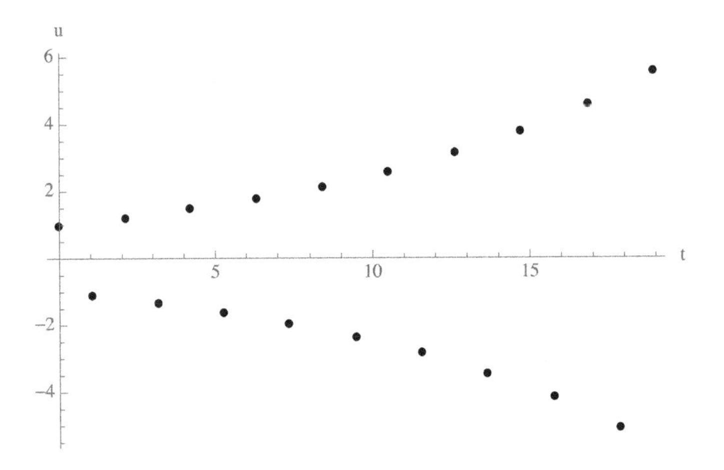

Figure 9.5 Numerical solution of (9.6) with $a = 2$ using the first-order explicit method with $\Delta t = 1.05$ and $t_{max} = 20$.

REMARK:

1. *The first-order explicit method is also called the "forward Euler method," "explicit Euler method," and "FTCS," which stands for "forward time, central space" when applied to partial differential equations.*
2. *Its global error is $O(\Delta t)$, that is, it is only first-order accurate.*
3. *As illustrated in this example, the first-order explicit method is typically only conditionally stable and results in very restrictive limitations on the time step necessary to maintain numerical stability. The time step restriction for stability is normally much more stringent than that required for accuracy.*
4. *Note that there is no difficulty if $f(t,u)$ is nonlinear as it is on the right-hand side and is evaluated at the previous time step.*

9.2.2 First-Order Implicit Method

In order to improve on the numerical stability properties of the algorithm, let us reconsider the initial-value problem (9.3) and (9.4) using a backward-difference approximation for the time derivative as follows:

$$\frac{u^{(n+1)} - u^{(n)}}{\Delta t} + O(\Delta t) = f\left(t^{(n+1)}, u^{(n+1)}\right).$$

Note that the function evaluations are now performed at the current time step, where the differential equation is now approximated as indicated in Figure 9.6. Rearranging so that the unknowns are on the left-hand side yields

$$u^{(n+1)} - \Delta t\, f\left(t^{(n+1)}, u^{(n+1)}\right) = u^{(n)} + O(\Delta t^2), \tag{9.10}$$

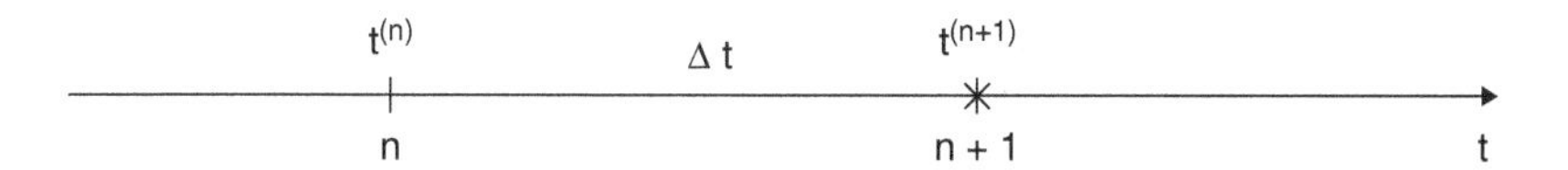

Figure 9.6 Finite-difference stencil for the first-order implicit method. The "×" marks the location where the differential equation is approximated.

which is now an implicit expression for $u^{(n+1)}$ at the current times step. That is, the unknown $u^{(n+1)}$ appears more than once in the equation and must be solved for implicitly.

Example 9.4 Apply the first-order implicit method to the first-order, linear initial-value problem considered in the previous section and evaluate numerical stability. The differential equation and initial condition are

$$\frac{du}{dt} = -au, \quad u(0) = u_0, \quad a > 0.$$

Thus, $f(t,u) = -au$.

Solution

Applying (9.10), the first-order implicit method leads to

$$u^{(n+1)} - \Delta t\, f\left(t^{(n+1)}, u^{(n+1)}\right) = u^{(n)}$$

$$u^{(n+1)} + a\Delta t u^{(n+1)} = u^{(n)}$$

$$(1 + a\Delta t)u^{(n+1)} = u^{(n)}$$

$$\therefore u^{(n+1)} = \frac{1}{(1 + a\Delta t)} u^{(n)}.$$

Compare this with (9.8) for the first-order explicit method. To maintain numerical stability now requires that the gain be such that

$$|G| = \left| \frac{1}{1 + a\Delta t} \right| \leq 1,$$

or

$$-1 \leq \frac{1}{1 + a\Delta t} \leq 1.$$

Note that $1 + a\Delta t > 1$; therefore, $1/\left(1 + a\Delta t\right)$ is always positive and less than one. As a result, the first-order implicit method applied to this differential equation is *unconditionally stable*. That is, there are no limitations on the size of the time step in order to maintain numerical stability of the algorithm. Obviously, accuracy is still influenced by the time step.

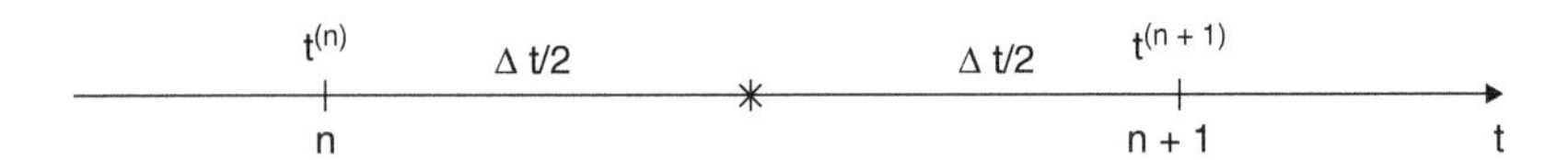

Figure 9.7 Finite-difference stencil for Crank–Nicolson method. The "×" marks the location where the differential equation is approximated.

REMARKS:

1. *The first-order implicit method is also called the "backward Euler method" or the "implicit Euler method."*
2. *The global error is $O(\Delta t)$, that is, it is first-order accurate.*
3. *If $f(t,u)$ is nonlinear in $u(t)$, then (9.10) is a nonlinear algebraic equation to be solved for $u^{(n+1)}$ using the methods in Chapter 12.*
4. *The first-order implicit method is stable for all Δt; therefore, it is* unconditionally stable. *The time step Δt can be chosen solely for accuracy, without regard for numerical stability issues.*
5. *As is typical of implicit methods, they improve numerical stability properties, such that one can take larger Δt, but they require more computation per time step. In the aggregate, the larger time steps allowed for by implicit methods more than make up for the additional computation per time step. Consequently, they are the preferred method for most initial-value problems*

9.2.3 Crank–Nicolson Method

In order to improve the accuracy of the algorithm from first order to second order, let us use a central, rather than forward or backward, difference as follows:

$$\frac{u^{(n+1)} - u^{(n)}}{\Delta t} + O(\Delta t^2) = \frac{1}{2}\left[f\left(t^{(n)}, u^{(n)}\right) + f\left(t^{(n+1)}, u^{(n+1)}\right)\right],$$

where the right-hand side is the average of $f(t,u)$ at the previous and current time steps. In this way, the differential equation is approximated at the midpoint between time steps as indicated in Figure 9.7, such that the average between function evaluations at the previous and current time steps provide an approximation at the midpoint. Because it is approximated at the midpoint, the time derivative is a second-order accurate central-difference approximation.

Once again, rearranging so that the unknowns are on the left-hand side yields

$$u^{(n+1)} - \frac{\Delta t}{2} f\left(t^{(n+1)}, u^{(n+1)}\right) = u^{(n)} + \frac{\Delta t}{2} f\left(t^{(n)}, u^{(n)}\right) + O(\Delta t^3), \tag{9.11}$$

which is again an implicit expression for $u^{(n+1)}$ at the current time step.

Example 9.5 Apply the Crank–Nicolson method to the first-order, linear initial-value problem considered in the previous two sections and evaluate numerical stability, but allow for $a = a_R + \mathrm{i} a_I$ to be complex. The differential equation and initial condition are

$$\frac{du}{dt} = -au, \quad u(0) = u_0, \quad a_R > 0. \tag{9.12}$$

Thus, $f(t,u) = -au$.

Solution

Applying (9.11), the Crank–Nicolson method applied to (9.12) leads to

$$u^{(n+1)} - \frac{1}{2}\Delta t\, f\left(t^{(n+1)}, u^{(n+1)}\right) = u^{(n)} + \frac{1}{2}\Delta t\, f\left(t^{(n)}, u^{(n)}\right)$$

$$u^{(n+1)} + \frac{1}{2}a\Delta t u^{(n+1)} = u^{(n)} - \frac{1}{2}a\Delta t u^{(n)}$$

$$\left(1 + \frac{1}{2}a\Delta t\right) u^{(n+1)} = \left(1 - \frac{1}{2}a\Delta t\right) u^{(n)}$$

$$\therefore u^{(n+1)} = \frac{1 - \frac{1}{2}a\Delta t}{1 + \frac{1}{2}a\Delta t} u^{(n)}.$$

Recalling that $a = a_R + \mathrm{i}a_I$ is complex, the gain is given by

$$G = \frac{1 - \frac{1}{2}a\Delta t}{1 + \frac{1}{2}a\Delta t} = \frac{1 - \frac{1}{2}a_R\Delta t - \mathrm{i}\frac{1}{2}a_I\Delta t}{1 + \frac{1}{2}a_R\Delta t + \mathrm{i}\frac{1}{2}a_I\Delta t} = \frac{re^{\mathrm{i}\theta}}{\rho e^{\mathrm{i}\phi}} = \frac{r}{\rho}e^{\mathrm{i}(\theta-\phi)}, \tag{9.13}$$

where

$$r = \left[\left(1 - \frac{1}{2}a_R\Delta t\right)^2 + \left(\frac{1}{2}a_I\Delta t\right)^2\right]^{1/2}, \quad \rho = \left[\left(1 + \frac{1}{2}a_R\Delta t\right)^2 + \left(\frac{1}{2}a_I\Delta t\right)^2\right]^{1/2}.$$

For numerical stability, $|G| \leq 1$, which in this case requires that $|G| = \frac{r}{\rho} \leq 1$. Because $a_R > 0$ and $r < \rho$, the Crank–Nicolson method is *unconditionally stable* for this case. Consequently, there is no restriction on the time step for numerical stability, which is true for the first-order implicit method as well; however, the Crank–Nicolson method is second-order accurate in time.

REMARKS:

1. *The implicit Crank–Nicolson method is also called the "trapezoid method."*
2. *The global error is $O(\Delta t^2)$, that is, it is second-order accurate.*
3. *If $f(t,u)$ is nonlinear in $u(t)$, then (9.11) is a nonlinear algebraic equation to be solved for $u^{(n+1)}$ (see Chapter 12). Therefore, implicit methods are ideal for linear initial-value problems, but nonlinear ones are better treated using predictor-corrector methods, such as Runge–Kutta methods, or multistep methods as discussed in Section 9.3.*
4. *The Crank–Nicolson method is* unconditionally stable *for all Δt as shown in the previous example.*
5. *The first-order explicit, first-order implicit, and Crank–Nicolson methods can all be applied to systems of first-order initial-value problems and extended to partial differential equations as shown in subsequent sections.*

9.2.4 Higher-Order and Systems of Ordinary Differential Equations

Here, we discuss how to apply the methods developed for a single first-order initial-value problem to higher-order and systems of initial-value problems. Consider a system of N first-order ordinary differential equations

$$\begin{aligned} \frac{du_1}{dt} &= f_1(t, u_1, u_2, \ldots, u_N), \\ \frac{du_2}{dt} &= f_2(t, u_1, u_2, \ldots, u_N), \\ &\vdots \\ \frac{du_N}{dt} &= f_N(t, u_1, u_2, \ldots, u_N), \end{aligned} \tag{9.14}$$

with the N initial conditions

$$u_1(0) = A_1, \quad u_2(0) = A_2, \quad \ldots, \quad u_N(0) = A_N. \tag{9.15}$$

Thus, there is one independent variable t and N dependent variables $u_1(t)$, $u_2(t)$, ..., $u_N(t)$.

Note that higher-order ordinary differential equations can be converted to a system of first-order differential equations as described in Section 2.6.2. Consider the Nth-order ordinary differential equation

$$\frac{d^N u}{dt^N} = f\left(t, u, \frac{du}{dt}, \frac{d^2u}{dt^2}, \ldots, \frac{d^{N-1}u}{dt^{N-1}}\right), \tag{9.16}$$

with the initial conditions

$$u(0) = A_1, \quad \frac{du}{dt}(0) = A_2, \quad \ldots, \quad \frac{d^{N-1}u}{dt^{N-1}}(0) = A_N. \tag{9.17}$$

To convert to a system of first-order equations, define a set of N new dependent variables according to

$$\begin{aligned} u_1 &= u, \\ u_2 &= \frac{du}{dt}, \\ &\vdots \\ u_N &= \frac{d^{N-1}u}{dt^{N-1}}. \end{aligned} \tag{9.18}$$

Differentiating each of equations (9.18) and writing in terms of the new variables yields

$$\begin{aligned}
\frac{du_1}{dt} &= \frac{du}{dt} &= u_2, \\
\frac{du_2}{dt} &= \frac{d^2u}{dt^2} &= u_3, \\
&\vdots \\
\frac{du_N}{dt} &= \frac{d^Nu}{dt^N} = f\left(t, u, \frac{du}{dt}, \ldots, \frac{d^{N-1}u}{dt^{N-1}}\right) = f\left(t, u_1, u_2, \ldots, u_N\right),
\end{aligned} \tag{9.19}$$

where the last equation comes from the original Nth-order equation (9.16). Thus, in the system of first-order differential equations (9.14), the right-hand side functions are

$$\begin{aligned}
f_1 &= u_2, \\
f_2 &= u_3, \\
&\vdots \\
f_{N-1} &= u_N, \\
f_N &= f\left(t, u_1, u_2, \ldots, u_N\right).
\end{aligned}$$

In terms of the new variables, the initial conditions are then

$$u_1(0) = A_1, \quad u_2(0) = A_2, \quad \ldots, \quad u_N(0) = A_N. \tag{9.20}$$

Therefore, (9.19) and (9.20) are a system of first-order ordinary differential equations in the form of (9.14) and (9.15).

Example 9.6 Apply the first-order explicit method to the following second-order initial-value problem:

$$\frac{d^2u}{dt^2} = -u(t), \quad u(0) = 0, \quad \dot{u}(0) = 1. \tag{9.21}$$

Note that the exact solution is given by

$$u(t) = \sin(t).$$

Solution

Begin by writing the second-order equation as a system of two first-order equations by letting

$$\begin{aligned}
u_1(t) &= u(t), \\
u_2(t) &= \dot{u}(t).
\end{aligned}$$

Differentiating, substituting the differential equation, and writing in terms of the new variables yields

$$\begin{aligned}\dot{u}_1 &= \dot{u} = u_2,\\ \dot{u}_2 &= \ddot{u} = -u(t) = -u_1(t).\end{aligned}$$

In terms of the new variables, the initial conditions are

$$u_1(0) = 0, \quad u_2(0) = 1. \tag{9.22}$$

Again, we have a system of two first-order ordinary differential equations that is equivalent to the one second-order ordinary differential equation (9.21). In matrix form, it is

$$\begin{bmatrix} \dot{u}_1(t) \\ \dot{u}_2(t) \end{bmatrix} = \begin{bmatrix} 0 & 1 \\ -1 & 0 \end{bmatrix} \begin{bmatrix} u_1(t) \\ u_2(t) \end{bmatrix}, \quad \begin{bmatrix} u_1(0) \\ u_2(0) \end{bmatrix} = \begin{bmatrix} 0 \\ 1 \end{bmatrix}.$$

Thus, we have the two first-order equations

$$\dot{u}_1 = u_2, \quad \dot{u}_2 = -u_1.$$

Recall that for the first-order explicit method, the equation is approximated at $t^{(n)}$ using a forward difference for the time derivatives. Applying the first-order explicit method from Section 9.2.1 to these two equations gives ($f_1 = u_2$, $f_2 = -u_1$)

$$\frac{u_1^{(n+1)} - u_1^{(n)}}{\Delta t} = u_2^{(n)} \quad \Rightarrow \quad u_1^{(n+1)} = u_1^{(n)} + \Delta t\, u_2^{(n)},$$

and

$$\frac{u_2^{(n+1)} - u_2^{(n)}}{\Delta t} = -u_1^{(n)} \quad \Rightarrow \quad u_2^{(n+1)} = u_2^{(n)} - \Delta t\, u_1^{(n)}.$$

Therefore, the solution for each successive time step $n = 0, 1, 2, 3, \ldots$ is obtained by solving these two explicit expressions in terms of information from the previous time step starting with the initial conditions (9.22), which are

$$A_1 = 0, \quad A_2 = 1.$$

The solution will be only first-order accurate and only conditionally stable.

Example 9.7 Apply the first-order implicit method to the same second-order initial-value problem from the previous example.

Solution

Recall that the equivalent first-order system of equations is given by

$$\dot{u}_1 = u_2, \quad \dot{u}_2 = -u_1.$$

In the first-order implicit method, the equation is approximated at $t^{(n+1)}$ using a backward difference for the time derivative.

Applying the first-order implicit method from Section 9.2.2 to these two equations gives

$$\frac{u_1^{(n+1)} - u_1^{(n)}}{\Delta t} = u_2^{(n+1)} \quad \Rightarrow \quad u_1^{(n+1)} - \Delta t\, u_2^{(n+1)} = u_1^{(n)},$$

and

$$\frac{u_2^{(n+1)} - u_2^{(n)}}{\Delta t} = -u_1^{(n+1)} \quad \Rightarrow \quad \Delta t\, u_1^{(n+1)} + u_2^{(n+1)} = u_2^{(n)},$$

or in matrix form

$$\begin{bmatrix} 1 & -\Delta t \\ \Delta t & 1 \end{bmatrix} \begin{bmatrix} u_1^{(n+1)} \\ u_2^{(n+1)} \end{bmatrix} = \begin{bmatrix} u_1^{(n)} \\ u_2^{(n)} \end{bmatrix}.$$

Consequently, the solution for each successive time step is obtained by solving this system of equations. Whereas explicit methods provide uncoupled explicit expressions to evaluate at each time step, implicit methods require solution of a system of equations to obtain the solution at each time step. The solution will be only first-order accurate, but the method is unconditionally stable.

Example 9.8 Apply the Crank–Nicolson method to the same second-order initial-value problem from the previous two examples.

Solution

Recall that the equivalent first-order system of equations is given by

$$\dot{u}_1 = u_2, \quad \dot{u}_2 = -u_1.$$

Applying the Crank–Nicolson method from Section 9.2.3 to these two equations, whereby the equations are approximated at the midpoint between $t^{(n)}$ and $t^{(n+1)}$ using a central difference, gives

$$\frac{u_1^{(n+1)} - u_1^{(n)}}{\Delta t} = \frac{u_2^{(n)} + u_2^{(n+1)}}{2} \quad \Rightarrow \quad u_1^{(n+1)} - \frac{\Delta t}{2} u_2^{(n+1)} = u_1^{(n)} + \frac{\Delta t}{2} u_2^{(n)},$$

and

$$\frac{u_2^{(n+1)} - u_2^{(n)}}{\Delta t} = -\frac{u_1^{(n)} + u_1^{(n+1)}}{2} \quad \Rightarrow \quad \frac{\Delta t}{2} u_1^{(n+1)} + u_2^{(n+1)} = -\frac{\Delta t}{2} u_1^{(n)} + u_2^{(n)},$$

or in matrix form

$$\begin{bmatrix} 1 & -\frac{\Delta t}{2} \\ \frac{\Delta t}{2} & 1 \end{bmatrix} \begin{bmatrix} u_1^{(n+1)} \\ u_2^{(n+1)} \end{bmatrix} = \begin{bmatrix} 1 & \frac{\Delta t}{2} \\ -\frac{\Delta t}{2} & 1 \end{bmatrix} \begin{bmatrix} u_1^{(n)} \\ u_2^{(n)} \end{bmatrix}.$$

As for the first-order implicit method, a system of equations must be solved at each time step in order to march the solution forward in time. The method is now second-order accurate but maintains unconditional stability.

REMARKS:

1. *Observe that implicit methods applied to systems of first-order equations require solution of a system of algebraic equations to advance each time step. If the original differential equation is linear, these algebraic equations are also linear.*
2. *Single-step methods are self-starting, and the time step Δt can be adjusted as the calculation progresses.*

Example 9.9 Consider the forced spring–mass system with drag from Section 7.1.1 governed by the second-order initial-value problem

$$\frac{d^2u}{dt^2} = -g - \frac{F}{m}\cos(\omega t) - \frac{k}{m}u - \frac{c}{m}\frac{du}{dt},$$

where the right-hand side is $f(t, u, \dot{u})$. The initial conditions are

$$u(0) = u_0, \quad \dot{u}(0) = v_0.$$

Solution

Because this is a second-order differential equation with $N = 2$, define the new variables

$$u_1 = u,$$
$$u_2 = \dot{u}.$$

Differentiating and writing in terms of the new variables gives

$$\dot{u}_1 = \dot{u} = u_2,$$
$$\dot{u}_2 = \ddot{u} = -g - \frac{F}{m}\cos(\omega t) - \frac{k}{m}u - \frac{c}{m}\dot{u} = -g - \frac{F}{m}\cos(\omega t) - \frac{k}{m}u_1 - \frac{c}{m}u_2.$$

This is a system of two equations for the two unknowns $u_1(t)$ and $u_2(t)$. The initial conditions are

$$u_1(0) = u_0, \quad u_2(0) = v_0.$$

Upon solving the preceding system of first-order differential equations for $u_1(t)$ and $u_2(t)$ using one of the previously mentioned explicit or implicit methods, the original variable is then simply $u(t) = u_1(t)$, which represents the position of the mass as a function of time. A sample solution is given in Figure 9.8.

For an additional example, recall the two-mass, three-spring system from Section 2.6.3. The equations of motion derived there consist of two second-order ordinary differential equations, which were converted into four first-order differential equations. Once again, either of the explicit or implicit methods could be applied to obtain the numerical solution of this linear system.

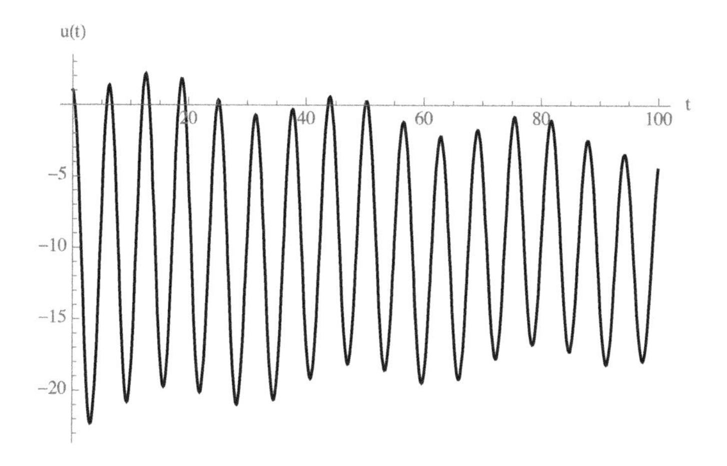

Figure 9.8 Numerical solution of the forced spring–mass system with drag for $g = 9.81$, $m = 1, k = 1, c = 0.01, F = 1, \omega = 0.2, u_0 = 1$, and $v_0 = 0$.

9.3 Additional Methods for Ordinary Differential Equations

Recall that for the single-step methods considered thus far, explicit methods are only first-order accurate and have poor numerical stability properties. However, they are especially easy to implement, as they do not require solution of a system of equations, and can be easily adapted to nonlinear differential equations. Implicit methods, on the other hand, can be second-order accurate and have improved numerical stability properties. However, they require more computation per time step than explicit methods, and they are difficult to implement for nonlinear equations.

Based on the advantages of explicit and implicit methods, one may articulate an initial-value problem wishlist that maintains the favorable accuracy and stability properties of an implicit method but with the ease of implementation and ability to easily adapt to nonlinear equations of an explicit method. There are two classes of methods that satisfy our wishlist:

1. *Predictor-corrector methods* consist of the following two steps:
   (i) *Predictor* – use an explicit method to "predict" the solution at the current time step.
   (ii) *Corrector* – use an implicit method to "correct" the predicted solution at the current time step.
2. *Multistep methods* take advantage of the known solutions at multiple previous time steps to approximate the solution at the current time step. Recall that single-step methods only use the solution at the previous time step.

We only consider a brief overview of these methods here as they only apply to ordinary differential equations and cannot be extended to partial differential equations (unless they can be converted to ordinary differential equations).

9.3.1 Predictor-Corrector Methods

A simple predictor-corrector method is known as *Heun's method*, which consists of the two steps:

(i) *Predictor* – use the first-order explicit method to "predict" the solution at the current step as follows:

$$\hat{u}^{(n+1)} = u^{(n)} + \Delta t\, f\left(t^{(n)}, u^{(n)}\right).$$

(ii) *Corrector* – use the Crank–Nicolson method with the predicted value $\hat{u}^{(n+1)}$ to "correct" $u^{(n+1)}$ at the current time step:

$$u^{(n+1)} = u^{(n)} + \frac{\Delta t}{2} f\left(t^{(n)}, u^{(n)}\right) + \frac{\Delta t}{2} f\left(t^{(n+1)}, \hat{u}^{(n+1)}\right).$$

Observe that because $\hat{u}^{(n+1)}$ is known, it moves to the right-hand side of the difference equation, creating an explicit expression for $u^{(n+1)}$ at the current time step.

The most popular predictor-corrector-type methods are the *Runge–Kutta methods*, which provide higher-order accuracy than the underlying explicit method alone. Unlike typical predictor-corrector methods, Runge–Kutta methods involve function evaluations for the right-hand side $f(t,u)$ of the ordinary differential equation(s) at intermediate points within each time step. For example, the *second-order Runge–Kutta method* requires a function evaluation at the midpoint between the previous and current time steps. Increasing the order of the approximation requires additional function evaluations for each time step. Runge–Kutta methods are typically only conditionally stable, but the time step limitations are much less restrictive than for the first-order explicit method alone. Although there are more calculations per time step than the first-order explicit method, therefore, larger time steps can be taken owing to both better accuracy and less restrictive stability criterion. The fourth-order Runge–Kutta method is one of the most widely used general-purpose initial-value problem algorithms for ordinary differential equations.

9.3.2 Multistep Methods

All of the methods discussed thus far are single-step methods as they only require information from the previous time step – and possibly intermediate points – to obtain the solution at the current time step. Multistep methods take advantage of solutions already obtained for multiple previous time steps. Because solutions at multiple time steps are required, each must have the same time step Δt, and a *starting method* is required to solve the first few iterations before the multistep method can be initiated. They can be explicit (Adams–Bashforth) or implicit (Adams–Moulton). For example, the fourth-order Adams–Bashforth method uses values of the right-hand-side functions at the current and three previous time steps.

Whereas the fourth-order Runge–Kutta method requires three new function evaluations of the right-hand-side function for each time step, the fourth-order Adams–Bashforth method only requires one new function evaluation (at the current time step); the others have already been calculated at previous time steps. However, Runge–Kutta methods are more stable than Adams–Bashforth methods, thereby allowing for larger time steps. Adams–Moulton methods are implicit; therefore, they have better stability properties than the explicit Adams–Bashforth methods.

Multistep methods can be implemented in a predictor-corrector framework as well. The explicit predictor step would then use a multistep Adams–Bashforth method. The implicit corrector step would then use a multistep Adams–Moulton method. This approach takes advantage of the simplicity and ability to treat nonlinear equations of explicit methods and the stability properties of implicit methods. Because predictor-corrector and multistep methods cannot be extended to general partial differential equations, they will not be discussed further. The reader will find more details on these methods in any undergraduate numerical methods book.

9.4 Partial Differential Equations

Whereas the elliptic partial differential equations that were the subject of the previous chapter are boundary-value problems, parabolic and hyperbolic partial differential equations are initial-value problems. Parabolic equations have one preferred direction of propagation of the solution, which is usually time. The canonical parabolic partial differential equation is the one-dimensional, unsteady diffusion equation

$$\frac{\partial u}{\partial t} = \alpha \frac{\partial^2 u}{\partial x^2}, \tag{9.23}$$

where $\alpha > 0$ is the diffusivity of the material through which diffusion is occurring. In various fields, the dependent variable $u(x,t)$ represents different quantities. Examples include temperature in heat conduction, velocity in momentum diffusion, and concentration in mass diffusion. This and its extensions to multidimensions are the primary focus of this chapter. Hyperbolic equations, such as the wave equation

$$\frac{\partial^2 u}{\partial t^2} = \sigma^2 \frac{\partial^2 u}{\partial x^2}, \tag{9.24}$$

are considered in Section 9.11. Recall that $u(x,t)$ is the amplitude of the wave, and σ is the wave speed in the medium.

Because parabolic problems are initial-value problems, we need to develop numerical methods that "march" in the preferred direction (time) in a step-by-step manner. Unlike elliptic problems, for which the approximate numerical solution must be stored throughout the entire domain, in parabolic problems, it is only necessary to store in memory the approximate solution at the current and previous time steps. Solutions obtained at earlier time steps can be saved to permanent storage, but do not need to be retained in memory.

Consider the general linear, one-dimensional, unsteady equation

$$\frac{\partial u}{\partial t} = a(x,t)\frac{\partial^2 u}{\partial x^2} + b(x,t)\frac{\partial u}{\partial x} + c(x,t)u + d(x,t), \tag{9.25}$$

which is parabolic forward in time for $a(x,t) > 0$. For the one-dimensional, unsteady diffusion equation (9.23), $a = 1$, and $b = c = d = 0$. Techniques developed for the canonical equation (9.23) can be used to solve the more general equation (9.25).

There are two basic techniques for numerically solving parabolic problems:

1. *Method of lines*: Discretize the spatial derivatives to reduce the partial differential equation in space and time to a set of ordinary differential equations in time and solve using predictor-corrector, Runge–Kutta, et cetera.
2. *Marching methods*: Discretize in both space and time.
   (i) *Explicit methods* – obtain a single equation for $u(x,t)$ at each grid point.
   (ii) *Implicit methods* – obtain a set of algebraic equations for $u(x,t)$ at all grid points for the current time step.

When faced with solving new, more difficult, problems, it is natural to first consider whether existing methods for simpler problems can be revised or generalized to solve the more complicated one. This is the approach taken in the method of lines, which harkens back to a time when initial-value problem solvers for ordinary differential equations were highly developed, but those for partial differential equations were not.[2] At that time, it was common to convert partial differential equations into sets of first-order ordinary differential equations so that existing methods for such problems could be applied to partial differential equations. Interestingly, the "old fashioned" method of lines is very similar to applying Galerkin projection, which is a popular method used to convert partial differential equations into so-called reduced-order models that only require solution of a system of first-order ordinary differential equations for the time-dependent coefficients in the expansion of the solution in terms of basis functions. For much more on Galerkin projection and related methods, see Chapter 13, and for more on the method of lines, see Moin (2010).

Because of their generality, our focus will be on marching methods. Explicit marching methods are designed to have a single unknown when the difference equation is applied at each grid point, which allows for the solution to be updated explicitly at each point in terms of the approximate solution at surrounding points (similar to the Gauss–Seidel method). Implicit marching methods, on the other hand, result in multiple unknowns when applied at each grid point, which requires solution of a system of algebraic equations to obtain the solution at each time step (similar to the ADI method).

[2] This was not because we did not know how to solve partial differential equations, but because the computational resources (both computational power and memory) were so limited.

9.5 Explicit Methods

In explicit methods, the spatial derivatives are all evaluated at the previous time level(s) resulting in a single unknown at the current time level $u_i^{(n+1)} = u(x_i, t_{n+1})$ on the left-hand side of the difference equation.

9.5.1 First-Order Explicit Method

As illustrated in Figure 9.9, second-order accurate central differences are used for the spatial derivatives, and a first-order accurate forward difference is used for the temporal derivative. For each of the methods, take particular note of the location in the grid at which the equation is approximated; these are the same as for ordinary initial-value problems. Applied to the one-dimensional, unsteady diffusion equation, we have the first-order accurate forward difference for the time derivative

$$\frac{\partial u}{\partial t} = \frac{u_i^{(n+1)} - u_i^{(n)}}{\Delta t} + O(\Delta t),$$

and the second-order accurate central difference for the spatial derivatives at the previous, that is, nth, time level, at which the approximate solution is known

$$\frac{\partial^2 u}{\partial x^2} = \frac{u_{i+1}^{(n)} - 2u_i^{(n)} + u_{i-1}^{(n)}}{\Delta x^2} + O(\Delta x^2).$$

Observe that it is not clear whether the time derivative is being approximated as a forward or backward difference; it is essential to be clear on the point in the grid at which the differential equation is being approximated. Substituting into (9.23) yields

$$\frac{u_i^{(n+1)} - u_i^{(n)}}{\Delta t} + O(\Delta t) = \alpha \frac{u_{i+1}^{(n)} - 2u_i^{(n)} + u_{i-1}^{(n)}}{\Delta x^2} + O(\Delta t^2),$$

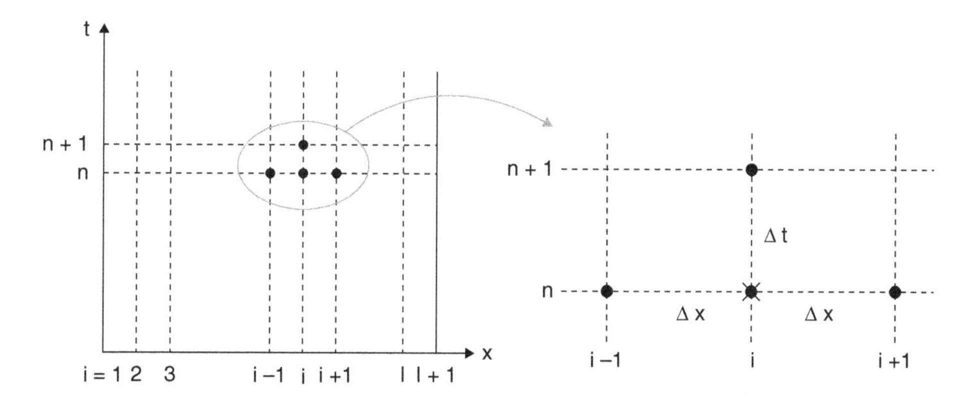

Figure 9.9 Schematic of the first-order explicit method for $u(x,t)$. The "×" marks the location where the differential equation is approximated.

and solving for the only unknown $u_i^{(n+1)}$ gives the explicit expression

$$u_i^{(n+1)} = u_i^{(n)} + \frac{\alpha \Delta t}{\Delta x^2}\left(u_{i+1}^{(n)} - 2u_i^{(n)} + u_{i-1}^{(n)}\right),$$

or

$$u_i^{(n+1)} = (1 - 2s)\, u_i^{(n)} + s\left(u_{i+1}^{(n)} + u_{i-1}^{(n)}\right), \quad i = 1, \ldots, I + 1, \tag{9.26}$$

where $s = \alpha \Delta t / \Delta x^2$.

REMARKS:

1. *Equation (9.26) is an explicit expression for $u_i^{(n+1)}$ at the $(n+1)$st time step in terms of $u_i^{(n)}$ at the nth time step. Each time step consists of a single sweep for $i = 1, \ldots, I+1$ at each time step $n+1, n = 0, 1, 2, \ldots$.*
2. *The method is second-order accurate in space and first-order accurate in time.*
3. *The time steps Δt may be varied from one time step to the next.*
4. *As for ordinary differential equations, it will be shown in Section 9.6 that there are restrictions on Δt and Δx for the first-order explicit method applied to the one-dimensional, unsteady diffusion equation to remain stable. We say that the method is* conditionally stable, *because for the numerical method to remain stable requires that*

$$s = \frac{\alpha \Delta t}{\Delta x^2} \leq \frac{1}{2},$$

which is very restrictive. If $s > \frac{1}{2}$, then the method is unstable, and errors in the solution grow to become unbounded as time proceeds.

5. *In choosing the time step, we must always be concerned with the temporal resolution from the accuracy point of view as well as from a numerical stability point of view.*

9.5.2 Richardson Method

The *Richardson method* seeks to improve on the temporal accuracy of the first-order explicit method by using a second-order accurate central difference for the time derivative as illustrated in Figure 9.10. Because the differential equation is approximated at the previous time step, the central difference involves the current time step along with the previous two. Thus, the one-dimensional, unsteady diffusion equation becomes

$$\frac{u_i^{(n+1)} - u_i^{(n-1)}}{2\Delta t} + O(\Delta t^2) = \alpha \frac{u_{i+1}^{(n)} - 2u_i^{(n)} + u_{i-1}^{(n)}}{\Delta x^2} + O(\Delta x^2),$$

or

$$u_i^{(n+1)} = u_i^{(n-1)} + 2s\left(u_{i+1}^{(n)} - 2u_i^{(n)} + u_{i-1}^{(n)}\right), \quad i = 1, \ldots, I + 1. \tag{9.27}$$

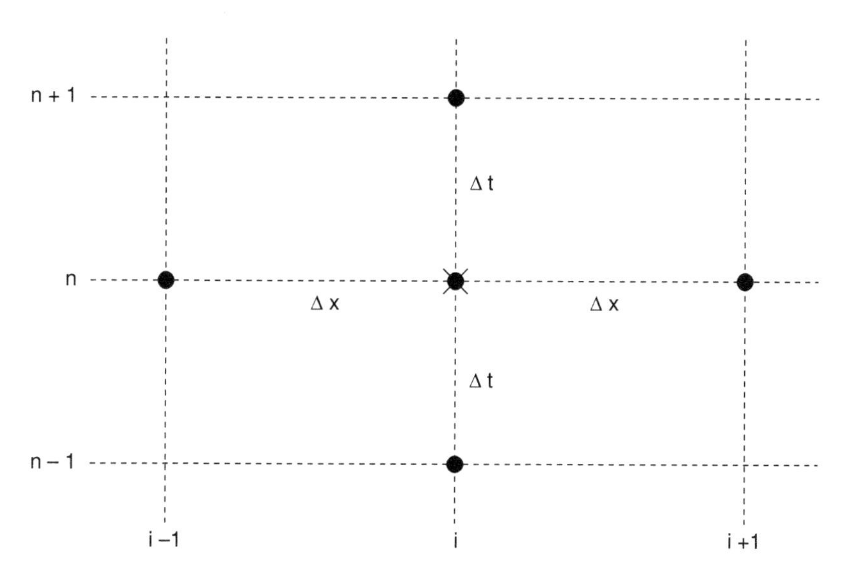

Figure 9.10 Schematic of the Richardson method for $u(x,t)$. The "×" marks the location where the differential equation is approximated.

REMARKS:

1. *The Richardson method is second-order accurate in both space and time.*
2. *We must keep Δt constant as we move from time step to time step, and a starting method is required as we need u_i at the two previous time steps.*
3. *The method is* unconditionally unstable *for $s > 0$; therefore, it is not used.*

9.5.3 DuFort–Frankel Method

In order to maintain second-order accuracy in time, but improve numerical stability, let us modify the Richardson method by taking an average between time levels for $u_i^{(n)}$. To devise such an approximation, consider the Taylor series approximation at t_{n+1} about t_n

$$u_i^{(n+1)} = u_i^{(n)} + \Delta t \left(\frac{\partial u}{\partial t}\right)_i^{(n)} + \frac{\Delta t^2}{2}\left(\frac{\partial^2 u}{\partial t^2}\right)_i^{(n)} + \cdots .$$

Similarly, consider the Taylor series approximation at t_{n-1} about t_n

$$u_i^{(n-1)} = u_i^{(n)} - \Delta t \left(\frac{\partial u}{\partial t}\right)_i^{(n)} + \frac{\Delta t^2}{2}\left(\frac{\partial^2 u}{\partial t^2}\right)_i^{(n)} + \cdots .$$

Adding these Taylor series together gives

$$u_i^{(n+1)} + u_i^{(n-1)} = 2u_i^{(n)} + \Delta t^2 \left(\frac{\partial^2 u}{\partial t^2}\right)_i^{(n)} + \ldots,$$

and solving for $u_i^{(n)}$ leads to

$$u_i^{(n)} = \frac{1}{2}\left(u_i^{(n+1)} + u_i^{(n-1)}\right) - \frac{1}{2}\Delta t^2 \left(\frac{\partial^2 u}{\partial t^2}\right)_i^{(n)} + \cdots. \tag{9.28}$$

Therefore, averaging across time levels in this manner (with uniform time step Δt) is $O(\Delta t^2)$ accurate.

Substituting the time average (9.28) into the difference equation (9.27) from Richardson's method gives

$$u_i^{(n+1)} = u_i^{(n-1)} + 2s\left[u_{i+1}^{(n)} - \left(u_i^{(n+1)} + u_i^{(n-1)}\right) + u_{i-1}^{(n)}\right],$$

or

$$(1+2s)u_i^{(n+1)} = (1-2s)u_i^{(n-1)} + 2s\left(u_{i+1}^{(n)} + u_{i-1}^{(n)}\right).$$

Solving for the single unknown yields the explicit expression

$$u_i^{(n+1)} = \frac{1-2s}{1+2s}u_i^{(n-1)} + \frac{2s}{1+2s}\left(u_{i+1}^{(n)} + u_{i-1}^{(n)}\right), \quad i = 1, \ldots, I+1. \tag{9.29}$$

However, let us consider the *consistency* of this approximation as discussed in Section 7.1.2. Including the truncation errors of each approximation, the one-dimensional, unsteady diffusion equation is approximated as in the Richardson method – with central differences for the temporal and spacial derivatives – in the form

$$\begin{aligned}\frac{\partial u}{\partial t} - \alpha\frac{\partial^2 u}{\partial x^2} = {} & \frac{u_i^{(n+1)} - u_i^{(n-1)}}{2\Delta t} - \frac{\Delta t^2}{6}\left(\frac{\partial^3 u}{\partial t^3}\right)_i^{(n)} + \cdots \\ & -\alpha\frac{u_{i+1}^{(n)} - 2u_i^{(n)} + u_{i-1}^{(n)}}{\Delta x^2} + \frac{\alpha\,\Delta x^2}{12}\left(\frac{\partial^4 u}{\partial x^4}\right)_i^{(n)} + \cdots.\end{aligned}$$

Substituting the time-averaging equation (9.28) for $u_i^{(n)}$ to implement the DuFort–Frankel method leads to

$$\begin{aligned}\frac{\partial u}{\partial t} - \alpha\frac{\partial^2 u}{\partial x^2} = {} & \frac{u_i^{(n+1)} - u_i^{(n-1)}}{2\Delta t} - \frac{\alpha}{\Delta x^2}\left[u_{i+1}^{(n)} + u_{i-1}^{(n)} - \left(u_i^{(n+1)} + u_i^{(n-1)}\right)\right] \\ & -\frac{\alpha\,\Delta t^2}{\Delta x^2}\left(\frac{\partial^2 u}{\partial t^2}\right)_i^{(n)} - \frac{\Delta t^2}{6}\left(\frac{\partial^3 u}{\partial t^3}\right)_i^{(n)} + \frac{\alpha\,\Delta x^2}{12}\left(\frac{\partial^4 u}{\partial x^4}\right)_i^{(n)} + \cdots.\end{aligned}$$

For consistency, all of the truncation error terms on the second line must go to zero as $\Delta t \to 0$ and $\Delta x \to 0$. That is, the discrete difference equation must reduce to the continuous differential equation as $\Delta x, \Delta t \to 0$. Although the second and third truncation error terms do so, the first term requires that $\Delta t \to 0$ faster than $\Delta x \to 0$, in which case $\Delta t \ll \Delta x$, for consistency. Because this is not the case in general, the DuFort–Frankel method is considered *inconsistent*.

REMARKS:

1. *The method is second-order accurate in both space and time.*
2. *We must keep Δt constant, and a starting method is necessary because the finite-difference equation involves two time levels.*
3. *The method is* unconditionally stable *for any $s = \alpha \Delta t / \Delta x^2$.*
4. *The method is inconsistent; therefore, it is not used.*

9.6 Numerical Stability Analysis

Whereas our concern is iterative convergence rate for iterative solvers applied to elliptic boundary-value problems, which is determined by the spectral radius of the iteration matrix, the concern is numerical stability in parabolic problems. In particular, the issue is how the numerical time-marching scheme handles the inevitable small round-off errors that are inherent to all numerical calculations. These errors effectively act as disturbances in the solution, and the question is what happens to these disturbances as the solution progresses in time? If the small errors decay in time and are damped out, then the numerical solution is *stable*. If the small errors grow in time and are amplified, then the numerical solution is *unstable*.

There are three general techniques commonly used to test numerical methods applied to parabolic partial differential equations for stability:

1. *Matrix method*: The matrix method is the most rigorous as it evaluates stability of the entire scheme, including treatment of boundary conditions. However, it is more difficult to perform because it involves determining the eigenvalues of a large matrix.
2. *Von Neumann method (Fourier analysis)*: The von Neumann method is only applicable to linear initial-value problems with constant coefficients; nonlinear problems must first be linearized. Therefore, it is more restrictive than the matrix method. In addition, it does not account for boundary conditions. As such, it often provides useful guidance, but it is not always conclusive for complicated problems. Nevertheless, it is the most commonly used as it is significantly easier to evaluate stability than using the more rigorous matrix method.
3. *Modified wavenumber analysis*: Extend numerical stability results for ordinary differential equations in time to partial differential equations in time and space.

9.6.1 Matrix Method

Let us denote the exact solution of the difference equation at $t = t_n$ by $\hat{u}_i^{(n)}$; then the error is

$$e_i^{(n)} = u_i^{(n)} - \hat{u}_i^{(n)}, \quad i = 2, \ldots, I, \tag{9.30}$$

where $u_i^{(n)}$ is the approximate numerical solution at $t = t_n$. In order to illustrate the matrix method, consider the first-order explicit method applied to the unsteady diffusion equation given by (9.26), which is repeated here:

$$u_i^{(n+1)} = (1 - 2s)u_i^{(n)} + s\left(u_{i+1}^{(n)} + u_{i-1}^{(n)}\right), \tag{9.31}$$

where $s = \alpha \Delta t / \Delta x^2$. Suppose that we have Dirichlet boundary conditions at $x = a$ and $x = b$. Both $u_i^{(n)}$ and $\hat{u}_i^{(n)}$ satisfy (9.31) (even though $u_i^{(n)} \neq \hat{u}_i^{(n)}$); thus, the error satisfies the same equation

$$e_i^{(n+1)} = (1 - 2s)e_i^{(n)} + s\left(e_{i+1}^{(n)} + e_{i-1}^{(n)}\right), \quad i = 2, \ldots, I. \tag{9.32}$$

Note that if u is specified at the boundaries, then the error is zero there, that is, $e_1^{(n)} = e_{I+1}^{(n)} = 0$. Equation (9.32) may be written in matrix form as

$$\mathbf{e}^{(n+1)} = \mathbf{A}\mathbf{e}^{(n)}, \quad n = 0, 1, 2, \ldots. \tag{9.33}$$

Thus, we imagine performing a matrix multiply to advance each time step in a manner similar to that for the matrix form of iterative methods. For the problem under consideration, the $(I - 1) \times (I - 1)$ matrix $\mathbf{A}$ and the $(I - 1) \times 1$ vector $\mathbf{e}^{(n)}$ are

$$\mathbf{A} = \begin{bmatrix} 1-2s & s & 0 & \cdots & 0 & 0 \\ s & 1-2s & s & \cdots & 0 & 0 \\ 0 & s & 1-2s & \cdots & 0 & 0 \\ \vdots & \vdots & \vdots & & \vdots & \vdots \\ 0 & 0 & 0 & \cdots & 1-2s & s \\ 0 & 0 & 0 & \cdots & s & 1-2s \end{bmatrix}, \quad \mathbf{e}^{(n)} = \begin{bmatrix} e_2^{(n)} \\ e_3^{(n)} \\ e_4^{(n)} \\ \vdots \\ e_{I-1}^{(n)} \\ e_I^{(n)} \end{bmatrix}.$$

The numerical method is stable if the eigenvalues λ_i of the matrix $\mathbf{A}$ are such that

$$|\lambda_i| \leq 1, \quad i = 1, \ldots, I - 1, \tag{9.34}$$

that is, the spectral radius is such that $\rho \leq 1$, in which case the error will not grow beyond that owing to truncation error during each time step. This is based on the same arguments as for iterative convergence in Section 6.4.2. Note that it only takes one eigenvalue to ruin stability for everyone! Because $\mathbf{A}$ for this case is a tridiagonal Toeplitz matrix, having constant elements along each diagonal, the eigenvalues are (see Section 6.3.3)

$$\lambda_i = 1 - (4s)\sin^2\left(\frac{i\pi}{2I}\right), \quad i = 1, \ldots, I - 1. \tag{9.35}$$

Note that there are $(I - 1)$ eigenvalues of the $(I - 1) \times (I - 1)$ matrix $\mathbf{A}$. For (9.34) to hold,

$$-1 \leq 1 - (4s)\sin^2\left(\frac{i\pi}{2I}\right) \leq 1.$$

The right inequality is true for all i, because $s > 0$ and $\sin^2(\cdot) > 0$. The left inequality is true if

$$1 - (4s)\sin^2\left(\frac{i\pi}{2I}\right) \geq -1$$

$$-(4s)\sin^2\left(\frac{i\pi}{2I}\right) \geq -2$$

$$s\sin^2\left(\frac{i\pi}{2I}\right) \leq \frac{1}{2}.$$

Because $0 \leq \sin^2\left(\frac{i\pi}{2I}\right) \leq 1$, this is true for all i if

$$s = \frac{\alpha \Delta t}{\Delta x^2} \leq \frac{1}{2}.$$

Thus, the first-order explicit method applied to the unsteady diffusion equation with Dirichlet boundary conditions is stable for $s \leq 1/2$, and we say that the method is only *conditionally stable.*

REMARKS:

1. *Whereas here we only needed eigenvalues for a tridiagonal matrix, in general the matrix method requires determination of the eigenvalues of a full* $(I+1) \times (I+1)$ *matrix.*
2. *The effects of different boundary conditions are reflected in* $\mathbf{A}$ *and, therefore, the resulting eigenvalues.*
3. *Because the eigenvalues depend on the number of grid points* $I+1$*, stability is influenced by the grid size* Δx*.*

The matrix method for evaluating numerical stability is the same as that used to obtain convergence properties of iterative methods for elliptic problems. Recall that the spectral radius $\rho(I, J)$ is the modulus of the largest eigenvalue of the iteration matrix, and $\rho \leq 1$ for an iterative method to converge. In iterative methods, however, we are concerned not only with whether they will converge or not, but the rate at which they converge. For example, recall that the Gauss–Seidel method converges twice as fast as the Jacobi method. Consequently, we seek to devise algorithms that minimize the spectral radius for maximum convergence rate. For parabolic problems, however, we are only concerned with stability in an absolute sense. There is no such thing as a time-marching method that is "more stable" than another stable one; we only care about whether the spectral radius is less than or equal to one – not how much less than one. Because of this, it is often advantageous to use a time-marching (parabolic) scheme for solving steady (elliptic) problems owing to the less restrictive stability criterion. This is sometimes referred to as the *pseudotransient method.* If we seek to solve the elliptic Laplace equation

$$\nabla^2 u = 0,$$

for example, it may be more computationally efficient to solve its unsteady parabolic counterpart

$$\frac{\partial u}{\partial t} = \nabla^2 u$$

until $\partial u/\partial t \to 0$, which corresponds to the steady solution.

9.6.2 Von Neumann Method (Fourier Analysis)

If the difference equation is linear, then we can take advantage of superposition of solutions using Fourier analysis. The error is regarded as a superposition of Fourier modes in the spatial coordinate x, and the linearity allows us to evaluate stability of each mode individually; if each mode is stable, then the linear superposition of the modes is also stable.

We expand the error at each time level as a Fourier series (see Section 3.2). It is then determined if the individual Fourier modes decay or amplify in time. Expanding the error at $t = 0$ ($n = 0$), we have

$$e(x,0) = \sum_{m=1}^{I-1} e_m(x,0) = \sum_{m=1}^{I-1} a_m(0)e^{\mathrm{i}\theta_m x}, \tag{9.36}$$

where $0 \leq x \leq 1$, $a_m(0)$ are the amplitudes of the Fourier modes $\theta_m = m\pi$, $t = 0$, and $\mathrm{i} = \sqrt{-1}$ is the imaginary number. At a later time $t = t_n$, the error is expanded as follows:

$$e(x,t_n) = \sum_{m=1}^{I-1} e_m(x,t_n) = \sum_{m=1}^{I-1} a_m(t_n)e^{\mathrm{i}\theta_m x}. \tag{9.37}$$

Note that the Fourier series handles the spatial dependence, while the coefficients $a_m(t_n)$ handle the temporal behavior. We want to determine how $a_m(t_n)$ behaves with time for each Fourier mode m. To do this, define the gain of mode m as

$$G_m(x,t_n) = \frac{e_m(x,t_n)}{e_m(x,t_n - \Delta t)} = \frac{a_m(t_n)}{a_m(t_n - \Delta t)},$$

which is the amplification factor for the mth mode during one time step. Hence, the error will not grow, and the method is stable, if $|G_m| \leq 1$ for all m. If it takes n time steps to get to time $t = t_n$, then the amplification after n time steps is

$$(G_m)^n = \frac{a_m(t_n)}{a_m(t_n - \Delta t)} \frac{a_m(t_n - \Delta t)}{a_m(t_n - 2\Delta t)} \cdots \frac{a_m(\Delta t)}{a_m(0)} = \frac{a_m(t_n)}{a_m(0)},$$

where $(G_m)^n$ is the nth power of G_m.

The von Neumann method consists of seeking a solution of the form (9.37), with $a_m(t_n) = (G_m)^n a_m(0)$, of the error equation corresponding to the difference equation. That is, the error at time $t = t_n$ is expanded as the Fourier series

$$e(x,t_n) = \sum_{m=1}^{I-1} (G_m)^n a_m(0) e^{\mathrm{i}\theta_m x}, \quad \theta_m = m\pi. \tag{9.38}$$

For the first-order explicit method, the error equation (9.32) is

$$e_i^{n+1} = (1-2s)\, e_i^n + s\left(e_{i+1}^n + e_{i-1}^n\right), \quad i = 2, \ldots, I. \tag{9.39}$$

This equation is linear; therefore, each mode m must satisfy the equation independently. Thus, substituting (9.38) into (9.39) gives (after canceling $a_m(0)$ in each term)

$$(G_m)^{n+1} e^{\mathrm{i}\theta_m x} = (1-2s)\,(G_m)^n e^{\mathrm{i}\theta_m x} + s(G_m)^n \left[e^{\mathrm{i}\theta_m (x+\Delta x)} + e^{\mathrm{i}\theta_m (x-\Delta x)}\right],$$

$m = 1, \ldots, I-1$. Note that subscript i corresponds to x, $i+1$ to $x+\Delta x$, and $i-1$ to $x - \Delta x$. The summations are dropped because we can consider each Fourier mode independently of the others. Dividing by $(G_m)^n e^{\mathrm{i}\theta_m x}$, which is common in each term, we have

$$\begin{aligned} G_m &= (1-2s) + s\left(e^{\mathrm{i}\theta_m \Delta x} + e^{-\mathrm{i}\theta_m \Delta x}\right) \\ &= 1 - 2s\left[1 - \cos\left(\theta_m \Delta x\right)\right] & & \left[\cos(ax) = \tfrac{1}{2}\left(e^{\mathrm{i}ax} + e^{-\mathrm{i}ax}\right)\right] \\ G_m &= 1 - (4s)\sin^2\left(\frac{\theta_m \Delta x}{2}\right), \quad i = 1, \ldots, I-1. & & \left[\sin^2 x = \tfrac{1}{2}\left[1 - \cos(2x)\right]\right] \end{aligned}$$

For stability, $|G_m| \leq 1$; therefore,

$$-1 \leq 1 - (4s)\sin^2\left(\frac{\theta_m \Delta x}{2}\right) \leq 1 \quad \text{for all} \quad \theta_m = m\pi, \quad m = 1, \ldots, I-1.$$

The right inequality holds for all m and $s > 0$ because

$$0 \leq \sin^2\left(\frac{\theta_m \Delta x}{2}\right) \leq 1.$$

From the same arguments used in application of the matrix method, the left inequality holds if $s = \alpha \Delta t / \Delta x^2 \leq \frac{1}{2}$.

REMARKS:

1. *For the first-order explicit method applied to the unsteady diffusion equation, therefore, we obtain the same stability criterion as from the matrix method; this is not normally the case.*
2. $|G_m|$ *is the modulus; thus, if we have a complex gain,* $|G_m|$ *equals the square root of the sum of the squares of the real and imaginary parts.*
3. *The advantage of the von Neumann method is in the fact that we can evaluate stability of each individual Fourier mode separately, in which case it is not necessary to consider a large matrix problem to evaluate stability. This is only the case, however, for linear equations, which is the primary limitation of the approach.*
4. *Von Neumann analysis applies for linear differential equations having constant coefficients, and the boundary conditions are not accounted for. Insofar as boundary conditions may influence the stability properties of a numerical method, the von Neumann method only provides guidance for stability criterion.*

9.6.3 Modified Wavenumber Analysis

Modified wavenumber analysis allows one to extend the stability analysis techniques used for ordinary differential equations in time to partial differential equations in space and time. It is very similar to the von Neumann method and has the same limitations with regard to boundary conditions and linearity. As an example, consider application of the Crank–Nicolson method to the one-dimensional, unsteady diffusion equation

$$\frac{\partial u}{\partial t} = \alpha \frac{\partial^2 u}{\partial x^2}, \qquad \alpha > 0.$$

First, a central difference approximation is applied to the spatial derivative while leaving the time derivative as is for now as follows:

$$\frac{\partial u_i}{\partial t} = \alpha \frac{u_{i+1} - 2u_i + u_{i+1}}{\Delta x^2}. \tag{9.40}$$

Let

$$u_i(t, x_i) = \psi(t) e^{\mathrm{i}kx_i},$$

which is a Fourier mode with wavenumber k in space, and the temporal behavior is captured by $\psi(t)$. Substituting into (9.40) yields

$$\frac{d\psi}{dt} e^{\mathrm{i}kx_i} = \frac{\alpha}{\Delta x^2} \left[e^{\mathrm{i}k(x_i + \Delta x)} - 2e^{\mathrm{i}kx_i} + e^{\mathrm{i}k(x_i - \Delta x)} \right] \psi(t),$$

which is of the form

$$\frac{d\psi}{dt} + \hat{k}^2 \psi = 0, \tag{9.41}$$

where $\hat{k}$ is the modified wavenumber, and

$$\hat{k}^2 = -\frac{\alpha}{\Delta x^2} \left[e^{\mathrm{i}k\Delta x} + e^{-\mathrm{i}k\Delta x} - 2 \right] = \frac{2\alpha}{\Delta x^2} \left[1 - \cos\left(k\Delta x \right) \right] = \frac{4\alpha}{\Delta x^2} \sin^2\left(\frac{k\Delta x}{2} \right).$$

Observe that the ordinary differential equation (9.41) for $\psi(t)$ is the same as (9.12) with $a = \hat{k}^2$. Therefore, we can combine the gain obtained for the Crank–Nicolson method with the modified wavenumber analysis here for finite-difference schemes used for the spatial derivatives. In other words, the stability analysis of the unsteady, first-order ordinary differential equation accounts for the method by which the time marching is carried out, for example, explicit, implicit, or Crank–Nicolson, and the modified wavenumber analysis extends this to take into account the differencing scheme used for the spatial derivatives in the partial differential equation.

Therefore, to evaluate the stability of the one-dimensional, unsteady diffusion equation solved using the Crank–Nicolson approach, consider the gain determined in Example 9.5, and substitute $a = \hat{k}^2$ from earlier for a central-difference approximation for the spatial derivative. This yields

$$G = \frac{1 - \frac{1}{2}\hat{k}^2 \Delta t}{1 + \frac{1}{2}\hat{k}^2 \Delta t} = \frac{1 - \frac{2\alpha\Delta t}{\Delta x^2} \sin^2\left(\frac{k\Delta x}{2}\right)}{1 + \frac{2\alpha\Delta t}{\Delta x^2} \sin^2\left(\frac{k\Delta x}{2}\right)}.$$

Because sine squared is always positive, the numerator is smaller than the denominator for all α and k, in which case the gain is always less than unity, and the method is unconditionally stable. As you can see, the modified wavenumber allows us to build on known results for corresponding ordinary differential equations in time, thereby allowing us to evaluate numerical stability of various time-marching schemes to the same equation. However, the von Neumann method is normally preferred as it can be applied directly to the partial differential equation of interest. See Moin (2010) for more on modified wavenumber numerical stability analysis.

9.6.4 Numerical Versus Physical Stability

The focus of our discussion has been on stability of the numerical algorithms used to solve physical problems. However, some physical systems themselves are susceptible to physical instabilities as well. When such systems are solved numerically, therefore, it is important to be able to distinguish between physical instabilities, which are real, and numerical instabilities that are a consequence of a bad choice of numerical method or numerical parameters.

Physical instabilities are common in fluid dynamics and convective heat transfer contexts. In such flows, small disturbances caused by geometric imperfections, vibrations, or other disturbances affect the underlying base flow. In numerical solutions, on the other hand, small round-off errors act as disturbances in the flow. Consequently, the issue is, "What happens to small disturbances or errors as a real flow and/or numerical solution evolves in time?" If they decay, then it is stable, and the disturbances/errors are damped out. If they grow, then they are unstable, and the disturbances/errors are amplified.

In computational fluid dynamics (CFD) and heat transfer, therefore, there are two possible sources of instability: (1) hydrodynamic instability, in which the flow itself is inherently unstable; and (2) numerical instability, in which the numerical algorithm magnifies the small errors. It is important to realize that the first is real and physical, whereas the second is not physical and signals that the time step needs to be reduced or a new numerical method is needed. The difficulty in computational fluid dynamics and heat transfer is that both are typically manifest in similar ways, that is, in the form of oscillatory solutions as illustrated in Example 9.3; therefore, it is often difficult to determine whether oscillatory numerical solutions are a result of a numerical or hydrodynamic instability.

Although beyond the scope of this book, stability analysis of fluid flows is a fascinating, rich, and difficult topic for which whole books have been written (see, for example, Drazin, 2002, and Charru, 2011). Such analysis is often quite difficult to perform, particularly for realistic flow situations. In addition, several assumptions must typically be made in order to make the analysis palatable, but it can provide conclusive evidence for hydrodynamic instability, particularly if confirmed by analytical or numerical results. In numerical stability analysis, on the other hand, we often only have guidance that is not always conclusive for complicated problems.

For example, recall that von Neumann analysis does not account for the influence of boundary conditions on numerical stability.

The interplay between these two subjects in fluid dynamics and heat transfer is exacerbated by the fact that typically these topics are treated separately: hydrodynamic stability in books on theory, and numerical stability in books on CFD. In practice, however, this distinction is often not so clear. Just because a numerical solution does not become oscillatory does not mean that no physical instabilities are present! For example, unsteady calculations are required to reveal a physical or numerical instability; that is, steady calculations suppress such instabilities as the issue numerically in steady problems is convergence, not numerical stability. In addition, there may not be sufficient resolution to reveal the unstable modes. This is because smaller grids allow for additional modes that may have different, often faster, growth rates. Consequently, it is generally not possible to obtain grid-independent solutions when hydrodynamic instabilities are present.

9.7 Implicit Methods

Having a better appreciation for numerical stability, let us return to implicit methods as applied to partial differential equations. Although explicit methods are very straightforward to implement, they are notoriously prone to numerical stability problems, and the situation generally gets worse as the equations get more complicated. We can improve dramatically on the stability properties of explicit methods by solving implicitly for more information at the current time level, which leads to a matrix problem to solve at each time step.

9.7.1 First-Order Implicit Method

Recall that in the first-order explicit method, a central difference approximation is used for spatial derivatives on the nth (previous) time level, and a first-order *forward* difference is used for the time derivative. In the first-order implicit method, we instead apply a central-difference approximation for the spatial derivatives on the $(n+1)$st (current) time level, and a first-order *backward* difference for the time derivative as illustrated in Figure 9.11.

Application of the first-order implicit method to the one-dimensional, unsteady diffusion equation yields

$$\frac{u_i^{(n+1)} - u_i^{(n)}}{\Delta t} + O(\Delta t) = \alpha \frac{u_{i+1}^{(n+1)} - 2u_i^{(n+1)} + u_{i-1}^{(n+1)}}{\Delta x^2} + O(\Delta x^2).$$

Collecting the unknowns on the left-hand side, we have the difference equation

$$s u_{i+1}^{(n+1)} - (1+2s)\, u_i^{(n+1)} + s u_{i-1}^{(n+1)} = -u_i^{(n)}, \quad i = 2, \ldots, I \tag{9.42}$$

applied at each time step. This is a *tridiagonal system* for $u_i^{(n+1)}$ at the current time level. Observe that it is strictly diagonally dominant.

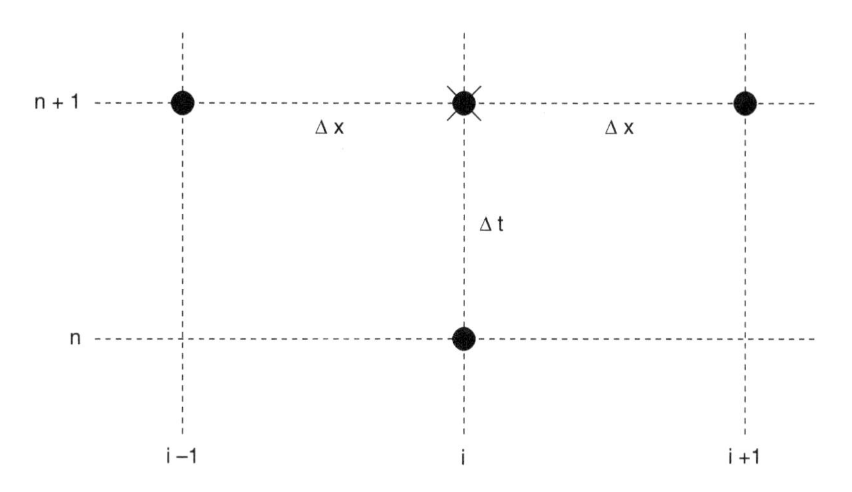

Figure 9.11 Schematic of the first-order implicit method for $u(x,t)$. The "×" marks the location where the differential equation is approximated.

Let us consider a von Neumann stability analysis. The error satisfies the finite-difference equation (9.42) according to

$$se_{i+1}^{(n+1)} - (1+2s)\,e_i^{(n+1)} + se_{i-1}^{(n+1)} = -e_i^{(n)}. \tag{9.43}$$

Again, we expand the error at time $t = t_n$ according to (9.38) as follows:

$$e(x,t_n) = \sum_{m=1}^{I-1} (G_m)^n a_m(0) e^{\mathrm{i}\theta_m x}, \quad \theta_m = m\pi. \tag{9.44}$$

Substituting into (9.43) for the error gives (canceling $a_m(0)$ in each term)

$$\begin{aligned} s(G_m)^{n+1} e^{\mathrm{i}\theta_m(x+\Delta x)} - (1+2s)\,(G_m)^{n+1} e^{\mathrm{i}\theta_m x} & \\ +s(G_m)^{n+1} e^{\mathrm{i}\theta_m(x-\Delta x)} &= -(G_m)^n e^{\mathrm{i}\theta_m x} \\ \left[s\left(e^{\mathrm{i}\theta_m \Delta x} + e^{-\mathrm{i}\theta_m \Delta x}\right) - (1+2s)\right] G_m &= -1 \\ \left[(2s)\cos\left(\theta_m \Delta x\right) - (1+2s)\right] G_m &= -1 \\ \left\{1 + 2s\left[1 - \cos\left(\theta_m \Delta x\right)\right]\right\} G_m &= 1. \end{aligned}$$

Thus, the gain of mode m is

$$G_m = \left[1 + (4s)\sin^2\left(\frac{\theta_m \Delta x}{2}\right)\right]^{-1}, \quad m = 1, \ldots, I-1.$$

The method is stable if $|G_m| \leq 1$ for all m. Therefore, noting that

$$1 + (4s)\sin^2\left(\frac{\theta_m \Delta x}{2}\right) > 1$$

for all θ_m and $s > 0$, the method is *unconditionally stable*.

REMARKS:

1. *Although second-order accurate in space, the first-order implicit method is only first-order accurate in time.*
2. *There are more computations required per time step than for explicit methods. However, one can typically use much larger time steps because they are only limited by temporal resolution requirements, not numerical stability. Specifically, for explicit methods, we must choose the time step* Δt *for both* accuracy *and* stability, *whereas for implicit methods, we must only be concerned with* accuracy.

9.7.2 Crank–Nicolson Method

Although the first-order implicit method dramatically improves the stability properties as compared to the explicit methods, it is only first-order accurate in time. Because we prefer second-order accuracy in time, consider approximating the equation midway between time levels as shown in Figure 9.12. This is known as the *Crank–Nicolson method.*

For the one-dimensional, unsteady diffusion equation, this is accomplished as follows:

$$\frac{u_i^{(n+1)} - u_i^{(n)}}{\Delta t} + O(\Delta t^2) = \frac{\alpha}{2}\left(\frac{\partial^2 u^{(n+1)}}{\partial x^2} + \frac{\partial^2 u^{(n)}}{\partial x^2}\right) + O(\Delta t^2),$$

which becomes

$$\frac{u_i^{(n+1)} - u_i^{(n)}}{\Delta t} = \frac{\alpha}{2}\left(\frac{u_{i+1}^{(n+1)} - 2u_i^{(n+1)} + u_{i-1}^{(n+1)}}{\Delta x^2} + \frac{u_{i+1}^{(n)} - 2u_i^{(n)} + u_{i-1}^{(n)}}{\Delta x^2}\right).$$

Later in this section, we show that averaging across time levels in this manner is indeed second-order accurate in time. Rearranging the difference equation with the unknowns on the left-hand side and the knowns on the right-hand side as usual leads to the tridiagonal form

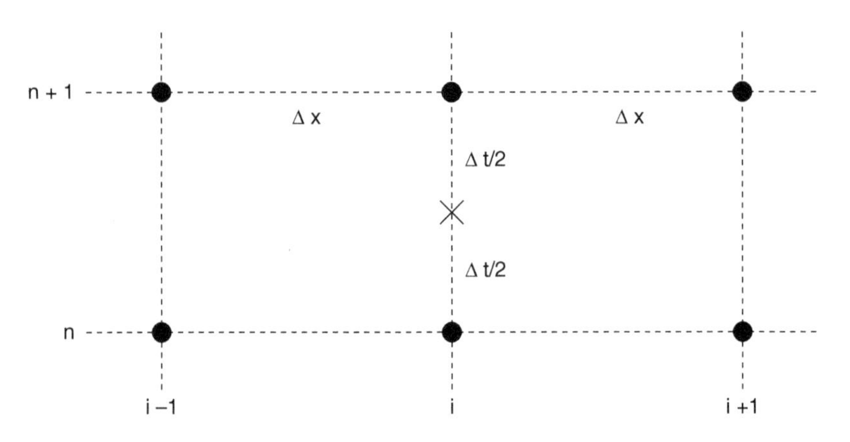

Figure 9.12 Schematic of the Crank–Nicolson method for $u(x,t)$. The "×" marks the location where the differential equation is approximated.

$$su_{i+1}^{(n+1)} - 2(1+s)u_i^{(n+1)} + su_{i-1}^{(n+1)} = -su_{i+1}^{(n)} - 2(1-s)u_i^{(n)} - su_{i-1}^{(n)}, \quad i = 2, \dots, I, \tag{9.45}$$

which we solve for $u_i^{(n+1)}$ at the current time level. As with the first-order implicit method, this results in a tridiagonal system of equations, but with more calculations required to evaluate the right-hand-side values.

REMARKS:

1. *The Crank–Nicolson method is second-order accurate in both space and time.*
2. *It is unconditionally stable for all s.*
3. *Derivative boundary conditions would be applied at the current time level.*
4. *The Crank–Nicolson method is sometimes called the* trapezoidal method.
5. *Because of its accuracy and stability properties, it is a very popular scheme for parabolic problems.*
6. *Observe that the approximations of the time derivatives for the first-order explicit, first-order implicit, and Crank–Nicolson methods all appear to be the same; however, they are actually forward-, backward-, and central-difference approximations, respectively, owing to the point in the grid at which the equation is approximated.*
7. *When nonlinear terms are included in unsteady parabolic equations, explicit methods lead to very restrictive time steps in order to maintain numerical stability. Implicit methods improve on the stability properties but require iteration at each time step in order to update the terms that make the equation nonlinear. This will be considered in Section 9.9.*

When describing the DuFort–Frankel method, it was established that averaging a variable across time levels is second-order accurate in time. Here, we again confirm this, and in doing so, we will illustrate a method to determine the accuracy of a specified finite-difference approximation.

As illustrated in Figure 9.13, consider averaging a quantity $u(x,t)$ midway between time levels as follows:

$$u_i^{(n+1/2)} = \frac{1}{2}\left(u_i^{(n+1)} + u_i^{(n)}\right) + T.E. \tag{9.46}$$

We seek an expression of the form $u_i^{(n+1/2)} = \hat{u} + T.E.$ to obtain the truncation error $T.E.$, where $\hat{u}$ is the exact value of $u_i^{(n+1/2)}$ midway between time levels. Let us expand each term in the expression (9.46) as a Taylor series about $(x_i, t_{n+1/2})$ as follows:

$$u_i^{(n+1)} = \sum_{k=0}^{\infty} \frac{1}{k!}\left(\frac{\Delta t}{2} D_t\right)^k \hat{u},$$

$$u_i^{(n)} = \sum_{k=0}^{\infty} \frac{1}{k!}\left(-\frac{\Delta t}{2} D_t\right)^k \hat{u} = \sum_{k=0}^{\infty} \frac{(-1)^k}{k!}\left(\frac{\Delta t}{2} D_t\right)^k \hat{u},$$

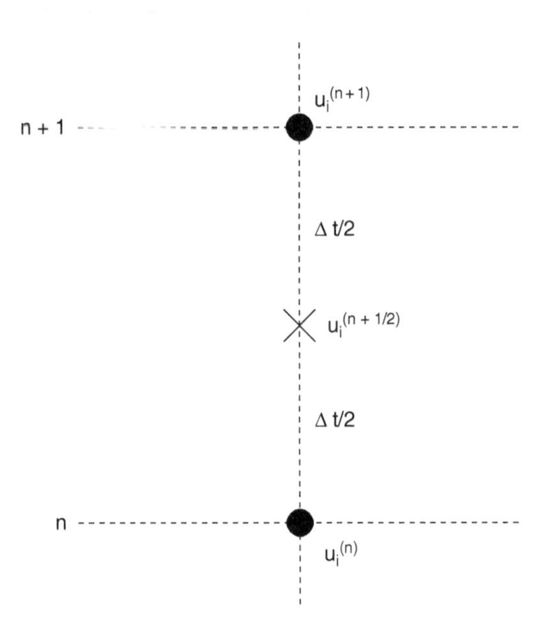

Figure 9.13 Schematic for averaging across time levels.

where $D_t = \partial/\partial t$. Substituting these expansions into (9.46) gives

$$u_i^{(n+1/2)} = \frac{1}{2}\sum_{k=0}^{\infty}\frac{1}{k!}\left[1+(-1)^k\right]\left(\frac{\Delta t}{2}D_t\right)^k \hat{u}.$$

Noting that

$$1+(-1)^k = \begin{cases} 0, & k = 1,3,5,\ldots \\ 2, & k = 0,2,4,\ldots \end{cases},$$

we can rewrite this expansion as (letting $k = 2m$)

$$u_i^{(n+1/2)} = \sum_{m=0}^{\infty}\frac{1}{(2m)!}\left(\frac{\Delta t}{2}D_t\right)^{2m} \hat{u}.$$

The first three terms of this expansion are

$$u_i^{(n+1/2)} = \hat{u} + \frac{1}{2!}\left(\frac{\Delta t}{2}D_t\right)^2 \hat{u} + \frac{1}{4!}\left(\frac{\Delta t}{2}D_t\right)^4 \hat{u} + \cdots.$$

The first term in the expansion ($m = 0$) is the exact value, and the second term ($m = 1$) gives the truncation error of the approximation. This truncation error term is

$$\frac{1}{8}\Delta t^2\frac{\partial^2\hat{u}}{\partial t^2},$$

where we note that the coefficient is different from that obtained in (9.28) because the averaging is carried out across two time steps as compared to only one here. Consequently, the approximation (9.46) is second-order accurate, such that

$$u_i^{(n+1/2)} = \frac{1}{2}\left(u_i^{(n+1)} + u_i^{(n)}\right) + O(\Delta t^2).$$

This confirms that averaging across time levels gives an $O(\Delta t^2)$ approximation of $u(x,t)$ at the midtime level $t_{n+1/2}$.

9.8 Boundary Conditions – Special Cases

In general, boundary conditions are treated as for boundary-value problems as discussed in Section 8.6. However, there is one eventuality that has not yet been addressed – when there is a singularity at a boundary. Such singularities do not normally arise because of the physics of the problem but rather its numerical representation. For example, we have been considering the one-dimensional, unsteady diffusion equation in Cartesian coordinates as our model equation throughout this chapter. Instead, suppose that we wish to solve this common model problem in cylindrical coordinates, for which the equation is given by ($\alpha = 1$)

$$\frac{\partial u}{\partial t} = \frac{1}{r}\frac{\partial}{\partial r}\left(r\frac{\partial u}{\partial r}\right), \quad 0 \le r \le R. \tag{9.47}$$

This equation arises from the full unsteady diffusion equation in cylindrical coordinates for axisymmetric problems in which there is no dependence on the azimuthal direction θ, and $\partial u/\partial r = 0$ at $r = 0$. Observe that at the left boundary $r = 0$, there is a singularity owing to the $1/r$ factor. Naively, we may simply remove the singularity by multiplying both sides by r, but then the time derivative is gone, and we have altered the form of the parabolic equation. Instead, we must be more creative; when in doubt, write a Taylor series! The idea is as follows: when there is a singularity *at* a point in a function or equation, write a Taylor series of the function *near* the point to determine how it behaves in the vicinity of the singular point.

The Taylor series with respect to r of the dependent variable $u(r,t)$ near $r = 0$ is given by

$$u(r,t) = u(0,t) + r\left.\frac{\partial u}{\partial r}\right|_{r=0} + \frac{1}{2}r^2\left.\frac{\partial^2 u}{\partial r^2}\right|_{r=0} + \ldots,$$

but $\partial u/\partial r = 0$ at $r = 0$; thus, the Taylor series reduces to

$$u(r,t) = u(0,t) + \frac{1}{2}r^2\left.\frac{\partial^2 u}{\partial r^2}\right|_{r=0} + \cdots. \tag{9.48}$$

Differentiating leads to

$$\frac{\partial u}{\partial r} = \overbrace{\left.\frac{\partial u}{\partial r}\right|_{r=0}}^{0} + r\left.\frac{\partial^2 u}{\partial r^2}\right|_{r=0} + \overbrace{\frac{1}{2}r^2\left.\frac{\partial^3 u}{\partial r^3}\right|_{r=0}}^{0} + \cdots.$$

The first term vanishes as before owing to axisymmetry, and the third term is neglected because it is quadratic in r, which is assumed small. Near the origin, therefore, the right-hand side of (9.47) becomes

$$\begin{aligned}
\frac{1}{r}\frac{\partial}{\partial r}\left(r\frac{\partial u}{\partial r}\right) &= \frac{1}{r}\frac{\partial}{\partial r}\left(r^2\left.\frac{\partial^2 u}{\partial r^2}\right|_{r=0}+\cdots\right) \\
&= \frac{1}{r}\left(2r\left.\frac{\partial^2 u}{\partial r^2}\right|_{r=0}+\cancelto{0}{r^2\left.\frac{\partial^3 u}{\partial r^3}\right|_{r=0}}+\cdots\right) \\
\frac{1}{r}\frac{\partial}{\partial r}\left(r\frac{\partial u}{\partial r}\right) &= 2\left.\frac{\partial^2 u}{\partial r^2}\right|_{r=0}+\cdots.
\end{aligned} \tag{9.49}$$

Writing the Taylor series (9.48) at $r = \Delta r$, corresponding to the first grid point in the interior at $i = 2$ from the left boundary at $i = 1$, gives

$$u_2 = u_1 + \frac{1}{2}(\Delta r)^2\left.\frac{\partial^2 u}{\partial r^2}\right|_{r=0}+\cdots.$$

Solving for $\partial^2 u/\partial r^2\big|_{r=0}$ yields

$$\left.\frac{\partial^2 u}{\partial r^2}\right|_{r=0} = 2\frac{u_2 - u_1}{(\Delta r)^2}+\cdots.$$

Substituting into (9.49) results in

$$\frac{1}{r}\frac{\partial}{\partial r}\left(r\frac{\partial u}{\partial r}\right) = 4\frac{u_2 - u_1}{(\Delta r)^2}+\cdots. \tag{9.50}$$

This is the approximation used for the spatial derivative on the right-hand side of (9.47) at the left boundary corresponding to $i = 1$ $(r = 0)$, where the singularity in the coordinate system occurs. All interior points $i = 2, 3, \ldots$ are treated as before.

9.9 Treatment of Nonlinear Convection Terms

As with elliptic partial differential equations, which correspond to steady problems, it is essential in fluid dynamics, heat transfer, and mass transfer to be able to handle nonlinear convective terms in unsteady parabolic contexts. Consider the one-dimensional, unsteady Burgers equation, which is a one-dimensional, unsteady diffusion equation with a convection term, given by

$$\frac{\partial u}{\partial t} = \nu\frac{\partial^2 u}{\partial x^2} - u\frac{\partial u}{\partial x}, \tag{9.51}$$

where $u(x,t)$ is the velocity in the x direction, and ν is the viscosity. Let us consider how the nonlinear convection term $u\partial u/\partial x$ is treated in the various schemes.

9.9.1 First-Order Explicit Method

Approximating spatial derivatives at the previous time level, and using a forward difference in time, we have

$$\frac{u_i^{(n+1)} - u_i^{(n)}}{\Delta t} = \nu \frac{u_{i+1}^{(n)} - 2u_i^{(n)} + u_{i-1}^{(n)}}{\Delta x^2} - u_i^{(n)} \frac{u_{i+1}^{(n)} - u_{i-1}^{(n)}}{2\Delta x} + O(\Delta x^2, \Delta t),$$

where the $u_i^{(n)}$ factor in the convection term is known from the previous time level. Writing in explicit form leads to

$$u_i^{(n+1)} = \left(s - \frac{1}{2}C_i^{(n)}\right) u_{i+1}^{(n)} + (1-2s)\, u_i^{(n)} + \left(s + \frac{1}{2}C_i^{(n)}\right) u_{i-1}^{(n)},$$

where $s = \nu\Delta t/\Delta x^2$, and $C_i^{(n)} = u_i^{(n)}\Delta t/\Delta x$ is the Courant number, which is the basis for the Courant–Friedrichs–Lewy (CFL) criterion for numerical stability of flows with convection. Because the nonlinearity owing to the convective term is in a spatial derivative, which is approximated at the previous time step, there is no particular issue introduced when treating nonlinear convective terms using the first-order explicit method. For stability, however, this method requires that $2 \leq Re_{\Delta x} \leq 2/C_i^{(n)}$, where $Re_{\Delta x} = u_i^{(n)}\Delta x/\nu$ is the mesh Reynolds number. Not only is this stability criterion very restrictive, we do not know the velocity $u_i^{(n)}$ in the Courant number before solving the problem.

9.9.2 Crank–Nicolson Method

In the Crank–Nicolson method, all spatial derivatives are approximated at the midpoint between time steps as shown in Figure 9.12. Therefore, the one-dimensional, unsteady Burgers equation becomes

$$\begin{aligned}\frac{u_i^{(n+1)} - u_i^{(n)}}{\Delta t} &= \frac{\nu}{2}\left(\frac{\partial^2 u^{(n+1)}}{\partial x^2} + \frac{\partial^2 u^{(n)}}{\partial x^2}\right) - \frac{1}{2}u_i^{(n+1/2)}\left(\frac{\partial u^{(n+1)}}{\partial x} + \frac{\partial u^{(n)}}{\partial x}\right)\\ &= \frac{\nu}{2\Delta x^2}\left(u_{i+1}^{(n+1)} - 2u_i^{(n+1)} + u_{i-1}^{(n+1)} + u_{i+1}^{(n)} - 2u_i^{(n)} + u_{i-1}^{(n)}\right)\\ &\quad - \frac{u_i^{(n+1/2)}}{4\Delta x}\left(u_{i+1}^{(n+1)} - u_{i-1}^{(n+1)} + u_{i+1}^{(n)} - u_{i-1}^{(n)}\right) + O(\Delta x^2, \Delta t^2),\end{aligned} \tag{9.52}$$

where we average across time levels to obtain the velocity according to

$$u_i^{(n+1/2)} = \frac{1}{2}\left(u_i^{(n+1)} + u_i^{(n)}\right) + O(\Delta t^2). \tag{9.53}$$

This results in the implicit finite-difference system of equations

$$
\begin{aligned}
&-\left(s-\frac{1}{2}C_i^{(n+1/2)}\right)u_{i+1}^{(n+1)}+2(1+s)u_i^{(n+1)}-\left(s+\frac{1}{2}C_i^{(n+1/2)}\right)u_{i-1}^{(n+1)} \\
&=\left(s-\frac{1}{2}C_i^{(n+1/2)}\right)u_{i+1}^{(n)}+2(1-s)u_i^{(n)}+\left(s+\frac{1}{2}C_i^{(n+1/2)}\right)u_{i-1}^{(n)},
\end{aligned}
\tag{9.54}
$$

which is tridiagonal. Here, $C_i^{(n+1/2)} = u_i^{(n+1/2)}\Delta t/\Delta x$, but we do not know $u_i^{(n+1/2)}$ yet. Therefore, this procedure requires iteration at each time step:

1. Begin with $u_i^{(k)} = u_i^{(n)} (k = 0)$, that is, use the value of u_i from the previous time step as an initial guess at the current time step.
2. Increment $k = k + 1$. Compute an update for $u_i^{(k)} = u_i^{(n+1)}, i = 1, \dots, I + 1$, by solving the tridiagonal system of equations (9.54).
3. Update $u_i^{(n+1/2)} = \frac{1}{2}\left(u_i^{(n+1)} + u_i^{(n)}\right), i = 1, \dots, I + 1$.
4. Return to step 2 and repeat until $u_i^{(k)} = u_i^{(n+1)}$ converges for all i.

It typically requires less than ten iterations to converge at each time step; if more are required, then the time step Δt is too large.

REMARKS:

1. *In elliptic problems, we use Picard linearization to treat nonlinear convective terms because we only care about the accuracy of the final converged solution, whereas here we want an accurate solution at each time step, thereby requiring iteration.*
2. *When nonlinear convective terms are included in unsteady parabolic equations, explicit methods lead to very restrictive time steps in order to maintain numerical stability. Implicit methods improve on the stability properties but require iteration at each time step.*

9.9.3 Upwind-Downwind Differencing

As in the elliptic case, we can also use upwind-downwind differencing. Moreover, the inclusion of the temporal dimension provides additional flexibility in how the upwind-downwind differencing is carried out to obtain the approximation at the desired location. Keeping in mind that we desire to maintain a compact scheme, let us reconsider the first-order convective term in the Crank–Nicolson approximation

$$u^{(n+1/2)}\frac{\partial u^{(n+1/2)}}{\partial x}.$$

If $u^{(n+1/2)} > 0$, then we approximate $\partial u^{(n+1/2)}/\partial x$ as shown in Figure 9.14, in which case

$$\frac{\partial u}{\partial x} = \frac{1}{2}\left(\left.\frac{\partial u^{(n+1)}}{\partial x}\right|_{i-1/2} + \left.\frac{\partial u^{(n)}}{\partial x}\right|_{i+1/2}\right)$$

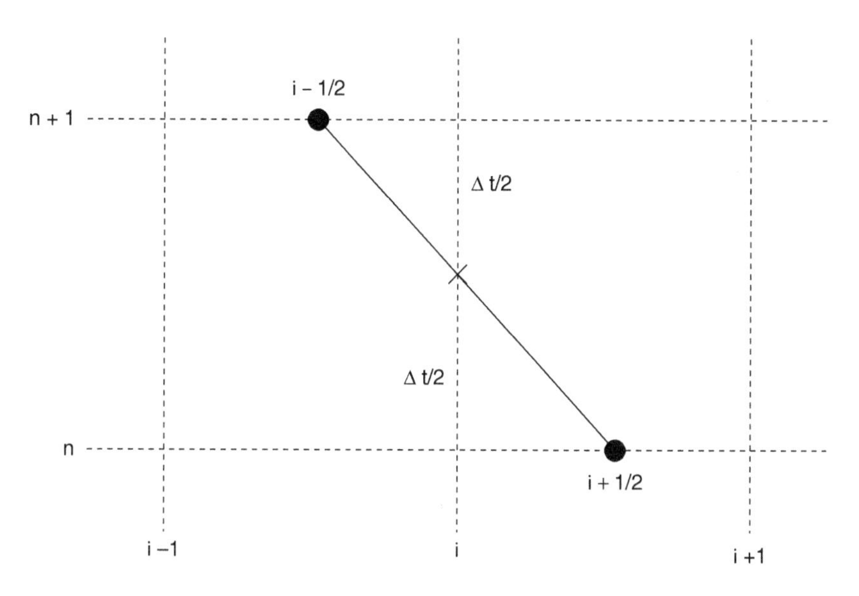

Figure 9.14 Crank–Nicolson with upwind-downwind differencing of the first-order convection term when $u^{(n+1/2)} > 0$.

is given by

$$\frac{\partial u}{\partial x} = \frac{1}{2}\left(\frac{u_i^{(n+1)} - u_{i-1}^{(n+1)}}{\Delta x} + \frac{u_{i+1}^{(n)} - u_i^{(n)}}{\Delta x}\right). \tag{9.55}$$

Although the finite-difference approximation at the current time level appears to be a backward difference and that at the previous time level appears to be a forward difference, they are in fact central differences evaluated at half-points in the grid.

If instead $u^{(n+1/2)} < 0$, then we approximate $\partial u^{(n+1/2)}/\partial x$ as shown in Figure 9.15, in which case

$$\frac{\partial u}{\partial x} = \frac{1}{2}\left(\left.\frac{\partial u^{(n+1)}}{\partial x}\right|_{i+1/2} + \left.\frac{\partial u^{(n)}}{\partial x}\right|_{i-1/2}\right)$$

is given by

$$\frac{\partial u}{\partial x} = \frac{1}{2}\left(\frac{u_{i+1}^{(n+1)} - u_i^{(n+1)}}{\Delta x} + \frac{u_i^{(n)} - u_{i-1}^{(n)}}{\Delta x}\right). \tag{9.56}$$

Substituting (9.55) and (9.56) into (9.52) in place of the central differences yields

$$u_i^{(n+1)} - u_i^{(n)} = \frac{1}{2}s\left(u_{i+1}^{(n+1)} - 2u_i^{(n+1)} + u_{i-1}^{(n+1)} + u_{i+1}^{(n)} - 2u_i^{(n)} + u_{i-1}^{(n)}\right)$$
$$- \frac{1}{2}C_i^{(n+1/2)}\begin{cases} u_i^{(n+1)} - u_{i-1}^{(n+1)} + u_{i+1}^{(n)} - u_i^{(n)}, & u_i^{(n+1/2)} > 0 \\ u_{i+1}^{(n+1)} - u_i^{(n+1)} + u_i^{(n)} - u_{i-1}^{(n)}, & u_i^{(n+1/2)} < 0 \end{cases}.$$

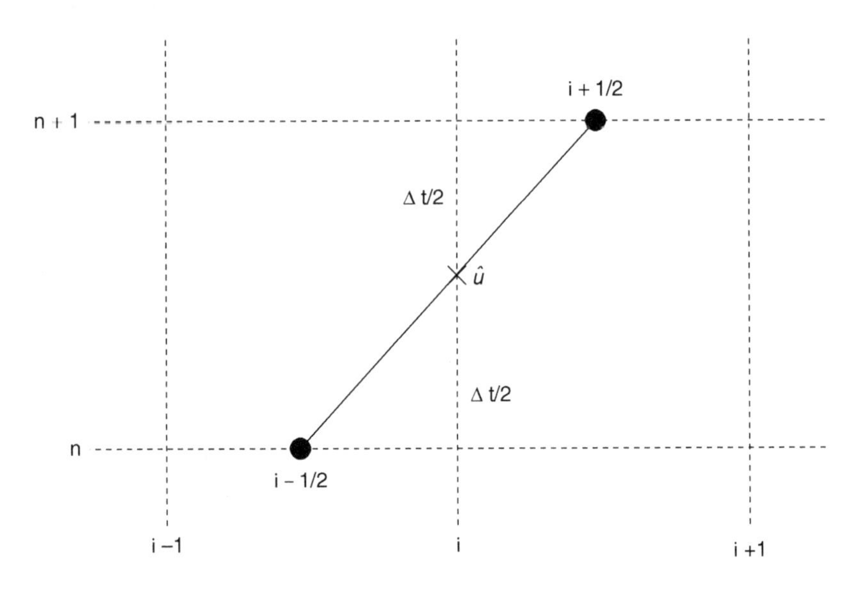

Figure 9.15 Crank–Nicolson with upwind-downwind differencing of the first-order convection term when $u^{(n+1/2)} < 0$.

This gives the tridiagonal problem

$$
\begin{aligned}
&- su_{i+1}^{(n+1)} + 2(1+s)u_i^{(n+1)} - su_{i-1}^{(n+1)} + C_i^{(n+1/2)} \left\{ \begin{array}{l} u_i^{(n+1)} - u_{i-1}^{(n+1)} \\ u_{i+1}^{(n+1)} - u_i^{(n+1)} \end{array} \right\} \\
&= su_{i+1}^{(n)} + 2(1-s)u_i^{(n)} + su_{i-1}^{(n)} - C_i^{(n+1/2)} \left\{ \begin{array}{l} u_{i+1}^{(n)} - u_i^{(n)} \\ u_i^{(n)} - u_{i-1}^{(n)} \end{array} \right\} \quad \begin{array}{l} , \quad u_i^{(n+1/2)} > 0 \\ , \quad u_i^{(n+1/2)} < 0 \end{array} .
\end{aligned}
\tag{9.57}
$$

Averaging across time levels in this manner leads to a difference equation (9.57) that is diagonally dominant for the one-dimensional, unsteady Burgers equation for all s and $C_i^{(n+1/2)}$ (note that $C_i^{(n+1/2)}$ may be positive or negative). Consequently, the iteration at each time step will converge. In addition, a numerical stability analysis would reveal that this method is *unconditionally stable*.

As for the Crank–Nicolson method for the case without upwind-downwind differencing, iteration at each time step is required owing to the nonlinear term $u\partial u/\partial x$, which may require under-relaxation on u_i for nonlinear equations; therefore,

$$
u_i^{(k+1)} \leftarrow (1-\omega)u_i^{(k)} + \omega u_i^{(k+1)}, \quad k = 0, 1, 2, \ldots,
$$

where $0 < \omega < 1$ for under-relaxation.

Let us determine the truncation error of the Crank–Nicolson scheme with upwind-downwind differencing as implemented here. We have used second-order accurate central differences for the $\partial u/\partial t$ and $\partial^2 u/\partial x^2$ terms; therefore, consider the $u^{(n+1/2)}\partial u^{(n+1/2)}\partial x$ term with $u^{(n+1/2)} < 0$ from (9.56), which yields

$$\frac{\partial u}{\partial x} = \frac{1}{2\Delta x}\left(u_{i+1}^{(n+1)} - u_i^{(n+1)} + u_i^{(n)} - u_{i-1}^{(n)}\right) + T.E. \tag{9.58}$$

We would like to determine the order of accuracy in the form of the truncation error $T.E.$ of this approximation. Because we already have the finite-difference approximation (9.58), we only need to obtain the truncation error. Therefore, the method introduced in Section 9.7.2 is appropriate. Here, $D_t = \partial/\partial t, D_x = \partial/\partial x$, and $\hat{u}$ is the exact value of $u_i^{(n+1/2)}$ midway between time levels as illustrated in Figure 9.15.

We seek an expression of the form

$$\frac{\partial u}{\partial x} = \frac{\partial \hat{u}}{\partial x} + T.E.$$

Expand each term in (9.58) as a two-dimensional Taylor series about $(x_i, t_{n+1/2})$ as follows:

$$\begin{aligned}
u_{i+1}^{(n+1)} &= \sum_{k=0}^{\infty} \frac{1}{k!}\left(\frac{\Delta t}{2}D_t + \Delta x D_x\right)^k \hat{u}, \\
u_i^{(n+1)} &= \sum_{k=0}^{\infty} \frac{1}{k!}\left(\frac{\Delta t}{2}D_t\right)^k \hat{u}, \\
u_i^{(n)} &= \sum_{k=0}^{\infty} \frac{1}{k!}\left(-\frac{\Delta t}{2}D_t\right)^k \hat{u} = \sum_{k=0}^{\infty} \frac{(-1)^k}{k!}\left(\frac{\Delta t}{2}D_t\right)^k \hat{u}, \\
u_{i-1}^{(n)} &= \sum_{k=0}^{\infty} \frac{1}{k!}\left(-\frac{\Delta t}{2}D_t - \Delta x D_x\right)^k \hat{u} = \sum_{k=0}^{\infty} \frac{(-1)^k}{k!}\left(\frac{\Delta t}{2}D_t + \Delta x D_x\right)^k \hat{u}.
\end{aligned}$$

Substituting these expansions into (9.58) produces

$$\begin{aligned}
\frac{\partial u}{\partial x} &= \frac{1}{2\Delta x}\sum_{k=0}^{\infty}\frac{1}{k!}\left\{\left[1-(-1)^k\right]\left(\frac{\Delta t}{2}D_t + \Delta x D_x\right)^k \right. \\
&\quad \left. + \left[-1+(-1)^k\right]\left(\frac{\Delta t}{2}D_t\right)^k\right\}\hat{u}, \\
\frac{\partial u}{\partial x} &= \frac{1}{2\Delta x}\sum_{k=0}^{\infty}\frac{1+(-1)^{k+1}}{k!}\left[\left(\frac{\Delta t}{2}D_t + \Delta x D_x\right)^k - \left(\frac{\Delta t}{2}D_t\right)^k\right]\hat{u}.
\end{aligned}$$

Note that

$$1+(-1)^{k+1} = \begin{cases} 0, & k = 0, 2, 4, \ldots \\ 2, & k = 1, 3, 5, \ldots \end{cases}.$$

Therefore, let $k = 2l+1$, in which case

$$\frac{\partial u}{\partial x} = \frac{1}{\Delta x}\sum_{l=0}^{\infty}\frac{1}{(2l+1)!}\left[\left(\frac{\Delta t}{2}D_t + \Delta x D_x\right)^{2l+1} - \left(\frac{\Delta t}{2}D_t\right)^{2l+1}\right]\hat{u}. \tag{9.59}$$

In order to treat the first term in square brackets, recall the *binomial theorem*

$$(a+b)^k = \sum_{m=0}^{k} \binom{k}{m} a^{k-m} b^m, \quad \binom{k}{m} = \frac{k!}{m!\,(k-m)!},$$

where $\binom{k}{m}$ are the binomial coefficients, and $0! = 1$. Application of the binomial theorem to the expression (9.59) leads to

$$\begin{aligned}
\frac{\partial u}{\partial x} &= \frac{1}{\Delta x} \sum_{l=0}^{\infty} \frac{1}{(2l+1)!} \left[\sum_{m=0}^{2l+1} \binom{2l+1}{m} \left(\frac{\Delta t}{2} D_t\right)^{2l+1-m} (\Delta x D_x)^m \right. \\
&\quad \left. - \left(\frac{\Delta t}{2} D_t\right)^{2l+1} \right] \hat{u}, \\
&= \frac{1}{\Delta x} \sum_{l=0}^{\infty} \frac{1}{(2l+1)!} \left[\sum_{m=1}^{2l+1} \binom{2l+1}{m} \left(\frac{\Delta t}{2} D_t\right)^{2l-m+1} (\Delta x D_x)^m \right. \\
&\quad \left. + \binom{2l+1}{0} \left(\frac{\Delta t}{2} D_t\right)^{2l+1} - \left(\frac{\Delta t}{2} D_t\right)^{2l+1} \right] \hat{u}, \quad \left[\binom{2l+1}{0} = 1 \right] \\
&= \sum_{l=0}^{\infty} \frac{1}{(2l+1)!} \sum_{m=1}^{2l+1} \frac{(2l+1)!}{m!\,(2l-m+1)!} \left(\frac{\Delta t}{2} D_t\right)^{2l-m+1} (\Delta x)^{m-1} (D_x)^m \hat{u}, \\
\frac{\partial u}{\partial x} &= \frac{\partial \hat{u}}{\partial x} + \sum_{l=1}^{\infty} \sum_{m=1}^{2l+1} \frac{1}{m!\,(2l-m+1)!} \left(\frac{\Delta t}{2} D_t\right)^{2l-m+1} (\Delta x)^{m-1} (D_x)^m \hat{u},
\end{aligned}$$

where the $\partial \hat{u}/\partial x$ term results from taking $l = 0, m = 1$ in the double summation. To obtain the truncation error, consider the $l = 1$ term for which $m = 1, 2, 3$ produces

$$\frac{1}{1!\,2!} \left(\frac{\Delta t}{2}\right)^2 D_t^2 D_x \hat{u} + \frac{1}{2!\,1!} \left(\frac{\Delta t}{2}\right) \Delta x D_t D_x^2 \hat{u} + \frac{1}{3!\,1!} \Delta x^2 D_x^3 \hat{u}.$$

Therefore, the truncation error is

$$O(\Delta t^2, \Delta t \Delta x, \Delta x^2).$$

Consequently, if $\Delta t < \Delta x$, then the approximation is $O(\Delta x^2)$ accurate, and if $\Delta t > \Delta x$, then the approximation is $O(\Delta t^2)$ accurate. This is significantly better than $O(\Delta x)$ or $O(\Delta t)$, but strictly speaking it is not $O(\Delta t^2, \Delta x^2)$. Note that the $O(\Delta t \Delta x)$ term arises owing to the diagonal averaging across time levels in the upwind-downwind differencing.

9.10 Multidimensional Problems

In order to illustrate the various methods for treating parabolic partial differential equations, we have focused on the spatially one-dimensional case. More likely, however, we are faced with solving problems that are two- or three-dimensional in space. Let us see how these methods extend to multidimensions.

Consider the two-dimensional, unsteady diffusion equation

$$\frac{\partial u}{\partial t} = \alpha\left(\frac{\partial^2 u}{\partial x^2} + \frac{\partial^2 u}{\partial y^2}\right), \quad u = u(x,y,t), \tag{9.60}$$

with initial condition

$$u(x,y,0) = u_0(x,y) \quad \text{at} \quad t = 0, \tag{9.61}$$

and boundary conditions (Dirichlet, Neumann, or Robin) on a closed contour C enclosing the domain. Observe that (9.60) is spatially elliptic in (x, y) and temporally parabolic in (t); therefore, it is solved using parabolic techniques adjusted for the spatial multidimensionality. As usual, we discretize the domain as shown in Figure 9.16, but now with t being normal to the (x, y) plane.

9.10.1 First-Order Explicit Method

Approximating (9.60) using a forward difference in time and central differences in space at the previous time level gives

$$\frac{u_{i,j}^{(n+1)} - u_{i,j}^{(n)}}{\Delta t} = \alpha\left(\frac{u_{i+1,j}^{(n)} - 2u_{i,j}^{(n)} + u_{i-1,j}^{(n)}}{\Delta x^2} + \frac{u_{i,j+1}^{(n)} - 2u_{i,j}^{(n)} + u_{i,j-1}^{(n)}}{\Delta y^2}\right)$$
$$+ O(\Delta x^2, \Delta y^2, \Delta t),$$

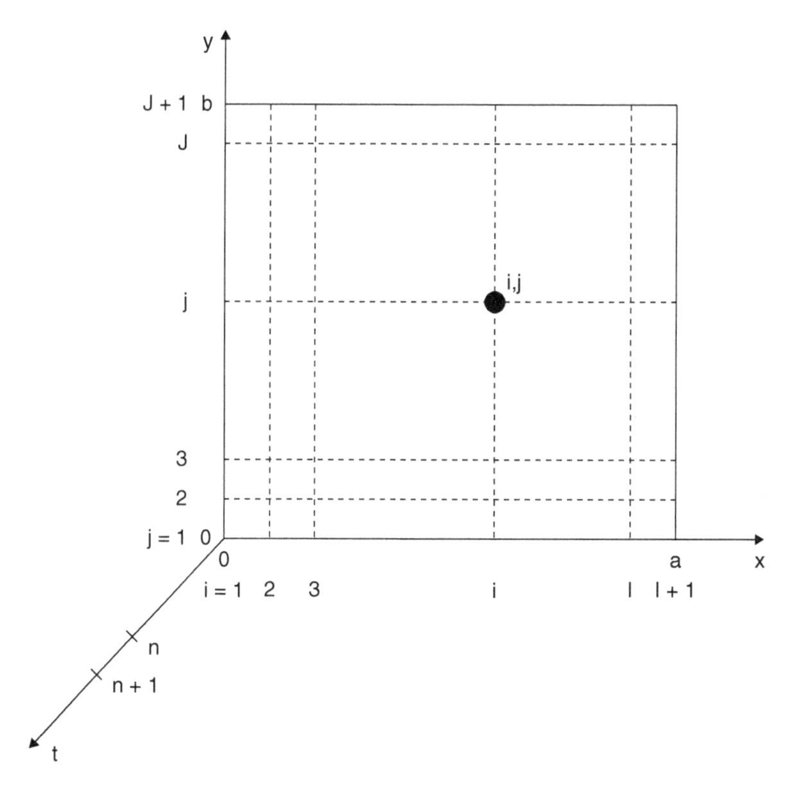

Figure 9.16 Discretization of two-dimensional domain for parabolic problems with $u(x, y, t)$.

where now $u_{i,j}^{(n)} = u(x_i, y_j, t_n)$. Therefore, solving for the only unknown gives the explicit expression

$$u_{i,j}^{(n+1)} = \left(1 - 2s_x - 2s_y\right) u_{i,j}^{(n)} + s_x \left(u_{i+1,j}^{(n)} + u_{i-1,j}^{(n)}\right) + s_y \left(u_{i,j+1}^{(n)} + u_{i,j-1}^{(n)}\right), \tag{9.62}$$

where $s_x = \alpha\Delta t/\Delta x^2$, and $s_y = \alpha\Delta t/\Delta y^2$. For numerical stability, a von Neumann stability analysis requires that

$$s_x + s_y \leq \frac{1}{2}.$$

Thus, for example, if $\Delta x = \Delta y$, in which case $s_x = s_y = s$, we must have

$$s \leq \frac{1}{4},$$

which is even more restrictive than for the one-dimensional, unsteady diffusion equation, which requires $s \leq \frac{1}{2}$ for stability. The three-dimensional case becomes even more restrictive, with $s \leq \frac{1}{6}$ for numerical stability.

9.10.2 First-Order Implicit Method

Applying a backward difference in time and central differences in space at the current time level leads to the implicit expression

$$\left(1 + 2s_x + 2s_y\right) u_{i,j}^{(n+1)} - s_x \left(u_{i+1,j}^{(n+1)} + u_{i-1,j}^{(n+1)}\right) - s_y \left(u_{i,j+1}^{(n+1)} + u_{i,j-1}^{(n+1)}\right) = u_{i,j}^{n}. \tag{9.63}$$

Observe that this produces a banded matrix with five unknowns, rather than a tridiagonal matrix as in the one-dimensional case.

REMARKS:

1. *For the two-dimensional, unsteady diffusion equation, the first-order implicit method is* unconditionally stable *for all s_x and s_y.*
2. *The usual Crank–Nicolson method could be used to obtain second-order accuracy in time. It produces a similar implicit equation as for the one-dimensional case, but with more terms on the right-hand side evaluated at the previous time step.*

9.10.3 ADI Method with Time Splitting

Rather than using the Crank–Nicolson method in its usual form to maintain second-order accuracy in time, it is more common to combine it with the *ADI method with time splitting*, which is often referred to as a *fractional-step method*. In this method, we split each time step into two half steps (or three for three dimensions), which results in two (or three) sets of tridiagonal problems per time step. In the two-dimensional case, we first solve implicitly for the terms associated with one coordinate direction followed by solving implicitly for the terms associated with the other coordinate direction.

For example, if we sweep along lines of constant y during the first half-time step, the difference equation with Crank–Nicolson becomes

$$\begin{aligned} &\frac{u_{i,j}^{(n+1/2)} - u_{i,j}^{(n)}}{\Delta t/2} \\ &= \alpha\left(\frac{u_{i+1,j}^{(n+1/2)} - 2u_{i,j}^{(n+1/2)} + u_{i-1,j}^{(n+1/2)}}{\Delta x^2} + \frac{u_{i,j+1}^{(n)} - 2u_{i,j}^{(n)} + u_{i,j-1}^{(n)}}{\Delta y^2}\right), \end{aligned} \tag{9.64}$$

where a central difference is used for the time derivative, and averaging across time levels is used for the spatial derivatives. Consequently, putting the unknowns on the left-hand side and the knowns on the right-hand side yields

$$\begin{aligned} &\frac{1}{2}s_x u_{i+1,j}^{(n+1/2)} - (1+s_x)\,u_{i,j}^{(n+1/2)} + \frac{1}{2}s_x u_{i-1,j}^{(n+1/2)} \\ &= -\frac{1}{2}s_y u_{i,j+1}^{(n)} - \left(1-s_y\right)u_{i,j}^{(n)} - \frac{1}{2}s_y u_{i,j-1}^{(n)}. \end{aligned} \tag{9.65}$$

Taking $i = 1,\dots,I+1$ leads to the tridiagonal problems (9.65) to be solved for $u_{i,j}^{(n+1/2)}$ at each $j = 1,\dots,J+1$, at the intermediate time level.

We then sweep along lines of constant x during the second half-time step using the difference equation in the form

$$\begin{aligned} &\frac{u_{i,j}^{(n+1)} - u_{i,j}^{(n+1/2)}}{\Delta t/2} \\ &= \alpha\left(\frac{u_{i+1,j}^{(n+1/2)} - 2u_{i,j}^{(n+1/2)} + u_{i-1,j}^{(n+1/2)}}{\Delta x^2} + \frac{u_{i,j+1}^{(n+1)} - 2u_{i,j}^{(n+1)} + u_{i,j-1}^{(n+1)}}{\Delta y^2}\right), \end{aligned} \tag{9.66}$$

which becomes

$$\begin{aligned} &\frac{1}{2}s_y u_{i,j+1}^{(n+1)} - \left(1+s_y\right)u_{i,j}^{(n+1)} + \frac{1}{2}s_y u_{i,j-1}^{(n+1)} \\ &= -\frac{1}{2}s_x u_{i+1,j}^{(n+1/2)} - (1-s_x)\,u_{i,j}^{(n+1/2)} - \frac{1}{2}s_x u_{i-1,j}^{(n+1/2)}. \end{aligned} \tag{9.67}$$

Taking $i = 1,\dots,I+1$ leads to the series of tridiagonal problems (9.67) to be solved for $u_{i,j}^{(n+1)}$ at each $j = 1,\dots,J+1$, at the current time level.

Note that this approach requires boundary conditions at the intermediate time level $(n+1/2)$ for (9.65). This is straightforward if the boundary condition does not change with time; however, a bit of care is required if this is not the case. For example, if the boundary condition at $x = 0$ is Dirichlet, but changing with time, as follows:

$$u(0,y,t) = a(y,t),$$

then

$$u_{1,j}^{(n)} = a_j^{(n)}.$$

Subtracting (9.66) from (9.64) gives

$$\frac{u_{i,j}^{(n+1/2)} - u_{i,j}^{(n)}}{\Delta t/2} - \frac{u_{i,j}^{(n+1)} - u_{i,j}^{(n+1/2)}}{\Delta t/2}$$
$$= \alpha \left(\frac{u_{i,j+1}^{(n)} - 2u_{i,j}^{(n)} + u_{i,j-1}^{(n)}}{\Delta y^2} - \frac{u_{i,j+1}^{(n+1)} - 2u_{i,j}^{(n+1)} + u_{i,j-1}^{(n+1)}}{\Delta y^2} \right),$$

and solving for the unknown at the intermediate time level results in

$$u_{i,j}^{(n+1/2)} = \frac{1}{2}\left(u_{i,j}^{(n)} + u_{i,j}^{(n+1)}\right)$$
$$+ \frac{1}{4}s_y \left[u_{i,j+1}^{(n)} - 2u_{i,j}^{(n)} + u_{i,j-1}^{(n)} - \left(u_{i,j+1}^{(n+1)} - 2u_{i,j}^{(n+1)} + u_{i,j-1}^{(n+1)} \right) \right].$$

Applying this equation at the boundary $x = 0$, leads to

$$u_{1,j}^{(n+1/2)} = \frac{1}{2}\left(a_j^{(n)} + a_j^{(n+1)}\right)$$
$$+ \frac{1}{4}s_y \left[a_{j+1}^{(n)} - 2a_j^{(n)} + a_{j-1}^{(n)} - \left(a_{j+1}^{(n+1)} - 2a_j^{(n+1)} + a_{j-1}^{(n+1)} \right) \right].$$

This provides the boundary condition for $u_{1,j}$ at the intermediate $(n+1/2)$ time level. Note that the first term on the right-hand side is the average of a at the (n)th and $(n+1)$st time levels, the second term is $\partial^2 a^{(n)}/\partial y^2$, and the third term is $\partial^2 a^{(n+1)}/\partial y^2$. Thus, if the boundary condition $a(t)$ does not depend on y, then $u_{1,j}^{(n+1/2)}$ is simply the average of $a^{(n)}$ and $a^{(n+1)}$.

REMARKS:

1. *The ADI method with time splitting is $O(\Delta x^2, \Delta y^2, \Delta t^2)$ accurate.*
2. *For stability, it is necessary to apply the von Neumann analysis at each half-time step and take the product of the resulting amplification factors, G_1 and G_2, to obtain G for the full time step. Such an analysis shows that the method is* unconditionally stable *for all s_x and s_y for the two-dimensional, unsteady diffusion equation.*
3. *In three dimensions, we require three fractional steps $(\Delta t/3)$ for each time step, and the method is only* conditionally stable, *where*

$$s_x, s_y, s_z \leq \frac{3}{2},$$

for stability ($s_z = \alpha\Delta t/\Delta z^2$). This is much less restrictive than for explicit methods.

9.10.4 Factored ADI Method

We can improve on the ADI method with time splitting using the *factored ADI method*. It provides a minor reduction in computational cost as well as improving on the stability properties for thee-dimensional cases. In addition, it can be extended naturally to the nonlinear convection case. Let us once again reconsider the two-dimensional, unsteady diffusion equation

$$\frac{\partial u}{\partial t} = \alpha \left(\frac{\partial^2 u}{\partial x^2} + \frac{\partial^2 u}{\partial y^2} \right), \tag{9.68}$$

and apply the Crank–Nicolson approximation

$$\frac{u_{i,j}^{(n+1)} - u_{i,j}^{(n)}}{\Delta t} = \frac{\alpha}{2} \left(\delta_x^2 u_{i,j}^{(n+1)} + \delta_x^2 u_{i,j}^{(n)} + \delta_y^2 u_{i,j}^{(n+1)} + \delta_y^2 u_{i,j}^{(n)} \right),$$

where δ^2 represents second-order central difference operators (as in Sections 8.7 and 8.9) as follows:

$$\delta_x^2 u_{i,j} = \frac{u_{i+1,j} - 2u_{i,j} + u_{i-1,j}}{\Delta x^2},$$

$$\delta_y^2 u_{i,j} = \frac{u_{i,j+1} - 2u_{i,j} + u_{i,j-1}}{\Delta y^2}.$$

Rewriting the difference equation with the unknowns on the left-hand side and the knowns on the right leads to

$$\left[1 - \frac{1}{2} \alpha \Delta t \left(\delta_x^2 + \delta_y^2 \right) \right] u_{i,j}^{(n+1)} = \left[1 + \frac{1}{2} \alpha \Delta t \left(\delta_x^2 + \delta_y^2 \right) \right] u_{i,j}^{(n)}. \tag{9.69}$$

We "factor" the difference operator on the left-hand side as follows:

$$\left[1 - \frac{1}{2} \alpha \Delta t \left(\delta_x^2 + \delta_y^2 \right) \right] \approx \left[1 - \frac{1}{2} \alpha \Delta t \delta_x^2 \right] \left[1 - \frac{1}{2} \alpha \Delta t \delta_y^2 \right], \tag{9.70}$$

where the first factor only involves the difference operator in the x-direction, and the second factor only involves the difference operator in the y-direction. Observe that the factored operator produces an extra term as compared to the unfactored operator, which is

$$\frac{1}{4} \alpha^2 \Delta t^2 \delta_x^2 \delta_y^2 = O(\Delta t^2).$$

Because this additional term is $O(\Delta t^2)$, the factorization (9.70) is consistent with the second-order accuracy in time of the Crank–Nicolson approximation. That is, while we are adding an additional error, it is of the same order as that inherent to the underlying discretization.

The factored form of (9.69) is

$$\left[1 - \frac{1}{2} \alpha \Delta t \delta_x^2 \right] \left[1 - \frac{1}{2} \alpha \Delta t \delta_y^2 \right] u_{i,j}^{(n+1)} = \left[1 + \frac{1}{2} \alpha \Delta t \left(\delta_x^2 + \delta_y^2 \right) \right] u_{i,j}^{(n)},$$

which can be solved in two steps by defining the intermediate variable

$$\hat{u}_{i,j} = \left[1 - \frac{1}{2} \alpha \Delta t \delta_y^2 \right] u_{i,j}^{(n+1)}. \tag{9.71}$$

The two-stage solution process is as follows:

1. Sweep along constant y-lines solving

$$\left[1-\frac{1}{2}\alpha\Delta t\delta_x^2\right]\hat{u}_{i,j}=\left[1+\frac{1}{2}\alpha\Delta t\left(\delta_x^2+\delta_y^2\right)\right]u_{i,j}^{(n)}, \tag{9.72}$$

   which produces a tridiagonal problem along each j for the intermediate variable $\hat{u}_{i,j}, i=1,\ldots,I+1$.
2. Sweep along constant x-lines solving (from (9.71))

$$\left[1-\frac{1}{2}\alpha\Delta t\delta_y^2\right]u_{i,j}^{(n+1)}=\hat{u}_{i,j}, \tag{9.73}$$

   which produces a tridiagonal problem along each i for $u_{i,j}^{(n+1)}, j=1,\ldots,J+1$ at the current time step. Note that the right-hand side of this equation is the intermediate variable solved for in (9.72).

REMARKS:

1. *The order of solution can be reversed, that is, we could define*

$$\hat{u}_{i,j}=\left[1-\frac{1}{2}\alpha\Delta t\delta_x^2\right]u_{i,j}^{(n+1)}$$

   instead of (9.71)*, and solve constant x-lines first. In addition, the method can be extended naturally to three dimensions, in which case there would be two intermediate variables.*
2. *Boundary conditions are required for the intermediate variable $\hat{u}_{i,j}$ in order to solve* (9.72)*. These are obtained from* (9.71) *applied at the boundaries (see Moin, 2010, and Fletcher, 1991a,b).*
3. *The factored ADI method is similar to the ADI method with time splitting, but we have an intermediate variable $\hat{u}_{i,j}$ rather than half time step $u_{i,j}^{(n+1/2)}$. Although they both require the same number of tridiagonal solves, the factored ADI method is somewhat faster as it only requires one evaluation of the spatial derivatives on the right-hand side per time step (for* (9.72)*) rather than two for the ADI method with time splitting (see* (9.65) *and* (9.67)*).*
4. *The factored ADI method is $O(\Delta x^2, \Delta y^2, \Delta t^2)$ accurate and is* unconditionally stable *– even for three-dimensional implementation of the unsteady diffusion equation.*
5. *If we have nonlinear convective terms, as in the unsteady Navier–Stokes or transport equation, for example, use upwind-downwind differencing as in Section 9.9.3. This algorithm is* compact *and (nearly) maintains second-order accuracy in space and time.*

Observe that numerical methods for parabolic problems do not require relaxation or acceleration parameters, as is often the case for elliptic solvers. Recall that the issue is numerical stability, not iterative convergence. As mentioned in Section 9.6.1, it is often advantageous to solve steady elliptic equations reframed as their unsteady parabolic counterparts using the *pseudotransient method*. In particular, you will recall that upwind-downwind differencing for elliptic equations, as discussed in Section 8.10.1, either requires sacrificing compactness for accuracy, by utilizing second-order accurate forward and backward differences, or accuracy for compactness, by utilizing first-order accurate forward and backward differences. In the case of parabolic equations, it is straightforward to develop a second-order accurate scheme that is also compact, as accomplished here.

Finally, the observant reader will have noticed that the ADI method for elliptic partial differential equations and first-order implicit, Crank–Nicolson, ADI with time splitting, and factored ADI methods for parabolic partial differential equations all boil down to a series of tridiagonal solves. These, of course, are carried out using the Thomas algorithm of Section 6.3.3. In fact, these methods are specifically designed to convert the solution of a differential equation into a series of tridiagonal solves precisely because of the unparalleled computational efficiency of the Thomas algorithm. Remember that it only requires $O(N)$ operations, where N is the number of points along constant x and y lines in a two-dimensional grid, for example.

9.11 Hyperbolic Partial Differential Equations

In addition to parabolic partial differential equations, hyperbolic equations are initial-value problems as well. However, they have two, rather than one, characteristic directions of propagation of solution. Let us briefly discuss numerical methods for their solution, which are similar to those for parabolic problems.

Recall that the canonical hyperbolic partial differential equation is the wave equation

$$\frac{\partial^2 u}{\partial t^2} = \sigma^2 \frac{\partial^2 u}{\partial x^2}, \tag{9.74}$$

where $u(x,t)$ is the amplitude of the wave, and σ is the wave speed in the medium. This second-order partial differential equation is equivalent to the two coupled first-order partial differential equations

$$\frac{\partial u}{\partial t} = \pm\sigma \frac{\partial v}{\partial x}, \tag{9.75}$$

$$\frac{\partial v}{\partial t} = \pm\sigma \frac{\partial u}{\partial x}. \tag{9.76}$$

To see this equivalence, differentiate (9.75) with respect to t to give

$$\frac{\partial^2 u}{\partial t^2} = \pm\sigma \frac{\partial^2 v}{\partial t \partial x}. \tag{9.77}$$

Similarly, differentiating (9.76) with respect to x gives

$$\frac{\partial^2 v}{\partial t \partial x} = \pm\sigma \frac{\partial^2 u}{\partial x^2}. \tag{9.78}$$

Table 9.1 Summary of numerical methods used for solving hyperbolic partial differential equations ("NL" = method is good for nonlinear equations).

Method	Type	Order of accuracy	Stability criterion	NL
First-order explicit	Explicit	$O(\Delta t, \Delta x^2)$	Unconditionally unstable	No
Lax	Explicit	$O(\Delta t, \Delta x^2)$	Stable for $\|C\| \leq 1$	Yes
Leap-frog	Explicit	$O(\Delta t^2, \Delta x^2)$	Stable for $\|C\| \leq 1$	No
Lax–Wendroff	Explicit	$O(\Delta t^2, \Delta x^2)$	Stable for $\|C\| \leq 1$	Yes
MacCormack	Explicit	$O(\Delta t^2, \Delta x^2)$	Stable for $\|C\| \leq 1$	Yes
First-order implicit	Implicit	$O(\Delta t, \Delta x^2)$	Unconditionally stable	No
Crank–Nicolson	Implicit	$O(\Delta t^2, \Delta x^2)$	Unconditionally stable	No
Beam–Warming	Implicit	$O(\Delta t^2, \Delta x^2)$	Unconditionally stable	Yes

Substituting (9.78) into the right-hand side of (9.77) leads to the second-order wave equation (9.74). Therefore, one can develop methods for the first-order wave equations (9.75) and (9.76) in order to solve the second-order equation (9.74). In addition, observe that these first-order wave equations resemble the one-dimensional, unsteady Euler equations for inviscid flow in fluid dynamics.

The numerical methods developed throughout this chapter for parabolic equations can be applied to hyperbolic equations as well. In addition, numerical stability analysis is carried out using the von Neumann method as before. Common methods used for hyperbolic equations are summarized in Table 9.1 along with pertinent characteristics. Interestingly, explicit methods are commonly used for hyperbolic problems, whereas they are rarely used for parabolic problems. This is because the numerical stability criteria are typically less stringent in practice for hyperbolic equations, and the convenience of using explicit methods often wins out over the improved stability properties of implicit methods. Recall that the Courant number is given by

$$C = \frac{\sigma \Delta t}{\Delta x},$$

where $|C| \leq 1$ is known as the Courant–Friedrichs–Lewy, or CFL, criterion. Note that the CFL criterion essentially says that the time step size Δt should be small enough so that a particle travels less than one grid size Δx during each time step.

REMARKS:

1. *The first-order explicit method applied to convection problems requires upwind-downwind differencing in order to make it stable, but then it is only first-order accurate in space.*
2. *In the Lax method, $u_i^{(n)}$ is replaced with the average $\frac{1}{2}\left(u_{i+1}^{(n)} + u_{i-1}^{(n)}\right)$ at the previous time level in order to stabilize the first-order explicit method.*
3. *The leap-frog method is analogous to the Richardson method for parabolic equations, as it uses a central difference in time. This means that the method uses three,*

rather than two, time levels during each step. This requires a starting method for the first time step.

4. *The Lax–Wendroff method only involves two time levels and is amenable for application to nonlinear equations.*
5. *The MacCormack method is a predictor-corrector algorithm that is good for nonlinear equations. It is the same as the Lax–Wendroff method when applied to the linear wave equation.*
6. *See Tannehill et al. (1997), Chung (2010), and Press et al. (2007) for more on numerical methods applied to hyperbolic partial differential equations.*

9.12 Coupled Systems of Partial Differential Equations

Numerical methods are a vast and rapidly evolving subject, and there are numerous additional topics that one must be aware of depending on their specific field of study. Here, we provide a brief introduction to solving coupled systems of partial differential equations, which is necessary in fluid dynamics and heat transfer, for example. In the next section, we introduce some issues related to implementing our numerical methods and algorithms on parallel computing architectures.

Until now, we have considered methods for solving single partial differential equations; but in many fields, we must solve systems of coupled equations for multiple dependent variables. In fluid dynamics, for example, we must solve the Navier–Stokes equations, which for two-dimensional, unsteady flow consists of two parabolic equations for the two velocity components and one elliptic equation for the pressure. There are two general methods for treating coupled equations numerically, *sequential* and *simultaneous*, or *coupled*, solutions.

In the sequential method, we iterate on each equation "sequentially" for its dominant variable, treating the other variables as known by using the most recent values. This requires one pass through the mesh for each equation at each iteration, and iteration until convergence at each time step if unsteady. If solving using Gauss–Seidel, for example, a sequential method would require one Gauss–Seidel expression for each of the N dependent variables that is updated in succession by sweeping through the grid once for each equation during each iteration or time step. Note that it may be necessary to use under-relaxation for convergence if nonlinearity is present. For unsteady problems, which are governed by parabolic partial differential equations, an outer loop is added to account for the time marching. The inner loop is performed to iteratively obtain the solution of the coupled equations at the current time step. Generally, there is a trade-off between the time step and the number of iterations required at each time step. Specifically, reducing the time step Δt reduces the number of inner loop iterations required for convergence. In practice, the time step should be small enough such that no more than ten to 20 iterations are required at each time step. The sequential method is most common because it is easiest to implement as it directly uses existing methods and codes for elliptic and parabolic equations. It is also straightforward to incorporate additional physics simply by adding the additional equation(s) into the sequential iterative process.

The simultaneous (or coupled) solution method combines the coupled equations into a single system of algebraic equations. If we have N dependent variables, it produces a block tridiagonal system of equations having $N \times N$ blocks that is solved for all the dependent variables simultaneously. In other words, the dependent variables at each grid point are stacked to form a single – very large – system of algebraic equations. See, for example, Vanka (1986) for an example utilizing such a coupled solution strategy.

9.13 Parallel Computing

The reader may have wondered when reading Chapters 8 and 9 why such a range of methods were described – usually progressing from simple, but inefficient or unstable, to more complicated, but efficient and stable. Was this for pedagogical or historical reasons? Why not simply cover the "best" algorithm for each case and not waste our time with less effective ones? Certainly, some of this was to help put the methods in some sort of context and help explain relative advantages and disadvantages. In addition, however, with the evolution of computer hardware technologies and programming paradigms, it is impossible to anticipate which methods and algorithms will be "best" on future computer architectures. This is particularly true for supercomputing architectures, which are highly and increasingly parallel, and programming methodologies. Therefore, it is essential that we be familiar with the full range of approaches and algorithms and be open to development of still more approaches that are optimized for new computer architectures.

As a simple example, consider the Jacobi and Gauss–Seidel methods for elliptic partial differential equations. While the traditional view – repeated in Chapter 8 – is that the Jacobi method should never be used because Gauss–Seidel is twice as fast and only requires half of the memory, the Jacobi method does have one advantage. With the large supercomputers available today, which have increasing numbers of CPUs (and GPUs), each with increasing numbers of cores, it is not hard to imagine a day when it would be possible to map each grid point in our computational domain to a single core on a supercomputer, thereby allowing for all of the points to be computed simultaneously in one single step. In this manner, the "sweeping through the grid" required for our iterative and time-marching algorithms would be replaced by a single parallel calculation at each grid point simultaneously on individual cores. While this would appear to lead to an astounding speedup in our iterative algorithms, Gauss–Seidel depends upon the sweeping sequence throughout the grid, thereby disallowing such extreme parallelism. The Jacobi method, on the other hand, has no such limitations. Despite its inherent disadvantages on serial computers, therefore, it may actually be preferred on some parallel architectures. Similarly, observe that the widely used Thomas algorithm is not amenable to parallelization. Thus, while we can compute multiple grid lines in parallel, using ADI, for example, we must perform each tridiagonal solve on a single core.

Let us begin by defining some common terms and acronyms used in parallel computing:

- **HPC:** High-performance computing.
- **Cloud (or grid) computing:** Parallel computing across a geographically distributed network via the internet; analogous to the electrical grid.
- **Parallel processing:** Multiple processors (CPUs and/or cores) perform operations simultaneously on different data elements. A parallel computer typically has multiple *nodes*, each of which has multiple *CPUs*, each of which has multiple *cores*, each of which can handle multiple *threads*.
- **Massively parallel:** Parallel computers involving thousands of processors and/or cores.
- **MPP:** Massively parallel processing.
- **SMP:** Symmetric multiprocessing; shared memory parallelism.
- **Shared memory parallelism:** All of the processors share and access the same memory.
- **Distributed memory parallelism:** All of the processors access their own dedicated memory.
- **OpenMP:** Most common function library used for parallelization on shared-memory computers.
- **MPI:** Message passing interface; most common library used for interprocessor communication on distributed memory computers.
- **SIMD:** Single instruction, multiple data; all processors perform the *same* instruction on different data elements.
- **MIMD:** Multiple instruction, multiple data; all processors perform *different* instructions on different data elements.
- **CPU:** Central processing unit; the "brains" of the computer that control all aspects of the hardware and perform the actual computations.
- **GPU:** Graphical processing unit; originally designed to operate graphics and displays; now used as coprocessors to CPUs. Highly optimized for vector and matrix operations.
- **Serial processing:** A computer code is executed line by line in sequence on one processor (or core).
- **Scalar processing:** A single processor or core performs a single calculation – instruction – on a single data element at a time.
- **Vector processing:** A single processor or core performs calculations on multiple data elements, that is, a vector, simultaneously (see SIMD).
- **Multicore CPUs:** A single chip with multiple cores.
- **Multithreading:** Each core may allow for multiple "threads" to be computed simultaneously.
- **Node:** A rack version of a PC (includes CPU(s), memory, I/O, etc.). The basic building block of a parallel cluster.

- **Cluster:** A parallel computer comprised of commodity, that is, off-the-shelf, hardware and open-source software, such as Linux.
- **Supercomputer:** The fastest computers available at a particular time. See the website www.top500.org for a list of the fastest 500 supercomputers in the world updated every six months.
- **Floating point number:** A real number in scientific notation containing a fixed number of significant figures, where the decimal point "floats." This is the way that most computers store real numbers.
- **Floating point operation:** An operation, such as addition, subtraction, multiplication, or division, on two floating point numbers.
- **FLOPS:** Floating point operations per second; measure of computer performance for numerical applications. Summary of FLOP prefixes:

MFLOPS	MegaFLOPS	Million (10^6) FLOPS
GFLOPS	GigaFLOPS	Billion (10^9) FLOPS
TFLOPS	TeraFLOPS	Trillion (10^{12}) FLOPS
PFLOPS	PetaFLOPS	10^{15} FLOPS
EFLOPS	ExaFLOPS	10^{18} FLOPS

- **Embarrassingly parallel:** Processors work in parallel with very little or no communication between them; for example, image processing and Seti@Home.
- **Coupled parallel**: Processors work in parallel but require significant communication between them; this is most common in engineering and science applications.

Roughly speaking, high-performance computing has evolved from supercomputers having moderate numbers of CPUs using shared-memory architectures to large clusters having many CPUs using distributed-memory architectures to massive clusters having many CPUs, each consisting of a number of cores and possibly GPUs, that are a hybrid distributed/shared-memory architecture, which is the most common currently.[3] While the scientist or engineer generally does not need to be concerned with hardware architectures when programming on serial computers, this is not the case for parallel computers. On distributed-memory architectures, for example, it is necessary to be

[3] One of several exceptions to this overall progression was the Connection Machine in the late 1980s and early 1990s. Although not a commercial success, it was a technological marvel for its time. While traditional supercomputers manufactured by Cray at the time consisted of tens of highly specialized CPUs, the Connection Machine by Thinking Machines Corporation was comprised of tens of thousands of off-the-shelf CPUs, the same ones found in contemporary personal computers. While surpassing traditional supercomputers in terms of theoretical peak speed – and costing substantially less – they required scientific computing aficionados to learn whole new algorithmic and programming paradigms using the SIMD approach. This very steep learning curve slowed adoption of this revolutionary new approach to high-performance computing in the scientific community. The concept of using off-the-shelf computing hardware in a massively parallel, distributed architecture is now the predominant supercomputing architecture – the primary difference being the adoption of open-source, rather than proprietary, software further decreasing the cost and increasing accessibility to a broader group of scientists and engineers.

concerned with where data are stored, which processors need that data, and how that data will be transferred to where they are needed. Likewise, how are the computations going to be distributed across the processing units? How is the computational grid mapped to the processors? Are there data dependency issues that must be addressed in which another processing unit needs the result of a calculation being done on another one; that is, can the calculations be carried out independently of one another on separate processing units? In order to make maximum use of the computational resources, how can an algorithm be designed that balances the load across all available processing units? Is the algorithm scalable on a particular architecture, that is, will doubling the number of processing units halve the total computational time (see Amdahl's law for a means to estimate speedups and scalability)? Such considerations are a moving target as computer architectures and programming methods and tools continue to evolve at a rapid pace. The primary difficulty for the programmer is that each computer architecture requires its own approach to programming, and not all algorithms are amenable to each.

In order to illustrate how difficult parallel programming is and how essential it is to pay attention to all of the details, let us consider *Amdahl's law*. It quantifies the speedup $S(N)$ of a code running on N processors as compared to the same code running on a single processor in a serial fashion. Amdahl's law is given by

$$S(N) = \frac{T(1)}{T(N)} = \frac{N}{N - (N-1)F},$$

where $T(N)$ is the computational time required using N processors, and F is the fraction of the serial run time that the code spends in the parallel portion. If $F = 1$, such that 100% of the code is executed in parallel, then $S(N) = N$ and an ideal linear speedup would be attained. Of course, this ideal can never be fully achieved. In reality, $F < 1$ and $S(N) < N$, leading to a sublinear speedup. Effectively then, the value of F limits the overall speedup that can be attained no matter how many processing units are available as follows. The maximum speedup possible for a given F with $N \to \infty$ is given by

$$\lim_{N\to\infty} S(N) = \lim_{N\to\infty} \frac{N}{N - (N-1)F} = \frac{1}{1-F} = \frac{1}{F_s},$$

where $F_s = 1 - F$ is the fraction of time in the serial code spent doing the serial portion. For example, if $F = 0.95$, then $F_s = 0.05$, and the maximum speedup is only $S(\infty) = 20$. That is, even if we have access to thousands or even millions of processors, our code will never run more than 20 times faster than the same code on a serial computer if "only" 95% of the code is run in parallel.

As discouraging as Amdahl's law is, it addresses the ideal case in which there is no additional overhead required to run the code in parallel. Of course, this is never the case as it ignores the additional overhead due to parallelization from interprocessor communication, synchronization of data between processors, and load imbalances across processing units. This additional overhead is addressed in the *modified Amdahl's law*, which is given by

$$S(N) = \frac{N}{N - (N-1)F + N^2 F_o},$$

where F_o is the fraction of time spent owing to parallel overhead, which further limits scalability.

9.14 Epilogue

The material in Part II is only the beginning. We have addressed general methods for numerically solving ordinary and partial differential equations. The real fun begins when these methods are adapted to field-specific settings such as CFD, computational solid mechanics, computational biology, et cetera. This is when the interplay between the physics and numerics comes to the fore, and numerical methods become a central tool to better understanding the physics of a system.

It is even becoming more plausible to combine various physics-based models to perform *multiphysics* simulations of very complex systems. For example, aircraft experience very large aerodynamic loads on their wings and other lifting surfaces. Of course, the aerospace engineer needs to simulate the detailed fluid mechanics around such surfaces in order to predict these loads. Her objective is to design the shapes of the aerodynamic surfaces and fuselage to maximize lift and minimize drag. However, she then also needs to account for these aerodynamic loads in order to address the internal aerostructures aspects of the plane's design. Traditionally, these simulations would have been performed separately, with the CFD simulations performed first in order to determine the aerodynamic loads, and then the aerostructures simulations second to combine these with other loads to design the structural aspects of the airplane. Increasingly, these simulations can be coupled to address fluid-structure interaction issues, such as flutter, that require two-way feedback between the simulations. When an optimization component is added to such a simulation – maximize lift, minimize drag, minimize structural loads, minimize weight, minimize flutter, maximize cargo capacity, et cetera – it is referred to as *multidisciplinary design optimization (MDO)*. The possibilities are truly endless!

Exercises

Unless stated otherwise, perform the exercises using hand calculations. A "user-defined function" is a standalone computer program to accomplish a task written in a programming language, such as Python, MATLAB, Mathematica, Fortran, or C++.

9.1 A simple pendulum is governed by the nonlinear differential equation

$$\ddot{\theta} + \frac{g}{\ell}\sin\theta = 0,$$

where g is acceleration owing to gravity, ℓ is the length of the pendulum, and $\theta(t)$ is the angle that the pendulum makes with the vertical. The initial conditions are

$$\theta(0) = 1, \quad \dot{\theta}(0) = 0.$$

(a) Convert the governing equation and initial conditions into a system of first-order differential equations.
(b) Obtain the system of algebraic equations to be solved during each time step using the first-order implicit method. Comment on how you would solve this system of equations (do not solve).

9.2 Consider the motion of a pendulum of length ℓ and mass m subject to damping. The governing equation for the angular displacement of the pendulum $\theta(t)$ is given by the nonlinear equation

$$\frac{d^2\theta}{dt^2} + \frac{c}{m\ell}\frac{d\theta}{dt} + \frac{g}{\ell}\sin\theta = 0,$$

where c is the damping coefficient, and g is the acceleration owing to gravity. The initial conditions are

$$\theta(0) = \pi/2, \quad \dot{\theta}(0) = 0.$$

(a) Convert the governing equation and initial conditions into a system of first-order differential equations.
(b) Write a user-defined function to solve the system from part (a) using the first-order explicit method. The parameters are as follows: $g = 9.81$, $\ell = 5$, $m = 1$, $c = 0.9$. Plot the solution of $\theta(t)$ for $0 \leq t \leq 25$. Determine the time step that you think provides an accurate solution; explain why you chose this time step.

9.3 The one-dimensional, unsteady diffusion equation given by (9.23) assumes that the material has homogeneous properties, such that the diffusivity α is a constant. If the diffusivity is instead a known function of x, such that $\alpha = \alpha(x)$, then the one-dimensional, unsteady diffusion equation is of the form

$$\frac{\partial u}{\partial t} = \frac{\partial}{\partial x}\left[\alpha(x)\frac{\partial u}{\partial x}\right].$$

Apply the first-order explicit method to obtain the expression for updating $u_i^{(n+1)} = u\left(x_i, t^{(n+1)}\right)$ at each successive time step.

9.4 Recall the three-spring, two-mass system considered in Section 2.6.3. Write a user-defined function to solve the system of four first-order ordinary differential equations using the first-order implicit method. Repeat the results shown in Figure 2.8.

9.5 Consider the one-dimensional, unsteady scalar-transport equation of the form

$$\frac{\partial u}{\partial t} = \alpha\frac{\partial^2 u}{\partial x^2} - c\frac{\partial u}{\partial x},$$

where α is the diffusivity, and c is the wave speed; both are constants. Using second-order accurate, central-difference approximations for all spatial derivatives, write down the difference equation in the form in which it would be solved using the following methods:

(a) First-order explicit method.
(b) Crank–Nicolson method.
(c) Which of these two methods would give a more accurate solution if the same time step is used? Why?
(d) Which method is likely to allow you to use larger time steps while maintaining numerical stability? Why?

9.6 Consider the unsteady, one-dimensional Burgers' equation

$$\frac{\partial u}{\partial t} = \nu \frac{\partial^2 u}{\partial x^2} - u \frac{\partial u}{\partial x},$$

where ν is the viscosity. Using central differences for all spatial derivatives, write down the difference equation in the form in which it would be solved for the first-order implicit method. Discuss how you would go about solving this equation.

9.7 Consider the general linear, one-dimensional, unsteady equation of the form

$$\frac{\partial u}{\partial t} = a(x,t)\frac{\partial^2 u}{\partial x^2} + b(x,t)\frac{\partial u}{\partial x} + c(x,t)u(x,t) + d(x,t),$$

where $a(x,t) > 0$. Using second-order accurate, central differences for all spatial derivatives, write down the difference equations in the form in which they would be solved based on the following methods:
(a) First-order explicit method.
(b) First-order implicit method.

9.8 Consider the general linear, one-dimensional, unsteady equation of the form

$$\frac{\partial u}{\partial t} = a(x)\frac{\partial^2 u}{\partial x^2} + b(x)\frac{\partial u}{\partial x},$$

where $a(x) > 0$. Using central differences for all spatial derivatives, write down the difference equations in the form in which they would be solved based on the following methods:
(a) DuFort–Frankel method.
(b) Crank–Nicolson method.

9.9 Consider the two-dimensional, unsteady diffusion equation

$$\frac{\partial u}{\partial t} = \alpha \left(\frac{\partial^2 u}{\partial x^2} + \frac{\partial^2 u}{\partial y^2} \right).$$

Describe how you would solve this equation using the factored-ADI method along with the first-order implicit method. Only consider the solution approach for interior points (do not be concerned with boundary conditions).

9.10 The one-dimensional, unsteady diffusion equation that follows is to be solved using the Crank–Nicolson method:

$$\frac{\partial u}{\partial t} = \alpha \frac{\partial^2 u}{\partial x^2}.$$

Perform a von Neumann stability analysis to determine the criterion necessary for stability of the numerical scheme.

9.11 The following one-dimensional, unsteady scalar-transport equation is to be solved using the first-order explicit method:

$$\frac{\partial \phi}{\partial t} = \alpha \frac{\partial^2 \phi}{\partial x^2} - u \frac{\partial \phi}{\partial x}.$$

Use second-order accurate, central differences for all spatial derivatives. Assuming that $u(x,t)$ is known, perform a von Neumann stability analysis to determine the criterion necessary for stability of the numerical scheme. Does the first-order convective term in this equation improve or diminish the numerical stability properties (compare with the first-order explicit method applied to the one-dimensional, unsteady diffusion equation)?

9.12 Consider the one-dimensional, unsteady scalar-transport equation

$$\frac{\partial \phi}{\partial t} = \alpha \frac{\partial^2 \phi}{\partial x^2} - u \frac{\partial \phi}{\partial x},$$

where the velocity u is a known constant throughout the domain, and the boundary conditions are $\phi(0,t) = 1, \phi(1,t) = 2$.

(a) Using second-order accurate, central differences for all spatial derivatives, write down the difference equation in the form in which it would be solved using the first-order explicit method.

(b) Determine the general requirement for stability of the numerical scheme from part (a) using the matrix method.

(c) From your results for part (b), determine the largest time step expected to produce a stable solution if $\alpha = 1$, $u = 1$, and the domain is divided into 10 equal intervals.

9.13 The two-dimensional, unsteady diffusion equation that follows is to be solved using the first-order explicit method:

$$\frac{\partial u}{\partial t} = \alpha \left(\frac{\partial^2 u}{\partial x^2} + \frac{\partial^2 u}{\partial y^2} \right).$$

Perform a von Neumann stability analysis to determine the criterion necessary for stability of the numerical scheme. Note that in two dimensions, the error is expanded as follows:

$$e(x,y,t) = \sum (G_m)^n \, e^{\mathrm{i}\theta_1 x + \mathrm{i}\theta_2 y}.$$

Does the two-dimensionality improve or diminish the numerical stability properties as compared with the first-order explicit method applied to the one-dimensional, unsteady diffusion equation?

9.14 The following one-dimensional wave equation is to be solved using an explicit method:

$$\frac{\partial u}{\partial t} + c \frac{\partial u}{\partial x} = 0,$$

where c is the constant wave speed.

(a) Perform a von Neumann stability analysis for the first-order explicit method to determine the criterion necessary for stability of the numerical scheme.
(b) To improve stability, Lax proposed to use an average for $u_i^{(n)}$ in the time derivative of the form

$$u_i^{(n)} = \frac{1}{2}\left(u_{i+1}^{(n)} + u_{i-1}^{(n)}\right).$$

Repeat the von Neumann stability analysis for the Lax method.

Part III

Least Squares and Optimization

10 Least-Squares Methods

> Neglect of mathematics works injury to all knowledge, since he who is ignorant of it cannot know the other sciences or the things of this world. And what is worse, men who are thus ignorant are unable to perceive their own ignorance and so do not seek a remedy. (Roger Bacon)

Having a good command of matrix methods from Part I, particularly the material on matrix calculus in Chapter 4, we are now ready to tackle some of the most important applications of these methods in optimization. This material builds on the fundamentals of least-squares methods, which is where we start our journey.

Typically, we first encounter least-squares methods in the context of curve fitting, which was introduced briefly in Section 1.1.1. The problem is posed as follows: a data set is comprised of a large number of data points relating two quantities, and a single best-fit line is sought that is the most faithful representation of the data. Posing such a curve-fitting problem in this way results in an overdetermined system of linear algebraic equations having more equations – one for each data point – than unknowns – two for the constants that define a straight line. Although such a system does not have a unique solution on its own, an optimization objective can be devised in which the distances between each point and the best-fit line is minimized such that it does lead to a unique solution. Such curve fitting techniques will be taken up in Chapter 11.

In addition to curve fitting, there are numerous additional optimization and control applications that result in the need to "solve" overdetermined, as well as underdetermined, systems of equations. Least-squares methods can be used directly to treat overdetermined and underdetermined systems of linear algebraic equations as discussed in Section 10.2. In addition, they are the basis for two additional numerical methods for approximating solutions of linear algebraic equations that are particularly well suited to solving systems of equations with large, sparse coefficient matrices. These are explored in Section 10.6.

10.1 Introduction to Optimization

All of the preceding scenarios are broadly classified as optimization problems. A "solution" of an overdetermined algebraic system is obtained by minimizing the residual of the solution in order to determine the solution vector that is closest to satisfying the equations. A "solution" of an underdetermined algebraic system is obtained by minimizing the distance of the solution vector from the origin (or some other point). The best fit line for a series of data is determined by minimizing the distance from the line to each of the data points. Terms like "best," "minimize," or "maximize" signal the need to formulate an optimization problem. Science and engineering are replete with such problems; they may be motivated by physical optimization principles, such as minimization of energy, or the desire to optimize some system or process in order to accomplish a desired task in an optimal manner. The fact that we will dedicate the entirety of Part III to optimization and related methods is a reflection of their importance and breadth of applications in science and engineering as well as a reflection of the wide variety of such problems and their many nuances.

All optimization and control problems consist of an *objective*, or *cost*, *function*[1] to be minimized (or maximized) that expresses the objective or goal of the optimization or control. Such algebraic or integral objective functions quantify the performance or cost of the process or system. Not only does an objective function specify the "objective" of the optimization or control, it is an "objective," as opposed to "subjective," measure of the system's performance. Most optimization and control problems also include *constraints* on the process or system behavior that must be adhered to in the optimization or control. Constraints may include a mathematical model for the process or system to be optimized, which is typically in the form of algebraic or differential equations that govern the system's behavior, as well as any limits on the system and its inputs or outputs. An optimal solution is one that minimizes (or maximizes) the objective function while adhering to all of the constraints. It tells us how the process or system should behave or evolve in order to fulfill the optimization objective.

Optimization problems with an algebraic objective function and algebraic constraint(s) are treated using differential calculus, while those with a variational functional (integral) objective function along with integral, algebraic, and/or differential constraints are handled using variational calculus. Consequently, optimization and control bring together and draw on both matrix and variational methods. The reader should read Part III of both Cassel (2013) and the present text in order to get a comprehensive treatment of methods used in optimization and control. See Table 10.1 for a classification of the various types of optimization and control problems in terms of the form of their objective function and constraints with references to where they are treated in this text or Cassel (2013).

Finally, we have been conditioned to assume that nonlinear problems are always harder to solve than linear ones. In optimization, however, the primary differentiator with respect to simplicity of the algorithm is whether it is a convex or nonconvex optimization problem. A convex optimization problem has an objective function and constraints that are bowl shaped and only have one global minimum (or maximum). A nonconvex optimization problem has an objective function and/or constraints that are not convex, in which case there may be several local minima. Clearly, to be convex, the problem must be nonlinear. Therefore, some – but certainly not all – nonlinear optimization problems are actually easier to solve than linear programming problems. For example, the quadratic programming problems introduced in Section

[1] Optimists call it an *objective* function; pessimists refer to it as a *cost* function.

Table 10.1 Classification of optimization problems with algebraic or integral objective functions and algebraic or differential constraints with section references in this text and Cassel (2013) (designated as "VMASE").

Type of optimization	Objective function	Constraint(s)	Section(s)
Linear programming (LP)	Algebraic linear	Algebraic linear equality/inequality	4.3.4, 12.7
Quadratic programming (QP)	Algebraic quadratic	Algebraic quadratic equality/inequality	4.3.3
Least squares (LS)	Algebraic quadratic	None	10.2, 11.1
Least squares (LS)	Algebraic quadratic	Algebraic linear equality	10.3, 10.4
Nonlinear programming (NP)	Algebraic nonlinear	None	10.5, 12.4
Nonlinear programming (NP)	Algebraic nonlinear	Algebraic nonlinear equality/inequality	12.5
Optimal control of discrete systems	Integral quadratic	Ordinary differential equation(s)	VMASE 10.4
Optimal control of continuous systems	Integral quadratic	Partial differential equation(s)	VMASE 10.5
State estimation – LS regression	Algebraic quadratic	None	11.9
State estimation for discrete systems	Integral quadratic	Linear ordinary differential equation(s)	VMASE 10.6.3
Image processing	Integral – Mumford & Shah	None	VMASE 11.1
Splines	Integral quadratic	Optional	11.6, VMASE 11.2
POD of discrete data	Algebraic quadratic	Algebraic quadratic – Normalization	13.7
POD of continuous data	Integral quadratic	Integral quadratic – Normalization	13.6
Numerical grid generation	Integral quadratic	None	VMASE 12.4

4.3.3 are convex and lead to solution of an eigenproblem for the optimal solution. In addition, least-squares methods are framed as quadratic optimization problems and take advantage of these properties of convex optimization.

10.2 Least-Squares Solutions of Algebraic Systems of Equations

Recall that our primary focus in Chapter 1 was on solving systems of equations $\mathbf{Au} = \mathbf{b}$ for which the rank of the coefficient matrix was equal to that of the augmented matrix and the number of unknowns, in which case there is a unique solution. This is the most common case in physical applications. By extension, the systems of linear algebraic equations that result from the application of numerical methods in Part II to physical problems also result in systems having a unique solution. As alluded to previously, however, there are many nonphysical settings for which $\mathbf{Au} = \mathbf{b}$ does not produce a unique solution, but where some form of a solution is still sought. This occurs either because we have more equations than unknowns, in which case the system is *overdetermined*, or fewer equations than unknowns, in which case it is *underdetermined*.[2] Using the principles developed in Chapter 4, which the reader may want to review before proceeding, we can formulate techniques for treating such overdetermined and underdetermined systems of linear algebraic equations.

10.2.1 Overdetermined Systems

An inconsistent system having more equations than unknowns is called overdetermined as there are more equations than necessary to determine a unique solution. Although there are no solutions that satisfy all of the equations, we may seek the "solution" that comes closest to doing so.

Consider a system of equations[3] $\mathbf{A}\hat{\mathbf{u}} = \mathbf{b}$ with $M > N$.[4] Although this system does not have a unique solution for $\hat{\mathbf{u}}$ in the usual sense, let us define the *residual* of a vector $\mathbf{u}$ according to[5]

$$\mathbf{r} = \mathbf{b} - \mathbf{Au}, \tag{10.1}$$

which, for a given coefficient matrix $\mathbf{A}$ and right-hand-side vector $\mathbf{b}$, is essentially a measure of how close the vector $\mathbf{u}$ is to satisfying the system of equations.[6] Hence, one could define the "solution" of an overdetermined system to be the vector $\mathbf{u}$ that comes closest to satisfying the system of equations, that is, the one with the smallest residual magnitude. As usual, we measure the size of vectors using the norm. Thus, we seek to minimize the square of the norm of the residual, which is the scalar quantity

$$J(\mathbf{u}) = \|\mathbf{r}\|^2 = \|\mathbf{b} - \mathbf{Au}\|^2. \tag{10.2}$$

This is the *objective function*. Because the norm involves the sum of squares, and we are seeking a minimum, this is known as the *least-squares method* of solution.

In order to minimize the objective function (10.2), we will need to differentiate with respect to the unknown solution $\mathbf{u}$. In preparation, let us expand the square of the norm of the residual as follows:

[2] Note that it is possible for an overdetermined ($M > N$) or underdetermined ($M < N$) system to have a unique solution if $\text{rank}(\mathbf{A}) = \text{rank}(\mathbf{A}_{aug}) = N$.

[3] As in Part II, we will use $\mathbf{u}$ generically for the dependent variable when the meaning is clear from the context, and $\hat{\mathbf{u}}$ for the exact solution and $\mathbf{u}$ for the approximate or least squares solution when it is necessary to distinguish between them throughout Part III.

[4] What follows also applies when $\text{rank}(\mathbf{A}) = \text{rank}(\mathbf{A}_{aug}) = N$ and a unique solution exists, as will become clear.

[5] Alternatively, the residual can be defined as $\mathbf{r} = \mathbf{Au} - \mathbf{b}$ without consequence.

[6] If the system did have an exact solution, then the residual of the exact solution would be zero.

$$\begin{aligned}
J(\mathbf{u}) &= \|\mathbf{r}\|^2 \\
&= \|\mathbf{b} - \mathbf{A}\mathbf{u}\|^2 \\
&= (\mathbf{b} - \mathbf{A}\mathbf{u})^T (\mathbf{b} - \mathbf{A}\mathbf{u}) \\
&= \left(\mathbf{b}^T - \mathbf{u}^T\mathbf{A}^T\right)(\mathbf{b} - \mathbf{A}\mathbf{u}) \\
&= \mathbf{b}^T\mathbf{b} - \mathbf{b}^T\mathbf{A}\mathbf{u} - \mathbf{u}^T\mathbf{A}^T\mathbf{b} + \mathbf{u}^T\mathbf{A}^T\mathbf{A}\mathbf{u} \\
J(\mathbf{u}) &= \mathbf{b}^T\mathbf{b} - \mathbf{b}^T\mathbf{A}\mathbf{u} - \left(\mathbf{b}^T\mathbf{A}\mathbf{u}\right)^T + \mathbf{u}^T\mathbf{A}^T\mathbf{A}\mathbf{u}.
\end{aligned}$$

Therefore, the objective function is quadratic – that is, second order – in $\mathbf{u}$ (see Section 2.3.3). Note that each of these four terms are scalars; thus, the transpose can be removed from the third term to yield

$$J(\mathbf{u}) = \mathbf{b}^T\mathbf{b} - 2\mathbf{b}^T\mathbf{A}\mathbf{u} + \mathbf{u}^T\mathbf{A}^T\mathbf{A}\mathbf{u}. \tag{10.3}$$

Because we seek a minimum, let us differentiate with respect to $\mathbf{u}$ and set equal to zero as follows. The first term is a constant and vanishes, while the final term is a quadratic with $\mathbf{A}^T\mathbf{A}$ being symmetric (see Section 2.4). Recall from Section 4.3.3 – as summarized in Section 4.4 – that the derivative of the quadratic is

$$\frac{d}{d\mathbf{u}}\left(\mathbf{u}^T\mathbf{A}^T\mathbf{A}\mathbf{u}\right) = 2\mathbf{A}^T\mathbf{A}\mathbf{u},$$

and we will show that

$$\frac{d}{d\mathbf{u}}\left(\mathbf{b}^T\mathbf{A}\mathbf{u}\right) = \mathbf{A}^T\mathbf{b}. \tag{10.4}$$

Therefore, differentiating equation (10.3) with respect to $\mathbf{u}$ yields

$$\frac{dJ}{d\mathbf{u}} = -2\mathbf{A}^T\mathbf{b} + 2\mathbf{A}^T\mathbf{A}\mathbf{u}.$$

Setting equal to zero gives

$$\mathbf{A}^T\mathbf{A}\mathbf{u} = \mathbf{A}^T\mathbf{b}. \tag{10.5}$$

Recall that if $\mathbf{A}$ is $M \times N$, then $\mathbf{A}^T\mathbf{A}$ is $N \times N$. Thus, the least-squares solution of the overdetermined system of equations $\mathbf{A}\hat{\mathbf{u}} = \mathbf{b}$ is equivalent to solving the square system of equations given by (10.5). If $\mathbf{A}^T\mathbf{A}$ is invertible, which is the case if the column vectors of $\mathbf{A}$ are linearly independent, then the solution $\mathbf{u}$ that minimizes the square of the residual is

$$\mathbf{u} = \left(\mathbf{A}^T\mathbf{A}\right)^{-1}\mathbf{A}^T\mathbf{b} = \mathbf{A}^+\mathbf{b}, \tag{10.6}$$

where $\mathbf{A}^+ = \left(\mathbf{A}^T\mathbf{A}\right)^{-1}\mathbf{A}^T$ is called the *pseudo-inverse* of the matrix $\mathbf{A}$, which is a generalization of the inverse of square matrices to rectangular matrices as introduced in Section 2.8.

If $\mathbf{A}$ is invertible, in which case it must be square, then $\mathbf{A}^+ = \left(\mathbf{A}^T\mathbf{A}\right)^{-1}\mathbf{A}^T = \mathbf{A}^{-1}\left(\mathbf{A}^T\right)^{-1}\mathbf{A}^T = \mathbf{A}^{-1}$, and the pseudo-inverse and regular inverse of $\mathbf{A}$ are the same. Therefore, the least-squares method for overdetermined systems is equivalent to solving the system $\mathbf{A}\hat{\mathbf{u}} = \mathbf{b}$ when rank($\mathbf{A}$) $= N$ and such a solution exists, in which case the residual is zero. If no such solution exists because there are more equations than unknowns, then the least-squares method produces the solution that minimizes the square of the norm of the residual. The pseudo-inverse, and thus the least-squares solution, is typically computed for large systems of equations using QR decomposition (see Section 6.5) or sometimes singular-value decomposition if ill conditioning is an issue (see Section 2.8).

REMARKS:

1. *We emphasize that the solution for* $\mathbf{u}$ *obtained from* (10.6) *is* not *the solution of the original overdetermined system of equations* $\mathbf{A}\hat{\mathbf{u}} = \mathbf{b}$*, which does not typically have a solution; instead, it is the solution that minimizes the residual of this inconsistent system of equations.*
2. *Observe from* (10.5) *that although* $\mathbf{A}\hat{\mathbf{u}} = \mathbf{b}$ *does not have a solution, premultiplying both sides by* $\mathbf{A}^T$ *gives the system of equations* $\mathbf{A}^T\mathbf{A}\mathbf{u} = \mathbf{A}^T\mathbf{b}$*, which does have a unique solution! Equation (10.5) is sometimes called a* normal system of equations *because the coefficient matrix* $\mathbf{A}^T\mathbf{A}$ *is normal (see Section 2.4).*
3. *In a sense, the least-squares solution* (10.6) *is more general than the methods considered in Chapter 1 as it includes the unique solution of* $\mathbf{A}\hat{\mathbf{u}} = \mathbf{b}$ *when* $\text{rank}(\mathbf{A}) = N$ *as well as solutions for overdetermined cases as well.*
4. *Geometrically, we can interpret the least-squares solution as follows:*
   - *The least-squares solution* $\mathbf{u}$ *is the linear combination of the columns of* $\mathbf{A}$ *that is closest to* $\mathbf{b}$*, such that taking* $\mathbf{A}\mathbf{u}$ *gives the vector that is closest to* $\mathbf{b}$.
   - *The vector* $\mathbf{A}\mathbf{u}$ *is the projection of* $\mathbf{b}$ *on the range of* $\mathbf{A}$.
5. *The following special cases are treated in the exercises: least squares with* $N = 1$ (10.6)*, least squares with an orthogonal coefficient matrix* (10.7)*, least squares for a system with weights on the equations* (10.8)*, least squares for a generalized quadratic objective function* (10.9)*, approximate the eigenvalues for given approximations of the eigenvectors from the Rayleigh quotient* (10.10)*, and determining the system matrix* $\mathbf{A}$ *for a given input* $\mathbf{u}$ *and output* $\mathbf{b}$ (10.12).
6. *For much more on the least-squares method and its applications, see Boyd and Vandenberghe (2018).*

Before doing an example, let us show that (10.4) holds. Expanding $\mathbf{b}^T\mathbf{A}\mathbf{u}$ gives the scalar expression

$$\begin{aligned}
\mathbf{b}^T\mathbf{A}\mathbf{u} &= \begin{bmatrix} b_1 & b_2 & \cdots & b_M \end{bmatrix} \begin{bmatrix} A_{11} & A_{12} & \cdots & A_{1N} \\ A_{21} & A_{22} & \cdots & A_{2N} \\ \vdots & \vdots & \ddots & \vdots \\ A_{M1} & A_{M2} & \cdots & A_{MN} \end{bmatrix} \begin{bmatrix} u_1 \\ u_2 \\ \vdots \\ u_N \end{bmatrix} \\
&= (b_1 A_{11} + b_2 A_{21} + \cdots + b_M A_{M1}) u_1 \\
&\quad + (b_1 A_{12} + b_2 A_{22} + \cdots + b_M A_{M2}) u_2 \\
&\quad \vdots \\
&\quad + (b_1 A_{1N} + b_2 A_{2N} + \cdots + b_M A_{MN}) u_N.
\end{aligned}$$

Differentiating with respect to each of the variables $u_i, i = 1, 2, \dots, N$, we have

$$\frac{\partial}{\partial u_1}\left(\mathbf{b}^T\mathbf{A}\mathbf{u}\right) = b_1A_{11} + b_2A_{21} + \cdots + b_MA_{M1},$$

$$\frac{\partial}{\partial u_2}\left(\mathbf{b}^T\mathbf{A}\mathbf{u}\right) = b_1A_{12} + b_2A_{22} + \cdots + b_MA_{M2},$$

$$\vdots$$

$$\frac{\partial}{\partial u_N}\left(\mathbf{b}^T\mathbf{A}\mathbf{u}\right) = b_1A_{1N} + b_2A_{2N} + \cdots + b_MA_{MN}.$$

A careful examination of this result will reveal that in matrix form

$$\frac{d}{d\mathbf{u}}\left(\mathbf{b}^T\mathbf{A}\mathbf{u}\right) = \mathbf{A}^T\mathbf{b}.$$

Example 10.1 Determine the least-squares solution of the overdetermined system of linear algebraic equations $\mathbf{A}\hat{\mathbf{u}} = \mathbf{b}$, where

$$\mathbf{A} = \begin{bmatrix} 2 & 0 \\ -1 & 1 \\ 0 & 2 \end{bmatrix}, \quad \mathbf{b} = \begin{bmatrix} 1 \\ 0 \\ -1 \end{bmatrix}.$$

Solution

To evaluate (10.6), observe that

$$\mathbf{A}^T\mathbf{A} = \begin{bmatrix} 2 & -1 & 0 \\ 0 & 1 & 2 \end{bmatrix}\begin{bmatrix} 2 & 0 \\ -1 & 1 \\ 0 & 2 \end{bmatrix} = \begin{bmatrix} 5 & -1 \\ -1 & 5 \end{bmatrix},$$

which is symmetric as expected. Then the inverse is

$$\left(\mathbf{A}^T\mathbf{A}\right)^{-1} = \frac{1}{24}\begin{bmatrix} 5 & 1 \\ 1 & 5 \end{bmatrix}.$$

Also

$$\mathbf{A}^T\mathbf{b} = \begin{bmatrix} 2 & -1 & 0 \\ 0 & 1 & 2 \end{bmatrix}\begin{bmatrix} 1 \\ 0 \\ -1 \end{bmatrix} = \begin{bmatrix} 2 \\ -2 \end{bmatrix}.$$

Thus, the least-squares solution is

$$\mathbf{u} = \mathbf{A}^+\mathbf{b} = \left(\mathbf{A}^T\mathbf{A}\right)^{-1}\mathbf{A}^T\mathbf{b} = \frac{1}{24}\begin{bmatrix} 5 & 1 \\ 1 & 5 \end{bmatrix}\begin{bmatrix} 2 \\ -2 \end{bmatrix} = \frac{1}{24}\begin{bmatrix} 8 \\ -8 \end{bmatrix} = \frac{1}{3}\begin{bmatrix} 1 \\ -1 \end{bmatrix}.$$

This is the vector $\mathbf{u}$ that comes closest to satisfying the original overdetermined system of equations $\mathbf{A}\hat{\mathbf{u}} = \mathbf{b}$ in a least-squares sense.

We will encounter the versatile least-squares method in several contexts throughout the remainder of the text, such as in curve fitting in Chapter 11. Before we proceed, however, let us see how the least-squares method for an $M \times N$ matrix $\mathbf{A}$ can be reframed in terms of the QR and singular-value decompositions. Recall from (10.6) that the least-squares method for overdetermined systems of equations results in needing to evaluate

$$\mathbf{u} = \mathbf{A}^+\mathbf{b},$$

where $\mathbf{A}^+$ is the pseudo-inverse of matrix $\mathbf{A}$, which is defined by

$$\mathbf{A}^+ = \left(\mathbf{A}^T\mathbf{A}\right)^{-1}\mathbf{A}^T.$$

Observe that given the QR decomposition $\mathbf{A} = \mathbf{QR}$, where $\mathbf{Q}$ is $M \times N$ and $\mathbf{R}$ is $N \times N$, we can rewrite the pseudo-inverse in the form

$$\begin{aligned}
\mathbf{A}^+ &= \left(\mathbf{A}^T\mathbf{A}\right)^{-1}\mathbf{A}^T \\
&= \left[(\mathbf{QR})^T\mathbf{QR}\right]^{-1}(\mathbf{QR})^T \\
&= \left(\mathbf{R}^T\mathbf{Q}^T\mathbf{QR}\right)^{-1}\mathbf{R}^T\mathbf{Q}^T \\
&= \left(\mathbf{R}^T\mathbf{R}\right)^{-1}\mathbf{R}^T\mathbf{Q}^T \\
&= \mathbf{R}^{-1}\left(\mathbf{R}^T\right)^{-1}\mathbf{R}^T\mathbf{Q}^T \\
\mathbf{A}^+ &= \mathbf{R}^{-1}\mathbf{Q}^T,
\end{aligned}$$

where we have used the facts that $\mathbf{Q}^T\mathbf{Q} = \mathbf{I}$ and $\left(\mathbf{R}^T\right)^{-1}\mathbf{R}^T = \mathbf{I}$. Recall that $\mathbf{R}$ is right triangular; therefore, its inverse is relatively easy to compute. In addition, observe that the system of equations in terms of the QR decomposition is now

$$\mathbf{Ru} = \mathbf{Q}^T\mathbf{b},$$

which has a triangular coefficient matrix. Consequently, the least-squares solution for $\mathbf{u}$ can be determined by simple backward substitution. This will be taken advantage of in the generalized minimum residual (GMRES) method to be discussed in Section 10.7, which is a popular numerical method for solving large, sparse systems of equations.

Recall from (2.51) that we can alternatively determine the pseudo-inverse from a singular-value decomposition of matrix $\mathbf{A} = \mathbf{U\Sigma V}^T$ according to

$$\mathbf{A}^+ = \mathbf{V\Sigma}^{-1}\mathbf{U}^T.$$

Consequently, we have three equivalent ways of writing the pseudo-inverse of a matrix $\mathbf{A}$ in terms of the original matrix $\mathbf{A}$, its QR decomposition, and its SVD:

$$\mathbf{A}^+ = \left(\mathbf{A}^T\mathbf{A}\right)^{-1}\mathbf{A}^T = \mathbf{R}^{-1}\mathbf{Q}^T = \mathbf{V\Sigma}^{-1}\mathbf{U}^T. \tag{10.7}$$

Therefore, the overdetermined least-squares problem (10.6) can be solved using QR decomposition or SVD, both of which avoid the need to invert a large dense matrix. Because least-squares algorithms are dominated by the computation of the QR decomposition or SVD, both methods are approximately equivalent when $M \gg N$, that is, systems having far more equations than unknowns, whereas QR decomposition is generally more efficient when $M \sim N$. Therefore, QR decomposition is typically used in all cases unless the problem is ill conditioned, in which case SVD may be more numerically stable.

10.2.2 Underdetermined Systems

Can we use a similar technique to obtain a unique solution for systems that have an infinite number of solutions? Such systems are called underdetermined as there are an insufficient number of equations to determine a unique solution. In other words, there are more unknowns than equations. We can use the least-squares method as in the previous section, but with two key modifications.

Consider a system of equations $\mathbf{A}\hat{\mathbf{u}} = \mathbf{b}$ for which $M < N$. Because this system will typically have an infinite number of solutions, the task now is to determine the unique solution that also satisfies an additional criteria. Specifically, we seek the solution $\mathbf{u}$ that is closest to the origin. We again use the norm to quantify this distance. Thus, we seek to minimize the square of the norm of $\mathbf{u}$

$$J(\mathbf{u}) = \|\mathbf{u}\|^2, \tag{10.8}$$

which is now our objective function. Of all the possible solutions $\hat{\mathbf{u}}$, we seek the one $\mathbf{u}$ that satisfies the original system of equations $\mathbf{A}\hat{\mathbf{u}} = \mathbf{b}$ *and* minimizes the objective function. Using the Lagrange multiplier method, we form the augmented function

$$\tilde{J}(\mathbf{u}) = \|\mathbf{u}\|^2 + \mathbf{\Lambda}^T (\mathbf{b} - \mathbf{A}\mathbf{u}), \tag{10.9}$$

where $\mathbf{\Lambda}$ is the vector of Lagrange multipliers, and $\mathbf{b} - \mathbf{A}\mathbf{u}$ is the residual. Differentiating with respect to $\mathbf{u}$ yields

$$\frac{d\tilde{J}}{d\mathbf{u}} = 2\mathbf{u} - \mathbf{A}^T\mathbf{\Lambda}.$$

Setting the derivative equal to zero, we can write

$$\mathbf{u} = \frac{1}{2}\mathbf{A}^T\mathbf{\Lambda}. \tag{10.10}$$

Substituting (10.10) into the original system of equations $\mathbf{A}\mathbf{u} = \mathbf{b}$ gives

$$\frac{1}{2}\mathbf{A}\mathbf{A}^T\mathbf{\Lambda} = \mathbf{b}.$$

If $\mathbf{A}\mathbf{A}^T$ is invertible, then

$$\mathbf{\Lambda} = 2\left(\mathbf{A}\mathbf{A}^T\right)^{-1}\mathbf{b}.$$

Substituting into (10.10) eliminates the Lagrange multiplier and provides the least-squares solution

$$\mathbf{u} = \mathbf{A}^T\left(\mathbf{A}\mathbf{A}^T\right)^{-1}\mathbf{b}. \tag{10.11}$$

This is the $\mathbf{u}$ that satisfies $\mathbf{A\hat{u}} = \mathbf{b}$ and minimizes $J(\mathbf{u}) = \|\mathbf{u}\|^2$. Observe the subtle, but important, differences between this result and (10.6) for the overdetermined system.

As with the least-squares method applied to overdetermined systems of equations, the least-squares solution (10.11) takes on a particularly nice form in terms of a QR decomposition. Let $\mathbf{A}^T = \mathbf{QR}$, such that $\mathbf{A} = \mathbf{R}^T\mathbf{Q}^T$. Then (10.11) becomes

$$\begin{aligned}
\mathbf{u} &= \mathbf{A}^T\left(\mathbf{AA}^T\right)^{-1}\mathbf{b} \\
&= \mathbf{QR}\left(\mathbf{R}^T\mathbf{Q}^T\mathbf{QR}\right)^{-1}\mathbf{b} \\
&= \mathbf{QR}\left(\mathbf{R}^T\mathbf{R}\right)^{-1}\mathbf{b} \\
&= \mathbf{QRR}^{-1}\left(\mathbf{R}^T\right)^{-1}\mathbf{b} \\
\mathbf{u} &= \mathbf{Q}\left(\mathbf{R}^T\right)^{-1}\mathbf{b}.
\end{aligned}$$

Let us set $\mathbf{v} = \left(\mathbf{R}^T\right)^{-1}\mathbf{b}$, which gives the left triangular system of linear algebraic equations

$$\mathbf{R}^T\mathbf{v} = \mathbf{b}. \tag{10.12}$$

As a result, once the QR decomposition of $\mathbf{A}^T$ has been obtained, one can solve (10.12) for the intermediate variable $\mathbf{v}$ using simple forward substitution, and the least-squares solution is then

$$\mathbf{u} = \mathbf{Qv},$$

which only requires a matrix-vector multiplication to obtain the least-squares solution $\mathbf{u}$. As for overdetermined systems, QR decomposition allows us to avoid evaluating the inverse of a large, dense matrix in (10.11).

Recall that in linear programming considered in Section 4.3.4, the system of linear algebraic constraints is an underdetermined system. However, note that the least-squares objective function used here is quadratic, not linear, which is why the usual method of determining an extremum works here but not for linear programming.

10.3 Least-Squares with Constraints

The least-squares solution for underdetermined systems is a constrained minimization problem. As we will see when applying least squares to optimization problems in later chapters, it is very often the case that the optimization is subject to various constraints. These constraints may be on the behavior of the system itself or the values of inputs to and outputs from the system. Most of the mathematical machinery required to handle such constraints has already been discussed in Section 4.3 and the last section using Lagrange multipliers. Just as the objective function may take on various forms, such as linear, quadratic, or nonlinear, the constraints have similar possibilities. In addition, however, the constraints may be of either equality or inequality type. We will focus on algebraic equality constraints in this chapter and leave inequality and other forms of constraints to Chapter 12 when the need arises.

Let us address the case when a least-squares solution is sought for an overdetermined system of algebraic equations subject to a set of linear constraints on the solution. As before, the objective function is to minimize the residual as follows:

$$J(\mathbf{u}) = \|\mathbf{r}\|^2 = \|\mathbf{b} - \mathbf{A}\mathbf{u}\|^2, \tag{10.13}$$

but now subject to the constraint that the least-squares solution $\mathbf{u}$ must satisfy the linear equality constraints

$$\mathbf{C}\mathbf{u} = \mathbf{d}. \tag{10.14}$$

Note that the solution $\mathbf{u}$ *must* satisfy the constraint (10.14) while minimizing the objective function (10.13). In other words, of all the possible values of $\mathbf{u}$ in the feasible space that satisfies (10.14), we seek the one that also minimizes the objective function (10.13). If you are not sure whether a quantity or expression should be included in the objective function or as a constraint, simply ask yourself whether the expression must be satisfied exactly, in which case it should be imposed as a constraint, or whether it is a quantity that is to be as small (or large) as possible, in which case it should be incorporated into the objective function.

If the rank of $\mathbf{A}$ is equal to the number of variables N in the solution vector $\mathbf{u}$ (and columns of $\mathbf{A}$), then there is a unique solution for the system $\mathbf{A}\mathbf{u} = \mathbf{b}$, and the residual is zero. This solution is unlikely to also satisfy the constraint equations (10.14). Likewise, if the rank of $\mathbf{C}$ is equal to the number of variables in $\mathbf{u}$ (and columns of $\mathbf{C}$), then there is a unique solution for $\mathbf{u}$, and there is no room for optimization via the least-squares objective function (10.13). Therefore, the case of interest in least-squares optimization is when rank($\mathbf{A}$) $> N$, such that $\mathbf{A}\mathbf{u} = \mathbf{b}$ is overdetermined, and rank($\mathbf{C}$) $< N$, such that $\mathbf{C}\mathbf{u} = \mathbf{d}$ is underdetermined. The *feasible space* is the (infinite) set of solutions that satisfy the constraint equations $\mathbf{C}\mathbf{u} = \mathbf{d}$, and the optimal least-squares solution is the point in the feasible space that minimizes the objective function (10.13).

In general, there are three ways to deal with such linear constraints:

1. Eliminate the constraints by substituting them directly into the objective function to produce an equivalent unconstrained problem. Because some of the variables will have been eliminated, the new unconstrained system is smaller than the original one. However, whereas $\mathbf{A}$ and $\mathbf{C}$ may be sparse, the new system likely is not.
2. Perform a descent method, such as steepest descent or Newton's method, but with a check to be sure that the algorithm remains in the feasible space of possible optimal solutions.
3. Reformulate the least-squares problem by augmenting the objective function with the constraints using Lagrange multipliers.

The first option is not always possible or computationally expedient, while the second option creates a much more complicated numerical algorithm. Therefore, let us consider the generally applicable Lagrange multiplier approach.

The objective function (10.13) is the same as (10.2) for the overdetermined case, which expands to give the objective function in the form of (10.3). The constraints

(10.14) are appended to the objective function J using Lagrange multipliers to form the augmented objective function $\tilde{J}$. Each constraint requires its own Lagrange multiplier, so if $\mathbf{C}$ is $P \times N$, such that $\mathbf{d}$ is $P \times 1$, then we need P Lagrange multipliers. These will be designated by the $P \times 1$ vector $\mathbf{\Lambda} = \begin{bmatrix} \Lambda_1 & \Lambda_2 & \cdots & \Lambda_P \end{bmatrix}^T$. The augmented objective function is then[7]

$$\tilde{J}(\mathbf{u}, \mathbf{\Lambda}) = \mathbf{b}^T\mathbf{b} - 2\mathbf{b}^T\mathbf{A}\mathbf{u} + \mathbf{u}^T\mathbf{A}^T\mathbf{A}\mathbf{u} + \mathbf{\Lambda}^T\left(\mathbf{d} - \mathbf{C}\mathbf{u}\right). \tag{10.15}$$

Observe that our augmented objective function is a function of the variables in $\mathbf{u}$ and the Lagrange multipliers $\mathbf{\Lambda}$. Therefore, minimization requires differentiation with respect to both sets of variables. Differentiating with respect to the Lagrange multipliers simply returns the original constraints (10.14), which provides no new information. As in Sections 4.3 and 4.4, differentiating the augmented objective function (10.15) with respect to $\mathbf{u}$ yields

$$\frac{d\tilde{J}}{d\mathbf{u}} = -2\mathbf{A}^T\mathbf{b} + 2\mathbf{A}^T\mathbf{A}\mathbf{u} - \mathbf{C}^T\mathbf{\Lambda},$$

where the first two terms are as before, and the final term arises from augmentation with the constraint. Setting these derivatives equal to zero results in an equivalent unconstrained least-squares optimization problem

$$2\mathbf{A}^T\mathbf{A}\mathbf{u} - \mathbf{C}^T\mathbf{\Lambda} = 2\mathbf{A}^T\mathbf{b}. \tag{10.16}$$

This system of linear algebraic equations is known as *optimality conditions* as they relate the Lagrange multipliers to the optimal solution $\mathbf{u}$, and these equations replace the normal equations (10.5) for the unconstrained least-squares problem.

We can combine the optimality conditions with the constraints (10.14) to form a large system of equations written in block form as follows:

$$\begin{bmatrix} 2\mathbf{A}^T\mathbf{A} & -\mathbf{C}^T \\ \mathbf{C} & \mathbf{0} \end{bmatrix} \begin{bmatrix} \mathbf{u} \\ \mathbf{\Lambda} \end{bmatrix} = \begin{bmatrix} 2\mathbf{A}^T\mathbf{b} \\ \mathbf{d} \end{bmatrix}. \tag{10.17}$$

The solution of these so called *Karush, Kuhn, and Tucker*, or *KKT*, *equations* would provide the optimal least-squares solution $\mathbf{u}$ – along with the Lagrange multipliers $\mathbf{\Lambda}$ – that minimizes the objective function (10.13) and satisfies the constraints (10.14). Note that there are $N + P$ KKT equations, and if you check the sizes of each block in the coefficient matrix, you will see that it is a $(N+P)\times(N+P)$ square matrix. Therefore, the KKT equations have a unique solution if the coefficient matrix is invertible. For the special case of the KKT equations when $\mathbf{A} = \mathbf{I}$, see Exercise 10.11.

As with unconstrained least-squares problems, the KKT equations can be solved using the QR decomposition of the coefficient matrix under most circumstances unless the problem turns out to be ill conditioned, in which case SVD should be used. A particularly elegant algorithm based on QR decomposition is provided in section 16.3 of Boyd and Vandenberghe (2018).

[7] The augmented function is sometimes called the "Lagrangian" owing to its similarity to the Lagrangian in Hamiltonian mechanics (see, for example, Cassel, 2013).

Note that the underdetermined case considered in Section 10.2.2 is a special case of this constrained least-squares problem, in which $\mathbf{A} = \mathbf{I}$ and $\mathbf{b} = \mathbf{0}$ in the residual. In the notation of this section, therefore, the least-squares solution (10.11) for the underdetermined problem is given by

$$\mathbf{u} = \mathbf{C}^T \left(\mathbf{C}\mathbf{C}^T\right)^{-1} \mathbf{d}, \tag{10.18}$$

where the system of equations $\mathbf{A}\mathbf{u} = \mathbf{b}$ that must be satisfied is now the constraint $\mathbf{C}\mathbf{u} = \mathbf{d}$.

Finally, some terminology. The optimization problem as formulated in the objective function (10.13) subject to the constraint (10.14) is sometimes referred to as the *primal problem*. The necessary conditions for a minimum given by the KKT equations (10.17) is then referred to as the *adjoint*, or *dual*, *problem*. In this structure, the Lagrange multipliers used to enforce the constraints are also called the *adjoint*, or *dual*, *variables* as they are the variables being solved for in the adjoint or dual problem. Because the primal and adjoint formulations are related mathematically, the adjoint problem can be used to expose features of the primal optimization problem as well as seek its solution. You will notice that in the present case, the adjoint problem is a system of linear algebraic equations as compared to the primal problem, which involves a quadratic objective function. Therefore, we can choose to solve whichever problem is easier, which in this case are the linear algebraic KKT equations for the adjoint problem.

Because the adjoint problem is a necessary, but not sufficient, condition for the optimal solution of the primal problem in the general case, the solution of the adjoint problem, which is often easier, may not provide the true optimal solution in all cases. If the original optimization problem – the primal problem – is convex, such that there is a single global minimum, then the solution of the adjoint problem will give the optimal solution of the primal problem as long as the constraints are well behaved. The case considered in this section, with a quadratic (convex) objective function and linear constraints, is such that the solutions to the adjoint and primal problems will be the same. For more on the primal-adjoint formulation of optimization problems, see Calafiore and El Ghaoui (2014).

10.4 Least-Squares with Penalty Functions

In addition to constraints, another common feature of least-squares problems is the desire to regulate the amount of control input required to achieve the optimization objective. Control regulation can be accomplished by adding a penalty term to the objective function as follows:

$$J(\mathbf{u}) = \gamma^2 \|\mathbf{r}\|^2 + \phi(\mathbf{u}) = \gamma^2 \|\mathbf{b} - \mathbf{A}\mathbf{u}\|^2 + \phi(\mathbf{u}). \tag{10.19}$$

The penalty term $\phi(\mathbf{u})$ is in some way related to the amount of control input required to achieve the desired objective. In this formulation of the objective function, the first term is called the *performance measure*, and the second term is the *penalty function*. In a control context, we can interpret the objective function in the following manner. The performance objective is to determine the input $\mathbf{u}$ that when operated on by $\mathbf{A}$, which represents the system behavior, gives a result that is as close as possible to the target behavior given by $\mathbf{b}$ in a least-squares sense. However, this is not to be done with unlimited control input; therefore, a penalty function is included such that the minimization leads to the best solution behavior that also minimizes the control input. The weight coefficient γ^2 determines the relative importance of the performance measure and the penalty function. A large weight coefficient implies that the control input is inexpensive and the performance measure is to be emphasized, whereas a small weight coefficient implies that the control input is expensive and the penalty function is to be emphasized.

A common choice for the penalty function is the L_2-norm of the input $\mathbf{u}$, which often represents some form of energy. Therefore, we seek to minimize the energy input to the system, and the objective function is of the form

$$J(\mathbf{u}) = \gamma^2 \|\mathbf{r}\|^2 + \|\mathbf{u}\|^2 = \gamma^2 \|\mathbf{b} - \mathbf{A}\mathbf{u}\|^2 + \|\mathbf{u}\|^2. \tag{10.20}$$

This is often referred to as the *linear-quadratic-regulator*, or *LQR*, *problem*. The system is *linear*, while the performance measure is *quadratic*, and the objective function includes a *regulator*. One can think of this as *multi-objective least squares* as there are more than one objective included in the objective function. See chapter 10 of Cassel (2013) to see how the LQR problem is formulated using variational methods.

10.5 Nonlinear Objective Functions

The least-squares problems considered thus far have been motivated by our interest in "solving" systems of linear algebraic equations even when a unique solution does not exist. In these cases, the residual vector arises from a linear system of algebraic equations; minimizing the norm of the residual produces a quadratic objective function. This is the basis for the standard least-squares problem.

In general optimization and control problems, however, the objective function can take on many different forms. As before, we wish to determine the values of the solution vector $\mathbf{u} = \begin{bmatrix} u_1 & u_2 & \cdots & u_N \end{bmatrix}^T$, but now such that the vector of general nonlinear functions $\mathbf{f}(\mathbf{u})$ is minimized. We could consider the function $\mathbf{f}(\mathbf{u}) = \begin{bmatrix} f_1(\mathbf{u}) & f_2(\mathbf{u}) & \cdots & f_M(\mathbf{u}) \end{bmatrix}^T$ to be a residual that we would like to minimize and seek a least-squares solution by minimizing the objective function

$$J(\mathbf{u}) = \|\mathbf{f}(\mathbf{u})\|^2 = f_1^2(u_1, u_2, \dots, u_N) + f_2^2(u_1, u_2, \dots, u_N) + \cdots + f_M^2(u_1, u_2, \dots, u_N). \tag{10.21}$$

Observe that $\mathbf{f}$ is a vector consisting of M functions of the N dependent, or design, variables. If $\mathbf{f}(\mathbf{u}) = \mathbf{r} = \mathbf{b} - \mathbf{A}\mathbf{u}$, then we have the standard least-squares problem from Sections 10.2 through 10.4. Alternatively, if we only have a single objective function $f(\mathbf{u})$, we could seek the optimal solution $\mathbf{u}$ that minimizes the single objective function directly, but not in a least-squares sense, in which case the objective function is given by

$$J(\mathbf{u}) = f(\mathbf{u}) = f(u_1, u_2, \ldots, u_N). \tag{10.22}$$

If constraints are imposed on the optimization, they are incorporated using Lagrange multipliers as before.

In order to produce the necessary conditions for a minimum in the objective function $J(\mathbf{u})$, we take the partial derivatives of the objective function with respect to each of the design variables u_n, $n = 1, 2, \ldots, N$. As you might imagine, optimization problems with nonlinear, but not quadratic, objective functions opens up a broader set of possible outcomes than the quadratic cases considered thus far, and it also leads to additional subtleties in the development of algorithms for solving such optimization problems. Optimization problems with nonlinear objective functions with or without constraints will be considered further in Chapter 12.

10.6 Conjugate-Gradient Method

In Section 6.4, we considered simple iterative methods, such as the Jacobi, Gauss–Seidel, and SOR methods, for solving systems of linear algebraic equations. Although these methods are very simple to implement, they typically require very large numbers of iterations to converge to an acceptable approximation of the solution. Having covered least-squares methods, it is now possible to approach solving systems of linear algebraic equations using optimization methods. This leads to two very popular algorithms that are particularly well suited to obtaining approximate solutions to systems of equations with large, sparse coefficient matrices. In the conjugate-gradient method, the norm of the error is minimized, while in the generalized minimum residual method, the norm of the residual is minimized. The former applies to symmetric, positive-definite matrices, while the latter applies to any nonsingular matrix. While the computations required during each iteration are significantly increased for these methods as compared to Jacobi, Gauss–Seidel, and SOR, the number of iterations until convergence is dramatically reduced, particularly for sparse matrices.

It turns out that when the coefficient matrix in a system of linear algebraic equations is symmetric and positive definite (all eigenvalues are positive), a particularly efficient algorithm can be devised that can be used to solve systems of linear algebraic equations and nonlinear optimization problems (see Section 12.5). In fact, it can lead to the exact solution if carried out for N steps using exact arithmetic for an $N \times N$ system of equations. While formulated as a direct method, it is more often implemented as an iterative method as it typically converges to an accurate approximate solution within fewer than N iterations.

Operationally, it only requires evaluation of the product of the coefficient matrix and a vector along with some straightforward vector operations during each iteration. Therefore, it is particularly effective for systems having sparse coefficient matrices as these operations can be done very efficiently. While being very efficient, the conjugate-gradient method is far more complex than our previous iterative methods for solving systems of linear algebraic equations; it takes advantage of much of the linear algebra that we learned in Part I. In the end, however, the method is quite straightforward to implement and again is very computationally efficient, making navigating the following developments worth our while.

The basic premise of the conjugate-gradient method is that the primary task of solving a system of algebraic equations $\mathbf{A}\hat{\mathbf{u}} = \mathbf{b}$, where $\hat{\mathbf{u}}$ is the exact solution, can be converted into an optimization problem, opening up a new class of solution techniques. At first, this may seem like an unnecessary complication; however, it leads to a particularly efficient solution strategy for systems comprised of symmetric, positive-definite matrices.

10.6.1 Equivalent Quadratic Optimization Problem

In the conjugate-gradient method, the error $\mathbf{e}$ is minimized as follows:

$$J_{CG} = \|\mathbf{e}\|_{\mathbf{A}}^2 = \|\mathbf{u} - \hat{\mathbf{u}}\|_{\mathbf{A}}^2,$$

where $\mathbf{u}$ is the approximate numerical solution. This is known as the $\mathbf{A}$-*norm* and is given by

$$J_{CG} = \mathbf{e}^T\mathbf{A}\mathbf{e} = \left(\mathbf{u} - \hat{\mathbf{u}}\right)^T \mathbf{A}\left(\mathbf{u} - \hat{\mathbf{u}}\right),$$

which is a quadratic for symmetric $\mathbf{A}$. Expanding the objective function yields

$$J_{CG}(\mathbf{u}) = \mathbf{u}^T\mathbf{A}\mathbf{u} - \mathbf{u}^T\mathbf{A}\hat{\mathbf{u}} - \hat{\mathbf{u}}^T\mathbf{A}\mathbf{u} + \hat{\mathbf{u}}^T\mathbf{A}\hat{\mathbf{u}}.$$

The middle two terms can be expressed as $\langle\mathbf{u}, \mathbf{A}\hat{\mathbf{u}}\rangle$ and $\langle\hat{\mathbf{u}}, \mathbf{A}\mathbf{u}\rangle$, respectively. These inner products are equivalent if $\mathbf{A}$ is symmetric (see Section 3.3.1); therefore,

$$J_{CG}(\mathbf{u}) = \mathbf{u}^T\mathbf{A}\mathbf{u} - 2\mathbf{u}^T\mathbf{A}\hat{\mathbf{u}} + \hat{\mathbf{u}}^T\mathbf{A}\hat{\mathbf{u}}.$$

In addition, the exact solution $\hat{\mathbf{u}}$ of the original system of algebraic equations satisfies $\mathbf{A}\hat{\mathbf{u}} = \mathbf{b}$. Therefore, the objective function becomes

$$J_{CG}(\mathbf{u}) = \mathbf{u}^T\mathbf{A}\mathbf{u} - 2\mathbf{u}^T\mathbf{b} + \hat{\mathbf{u}}^T\mathbf{b}. \tag{10.23}$$

Minimization is with respect to the approximate solution $\mathbf{u}$; therefore, the last term is a constant with respect to the minimization. As a result, this is equivalent to minimizing the function

$$f(\mathbf{u}) = \frac{1}{2}\mathbf{u}^T\mathbf{A}\mathbf{u} - \mathbf{u}^T\mathbf{b} \tag{10.24}$$

with respect to the approximate solution $\mathbf{u}$ that is being sought. From Section 4.4, this minimization with symmetric $\mathbf{A}$ requires that the gradient of the function vanish as follows:

$$\nabla f(\mathbf{u}) = \mathbf{A}\mathbf{u} - \mathbf{b} = \mathbf{0}, \tag{10.25}$$

which of course is the case if the system of equations is satisfied exactly by $\mathbf{u}$. This is the same as taking the partial derivative of $f(u_1, u_2, \ldots, u_n, \ldots, u_N)$ with respect to each of the variables u_n and setting equal to zero. Recall that a vanishing gradient is a necessary condition for an extremum in a function. Moreover, note that the Hessian is given by

$$\mathbf{H}(\mathbf{u}) = \boldsymbol{\nabla}\left[\boldsymbol{\nabla} f(\mathbf{u})\right] = \mathbf{A},$$

which is positive definite. Recall from Section 4.3.1 that this provides a sufficient condition for a minimum.

We have shown that solving the system of linear algebraic equations $\mathbf{A}\hat{\mathbf{u}} = \mathbf{b}$, where $\mathbf{A}$ is symmetric and positive definite, is equivalent to minimizing the quadratic function (10.24), where the optimal solution $\mathbf{u}$ approximates the exact solution $\hat{\mathbf{u}}$. Consequently, we have a quadratic programming problem (see Section 4.3.3), which is convex and only has a single minimum.

Two algorithms for solving the quadratic programming problem are discussed: steepest descent and conjugate gradient. The method of steepest descent is simpler, but slower, and the conjugate-gradient method is more involved, but faster. Both involve proceeding down the quadratic bowl toward the global minimum by reducing the residual in a step-by-step fashion. Similar to the Arnoldi method for approximating eigenvalues in Section 6.5.3, these methods are direct methods for obtaining solutions of linear systems of algebraic equations. Thus, each step in the process will be denoted by subscript k.[8] However, they are implemented in an iterative fashion and terminated when the solution is "close enough" to the true minimum. As such, we will present them as iterative methods with the understanding that they never require more than N "iterations."

We define the residual in the usual way

$$\mathbf{r}_k = \mathbf{b} - \mathbf{A}\mathbf{u}_k, \tag{10.26}$$

where $\mathbf{u}_k$ is the approximate solution at the kth iteration, and we seek successive approximations for $\mathbf{u}_k$ that progressively drive the residual toward zero. Observe from (10.25) and (10.26) that the residual is the negative of the gradient of $f(\mathbf{u})$ according to

$$\mathbf{r}_k = -\boldsymbol{\nabla} f\left(\mathbf{u}_k\right). \tag{10.27}$$

The iterative process is defined as follows:

$$\mathbf{u}_{k+1} = \mathbf{u}_k + \alpha_k \mathbf{p}_k, \quad k = 0, 1, 2, \ldots, \tag{10.28}$$

where $\mathbf{p}_k$ is the search direction at each iteration, and α_k is how far to proceed – the step size – in that direction. Once the search direction $\mathbf{p}_k$ is determined at each step, the distance α_k is determined by minimizing $f(\mathbf{u}_{k+1}) = f(\mathbf{u}_k + \alpha_k \mathbf{p}_k)$. The methods of steepest descent and conjugate gradient differ in how the search direction and step size are chosen.

[8] Recall from Part II that we denote iteration numbers in numerical methods by superscripts in parentheses.

10.6.2 Method of Steepest Descent

Because the function for which the minimum is sought is a bowl-shaped quadratic, a logical strategy would be as follows:

1. Choose an initial guess $\mathbf{u}_0$ for $k = 0$.
2. Identify the direction $\mathbf{p}_k$ of steepest descent.
3. Proceed a distance α_k in that direction until a minimum is reached, such that the function no longer descends.
4. Increment k and repeat steps 2 and 3 until a local minimum is reached.

The direction $\mathbf{p}_k$ of steepest descent corresponds to the negative of the gradient of $f(\mathbf{u}_k)$. Therefore, from (10.27), the search direction in such an algorithm is given by

$$\mathbf{p}_k = -\nabla f(\mathbf{u}_k) = \mathbf{r}_k. \tag{10.29}$$

Consequently, the residual vector is in the direction of steepest descent. To see the consequences of this, consider two simple cases. If the contours of constant f are drawn, the negative gradient will always be perpendicular to the contour passing through the given point. For circular contours, this is always pointing toward the minimum at the bottom of the circular bowl. For contours that are shaped like elongated ellipses, however, if the point is far from a minor or major axis, the normal to a local contour, while pointing down the hill, could point in a direction that is nearly perpendicular to the path directly from that point to the minimum at the bottom of the bowl. Consequently, although the negative gradient vector is always pointing in the direction of steepest descent locally, it may or may not form a path leading directly toward the sought-after minimum.

Note that the minimization problem in step 3 is one dimensional in the direction of the negative of the gradient. More specifically, the step size α_k is determined by setting the directional derivative along the search direction to zero. This is a valid approach and is generally very robust in identifying a local minimum; it is called the *method of steepest descent* because of the way that the search direction is selected for each iteration. However, a closer look reveals that step 2 will always result in a right-angle change of direction at each iteration – it just requires determining whether to turn left or right. This would be like walking generally in the northeast direction in a city with north-south and east-west only streets. It hardly would produce the path having the shortest distance, which is a straight line.

Let us determine the step size α_k in the method of steepest descent. The step size is sought for which the directional derivative along $\mathbf{p}_{k+1}$ is zero (see Section 4.1.2), corresponding to a minimum in f along the search direction. Therefore,

$$\mathbf{p}_{k+1}^T \nabla f_k = 0.$$

The best estimate for the gradient of f is from the previous iteration, while the search direction corresponds to the residual from the current iteration. Therefore, from (10.29), setting the directional derivative to zero is equivalent to

$$\mathbf{p}_{k+1}^T \mathbf{p}_k = \mathbf{r}_{k+1}^T \mathbf{r}_k = 0, \tag{10.30}$$

which shows that the search direction and residual for each iteration is always orthogonal to that in the previous iteration.

Let us consider (10.28) relating the new approximate solution to the previous approximate solution and the search direction. Premultiplying by $\mathbf{A}$ and subtracting $\mathbf{b}$ from both sides yields

$$\mathbf{A}\mathbf{u}_{k+1} - \mathbf{b} = \mathbf{A}\mathbf{u}_k - \mathbf{b} + \alpha_k \mathbf{A}\mathbf{p}_k,$$

but from the definition of the residual (10.26), this becomes

$$\mathbf{r}_{k+1} = \mathbf{r}_k - \alpha_k \mathbf{A}\mathbf{p}_k, \tag{10.31}$$

or in the method of steepest descent, because $\mathbf{p}_k = \mathbf{r}_k$,

$$\mathbf{r}_{k+1} = \mathbf{r}_k - \alpha_k \mathbf{A}\mathbf{r}_k.$$

Premultiplying by $\mathbf{r}_k^T$ gives

$$\mathbf{r}_k^T \mathbf{r}_{k+1} = \mathbf{r}_k^T \mathbf{r}_k - \alpha_k \mathbf{r}_k^T \mathbf{A}\mathbf{r}_k.$$

Because of orthogonality of successive residuals, the left-hand side vanishes, and we can solve for the step size

$$\alpha_k = \frac{\mathbf{r}_k^T \mathbf{r}_k}{\mathbf{r}_k^T \mathbf{A}\mathbf{r}_k}. \tag{10.32}$$

Summarizing the method of steepest descent, the search direction in step 2 corresponds to the residual at the current iteration, the step size is given by (10.32), and the next approximation for the solution is given by (10.28). As implied earlier, the method of steepest descent converges slowly owing to the right-angle changes in the search direction during each iteration; therefore, it is rarely used in practice. See Section 12.5 for more details on the method of steepest descent as applied to optimization problems.

10.6.3 Conjugate-Gradient Method

Instead of the method of steepest descent's right-angle turns at each iteration, we would prefer a path more along the lines of that taken by a ball rolling down the quadratic bowl directly toward the local minimum. How can an algorithm be devised that more closely adheres to such a path? This is the *conjugate-gradient method*. The name of the method arises from the central role played by choosing the appropriate search directions $\mathbf{p}_k$ along which to proceed during the iterative process in seeking a minimum of (10.24).

Recall that two vectors are orthogonal if their inner product is zero, such that

$$\langle \mathbf{p}_k, \mathbf{p}_{k+1} \rangle = \mathbf{p}_k^T \mathbf{p}_{k+1} = 0.$$

As given in (10.30), this is how successive search direction vectors are related in the method of steepest descent. Therefore, the path toward the minimum is comprised of a series of right-angle turns, which is clearly not the ideal path to proceed down a convex bowl toward a minimum as it utilizes the same two search directions in the sequence of iterations. Instead, it is better to use *conjugate* paths.

Two vectors are said to be conjugate with respect to the symmetric, positive-definite matrix $\mathbf{A}$ if

$$\langle \mathbf{p}_k, \mathbf{A}\mathbf{p}_{k+1} \rangle = \mathbf{p}_k^T \mathbf{A}\mathbf{p}_{k+1} = 0.$$

Rather than $\mathbf{p}_k$ being orthogonal to $\mathbf{p}_{k+1}$, as in the method of steepest descent, therefore, it is orthogonal to $\mathbf{A}\mathbf{p}_{k+1}$. Conjugate vectors are linearly independent (for nonsingular $\mathbf{A}$) and as different as possible from all previous search directions. Instead of proceeding down the gradient at the current point, we devise a scheme in which each search direction is conjugate with respect to $\mathbf{A}$ to the previous one. From (10.27), the residual is still equal to the negative of the gradient of f; however, this no longer corresponds to the search direction $\mathbf{p}_k$ as in the method of steepest descent.

Let us determine the step size α_k in the conjugate-gradient method. The step size is sought for which the directional derivative along $\mathbf{p}_k$ is zero, corresponding to a minimum in f along the search direction. Therefore,

$$\mathbf{p}_k^T \nabla f_{k+1} = (\nabla f_{k+1})^T \mathbf{p}_k = 0.$$

Because $\mathbf{r} = -\nabla f$, this is equivalent to

$$\mathbf{r}_{k+1}^T \mathbf{p}_k = 0. \tag{10.33}$$

Therefore, the residual vector is orthogonal to the search direction; in fact, each residual vector is orthogonal to all previous search directions. In this way, each new residual and orthogonal search direction is linearly independent with all the previous ones. Note that if each residual is orthogonal to all previous search directions, it is necessarily orthogonal to all previous residuals as well. Therefore,

$$\mathbf{r}_{k+1}^T \mathbf{r}_k = 0. \tag{10.34}$$

As in the method of steepest descent, the residual at the next iteration is a linear combination of the previous residual and $\mathbf{A}\mathbf{p}_k$ according to

$$\mathbf{r}_{k+1} = \mathbf{r}_k - \alpha_k \mathbf{A}\mathbf{p}_k. \tag{10.35}$$

Premultiplying by $\mathbf{r}_k^T$ gives

$$\mathbf{r}_k^T \mathbf{r}_{k+1} = \mathbf{r}_k^T \mathbf{r}_k - \alpha_k \mathbf{r}_k^T \mathbf{A}\mathbf{p}_k.$$

The left-hand side vanishes owing to orthogonality of successive residuals according to (10.34); therefore, solving for the step size yields

$$\alpha_k = \frac{\mathbf{r}_k^T \mathbf{r}_k}{\mathbf{r}_k^T \mathbf{A}\mathbf{p}_k}. \tag{10.36}$$

Observe that if the search direction is equal to the residual, then this expression becomes the corresponding one for the method of steepest descent given by (10.32).

Let us define the search direction at the next iteration as a linear combination of that at the current iteration and the residual according to

$$\mathbf{p}_{k+1} = \mathbf{r}_{k+1} + \beta_k \mathbf{p}_k. \tag{10.37}$$

This expression can be used to obtain the step size for the search direction β_k as well as simplify our expression for the step size (10.36). Begin by premultiplying the matrix $\mathbf{A}$ to give

$$\mathbf{A}\mathbf{p}_{k+1} = \mathbf{A}\mathbf{r}_{k+1} + \beta_k \mathbf{A}\mathbf{p}_k. \tag{10.38}$$

Let us premultiply (10.38) by $\mathbf{p}_{k+1}^T$ to yield

$$\mathbf{p}_{k+1}^T \mathbf{A}\mathbf{p}_{k+1} = \mathbf{p}_{k+1}^T \mathbf{A}\mathbf{r}_{k+1} + \beta_k \mathbf{p}_{k+1}^T \mathbf{A}\mathbf{p}_k.$$

The last term vanishes because the search direction vectors are conjugate with respect to $\mathbf{A}$. Therefore, shifting the iteration index by negative one, taking the transpose (and recalling that $\mathbf{A} = \mathbf{A}^T$), and substituting into (10.36) gives the step size

$$\alpha_k = \frac{\mathbf{r}_k^T \mathbf{r}_k}{\mathbf{p}_k^T \mathbf{A}\mathbf{p}_k}. \tag{10.39}$$

Now, let us obtain an expression for β_k. Returning to (10.38), premultiplying by $\mathbf{p}_k^T$ leads to

$$\mathbf{p}_k^T \mathbf{A}\mathbf{p}_{k+1} = \mathbf{p}_k^T \mathbf{A}\mathbf{r}_{k+1} + \beta_k \mathbf{p}_k^T \mathbf{A}\mathbf{p}_k.$$

Because $\mathbf{p}_k$ and $\mathbf{p}_{k+1}$ are conjugate with respect to $\mathbf{A}$, the left-hand side is zero, and solving for β gives

$$\beta_k = -\frac{\mathbf{p}_k^T \mathbf{A}\mathbf{r}_{k+1}}{\mathbf{p}_k^T \mathbf{A}\mathbf{p}_k}. \tag{10.40}$$

Premultiplying $\mathbf{r}_{k+1}^T$ on both sides of (10.35) yields

$$\mathbf{r}_{k+1}^T \mathbf{r}_{k+1} = \mathbf{r}_{k+1}^T \mathbf{r}_k - \alpha_k \mathbf{r}_{k+1}^T \mathbf{A}\mathbf{p}_k.$$

But the first term on the right-hand side is zero owing to orthogonality of residuals according to (10.34); therefore, taking the transpose of the equation yields

$$\mathbf{r}_{k+1}^T \mathbf{r}_{k+1} = -\alpha_k \mathbf{p}_k^T \mathbf{A}\mathbf{r}_{k+1},$$

where the fact that $\mathbf{A}^T = \mathbf{A}$ has been used. Observe that the right-hand side is the numerator in the expression (10.40) for β_k; therefore, substituting gives

$$\beta_k = \frac{1}{\alpha_k} \frac{\mathbf{r}_{k+1}^T \mathbf{r}_{k+1}}{\mathbf{p}_k^T \mathbf{A}\mathbf{p}_k}. \tag{10.41}$$

Substituting the expression (10.39) for α_k leads to the result

$$\beta_k = \frac{\mathbf{r}_{k+1}^T \mathbf{r}_{k+1}}{\mathbf{r}_k^T \mathbf{r}_k}. \tag{10.42}$$

Given the preceding analysis, we are now ready to formulate the conjugate-gradient algorithm:

1. Initial guess ($k = 0$):
   - Solution: $\mathbf{u}^{(0)}$,
   - Residual: $\mathbf{r}^{(0)} = \mathbf{b} - \mathbf{A}\mathbf{u}^{(0)}$,
   - Search direction: $\mathbf{p}^{(0)} = \mathbf{r}^{(0)}$.
2. Do while $\|\mathbf{r}_k\| <$ tolerance and $k < N$:
   (i) Determine the step size from (10.39):
$$\alpha_k = \frac{\mathbf{r}_k^T \mathbf{r}_k}{\mathbf{p}_k^T \mathbf{A} \mathbf{p}_k}.$$
   (ii) Update the approximation of the solution from (10.28):
$$\mathbf{u}_{k+1} = \mathbf{u}_k + \alpha_k \mathbf{p}_k.$$
   (iii) Update the residual from (10.35):
$$\mathbf{r}_{k+1} = \mathbf{r}_k - \alpha_k \mathbf{A} \mathbf{p}_k.$$
   (iv) Calculate the step size for the search direction from (10.42):
$$\beta_k = \frac{\mathbf{r}_{k+1}^T \mathbf{r}_{k+1}}{\mathbf{r}_k^T \mathbf{r}_k}.$$
   (v) Determine the new search direction from (10.37):
$$\mathbf{p}_{k+1} = \mathbf{r}_{k+1} + \beta_k \mathbf{p}_k.$$
   (vi) Increment $k = k + 1$.
3. End do.

The power of the conjugate-gradient method is in the fact that each iteration produces additional information, along with all of the previous iterations, for locating an estimate of the local minimum. The method of steepest descent only uses information from the previous iteration to compute the current iteration, and all previous information is lost. The conjugate-gradient method accumulates knowledge about the function during each successive iteration and uses it during the next one in order to provide the best approximation for the local minimum. It is in this way that the conjugate-gradient method can lead to the exact minimum in only N "iterations" if exact arithmetic is used, because after N steps it has explored all possible directions. In this way, it can be considered a direct method for solving a system of algebraic equations with a symmetric, positive definite coefficient matrix. Because each step produces an improved estimate of the solution $\mathbf{u}$, however, it is normally implemented as an iterative method by imposing our usual iterative convergence criterion to terminate the iterations when the residual is sufficiently small.

A careful look at this algorithm will reveal that it only requires one matrix-vector product $\mathbf{A}\mathbf{p}$ in steps (i) and (iii) along with some straightforward vector additions and inner products. Although the analysis and algorithm for the conjugate-gradient method are more involved than the method of steepest descent, therefore, the computational time is still dominated by the single matrix-vector multiplication. The additional vector operations in the conjugate-gradient method only add a small amount of computational time to each iteration, but it requires no more than N iterations to converge. Furthermore, when used to solve linear algebraic problems as here, the conjugate-gradient method always converges to the single minimum corresponding to the unique solution of the system of algebraic equations regardless of the initial guess. As a result, the conjugate-gradient method is the preferred technique for solving linear systems of algebraic equations with symmetric, positive definite coefficient matrices and general nonlinear optimization problems (see Section 12.5), particularly when the coefficient matrix is sparse.

REMARKS:

1. *Observe from* (10.29) *that step (v) indicates that the conjugate-gradient method leads to an adjustment of the search direction as compared to the steepest descent method, for which* $\mathbf{p}_k = \mathbf{r}_k$. *This adjustment results in search directions that are conjugate rather than orthogonal.*
2. *Note that at no point during these developments have we mentioned "conjugate gradients." Therefore, the name is a bit misleading; it is the search directions that are conjugate.*

10.7 Generalized Minimum Residual (GMRES) Method

Although the conjugate-gradient method is very efficient for solving large, sparse systems of linear algebraic equations, it is limited to those having coefficient matrices that are symmetric and positive definite. Like the conjugate-gradient method, the *generalized minimum residual* (GMRES) *method*[9] converts solving a system of linear equations into an optimization problem. Unlike the conjugate-gradient method, however, it can be used to solve any linear system of algebraic equations with a nonsingular coefficient matrix. Recall that in the conjugate-gradient method, it is the error between the exact and numerical solutions that is minimized. Introduced by Saad and Schultz (1986), in GMRES, it is the residual of the numerical solution that is minimized as suggested by its name.

Once again, we seek an approximate solution of the system of linear algebraic equations $\mathbf{A}\hat{\mathbf{u}} = \mathbf{b}$, where $\mathbf{A}$ is the $N \times N$ coefficient matrix, and $\hat{\mathbf{u}}$ and $\mathbf{b}$ are $N \times 1$ vectors. The squared norm of the residual of a numerical approximation $\mathbf{u}$ to the solution is given by

[9] It is called the "generalized" minimum residual method because it is a generalization of an earlier method, called the *minimum residual method*, that was developed for solving systems of linear algebraic equations with a symmetric coefficient matrix. This method has been supplanted by the conjugate-gradient method considered in the last section.

$$J_{GMRES} = \|\mathbf{r}\|^2 = \|\mathbf{b} - \mathbf{A}\mathbf{u}\|^2, \tag{10.43}$$

which is to be minimized.

Although minimization of the residual in the form (10.43) using the least-squares method does not hold any particular inherent advantage over previously discussed methods, it can be recast into an objective function that can be efficiently solved using the Arnoldi method discussed in Section 6.5.3. The optimization problem is solved using the least-squares method during each iteration, and an algorithm is sought that iteratively reduces the residual until it is smaller than some tolerance value. This is done by forming a sequence of approximate solutions $\mathbf{u}_k$ using the Arnoldi method that get progressively closer to the exact solution. In fact, if exact arithmetic is used, the exact solution – with zero residual – is obtained when $k = N$. In practice, it is implemented as an iterative method that is terminated when the residual is sufficiently small.

To see how this is the case, let us write successive approximations of the solution as a linear combination of the orthonormal basis vectors

$$\mathbf{u}_k = a_1\mathbf{q}_1 + a_2\mathbf{q}_2 + \cdots + a_i\mathbf{q}_i + \cdots + a_k\mathbf{q}_k = \mathbf{Q}_k\mathbf{a}_k, \tag{10.44}$$

where $i = 1,2,\ldots,k$ and $k = 1,2,\ldots,N$, and

$$\mathbf{Q}_k = \begin{bmatrix} \vdots & \vdots & & \vdots & & \vdots \\ \mathbf{q}_1 & \mathbf{q}_2 & \cdots & \mathbf{q}_i & \cdots & \mathbf{q}_k \\ \vdots & \vdots & & \vdots & & \vdots \end{bmatrix}$$

is an orthogonal matrix. Here, the vectors $\mathbf{q}_i, i = 1,2,\ldots,k$ are the orthonormal basis vectors of the Krylov subspace

$$\mathcal{K}_k(\mathbf{A},\mathbf{q}_1) = \text{span}\left\{\mathbf{q}_1, \mathbf{A}\mathbf{q}_1, \mathbf{A}^2\mathbf{q}_1, \ldots, \mathbf{A}^{k-1}\mathbf{q}_1\right\},$$

which could be obtained using Gram–Schmidt orthogonalization, for example. In terms of the orthogonal matrix $\mathbf{Q}_k$ and vector of coefficients $\mathbf{a}_k$, the objective function (10.43) at step k can be written in the form

$$J_{GMRES} = \|\mathbf{r}_k\|^2 = \|\mathbf{b} - \mathbf{A}\mathbf{Q}_k\mathbf{a}_k\|^2.$$

Recall from Section 6.5 the similarity transformation $\mathbf{A} = \mathbf{Q}\mathbf{H}\mathbf{Q}^T$, where $\mathbf{H}$ is a Hessenberg matrix; therefore, $\mathbf{A}\mathbf{Q} = \mathbf{Q}\mathbf{H}$, and we have

$$J_{GMRES} = \|\mathbf{r}_k\|^2 = \|\mathbf{b} - \mathbf{Q}_k\mathbf{H}_k\mathbf{a}_k\|^2.$$

Also recall from Section 2.4 that the L_2-norm (length) of a vector is not affected by multiplication by an orthogonal matrix. Therefore, this objective function can be written

$$J_{GMRES} = \|\mathbf{r}_k\|^2 = \|\mathbf{Q}_k^T\mathbf{b} - \mathbf{Q}_k^T\mathbf{Q}_k\mathbf{H}_k\mathbf{a}_k\|^2,$$

but $\mathbf{Q}^T\mathbf{Q} = \mathbf{I}$ owing to orthogonality, in which case

$$J_{GMRES} = \|\mathbf{r}_k\|^2 = \|\mathbf{Q}_k^T\mathbf{b} - \mathbf{H}_k\mathbf{a}_k\|^2. \tag{10.45}$$

Now consider the matrix-vector product

$$\mathbf{Q}_k^T\mathbf{b} = \begin{bmatrix} \cdots & \mathbf{q}_1^T\mathbf{b} & \cdots \\ \cdots & \mathbf{q}_2^T\mathbf{b} & \cdots \\ & \vdots & \\ \cdots & \mathbf{q}_k^T\mathbf{b} & \cdots \end{bmatrix}.$$

If we set

$$\mathbf{q}_1 = \frac{\mathbf{b}}{\|\mathbf{b}\|}, \tag{10.46}$$

which is the normalized right-hand-side vector, then because all of the remaining orthonormal vectors $\mathbf{q}_j$ are orthogonal to $\mathbf{q}_1$, we have that $\mathbf{q}_j^T\mathbf{b} = 0$ for $j \geq 2$, and

$$\mathbf{Q}_k^T\mathbf{b} = \frac{\mathbf{b}^T\mathbf{b}}{\|\mathbf{b}\|}\mathbf{e}_k = \frac{\|\mathbf{b}\|^2}{\|\mathbf{b}\|}\mathbf{e}_k = \|\mathbf{b}\|\mathbf{e}_k,$$

where

$$\mathbf{e}_k = \begin{bmatrix} 1 \\ 0 \\ \vdots \\ 0 \end{bmatrix}$$

is a $k \times 1$ vector. Then the objective function (10.45) becomes

$$J_{GMRES} = \|\mathbf{r}_k\|^2 = \big\| \|\mathbf{b}\|\mathbf{e}_k - \mathbf{H}_k\mathbf{a}_k \big\|^2. \tag{10.47}$$

How is the least-squares problem for the objective function (10.47) better than that in (10.43)? Observe that in (10.43), the matrix $\mathbf{A}$ is $N \times N$ and the right-hand-side vector $\mathbf{b}$ is $N \times 1$. However, in (10.47), the Hessenberg matrix $\mathbf{H}_k$ is only $k \times k$ and the vector $\mathbf{e}_k$ is $k \times 1$. At each step k of the algorithm, only a $k \times k$ least-squares problem needs to be solved, and typically the iterations will terminate for $k \ll N$, particularly for sparse matrices.

Not only is the least-squares problem significantly smaller, it is particularly straightforward to solve. Recall that we said we would minimize the residual objective function by solving the least-squares problem. Let us see how straightforward this is for (10.47). If we had the solution of the system of linear algebraic equations

$$\mathbf{H}_k\mathbf{a}_k = \|\mathbf{b}\|\mathbf{e}_k, \tag{10.48}$$

the residual would be zero, and the objective function would be minimized for each $\mathbf{H}_k$. Normally, we are seeking to minimize the residual in order to avoid solving such a system of algebraic equations. Here, however, the Hessenberg coefficient matrix only has one nonzero lower diagonal; therefore, zeroing this lower diagonal would produce a system with a right triangular matrix that can be solved by simple backward substitution. This can be accomplished using the QR decomposition of the Hessenberg matrix

$$\mathbf{H}_k = \mathbf{Q}_k \mathbf{R}_k,$$

where $\mathbf{R}$ is a right triangular matrix. Substituting into equation (10.48) and premultiplying $\mathbf{Q}_k^T$ gives

$$\mathbf{Q}_k^T \mathbf{Q}_k \mathbf{R}_k \mathbf{a}_k = \|\mathbf{b}\| \mathbf{Q}_k^T \mathbf{e}_k,$$

but $\mathbf{Q}_k^T \mathbf{Q}_k = \mathbf{I}_k$; thus, we have the system of equations

$$\mathbf{R}_k \mathbf{a}_k = \|\mathbf{b}\| \mathbf{Q}_k^T \mathbf{e}_k. \tag{10.49}$$

Because $\mathbf{R}_k$ is right triangular, this system can be solved using simple backward substitution to produce the solution $\mathbf{a}_k$ that minimizes the residual objective function. Once this least-squares problem is solved for the coefficients $\mathbf{a}_k$ in the linear combination (10.44), the kth approximation of the solution to the original system of algebraic equations is given by

$$\mathbf{u}_k = \mathbf{Q}_k \mathbf{a}_k. \tag{10.50}$$

While typical iterative methods require the same amount of computational effort during each iteration, Arnoldi-based methods, such as GMRES, are such that each iteration requires solution of a matrix problem that increases in size by one each time. Therefore, more computational effort, both operations and storage, is expended each successive iteration. One common remedy is to restart the Arnoldi iteration after a certain number of steps using the improved initial guess as the starting value in building the basis vectors for the new Krylov subspace. While reducing computational and memory requirements, this approach can lead to convergence problems.

The GMRES algorithm is very similar to the Arnoldi method used to obtain the eigenvalues of a matrix. However, instead of computing the eigenvalues of a Hessenberg matrix at each iteration, a least-squares solution of the Hessenberg matrix is obtained in order to minimize the residual objective function as described earlier. In the GMRES algorithm, we form the orthonormal projection matrix $\mathbf{Q}$ using the following step-by-step direct method that produces a Hessenberg matrix $\mathbf{H}$, the least-squares solution of which approximates the solution of the linear system of algebraic equations $\mathbf{A}\hat{\mathbf{u}} = \mathbf{b}$:

1. Specify the starting Arnoldi vector $\mathbf{q}_0 = \mathbf{b}$.
2. Normalize: $\mathbf{q}_1 = \mathbf{q}_0 / \|\mathbf{q}_0\|$.
3. Set $\mathbf{Q}_1 = \mathbf{q}_1$.
4. Set $k = 1$.
5. Do while $\|\mathbf{r}_k\| < \text{tolerance}$ and $k < N$:
   (i) Increment $k = k + 1$.
   (ii) Multiply $\mathbf{q}_k = \mathbf{A}\mathbf{q}_{k-1}$.
   (iii) Orthonormalize $\mathbf{q}_k$ against $\mathbf{q}_1, \mathbf{q}_2, \ldots, \mathbf{q}_{k-1}$.
   (iv) Append $\mathbf{q}_k$ to $\mathbf{Q}_{k-1}$ to form $\mathbf{Q}_k$.

(v) Form the Hessenberg matrix $\mathbf{H}_k = \mathbf{Q}_k^T \mathbf{A} \mathbf{Q}_k$.
(vi) Use least-squares with a QR decomposition of $\mathbf{H}_k$ to determine the vector $\mathbf{a}_k$ that minimizes $J_{GMRES} = \left\| \|\mathbf{b}\| \mathbf{e}_k - \mathbf{H}_k \mathbf{a}_k \right\|^2$.
(vii) Compute an approximate solution $\mathbf{u}_k = \mathbf{Q}_k \mathbf{a}_k$.

6. End do.

At each step k, an $N \times k$ orthogonal matrix $\mathbf{Q}_k$ is produced, the columns of which form an orthonormal basis for the Krylov subspace $\mathcal{K}_k(\mathbf{A}, \mathbf{q}_0)$. Using the projection matrix $\mathbf{Q}_k$, we transform $\mathbf{A}$ to produce a $k \times k$ Hessenberg matrix $\mathbf{H}_k$, which is an orthogonal projection of $\mathbf{A}$ onto the Krylov subspace $\mathcal{K}_k$. As for the Arnoldi method, if $\mathbf{A}$ is symmetric, then the Hessenberg matrix $\mathbf{H}$ is actually a tridiagonal matrix, thereby making the calculations somewhat easier and more efficient.

Observe that the only requirement for GMRES to work is that the coefficient matrix $\mathbf{A}$ be nonsingular, and there is no requirement for diagonal dominance as with traditional iterative methods. Although the GMRES method can be applied to any nonsingular system of equations, however, the Arnoldi algorithm is most efficient for sparse matrices. Therefore, GMRES is particularly efficient for solving large systems of linear algebraic equations with sparse coefficient matrices.

10.8 Summary of Krylov-Based Methods

The Arnoldi, conjugate-gradient, and GMRES methods are unified within the framework of Krylov subspaces. Recall from (10.26) and (10.43) that for the conjugate-gradient and GMRES methods, the residual for the initial guess $\mathbf{u}_0$ is defined by

$$\mathbf{r}_0 = \mathbf{b} - \mathbf{A}\mathbf{u}_0.$$

The approximate solution at the kth step of the algorithm then lies in the vector space given by

$$\mathbf{u}_0 + \mathcal{K}_k(\mathbf{A}, \mathbf{r}_0),$$

where the Krylov subspace is defined by

$$\mathcal{K}_k(\mathbf{A}, \mathbf{r}_0) = \text{span}\left\{\mathbf{r}_0, \mathbf{A}\mathbf{r}_0, \mathbf{A}^2\mathbf{r}_0, \ldots, \mathbf{A}^{k-1}\mathbf{r}_0\right\}.$$

The conjugate-gradient and GMRES methods correspond to imposing different conditions on the residual. In the conjugate-gradient method, the residual is orthogonal to the Krylov subspace, whereas for the GMRES method, the residual is minimized over the Krylov subspace. Therefore, we can summarize Krylov-based algorithms as in Table 10.2.

It is not possible to formally compare the convergence rates of the conjugate-gradient and GMRES methods with the standard iterative methods, such as Jacobi, Gauss–Seidel, and SOR discussed in Section 6.4. In practice, however, they are found to be quite computationally efficient, and their advantages really shine when applied to systems with large sparse coefficient matrices. In addition, the more sparse the

Table 10.2 Krylov-based methods for solving systems of linear algebraic equations and algebraic eigenproblems.

To solve **systems of linear algebraic equations**:		
Form of matrix **A**	Krylov-based method	Section
Symmetric, positive-definite	Conjugate-gradient	10.6
Nonsymmetric	GMRES	10.7

To solve **algebraic eigenproblems**:		
Form of matrix **A**	Krylov-based method	Section
Symmetric	Lanczos	6.5.3
Nonsymmetric	Arnoldi	6.5.3

coefficient matrix **A** is, the smaller k can be to give an accurate approximation of the solution $\hat{\mathbf{u}}$. See Trefethen and Bau (1997), Demmel (1997), and Golub and Van Loan (2013) for further discussion of the conjugate-gradient, Arnoldi, and GMRES methods as well as preconditioning techniques that can be used to speed iterative convergence by reducing the number of iterations required for ill-conditioned matrices.

Exercises

Unless stated otherwise, perform the exercises using hand calculations. Exercises to be completed using "built-in functions" should be completed using the built-in functions within Python, MATLAB, or Mathematica for the vector and matrix operation(s).

10.1 Consider the following three equations for two unknowns:

$$\begin{aligned} 2u_1 &= 1, \\ -u_1 + u_2 &= 0, \\ 2u_2 &= -1. \end{aligned}$$

(a) Directly determine the least-squares solution of this overdetermined system by minimizing the square of the norm of the residual for this system of linear algebraic equations (do not use the general results from the least-squares method)

$$\begin{aligned} J(\mathbf{u}) = \|\mathbf{r}\|^2 = \|\mathbf{b} - \mathbf{A}\mathbf{u}\|^2 &= \sum_{i=1}^{M} \left(b_i - \mathbf{a}_i^T \mathbf{u}\right)^2 \\ &= (1 - 2u_1)^2 + (0 + u_1 - u_2)^2 + (-1 - 2u_2)^2, \end{aligned}$$

where $\mathbf{a}_i, i = 1, \ldots, M$ are the row vectors of the $M \times N$ matrix **A**.

(b) Show that the least-squares solution from part (a) does not satisfy the system of equations $\mathbf{A}\hat{\mathbf{u}} = \mathbf{b}$ by calculating the residual vector

$$\mathbf{r} = \mathbf{b} - \mathbf{A}\mathbf{u},$$

and the value of the objective function

$$J(\mathbf{u}) = \|\mathbf{r}\|^2.$$

10.2 Recall that the least-squares method for overdetermined systems of linear algebraic equations applies to square systems having a unique solution as well. Obtain the unique solution for the 3×3 system of linear algebraic equations considered in Examples 1.4, 1.8, 1.9, and 2.3 using the least-squares method.

10.3 Consider the matrix and vector

$$\mathbf{A} = \begin{bmatrix} 2 & 1 \\ 1 & 2 \\ 1 & 1 \end{bmatrix}, \quad \mathbf{b} = \begin{bmatrix} 3 \\ 3 \\ 1 \end{bmatrix}.$$

(a) Confirm that the system of linear algebraic equations $\mathbf{A}\hat{\mathbf{u}} = \mathbf{b}$ is inconsistent and has no solution(s).

(b) Obtain the least-squares solution for the overdetermined system $\mathbf{A}\mathbf{u} = \mathbf{b}$.

(c) Check your result for part (b) using built-in functions for the QR decomposition and the SVD of $\mathbf{A}$.

10.4 Solve the overdetermined system in Exercise 1.18 using the least-squares method.

10.5 Recall the truss structure in Example 1.2, which resulted in ten equations for seven unknowns; therefore, it is an overdetermined system. However, we expect a unique solution for such a truss problem. Reconcile this apparent discrepancy, and determine the unique solution using the least-squares method. Use built-in functions to carry out the steps.

10.6 Consider the least-squares problem $\mathbf{A}\mathbf{u} = \mathbf{b}$ for the special case when $N = 1$, in which case matrix $\mathbf{A} = \mathbf{a}$ is an $M \times 1$ vector. In this case, the solution u is a scalar. Determine the general solution for $\mathbf{a}u = \mathbf{b}$ using the least-squares method, and show that it is given by

$$u = \frac{\langle \mathbf{a}, \mathbf{b} \rangle}{\|\mathbf{a}\|^2} = \frac{\|\mathbf{b}\| \cos\theta}{\|\mathbf{a}\|},$$

where θ is the angle between the vectors $\mathbf{a}$ and $\mathbf{b}$.

10.7 Consider the least-squares problem $\mathbf{A}\mathbf{u} = \mathbf{b}$ for the special case when the $M \times N$ $(M > N)$ matrix $\mathbf{A}$ has orthonormal column vectors, in which case we denote it by $\mathbf{A} = \mathbf{Q}$. Determine the general solution for $\mathbf{Q}\mathbf{u} = \mathbf{b}$ using the least-squares method. Observe how much less work this is to compute than the general approach given by (10.6).

10.8 In the standard least-squares method, each equation in the system is weighted equally. In some cases, one may have more confidence in the validity of some equations than others; this information can be incorporated by having larger weights associated with those equations for which confidence is high and lower weights for those for which confidence is low. The objective function in the standard least-squares method is defined by

$$J(\mathbf{u}) = \|\mathbf{r}\|^2 = \|\mathbf{b} - \mathbf{A}\mathbf{u}\|^2 = \sum_{i=1}^{M} \left(b_i - \mathbf{a}_i^T \mathbf{u}\right)^2,$$

where $\mathbf{a}_i, i = 1, \ldots, M$ are the row vectors of the $M \times N$ matrix $\mathbf{A}$. In the weighted least-squares method, positive weights $w_i, i = 1, \ldots, M$ for each of the M equations are incorporated into the objective function as follows:

$$J_w(\mathbf{u}) = \sum_{i=1}^{M} w_i \left(b_i - \mathbf{a}_i^T \mathbf{u}\right)^2.$$

Show that the objective function for the weighted least-squares method can be written in the matrix form

$$J_w(\mathbf{u}) = \|\mathbf{W}\left(\mathbf{b} - \mathbf{A}\mathbf{u}\right)\|^2,$$

where $\mathbf{W}$ is a diagonal matrix with the weights $w_i, i = 1, \ldots, M$ along its diagonal. Therefore, the weighted least-squares method is equivalent to a standard least-squares method of the form

$$J_w(\mathbf{u}) = \|\mathbf{d} - \mathbf{B}\mathbf{u}\|^2,$$

where $\mathbf{B} = \mathbf{W}\mathbf{A}$ and $\mathbf{d} = \mathbf{W}\mathbf{b}$. Modify the least-squares result (10.6) for the weighted case.

10.9 Consider the generalized least-squares problem for the objective function

$$J(\mathbf{u}) = \|\mathbf{b} - \mathbf{A}\mathbf{u}\|^2 + \mathbf{c}^T \mathbf{u} + \mathbf{d},$$

where $\mathbf{A}$, $\mathbf{b}$, $\mathbf{c}$, and $\mathbf{d}$ are known. Determine the least-squares solution for $\mathbf{u}$ and compare with the result (10.6) for the standard least-squares problem.

10.10 The focus in this chapter has been on least-squares solutions of linear algebraic systems of equations. Let us consider the algebraic eigenproblem $\mathbf{A}\hat{\mathbf{u}} = \lambda\hat{\mathbf{u}}$, where $\mathbf{A}$ is real and symmetric. Suppose that we have approximations of the eigenvectors $\mathbf{u}$ from some numerical algorithm, and we seek the associated eigenvalues λ. Because the eigenvectors are only approximate, they do not satisfy the eigenproblem exactly. However, similar to the least-squares method applied to linear systems of equations, let us define an objective function as the residual defined by

$$J(\lambda) = \|\mathbf{r}_\lambda\|^2 = \|\mathbf{A}\mathbf{u} - \lambda\mathbf{u}\|^2.$$

For the exact eigenvalues and eigenvectors, the residual would be zero. For the approximate eigenvectors, we seek to minimize the objective function using the least-squares method to obtain the eigenvalues. Show that minimizing the objective function $J(\lambda)$ results in the estimate for the eigenvalues

$$\lambda = \frac{\mathbf{u}^T \mathbf{A}\mathbf{u}}{\mathbf{u}^T \mathbf{u}}.$$

This is called the *Rayleigh quotient*. Because we would like to apply the result even when exact eigenvectors are not known, do not use the original eigenproblem $\mathbf{A}\hat{\mathbf{u}} = \lambda\hat{\mathbf{u}}$ in your derivation. It is straightforward to show that the Rayleigh quotient holds for exact eigenvalues and eigenvectors; its real value is in the fact that the least-squares method shows that it also holds for approximations of the eigenvectors, thereby leading to corresponding approximations of the eigenvalues.

10.11 Recall that in the constrained least-squares problem, we seek to minimize the residual

$$J(\mathbf{u}) = \|\mathbf{r}\|^2 = \|\mathbf{b} - \mathbf{A}\mathbf{u}\|^2$$

subject to the constraint

$$\mathbf{C}\mathbf{u} = \mathbf{d}.$$

Let us consider the special case when $\mathbf{A} = \mathbf{I}$, in which case the objective is to minimize the distance between $\mathbf{u}$ and $\mathbf{b}$ as follows:

$$J(\mathbf{u}) = \|\mathbf{b} - \mathbf{u}\|^2.$$

Show that the KKT equations simplify to give the least-squares solution

$$\mathbf{u} = \mathbf{b} + \mathbf{C}^T \left(\mathbf{C}\mathbf{C}^T\right)^{-1} (\mathbf{d} - \mathbf{C}\mathbf{b}).$$

Hint: Follow the logic in Section 10.2.2 for the underdetermined case.

10.12 Throughout the text, we have considered the system of linear algebraic equations $\mathbf{A}\hat{\mathbf{u}} = \mathbf{b}$, where $\mathbf{A}$ and $\mathbf{b}$ are known, and $\hat{\mathbf{u}}$ is sought. Let us instead imagine that we have a system with an unknown behavior as represented by the matrix $\mathbf{A}$. Instead, we have a series of input data $\mathbf{u}_i, i = 1, \ldots, N$ and the corresponding output data $\mathbf{b}_i, i = 1, \ldots, N$ for the system, and we seek the system matrix $\mathbf{A}$. In a least-squares context, this could be formulated as seeking the matrix $\mathbf{A}$ that minimizes the objective function

$$J(\mathbf{A}) = \sum_{i=1}^{N} \|\mathbf{b}_i - \mathbf{A}\mathbf{u}_i\|^2 = \|\mathbf{B} - \mathbf{A}\mathbf{U}\|^2,$$

where the known input and output vectors $\mathbf{u}_i$ and $\mathbf{b}_i$ are the columns of matrices $\mathbf{U}$ and $\mathbf{B}$, respectively. Show that the least-squares solution is given by

$$\mathbf{A} = \mathbf{B}\mathbf{U}^+.$$

11 Data Analysis: Curve Fitting and Interpolation

> Can one think that because we are engineers, beauty does not preoccupy us or that we do not try to build beautiful, as well as solid and long-lasting, structures? Aren't the genuine functions of strength always in keeping with unwritten conditions of harmony? (Gustave Eiffel)

Science and engineering are all about data that quantify the relationships between various physical quantities of interest. Such data may be obtained experimentally (empirically) or numerically. The data can then be used to develop, refine, and confirm mathematical models of physical phenomena. In many cases, the goal is to develop a simple functional representation of a complex or noisy data set; this is the goal of *curve fitting*. In other instances, the objective is to provide a means to infer data for values of the variables that have not been directly measured; this is accomplished using *interpolation*. More specifically, interpolation involves estimating values between known data points,[1] and curve fitting is used to determine the "best-fit" curve representing a discrete data set. In general, interpolation is preferred when one has a high degree of confidence in the accuracy of the data, whereas curve fitting is the typical choice when the data are noisy and inaccurate. Curve fitting and interpolation in their various forms are considered in turn, and it is shown how they can be framed as overdetermined systems that can be solved using the least-squares method introduced in Chapter 10.

11.1 Linear Regression

When analyzing the data obtained from real systems, there is an underlying assumption that the actual behavior of the system conforms to some prescribed functional behavior. When we solve differential equations governing physical systems, the solution provides this functional form, which is adhered to exactly. When we take experimental measurements of such a system, however, we introduce experimental noise and measurement errors into the data. In order to extract the underlying functional behavior from such a set of data requires some form of a curve fit that best represents the data. The most common form of such a curve fitting is linear regression, in which the data are assumed to exhibit a linear behavior.

[1] Estimating values outside the range of the known data points is known as *extrapolation*.

11.1.1 Linear Least-Squares Regression

When it is believed that the underlying behavior of a system or data set is linear, the line that best represents the data set is sought. Only two points are required to uniquely define a straight line; any additional points result in an overdetermined system having too many points to obtain a unique line, as discussed in Section 10.2.1. Therefore, we formulate a least-squares optimization problem that computes the best-fit line to the data points. This best-fit *trial function* is given by the equation

$$f(x) = a_0 + a_1 x, \tag{11.1}$$

such that the basis functions are 1 and x. For example, suppose that we have the $N = 19$ data points illustrated in Figure 11.1 identified by their x and y coordinates $(x_i, y_i), i = 1, \ldots, N$. We seek the coefficients a_0 and a_1 in the trial function that best fits the data.

In order to quantify the error between the trial function (11.1) and the data points, the *residual* is defined for each data point (x_i, y_i) as follows:

$$r_i = y_i - f(x_i) = y_i - (a_0 + a_1 x_i), \quad i = 1, \ldots, N, \tag{11.2}$$

where $f(x)$ is the best-fit line being sought. This is illustrated in Figure 11.2. Note that implicit in this definition of the residual is that there are only errors in the dependent variable y, not the independent variable x. Using the individual residuals for each point, the total error can be defined as

$$E = \sum_{i=1}^{N} r_i^2 = \sum_{i=1}^{N} \left[y_i - f(x_i)\right]^2 = \sum_{i=1}^{N} \left[y_i - (a_0 + a_1 x_i)\right]^2,$$

such that the error is the sum of the squares of the residuals. The residuals are squared so that positive and negative residuals do not cancel one another and are treated the

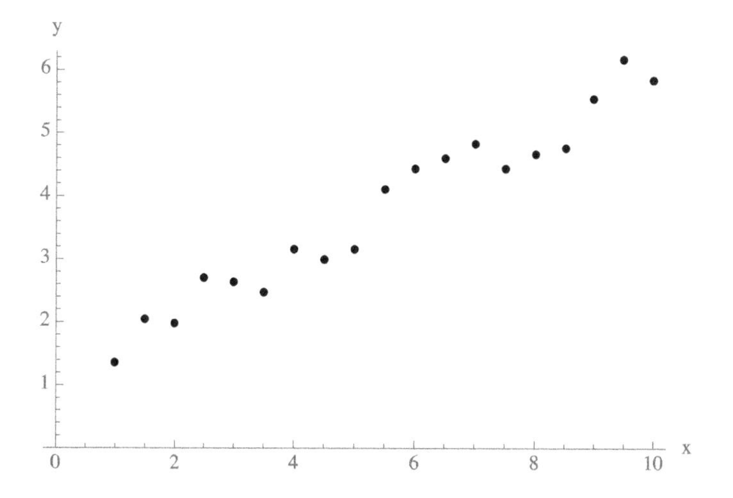

Figure 11.1 $N = 19$ data points.

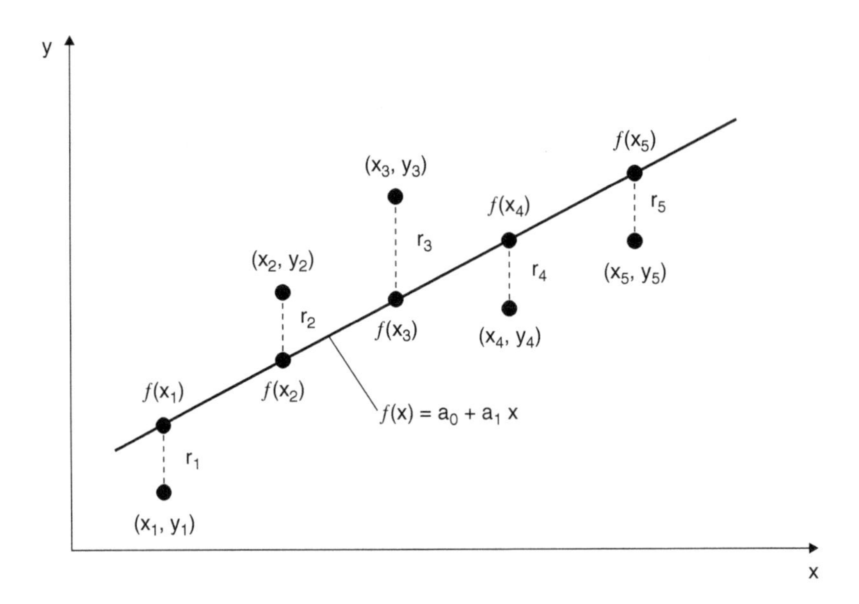

Figure 11.2 Data points, residuals, and best-fit line in linear regression.

same.[2] The best-fit line is given by the constants a_0 and a_1 that minimize the total error E. Therefore, it is referred to as *linear least-squares regression* as we are minimizing the squares of the residuals ("least squares") in order to produce the best-fit line ("linear").

To minimize an algebraic function, we differentiate with respect to the variable(s) being sought and set equal to zero. Note that the data points $(x_i, y_i), i = 1, \ldots, N$ are known, and we seek the coefficients a_0 and a_1 in the best-fit line. Thus, we have

$$\frac{\partial E}{\partial a_0} = -2\sum_{i=1}^{N}\left[y_i - a_0 - a_1 x_i\right] = 0,$$

$$\frac{\partial E}{\partial a_1} = -2\sum_{i=1}^{N}\left[y_i - a_0 - a_1 x_i\right]x_i = 0.$$

Note that $\sum_{i=1}^{N} a_0 = N a_0$; therefore, these two equations can be rewritten in the form

$$\begin{aligned} N a_0 + \left(\sum_{i=1}^{N} x_i\right) a_1 &= \sum_{i=1}^{N} y_i, \\ \left(\sum_{i=1}^{N} x_i\right) a_0 + \left(\sum_{i=1}^{N} x_i^2\right) a_1 &= \sum_{i=1}^{N} x_i y_i, \end{aligned}$$

[2] Squaring the residual is the lowest power that eliminates the influence of the sign. The absolute value would also render the error positive, but the absolute value is not continuous and would lead to difficulties in developing an algorithm.

which is a 2×2 system of linear algebraic equations for the coefficients a_0 and a_1. It can be written more compactly in matrix form as

$$\begin{bmatrix} N & S_x \\ S_x & S_{xx} \end{bmatrix} \begin{bmatrix} a_0 \\ a_1 \end{bmatrix} = \begin{bmatrix} S_y \\ S_{xy} \end{bmatrix},$$

where

$$S_x = \sum_{i=1}^{N} x_i, \quad S_y = \sum_{i=1}^{N} y_i, \quad S_{xy} = \sum_{i=1}^{N} x_i y_i, \quad S_{xx} = \sum_{i=1}^{N} x_i^2.$$

Solving this 2×2 system of linear algebraic equations, which are often referred to as the *normal equations*, gives the coefficients in the linear trial function $f(x) = a_0 + a_1 x$ as

$$a_0 = \frac{S_{xx} S_y - S_{xy} S_x}{N S_{xx} - (S_x)^2}, \tag{11.3}$$

and

$$a_1 = \frac{N S_{xy} - S_x S_y}{N S_{xx} - (S_x)^2}. \tag{11.4}$$

The quality of the least-squares curve fit is given by the *correlation*, or *cross-correlation, coefficient*, which is given by

$$R = \frac{N S_{xy} - S_x S_y}{\sqrt{N S_{xx} - (S_x)^2}\sqrt{N S_{yy} - (S_y)^2}}, \tag{11.5}$$

where

$$S_{yy} = \sum_{i=1}^{N} y_i^2.$$

A perfect correlation, in which all of the data points lie exactly on the linear regression line, corresponds to $R = 1$, and the closer to one the better.

For our sample data exhibited in Figure 11.1, the best-fit line calculated using (11.3) and (11.4) is given by

$$f(x) = 1.24508 + 0.440729x,$$

and superimposed on the data in Figure 11.3. The correlation coefficient for this case is given by $R = 0.982941$, which signifies a very good fit to the data.

REMARKS:

1. *Note that the residual is defined by the vertical distance from the data point to the linear regression line, not the perpendicular distance. This reflects the assumption that the errors in the data are in the dependent variable y, not the independent variable x.*
2. *The normal equations are* symmetric *and, therefore,* normal *in the sense of Section 2.4; this is why they are called the* normal equations.

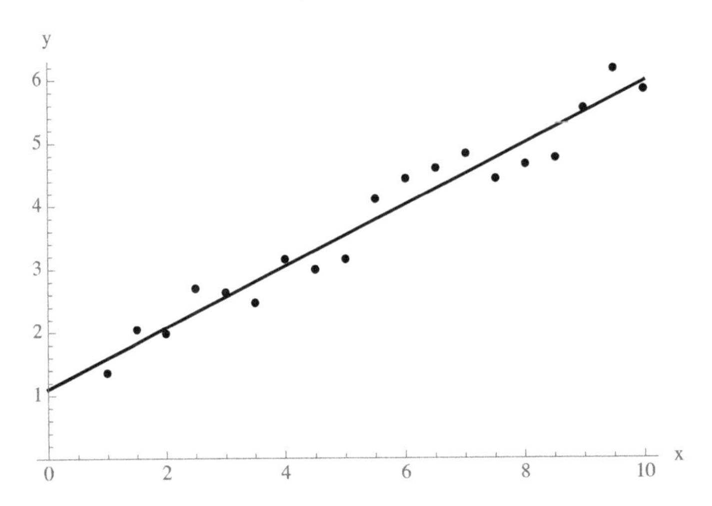

Figure 11.3 Best-fit linear regression line superimposed on data points shown in Figure 11.1.

3. *Linear (and polynomial) least-squares regression can be considered an optimization problem where the total error E is the objective function to be minimized to determine the best-fit line that provides the "optimal" curve fit.*
4. *See Section 11.3 for an alternative way to formulate linear regression problems directly as an overdetermined system of equations as in Section 10.2.1. In addition, see Section 11.8, which shows how to perform this same linear regression using just the largest singular value from a singular-value decomposition.*

11.1.2 Linearizing Certain Nonlinear Data

Obviously, not all data are well represented by a linear relationship. In some cases, however, we can represent nonlinear equations as linear ones, thereby allowing us to use linear regression. Consider, for example, the data shown in Figure 11.4. Looking at the data leads one to think that the underlying functional behavior may correspond to an exponential function. Consider an exponential relationship of the form

$$y = c_0 e^{c_1 x}. \tag{11.6}$$

Taking the natural logarithm of both sides gives

$$\ln y = \ln\left(c_0 e^{c_1 x}\right) = \ln c_0 + \ln\left(e^{c_1 x}\right),$$

which yields

$$\ln y = \ln c_0 + c_1 x. \tag{11.7}$$

Thus, a plot of $\ln y$ versus x would be a straight line with slope c_1 and intercept $\ln c_0$. Consequently, linear regression could be used to obtain c_1 and $\ln c_0$ for the best-fit line on a $\ln y$ versus x plot.

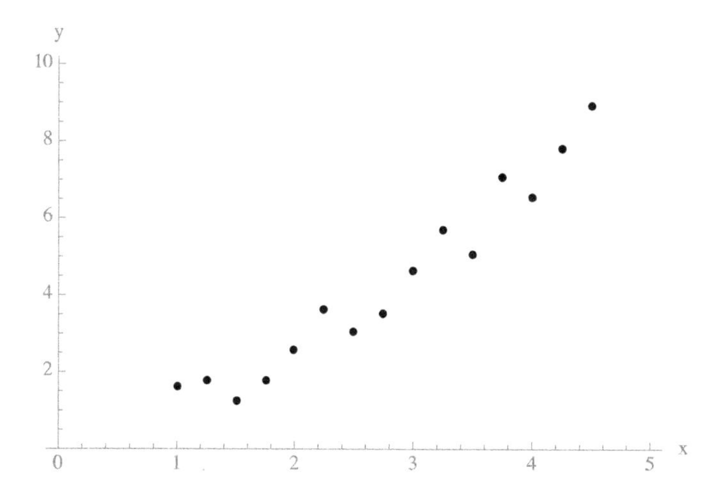

Figure 11.4 $N = 17$ data points.

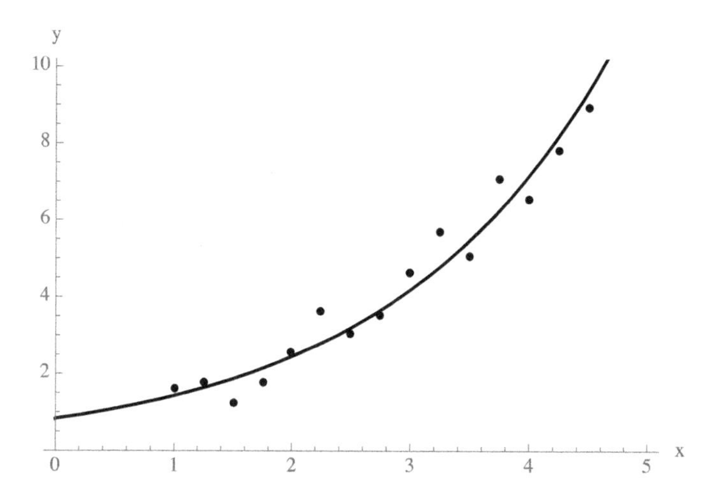

Figure 11.5 Exponential curve fit to the data points shown in Figure 11.4.

For the data given in Figure 11.4, the best-fit exponential curve is shown in Figure 11.5 and is given by

$$y = 1.09468e^{0.48215x}.$$

Observe from Figure 11.6 that the curve fit is a straight line when plotted on a $\ln y$ versus x plot.

Similarly, suppose that the exponential relationship is of the form

$$y = c_0 10^{c_1 x}. \tag{11.8}$$

Taking the base-10 logarithm of both sides gives

$$\log y = \log\left(c_0 10^{c_1 x}\right) = \log c_0 + \log\left(10^{c_1 x}\right),$$

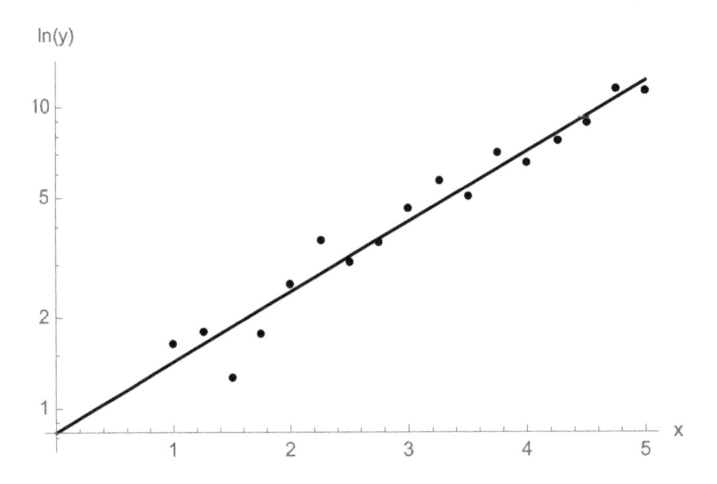

Figure 11.6 Exponential curve fit to the data points shown in Figure 11.4 on a $\ln y$ versus x plot.

which yields

$$\log y = \log c_0 + c_1 x. \tag{11.9}$$

Now a plot of $\log y$ versus x would be a straight line with slope c_1 and intercept $\log c_0$. Consequently, linear regression could be used to obtain c_1 and $\log c_0$ for the best-fit line on a $\log y$ versus x plot.

Finally, suppose that the data are expected to be represented by a power function of the form

$$y = c_0 x^{c_1}. \tag{11.10}$$

Again, taking the natural logarithm of both sides gives

$$\ln y = \ln\left(c_0 x^{c_1}\right) = \ln c_0 + \ln\left(x^{c_1}\right),$$

which yields

$$\ln y = \ln c_0 + c_1 \ln x. \tag{11.11}$$

In this case, a plot of $\ln y$ versus $\ln x$ would be a straight line with slope c_1 and intercept $\ln c_0$. Again, linear regression could be used to obtain c_1 and $\ln c_0$ for the best-fit line on a $\ln y$ versus $\ln x$ plot. Note that base-10 logarithms could also be used in this case.

Example 11.1 Determine the best-fit curve of the form

$$y(x) = c_0 x^{c_1}$$

for the following data:

i	1	2	3	4
x_i	2.0	3.0	4.0	5.0
y_i	4.0	9.0	16.0	25.0

Solution

In order to convert the curve-fit problem into a linear-regression problem, take the natural logarithm to obtain

$$\ln y = \ln c_0 + \ln \left(x^{c_1}\right),$$

or

$$\ln y = \ln c_0 + c_1 \ln x,$$

or

$$\hat{y}(x) = a_0 + a_1 \hat{x},$$

which is now a linear regression problem for the constants a_0 and a_1, where

$$\hat{y} = \ln y, \quad a_0 = \ln c_0, \quad \hat{x} = \ln x, \quad a_1 = c_1.$$

Taking the natural logarithms of the values of x_i and y_i in the provided data yields

i	1	2	3	4
$\hat{x}_i = \ln x_i$	0.6931	1.0986	1.3863	1.6094
$\hat{y}_i = \ln y_i$	1.3863	2.1972	2.7726	3.2189

For linear regression of the data, we need to solve the system of equations

$$\begin{bmatrix} N & \sum \hat{x}_i \\ \sum \hat{x}_i & \sum \hat{x}_i^2 \end{bmatrix} \begin{bmatrix} a_0 \\ a_1 \end{bmatrix} = \begin{bmatrix} \sum \hat{y}_i \\ \sum \hat{x}_i \hat{y}_i \end{bmatrix}.$$

Evaluating each sum yields

$$\begin{bmatrix} 4 & 4.78749 \\ 4.78749 & 6.1995 \end{bmatrix} \begin{bmatrix} a_0 \\ a_1 \end{bmatrix} = \begin{bmatrix} 9.57498 \\ 12.399 \end{bmatrix}.$$

Solving gives the solution (if exact arithmetic is used)

$$a_0 = 0, \quad a_1 = 2;$$

therefore,

$$c_0 = e^{a_0} = 1, \quad c_1 = a_1 = 2.$$

Thus, the best-fit curve to the original data is

$$y(x) = 1x^2,$$

which is, in fact, an exact representation of the original data in this case.

11.2 Polynomial Regression

For situations in which none of the aforementioned curve fits are satisfactory, we may use polynomial least-squares regression, which is a generalization of linear regression. For example, let us consider the data plotted in Figure 11.7. We use the same definition of the residual as before, that is,

$$r_i = y_i - f(x_i), \quad i = 1, \ldots, N, \tag{11.12}$$

but now the trial function $f(x)$ is a polynomial of Mth-order of the form

$$f(x) = a_0 + a_1 x + a_2 x^2 + \cdots + a_M x^M. \tag{11.13}$$

Now the basis functions are integer powers of x of the form $x^0 = 1$, $x^1 = x$, x^2, …, x^M. Note that M is the order of the polynomial, and N is the number of data points. As before, the total error is the sum of the squares of the residuals for each point according to

$$E = \sum_{i=1}^{N} r_i^2 = \sum_{i=1}^{N} \left[y_i - f(x_i)\right]^2 = \sum_{i=1}^{N} \left[y_i - \left(a_0 + a_1 x_i + a_2 x_i^2 + \cdots + a_M x_i^M\right)\right]^2.$$

Again, we seek the constants a_0, a_1, …, a_M that minimize the total error. Thus, differentiating with respect to each of the coefficients and setting equal to zero gives

$$\frac{\partial E}{\partial a_0} = -2\sum_{i=1}^{N} \left[y_i - a_0 - a_1 x_i - a_2 x_i^2 - \cdots - a_M x_i^M\right] = 0,$$

$$\frac{\partial E}{\partial a_1} = -2\sum_{i=1}^{N} \left[y_i - a_0 - a_1 x_i - a_2 x_i^2 - \cdots - a_M x_i^M\right] x_i = 0,$$

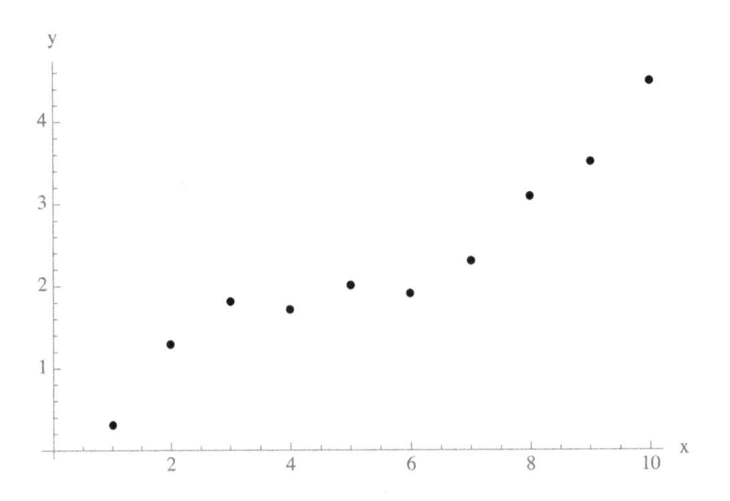

Figure 11.7 $N = 10$ data points.

$$\frac{\partial E}{\partial a_2} = -2\sum_{i=1}^{N}\left[y_i - a_0 - a_1x_i - a_2x_i^2 - \cdots - a_Mx_i^M\right]x_i^2 = 0,$$

$$\vdots$$

$$\frac{\partial E}{\partial a_M} = -2\sum_{i=1}^{N}\left[y_i - a_0 - a_1x_i - a_2x_i^2 - \cdots - a_Mx_i^M\right]x_i^M = 0.$$

This is a set of $M+1$ linear algebraic equations for the $M+1$ unknown constants a_0, $a_1, \ldots, a_M$. In matrix form, it is

$$\begin{bmatrix} N & \sum x_i & \sum x_i^2 & \cdots & \sum x_i^M \\ \sum x_i & \sum x_i^2 & \sum x_i^3 & \cdots & \sum x_i^{M+1} \\ \sum x_i^2 & \sum x_i^3 & \sum x_i^4 & \cdots & \sum x_i^{M+2} \\ \vdots & \vdots & \vdots & \ddots & \vdots \\ \sum x_i^M & \sum x_i^{M+1} & \sum x_i^{M+2} & \cdots & \sum x_i^{2M} \end{bmatrix} \begin{bmatrix} a_0 \\ a_1 \\ a_2 \\ \vdots \\ a_M \end{bmatrix} = \begin{bmatrix} \sum y_i \\ \sum x_i y_i \\ \sum x_i^2 y_i \\ \vdots \\ \sum x_i^M y_i \end{bmatrix}, \quad (11.14)$$

where all of the summations are from $i = 1, \ldots, N$.

For the data displayed in Figure 11.7, the best-fit cubic polynomial, with $M = 3$, is shown in Figure 11.8 and is given by

$$f(x) = -0.693333 + 1.3488x - 0.242716x^2 + 0.0160645x^3.$$

REMARKS:

1. *Clearly, increasing the order of the polynomial increases the size of the matrix to solve for its coefficients.*
2. *For* linear regression, *set $M = 1$.*
3. *For* polynomial regression, *set $M < N - 1$, where N is the number of data points.*
4. *For* polynomial interpolation, *set $M = N - 1$, which is discussed in Section 11.5.*

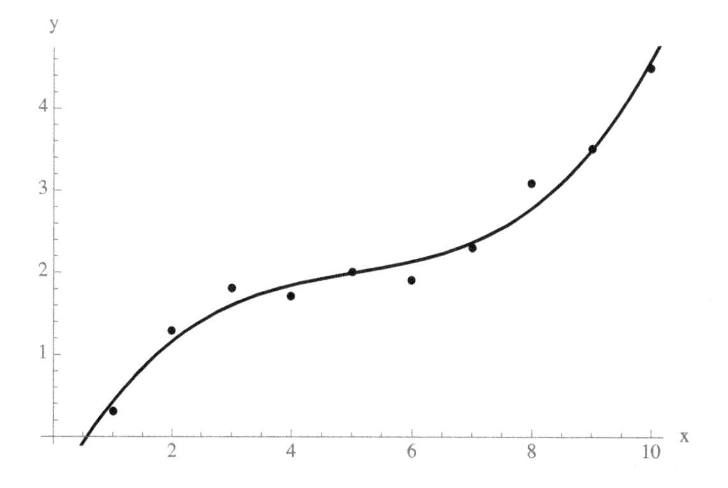

Figure 11.8 Best-fit cubic polynomial to data points shown in Figure 11.7.

11.3 Least-Squares Regression as an Overdetermined System

Let us reconsider the linear-regression problem from the point of view of an overdetermined system of linear algebraic equations and apply the least-squares method of Section 10.2.1. In an ideal situation, each of the N data points would lie on the curve $f(x) = a_0 + a_1 x$, such that

$$\begin{aligned} a_0 + a_1 x_1 &= y_1, \\ a_0 + a_1 x_2 &= y_2. \\ &\vdots \\ a_0 + a_1 x_N &= y_N. \end{aligned}$$

In matrix form, this can be written

$$\begin{bmatrix} 1 & x_1 \\ 1 & x_2 \\ \vdots & \vdots \\ 1 & x_N \end{bmatrix} \begin{bmatrix} a_0 \\ a_1 \end{bmatrix} = \begin{bmatrix} y_1 \\ y_2 \\ \vdots \\ y_N \end{bmatrix}, \tag{11.15}$$

or simply

$$\mathbf{\Phi a} = \mathbf{y}.$$

Observe that the basis functions are contained in the matrix $\mathbf{\Phi}$. Premultiplying by $\mathbf{\Phi}^T$ on both sides gives

$$\mathbf{\Phi}^T \mathbf{\Phi a} = \mathbf{\Phi}^T \mathbf{y},$$

or

$$\begin{bmatrix} 1 & 1 & \cdots & 1 \\ x_1 & x_2 & \cdots & x_N \end{bmatrix} \begin{bmatrix} 1 & x_1 \\ 1 & x_2 \\ \vdots & \vdots \\ 1 & x_N \end{bmatrix} \begin{bmatrix} a_0 \\ a_1 \end{bmatrix} = \begin{bmatrix} 1 & 1 & \cdots & 1 \\ x_1 & x_2 & \cdots & x_N \end{bmatrix} \begin{bmatrix} y_1 \\ y_2 \\ \vdots \\ y_N \end{bmatrix}.$$

Multiplying on the left and right sides yields the 2×2 system of linear algebraic equations

$$\begin{bmatrix} N & \sum x_i \\ \sum x_i & \sum x_i^2 \end{bmatrix} \begin{bmatrix} a_0 \\ a_1 \end{bmatrix} = \begin{bmatrix} \sum y_i \\ \sum x_i y_i \end{bmatrix}$$

for the coefficients a_0 and a_1 in the best-fit line. Observe that this is the same result for the normal equations as obtained by minimizing the total error.

The matrix form (11.15) suggests an alternative technique for obtaining the a_0 and a_1 coefficients in the best-fit line. Note that only when $N = 2$ does (11.15) give a system of algebraic equations that produces a unique solution, for which the line passes through both of the points. For $N > 2$, this system of equations is overdetermined with there being more equations than unknowns. In other words, there is no single line that passes through all of the points in the general case.

Such overdetermined systems were solved in Section 10.2.1 using the least-squares method by minimizing the square of the norm of the residual

$$\mathbf{r} = \mathbf{y} - \mathbf{\Phi a}.$$

From (10.6), the solution was found to be

$$\mathbf{a} = \mathbf{\Phi}^{+}\mathbf{y}, \tag{11.16}$$

where $\mathbf{\Phi}^{+}$ is the pseudo-inverse of $\mathbf{\Phi}$, which is given by

$$\mathbf{\Phi}^{+} = \left(\mathbf{\Phi}^T\mathbf{\Phi}\right)^{-1}\mathbf{\Phi}^T.$$

Generalizing (11.15), the general Mth-order polynomial regression problem is then of the form

$$\begin{bmatrix} 1 & x_1 & x_1^2 & \cdots & x_1^M \\ 1 & x_2 & x_2^2 & \cdots & x_2^M \\ & \vdots & & \ddots & \vdots \\ 1 & x_N & x_N^2 & \cdots & x_N^M \end{bmatrix} \begin{bmatrix} a_0 \\ a_1 \\ \vdots \\ a_M \end{bmatrix} = \begin{bmatrix} y_1 \\ y_2 \\ \vdots \\ y_N \end{bmatrix}, \tag{11.17}$$

which is still of the form

$$\mathbf{\Phi a} = \mathbf{y}.$$

The coefficient matrix in (11.17) is known as a *Vandermonde matrix*. Note that $\mathbf{\Phi}$ is $N \times (M+1)$, $\mathbf{a}$ is $(M+1) \times 1$, and $\mathbf{y}$ is $N \times 1$. As in the linear-regression case, the least-squares polynomial-regression problem is obtained in terms of the pseudo-inverse using equation (11.16).

As in Section 10.2.1, there are three methods for solving the overdetermined least-squares problem (11.17) that arises in polynomial regression:

1. Solve the preceding normal equations via

$$\mathbf{a} = \left(\mathbf{\Phi}^T\mathbf{\Phi}\right)^{-1}\mathbf{\Phi}^T\mathbf{y}.$$

   Because $\mathbf{\Phi}^T\mathbf{\Phi}$ is positive definite, this can be done efficiently using Cholesky decomposition (see Section 6.3.2).
2. Obtain the QR decomposition $\mathbf{\Phi} = \mathbf{QR}$ and evaluate

$$\mathbf{a} = \mathbf{R}^{-1}\mathbf{Q}^T\mathbf{y}.$$

3. Obtain the singular-value decomposition $\mathbf{\Phi} = \mathbf{U\Sigma V}^T$ and evaluate

$$\mathbf{a} = \mathbf{V\Sigma}^{-1}\mathbf{U}^T\mathbf{y}.$$

It turns out that the condition number of $\mathbf{\Phi}$ is such that

$$\kappa\left(\mathbf{\Phi}^T\mathbf{\Phi}\right) = \left[\kappa\left(\mathbf{\Phi}\right)\right]^2.$$

Thus, if the condition number for matrix $\mathbf{\Phi}$ is large, then the corresponding condition number for its product with its transpose is significantly larger. Consequently, the first method is rarely used for large data sets. Instead, QR decomposition is typically the most efficient technique unless the system is ill conditioned, in which case SVD is a better choice to reduce the effects of round-off errors in the calculations. In the case of linear regression, we have the solution of the normal equations in the explicit form (11.3) and (11.4).

Least-squares regression suggests some ideas for improvement in more complex settings involving more complicated systems and larger quantities of data. Specifically, least-squares regression requires us to preselect the basis functions for the curve-fitting procedure. What if the "best" basis functions for developing a simplified model of a system could be determined directly from the data? This is the idea behind proper-orthogonal decomposition (POD) and reduced-order modeling (ROM), which will be considered in detail in Chapter 13. We will also revisit least-squares curve fitting reframed as a state-estimation problem in Section 11.9.

11.4 Least Squares with Orthogonal Basis Functions – Fourier Series

Thus far, our trial functions for curve fitting have been comprised of nonnegative powers of x. In principle, any functions could be used as basis functions. For example, recall from Chapter 3 that Fourier series are particularly convenient to work with given the fact that any piecewise continuous function in a prescribed interval can be expressed to an arbitrary level of accuracy using as many terms as necessary in a Fourier series. Such Fourier series do not suffer from the increasingly oscillatory behavior as observed with higher-order polynomials. In contrast, curve fitting using Fourier series gets increasingly accurate throughout the entire domain as the number of Fourier basis functions increases. They have the added benefit of being mutually orthogonal over the interval $0 \leq x \leq 2\pi$ with respect to the weight function $w(x) = 1$.

While Fourier series were previously obtained as the orthogonal eigenfunctions of certain differential operators, there is no reason why they could not also be used in curve-fitting contexts as well to represent general behavior exhibited by various data sets.[3] Specifically, can truncated Fourier series be used to accurately represent discrete experimental or numerical data? Remarkably, determination of the coefficients in the Fourier series turns out to be a least-squares problem, just as for linear and polynomial regression.

Let us generalize the least-squares method applied previously to polynomial curve fitting to allow for more general basis functions. The residual is defined by

$$r_i = y_i - f(x_i), \quad i = 1, \dots, N, \tag{11.18}$$

[3] The reader may wonder why we focus on Fourier series when other functions, such as Chebyshev and Legendre polynomials or Bessel functions, have similar properties. In fact, we will revisit this issue when we discuss spectral numerical methods in Section 13.2.5.

where $f(x)$ is the trial function, and (x_i, y_i) are the N data points to be curve fit. The total error is defined in a least-squares sense by

$$E = \sum_{i=1}^{N} r_i^2 = \sum_{i=1}^{N} \left[y_i - f(x_i)\right]^2. \tag{11.19}$$

The trial function is of the general form

$$f(x) = \sum_{j=0}^{M} a_j \phi_j(x), \tag{11.20}$$

where for polynomial regression, for example,

$$\phi_j(x) = x^j, \quad j = 0, \ldots, M,$$

and $M = 1$ for linear regression. Substituting the trial function (11.20) into the total error (11.19) gives

$$E = \sum_{i=1}^{N} \left[y_i - \sum_{j=0}^{M} a_j \phi_j(x_i) \right]^2. \tag{11.21}$$

Minimizing the error requires that

$$\frac{\partial E}{\partial a_k} = 0, \quad k = 0, \ldots, M,$$

which results in

$$\frac{\partial E}{\partial a_k} = -2 \sum_{i=1}^{N} \left[y_i - \sum_{j=0}^{M} a_j \phi_j(x_i) \right] \phi_k(x_i) = 0, \quad k = 0, \ldots, M,$$

or

$$\sum_{i=1}^{N} \left[y_i \phi_k(x_i) - \sum_{j=0}^{M} a_j \phi_j(x_i) \phi_k(x_i) \right] = 0, \quad k = 0, \ldots, M. \tag{11.22}$$

These $M + 1$ coupled linear algebraic equations are solved for the coefficients a_j, $j = 0, \ldots, M$ in the trial function. For polynomial basis functions, this results in the normal equations as obtained previously for the coefficients.

If instead of powers of x, trigonometric functions are used for the basis functions, their orthogonality over the interval $0 \le x \le 2\pi$ can be exploited as follows to obtain a particularly straightforward curve-fit solution. If the basis functions $\phi_j(x_i)$ are orthonormal (orthogonal and normalized to length one), in which case

$$\langle \phi_j(x_i), \phi_k(x_i) \rangle = \delta_{j,k},$$

then (11.22) simplifies to

$$\sum_{i=1}^{N} \left[y_i \phi_k(x_i) - a_k \right] = 0, \quad k = 0, \ldots, M.$$

Recall that $\sum_{i=1}^{N} a_k = N a_k$; therefore, this gives an explicit expression for the trial function coefficients in the form

$$a_k = \frac{1}{N} \sum_{i=1}^{N} y_i \phi_k(x_i), \quad k = 0, \ldots, M. \tag{11.23}$$

That is, the normal equations are diagonal and can be solved in an uncoupled fashion without the need to solve a coupled system of equations! This is because of the orthogonality of the basis functions.

For example, say that we have N equally spaced data points x_i in the interval $0 \leq x \leq 2\pi$, and we select basis functions $\phi_k(x)$ that are 2π-periodic of the form $\cos(0x) = 1$, $\cos(x)$, $\cos(2x), \ldots$, $\cos(kx), \ldots$, $\sin(x)$, $\sin(2x), \ldots$, $\sin(kx), \ldots$. After normalizing the already orthogonal basis functions, the trial function is given by

$$f(x) = \frac{a_0}{\sqrt{2\pi}} + \frac{1}{\sqrt{\pi}} \sum_{m=1}^{M} \left[a_m \cos(mx) + b_m \sin(mx) \right]. \tag{11.24}$$

We cannot have more coefficients in the trial function than data points; therefore, the Fourier series (11.24) must be such that $2M+1 \leq N$. Consequently, a discrete Fourier series is essentially a weighted linear combination of sine and cosine functions having various frequencies. From (11.23), the coefficients in the Fourier series (11.24) are then given by

$$\begin{aligned} a_0 &= \frac{1}{N\sqrt{2\pi}} \sum_{i=1}^{N} y_i, \\ a_m &= \frac{1}{N\sqrt{\pi}} \sum_{i=1}^{N} y_i \cos(mx_i), \quad k = 0, \ldots, M, \\ b_m &= \frac{1}{N\sqrt{\pi}} \sum_{i=1}^{N} y_i \sin(mx_i), \quad k = 0, \ldots, M. \end{aligned}$$

Observe that these are simply discrete versions of the continuous Fourier coefficients obtained in Section 3.1.3.

REMARKS:

1. *As with any type of trial function, a Fourier series representation improves as M is increased and more frequencies, or modes, are included in the curve fit. Moreover, Fourier series, do not suffer the plight of polynomials that can become oscillatory as their degree is increased.*
2. *There is no need to solve an entire system of equations for the coefficients in the Fourier series, as with polynomial regression and splines, owing to orthogonality of sines and cosines.*
3. *Once we have a Fourier series representation of a data set, we can use discrete Fourier transforms to quantify the dominant frequencies contained within a data set.*

4. *This approach with Fourier series is the basis for traditional signal and image processing and data compression techniques.*
5. *For much more on continuous and discrete Fourier series and Fourier transforms, including fast Fourier transforms, see Press et al. (2007).*

In many cases, the reason for seeking an accurate curve fit of a data set is in order to perform further analysis on that data. For example, image and signal processing are largely focused on performing Fourier, or spectral, analysis on data sets that are periodic or exhibit oscillations. To illustrate, let us consider a sine wave comprised of two frequencies as follows

$$f(x) = \sin(x) + \frac{3}{4}\sin(6x).$$

The two frequencies, which have different amplitudes, are one and six; this is essentially a two-term Fourier series representation of the function. With $N = 401$ equally spaced points over the total period $0 \leq x \leq 5\pi$, the data are shown in Figure 11.9. Note how superposition of the two frequencies is evident. Taking the fast Fourier transform (FFT) of this discrete data set, which transforms the data to the frequency domain, and performing the inverse Fourier transform reproduces the original data set precisely.

Now let us analyze the frequency information in the original data set, which contains two frequencies. The power spectrum is the square of the modulus of the FFT of the data. It indicates the power contained in each frequency within the data set. Specifically, the power of a discrete distribution is calculated by summing the squares of the Fourier coefficients associated with each frequency. Observe in Figure 11.10 that the power spectrum is symmetric about its midpoint owing to aliasing, and it clearly delineates the two frequencies contained in the data.

In order to plot the power spectrum with respect to frequency, we adjust the abscissa to give Figure 11.11 (and only plot the frequencies $0 < \omega < 10$). Thus, we can

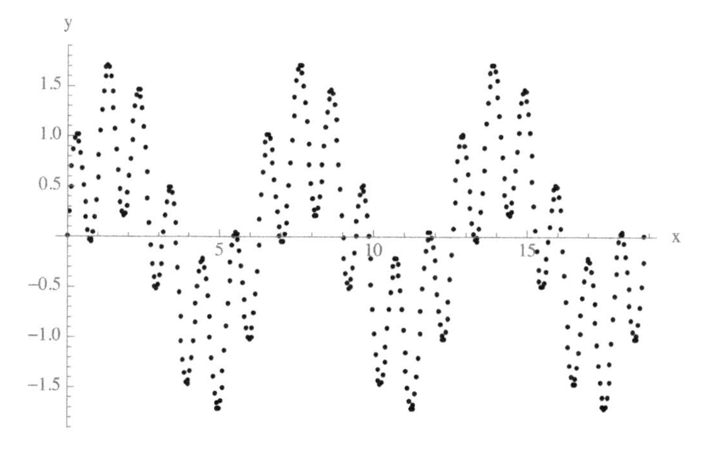

Figure 11.9 Plot of two-frequency wave with $N = 401$ equally spaced points.

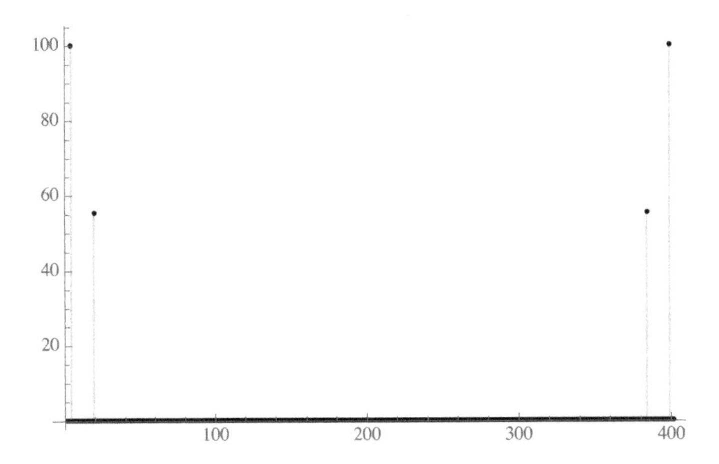

Figure 11.10 Plot of the power spectrum of the two-frequency wave with $N = 401$ equally spaced points with respect to the index: $1, \dots, N$.

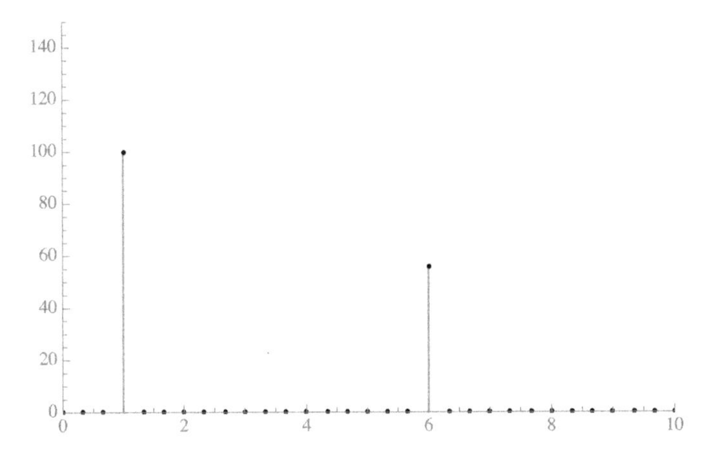

Figure 11.11 Zoomed plot of the power spectrum of the two-frequency wave with respect to frequency.

clearly see that the discrete Fourier transform picks out the two frequencies one and six contained in the original discrete data set and indicates that frequency one contains more energy (power) than that corresponding to a frequency of six.

For a more interesting example, let us consider the Wolf sunspot number for solar activity. It has been observed that there is a relatively periodic occurrence of peaks in solar activity, which coincides with increases in the number of sunspots and solar flares on the surface of the sun. These peaks in solar activity impact the Earth's upper atmosphere, called the ionosphere, and can disrupt communications, damage satellite electronics, and influence GPS signals.

In order to illustrate Fourier analysis of a time series of data, let us consider the frequency of sunspots that are generated by the sun to see if there is any particular pattern to their occurrence. Johann Wolf devised the *sunspot number* in 1848 to

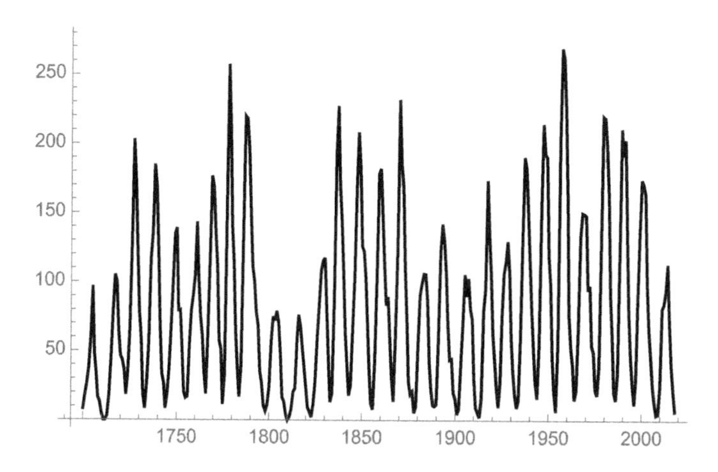

Figure 11.12 Time series of Wolf sunspot number annually from 1700 to 2018.

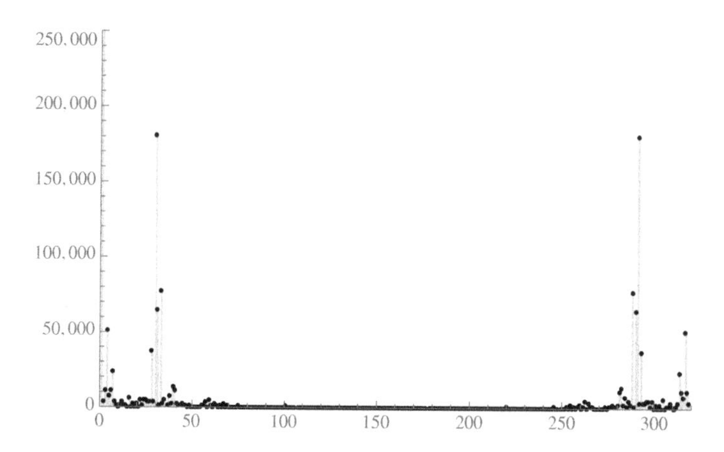

Figure 11.13 Power spectrum of sunspot number data.

quantify solar activity. It involves counting the number of sunspots and groups of sunspots – assumed to be equivalent to ten sunspots – that appear on the surface of the sun. Such data extend all the way back to 1700; therefore, they provide an interesting data set to analyze for historical features. In the mid-1800s, Wolf estimated that peaks in natural solar activity occurred with a period of 11.1 years. Let us see if this estimate holds for a longer series of data that extends from 1700 to 2018.

The Wolf sunspot number is plotted in Figure 11.12 for each year from 1700 to 2018. It is not possible from this plot itself to accurately determine the typical period of solar activity. However, we can take the discrete Fourier transform of the data in order to extract frequency information inherent in the discrete data. Taking the discrete Fourier transform gives the power spectrum shown in Figure 11.13. Let us adjust the abscissa to show the period (reciprocal of frequency) and limit the range in order to reveal the dominant period(s); this is shown in Figure 11.14. Observe that there

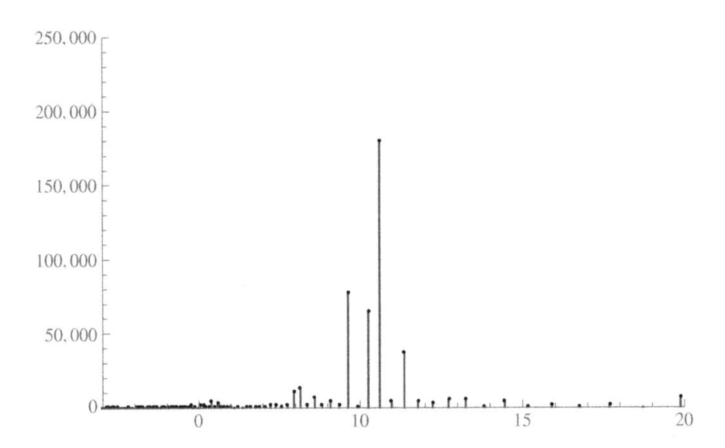

Figure 11.14 Zoomed power spectrum of sunspot number data with respect to frequency.

are several dominant peaks between a period of 9.5 to 11.5 years, but that the most dominant peak is indeed near 11 years, as observed by Wolf.

11.5 Polynomial Interpolation

Although closely related to curve fitting, interpolation has a fundamentally different objective. In curve fitting, the best-fit curve is sought that represents the data. In general, none of the data points fall directly on the regression curve. In interpolation, by contrast, the interpolating function passes through all of the data points and provides an estimate between data points.

Recall from Section 11.2 that increasing the order (degree) of the polynomial in polynomial regression leads to a better curve-fit representation of the data. In fact, a polynomial of degree $M = N - 1$, where N is the number of data points, produces a trial function $f(x)$ that passes through all of the points. Consequently, setting $M = N - 1$ corresponds to *polynomial interpolation*, while $M < N - 1$ corresponds to *polynomial regression* curve fitting. Recall that interpolation is preferred when one has a high degree of confidence in the accuracy of the data, whereas curve fitting is the typical choice when the data are noisy and inaccurate. Because interpolating functions pass through all of the data points, they allow for estimation of the function for values of x that are between data points.

The primary shortcoming of polynomial interpolation is that it often produces highly oscillatory polynomials for large N. For example, let us revisit the data shown in Figure 11.7. Because there are $N = 10$ points, polynomial interpolation seeks a polynomial of degree $M = N - 1 = 9$. The procedure is as shown for polynomial regression, and the result is shown in Figure 11.15. As you can see, the interpolating polynomial passes through all $N = 10$ of the data points and allows us to estimate values of y for values of x between data points. However, looking more closely

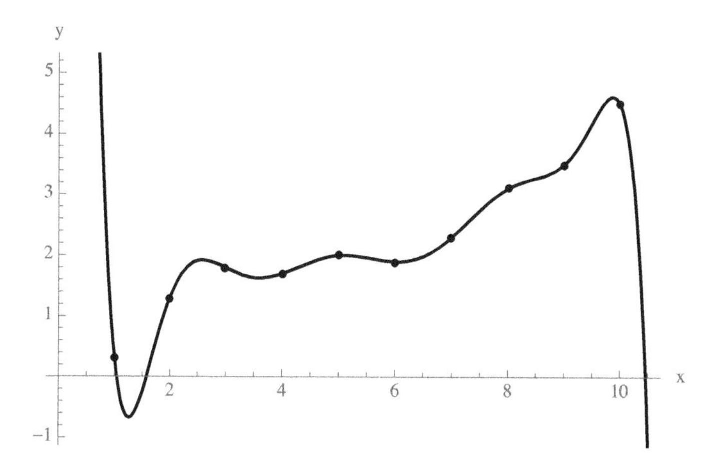

Figure 11.15 Polynomial interpolating function with $M = 9$ for data points shown in Figure 11.7.

at the interpolation polynomial, we see that it does not provide a good estimate for values of x between the pairs of points at either end of the data set. This is a primary disadvantage of interpolating polynomials, and it becomes more pronounced as the degree of the polynomial increases, that is, as more data points are included. Therefore, it is generally not used for N greater than about ten or so as estimated values between data points can be highly inaccurate. This deficiency is remedied using spline interpolation.

11.6 Spline Interpolation

Rather than producing a single high-order interpolating polynomial, we may instead construct a large number of low-order polynomials that each approximate a small portion of the overall data. These are called *splines* after the draftsmen's device used to draw smooth curves between a series of noncollinear points. A linear spline, for example, would simply consist of a series of straight lines connecting each pair of successive points as shown in Figure 11.16. While being simple to construct, linear splines have discontinuities in the first derivatives – slopes – at the points common to two successive splines. Quadratic splines have continuous derivatives but discontinuous second derivatives – curvatures – at the points between adjacent splines.

The sweet spot is *cubic splines*, which have continuous first and second derivatives, but no difficulties with oscillations as for high-order polynomials. Cubic splines consist of third-order polynomials between each pair of data points as illustrated in Figure 11.17. The cubic spline

$$f_i(x) = a_i + b_i\,(x - x_i) + c_i\,(x - x_i)^2 + d_i\,(x - x_i)^3, \quad i = 1, \ldots, N - 1 \tag{11.25}$$

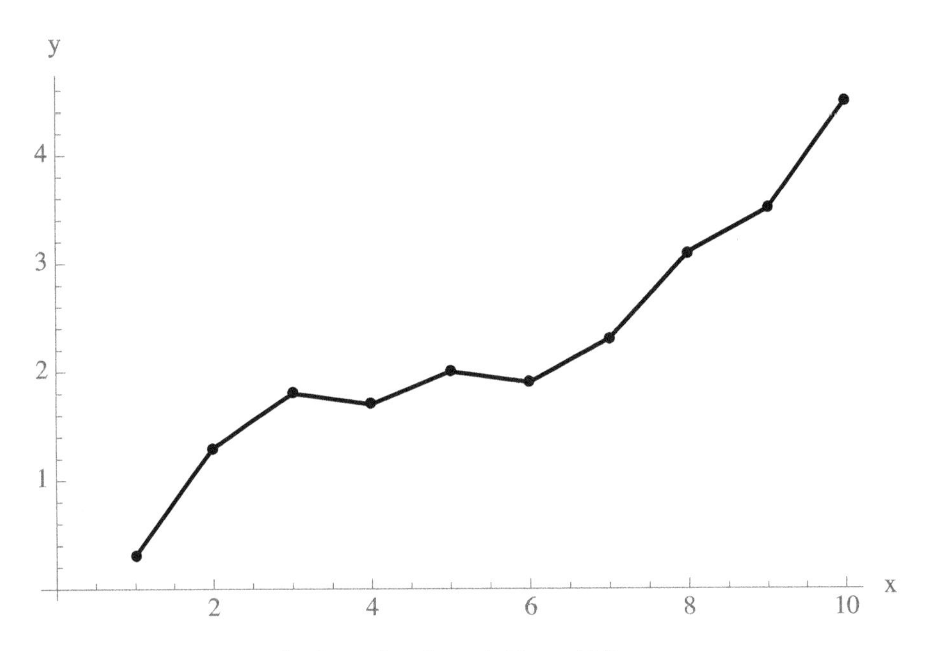

Figure 11.16 Linear spline for data points shown in Figure 11.7.

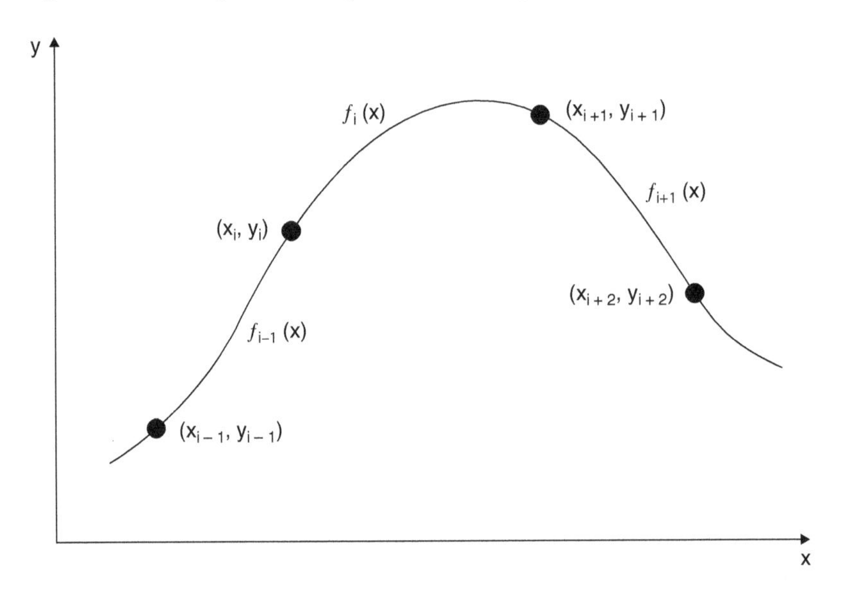

Figure 11.17 Schematic of cubic splines $f_i(x)$ between each pair of data points (x_i, y_i).

connects the points (x_i, y_i) and (x_{i+1}, y_{i+1}). In order to determine the constants a_i, b_i, c_i, and d_i for each $f_i(x), i = 1, \ldots, N-1$, we impose three conditions:

1. Endpoints of each successive cubic polynomial must match with the adjacent cubics and the data points:

$$f_i(x_i) = f_{i-1}(x_i) = y_i, \quad f_i(x_{i+1}) = f_{i+1}(x_{i+1}) = y_{i+1}.$$

2. First derivatives – slopes – must match at the endpoints of each cubic polynomial:

$$f_i'(x_i) = f_{i-1}'(x_i), \quad f_i'(x_{i+1}) = f_{i+1}'(x_{i+1}).$$

3. Second derivatives – curvatures – must match at the endpoints of each cubic polynomial:

$$f_i''(x_i) = f_{i-1}''(x_i), \quad f_i''(x_{i+1}) = f_{i+1}''(x_{i+1}).$$

Let us define the nonuniform spacing between data points by

$$h_i = x_{i+1} - x_i.$$

After a significant amount of algebra, applying these three conditions to (11.25) leads to the requirement that the coefficients c_i in the cubic polynomials satisfy the system of equations

$$h_{i-1}c_{i-1} + 2\left(h_{i-1} + h_i\right)c_i + h_ic_{i+1} = 3\left(\frac{y_{i+1} - y_i}{h_i} - \frac{y_i - y_{i-1}}{h_{i-1}}\right). \quad (11.26)$$

Given the values of y_i and h_i from the data, this results in a tridiagonal system of algebraic equations for the $c_i, i = 1, \ldots, N-1$ coefficients that can be solved using the Thomas algorithm from Section 6.3.3. Given these solutions for $c_i, i = 1, \ldots, N-1$, the other coefficients are given by

$$\begin{aligned} a_i &= y_i \\ d_i &= \frac{c_{i+1} - c_i}{3h_i}, \\ b_i &= \frac{y_{i+1} - y_i}{h_i} - c_ih_i - d_ih_i^2, \end{aligned}$$

for $i = 1, \ldots, N-1$, which nearly specifies the cubic polynomials given by equation (11.25) for nonuniformly spaced data points. Two additional conditions are required at the two endpoints of the data set in order to close the problem such that it has a unique solution. For example, one could specify the slopes of the splines at the endpoints as if they were clamped. Alternatively, a *natural spline* occurs if the curvatures are specified to be zero at the endpoints. For example, the cubic spline with natural splines at the endpoints for the data shown in Figure 11.7 is plotted in Figure 11.18. Another common condition is known as the *not-a-knot spline*,[4] in which the third derivatives are also specified to be continuous at the first interior points from the ends. See section 11.2 of Cassel (2013) for more on splines.

11.7 Curve Fitting and Interpolation of Multidimensional Data

Our focus has been on curve fitting and interpolation of one-dimensional data, in which the data are only a function of one independent variable $y(x)$. For multidimensional data sets, it is not difficult to imagine some relatively straightforward

[4] In splines, a *knot* is a point that the spline must pass through.

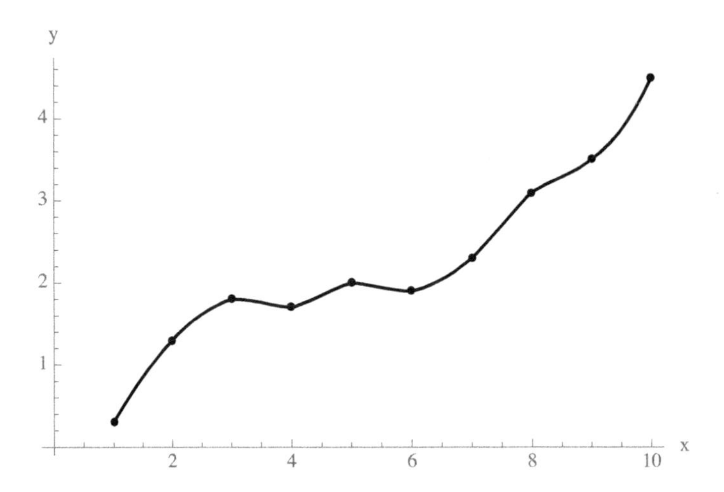

Figure 11.18 Cubic-spline interpolation for data points shown in Figure 11.7.

extensions of the methods discussed thus far. For example, let us consider *bilinear regression*, in which the best-fit function $y(x_1, x_2)$ is a function of two independent variables. The best-fit surface, which is now a plane, is of the form

$$y(x_1, x_2) = a_0 + a_1 x_1 + a_2 x_2.$$

The constants a_0, a_1, and a_2 are obtained by solving the following system of equations, which are obtained from a least-squares regression as before,

$$\begin{bmatrix} N & \sum_{i=1}^{N} x_{1i} & \sum_{i=1}^{N} x_{2i} \\ \sum_{i=1}^{N} x_{1i} & \sum_{i=1}^{N} x_{1i}^2 & \sum_{i=1}^{N} x_{1i} x_{2i} \\ \sum_{i=1}^{N} x_{2i} & \sum_{i=1}^{N} x_{1i} x_{2i} & \sum_{i=1}^{N} x_{2i}^2 \end{bmatrix} \begin{bmatrix} a_0 \\ a_1 \\ a_2 \end{bmatrix} = \begin{bmatrix} \sum_{i=1}^{n} y_i \\ \sum_{i=1}^{n} x_{1i} y_i \\ \sum_{i=1}^{n} x_{2i} y_i \end{bmatrix}. \tag{11.27}$$

Observe that the coefficient matrix remains symmetric as in the one-dimensional case.

Example 11.2 Consider the flow of a fluid through an inclined pipe. A series of experiments are done to measure the volume flow rate Q for a range of pipe diameters D and slopes S. The results are given in the following table:

i	1	2	3	4	5	6	7	8	9
D (ft)	1	2	3	1	2	3	1	2	3
S (ft/ft)	0.001	0.001	0.001	0.01	0.01	0.01	0.05	0.05	0.05
Q (ft^3/s)	1.4	8.3	24.2	4.7	28.9	84.0	11.1	69.0	200.0

Such data are expected to be well approximated by the two-dimensional power function

$$Q(D, S) = Q_0 D^{a_1} S^{a_2}. \tag{11.28}$$

Use bilinear regression to determine the constants Q_0, a_1, and a_2 that provide the best fit to the given data.

Solution

Taking the base-10 logarithm[5] of both sides of (11.28) yields

$$\log_{10} Q = \log_{10} Q_0 + \log_{10} D^{a_1} + \log_{10} S^{a_2},$$

or

$$\log_{10} Q = \log_{10} Q_0 + a_1 \log_{10} D + a_2 \log_{10} S,$$

which can be written in the form

$$y(x_1, x_2) = a_0 + a_1 x_1 + a_2 x_2,$$

where

$$y = \log_{10} Q, \quad a_0 = \log_{10} Q_0, \quad x_1 = \log_{10} D, \quad x_2 = \log_{10} S.$$

Note from the second of these relationships that $Q_0 = 10^{a_0}$. Rewriting the data in terms of these new variables gives

i	1	2	3	4	5	6	7	8	9
x_1	0	0.301	0.477	0	0.301	0.477	0	0.301	0.477
x_2	−3	−3	−3	−2	−2	−2	−1.301	−1.301	−1.301
y	0.146	0.919	1.383	0.672	1.460	1.924	1.045	1.838	2.301

Then the coefficients a_0, a_1, and a_2 are obtained by solving the system of equations (11.27), which for the data given is

$$\begin{bmatrix} 9 & 2.3344 & -18.9031 \\ 2.3344 & 0.9548 & -4.9031 \\ -18.9031 & -4.9031 & 44.078 \end{bmatrix} \begin{bmatrix} a_0 \\ a_1 \\ a_2 \end{bmatrix} = \begin{bmatrix} 11.6915 \\ 3.9462 \\ -22.2077 \end{bmatrix}.$$

Solving this linear system of algebraic equations yields

$$a_0 = 1.74797, \quad a_1 = 2.61584, \quad a_2 = 0.53678.$$

Therefore,

$$Q_0 = 10^{a_0} = 55.9719.$$

The resulting best fit curve (11.28) is then

$$Q(D, S) = 55.9719 D^{2.61584} S^{0.53678}.$$

[5] As in Example 11.1, we could also use the natural logarithm.

With regard to interpolation between data points, various methods are available depending upon the situation. In numerical methods, for example, it is common for the results to be available on a regular grid throughout the domain with small distances between the grid points. In such situations, bilinear interpolation is typically sufficient to calculate values of the dependent variable(s) between grid points. Higher-order functions, for example quadratic polynomials, can also be used if necessary, just as in finite-element methods.

If the data are not available on a regular grid, but rather at points scattered throughout the domain, then a more robust technique is called for. A popular method in such situations is to utilize *radial basis functions*. As the name suggests, basis functions are used that approximate the behavior of the data in a radially axisymmetric fashion about each data point. In other words, the functional behavior surrounding each point is assumed to be the same in all directions. The interpolating function is then the linear combination of all the radial basis functions corresponding to each of the N data points $\mathbf{x}_n$ according to

$$y(x_1, x_2) = \sum_{n=0}^{N-1} a_n \phi(r) = \sum_{n=0}^{N-1} a_n \phi(|\mathbf{x} - \mathbf{x}_n|).$$

The a_n coefficients in the linear combination are then obtained by enforcing that the interpolating function exactly match the value of y at each data point. This leads to an $N \times N$ system of algebraic equations for the coefficients. There are various possibilities for the radial basis functions that specify how the function behaves around each data point; for example, Gaussian and spline functions are useful in certain circumstances. Each one has advantages and disadvantages in various situations. The *Gaussian radial basis function* is given by

$$\phi(r) = \exp\left(-\frac{r^2}{2r_0^2}\right),$$

where r_0 is a user-selected scale factor. The scale factor is chosen to be between the typical distance between data points and the overall size of the features in the data that are being approximated. The most common choice is called the *multiquadric radial basis function*, which has the form

$$\phi(r) = \sqrt{r^2 + r_0^2}.$$

For more on radial basis functions and their applications, see Press et al. (2007) and Fasshauer (2007).

11.8 Linear Regression Using Singular-Value Decomposition

Let us return to the original data set shown in Figure 11.1 with $N = 19$ data points. Here, we show how singular-value decomposition (SVD) from Section 2.8 can be used to reconstruct the data using various levels of fidelity. This provides an introduction to how SVD can be used in data-reduction techniques.

In order to perform the SVD, we form the data matrix consisting of the coordinates of each data point in the form

$$\begin{bmatrix} x_1 & x_2 & \cdots & x_N \\ y_1 & y_2 & \cdots & y_N \end{bmatrix},$$

where here $N = 19$. The SVD is actually performed on the mean-adjusted data matrix by subtracting off the mean of each row; this is discussed in more detail in Section 13.7.2. With the data matrix being 2×19, the resulting matrix $\mathbf{U}$ of left singular vectors is 2×2 and the matrix $\mathbf{V}$ of right singular vectors is 19×19. The diagonal matrix $\mathbf{\Sigma}$ of singular values is 2×19 with the two singular values on its diagonal.

Recall from equation (2.50) that the mean-adjusted data matrix can be reconstructed for our $M = 2$ case using the relationship

$$\mathbf{A} = \sigma_1 \mathbf{u}_1 \mathbf{v}_1^T + \sigma_2 \mathbf{u}_2 \mathbf{v}_2^T,$$

where σ_n are the singular values, $\mathbf{u}_n$ are the left singular vectors (columns of $\mathbf{U}$), and $\mathbf{v}_n$ are the right singular vectors (columns of $\mathbf{V}$) of the mean-adjusted data matrix $\mathbf{A}$. For a full reconstruction of the original data, we use both singular values σ_1 and σ_2 and corresponding left/right singular vectors to exactly reproduce the original data points shown in Figure 11.1. The best-fit line via linear regression corresponds to using only the first – largest – singular value σ_1 and the corresponding left/right singular vectors, which are plotted as the open circles in Figure 11.19; the figure also includes the original data along with the linear regression line. As you can see, the linear regression formula produces a line that passes exactly through the points corresponding to the first principal component produced by the SVD. Therefore, linear

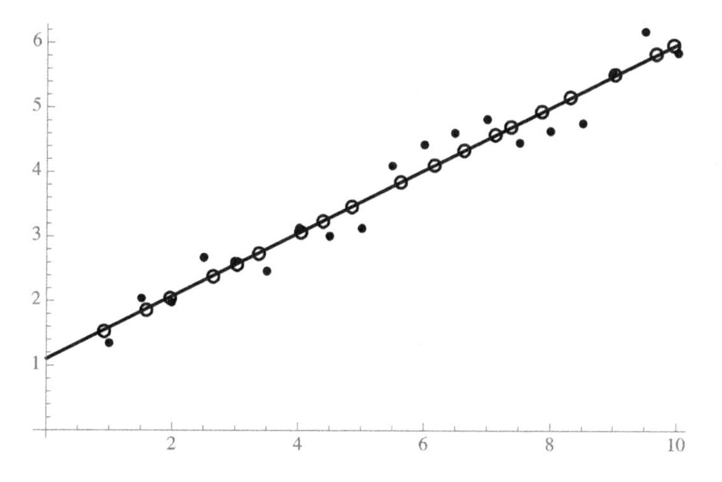

Figure 11.19 Original data points shown in Figure 11.1 (dots), linear regression line from Figure 11.3 (solid line), and data reconstruction using only the first principal component from the SVD (open circles).

regression corresponds to the first principle component of the data matrix. This is a simple example of data reduction using SVD; we will consider more general scenarios in Chapter 13.

11.9 Least-Squares Regression as State Estimation

Recall that the objective of least-squares regression curve fitting is to take a set of measured data points for a system and determine the best-fit curve that minimizes the error between the curve and the data points. In the vocabulary of state estimation, the data points are output *measurements* or *observations* of the state. For real systems that produce such data, there are inaccuracies that are present owing to inherent errors in the measurement techniques used and/or random noise in those measurements. Therefore, one would like to develop a representation of the state that is as faithful to the measured data as possible while adhering to a prescribed functional form of the model. Such an optimal representation that takes into account both the measured data as well as a system model could then be used for further analysis or prediction.

Let us consider a system that has some sort of u versus x behavior. In a real system, we do not have the actual $u(x)$ functional form that relates the dependent variable u to the independent variable x. However, suppose that we have two pieces of information about the system: (1) we have a general model of the system in the form of an Mth-order polynomial with unknown coefficients, and (2) a set of N output measurements from the system that give some representation of the system's behavior. These measured values include errors and noise as mentioned earlier. Therefore, we will denote them by $\mathbf{v}$ in order to distinguish them from the estimated system states $\hat{\mathbf{u}}$ that we seek. In addition, the system states predicted by our system model will also differ from the actual state owing to imperfections in our model; we will denote the predicted values of the state by $\mathbf{u}$.

In order for the output measurements to be useful, we need to know how they are related to the state vector, because the measurements $\mathbf{v}$ may not be of the state vector $\mathbf{u}$ directly. There are several reasons why this may be the case. First, the sensors used to make the measurements may be inaccurate owing to noise. Second, the locations where the measurements are taken may differ from those where the state variables are sought. Third, the measurements may be of quantities different from those corresponding to the state variables. Instead, ancillary quantities are measured from which the state of the system can be *estimated*. For example, your car does not directly measure the speed of the vehicle as displayed on your speedometer. Instead, it measures the rate of rotation of your wheels, which along with the circumference of the wheels can be used to calculate an estimate of the speed. The known relationship between the output measurements and the sought-after estimated state vector are given by the *observation model*

$$\mathbf{v} = \mathbf{C}\hat{\mathbf{u}}. \tag{11.29}$$

In the case of least-squares polynomial regression, it is assumed that the measurements directly correspond to the state values, such that $\mathbf{C} = \mathbf{I}$.

Let us suppose that we have N output measurements (x_i, v_i) from the system. In addition, we have reason to believe that the u versus x behavior of the system is in the form of an Mth-order polynomial, where $M \leq N$. The system model is in the form of the trial function

$$u(x) = a_0 + a_1 x + a_2 x^2 + \cdots + a_M x^M, \tag{11.30}$$

where we need to *estimate* the coefficients $a_i, i = 0, 1, \ldots, M$. Specifically, we seek the vector $\mathbf{a} = \begin{bmatrix} a_0 & a_1 & \cdots & a_M \end{bmatrix}^T$ for which $u(x)$ provides the optimal polynomial curve fit to the data. Note that v denotes measured data values, while u denotes the state values as estimated by the system model. That is, they are both different, but related, representations of the same information about the system. In matrix form, (11.30) is of the form

$$\mathbf{\Phi a} = \mathbf{u}, \tag{11.31}$$

or

$$\begin{bmatrix} 1 & x_1 & x_1^2 & \cdots & x_1^M \\ 1 & x_2 & x_2^2 & \cdots & x_2^M \\ \vdots & \vdots & \vdots & \ddots & \vdots \\ 1 & x_N & x_N^2 & \cdots & x_N^M \end{bmatrix} \begin{bmatrix} a_0 \\ a_1 \\ \vdots \\ a_M \end{bmatrix} = \begin{bmatrix} u_1 \\ u_2 \\ \vdots \\ u_N \end{bmatrix}. \tag{11.32}$$

The $M + 1$ columns of $\mathbf{\Phi}$ are the basis functions being used in the regression and represent the terms in the Mth-order polynomial trial function. The measured values of $\mathbf{u}$, which we are denoting by $\mathbf{v}$, do not lie on this polynomial exactly; they differ by the residual

$$\mathbf{r} = \hat{\mathbf{u}} - \mathbf{u} = \mathbf{v} - \mathbf{u}, \tag{11.33}$$

where the fact that $\mathbf{v} = \mathbf{C}\hat{\mathbf{u}} = \hat{\mathbf{u}}$, with $\mathbf{C} = \mathbf{I}$, for the least-squares regression has been used.

We seek the optimal coefficients $\mathbf{a}$ for the best-fit polynomial given by the trial function (11.30) to the measured data. This is accomplished by minimizing the following objective function, which is the square of the norm of the residual (11.33)

$$J = \|\mathbf{r}\|^2 = \|\mathbf{v} - \mathbf{u}\|^2 = (\mathbf{v} - \mathbf{u})^T (\mathbf{v} - \mathbf{u}). \tag{11.34}$$

A weight matrix could be included to emphasize measurements that are more accurate than others. Multiplying out the objective function gives

$$J = \mathbf{v}^T\mathbf{v} - \mathbf{u}^T\mathbf{v} - \mathbf{v}^T\mathbf{u} + \mathbf{u}^T\mathbf{u}.$$

Substituting (11.31) yields

$$J = \mathbf{v}^T\mathbf{v} - \mathbf{a}^T\mathbf{\Phi}^T\mathbf{v} - \mathbf{v}^T\mathbf{\Phi a} + \mathbf{a}^T\mathbf{\Phi}^T\mathbf{\Phi a}.$$

The measurements $\mathbf{v}$ and basis functions $\mathbf{\Phi}$ are known; therefore, we seek the vector of coefficients $\mathbf{a}$ by minimizing J. Differentiating the objective function J with respect to $\mathbf{a}$ and setting equal to zero gives

$$-\mathbf{\Phi}^T\mathbf{v} - \mathbf{\Phi}^T\mathbf{v} + 2\mathbf{\Phi}^T\mathbf{\Phi}\mathbf{a} = \mathbf{0}.$$

Observe that this means that

$$\mathbf{\Phi}^T\mathbf{\Phi}\mathbf{a} = \mathbf{\Phi}^T\mathbf{v}.$$

Recall that the product of a matrix and its transpose is always square, thus, if it is invertible, we can solve for the coefficients

$$\mathbf{a} = \left(\mathbf{\Phi}^T\mathbf{\Phi}\right)^{-1}\mathbf{\Phi}^T\mathbf{v} = \mathbf{\Phi}^+\mathbf{v}, \tag{11.35}$$

where $\mathbf{\Phi}^+$ is the pseudo-inverse of $\mathbf{\Phi}$ as introduced in Section 2.8. In this way, least-squares curve fitting can be couched in terms of least-squares state estimation.

One can think of the resulting trial function (11.30) with coefficients (11.35) as a simple reduced-order model of the system – an Mth-order polynomial that provides the optimal representation of $N \geq M$ data points. We will return to this idea in Chapter 13, where we develop a more general method for representing complex data sets in a manner that is amenable for use in closed-loop, feedback control algorithms.

Finally, let us also show that least-squares curve fitting can be expressed within the method of weighted residuals framework introduced in Section 7.3. Recall that our residual is defined here by

$$\mathbf{r} = \hat{\mathbf{u}} - \mathbf{u} = \mathbf{v} - \mathbf{u} = \mathbf{v} - \mathbf{\Phi}\mathbf{a}. \tag{11.36}$$

In the method of weighted residuals, we orthogonally project the residual onto a set of weight vectors $\mathbf{w}_i, i = 1, 2, \ldots, M$, such that

$$\langle \mathbf{r}, \mathbf{w}_i \rangle = 0, \quad i = 1, 2, \ldots, M,$$

which is equivalent to

$$\mathbf{W}^T\mathbf{r} = \mathbf{0},$$

where the matrix $\mathbf{W}$ has the weight vectors $\mathbf{w}_i$ as its columns. Substituting the definition of the residual (11.36) gives

$$\mathbf{W}^T\mathbf{\Phi}\mathbf{a} = \mathbf{W}^T\mathbf{v}. \tag{11.37}$$

The matrix product $\mathbf{W}^T\mathbf{\Phi}$ is square. If it is also invertible, then

$$\mathbf{a} = \left(\mathbf{W}^T\mathbf{\Phi}\right)^{-1}\mathbf{W}^T\mathbf{v}. \tag{11.38}$$

In the Galerkin method, the weight vectors are equal to the basis vectors, in which case

$$\mathbf{W} = \mathbf{\Phi},$$

and (11.38) becomes

$$\mathbf{a} = \left(\mathbf{\Phi}^T \mathbf{\Phi}\right)^{-1} \mathbf{\Phi}^T \mathbf{v} = \mathbf{\Phi}^+ \mathbf{v}, \tag{11.39}$$

which is the same as that obtained previously using least-squares estimation techniques. Recall that in the method of weighted-residuals context, the least-squares method can be expressed with the weight vectors being

$$\mathbf{w}_i = -\frac{\partial \mathbf{r}}{\partial a_i} = \boldsymbol{\phi}_i,$$

which is the ith column of $\mathbf{\Phi}$. In the linear case, therefore, the least-squares regression for overdetermined linear systems of equations is equivalent to the Galerkin method, which is equivalent to least-squares state estimation. We will utilize this method of weighted residuals framework again in Chapter 13 in the context of Galerkin projection.

The least-squares approach used here is an example of data-driven *system identification*. We specify the possible general forms of the basis vectors and determine how much of each basis vector to use in order to accurately represent the data from the system. See Section 13.9 for more on system identification. More generally, state estimation is an essential aspect of optimal control. It is discussed in this context in section 10.6.3 of Cassel (2013).

11.10 Definitions of the Residual

The observant reader will have noticed – and may be frustrated by – the fact that a different definition of the residual has been introduced in Section 11.9 as compared to those encountered thus far. We first encountered the residual as a proxy for the error in Section 6.2.2, where it was defined for a linear system of algebraic equations by

$$\mathbf{r}_m = \mathbf{b} - \mathbf{A}\mathbf{u}. \tag{11.40}$$

The subscript m designates that this is appropriate for model-driven contexts when a mathematical model is available. Defined in this way, the residual is a measure of how "incorrect" a candidate solution $\mathbf{u}$ is with respect to solving the system of equations $\mathbf{A}\hat{\mathbf{u}} = \mathbf{b}$; the residual vanishes if $\mathbf{u}$ is the exact solution $\hat{\mathbf{u}}$ of the system. The continuous counterpart, first encountered in Section 7.3 in our discussion of spectral numerical methods, is given by

$$r_m = f - \mathcal{L}u. \tag{11.41}$$

Again, the residual quantifies how well the approximate solution $u(x)$ satisfies the differential equation $\mathcal{L}\hat{u} = f$. Observe that in most cases, it does not matter if we define the residual as $\mathbf{r} = \mathbf{b} - \mathbf{A}\mathbf{u}$ or $\mathbf{r} = \mathbf{A}\mathbf{u} - \mathbf{b}$. This is particularly the case if it is the norm squared of the residual that is to be minimized as the norm squared of both forms are equivalent. If the norm itself is used, as in multigrid methods in Section 8.8, the alternative definitions will simply result in a sign change. Therefore, many authors take advantage of this and use the definition that produces a desirable final form of the equation or algorithm. In fact, we will typically use the forms given in (11.40) and (11.41).

So why did we define the residual as the difference between an actual and approximate solution in least-squares state estimation in Section 11.9? The simple answer is that the usual definition was not possible because a system of linear algebraic equations was not available for the regression case. In least-squares regression, we only have data – there is no mathematical model available. As will be seen in Section 13.7 on proper-orthogonal decomposition, when this is the case, we define the residual as the difference between the approximate and actual values of the solution in the form

$$\mathbf{r}_d = \mathbf{u} - \hat{\mathbf{u}}. \tag{11.42}$$

The subscript d designates that this definition is appropriate for data-driven contexts, for which a mathematical model is not available.

While one may wonder whether these definitions of the residual are actually equivalent, it is straightforward to show that this is not the case, even for $\mathbf{A}$ square and invertible. To show this, let us define the residual as for model-driven contexts by (11.40). If $\hat{\mathbf{u}}$ is the solution of the system of linear algebraic equations $\mathbf{A}\hat{\mathbf{u}} = \mathbf{b}$, then

$$\mathbf{r}_d = \mathbf{u} - \hat{\mathbf{u}} = \mathbf{u} - \mathbf{A}^{-1}\mathbf{b}.$$

Premultiplying $\mathbf{A}$ on both sides yields

$$\mathbf{A}\mathbf{r}_d = \mathbf{A}\mathbf{u} - \mathbf{b},$$

which when compared with the definition (11.40) shows that

$$\mathbf{A}\mathbf{r}_d = -\mathbf{r}_m,$$

and the two definitions of the residual differ by a factor of the coefficient matrix $\mathbf{A}$.

In summary, the form (11.40) or (11.41) is used to define the residual when a mathematical model of the system is available; therefore, it is for *model-driven methods*. The form (11.42) is used when only raw data are available with no mathematical model; therefore, it is for *data-driven methods*. The least-squares methodology is used in both cases, but it is applied to different definitions of the residual appropriate to the context.

Exercises

Unless stated otherwise, perform the exercises using hand calculations. Exercises to be completed using "built-in functions" should be completed using the built-in functions within Python, MATLAB, or Mathematica for the vector and matrix operation(s) or algorithm. A "user-defined function" is a standalone computer program to accomplish a task written in a programming language, such as Python, MATLAB, Mathematica, Fortran, or C++.

11.1 Consider the following data:

x	1	2	3	4	5
y	0.7	2.2	2.8	4.4	4.9

Perform a linear regression of this data set using the normal equations to obtain the coefficients in the best-fit linear trial function $f(x) = a_0 + a_1 x$.

11.2 Consider the following data:

x	−7	−5	−1	0	2	5	6
y	15	12	5	2	0	−5	−9

(a) Perform a linear regression of this data set using the normal equations to obtain the coefficients in the best-fit linear trial function $f(x) = a_0 + a_1 x$.

(b) Check your results for the coefficients in the trial function using a built-in function.

(c) Plot the data points as dots and the best-fit line as a solid line on the same figure.

11.3 Using linear least-squares regression, determine the best-fit curve of the form

$$u(x) = c_0 e^{c_1 x}$$

for the following data:

i	1	2	3	4
x	0	2	4	6
u	5.0	3.7	2.7	2.0

11.4 Recall that the coefficients in an exponential function

$$y = c_0 e^{c_1 x}$$

can be determined by converting the exponential into a linear regression problem.

(a) Write a user-defined function that takes two vectors as its input arguments that contain the values of x_i and y_i $i = 1, \ldots, N$ in the data set, performs the linear regression, and returns the values of the coefficients c_0 and c_1.

(b) Consider the following data giving the population of the Earth in billions for selected years between 1850 and 2000:

Year	1850	1900	1950	1980	2000
Population	1.3	1.6	3.0	4.4	6.0

Assuming that this population growth data follow roughly an exponential growth curve, use your user-defined function from part (a) to determine the best-fit values of the coefficients c_0 and c_1 in the exponential function.

(c) Use a built-in function to determine the coefficients in the exponential function directly, that is, without using linear regression, and compare with your result in part (b).

(d) Plot the data and your best-fit curve. Using the best-fit curve (from part (b) or (c)), estimate the population in the year 1970.

11.5 A specimen is placed in an axial-load test machine. The applied tensile force F on the specimen leads to the length of the specimen becoming L. The axial stress σ_x and strain ϵ_x are defined by

$$\sigma_x = \frac{F}{A_0}\frac{L}{L_0}, \quad \text{and} \quad \epsilon_x = \ln\frac{L}{L_0},$$

where A_0 is the original unloaded cross-sectional area of the specimen, and L_0 is the original unloaded length of the specimen. Beyond the yield point of the stress–strain curve, the relationship between stress and strain can often be modeled by the relationship

$$\sigma_x = K\epsilon_x^m.$$

The values of force F and length L measured in an experiment are given by

F(kN)	24.6	29.3	31.5	33.3	34.8	35.7	36.6
L (mm)	12.58	12.82	12.91	12.95	13.05	13.21	13.35

F(kN)	37.5	38.8	39.6	40.4
L (mm)	13.49	14.08	14.21	14.48

The initial cross-sectional area is $A_0 = 1.25 \times 10^{-4}\ m^2$, and the original specimen length is $L_0 = 0.0125\ m$. Use a built-in function to determine the best-fit coefficients K and m in the trial function. Plot the data and your best-fit curve.

11.6 Determine the best-fit parabola $y(x) = a_0 + a_1x + a_2x^2$ for the data set:

x	−3	0	2	4
y	3	1	1	3

Plot the parabola and the data points.

11.7 Determine the best-fit cubic polynomial $y(x) = a_0 + a_1x + a_2x^2 + a_3x^3$ for the data set:

x	−1.0	−0.8	−0.6	−0.4	−0.2	0.0	0.2	0.4	0.6	0.8	1.0
y	0.05	0.08	0.14	0.23	0.35	0.50	0.65	0.77	0.86	0.92	0.95

Plot the cubic polynomial and the data points.

11.8 Values of specific enthalpy h of an Argon plasma in equilibrium as a function of temperature are given by the following:

T(K)	5,000	7,500	10,000	12,500	15,000	17,500	20,000
h (MJ/kg)	3.3	7.5	41.8	51.8	61	101.1	132.9

T(K)	22,500	25,000	27,500	30,000
h (MJ/kg)	145.5	171.4	225.8	260.9

Determine the cubic spline interpolating function for the data using a built-in function. Using this spline function, interpolate the value of h for $T = 13,000\ K$.

11.9 Determine the best-fit two-dimensional plane $y(x_1, x_2) = a_0 + a_1x_1 + a_2x_2$ for the data set:

x_1	1	2	2	5	4	5
x_2	2	3	1	5	6	4
y	2	2	3	3	5	6

11.10 Determine the best-fit two-dimensional plane $y(x_1, x_2) = a_0 + a_1x_1 + a_2x_2$ for the data set:

x_1	10	10	20	50	60	60
x_2	5	45	25	25	5	45
y	50	40	36	32	32	19

11.11 An algorithm is sought for determining the best-fit circle

$$(x_i - a_1)^2 + (y_i - a_2)^2 = R^2$$

for $N \geq 3$ given points (x_i, y_i), $i = 1, \ldots, N$. Recall that three points uniquely define a circle; therefore, more points would lead to an overdetermined nonlinear system of equations for the parameters a_1, a_2, and R.

(a) Show that by defining the additional parameter

$$a_3 = R^2 - a_1^2 - a_2^2, \tag{11.43}$$

a linear system of algebraic equations can be obtained for a_1, a_2, and a_3. The solution for this (typically) overdetermined system can then be used to obtain R from (11.43), thereby giving the parameters a_1, a_2, and R defining the best-fit circle.

(b) Test the algorithm by determining the "best-fit" circle for the three points $(1,4)$, $(3,2)$, and $(1,0)$.

12 Optimization and Root Finding of Algebraic Systems

> Mathematics is the instrument by which the engineer tunnels our mountains, bridges our rivers, constructs our aqueducts, erects our factories, and makes them musical by the busy hum of spindles. Take away the results of the reasoning of mathematics, and there would go with it nearly all the material achievements which give convenience and glory to modern civilization. (Edward Brooks, *Mental Science and Culture*, 1891)

12.1 Introduction

In the previous chapter, we considered a specific optimization problem – that of determining the best-fit curve for a particular data set by minimizing the distance between the data points and the best-fit curve. In this chapter, we extend these methods to include a much broader set of scenarios. In order to build our intuition in optimization, and the related topic of root finding, let us first consider the case with a single nonlinear function of a single variable $f(x)$. For such a scenario, it is straightforward to see graphically the important features that may correspond to solutions of the root finding or optimization problem.

A root of a function corresponds to the location where that function is zero, while in general, an optimal solution corresponds to a minimum or maximum in the objective function. We will consider minimums throughout with the understanding that simply taking the negative of the objective function changes a maximum to a minimum. Figure 12.1 illustrates a function $f(x)$ and the various points of interest that could appear on such a function. The *roots*, or *zeros*, are values of the independent variable x at which $f(x) = 0$. A point where $f'(x) = 0$ is called a *stationary point*. For such a stationary point also to be a *local minimum* requires the curvature to be positive, such that $f''(x) > 0$. Similarly, for a stationary point to be a *local maximum* requires $f''(x) < 0$, that is, to have negative curvature at the stationary point. A function may have multiple local minimums and/or maximums; the lowest local minimum and highest local maximum are called *global minimums* and *maximums*.

Note that a minimum or maximum – or any stationary point – of $f(x)$ is a root of $f'(x) = 0$, that is, a point having zero slope. Therefore, identifying minimums and maximums for optimal solutions is closely related to calculating the roots of a function. In fact, let us first develop algorithms for approximating the roots of a

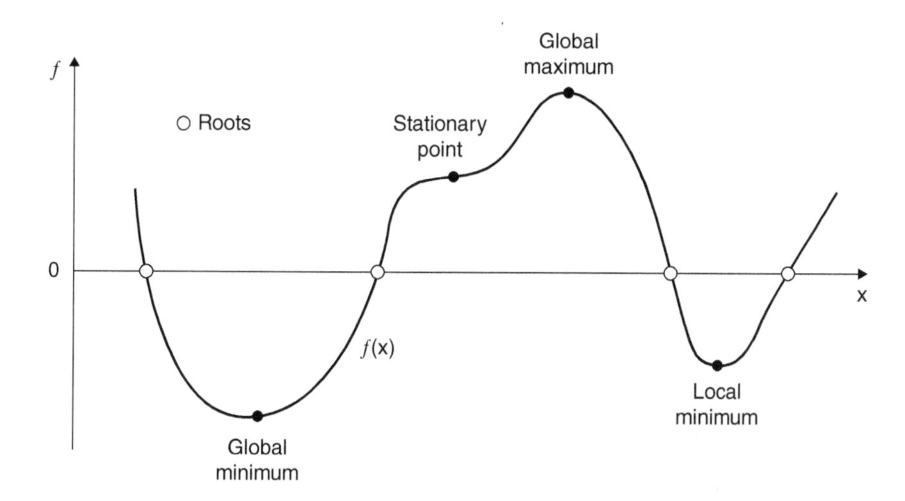

Figure 12.1 A one-dimensional function illustrating roots, stationary points, and minimums and maximums.

nonlinear equation $f(x)$, then we will return to optimization and build on these root-finding techniques.

12.2 Nonlinear Algebraic Equations – Root Finding

12.2.1 Introduction

In earlier chapters, we primarily considered *linear* algebraic equations for which the relationships between the variables are linear, that is, represented by straight lines. For example,

$$f(x) = mx + b.$$

The value of x at which $f(x) = 0$ is given by

$$x = -\frac{b}{m}.$$

Such "zeros" of a function are called its *roots*. Here, we turn our attention to *nonlinear* algebraic equations. For example, recall the quadratic polynomial

$$f(x) = ax^2 + bx + c.$$

The solutions, zeros, or roots, at which $f = 0$, that is, where the function intersects the x-axis, are given by the quadratic formula

$$x = \frac{-b \pm \sqrt{b^2 - 4ac}}{2a}.$$

A quadratic polynomial is one of the few nonlinear equations for which the roots may be obtained exactly using an analytical expression. Here, we seek general iterative numerical methods that can be used to approximately determine the root(s) of any nonlinear function $f(x)$.

Let us consider the example of the relationship between stress (force per unit area) and strain (elongation per unit length) in materials. Many materials behave according to linear *constitutive relations*. The most common by far is Hooke's law, which prescribes a linear relationship between stress and strain. Hooke's law for axial strain only in the x-direction is given by

$$\sigma_x = E\epsilon_x = E\frac{du}{dx},$$

where σ_x is the axial stress; $\epsilon_x = du/dx$ is the axial strain; E is Young's modulus, which is a property of the material; and $u(x)$ is the axial displacement of the material. For this linear case, it is easy to switch between stress σ and strain ϵ or displacement u when necessary.

However, some materials have a nonlinear constitutive relationship or behave nonlinearly for large strains. For the sake of discussion, let us suppose that we have data representing the forces F required to stretch a rubber band to displacements u and that this nonlinear relationship between force and axial displacement can be written as a cubic polynomial of the form

$$F(u) = au^3 + bu^2 + cu + d,$$

where the coefficients a, b, c, and d could be determined using curve fitting from experimental data, for example (see Chapter 11). Then, if the displacement u is given, one could obtain the force F explicitly by evaluating this equation. On the other hand, if the applied force $F = P$ is given and we seek the corresponding displacement u, we must determine the root(s) of the nonlinear cubic equation

$$F(u) - P = 0,$$

or

$$au^3 + bu^2 + cu + d - P = 0.$$

There are two general approaches to solving for the roots of nonlinear equations:

1. Open methods:
   - Newton's method (also called the Newton–Raphson method)
   - Secant method
2. Bracketing methods:
   - Bisection method
   - Regula falsi method (also called the false-position method)

Each method has its advantages and disadvantages. For example, bracketing methods tend to be more robust as long as the starting values indeed bracket a root, while open methods tend to converge to the root faster, that is, with fewer iterations, than bracketing methods, but only if one is close enough to the root. Here, however, we focus on Newton's method because it can be naturally extended to systems of nonlinear equations and draws on matrix methods. In addition, it exhibits very fast convergence – when it works.

12.2.2 Newton's Method

The basic idea of Newton's method[1] is to linearize the nonlinear function $f(x) = 0$ by approximating it locally for each iteration as a straight line. To start, consider the Taylor series of the nonlinear function $f(x)$ near a point $x^{(k)}$

$$f(x) = f\left(x^{(k)}\right) + \Delta x^{(k)} f'\left(x^{(k)}\right) + \left[\Delta x^{(k)}\right]^2 \frac{f''\left(x^{(k)}\right)}{2!} + \dots,$$

where $\Delta x^{(k)} = x - x^{(k)}$. If x is sufficiently close to the point $x^{(k)}$ about which the Taylor series is written, that is, if $\Delta x^{(k)}$ is small enough such that the second-order (quadratic) term $\left[\Delta x^{(k)}\right]^2$ is *very* small, then the series can be truncated after the $O\left[\Delta x^{(k)}\right]$ term. Because the first term that has been truncated is second order, this gives the *truncation error* $E\left(x^{(k)}\right)$ as

$$E\left(x^{(k)}\right) = O\left[\Delta x^{(k)}\right]^2.$$

Note that all higher-order terms that have been truncated are even smaller than the quadratic term. The truncation error also indicates the convergence rate, which tells us how fast the iterative process will converge to the correct solution. Therefore, we say that Newton's method has a "quadratic," or "second-order," convergence rate.

We are seeking a solution of $f(x) = 0$; therefore, setting the Taylor series equal to zero, truncating, and solving for $\Delta x^{(k)}$ in the second term yields

$$\Delta x^{(k)} = -\frac{f\left(x^{(k)}\right)}{f'\left(x^{(k)}\right)}. \tag{12.1}$$

Given an estimate of the root $x^{(k)}$, (12.1) gives the adjustment $\Delta x^{(k)} = x^{(k+1)} - x^{(k)}$ that provides a better estimate $x^{(k+1)}$ of where $f(x) = 0$. This is done by extending the tangent line at $x^{(k)}$ to the x-axis in order to provide the next estimate $x^{(k+1)}$ for the root. In this way, we see that Newton's method converts a nonlinear equation $f(x)$ into a series of linear equations for Δx upon which we iterate to obtain an approximation of the root.

Newton's method then consists of the iterative procedure as illustrated in Figure 12.2 and enumerated here:

1. Choose the initial guess ($k = 0$): $x = x^{(0)}$.
2. Determine the change in the estimate of the root $\Delta x^{(k)}$ from (12.1).
3. Update the approximation of the root: $x^{(k+1)} = x^{(k)} + \Delta x^{(k)}$.

[1] This method is also known as the Newton–Raphson method as Joseph Raphson published the method 20 years before Newton did despite the fact that Newton devised the method 20 years prior to Raphson's publishing it.

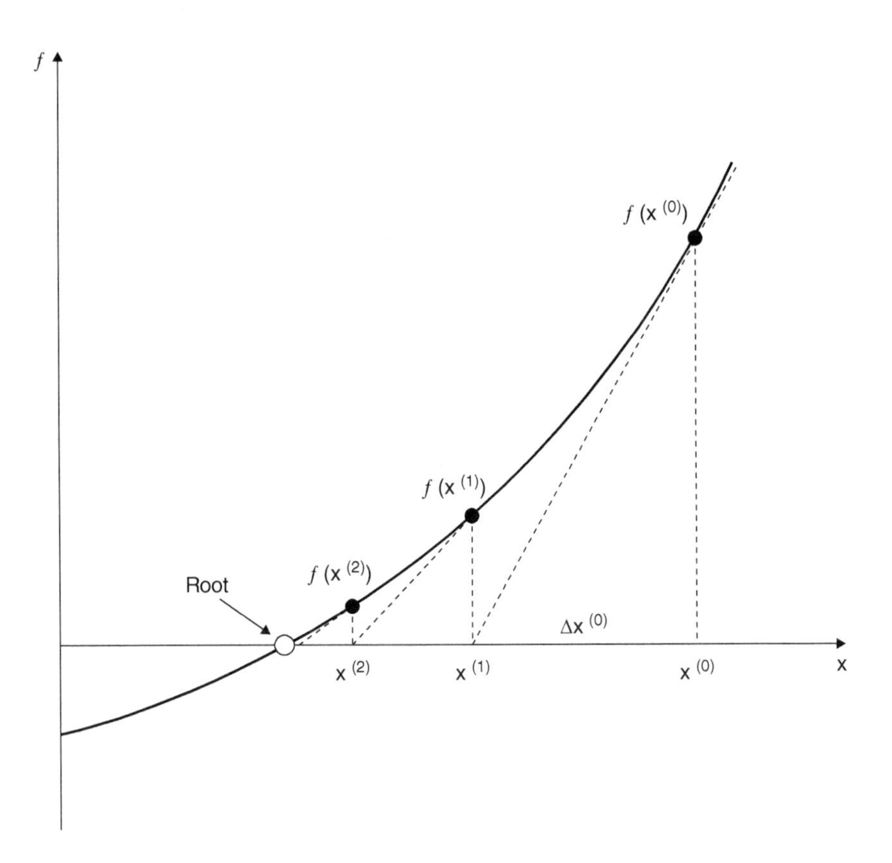

Figure 12.2 Schematic of the iterative process by which Newton's method approaches a root.

4. Increment the iteration number: $k = k + 1$.
5. Repeat steps 2 through 4 iteratively until $\Delta x^{(k)}$ and or $f\left(x^{(k)}\right)$ converges to zero, or in practice is less than some small tolerance: $k = 1, 2, 3, \ldots$

REMARKS:

1. *Newton's method requires evaluation of the derivative of the function at each estimate of the root.*
2. *Newton's method requires an initial guess $x^{(0)}$ near to the root being sought. If there are multiple roots of $f(x)$, the choice of initial guess determines which of the roots is approximated.*
3. *Newton's method exhibits a quadratic (second-order) convergence rate, which is very fast.*
4. *It is common to use other, more robust, bracketing methods, such as bisection, to obtain the initial guess for Newton's method, which then converges rapidly to the actual root.* Brent's method *is one such algorithm.*
5. *Observe from* (12.1) *that there is a problem if $f'\left(x^{(k)}\right) \approx 0$, in which case Newton's method may "shoot" past the sought-after root owing to the very shallow slope, thereby producing a large $\Delta x^{(k)}$.*

6. *We do not actually calculate the truncation error $E(x)$; it simply indicates the convergence rate of the algorithm.*
7. *Newton's method is the standard method used by mathematical software, such as MATLAB and Mathematica.*
8. *Recall that the methods of steepest descent and conjugate gradients from Section 10.6 are Newton-like methods that consist of determining the direction to proceed and the distance to proceed in that direction.*

12.2.3 Physical Examples

Before seeing how Newton's method can be extended to systems of nonlinear algebraic equations, let us consider two physical examples.

Equations of State

In thermodynamics, we use *equations of state* to relate pressure P, temperature T, and volume V of gases. The most common equation of state is that for an *ideal gas*, which is given by

$$PV = n\hat{R}T \quad \text{or} \quad P\hat{v} = \hat{R}T, \tag{12.2}$$

where P is the absolute pressure, $\hat{v} = V/n$ is the molal volume (volume per mole), T is the absolute temperature, and $\hat{R} = 0.082054\ L \cdot atm/(mol \cdot K)$ is the universal gas constant. The ideal gas equation of state is accurate for most gases at low temperatures and pressures. Observe that using the ideal gas equation of state, the molal volume $\hat{v}$ can be determined explicitly for a given pressure P and temperature T from

$$\hat{v} = \frac{\hat{R}T}{P}. \tag{12.3}$$

An alternative, but more complicated, equation of state is the *van der Waals* equation of state, which is accurate over a wider range of pressures and temperatures than the ideal gas equation of state. It is given by

$$\left(P + \frac{a}{\hat{v}^2}\right)(\hat{v} - b) = \hat{R}T, \tag{12.4}$$

where a and b are empirical constants for each gas. Observe that when $a = b = 0$, the van der Waals equation of state reduces to that for an ideal gas. The term involving the parameter a accounts for the higher intramolecular forces between molecules as the volume decreases. The term involving the parameter b accounts for the fact that the gas molecules can only be compressed so much. For a given gas, with its values of a and b, at specified temperature T and pressure P, the van der Waals equation of state cannot be solved explicitly for the molal volume $\hat{v}$ as is the case for the ideal gas equation of state. Instead, it is necessary to determine the root $\hat{v}$ for given P and T from the nonlinear algebraic equation

$$f(\hat{v}) = \left(P + \frac{a}{\hat{v}^2}\right)(\hat{v} - b) - \hat{R}T = 0. \tag{12.5}$$

Table 12.1 Molal volume approximated using the ideal gas and van der Waals equations of state for $T = 300\,K$ and various pressures.

Pressure (atm)	Ideal Gas	Van der Waals	% difference
1	24.6162	24.5126	0.42
10	2.46162	2.3545	4.5
100	0.246162	0.079511	210

Using Newton's method requires the derivative of $f(\hat{v})$ with respect to $\hat{v}$, which is given by

$$\frac{df}{d\hat{v}} = f'(\hat{v}) = -\frac{2a}{\hat{v}^3}(\hat{v} - b) + \left(P + \frac{a}{\hat{v}^2}\right) = P - \frac{a}{\hat{v}^2} + \frac{2ab}{\hat{v}^3}. \tag{12.6}$$

The Newton iteration equation for the change in molal volume is

$$\Delta\hat{v}^{(k)} = -\frac{f(\hat{v}^{(k)})}{f'(\hat{v}^{(k)})}, \tag{12.7}$$

which is used to update the approximation of the molal volume using

$$\hat{v}^{(k+1)} = \hat{v}^{(k)} + \Delta\hat{v}^{(k)}. \tag{12.8}$$

A logical initial guess for the van der Waals solution is that given by the ideal equation of state for the same pressure and temperature as given by (12.3). See Table 12.1 for a comparison of the molal volume for carbon dioxide ($a = 3.592$ and $b = 0.04267$) at $T = 300\,K$ and $P = 1, 10$, and $100\,atm$ using the ideal gas and van der Waals equations of state. These results show that the ideal gas equation of state is not appropriate for carbon dioxide at high pressures (they differ by approximately 10% at $P = 20\,atm$ – not shown).

Static Nonlinear Springs

Consider the static equilibrium of the nonlinear spring–mass system shown in Figure 12.3. Recall that for a linear spring, the force in the spring is equal to the spring constant multiplied by the elongation of the spring. The nonlinear springs here are instead proportional to the square of the elongation, making them nonlinear. Summing the vertical forces for free-body diagrams of the two masses leads to the two nonlinear algebraic equations

$$\begin{aligned} m_1 g + K_2(x_2 - x_1)^2 - K_1 x_1^2 &= 0, \\ m_2 g - K_2(x_2 - x_1)^2 &= 0, \end{aligned} \tag{12.9}$$

that must be satisfied for the spring–mass system to be in static equilibrium. Thus, we have a system of two coupled nonlinear algebraic equations for the positions x_1 and x_2. Consequently, we need to extend Newton's method to cases involving systems of equations.

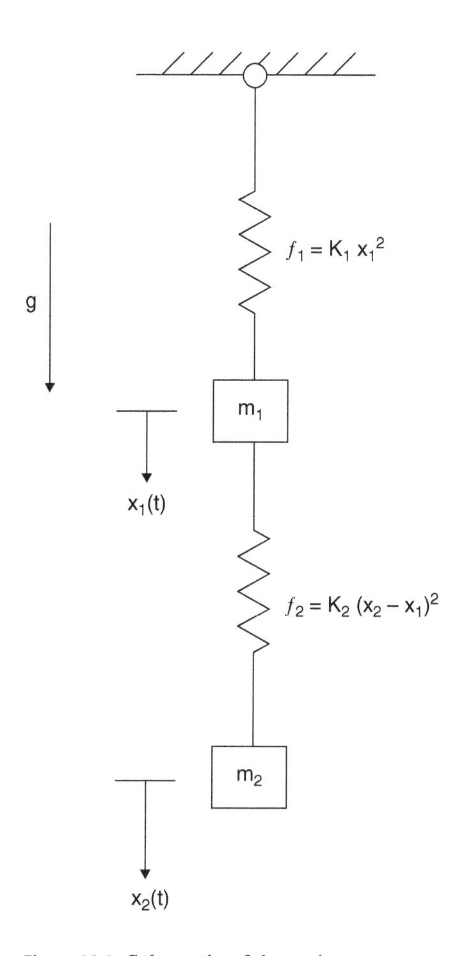

Figure 12.3 Schematic of the spring–mass system with two nonlinear springs.

12.2.4 Systems of Nonlinear Equations

Consider the system of N nonlinear equations in N variables

$$\begin{aligned} f_1(x_1, x_2, \ldots, x_N) &= 0, \\ f_2(x_1, x_2, \ldots, x_N) &= 0, \\ &\vdots \\ f_N(x_1, x_2, \ldots, x_N) &= 0. \end{aligned} \tag{12.10}$$

Following Newton's method for a single nonlinear equation, the multidimensional Taylor series expansion of the first of equations (12.10) is given by

$$f_1\left(x_1, x_2, \ldots, x_N\right) = f_1\left(x_1^{(k)}, x_2^{(k)}, \ldots, x_N^{(k)}\right) + \Delta x_1^{(k)} \left.\frac{\partial f_1}{\partial x_1}\right|_{x_1^{(k)},\ldots,x_N^{(k)}}$$

$$+ \Delta x_2^{(k)} \left.\frac{\partial f_1}{\partial x_2}\right|_{x_1^{(k)},\ldots,x_N^{(k)}} + \cdots + \Delta x_N^{(k)} \left.\frac{\partial f_1}{\partial x_N}\right|_{x_1^{(k)},\ldots,x_N^{(k)}} + \ldots,$$

where $\Delta x_n^{(k)} = x_n - x_n^{(k)}, n = 1, \dots, N$. Similar Taylor series expansions are also formed for f_2 through f_N as well. As in the one-dimensional case, the Taylor series can be truncated if the step sizes $\Delta x_n^{(k)}$ are small.

Setting each of the Taylor series equal to zero, truncating quadratic and higher-order terms (not shown), and expressing in matrix form, with $f_n = f_n\left(x_1^{(k)}, x_2^{(k)}, \dots, x_N^{(k)}\right)$, $n = 1, \dots, N$, these equations are

$$\mathbf{A}^{(k)}\mathbf{\Delta x}^{(k)} = -\mathbf{f}^{(k)},$$

or

$$\begin{bmatrix} \frac{\partial f_1}{\partial x_1} & \frac{\partial f_1}{\partial x_2} & \cdots & \frac{\partial f_1}{\partial x_N} \\ \frac{\partial f_2}{\partial x_1} & \frac{\partial f_2}{\partial x_2} & \cdots & \frac{\partial f_2}{\partial x_N} \\ \vdots & \vdots & \ddots & \vdots \\ \frac{\partial f_N}{\partial x_1} & \frac{\partial f_N}{\partial x_2} & \cdots & \frac{\partial f_N}{\partial x_N} \end{bmatrix} \begin{bmatrix} \Delta x_1^{(k)} \\ \Delta x_2^{(k)} \\ \vdots \\ \Delta x_N^{(k)} \end{bmatrix} = - \begin{bmatrix} f_1 \\ f_2 \\ \vdots \\ f_N \end{bmatrix}, \tag{12.11}$$

where both $\mathbf{A}^{(k)}$ and $\mathbf{f}^{(k)}$ are known when evaluated at the current iteration $\left(x_1^{(k)}, x_2^{(k)}, \dots, x_N^{(k)}\right)$.

Consequently, the linear system of algebraic equations (12.11) is solved in step 2 of Newton's method for the adjustments $\Delta x_n^{(k)}$. Step 3 then becomes

$$x_n^{(k+1)} = x_n^{(k)} + \Delta x_n^{(k)}, \quad n = 1, 2, \dots, N.$$

We again repeat steps 2 and 3 iteratively until the algorithm converges to an approximation of the root noting that the linear system of equations (12.11) must be solved for each iteration. The determinant of the coefficient matrix $\mathbf{A}$ containing the derivatives is called the *Jacobian*. In order for this system of equations to have a unique solution, the Jacobian must be nonzero. Again, we see that Newton's method converts a system of nonlinear equations to a system of linear equations for Δx_n upon which we iterate to obtain an approximation of the root.

Returning to the static nonlinear spring–mass system shown in Figure 12.3, where $N = 2$, we have the two nonlinear algebraic functions

$$f_1(x_1, x_2) = m_1 g + K_2(x_2 - x_1)^2 - K_1 x_1^2,$$
$$f_2(x_1, x_2) = m_2 g - K_2(x_2 - x_1)^2.$$

In order to apply Newton's method, we require the partial derivatives in matrix $\mathbf{A}$:

$$\frac{\partial f_1}{\partial x_1} = -2K_2(x_2 - x_1) - 2K_1 x_1,$$

$$\frac{\partial f_1}{\partial x_2} = 2K_2(x_2 - x_1),$$

$$\frac{\partial f_2}{\partial x_1} = 2K_2(x_2 - x_1),$$

$$\frac{\partial f_2}{\partial x_2} = -2K_2(x_2 - x_1).$$

Then at each iteration, we solve the following system of linear equations for $\Delta x_n^{(k)}$:

$$\begin{bmatrix} \frac{\partial f_1}{\partial x_1} & \frac{\partial f_1}{\partial x_2} \\ \frac{\partial f_2}{\partial x_1} & \frac{\partial f_2}{\partial x_2} \end{bmatrix} \begin{bmatrix} \Delta x_1^{(k)} \\ \Delta x_2^{(k)} \end{bmatrix} = \begin{bmatrix} -f_1 \\ -f_2 \end{bmatrix},$$

where the coefficient matrix $\mathbf{A}^{(k)}$ and right-hand-side vector $\mathbf{f}^{(k)}$ are evaluated for $x_1^{(k)}$ and $x_2^{(k)}$ at each iteration. For example, consider the case with

$$m_1 = m_2 = 1, \quad K_1 = K_2 = 1, \quad g = 9.81.$$

Starting with the initial guess

$$x_1^{(0)} = 1, \quad x_2^{(0)} = 2,$$

the first iteration ($k = 0$) leads to the system of linear equations

$$\begin{bmatrix} -4 & 2 \\ 2 & -2 \end{bmatrix} \begin{bmatrix} \Delta x_1^{(0)} \\ \Delta x_2^{(0)} \end{bmatrix} = \begin{bmatrix} -9.81 \\ -8.81 \end{bmatrix}.$$

We solve this system of equations for $\Delta x_1^{(0)}$ and $\Delta x_2^{(0)}$ and update the approximations to the roots according to

$$x_1^{(1)} = x_1^{(0)} + \Delta x_1^{(0)},$$

$$x_2^{(1)} = x_2^{(0)} + \Delta x_2^{(0)}.$$

The solutions for $\Delta x_1^{(0)}$ and $\Delta x_2^{(0)}$ as well as $x_1^{(1)}$ and $x_1^{(1)}$ for the first iteration are given in Table 12.2 along with the next three iterations. For comparison, the following are the exact solutions:

Table 12.2 First four iterations of Newton's method for the nonlinear spring–mass system shown in Figure 12.3.

k	$\Delta x_1^{(k)}$	$\Delta x_2^{(k)}$	$x_1^{(k+1)}$	$x_2^{(k+1)}$
0	9.31	13.715	10.31	15.715
1	−4.2035	−5.9985	6.1065	9.7165
2	−1.44677	−1.89304	4.65974	7.82346
3	−0.224598	−0.256073	4.43514	7.56739

$$x_1 = \sqrt{\frac{(m_1+m_2)g}{K_1}} = 4.42945,$$

$$x_2 = x_1 + \sqrt{\frac{m_2 g}{K_2}} = 7.56154.$$

Observe that the solution that results after only four iterations of Newton's method agrees quite well with the exact solution.

12.3 Optimization

One of the most powerful and practical uses of matrix and numerical methods is to provide a formal procedure for determining the optimal, or best, solution of a system. That is, how can the system be best utilized to meet some performance or design objective? In optimization of systems governed by algebraic equations, the objective is to determine the "best" means of accomplishing a task that can be captured by a particular set of linear or nonlinear algebraic equations. Typically, such an objective is accomplished by minimizing or maximizing an algebraic *objective function* denoted by $J = f(x_1, x_2, \ldots, x_N)$ that encapsulates the desired performance of the system. The variables $x_1, x_2, \ldots, x_N$ that are sought to minimize the objective function are called the *design*, or *control*, *variables.*

An example of this that has already been encountered is curve fitting using polynomial regression in Chapter 11. We determine the best-fit polynomial that minimizes the total error, which is the sum of the squares of the residuals. The residuals are the vertical distances between the best-fit polynomial and each data point. The total error is the objective function, and the coefficients in the best-fit trial function are the design variables. In this way, curve fitting is essentially an optimization problem in which the optimal curve that fits a data set is sought. More generally, all least-squares problems are optimization problems that seek to minimize an algebraic objective function. In overdetermined systems, the objective function being minimized is the residual of the algebraic system of equations. In underdetermined systems, the objective function being minimized is a function representing the distance of the solution from the origin.

In a topic as broad as optimization, there are numerous possible settings and scenarios to be considered. For example, one could have a single objective function $[J = f(x)]$ or multi-objective functions $[J_1 = f_1(x), J_2 = f_2(x), \ldots, J_M = f_M(x)]$. The objective function(s) could be a function of a single design variable (x) or several $(x_1, x_2, \ldots, x_N)$. The optimization could be unconstrained or subject to algebraic constraints. The objective function(s) and/or constraint(s) could be linear or nonlinear. The constraints could be equality and/or inequality in nature. Finally, the objective function and/or constraints can be continuous or discrete functions.

Although we normally consider linear problems first before tackling nonlinear cases, here we will actually do the opposite as optimization of nonlinear problems follows directly from the calculus of extrema in Chapter 4 and the root-finding methods developed earlier in this chapter. Optimization of linear problems, on the other hand, requires a completely different approach as they do not succumb to such general methods as discussed in Section 4.3.4. Instead, specialized methods must be developed that deal with such problems and the various types of constraints that are inherent to linear optimization problems. Treating nonlinear optimization will allow us to build on standard methods and ease into the issue of enforcing constraints on the optimization process.

Finally, note that optimization is often called "programming" for historical reasons. This is because some of the original algorithms were developed during World War II for military optimization problems having to do with supply chain and logistics "programs" as they were called in the military. It has nothing to do with computer programming as we think of it today.

12.4 Nonlinear Unconstrained Optimization

The necessary and sufficient conditions for extrema of algebraic functions with and without constraints is discussed in Sections 4.3 and Chapter 10, where least-squares methods and quadratic programming are described in a general context. This material provides the basis for what is to come and should be reviewed in detail. Essentially, extrema of a nonlinear function are found by taking the derivative of the algebraic equation(s) with respect to each of the independent variables and setting equal to zero. The values of the independent variables for which this condition is satisfied represent the optimal solution. Recall from the introduction to this chapter that optimization of a nonlinear algebraic function $f(x)$, which requires locating minimums or maximums, simply requires determining the root(s) of its derivative $f'(x)$.

Let us consider some examples in which the objective function $J = f(x)$ or $J = f(x_1,x_2,\ldots,x_N)$ is a single nonlinear algebraic expression of one or more design variables. For simple objective functions, differentiating the objective function with respect to each of the design variables and setting equal to zero to determine the optimal solution directly is straightforward and produces the exact solution. If necessary, second derivatives can be evaluated at extrema points to ensure that they are the minimum or maximum that we seek. In more complex scenarios, Newton's iterative method can be used as a numerical method to approximate the root(s) of $f'(x_1,x_2,\ldots,x_N)$ in order to evaluate the optimal solution.

12.4.1 One-Dimensional Objective Function

Let us first consider a case involving a single objective function that is a function of only one variable.

Example 12.1 The fuel economy of your car varies quite substantially with speed. For example, the data measured for a particular sedan are shown in Figure 12.4. Develop a polynomial curve fit that accurately represents such data, and determine the speed at which the maximum fuel economy for this automobile occurs.

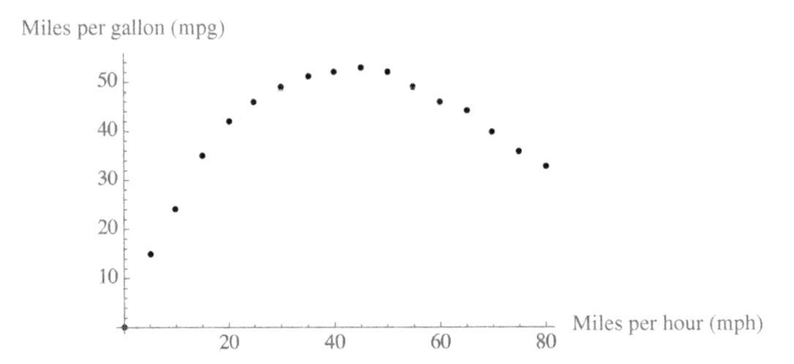

Figure 12.4 Fuel economy as a function of speed measured for a particular sedan.

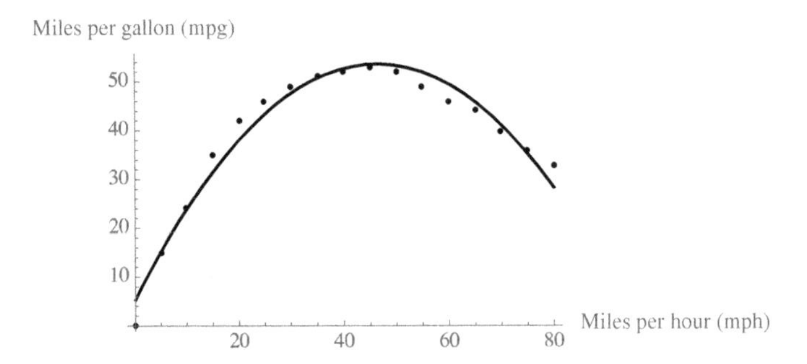

Figure 12.5 Quadratic curve fit to fuel economy data.

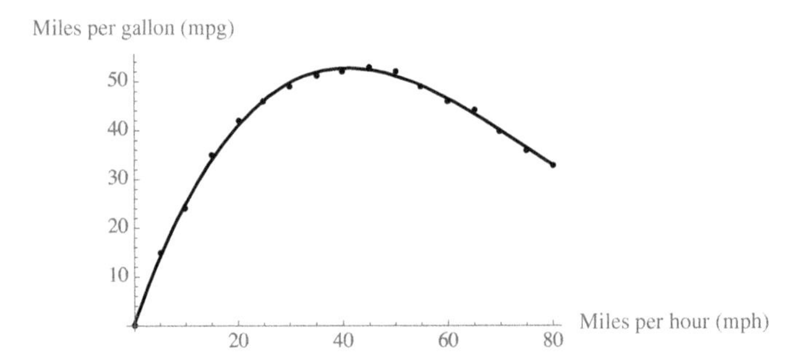

Figure 12.6 Cubic curve fit to fuel economy data.

Solution

A quadratic curve fit is shown in Figure 12.5. Observe that a simple parabola does not provide a good representation of the experimental data. A cubic polynomial curve fit is shown in Figure 12.6 and given by

$$J = f(x) = 0.00022291x^3 - 0.0491796x^2 + 2.91153x + 0.653251.$$

This produces a reasonably good representation of the data. Thus, we will use it to determine the speed at which the maximum fuel economy occurs.

In some cases, we can simply differentiate the objective function $J = f(x)$ with respect to the design variable x, set equal to zero, and solve for the extremum. Here, for example, the derivative of the objective function produces a quadratic function, for which the quadratic formula provides the roots. Unfortunately, this is only the case for simple objective functions having a small number of extrema. Instead, we need a more general procedure. Recall that an extremum of a function is equivalent to the root of the first derivative of that function. Therefore, let us differentiate our cubic curve fit as follows:

$$f'(x) = 0.000668731x^2 - 0.0983591x + 2.91153,$$

which gives a quadratic polynomial. We can then use a root-finding technique to approximate the root(s) of $f'(x) = 0$, which corresponds to an extremum of the objective function $f(x)$.

Recall that each iteration of the standard Newton's method for root finding in $f(x)$ is given by

$$x^{(k+1)} = x^{(k)} + \Delta x^{(k)}, \quad \text{where} \quad \Delta x^{(k)} = -\frac{f\left(x^{(k)}\right)}{f'\left(x^{(k)}\right)}.$$

Because we must differentiate the objective function to convert an extremum to a root, that is, we seek points where $f'(x) = 0$, an iteration of Newton's method in optimization contexts becomes

$$x^{(k+1)} = x^{(k)} + \Delta x^{(k)}, \quad \text{where} \quad \Delta x^{(k)} = -\frac{f'\left(x^{(k)}\right)}{f''\left(x^{(k)}\right)}. \tag{12.12}$$

This is equivalent to writing a Taylor series expansion for $f(x)$ in which we retain second-order derivatives and setting the derivative of the Taylor series equal to zero. Newton's method requires analytical evaluation of the first and second derivatives of the objective function. This is not always practical in large-scale, multidimensional problems.

Because a quadratic polynomial has two roots, we must choose an initial guess that is within the range of our data and near the maximum in order to capture the proper root that corresponds to the desired maximum in the objective function. This leads to the result that $x = 41.0677$ mph corresponds to the maximum fuel economy for this vehicle. Recall that in general, we seek minimums of the objective function. Therefore, we could have multiplied our curve fit by negative one and then sought a minimum instead of a maximum. Obviously, we would end up with the same result.

12.4.2 Multidimensional Objective Function

Next we consider an example for which the single objective function is a function of multiple design variables. For example, a two-dimensional objective function may have the surface plot shown in Figure 12.7. This objective function has two local minimums (one of which is global) and a single global maximum. Typically, our

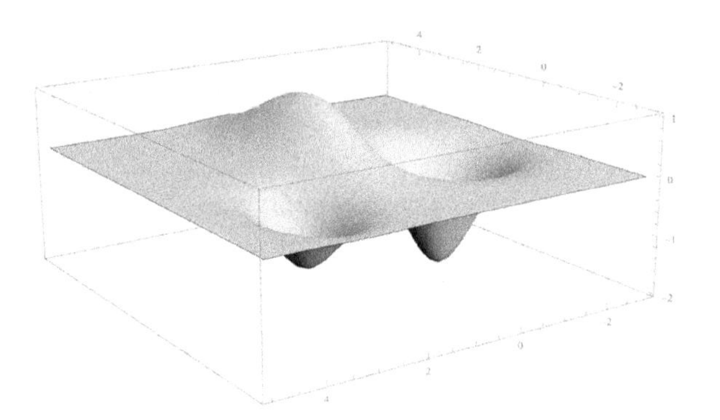

Figure 12.7 Surface plot of objective function with two design variables.

objective in the optimization procedure would be to identify the values of the design variables that correspond to the global minimum (maximum) in the objective function. Let us illustrate this with an objective function involving three design variables.

Example 12.2 Obtain the location (x_1, x_2, x_3) at which a minimum value of the function

$$f(x_1, x_2, x_3) = \frac{f_1(x_1, x_2, x_3)}{f_2(x_1, x_2, x_3)} = \frac{x_1^2 + 6x_1x_2 + 4x_2^2 + x_3^2}{x_1^2 + x_2^2 + x_3^2}$$

occurs.

Solution

At a point (x_1, x_2, x_3) where $f(x_1, x_2, x_3)$ has zero slope, we have

$$\frac{\partial f}{\partial x_n} = 0, \quad n = 1, 2, 3.$$

With $f = f_1/f_2$, this requires that (using the quotient rule)

$$\frac{1}{f_2}\left(\frac{\partial f_1}{\partial x_n} - \frac{f_1}{f_2}\frac{\partial f_2}{\partial x_n}\right) = 0, \quad n = 1, 2, 3,$$

or letting $\lambda = f = f_1/f_2$ (and canceling $1/f_2$)

$$\frac{\partial f_1}{\partial x_n} - \lambda\frac{\partial f_2}{\partial x_n} = 0, \quad n = 1, 2, 3. \tag{12.13}$$

Recall that

$$f_1 = x_1^2 + 6x_1x_2 + 4x_2^2 + x_3^2, \quad f_2 = x_1^2 + x_2^2 + x_3^2.$$

Therefore, the partial derivatives in (12.11) are given by

$$\frac{\partial f_1}{\partial x_1} = 2x_1 + 6x_2, \quad \frac{\partial f_2}{\partial x_1} = 2x_1,$$
$$\frac{\partial f_1}{\partial x_2} = 6x_1 + 8x_2, \quad \frac{\partial f_2}{\partial x_2} = 2x_2,$$
$$\frac{\partial f_1}{\partial x_3} = 2x_3, \quad \frac{\partial f_2}{\partial x_3} = 2x_3.$$

Substituting into (12.13) gives the following three equations for the design variables

$$2x_1 + 6x_2 - \lambda 2x_1 = 0,$$
$$6x_1 + 8x_2 - \lambda 2x_2 = 0,$$
$$2x_3 - \lambda 2x_3 = 0,$$

or canceling 2 from each equation and rearranging gives the eigenproblem

$$\begin{bmatrix} 1 & 3 & 0 \\ 3 & 4 & 0 \\ 0 & 0 & 1 \end{bmatrix} \begin{bmatrix} x_1 \\ x_2 \\ x_3 \end{bmatrix} = \lambda \begin{bmatrix} x_1 \\ x_2 \\ x_3 \end{bmatrix}.$$

Interestingly, we were able to convert this nonlinear optimization problem into an eigenproblem, which is a linear system of algebraic equations. The eigenvalues are possible values of the objective function, and the eigenvectors are the corresponding values of the design variables. This is not normally the case. The relevant eigenvalue/vector pairing that produces the minimum in the objective function is simply determined from the eigenvalues.

For a nontrivial solution, the required eigenvalues are (from MATLAB or Mathematica, for example)

$$\lambda_1 = 1, \quad \lambda_2 = \frac{1}{2}(5 + 3\sqrt{5}) = 5.8541, \quad \lambda_3 = \frac{1}{2}(5 - 3\sqrt{5}) = -0.8541,$$

and the corresponding eigenvectors are

$$\mathbf{x}_1 = \begin{bmatrix} 0 \\ 0 \\ 1 \end{bmatrix}, \quad \mathbf{x}_2 = \begin{bmatrix} -\frac{4}{3} + \frac{1}{6}(5 + 3\sqrt{5}) \\ 1 \\ 0 \end{bmatrix}, \quad \mathbf{x}_3 = \begin{bmatrix} -\frac{4}{3} + \frac{1}{6}(5 - 3\sqrt{5}) \\ 1 \\ 0 \end{bmatrix}.$$

Of the three eigenvalues (objective function values), $f = \lambda_3 = \frac{1}{2}(5 - 3\sqrt{5})$ has the minimum value, which occurs for the eigenvector (design variables)

$$\begin{bmatrix} x_1 \\ x_2 \\ x_3 \end{bmatrix} = \mathbf{x}_3 = \begin{bmatrix} -\frac{4}{3} + \frac{1}{6}(5 - 3\sqrt{5}) \\ 1 \\ 0 \end{bmatrix}.$$

As you can see from just the first two examples, optimization draws from numerous areas of mathematics and numerical methods, including differential calculus, eigenproblems, curve fitting to obtain the objective function from empirical data, and root finding to determine the value(s) of the design variable(s) that lead to an extremum in the objective function

12.5 Numerical Methods

In order to solve more complicated nonlinear optimization problems, numerical methods are required. Newton's method for root finding and the steepest descent and conjugate-gradient methods described in Section 10.6 can be adapted for such problems.

12.5.1 Method of Steepest Descent

Looking at the plot of the two-dimensional objective function in Figure 12.7 suggests another approach for objective functions of multiple design variables. If a minimum is sought, then one could imagine starting somewhere in that minimum's "bowl" and proceeding down the bowl until the minimum is reached. This is essentially a multidimensional version of a Newton-like method consisting of two steps:

1. Determine the "best" direction to proceed down the bowl (given by the slope of the tangent line in Newton's method).
2. Determine how far to proceed in that direction (Δx in Newton's method).

If, for example, the objective function is given by $J = f(x_1, x_2, \ldots, x_n, \ldots, x_N)$, the extrema are the roots of the system of equations

$$\frac{\partial f}{\partial x_1} = 0, \quad \frac{\partial f}{\partial x_2} = 0, \quad \ldots \quad \frac{\partial f}{\partial x_n} = 0, \quad \ldots \quad \frac{\partial f}{\partial x_N} = 0.$$

Note that this is equivalent to setting the gradient of the function equal to zero as follows:

$$\boldsymbol{\nabla} f = \begin{bmatrix} \frac{\partial f}{\partial x_1} & \frac{\partial f}{\partial x_2} & \cdots & \frac{\partial f}{\partial x_n} & \cdots & \frac{\partial f}{\partial x_N} \end{bmatrix}^T = \mathbf{0}.$$

In the *method of steepest descent*, one simply proceeds down in the direction that has the largest negative slope during each iteration until a minimum in the function is reached. Recall the two steps that comprise each iteration:

1. Determine the direction to proceed: at a given point, the direction of steepest descent, which is the negative of the gradient, is determined. For the kth iteration, the direction of steepest descent $\mathbf{p}_k$ is

$$\mathbf{p}^{(k)} = -\boldsymbol{\nabla} f\left(\mathbf{x}^{(k)}\right).$$

2. Determine how far to proceed: we then proceed along a straight line in the direction of steepest descent as long as the objective function is decreasing. Therefore, the step size

$$\Delta \mathbf{x}^{(k)}$$

is chosen as the distance between the current starting point and the minimum along the straight descent path. Given the step sizes, we update the values of the design variables at the current iteration according to

$$\mathbf{x}^{(k+1)} = \mathbf{x}^{(k)} + \Delta \mathbf{x}^{(k)}.$$

Once the location of the next point in the design variables is determined, the next iteration is initiated from this point.

REMARKS:

1. *Essentially, the method of steepest descent converts the multidimensional descent problem into a series of one-dimensional minimum-seeking steps, one for each iteration of the algorithm.*
2. *If you imagine hiking into a valley using this procedure, while you will likely reach the lowest elevation in the valley eventually, it is easy to see how such a path would likely be somewhat meandering and not the shortest path toward the minimum (see Section 10.6 for more discussion of this).*
3. *The method of steepest descent only has a first-order convergence rate. Newton's method, to be discussed next, has a second-order convergence rate.*

12.5.2 Newton's Method

As in the method of steepest descent, a single objective function of N design variables can be converted into a system of N nonlinear equations by evaluating the gradient, which is the derivative of the objective function with respect to each of the design variables, and setting equal to zero. This system of nonlinear equations then can be solved using Newton's method as formulated in Section 12.2.4. Alternatively, we can evaluate the multidimensional Taylor series of the objective function in vector form first and then evaluate its gradient to produce the system of linear algebraic equations to solve for the step sizes. Let us pursue this latter approach.

Suppose that we have the single objective function as a function of N design variables

$$J = f(x_1, x_2, \dots, x_N), \tag{12.14}$$

or $J = f(\mathbf{x})$ for short. In vector form, the Taylor series of the objective function about a point $\mathbf{x}^{(k)}$ is given by

$$f(\mathbf{x}) = f^{(k)} + \left[\nabla f^{(k)}\right]^T \Delta \mathbf{x}^{(k)} + \frac{1}{2}\left[\Delta \mathbf{x}^{(k)}\right]^T \nabla \left[\nabla f^{(k)}\right]^T \Delta \mathbf{x}^{(k)} + \dots, \tag{12.15}$$

where each of the following are evaluated for the design variables at the kth iteration:

$$
\begin{aligned}
f^{(k)} &= f\left(\mathbf{x}^{(k)}\right) &&= \text{objective function},\\
\boldsymbol{\nabla} f^{(k)} &= \boldsymbol{\nabla} f\left(\mathbf{x}^{(k)}\right) &&= \text{gradient of the objective function},\\
\Delta \mathbf{x}^{(k)} &= \mathbf{x} - \mathbf{x}^{(k)} &&= \text{step sizes for iteration},\\
\boldsymbol{\nabla}\left[\boldsymbol{\nabla} f^{(k)}\right]^T &= \mathbf{H}^{(k)} &&= \text{Hessian matrix}.
\end{aligned}
$$

The $N \times N$ Hessian matrix consists of all possible second derivatives of the objective function and is given by

$$
\mathbf{H}^{(k)} = \boldsymbol{\nabla}\left[\boldsymbol{\nabla} f^{(k)}\right]^T = \begin{bmatrix}
\dfrac{\partial^2 f}{\partial x_1^2} & \dfrac{\partial^2 f}{\partial x_1 \partial x_2} & \cdots & \dfrac{\partial^2 f}{\partial x_1 \partial x_N} \\
\dfrac{\partial^2 f}{\partial x_2 \partial x_1} & \dfrac{\partial^2 f}{\partial x_2^2} & \cdots & \dfrac{\partial^2 f}{\partial x_2 \partial x_N} \\
\vdots & \vdots & \ddots & \vdots \\
\dfrac{\partial^2 f}{\partial x_N \partial x_1} & \dfrac{\partial^2 f}{\partial x_N \partial x_2} & \cdots & \dfrac{\partial^2 f}{\partial x_N^2}
\end{bmatrix}.
$$

Observe that if all of the derivatives of f are continuous, then the Hessian matrix is symmetric. Using this Hessian matrix, the *second-derivative test* can be used to determine what type of extrema exists at each stationary point as follows (see Section 4.3.1):

- If the Hessian matrix is negative definite (all eigenvalues are negative), then the function $f(x_1, \ldots, x_N)$ is concave and has a single local maximum.
- If the Hessian matrix is positive definite (all eigenvalues are positive), then the function $f(x_1, \ldots, x_N)$ is convex and has a single local minimum.
- If the Hessian matrix has both positive and negative eigenvalues, then the function $f(x_1, \ldots, x_N)$ has a saddle point.
- If the Hessian matrix is semidefinite (at least one eigenvalue is zero), then the second-derivative test is inconclusive.

Returning to the objective function (12.14), at an extremum point

$$
\frac{\partial f(\mathbf{x})}{\partial x_n} = 0, \quad n = 1, 2, \ldots, N,
$$

which is equivalent to setting the gradient of the objective function equal to zero according to

$$
\boldsymbol{\nabla} f = \begin{bmatrix} \dfrac{\partial f}{\partial x_1} & \dfrac{\partial f}{\partial x_2} & \cdots & \dfrac{\partial f}{\partial x_n} & \cdots & \dfrac{\partial f}{\partial x_N} \end{bmatrix}^T = \mathbf{0}.
$$

Evaluating the gradient, with its partial derivatives with respect to the design variables $\mathbf{x}$, of the Taylor series (12.15) for $f(\mathbf{x})$ and setting equal to zero gives

$$
\boldsymbol{\nabla} f = \boldsymbol{\nabla} f^{(k)} + \mathbf{H}^{(k)} \Delta \mathbf{x}^{(k)} = \mathbf{0}. \tag{12.16}
$$

If the Hessian matrix is not singular, then (12.16) can be solved for the step size at each iteration from the following system of linear algebraic equations:

$$\mathbf{H}^{(k)}\Delta\mathbf{x}^{(k)} = -\boldsymbol{\nabla} f^{(k)}. \tag{12.17}$$

When written out, this system of equations is given by

$$\begin{bmatrix} \frac{\partial^2 f}{\partial x_1^2} & \frac{\partial^2 f}{\partial x_1 \partial x_2} & \cdots & \frac{\partial^2 f}{\partial x_1 \partial x_N} \\ \frac{\partial^2 f}{\partial x_2 \partial x_1} & \frac{\partial^2 f}{\partial x_2^2} & \cdots & \frac{\partial^2 f}{\partial x_2 \partial x_N} \\ \vdots & \vdots & \ddots & \vdots \\ \frac{\partial^2 f}{\partial x_N \partial x_1} & \frac{\partial^2 f}{\partial x_N \partial x_2} & \cdots & \frac{\partial^2 f}{\partial x_N^2} \end{bmatrix} \begin{bmatrix} \Delta x_1^{(k)} \\ \Delta x_2^{(k)} \\ \vdots \\ \Delta x_N^{(k)} \end{bmatrix} = - \begin{bmatrix} \frac{\partial f}{\partial x_1} \\ \frac{\partial f}{\partial x_2} \\ \vdots \\ \frac{\partial f}{\partial x_N} \end{bmatrix}.$$

As you can see, Newton's method for optimization requires evaluation of both the gradient (first derivatives) and Hessian (second derivatives) of the objective function at each iteration. Compare this with (12.11) for the traditional Newton's method applied to root finding, in which we only require evaluation of the function and its gradient at each iteration. In this way, we are essentially applying the standard Newton's method for root finding to the gradient of the objective function. Whereas the standard Newton's method for root finding approximates the function locally by a straight line (or plane in multidimensions) at each iteration, that for optimization approximates the function locally at each iteration as a quadratic function (parabola or paraboloid). That is, a parabolic function locally approximating $f(\mathbf{x})$ corresponds to a linear approximation for its gradient $\boldsymbol{\nabla} f(\mathbf{x})$.

As with the standard Newton's method, the gradient and Hessian are evaluated at the current iteration in order to obtain the step sizes $\Delta x_n^{(k)}, n = 1, \dots, N$ to the next iteration. The updated values of the design variables at the next iteration are then obtained using

$$\mathbf{x}^{(k+1)} = \mathbf{x}^{(k)} + \Delta\mathbf{x}^{(k)}. \tag{12.18}$$

Equations (12.17) and (12.18) are the multidimensional version of (12.12).

REMARKS:

1. *As for root finding, use of Newton's method for optimization exhibits a quadratic – second-order – convergence rate if the initial guess is sufficiently close to the minimum being sought.*
2. *The inverse of the Hessian matrix is required to solve the system* (12.17). *The computational cost of computing this inverse is high for very large problems with numerous design variables. This issue is addressed in* quasi-Newton methods *that approximate the Hessian matrix using only first-order, rather than second-order, partial derivatives. As the iterations proceed, an improved approximation of the*

true Hessian matrix is obtained without the need to evaluate the second derivatives and the inverse.

3. *Using terminology from dynamical systems (see Chapter 5), the basin of attraction of each root of a function as determined by Newton's method can be quite interesting. For example, the basins of attraction of each of the three roots of the complex cubic function $f(z) = z^3 - 1$ lead to a structure known as the* Julia fractal set. *Not only is it quite surprising, it is very beautiful (a simple internet search will reveal many artistic renditions of the Julia set, where each color represents the basis of attraction of a single root).*

12.5.3 Conjugate-Gradient Method

Whereas the conjugate-gradient method was applied to a system of equations with symmetric, positive-definite coefficient matrices in Section 10.6, which corresponds to a quadratic programming problem, it can be applied to more general nonlinear optimization problems by approximating the objective function locally as a quadratic. Of course, nonlinear optimization problems often are not quadratic with a single minimum; however, a general nonlinear objective function with multiple local minima can be locally approximated as a quadratic function in the vicinity of a local minimum for the purpose of determining the minimizing direction to proceed at each iteration. In this way, we can take advantage of the mathematical properties of determining minimums of convex functions locally as we proceed through the function toward a local minimum. It is essentially a multidimensional version of Newton's method. Like Newton's method, therefore, convergence is very fast if a good initial guess is available, and the choice of initial guess will determine which local minimum is approximated. Unlike Newton's method, however, there is no need to invert a Hessian matrix during each step of the algorithm. Some of the details must be modified when applied to nonlinear optimization problems as compared to linear conjugate-gradient methods. The most common conjugate-gradient algorithm for optimization problems is due to Fletcher and Reeves (1964).

12.5.4 Gradient-Based and Gradient-Free Methods

You will have noticed a common theme in the optimization methods covered in this chapter for nonlinear objective functions: they all require evaluation of the first derivative(s), or gradient, and in some cases the second derivative(s), or Hessian, of the objective function(s). This is the case for Newton's method, the method of steepest descent, and the conjugate-gradient method. These gradients and Hessians, which are either evaluated analytically or numerically, guide the methods in their methodical search for an extremum. Such approaches are called *gradient-based methods*. As emphasized throughout, they can be very effective if a good initial guess is available that is near the extremum being sought. For functions with a single extremum, such as convex functions, gradient-based methods will locate the global extremum very efficiently.

For objective functions containing multiple local extrema, however, it may be difficult to locate the global extremum that is typically the objective of the optimization problem. In addition, function and derivative evaluations can dominate the computational time in some algorithms. Finally, there has been an underlying assumption that our objective functions are smooth, such that their gradient and Hessian calculations are smooth as well, which is often not the case. Noise in the data can lead to numerous small apparent extrema in the function that are not truly present in the underlying system behavior. When this is the case, the gradient and Hessian are even less smooth than the underlying function and lead to great difficulty in locating the true extremum(s) of the function. Owing to one or more of these reasons, there is a need for *gradient-free*, or *derivative-free*, optimization algorithms. These algorithms are very different from those outlined earlier and draw on different mathematics in their search for global extrema. These include genetic algorithms, which consist of a search technique inspired by biological evolution to determine the minimum or maximum, as well as adjoint methods, which are covered in chapter 10 of Cassel (2013).

12.6 Nonlinear Constrained Optimization

In addition to the objective function(s), an optimization problem may have one or both of two types of constraints: (1) equality constraints that specify an algebraic relationship between the design variables that must be satisfied exactly by the system and/or (2) inequality constraints that specify limits on the values of the design variables that must be honored by the system.

12.6.1 Optimization Subject to Equality Constraints

When the optimization problem involves algebraic equality constraints on the minimization (maximization) of the objective function, the method of Lagrange multipliers is used. It allows for the constrained minimization (maximization) problem to be converted to an equivalent unconstrained extremum problem that can be solved in the usual manner. If the number of equality constraints is equal to the number of design variables, then there is a unique solution of the constrained system, and there is no room for optimization. Consequently, we are interested in the case when there are fewer constraints than design variables. Let us illustrate the method of Lagrange multipliers for imposing equality constraints using a geometric problem.

Example 12.3 Determine the semimajor and semiminor axes of the ellipse centered at the origin defined by

$$(x_1 + x_2)^2 + 2(x_1 - x_2)^2 = 8,$$

which may be written

$$g(x_1, x_2) = 3x_1^2 - 2x_1x_2 + 3x_2^2 - 8.$$

Solution

To determine the semimajor (minor) axis, calculate the farthest (nearest) point on the ellipse from the origin. Therefore, we maximize (minimize) the distance objective function $J = f(x_1, x_2) = x_1^2 + x_2^2$, which is the square of the distance from the origin, subject to the constraint that the coordinates (x_1, x_2) be on the ellipse. Squaring the distance produces the same semiminor and semimajor axes but eases the analysis as negative distances and positive distances are treated the same. Clearly, without the constraint, minimizing and maximizing the distance objective function would result in the minimum being zero and the maximum being infinity.

As discussed in Section 4.3.2, the method of Lagrange multipliers involves defining an augmented function as follows:

$$\tilde{f}(\Lambda, x_1, x_2) = f(x_1, x_2) + \Lambda g(x_1, x_2),$$

where Λ is the Lagrange multiplier that is multiplied by the constraint that the extrema be on the ellipse. Determining the extremum(s) of the objective function subject to the constraint is equivalent to finding the extremum(s) of the augmented function subject to no constraint. Therefore, we apply the same process as before to the augmented function. To determine the extrema of the algebraic function $\tilde{f}(x_1, x_2)$, we evaluate

$$\frac{\partial \tilde{f}}{\partial x_1} = 0, \quad \frac{\partial \tilde{f}}{\partial x_2} = 0,$$

with

$$\tilde{f}(x_1, x_2) = x_1^2 + x_2^2 + \Lambda\left(3x_1^2 - 2x_1x_2 + 3x_2^2 - 8\right).$$

Evaluating the partial derivatives and setting equal to zero gives

$$\frac{\partial \tilde{f}}{\partial x_1} = 2x_1 + \Lambda\,(6x_1 - 2x_2) = 0,$$

$$\frac{\partial \tilde{f}}{\partial x_2} = 2x_2 + \Lambda\,(-2x_1 + 6x_2) = 0.$$

Thus, we have two equations for x_1 and x_2 given by

$$3x_1 - x_2 = \lambda_n x_1,$$
$$-x_1 + 3x_2 = \lambda_n x_2,$$

where $\lambda_n = -1/\Lambda$. This is an eigenproblem of the form $\mathbf{A}\mathbf{x}_n = \lambda_n \mathbf{x}_n$, where

$$\mathbf{A} = \begin{bmatrix} 3 & -1 \\ -1 & 3 \end{bmatrix}.$$

The eigenvalues of the symmetric matrix $\mathbf{A}$ are $\lambda_1 = 2$ and $\lambda_2 = 4$ with the corresponding eigenvectors

$$\mathbf{x}_1 = \begin{bmatrix} 1 \\ 1 \end{bmatrix}, \quad \mathbf{x}_2 = \begin{bmatrix} -1 \\ 1 \end{bmatrix}.$$

The two eigenvectors $\mathbf{x}_1$ and $\mathbf{x}_2$ (along with $-\mathbf{x}_1$ and $-\mathbf{x}_2$) give the directions of the semimajor and semiminor axes, which are along lines that bisect the first and third quadrants and second and fourth quadrants, respectively.

In order to determine which eigenvectors correspond to the semimajor and semiminor axes, we recognize that a point on the ellipse must satisfy

$$(x_1 + x_2)^2 + 2(x_1 - x_2)^2 = 8.$$

Considering $\mathbf{x}_1 = \begin{bmatrix} 1 & 1 \end{bmatrix}^T$, let us set $x_1 = c_1$ and $x_2 = c_1$. Substituting into the equation for the ellipse yields

$$4c_1^2 + 0 = 8,$$

in which case $c_1 = \pm\sqrt{2}$. Therefore, $x_1 = \sqrt{2}$ and $x_2 = \sqrt{2}$ or ($x_1 = -\sqrt{2}$ and $x_2 = -\sqrt{2}$), and the length of the corresponding axis is $\sqrt{x_1^2 + x_2^2} = 2$.

Similarly, considering $\mathbf{x}_2 = \begin{bmatrix} -1 & 1 \end{bmatrix}^T$, let us set $x_1 = -c_2$ and $x_2 = c_2$. Substituting into the equation for the ellipse yields

$$0 + 8c_2^2 = 8,$$

in which case $c_2 = \pm 1$. Therefore, $x_1 = -1$ and $x_2 = 1$ (or $x_1 = 1$ and $x_2 = -1$), and the length of the corresponding axis is $\sqrt{x_1^2 + x_2^2} = \sqrt{2}$. As a result, the eigenvector $\mathbf{x}_1$ corresponds to the semimajor axis, and $\mathbf{x}_2$ corresponds to the semiminor axis.

REMARKS:

1. *In the preceding example, both the objective function and constraints are quadratic functions. In such quadratic programming problems, the optimal solution for the augmented function always produces a generalized eigenproblem of the form*

$$\mathbf{A}\mathbf{x}_n = \lambda_n \mathbf{B}\mathbf{x}_n$$

to be solved for the Lagrange multipliers, which are the eigenvalues, and the corresponding optimal design variables, which are the eigenvectors (see Section 4.3.3). In the previous example, $\mathbf{B} = \mathbf{I}$, which corresponds to a regular eigenproblem.

2. *If there are multiple constraints*

$$g_1(x_1,\ldots,x_N) = c_1, \quad g_2(x_1,\ldots,x_N) = c_2, \quad \ldots, \quad g_S(x_1,\ldots,x_N) = c_S,$$

then each one gets its own Lagrange multiplier, such that the augmented objective function is of the form

$$\tilde{J} = J + \Lambda_1(g_1 - c_1) + \Lambda_2(g_2 - c_2) + \cdots + \Lambda_S(g_S - c_S).$$

3. *Observe that evaluating*

$$\frac{\partial \tilde{J}}{\partial \Lambda} = 0$$

simply produces the constraint equation.

Example 12.4 The maximum bending stress in a beam with a rectangular cross section is given by[2]

$$\sigma_{max} = \frac{My}{I},$$

where $M = 2,000$ lbf-in is the maximum bending moment, y is the vertical distance from the neutral axis to the outer edge of the rectangular beam, and I is the moment of inertia of the cross section about the neutral axis. Determine the width x_1 and height x_2 of the rectangular beam that fits within a circular shape of radius $R = 10$ in that minimizes the maximum bending stress σ_{max} (see Figure 12.8).

Solution

For a rectangular beam:

$$y = \frac{1}{2}x_2, \quad I = \frac{1}{12}x_1x_2^3.$$

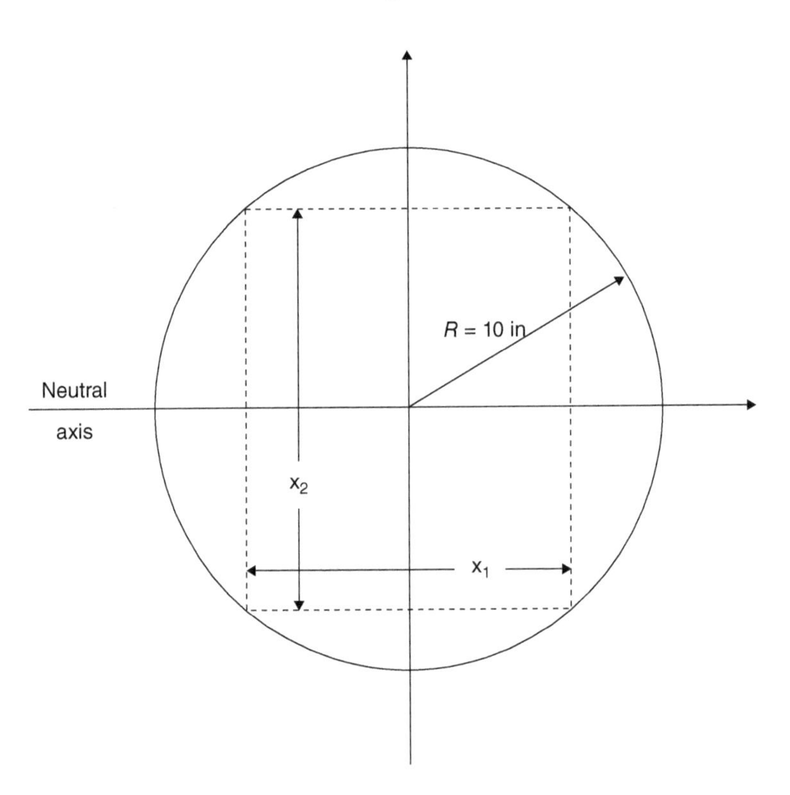

Figure 12.8 Cross section of rectangular beam cut from a circular cross section with radius 10 in.

[2] This example is adapted from Rao (2002).

Therefore, the objective function to be minimized is given by

$$J = \sigma_{max}(x_1, x_2) = \frac{2,000\left(\frac{1}{2}x_2\right)}{\frac{1}{12}x_1x_2^3} = \frac{12,000}{x_1x_2^2}.$$

The constraint is that the rectangular cross section fits within the circular shape. Therefore, from the Pythagorean theorem

$$g(x_1, x_2) = \left(\frac{x_1}{2}\right)^2 + \left(\frac{x_2}{2}\right)^2 = R^2,$$

or

$$g(x_1, x_2) = x_1^2 + x_2^2 - 400 = 0.$$

The augmented objective function is then

$$\tilde{J}(\Lambda, x_1, x_2) = \frac{12,000}{x_1x_2^2} + \Lambda\left(x_1^2 + x_2^2 - 400\right),$$

where Λ is the Lagrange multiplier. Setting the derivatives of the augmented function with respect to each of the design variables equal to zero yields

$$\frac{\partial \tilde{J}}{\partial x_1} = -\frac{12,000}{x_1^2x_2^2} + 2\Lambda x_1 = 0,$$

$$\frac{\partial \tilde{J}}{\partial x_2} = -\frac{24,000}{x_1x_2^3} + 2\Lambda x_2 = 0.$$

Because we really only care about determining the design variables x_1 and x_2, and not the Lagrange multiplier Λ, let us eliminate Λ by multiplying the first of these equations by x_2 and subtracting the second equation multiplied by x_1. This gives

$$-\frac{12,000}{x_1^2x_2^2}x_2 + \frac{24,000}{x_1x_2^3}x_1 = 0,$$

which after simplification leads to

$$x_2 = \sqrt{2}x_1.$$

Substituting this result into the constraint yields

$$x_1^2 + x_2^2 = 400,$$
$$x_1^2 + 2x_1^2 = 400,$$
$$3x_1^2 = 400,$$

$$\therefore x_1 = \sqrt{\frac{400}{3}} = \frac{20}{\sqrt{3}} = 11.547.$$

The corresponding value of x_2 is then

$$x_2 = \sqrt{2}x_1 = \sqrt{\frac{2}{3}}20 = 16.3299.$$

12.6.2 Optimization Subject to Inequality Constraints

Inequality constraints can be converted into equality constraints using *slack variables* and then solved using the Lagrange multiplier method. We will discuss and illustrate slack variables in the context of linear programming to be considered next.

12.7 Linear Programming – Formulation

There are three issues that often render linear programming problems more difficult to solve in practice than nonlinear programming problems: (1) because linear equations do not have finite extrema, they require constraints; (2) because they are typically "easier" to solve numerically, linear programming problems are often much larger than nonlinear ones, with many more design variables and constraints; and (3) linear programming introduces an excessive amount of new jargon. For these reasons, we have saved linear programming for last.

12.7.1 Standard Formulation

The general algebraic optimization problem is to determine the values of the *design*, or *control*, *variables* $x_1, x_2, \dots, x_n, \dots, x_N$ that minimizes the objective, or cost, function

$$J = f(x_1, x_2, \dots, x_n, \dots, x_N) \tag{12.19}$$

subject to the *equality constraint(s)*

$$g_m(x_1, x_2, \dots, x_n, \dots, x_N) = 0, \quad m = 1, \dots, M \tag{12.20}$$

and the *inequality constraint(s)*

$$h_s(x_1, x_2, \dots, x_n, \dots, x_N) \leq 0, \quad s = 1, \dots, S \tag{12.21}$$

and the *positivity constraints*

$$x_1 \geq 0, \quad x_2 \geq 0, \quad \dots \quad x_n \geq 0, \quad \dots \quad x_N \geq 0. \tag{12.22}$$

The constraints provide restrictions on the values that the design variables can possess. For example, a rocket engine has a minimum (zero) and maximum thrust, a company's profit cannot be larger than its revenue (unfortunately, it can be negative), and a manufacturer cannot produce a negative number of widgets. If the functions f, g, and h are all linear functions of the design variables, then it is called a *linear programming*, or *optimization*, *problem*. If any of these functions are nonlinear, it is a *nonlinear programming*, or *optimization*, *problem*. As a special case of nonlinear programming, if the objective function is quadratic and constraints linear, then it is a *quadratic programming problem*. Our primary task here is to develop methods for incorporating the various types of constraints into linear programming problems.

Let us express the design variables as the vector

$$\mathbf{x} = \begin{bmatrix} x_1 & x_2 & \dots & x_n & \dots & x_N \end{bmatrix}^T.$$

When f, g, and h are all linear, this linear programming problem can be expressed in the matrix form that requires minimizing the objective function

$$J(\mathbf{x}) = \mathbf{c}^T\mathbf{x} = c_1x_1 + c_2x_2 + \cdots + c_nx_n + \cdots + c_Nx_N, \tag{12.23}$$

subject to the *equality constraint(s)*

$$\mathbf{A}\mathbf{x} = \mathbf{b}, \tag{12.24}$$

and the *inequality constraint(s)*

$$\mathbf{C}\mathbf{x} \leq \mathbf{d}, \tag{12.25}$$

and the *positivity constraints*

$$x_1 \geq 0, \quad x_2 \geq 0, \quad \dots \quad x_n \geq 0, \quad \dots \quad x_N \geq 0. \tag{12.26}$$

The vector $\mathbf{c}$ contains the constants in the linear objective function, $\mathbf{A}$ is an $M \times N$ matrix, $\mathbf{b}$ is an $M \times 1$ vector, $\mathbf{C}$ is an $S \times N$ matrix, and $\mathbf{d}$ is an $S \times 1$ vector.

Note that in the linear case, two design variables simply denote a straight line and three or more design variables denote a plane surface for the objective function. If the objective function has a minimum, therefore, it must occur at infinity or be bounded by the equality, inequality, or positivity constraints. The number of degrees of freedom of the optimization problem is $N - M - S$. The typical optimization problem is underdetermined, such that $M + S \leq N$. If $M + S > N$, then the problem is overdetermined (see Section 10.2).

The set of values of the design variables that satisfy the constraints is called the *feasible space* or *feasible region*. The feasible point that also minimizes the objective function is the optimal solution. The feasible space in linear programming problems is (hopefully) a closed polygon or polyhedron. The feasible point that also minimizes the objective function is the optimal solution that we seek, and the optimal solution corresponds to one of the vertices of the polygon/polyhedron.

Comments on formulation:

1. In deciding whether an algebraic equation should be included in the objective/cost function or as a constraint, ask the following question: "Must the algebraic equation be identically satisfied by any optimal solution?" If so, include it as a constraint. If the algebraic equation represents the objective or cost to be minimized, in which case we seek its minimum value, then it is the objective/cost function.
2. Optimization problems are typically framed in terms of minimizing the objective function. If a problem is naturally posed as maximizing the objective function, simply change the sign of J in order to reframe it as a minimization problem.
3. If the inequality constraints are expressed in terms of "$\geq$", rather than "$\leq$", then multiply by -1.
4. If the lower limits of the design variables are not naturally zero, then simply shift the variables so that they are.

Example 12.5 Suppose that we have two train engines that incur different costs per car pulled as well as different car capacities. The cost is ten per car for train one and 15 per car for train two. We want to transport 100 total cars between the two trains, but train one can pull no more than 50 cars, and train two no more than 75 cars. The first train pulls x_1 cars, while the second pulls x_2 cars. Determine the number of cars for each train that minimizes the total cost.

Solution

Because trains cannot have negative cars, we have the *positivity constraints*

$$x_1 \geq 0, \quad x_2 \geq 0. \tag{12.27}$$

The maximum number of cars that can be pulled by each train engine form the *inequality constraints*

$$x_1 \leq 50, \quad x_2 \leq 75. \tag{12.28}$$

The number of cars must total 100; therefore, we have the *equality constraint*

$$x_1 + x_2 = 100. \tag{12.29}$$

These constraints define the region as shown in Figure 12.9 within the (x_1, x_2) domain that forms the *feasible space* within which the optimal solution resides. Observe that enforcement of each additional constraint further confines the feasible space within the (x_1, x_2) design space.

The cost function to be minimized reflects the total cost of the two trains to pull their respective cars. The *cost function* is given by

$$J(x_1, x_2) = 10x_1 + 15x_2. \tag{12.30}$$

Therefore, the linear programming problem is to minimize the total cost (12.30) while adhering to the constraints (12.27) through (12.29). In order to visualize the cost function on the feasible space, lines of constant $J(x_1, x_2)$ can be superimposed on the feasible space to get a graphical sense for which "corner" or vertex of the feasible space corresponds to the optimal solution. These will be parallel straight lines overlaid on the (x_1, x_2) feasibility plot. In this problem, the feasible space is a finite line segment; therefore, the optimal solution is the point A or B at either end of the line segment.

Because there are only two possible optimal solutions, we can simply determine which one produces the smallest value of the cost function:

$$A: \quad x_1 = 25, x_2 = 75 \quad \Rightarrow \quad J = 1{,}375$$

$$B: \quad x_1 = 50, x_2 = 50 \quad \Rightarrow \quad J = 1{,}250$$

Point B corresponds to a smaller value of the cost function; therefore, it is the optimal solution.

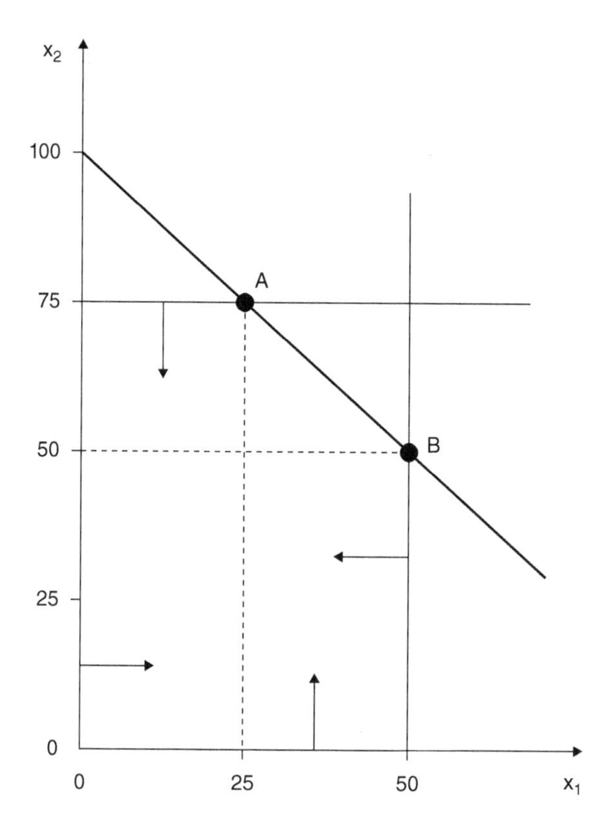

Figure 12.9 Feasible space for Example 12.5.

In general, there are four possibilities when solving linear programming problems:

1. A unique optimal solution occurs when the minimum in J occurs at a vertex (corner) of the feasible space.
2. An infinity of optimal solutions exist if the lines of constant J are parallel to one of the boundaries of the feasible space, that is, all of the feasible points along such a boundary have the same minimum value of J.
3. If the feasible space is not a closed polygon/polyhedron, then the minimum value of J occurs at infinity.
4. If the constraints are not consistent, such that there are no values of the design variables that are consistent with all of the constraints, then there is no feasible space or optimal solution at all.

As in the previous example, we will focus on the first possibility, because it is the most common and desirable.

In practical optimization problems, the number of design variables and constraints are typically very large; therefore, it is not possible to visualize the feasible space graphically. Instead, we will need a more algorithmic approach that sorts through the tens, hundreds, or even thousands of vertices in the feasible space to determine the optimal solution. The first step in devising such a method is to convert the inequality constraints into equality ones using slack variables.

12.7.2 Slack Variables for Inequality Constraints

The so-called *standard*, or *normal, form* of a linear programming problem consists of the linear algebraic objective function (12.23), the set of linear equality constraints (12.24), and the positivity constraints (12.26) that ensure that all of the design variables are nonnegative. Therefore, the inequality constraints (12.25) must be replaced with equivalent equality constraints. This is accomplished using *slack variables.*

Consider the following inequality constraint:

$$x_1 + 2x_2 \leq 5. \tag{12.31}$$

This is equivalent to requiring that

$$5 - x_1 - 2x_2 \geq 0.$$

Thus, we can define a new variable

$$s_1 = 5 - x_1 - 2x_2$$

that must be nonnegative. The larger s_1 is, the more "slack" is available in the quantity represented by this inequality constraint. If $s_1 = 0$, then there is no slack in that quantity, and there is just enough to satisfy the inequality constraint. Therefore, the inequality constraint (12.31) may be expressed as the equality constraint

$$x_1 + 2x_2 + s_1 = 5, \tag{12.32}$$

where

$$s_1 \geq 0. \tag{12.33}$$

The new variable s_1 that has been introduced is called a *slack variable* because it "takes up the slack" or quantifies the slack available in the variables between the two sides of the inequality constraint that is required to render it an equality constraint. In this way, the inequality constraint (12.31) is replaced by the equality constraint (12.32) and the requirement (12.33) that the slack variable be nonnegative.

One additional slack variable is introduced for each inequality constraint, thereby increasing the total number of variables accordingly. If there are S inequality constraints, then the total number of variables becomes $N + S$. The slack variables introduced for each of the inequality constraints become additional unknowns to be calculated as part of the solution procedure. The total number of equality constraints in the standard form is $M + S$, which is less than the total number of variables $N + S$ $(N > M)$. Therefore, the constraints form an underdetermined system having fewer equations than unknowns. The underdetermined system $\mathbf{Ax} = \mathbf{b}$ for the constraints has an infinity of solutions that correspond to the feasible space. Of the infinity of possible points in the feasible space, we seek the one that minimizes the objective function. Observe that in our train-car example, $N = 2$, $M = 1$, and $S = 2$.

REMARKS:

1. *The introduction of the slack variables does not change the objective function; the coefficients of the additional slack variables are simply zero.*
2. *In terms of the design and slack variables in standard form, the feasible space is bounded by the positivity constraints, such that one or more of the variables will be zero for the optimal solution.*

Example 12.6 Put the train-car problem in Example 12.5 into standard form.

Solution

Recall that the train-car linear programming problem consists of minimizing the cost function

$$J(x_1, x_2) = 10x_1 + 15x_2,$$

subject to the equality constraint

$$x_1 + x_2 = 100,$$

and the inequality constraints

$$x_1 \leq 50, \quad x_2 \leq 75, \tag{12.34}$$

and the positivity constraints

$$x_1 \geq 0, \quad x_2 \geq 0.$$

Introducing slack variables in order to convert the inequality constraints (12.34) into equality ones leads to the standard form of the optimization problem, which seeks to minimize

$$J(x_1, x_2) = 10x_1 + 15x_2, \tag{12.35}$$

subject to the equality constraints

$$\begin{aligned} x_1 + x_2 &= 100, \\ x_1 + s_1 &= 50, \\ x_2 + s_2 &= 75, \end{aligned} \tag{12.36}$$

and the positivity constraints

$$x_1 \geq 0, \quad x_2 \geq 0, \quad s_1 \geq 0, \quad s_2 \geq 0, \tag{12.37}$$

where now the design and slack variables are

$$\mathbf{x} = \begin{bmatrix} x_1 & x_2 & s_1 & s_2 \end{bmatrix}^T.$$

When expressed in standard form, the feasible space is bounded by the positivity constraints on both the design and slack variables as illustrated in Figure 12.10.

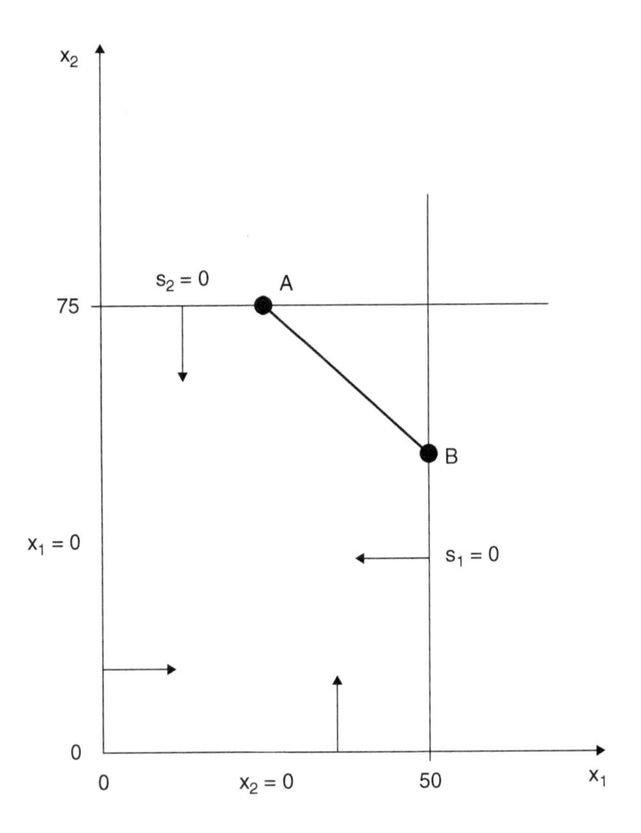

Figure 12.10 Feasible space in standard form for Example 12.6.

We can write the equality constraints (12.36) in the 3 × 4 matrix form $\mathbf{Ax} = \mathbf{b}$ as follows:

$$\begin{bmatrix} 1 & 1 & 0 & 0 \\ 1 & 0 & 1 & 0 \\ 0 & 1 & 0 & 1 \end{bmatrix} \begin{bmatrix} x_1 \\ x_2 \\ s_1 \\ s_2 \end{bmatrix} = \begin{bmatrix} 100 \\ 50 \\ 75 \end{bmatrix}. \tag{12.38}$$

Once the linear programming problem is posed in the standard form, we might expect to proceed as for nonlinear programming problems by differentiating the objective function with respect to each of the design and slack variables, setting equal to zero, and incorporating the equality constraints using Lagrange multipliers. From the cost function (12.35), however, differentiating with respect to x_1 and x_2 and setting equal to zero results in

$$10 = 0, \quad 15 = 0,$$

which clearly is not possible. Again, this is because linear functions only have extrema at infinity as discussed in Section 4.3.4. Therefore, we need a special approach for linear programming problems. This approach is called the *simplex method*.

12.8 Linear Programming – Simplex Method

The simplex method consists of combining the objective function and equality constraints of a linear programming problem formulated in standard form into an augmented matrix. Recall that the *standard*, or *normal, form* is of the following general matrix form. We seek to minimize the objective function

$$J(\mathbf{x}) = \langle \mathbf{c}, \mathbf{x} \rangle = \mathbf{c}^T \mathbf{x} = c_1 x_1 + c_2 x_2 + \cdots + c_n x_n + \cdots + c_N x_N \tag{12.39}$$

subject to the $M + S$ *equality constraint(s)*

$$\mathbf{A}\mathbf{x} = \mathbf{b}, \tag{12.40}$$

and the $N + S$ *positivity constraints*

$$x_1 \geq 0, \quad \ldots \quad x_N \geq 0, \quad s_1 \geq 0, \quad \ldots \quad s_S \geq 0. \tag{12.41}$$

Note that now the design and slack variables are

$$\mathbf{x} = \begin{bmatrix} x_1 & \ldots & x_N & s_1 & \ldots & s_S \end{bmatrix}^T,$$

and $\mathbf{A}$ is of size $(M + S) \times (N + S)$, and $\mathbf{b}$ is $(M + S) \times 1$ owing to conversion of the inequality into equality constraints and the introduction of the slack variables required to do so.

The simplex method recognizes that an extremum of a linear function subject to linear constraints must occur at one of the corner points of the feasible space as defined by the constraints. The corner points that are located in the feasible space are known as the *basic feasible solutions*. The basic feasible solution having the smallest value of the objective function is the *optimal solution* that we seek. Because the minimum occurs at a corner of the feasible space, we could simply evaluate the objective function at each of the corners and select the one having the minimum value as in Example 12.5. However, this is unnecessarily tedious, particularly for large problems with numerous corner points, many of which may not be feasible.

For a problem with $N + S$ variables and $M + S$ equality constraints, the number of basic feasible solutions is given by

$$\frac{(N + S)!}{(M + S)!\,(N - M)!}.$$

For example, if $N = 20$, $M = 5$, and $S = 5$, there are more than 3.2 million possible basic feasible solutions! Instead, the simplex method specifies a step-by-step procedure that progresses through the basic feasible solutions – the corner points – in such a way as to decrease the value of the objective function during each step and finally arrive at the optimal (minimum) solution.

The simplex algorithm is facilitated by the so-called *simplex tableau*, which is given by the augmented matrix problem

$$\begin{bmatrix} 1 & -\mathbf{c}^T \\ \mathbf{0} & \mathbf{A} \end{bmatrix} \begin{bmatrix} J \\ \mathbf{x} \end{bmatrix} = \begin{bmatrix} 0 \\ \mathbf{b} \end{bmatrix}. \tag{12.42}$$

The first equation defines the objective function (12.39) written in the form

$$J - \mathbf{c}^T\mathbf{x} = 0,$$

and the remaining equations correspond to the equality constraints (12.40). The vector $\mathbf{c}$ has $N+S$ elements and contains the coefficients in the objective function (recall that the coefficients of the slack variables are zero). The $(M+S)\times(N+S)$ matrix $\mathbf{A}$ and $(M+S)\times 1$ vector $\mathbf{b}$ contain the coefficients in the equality constraints along with the right-hand-side vector, respectively, for the problem expressed in standard form. Note that the $\mathbf{0}$ on the left-hand side is an $(M+S)\times 1$ vector (same size as $\mathbf{b}$), while that on the right-hand side is a scalar. The simplex tableau consists of $M+S+1$ equations for $N+S+1$ variables, which is underdetermined for $N > M$.

From our train-car example in standard form (12.35) through (12.37), the simplex tableau (12.42) is given by the system of equations

$$\begin{bmatrix} 1 & -10 & -15 & 0 & 0 \\ 0 & 1 & 1 & 0 & 0 \\ 0 & 1 & 0 & 1 & 0 \\ 0 & 0 & 1 & 0 & 1 \end{bmatrix} \begin{bmatrix} J \\ x_1 \\ x_2 \\ s_1 \\ s_2 \end{bmatrix} = \begin{bmatrix} 0 \\ 100 \\ 50 \\ 75 \end{bmatrix}. \tag{12.43}$$

This corresponds to the system of equations for the constraints (12.38) augmented with the objective function (12.35). This mathematical form is sometimes expressed in the tamer looking – and more informative – tabular form:

J	x_1	x_2	s_1	s_2	$\mathbf{b}$
1	−10	−15	0	0	0
0	1	1	0	0	100
0	1	0	1	0	50
0	0	1	0	1	75

The simplex tableau is generally quite sparse, particularly as the number of design and slack variables increases.

Linear programming methods typically result in fewer equations (objective function plus equality constraints) than total variables $(N+S+1)$; therefore, they are underdetermined (see Section 10.2.2). Consequently, the system of linear algebraic equations expressed in the simplex tableau does not have a unique solution, but rather an infinity of solutions. To determine the basic feasible solution that is optimal, note the following guiding principles:

1. Because the optimal solution occurs at one of the corners of the feasible space, the value of $(N+S)-(M+S) = N-M$ of the design and/or slack variables must be zero for the final optimal solution. For example, if there are three more variables than equations $(N-M=3)$, then three of the variables must be at their zero limit, corresponding to the positivity constraint, for the optimal point. The $N-M$

subset of design and slack variables that are zero at each basic feasible point are called *nonbasic variables*, while the remaining nonzero variables are called *basic variables*.

2. Essentially, the simplex method provides a step-by-step method for determining which of the variables are the basic and nonbasic variables that lead to the optimal solution corresponding to the minimum value of the objective function.
3. With the nonbasic variables set to zero, a unique solution can be obtained for the remaining basic variables and the objective function. Of all the possible combinations of basic and nonbasic variables, the solution having the minimum value of the objective function corresponds to the optimal solution.
4. Unless a particular choice of basic and nonbasic variables happens to correspond to the optimal solution, one or more of the variables will be negative and violate its positivity constraint.
5. Setting one of the negative variables equal to zero, that is, making it one of the nonbasic variables, will then decrease the value of the objective function $J(\mathbf{x})$.
6. Because setting the most negative variable to zero is likely to lead to the largest decrease in the objective function, that is the one typically chosen. This variable is sometimes referred to as the *entering variable*.
7. But how do we select the *leaving variable* that it replaces? The standard method is to determine where the constraint lines intersect with the line corresponding to the entering variable. This value is the ratio of the right-hand side of the constraint equation to the coefficient of the entering variable.
8. Using these principles, we cycle through the various possible nonbasic variables until we arrive at the optimal solution for which none of the variables are negative.

Example 12.7 Solve the train-car example in standard form (12.43) using the simplex method.

Solution

Recall that the simplex tableau in standard form is given by

$$\begin{bmatrix} 1 & -10 & -15 & 0 & 0 \\ 0 & 1 & 1 & 0 & 0 \\ 0 & 1 & 0 & 1 & 0 \\ 0 & 0 & 1 & 0 & 1 \end{bmatrix} \begin{bmatrix} J \\ x_1 \\ x_2 \\ s_1 \\ s_2 \end{bmatrix} = \begin{bmatrix} 0 \\ 100 \\ 50 \\ 75 \end{bmatrix}. \tag{12.44}$$

In this case, $N - M = 1$; therefore, there is one variable in the nonbasic set, which will be zero for the basic feasible solutions, and four variables in the basic set. We simply need to determine which of the design and slack variables lie in the basic and nonbasic subsets of variables for the optimal solution.

Recall that there are two possible solutions corresponding to points A and B in the feasible space shown in Figure 12.10. Point A corresponds to $s_2 = 0$, and point B corresponds to $s_1 = 0$. Let us consider each point in turn:

- **Point** A ($s_2 = 0$): Separating the nonbasic variable (s_2) from the basic variables (J, x_1, x_2, s_1) and their respective coefficients leads to

$$\begin{bmatrix} 1 & -10 & -15 & 0 \\ 0 & 1 & 1 & 0 \\ 0 & 1 & 0 & 1 \\ 0 & 0 & 1 & 0 \end{bmatrix} \begin{bmatrix} J \\ x_1 \\ x_2 \\ s_1 \end{bmatrix} + \begin{bmatrix} 0 \\ 0 \\ 0 \\ 1 \end{bmatrix} \cancelto{0}{[s_2]} = \begin{bmatrix} 0 \\ 100 \\ 50 \\ 75 \end{bmatrix}. \tag{12.45}$$

This is the simplex tableau (12.44) with the last column of the coefficient matrix and its variable separated out. In matrix form, this is

$$\mathbf{A}_B \mathbf{x}_B + \mathbf{A}_N \cancelto{0}{\mathbf{x}_N} = \mathbf{b},$$

where $\mathbf{x}_B$ and $\mathbf{A}_B$ correspond to the basic variables and their coefficients, and $\mathbf{x}_N$ and $\mathbf{A}_N$ correspond to the nonbasic variable and its coefficient. Observe that $\mathbf{A}_B \mathbf{x}_B = \mathbf{b}$ now consists of four equations for four unknowns and has a unique solution. It is

$$\begin{bmatrix} J \\ x_1 \\ x_2 \\ s_1 \end{bmatrix} = \mathbf{A}_B^{-1} \mathbf{b} = \begin{bmatrix} 1,375 \\ 25 \\ 75 \\ 25 \end{bmatrix}.$$

This result consists of $x_1 = 25$ cars on train one and $x_2 = 75$ cars on train two, with a total cost of $J = 1,375$ as determined in Example 12.5.

- **Point** B ($s_1 = 0$): Separating the nonbasic variable (s_1) from the basic variables (J, x_1, x_2, s_2) and their respective coefficients leads to

$$\begin{bmatrix} 1 & -10 & -15 & 0 \\ 0 & 1 & 1 & 0 \\ 0 & 1 & 0 & 0 \\ 0 & 0 & 1 & 1 \end{bmatrix} \begin{bmatrix} J \\ x_1 \\ x_2 \\ s_2 \end{bmatrix} + \begin{bmatrix} 0 \\ 0 \\ 1 \\ 0 \end{bmatrix} \cancelto{0}{[s_1]} = \begin{bmatrix} 0 \\ 100 \\ 50 \\ 75 \end{bmatrix}. \tag{12.46}$$

This is the simplex tableau (12.44) with the second-to-last column of the coefficient matrix and its variable separated out. Again, we have four remaining equations for four unknowns, which has the unique solution

$$\begin{bmatrix} J \\ x_1 \\ x_2 \\ s_2 \end{bmatrix} = \mathbf{A}_B^{-1} \mathbf{b} = \begin{bmatrix} 1,250 \\ 50 \\ 50 \\ 25 \end{bmatrix}.$$

This result consists of $x_1 = 50$ cars on train one and $x_2 = 50$ cars on train two with a total cost of $J = 1,250$ as determined in Example 12.5.

Once again, we see that point B leads to the smaller value of the objective function corresponding to 50 cars on each train.

The simplex method consists of the following general steps:

1. Reformulate the linear programming optimization problem in standard form. That is, if necessary:
   - Convert maximization of the objective function to minimization.
   - Convert "$\geq$" inequality constraints to "$\leq$" inequality constraints.
   - Introduce slack variables to convert the inequality constraints into equality ones.
2. Form the simplex tableau from the standard form.
3. Determine at which basic feasible solution to start the simplex algorithm.
4. Proceed from one basic feasible solution to the next along the boundary of the feasible space in such a way that the objective function J decreases (or at least does not increase) at each step until the optimal solution is reached. At each step, the entering and leaving variables must be selected for the set of basic variables that correspond to the new basic feasible solution.

For details of how the starting point is chosen and how the entering and leaving variables are selected at each step, see Press et al. (2007), Messac (2015), and Miller (2014). Let us put these steps all together in another example.

Example 12.8 Raw material in the amount of 77 units is delivered each week to a manufacturing plant. The plant has the ability to produce two different products from this raw material, and each requires different amounts of raw material, amounts of time to produce, have different onsite storage requirements, and produce different profits as follows. Product A requires seven units of raw material, takes ten hours per unit to produce, the plant can only store up to nine units per week, and the profit is 150 per unit. Product B requires 11 units of raw material, takes eight hours per unit to produce, the plant can only store up to six units per week, and the profit is 175 per unit. The plant can be in production up to 80 hours per week, and only one product can be produced at a time. Determine how much of each product to produce each week to maximize profits.

Solution

The design variables are x_1, which equals the units of product A produced each week, and x_2, which equals the units of product B produced each week. The objective function reflects the desire to maximize the profit each week, which is given by

$$\hat{J}(x_1, x_2) = 150x_1 + 175x_2. \tag{12.47}$$

The constraints are as follows. The total raw material used to produce each product must be less than that delivered each week according to

$$7x_1 + 11x_2 \leq 77. \tag{12.48}$$

The total time of production must not exceed 80 hours per week as follows:

$$10x_1 + 8x_2 \leq 80. \tag{12.49}$$

Onsite storage limits the units of each product that can be produced each week:

$$x_1 \leq 9, \quad x_2 \leq 6. \tag{12.50}$$

Finally, we have the positivity constraints

$$x_1 \geq 0, \quad x_2 \geq 0. \tag{12.51}$$

In order to put this linear programming problem in standard form, we pose the optimization problem in terms of minimization, let us take the negative of the objective function to replace (12.47) by

$$J(x_1, x_2) = -\hat{J}(x_1, x_2) = -150x_1 - 175x_2. \tag{12.52}$$

In addition, we introduce slack variables to convert each of the inequality to equality constraints as follows:

$$\begin{aligned} &7x_1 + 11x_2 + s_1 = 77, \\ &10x_1 + 8x_2 + s_2 = 80, \\ &x_1 + s_3 = 9, \\ &x_2 + s_4 = 6. \end{aligned} \tag{12.53}$$

Note that $N = 2$ (design variables), $M = 0$ (equality constraints), and $S = 4$ (inequality constraints) in the original formulation. Therefore, there are $N - M = 2$ variables in the nonbasic set; the remaining four variables are in the basic set. From these constraints, the feasible space is illustrated graphically in Figure 12.11. Note that the s_3 constraint is redundant, that is, it is outside the feasible space. The basic feasible solutions are given by the corner points of the feasible space: A, B, C, D, and E, where the corresponding nonbasic variables are given in Table 12.3.

From (12.52) and (12.53), the simplex tableau is given by

$$\begin{bmatrix} 1 & 150 & 175 & 0 & 0 & 0 & 0 \\ 0 & 7 & 11 & 1 & 0 & 0 & 0 \\ 0 & 10 & 8 & 0 & 1 & 0 & 0 \\ 0 & 1 & 0 & 0 & 0 & 1 & 0 \\ 0 & 0 & 1 & 0 & 0 & 0 & 1 \end{bmatrix} \begin{bmatrix} J \\ x_1 \\ x_2 \\ s_1 \\ s_2 \\ s_3 \\ s_4 \end{bmatrix} = \begin{bmatrix} 0 \\ 77 \\ 80 \\ 9 \\ 6 \end{bmatrix}. \tag{12.54}$$

Table 12.3 Possible basic feasible solutions for Example 12.8.

Basic feasible point	Nonbasic (zero) variables
A	x_1, x_2
B	x_2, s_2
C	s_1, s_2
D	s_1, s_4
E	x_1, s_4

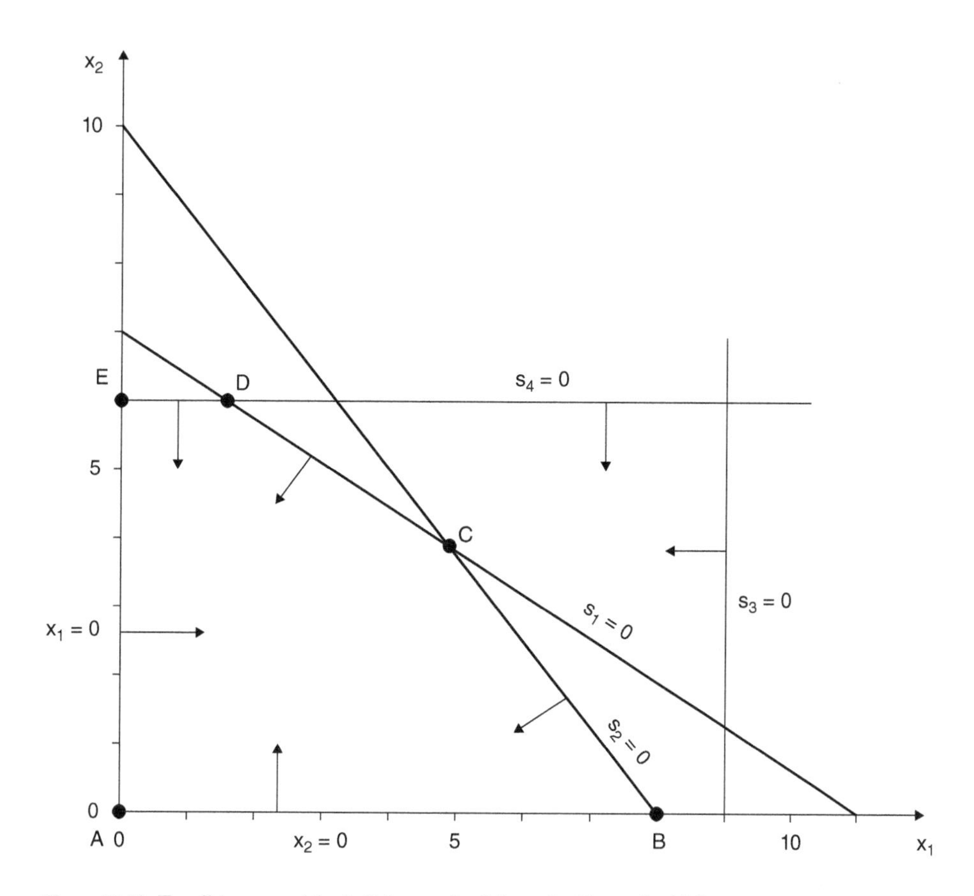

Figure 12.11 Feasible space (shaded) in standard form for Example 12.8.

If, for example, we start the simplex algorithm at point A, where $x_1 = 0$ and $x_2 = 0$ (the plant produces nothing), the value of the objective function is $J = 0$ (no profit is produced). The simplex algorithm then proceeds as shown in Table 12.4. The simplex algorithm determines the order in which to evaluate the basic feasible points such that the objective function has the greatest decrease in the objective function at each step. This leads from point A to B to C. Point C corresponds to the maximum profit per week of

$$\hat{J} = -J = 1{,}413.89.$$

REMARKS:

1. *A* simplex *is the simplest polygon or polyhedron in each dimension (a line segment in one dimension, a triangle in two dimensions, a tetrahedron in three dimensions, et cetera), and* tableau *is a fancy word for a table.*
2. *Linear programming codes are very complex as they must be able to handle a wide variety of scenarios and eventualities in optimization. For example, how to choose*

Table 12.4 Values of objective function J at each basic feasible solutions for Example 6.8.

Basic feasible point	Nonbasic (zero) variables	J
A	x_1, x_2	0
B	x_2, s_2	$-1,200$
C	s_1, s_2	$-1,413.89$

the starting basic feasible solution to start the simplex algorithm? How to choose the next basic feasible solution – and associated leaving and entering variables – at each step?

3. *For very large linear programming problems, the* interior-point method *can sometimes locate the optimal solution faster than the simplex method. While the simplex method only proceeds through the feasible points sequentially, that is, along the edges of the feasible space, the interior-point method has the freedom to cross the feasible space in search of the optimal solution.*

12.9 Optimal Control

Thus far in Part III, we have considered optimization in a general context as applied to all manner of systems, processes, and data analysis tasks. In *optimal control*, optimization methods are particularized to time-evolving systems that arise in engineering applications. It could be an actual mechanical or electrical system or a chemical, biological, financial, or manufacturing process. For example, imagine a system that has the capability to sense its current state and, based on the anticipated needs of the system, adjust its operation in real time in order to accomplish a predefined objective in the best way possible. Sounds complicated; must be expensive. This describes the thermostat in your home – particularly modern "smart" thermostats that learn your comings and goings as well as your heating and cooling preferences. The system itself has been designed to perform some task, and it is supplemented with sensors and actuators to sense the present state of the system (thermometer) and actuate devices that act to control the system (furnace and air conditioner) in order to achieve some objective (keep you comfortable). In optimal control parlance, we would refer to this as closed-loop feedback control.

Thus, optimal control is at the same time a special case of optimization and a generalization of it to the controlled manipulation of systems and processes to achieve a predefined objective. While the formulation of optimal control problems involves a subset of those for more general optimization problems, with an objective function and constraints, the application to physical systems and processes brings in a number of additional aspects that require our specific attention. While optimization and control are important areas in their own right, with numerous applications throughout science and engineering, these techniques also provide the foundation for reduced-order modeling methods to be presented in Chapter 13.

12.9.1 Introduction

In addition to the objective function or functional, which quantifies the control objective, the essential feature of optimal control problems is a mathematical model for the behavior of the system that is in the form of a differential, or algebraic, equation. A physical system that evolves in time is generically referred to as a *dynamical system*. The mathematical model of a discrete dynamical system is an (or system of) ordinary differential equation(s), while that of a continuous system is a (or system of) partial differential equation(s); see Chapter 5 for more on dynamical systems. Therefore, the primary distinguishing feature of optimal control, as compared to general optimization problems, is a differential equation that governs the behavior of the system known as the *state equation(s)*. This state equation is enforced as a constraint on the minimization of the algebraic or integral objective function that defines the aim of the control. In addition, optimal control introduces some additional aspects into the optimization framework. In particular, how do we use the sensor measurements to *estimate* the actual state of the system? How do we determine the actuator's operation that optimizes the objective?

Control theory and methods occupy entire books and are a rich source of mathematical and practical topics for further study, and they rely primarily on variational methods. Therefore, the reader is referred to chapter 10 of Cassel (2013) for a detailed treatment of optimal control applied to discrete and continuous systems governed by ordinary and partial differential state equations, respectively. For a unified classification of the full range of optimization and control problems addressable using matrix and variational methods, see Table 10.1.

12.9.2 Optimize-Then-Discretize Versus Discretize-Then-Optimize

Chapter 10 of Cassel (2013) follows an optimize-then-discretize approach to formulating optimal control problems. The control problem is fully formulated using continuous state and control functions; only after obtaining the state-adjoint coupling do we even consider how to discretize and approximately solve the resulting differential equations. This approach requires variational methods to obtain the adjoint equation to be solved backward in time for the adjoint or control variable.

Many espouse a discretize-then-optimize strategy instead. While there are philosophical arguments to be made for each, the primary advantage of this approach is a practical one. By first discretizing the state equations and objective functional before formulating the adjoint problem for the control, we have a system of linear algebraic equations rather than a system of first-order ordinary differential equations. As a result, there is no need to appeal to variational methods to formulate the control problem; all operations stay within the realm of matrix methods. In terms of obtaining the formulation, therefore, the advantage of the discretize-then-optimize approach is that we have the full arsenal of linear least-squares and linear algebra methods at our disposal. The advantage of the optimize-then-discretize strategy is that the variational approach naturally can be applied to nonlinear state equations, nonquadratic objective functionals, and nonhomogeneous boundary conditions.

Unfortunately, these two approaches do not commute. That is, when applied to the same objective functional and state equations, they do not produce the same resulting linear system to solve for the discretized state and adjoint/control variables. This is not even the case for LQ problems. The optimize-then-discretize approach is embraced here and in Cassel (2013) as (1) the computational grids can be tailored for solution of the state and adjoint equations separately and (2) the variational approach provides a great deal more flexibility in how the optimization algorithm is formulated and constructed.

In the discretize-then-optimize approach, if the system is linear, we simply discretize the unsteady state equation in time using the methods in Chapter 9 in order to produce the algebraic optimal control formulation. If the system is nonlinear, then the state equations are first linearized, for example, using Newton linearization (see Section 8.10), and then the appropriate numerical method is applied for initial-value problems in order to obtain the linear algebraic optimal control formulation in state-space form.

The continuous system of ordinary differential equations governing the state of the linear discrete system is given by

$$\dot{\mathbf{u}}(t) = \mathbf{A}\mathbf{u}(t) + \mathbf{B}\phi(t), \tag{12.55}$$

where $\mathbf{u}(t)$ is the state variable, and $\phi(t)$ is the control variable. A popular choice for solving such initial-value problems numerically would be one of the Runge–Kutta methods, but we will illustrate using the first-order explicit (Euler) method for simplicity (see Section 9.2). In the first-order explicit method, the differential equations are approximated at the previous time step, and a first-order accurate in time forward difference is used for the time derivative. This gives

$$\frac{\mathbf{u}^{(n+1)} - \mathbf{u}^{(n)}}{\Delta t} = \mathbf{A}\mathbf{u}^{(n)} + \mathbf{B}\phi^{(n)},$$

where Δt is the time step, and the superscripts represent the index for each successive time step. Solving for the state vector at the new time level yields

$$\mathbf{u}^{(n+1)} = (\mathbf{I} + \Delta t\mathbf{A})\,\mathbf{u}^{(n)} + \Delta t\mathbf{B}\phi^{(n)}. \tag{12.56}$$

This is a discretized version of the continuous differential state equations governing the discrete system. Clearly, the precise form of the discretized state equations (12.56) will depend upon the numerical method used in its discretization; however, it will be of the general form

$$\mathbf{u}^{(n+1)} = \hat{\mathbf{A}}\mathbf{u}^{(n)} + \hat{\mathbf{B}}\phi^{(n)}, \tag{12.57}$$

which is of very similar form as the continuous state equations (12.55) but with modified dynamic and input matrices.

Because the continuous time domain $0 \leq t \leq t_f$ has been discretized, the continuous integral in the objective functional must be exchanged for a discrete sum. Therefore, a quadratic integral objective functional of the form

$$J[\mathbf{u}(t), \boldsymbol{\phi}(t)] = \frac{1}{2}\int_0^{t_f} \left(\gamma^2 \mathbf{u}^T \mathbf{Q}\mathbf{u} + \boldsymbol{\phi}^T \mathbf{R}\boldsymbol{\phi}\right) dt$$

becomes the following algebraic objective function:

$$J[\mathbf{u}^{(n)}, \boldsymbol{\phi}^{(n)}] = \frac{1}{2}\sum_{n=0}^{N}\left[\gamma^2 \left(\mathbf{u}^{(n)}\right)^T \mathbf{Q}\mathbf{u}^{(n)} + \left(\boldsymbol{\phi}^{(n)}\right)^T \mathbf{R}\boldsymbol{\phi}^{(n)}\right], \tag{12.58}$$

where $n = N$ corresponds to $t = t_f$. Consequently, we wish to minimize the algebraic objective function (12.58) subject to the constraint that the system of linear algebraic equations (12.57) must be satisfied. Rather than being a variational LQ problem leading to the differential form of the Riccati equation, we have a linear least-squares LQ problem that leads to an algebraic form of the Riccati equation.

Likewise, the state-estimation problem discussed in section 10.6.3 of Cassel (2013) applied to this discretized formulation gives rise to the *Kalman filter*. It is very similar to the Kalman–Bucy filter but applied to discrete state, output, and input vectors, with the optimization procedure set up in algebraic (least squares), rather than variational, form. For a complete treatment of optimization and control from a discretize-then-optimize point of view, see Borrelli et al. (2017), for example.

Exercises

Unless stated otherwise, perform the exercises using hand calculations. Exercises to be completed using "built-in functions" should be completed using the built-in functions within Python, MATLAB, or Mathematica for the vector and matrix operation(s) or algorithm.

12.1 Consider the torsional bar shown in Figure 12.12. The rigid bar is massless and of length ℓ with a vertical force F at its end. A torsional spring at the bar's base has torsion given by $T = k\theta$, where θ is the angle of the bar from the vertical, and k is the torsional stiffness of the spring. The angle θ is sought where the system is in static equilibrium under the action of the force F and torsion T.

(a) Apply the principles of static equilibrium to a free-body diagram to show that the governing equation is given by

$$f(\theta) = k\theta - F\ell \sin\theta = 0.$$

Observe that this is a nonlinear equation for θ given k, F, and ℓ.

(b) Evaluate the first two iterations of Newton's method to obtain an estimate of θ for $k = 8$, $\ell = 10$, and $F = 1$ with an initial guess of $\theta^{(0)} = 0.9$ radians.

(c) Use a built-in root-finding function to obtain the fully converged root for θ and compare with your approximation from part (b).

12.2 Consider an object of mass m free falling under the action of its own weight $W = mg$ and resisted by drag $F_D = -cv$, where $v(t)$ is the velocity and $c > 0$.

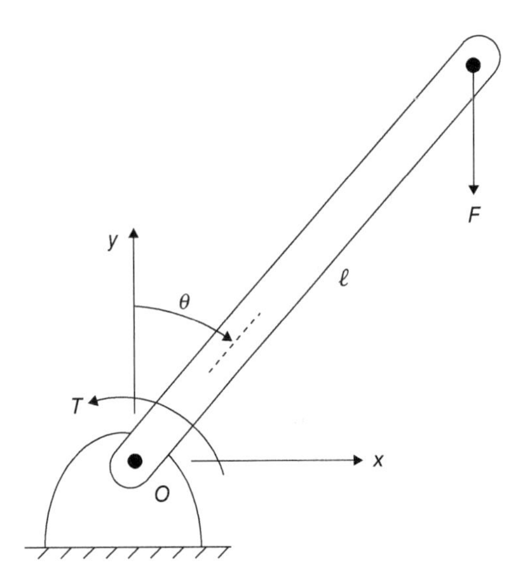

Figure 12.12 Schematic for Exercise 12.1.

(a) Apply Newton's second law to a free-body diagram to show that the governing equation is given by

$$\frac{dv}{dt} + \frac{c}{m} v = -g, \quad v(0) = 0,$$

where g is the acceleration owing to gravity.

(b) Show that the exact solution of this differential equation is given by

$$v(t) = \frac{mg}{c}\left(e^{-\frac{c}{m}t} - 1\right).$$

Therefore, given the mass m, acceleration owing to gravity g, and the drag constant c, one may determine the velocity $v(t)$ "explicitly" as a function of time.

(c) Suppose instead that one would like to determine the drag constant c given m, g, and the time t required for the object to reach a specified velocity v. In this case, it is not possible to solve explicitly for the drag constant c in terms of the remaining parameters. Instead, the nonlinear equation

$$f(c) = \frac{mg}{c}\left(e^{-\frac{c}{m}t} - 1\right) - v = 0$$

must be solved to determine the root c. Use a built-in root-finding function to obtain the fully converged root for c with $m = 1$, $g = 9.81$, $t = 2$, $v = -7.2$, and an appropriate initial guess.

12.3 In fluid mechanics, the Moody diagram for turbulent pipe flow provides a graphical relationship among the friction factor f, the Reynolds number Re, and the surface roughness ϵ/D. The Moody diagram can be approximated by the Colebrook equation

$$\frac{1}{\sqrt{f}} = -2.0 \log_{10}\left[\frac{\epsilon/D}{3.7} + \frac{2.51}{Re\sqrt{f}}\right].$$

For $Re = 2 \times 10^5$ and $\epsilon/D = 0.004$, solve the Colebrook equation for the corresponding friction factor f using a built-in root-finding function with an appropriate initial guess.

12.4 A sphere is initially at rest and dropped in still water. Suppose that the drag force f_D acting on the sphere can be approximated by the equation

$$f_D = \frac{1}{2}\rho A C_D V^2,$$

where ρ is the density of the water, A is the cross-sectional area of the sphere, C_D is the drag coefficient, and V is the velocity.

(a) Derive an expression for the terminal velocity of the sphere in the form $P(V) = f$, where f is a constant (hint: the sphere reaches terminal velocity when it stops accelerating).

(b) Obtain an approximation for the terminal velocity by performing two iterations of Newton's method. The mass of the sphere is $500\,kg$ and its diameter is $0.25\,m$. Take the gravitational constant to be $g = 9.81\,m/s^2$. Assume that $C_D = 1.0$, and $\rho = 1000\,kg/m^3$.

12.5 Consider the deflection $v(x)$ of a simply supported beam with constant cross section, length ℓ, and a linearly increasing load distribution having a maximum p_ℓ is given by

$$v(x) = \frac{p_\ell}{120EI\ell}\left(-x^5 + 2\ell^2 x^3 - \ell^4 x\right).$$

For a beam of length $\ell = 600\,cm$, Young's modulus $E = 50,000\,kN/cm^2$, moment of inertia $I = 30,000\,cm^4$, and maximum load $p_\ell = 2.5\,kN/cm$, perform the following:

(a) Plot the deflection curve.

(b) Determine the point x along the beam that has the maximum deflection. Is this value consistent with your plot in part (a)?

(c) Check the numerical value of your answer in part (b) using a built-in root-finding function.

12.6 A conically shaped pit is to be designed that will store $50\,m^3$ of liquid. It costs $\$100/m^3$ for excavation, $\$50/m^2$ to line the conical sides, and $\$25/m^2$ to cover the flat top of the tank.

(a) Set up the equation(s) necessary to determine the dimensions of the conical enclosure that minimizes its total cost.

(b) Obtain the optimal solution using a built-in root-finding function.

12.7 An airfoil experiences drag owing to friction from shear stresses along the surface, which is called friction drag, and form (or pressure) drag owing to the wake that forms behind a body moving through a fluid. The total drag force F_D on an airfoil can be approximated by the relationship

$$F_D = 0.01\sigma U^2 + \frac{0.95}{\sigma}\left(\frac{W}{U}\right)^2,$$

where W is the weight, U is the velocity, and σ is the ratio of the density of air between the flight altitude and sea level. Consider the case with $\sigma = 0.6$.

(a) Determine the velocity for which the drag is a minimum for a weight of $W = 16,000$.
(b) In order to see how sensitive the minimum drag (and associated velocity) is to changes in the plane's weight, plot the minimum F_D and the corresponding velocity U for weights between $12,000 \leq W \leq 20,000$.

12.8 Consider the following equations for a parabola and a circle:

$$x^2 - y = 0,$$

$$x^2 + y^2 = 1.$$

Perform one iteration of Newton's method for a system of nonlinear equations to estimate the location where the two curves intersect. Take the point $\left(x^{(0)}, y^{(0)}\right) = (0.5, 0.5)$ as the initial guess.

12.9 The point (x_0, y_0, z_0) on the plane

$$2x - y + z = 3$$

is sought that is closest to the origin. Determine this point by minimizing the square of the distance from the origin

$$J = f(x, y, z) = x^2 + y^2 + z^2$$

subject to an appropriate constraint.

12.10 Consider the two nonlinear algebraic equations

$$4x_1^2 - x_2^3 = -28,$$

$$3x_1^3 + 4x_2^2 = 145.$$

(a) Determine the general form of the system of linear algebraic equations (12.11) that is to be solved for $\Delta x_1^{(k)}$ and $\Delta x_2^{(k)}$ at each iteration of Newton's method.
(b) Using a built-in function, calculate the point where the nonlinear equations intersect. Use the initial guesses $x_1^{(0)} = 1$ and $x_2^{(0)} = 1$.

12.11 Consider a linear programming problem that seeks to minimize the objective function

$$f(x_1, x_2) = 4x_1 + 3x_2,$$

subject to the constraints

$$x_1 + 2x_2 \geq 2,$$

$$3x_1 + x_2 \geq 3,$$

$$5x_1 + 3x_2 \geq 6,$$

$$x_1 \geq 0, x_2 \geq 0.$$

(a) Convert the linear programming problem into its standard form.
(b) Draw the feasible space for this linear programming problem, and identify the basic feasible solutions.
(c) Determine which basic feasible solution corresponds to the optimal solution of the linear programming problem using the simplex tableau.
(d) Use a built-in optimization function to verify your solution.

12.12 A company manufactures three types of heaters, H_1, H_2, and H_3, using two machines, M_1 and M_2. The price charged for each of the three heaters is \$40, \$63, and \$88, respectively. The number of each type of heater, x_1, x_2, and x_3, respectively, is sought that can be manufactured per hour in order to maximize the company's hourly revenue. Each heater H_1 takes two minutes on machine M_1 and five minutes on machine M_2. Each heater H_2 takes eight minutes on machine M_1 and three minutes on machine M_2. Each heater H_3 takes four minutes on machine M_1 and six minutes on machine M_2. In order to maintain full employment in the factory, the total number of heaters manufactured each hour must be at least 25.

(a) Formulate the linear programming problem to be solved for the production amounts x_1, x_2, and x_3 to maximize hourly revenue for the three heaters.

(b) Convert the linear programming problem into its standard form and write down the simplex tableau.

(c) How many variables are in the basic and nonbasic sets for the optimal solution?

(d) Determine which basic feasible solution corresponds to the optimal solution of the linear programming problem using the simplex tableau.

(e) Use a built-in optimization function to verify your solution.

13 Data-Driven Methods and Reduced-Order Modeling

There is nothing more practical than a good theory.
(Various[1])

13.1 Introduction

As highlighted in Chapter 5, an important objective for scientists and engineers is to develop techniques for better understanding the behaviors inherent in solutions of complex, large-dimensional, linear and nonlinear systems. This includes determining the types of solutions that are permissible and their stability characteristics. The loosely defined collection of methods under consideration is generally referred to as dynamical systems theory and includes both qualitative and quantitative techniques that are applicable to the full range of fields involving time-evolving systems. As our ability to address increasingly large systems expands through advanced experimental measurements and numerical methods, and the sheer size of the resulting data sets grows rapidly, significant effort has been directed toward developing formal approaches to (1) extract important dynamic features that dominate the evolution of a system and (2) formulate simplified mathematical models that capture the essential features of a system's evolution. Both of these objectives are addressed in this chapter for discrete and continuous systems. The emphasis is on formal techniques that are firmly grounded in linear systems theory.

The dimensionality of a discrete system corresponds to its number of degrees of freedom, which corresponds to the number of dependent variables required to fix the state of the system. For discrete systems, this corresponds to the number of coupled second-order ordinary differential equations in time that must be solved. Continuous systems have an infinite number of degrees of freedom as they are spatially distributed and are governed by partial differential equations. For systems having a large (or infinite) number of degrees of freedom, or *order*, the computational time required to solve such systems using the methods given in Part II becomes very long, and the resulting solutions are often difficult to interpret. This has led to development of a series of techniques to reduce the order of the system in order to aid in both its solution and interpretation of the results. A *reduced-order model* (ROM), then, is a simplified

[1] This quote has been variously attributed to James Clerk Maxwell, Ludwig Boltzmann, Albert Einstein, Kurt Lewin, and even Leonid Brezhnev.

mathematical model of a system that captures the essential features of its solution. Such a ROM can be used to directly aid in interpretation of the solution, be used as the basis for stability analysis, or provide a compact mathematical model for use in optimization or feedback-control algorithms.

We have already encountered, or can easily imagine, some simple examples of reduced-order models. Perhaps the simplest example is linear, least-squares regression in Chapter 11. A data set involving a large number N of point measurements of some quantity that exhibits roughly linear behavior is characterized as a single, best-fit line. Thus, a single line, defined uniquely by only two points ($2 \ll N$), represents a large collection of individual points. More generally, an Nth-order interpolating polynomial can exactly pass through N data points. Polynomial regression using any order polynomial less than N will produce a reduced-order model of the full data set.

The stability analysis of discrete systems in Chapters 5 typically involves first isolating individual modes so that the most unstable, or least stable, modes can be identified. The Saltzman–Lorenz model considered in Section 5.6 is an example of an extreme reduced-order model in which the full nonlinear, infinite-dimensional Navier–Stokes partial differential equations are reduced to a system of three nonlinear ordinary differential equations for the Rayleigh–Bénard convection problem.

In addition to these examples, the careful reader of Part I will have observed our – sometimes not so subtle – theme concerning how solutions of various types of problems can be represented as a linear combination of basis vectors or functions in discrete and continuous cases, respectively. For example, recall that in Section 2.3.2 we expressed the solution of a system of linear algebraic equations having a real, symmetric coefficient matrix as a linear combination of the eigenvectors of the coefficient matrix. Why not simply truncate this expansion so that we approximate the solution based on just a subset of the eigenmodes? We could even be intelligent about the truncation process by only truncating those modes that have small coefficients. In Section 2.6, we expressed the solution of a system of first-order, linear ordinary differential equations in terms of a linear combination of the eigenvectors of the coefficient matrix. Again, we could truncate this expansion by eliminating terms with small coefficients. In Chapter 3, we represented the solution of a self-adjoint ordinary or partial differential equation as an expansion in terms of the orthonormal eigenfunctions of the differential operator. Because such systems are infinite dimensional, these expansions have an infinite number of terms and they cry out to be truncated after a finite number of terms.

While the preceding scenarios would indeed be examples of reduced-order models in the spirit of our objective, there are a number of questions that remain. How many basis vectors or functions should we retain? Which ones? For a given number of terms, is there a better, that is, more optimal, set of basis vectors or functions? In the case of eigenfunction solutions of differential equations, for example, we can only be assured that keeping all infinity of the terms will produce an exact solution; nothing can be said about their optimality when truncated. More importantly, such simple truncations do nothing to extend these techniques beyond the very limited cases addressed by their specific theories. What about systems of linear algebraic equations for which the coefficient matrix is not real and symmetric or differential equations having differential operators that are not self-adjoint? What about an experimental data set for which we do not have a precise mathematical model? We need a more general technique that allows us to apply the spirit of this approach to a broader range of problems. In addition, it would be beneficial if the outcome provided useful information about the solution behavior in the spirit of our emphasis in Chapter 5 on techniques for analyzing and interpreting solutions of complex problems.

Summarizing, the objective of a reduced-order model is to provide a simplified model that faithfully captures the dominant physics of the system that can be used to better understand a system's behavior via dynamical systems theory or incorporated into an optimization or feedback control setting. Specifically, reduced-order modeling consists of two steps:

1. Determine the "best" set of basis vectors or functions for a particular set of experimental or numerical data.
2. Use the basis vectors or functions to formulate a reduced-order mathematical model of the system having fewer degrees of freedom than the original system using Galerkin projection.

In some cases, we are only interested in the basis vectors or functions themselves as they encapsulate the most important features and behaviors in the system (step 1). In other cases, it is the ROM that is of interest as it can be used for further analysis using dynamical systems theory or as the basis for control of the system (step 2). In addition, while the basis vectors or functions depend upon the nature of the system being analyzed, such as whether it is linear or nonlinear, the general Galerkin projection approach applies in all cases. Therefore, we discuss it first as it provides the common framework for the methods to be discussed throughout the chapter and helps us understand how the basis vectors or functions determined in step 1 will be used.

13.2 Projection Methods for Continuous Systems

We begin by considering continuous systems for which variational methods provide the fundamental motivation and mathematical machinery. These methods are then adapted to discrete systems using vectors and matrices in the next section. One-dimensional systems are considered for now in order to keep the focus on the underlying concepts and methods. Although it may be tempting to jump directly to discussion of the discrete case, as this is the context in which these methods are generally applied, the reader will be rewarded with a broader and more unified framework for Galerkin projection and proper-orthogonal decomposition by first considering the continuous case.

13.2.1 Variational Methods Primer

Projection methods, and the closely associated method of weighted residuals, have their origins in variational methods (see, for example, Cassel 2013). The one-dimensional variational problem is stated as follows: determine the function $\hat{u}(x)$ that is an exact stationary function of the functional

$$I[\hat{u}(x)] = \int_{x_0}^{x_1} F[x, \hat{u}(x), \hat{u}'(x)]dx. \tag{13.1}$$

Taking the variation of the functional and setting equal to zero according to[2]

$$\delta I[\hat{u}] = \delta \int_{x_0}^{x_1} F[x, \hat{u}(x), \hat{u}'(x)]dx = 0 \tag{13.2}$$

leads to the necessary condition for an extremum given by the Euler equation

$$\frac{\partial F}{\partial \hat{u}} - \frac{d}{dx}\left(\frac{\partial F}{\partial \hat{u}'}\right) = 0. \tag{13.3}$$

For integrands of the form (13.1), this corresponds to a second-order ordinary differential equation for $\hat{u}(x)$ of the general form

$$\mathcal{L}\hat{u}(x) = f(x), \tag{13.4}$$

where in this case, the differential operator $\mathcal{L}$ may be linear or nonlinear.

In the *inverse problem*, we convert the differential Euler equation (13.4) back to its *proper variational form* (13.2). To do so, the inner product of the variation of the dependent variable is taken with the differential equation (13.4), that is, the differential equation is projected onto $\delta\hat{u}$ according to

$$\int_{x_0}^{x_1} \left(f - \mathcal{L}\hat{u}\right) \delta\hat{u}\, dx = 0, \tag{13.5}$$

which is known as the *reduced variational form.* Carrying out the necessary integration by parts allows one to return to the proper variational form (13.2) by extracting the variation δ out of the integrand if the differential operator $\mathcal{L}$ is self-adjoint and the boundary conditions are homogeneous.

If the Euler equation (13.3) or (13.4) cannot be solved exactly in closed form, we may obtain an approximate solution using one of the following approaches:

1. Solve the proper variational form (13.2) approximately using the Rayleigh–Ritz method or finite-element method based on it.
2. Solve the reduced variational form (13.5) using projection methods, such as Galerkin projection.
3. Solve the differential Euler equation using numerical methods, such as finite-difference methods, spectral methods, or finite-element methods, where the latter two are based on the Galerkin (or other weighted residual) approach.

[2] This is equivalent to multiplying the variation of the dependent variable and the differential equation and integrating over the domain.

A comprehensive arsenal of numerical methods for solving differential equations is provided in Part II. Here, we focus on projection-based methods that begin with the reduced variational form (13.5).

13.2.2 Rayleigh–Ritz Method

Lord Rayleigh in 1877 and Walther Ritz in 1909 devised a method for determining an approximation $u(x)$ to the exact stationary function $\hat{u}(x)$ as the linear combination of a set of basis functions of the form

$$u(x) = \phi_0(x)+\sum_{n=1}^{N} a_n\phi_n(x) = \phi_0(x) + a_1\phi_1(x) + \cdots + a_n\phi_n(x) + \cdots + a_N\phi_N(x). \tag{13.6}$$

This is referred to as a *trial function*. The linearly independent *basis functions* $\phi_n(x)$, $n = 0, \ldots, N$ that comprise the trial function account for the spatial dependence in the solution and are specified functions that satisfy the boundary conditions. Any nonhomogeneous boundary conditions are satisfied by $\phi_0(x)$, and the remaining basis functions are homogeneous at the boundaries of the domain. Substituting the approximate trial function (13.6) into the proper variational form (13.2) for $\hat{u}(x)$, integrating with respect to x, and taking the variation with respect to the unknown coefficients in the trial function leads to an $N \times N$ system of algebraic equations for the coefficients a_n, $n = 1, \ldots, N$. If the problem is unsteady, that is, evolves with time, then this process will lead to an $N \times N$ system of ordinary differential equations for the time-dependent coefficients $a_n(t)$, $n = 1, \ldots, N$. The solution produces the coefficients that for the given basis functions lead to the trial function that is closest to the exact stationary function $\hat{u}(x)$. This is the *Rayleigh–Ritz method* for approximating stationary functions of variational functionals.

Summarizing, the Rayleigh–Ritz method replaces solving the Euler differential equation (13.3) for the exact stationary function $\hat{u}(x)$ with solving a system of algebraic equations (or ordinary differential equations for unsteady problems) for the coefficients in the trial function $u(x)$ given by (13.6) that approximates this stationary function.

13.2.3 Galerkin Method

Because the trial function (13.6) is only an approximate solution, it does not satisfy the differential equation (13.4) exactly. As a proxy for the error of this approximate solution, which is not known in general given that we normally do not have the exact solution, we define the *residual* from the differential equation by

$$r(x) = f(x) - \mathcal{L}u(x). \tag{13.7}$$

The differential operator could be linear or nonlinear, steady or unsteady, and ordinary or partial; however, we will stay with our one-dimensional framework for the sake of simplicity at this point. Observe that if the approximation is the same as the exact solution, that is, $u(x) = \hat{u}(x)$, then the residual vanishes. The *reduced variational form* (13.5) can be written in terms of the trial function as

$$\int_{x_0}^{x_1} r(x)\delta u \, dx = 0. \tag{13.8}$$

Although the reduced variational form (13.5) is only useful as an exact representation of the so-called *weak form* of the differential equation, the analogous form (13.8) in terms of the trial function can always be viewed as the inner product of the residual with the variation of the trial function. In other words, this is a projection of the residual of the differential equation onto $\delta u(x)$, which because the inner product is equal to zero is an *orthogonal projection*.

Let us more closely examine what (13.8) is telling us. Because the basis functions $\phi_n(x)$, $n = 0, \ldots, N$ in the trial function (13.6) are specified and do not vary, and it is only the coefficients a_n, $n = 1, \ldots, N$ that vary, taking the variation of the trial function (13.6) and substituting into (13.8) yields

$$\int_{x_0}^{x_1} r(x) \left[\phi_1(x)\delta a_1 + \cdots + \phi_n(x)\delta a_n + \cdots + \phi_N(x)\delta a_N \right] dx = 0.$$

Because the coefficients are arbitrary, for this sum to vanish, the expression multiplying each variation must vanish. That is,

$$\int_{x_0}^{x_1} r(x)\phi_i(x) \, dx = 0, \quad i = 1, \ldots, N, \tag{13.9}$$

which simply specifies that the inner product of the residual with each basis function be zero, that is, $\langle r(x), \phi_i(x) \rangle = 0$ for $i = 1, \ldots, N$. Thus, the projection of the residual onto each of the basis functions is zero. The index is changed to i so that there is no confusion with the index n that identifies the basis functions in the residual $r(x)$. Note that each of the orthogonal projections (13.9) includes all coefficients a_n and basis functions $\phi_n(x)$, $n = 1, \ldots, N$ in the residual but only one of the basis functions $\phi_i(x)$. Evaluating these N definite integrals removes the dependence on the spatial coordinate x and leads to the $N \times N$ algebraic (or ordinary differential) equations for the coefficients in the trial function. This is called the *Galerkin method* after Boris Galerkin, who introduced it in 1915, and it is equivalent to the reduced variational form for trial functions of the given form with known basis functions.

There are clearly significant similarities between the Rayleigh–Ritz and Galerkin methods. In fact, they are equivalent for self-adjoint differential operators, for which a proper variational form exists. That is, integration by parts can be used to convert the reduced variational form (13.8) to the proper variational form (13.2) for application of the Rayleigh–Ritz method in the self-adjoint case. The difference is that the Rayleigh–Ritz method applies the trial function directly in the proper variational (weak) form of the problem, whereas the Galerkin method starts with the mathematically equivalent differential (strong) form. As a result, the Galerkin method is not limited to situations for which a proper variational form exists, and it can be applied to a much broader class of problems, including nonlinear ordinary and partial differential equations.

Recall from Section 3.3 that while any second-order, linear ordinary differential equation can be converted to a self-adjoint, that is, Sturm–Liouville, form, this is not the case for nonlinear ordinary differential equations or partial differential equations that do not have self-adjoint differential operators. Consequently, not all differential equations can be converted to an equivalent proper variational form via the inverse method, and the Rayleigh–Ritz approximation method is confined to a narrow class of problems. Just as with the Rayleigh–Ritz method, steady problems produce algebraic systems of equations for the constant coefficients, while unsteady problems produce a system of ordinary differential equations for the time-dependent coefficients in the Galerkin method.

The Galerkin method is the basis for numerical methods and ROM frameworks discussed further in this chapter. In ROM contexts, this is referred to as Galerkin projection. In numerical methods, it is simply referred to as the Galerkin method. It is helpful, however, to realize that all such methods trace their roots back to the inverse variational problem. The Galerkin method is particularly straightforward when the basis functions are mutually orthogonal, in which case all of the products of basis functions $\phi_n(x)\phi_i(x)$ in (13.9) vanish except when $n = i$. The advantages of this will be evident in the subsequent discussion.

13.2.4 Method of Weighted Residuals

The basic approach encapsulated in the Galerkin method can be generalized as the *method of weighted residuals* by replacing the basis functions $\phi_n(x)$ in the orthogonal projection (13.9) by a set of *weight* (test) *functions* $w_i(x)$, $i = 1, \ldots, N$ as follows:

$$\int_{x_0}^{x_1} r(x) w_i(x)\, dx = 0, \quad i = 1, \ldots, N. \tag{13.10}$$

This is then equivalent to setting the inner product of the residual, which contains the trial function, with each weight function to zero, that is, $\langle r(x), w_i(x) \rangle = 0$, $i = 1, \ldots, N$. Thus, it is an orthogonal projection of the residual onto a function space spanned by the weight functions. Alternatively, this can be viewed as averaging the residual – as a proxy for the error – in the weighted differential equation over the domain. Different weight functions can be chosen in order to arrive at various weighted residual methods. The Galerkin method, then, corresponds to the weight functions being specified to be the same as the basis functions in the trial function according to

$$w_i(x) = \phi_i(x) = \frac{\partial u}{\partial a_i}, \tag{13.11}$$

which is equivalent to differentiating the trial function with respect to each of its coefficients.

An alternative to the variationally motivated Rayleigh–Ritz and Galerkin methods arises from focusing on the residual (13.7) directly. Because the exact solution corresponds to a residual of zero, it makes sense that we could arrive at a good approximation to the solution of the differential equation by determining the trial function that minimizes some norm of the residual. Because the basis functions are already specified in the trial function, minimization requires determining the coefficients. If we do so in a mean square sense, this corresponds to minimizing the square of the L_2-norm of the residual $\langle r(x), r(x) \rangle$, or

$$\int_{x_0}^{x_1} r^2(x)dx.$$

The goal is then to determine the coefficients a_i, $i = 1, \dots, N$ in the trial function by minimizing this inner product with respect to the coefficients. Differentiating and setting equal to zero yields

$$\frac{\partial}{\partial a_i}\left[\int_{x_0}^{x_1} r^2(x)dx\right] = 0, \quad i = 1, \dots, N,$$

or

$$2\left[\int_{x_0}^{x_1} r(x)\frac{\partial r}{\partial a_i}dx\right] = 0, \quad i = 1, \dots, N.$$

This is known as the *least-squares method*[3] and is equivalent to a method of weighted residuals with the weight functions being

$$w_i(x) = -\frac{\partial r}{\partial a_i}.$$

Note the similarity between the least-squares and Galerkin methods. In the Galerkin method, it is the trial function, rather than the residual, that is differentiated with respect to each of the coefficients to produce the weight functions. Because the weight functions arise from the residual itself, they require evaluation of derivatives of the basis functions and are typically more complicated than simply using the basis functions as in the Galerkin method. While both Galerkin projection and least-squares methods can be couched in terms of the method of weighted residuals, they each have very different origins. Whereas Galerkin projection arises from a reduced variational inverse problem, least squares originates from an optimization problem. Because least squares does not originate from a variational principle, there is no assurance that minimizing the residual will result in the smallest error for all problems. Consequently, it has theoretical as well as practical disadvantages for continuous systems owing to the use of more complex weight functions as compared to Galerkin projection.

In addition to the Galerkin and least-squares methods, a third common weighted residual method is the *collocation method*. In this case, the weight functions are set equal to N Dirac delta functions centered at preselected collocation points x_i as follows:

$$w_i(x) = \hat{\delta}(x - x_i),$$

[3] Compare this with the least-squares method in Section 10.2.1, where the square of the residual of the discrete problem is minimized in order to obtain the best solution of an overdetermined system of equations.

such that the residual is equal to zero at each of the collocation points. The finite-volume method, which is a numerical method popular in fluid mechanics and heat transfer, is essentially the collocation method applied to each finite-volume subdomain. It is also useful in spectral numerical methods (see Section 7.3) applied to nonlinear differential equations.

The primary value of the method of weighted residuals in general, and the Galerkin method in particular, is that no proper variational form is required for the given differential equation; therefore, it is applicable to essentially any scenario involving ordinary and partial differential equations, including nonlinear equations. In the Rayleigh–Ritz, Galerkin, and other weighted residual methods, we choose the basis functions primarily for their ease of integration to produce the system of algebraic (or differential) equations for the coefficients in the trial function. Of the methods of weighted residuals, the Galerkin method is the one that is most faithful to the original variational underpinnings. Therefore, it is generally preferred unless there is some other reason to use an alternative set of weight functions.

The finite-element method (see Section 7.4) is an extension of the Rayleigh–Ritz and Galerkin methods in which the trial function is applied locally over a series of small subdomains, called elements, rather than globally over the entire domain. Because there are many small elements, low-order polynomials are typically used as the basis functions for each element covering its small subdomain. Therefore, there is a trade-off between having many elements with low-order basis functions and fewer elements with higher-order basis functions. The restriction to ordinary and partial differential operators that are self-adjoint was a major limitation and stunted development of early finite-element methods that were based on the Rayleigh–Ritz method. Most modern finite-element methods are based on the Galerkin method, which has enabled their application in nearly all areas of physics and engineering.

Within the Galerkin method, we may select the basis functions in two ways depending on our objective:

1. Preselect the basis functions for their orthogonality properties and ease of integration. This gives rise to *spectral numerical methods*, in which Fourier series or Chebyshev polynomials are typically used as the basis functions.
2. Calculate the basis functions from numerical or experimental data obtained from the system for a particular set of parameters. This is accomplished using *proper-orthogonal decomposition* and its extensions and is the basis of *reduced-order models* for the system's behavior.

Once the basis functions are selected or calculated, Galerkin projection is then used to determine the coefficients in the trial function. Each of these projection-based methods is considered in the next two sections.

13.2.5 Spectral Numerical Method

As discussed in Part II, the goal of numerical methods for differential equations is to seek an approximate solution of the differential equation

$$\mathcal{L}\hat{u}(x) = f(x).$$

Finite-difference, finite-volume, and finite-element methods involve *local* approximations of the solution leading to a large system of algebraic equations for the coefficients in each local approximation. Application of the method of weighted residuals, such as the Galerkin method, directly as a numerical technique to solve an ordinary or partial differential equation is known as *spectral numerical methods*[4] and involve *global* basis functions in the trial function (13.6) that each apply across the entire domain to form an approximate solution of the differential equation.

As introduced in Section 7.3, in spectral methods, the right-hand-side function $f(x)$ is expanded as a linear combination of the same basis functions as used for the solution according to

$$f(x) = \sum_{n=1}^{N} \hat{a}_n \phi_n(x) = \hat{a}_1 \phi_1(x) + \cdots + \hat{a}_n \phi_n(x) + \cdots + \hat{a}_N \phi_N(x). \qquad (13.12)$$

Taking the inner product of each basis function with this expansion for $f(x)$ produces the $\hat{a}_n$ coefficients in this expansion (see Section 3.2.2). This process is particularly straightforward if the basis functions are mutually orthogonal. Using the Galerkin method, for which $w_i(x) = \phi_i(x)$, substituting the trial function (13.6) and expansion (13.12) for $f(x)$ into the weighted residual expression (13.10) leads to the system of equations

$$\int_{x_0}^{x_1} \left\{ \sum_{n=1}^{N} \hat{a}_n \phi_n(x) - \mathcal{L}\left[\phi_0(x) + \sum_{n=1}^{N} a_n \phi_n(x) \right] \right\} \phi_i(x)\, dx = 0, \quad i = 1, \ldots, N. \qquad (13.13)$$

It is apparent that using mutually orthogonal basis functions is advantageous so that the integrals of products, that is, inner products, of basis functions all vanish except when there is a product of a basis function with itself when $n = i$. This eases formulation of the system of algebraic or ordinary differential equations to be solved for the coefficients in the trial function. The function $\phi_0(x)$ is defined in such as way as to satisfy any nonhomogeneous boundary conditions, while the remaining basis functions are then homogeneous at the boundaries of the domain. It is important to note that the basis functions are selected to be mutually orthogonal for convenience – they only need to be linearly independent.

The basis functions are specified to approximate the spatial dependence of the solution, and steady problems result in an $N \times N$ system of algebraic equations for the coefficients, while unsteady problems lead to an $N \times N$ system of ordinary differential equations for the time-dependent coefficients. For problems with periodic boundary conditions, trigonometric functions, such as Fourier series, are used for the basis functions. For nonperiodic problems, on the other hand, Chebyshev or Legendre

[4] The term *spectral* is often used synonymously with Fourier analysis. Indeed, certain spectral methods are based on the use of Fourier series as basis functions. When this is not the case, however, the term spectral method is still used.

polynomials are typically used owing to their mutual orthogonality and ease of integration. For nonlinear problems, collocation is typically used rather than the Galerkin method, and this is referred to as the *pseudospectral method.*

Once again, the primary distinguishing feature of spectral methods relative to other popular numerical methods is that the basis functions are applied globally, rather than locally, throughout the domain. They give highly accurate approximate solutions when the underlying solution is smooth, and it is for smooth solutions that *spectral convergence*[5] is achieved with increasing N. For solutions exhibiting large gradients somewhere in the domain, such as shockwaves or discontinuities, there is typically a dramatic increase in the number of basis functions that must be included in the trial function in order to maintain a certain level of accuracy; this is its primary deficiency. See Sections 7.1.3 and 7.3 for more details on spectral numerical methods, an example, and comparison with other numerical methods.

13.3 Galerkin Projection and Reduced-Order Modeling for Continuous Systems

Suppose that rather than determining a solution of the original differential equation, as with spectral methods, the objective is to develop a *reduced-order model* (ROM) of the governing equation. The ROM is a simplified model that faithfully captures the dominant physics of a system that can be used to better understand a system's behavior. This ROM could be used for further analysis using dynamical systems theory, or the low-dimensional model may be useful in optimization or feedback control algorithms.

In line with our theme of considering continuous models first, let us now consider the single partial differential equation

$$\mathcal{L}\hat{u}(x,t) = f(x),$$

which is subject to the necessary boundary and/or initial conditions, where now the dependent variable is $\hat{u}(x,t)$, and the right-hand side is some algebraic function $f(x)$.

Because the system is continuous and infinite dimensional, we would require an infinite number of basis functions in order to express the solution exactly as in Chapter 3. In the spectral numerical method, we utilize N basis functions in order to approximate this exact solution. Rather than requiring all N basis functions in the trial function, imagine that one could obtain an adequate representation of the system using only R basis functions in our expansion (13.6), where $R \ll N$ in order to dramatically reduce the order of the model. In the ROM, then, the system of algebraic or ordinary

[5] Spectral convergence corresponds to a convergence rate as the number of basis functions increases that decreases exponentially rather than algebraically. Exponential decay is faster than any algebraic power of the grid size. We call it spectral convergence because it is the rate of convergence as additional terms are added to a Fourier (spectral) series. Finite-difference methods exhibit algebraic convergence rates corresponding to some integer power (usually one or two) of the grid size.

differential equations now will be $R \times R$ for the time-dependent coefficients in the trial function[6]

$$\begin{aligned} u(x,t) &= \phi_0(x) + \sum_{n=1}^{R} a_n(t)\phi_n(x) \\ &= \phi_0(x) + a_1(t)\phi_1(x) + \cdots + a_n(t)\phi_n(x) + \cdots + a_R(t)\phi_R(x). \end{aligned} \tag{13.14}$$

Once again, we approximate the solution in terms of a linear combination of a finite number of basis functions that capture the spatial dependence of the solution. There is no fundamental difference between this trial function and (13.6). The use of R instead of N is simply emphasizing that the ROM often requires fewer basis functions than is required for its numerical solution.

As with spectral numerical methods, there is no mathematical requirement that the basis functions be mutually orthogonal; however, there are significant advantages if this is the case. As in spectral methods, we expand the forcing term $f(x)$ in terms of the same basis functions as the solution according to

$$f(x) = \sum_{n=1}^{R} \hat{a}_n\phi_n(x) = \hat{a}_1\phi_1(x) + \cdots + \hat{a}_n\phi_n(x) + \cdots + \hat{a}_R\phi_R(x), \tag{13.15}$$

where the $\hat{a}_n$ coefficients are determined as in Section 3.2.2 and for spectral methods. For example, if the basis functions are orthonormal, then the coefficients are given by

$$\hat{a}_n = \left\langle f(x), \phi_n(x) \right\rangle.$$

We define a *residual function* as in (13.7) according to

$$r(x,t) = f(x) - \mathcal{L}u(x,t). \tag{13.16}$$

Defining a set of *weight functions* $w_i(x), i = 1, \ldots, R$, we project the residual onto the weight functions and enforce their orthogonality according to

$$\langle r(x,t), w_i(x) \rangle = 0, \quad i = 1, \ldots, R.$$

Because we are dealing with functions, the inner products are given by the integrals

$$\int_{x_0}^{x_1} \left[f(x) - \mathcal{L}u(x,t) \right] w_i(x)\, dx = 0, \quad i = 1, \ldots, R, \tag{13.17}$$

which is over the spatial domain $x_0 \leq x \leq x_1$, and we have substituted the expression (13.16) for the residual function.

In the Galerkin method, we use the same expressions for the weight functions as we use for the basis functions, such that $w_i(x) = \phi_i(x)$. Substituting the expansions (13.14) and (13.15) for the solution $u(x,t)$ and right-hand-side function $f(x)$, the

[6] The observant reader will notice the similarity between this representation of the solution and that used for obtaining exact solutions of partial differential equations using separation of variables in Chapter 3. There it was a formal means to facilitate the analytical solution; here it is to split up the spatial and temporal dependence for similar reasons. However, because we do not actually seek the solution of a differential equation, but rather the basis functions for a particular set of data, we can apply this deconstruction to any problem, not just separable ones.

definite integrals (13.17) eliminate the explicit dependence on the spatial coordinate x and produce a system of R ordinary differential equations in time for the time-dependent coefficients $a_n(t), n = 1, \ldots, R$. For steady problems that do not depend on time, it produces a system of algebraic equations for the coefficients $a_n, n = 1, \ldots, R$. Observe that we have converted the continuous, infinite-dimensional partial differential equation $\mathcal{L}\hat{u} = f$ into a system of R first-order ordinary differential equations (or algebraic equations for a steady system). This is the reduced-order model.

Note that this is referred to as an orthogonal projection, not because the basis functions must be mutually orthogonal, but because orthogonality of the residual and basis functions is being enforced inherently in the method. This projection-based ROM framework can be applied to linear and nonlinear, autonomous and nonautonomous, ordinary and partial differential equations governing discrete (finite degrees of freedom) and continuous (infinite degrees of freedom) systems. Along with its ability to accommodate any set of linearly independent, or preferably orthogonal, basis functions, this generality is one of its greatest assets. Despite its generality and modern appeal, note how similar projection methods are to the original inverse variational problem.

REMARKS:

1. *The general Galerkin projection procedure can be summarized as follows:*
   (i) *Identify the basis functions to be used to represent the solution in the trial function and the weight functions. Recall that they are the same in the Galerkin method.*
   (ii) *Expand the solution and right-hand-side function in terms of a linear combination of the R basis functions.*
   (iii) *Substitute the expansions into the differential equation(s) and form the residual.*
   (iv) *Project the residual onto the set of weight functions by setting their inner products to zero to form the ROM.*
   (v) *Solve the R-dimensional ROM for the (time-dependent) coefficients in the trial function for the solution.*

   All but step (i) have been considered in this section. It remains to determine the optimal basis functions for a given set of data. Recall that in spectral numerical methods, no effort is made to choose the basis functions that are "optimal" for the particular problem.
2. *For the special case of homogeneous, linear ordinary differential equations governing steady problems, Galerkin projection produces a homogeneous system of linear algebraic equations for the constant coefficients, and the only unique solution is the trivial solution. Therefore, the nonhomogeneity must be moved from the boundary conditions to the equation itself via a transformation (see Section 3.3.3).*
3. *See section 3.2 of Cassel (2013) for more on Galerkin projection placed into the context of variational methods.*

Example 13.1 Consider the one-dimensional, unsteady diffusion equation

$$\frac{\partial \hat{u}}{\partial t} = \alpha \frac{\partial^2 \hat{u}}{\partial x^2}, \tag{13.18}$$

where $\alpha > 0$ is the diffusivity. It governs the temperature distribution $\hat{u}(x,t)$ owing to unsteady heat conduction in the one-dimensional domain $0 \leq x \leq \ell$. The initial condition is

$$\hat{u}(x,0) = u_0(x), \tag{13.19}$$

and the boundary conditions are

$$\hat{u}(0,t) = 0, \quad \hat{u}(\ell,t) = 0. \tag{13.20}$$

Determine the approximate solution using Galerkin projection with the basis functions[7]

$$\phi_n(x) = \sin\left(\frac{n\pi}{\ell}x\right), \quad n = 1, \ldots, R, \tag{13.21}$$

which all satisfy the boundary conditions and are mutually orthogonal over the interval $0 \leq x \leq \ell$.

Solution

We expand the solution in terms of the basis functions (13.21) as follows:

$$u(x,t) = \sum_{n=1}^{R} a_n(t)\phi_n(x), \tag{13.22}$$

where you will note that $\phi_0(x) = 0$ owing to the homogeneous boundary conditions (13.20). This is the trial function. In the general form $\mathcal{L}u = f$, the one-dimensional, unsteady diffusion equation (13.18) has

$$\mathcal{L} = \frac{\partial}{\partial t} - \alpha \frac{\partial^2}{\partial x^2}, \quad f(x) = 0.$$

Therefore, the residual is given by

$$r(x,t) = f - \mathcal{L}u = -\frac{\partial u}{\partial t} + \alpha \frac{\partial^2 u}{\partial x^2}.$$

In Galerkin projection, the weight functions are set equal to the basis functions (13.21), such that

$$w_i(x) = \phi_i(x) = \sin\left(\frac{i\pi}{\ell}x\right), \quad i = 1, \ldots, R.$$

[7] Observe that from Section 3.4.2, where the one-dimensional, unsteady diffusion equation is solved using separation of variables, these are the orthonormal eigenfunctions of the spatial differential operator for this equation with the given boundary conditions. We can choose any basis functions; we choose the eigenfunctions in order to see what happens in the context of Galerkin projection.

Consequently, projecting the residual onto the set of weight functions by setting their inner products to zero produces

$$\langle r(x,t), w_i(x)\rangle = 0, \quad i = 1, \dots, R,$$

or

$$\int_0^\ell (f(x) - \mathcal{L}u(x,t))\, w_i(x)\, dx = 0, \quad i = 1, \dots, R,$$

which for our problem is

$$\int_0^\ell \left[-\frac{\partial}{\partial t} + \alpha \frac{\partial^2}{\partial x^2} \right] \left[\sum_{n=1}^{R} a_n(t)\phi_n(x) \right] \phi_i(x)\, dx = 0, \quad i = 1, \dots, R.$$

Performing the differentiation inside the integral produces

$$\int_0^\ell \sum_{n=1}^{R} \left[-\dot{a}_n(t)\phi_n(x)\phi_i(x) + \alpha a_n(t)\phi_n''(x)\phi_i(x) \right] dx = 0, \quad i = 1, \dots, R,$$

where as usual the dot denotes differentiation with respect to time, and primes denote differentiation with respect to the spatial variable x. For substitution into the second term, differentiating the basis functions (13.21) twice yields

$$\phi_n''(x) = -\frac{n^2\pi^2}{\ell^2} \sin\left(\frac{n\pi}{\ell}x\right) = -\frac{n^2\pi^2}{\ell^2}\phi_n(x).$$

Substituting leads to

$$\sum_{n=1}^{R} \left[\dot{a}_n(t) + \frac{\alpha n^2\pi^2}{\ell^2} a_n(t) \right] \int_0^\ell \phi_n(x)\phi_i(x) dx = 0, \quad i = 1, \dots, R.$$

Because the basis functions are mutually orthogonal over the interval $0 \leq x \leq \ell$ in this case, the inner product of the basis functions with themselves all vanish except for the one term when $n = i$. Thus, we have the system of equations for the time-dependent coefficients

$$\dot{a}_i(t) = -\frac{\alpha i^2\pi^2}{\ell^2} a_i(t), \quad i = 1, \dots, R.$$

The solutions to these R equations are then

$$a_i(t) = c_i \exp\left(-\frac{\alpha i^2\pi^2}{\ell^2} t \right), \quad i = 1, \dots, R,$$

where the c_i coefficients are constants of integration. Therefore, the expansion (13.22) gives the solution

$$u(x,t) = \sum_{i=1}^{R} c_i \exp\left(-\frac{\alpha i^2\pi^2}{\ell^2} t \right) \sin\left(\frac{i\pi}{\ell} x \right). \tag{13.23}$$

Finally, the c_i coefficients are determined from the nonhomogeneous initial condition (13.19) as in Section 3.4.2, producing Fourier sine coefficients of the initial condition $u_0(x)$.

Observe that we have obtained a truncated version of the same expansion as (3.66), which was determined using separation of variables and an eigenfunction expansion. This is no surprise as we used the eigenfunctions of the differential operator as the basis functions in the Galerkin projection. Therefore, this exercise has shown that an eigenfunction expansion truncated to R terms is indeed a valid ROM in the sense of Galerkin projection as imagined in the introduction of this chapter.

The advantage of Galerkin projection is that it can be applied with any set of linearly independent basis functions and can be extended naturally to nonlinear partial differential equations that do not have self-adjoint differential operators. Unlike this example, however, the system of ordinary differential equations for the $a_i(t)$ coefficients generally will be fully coupled, and increasing R will alter *all* of the coefficients, not simply add additional terms in the expansion with the terms having smaller i remaining the same.

In this example, it is because we used the orthonormal eigenfunctions as the basis functions that the ordinary-differential equations for the time-dependent coefficients ended up being uncoupled. Despite the fact that using the eigenfunctions as the basis functions leads to such a satisfying result, the eigenfunctions of the differential operator do not necessarily produce an "optimal" set of basis functions for a given problem having R modes. This will be taken up in a later section.

Example 13.2 As in the previous example, let us once again consider the one-dimensional, unsteady diffusion equation

$$\frac{\partial \hat{u}}{\partial t} = \alpha \frac{\partial^2 \hat{u}}{\partial x^2}, \tag{13.24}$$

where $\alpha > 0$ is the diffusivity. The initial condition is

$$\hat{u}(x,0) = u_0(x), \tag{13.25}$$

and the boundary conditions are

$$\hat{u}(0,t) = 0, \quad \hat{u}(\ell,t) = 0, \tag{13.26}$$

which are the same as before. In addition, the same basis functions are used, which are taken to be the orthonormal eigenfunctions of the differential operator as follows:

$$\phi_n(x) = \sin\left(\frac{n\pi}{\ell}x\right), \quad n = 1, \ldots, R. \tag{13.27}$$

In this example, however, we repeat the projection process using the least-squares method to determine the solution.

Solution

We again expand the solution in terms of the basis functions (13.27) as follows:

$$u(x,t) = \sum_{n=1}^{R} a_n(t)\phi_n(x). \tag{13.28}$$

In the general form $\mathcal{L}\hat{u} = f$, the one-dimensional, unsteady diffusion equation (13.24) has

$$\mathcal{L} = \frac{\partial}{\partial t} - \alpha\frac{\partial^2}{\partial x^2}, \quad f(x) = 0.$$

Therefore, the residual is given by

$$r(x,t) = f - \mathcal{L}u = -\frac{\partial u}{\partial t} + \alpha\frac{\partial^2 u}{\partial x^2} = \left[-\frac{\partial}{\partial t} + \alpha\frac{\partial^2}{\partial x^2}\right]\left[\sum_{n=1}^{R} a_n(t)\phi_n(x)\right],$$

which gives

$$r(x,t) = -\sum_{n=1}^{R} \dot{a}_n(t)\phi_n(x) + \alpha\sum_{n=1}^{R} a_n(t)\phi_n''(x). \tag{13.29}$$

Thus far, everything is the same as in the previous example using Galerkin projection.

In the least-squares method, the weight functions are given by

$$w_i(x) = -\frac{\partial r}{\partial a_i} = \alpha\phi_i''(x), \quad i = 1, \dots, R. \tag{13.30}$$

Projecting the residual onto these weight functions and setting equal to zero, that is,

$$\langle r(x,t), w_i(x)\rangle = 0, \quad i = 1, \dots, R,$$

or

$$\alpha\int_0^\ell \sum_{n=1}^{R}\left[-\dot{a}_n(t)\phi_n(x)\phi_i''(x) + \alpha a_n(t)\phi_n''(x)\phi_i''(x)\right]dx = 0, \quad i = 1, \dots, R.$$

However, note that

$$\phi_n''(x) = -\frac{n^2\pi^2}{\ell^2}\sin\left(\frac{n\pi}{\ell}x\right) = -\frac{n^2\pi^2}{\ell^2}\phi_n(x).$$

Substituting this into the projection gives

$$\alpha\frac{i^2\pi^2}{\ell^2}\int_0^\ell \sum_{n=1}^{R}\left[\dot{a}_n(t)\phi_n(x)\phi_i(x) + \frac{\alpha n^2\pi^2}{\ell^2}a_n(t)\phi_n(x)\phi_i(x)\right]dx = 0, \quad i = 1, \dots, R.$$

This can be written in the form

$$\sum_{n=1}^{R}\left[\dot{a}_n(t) + \frac{\alpha n^2\pi^2}{\ell^2}a_n(t)\right]\int_0^\ell \phi_n(x)\phi_i(x)dx = 0, \quad i = 1, \dots, R,$$

which is the same as for the Galerkin method. Therefore, the same expansion is obtained using the Galerkin and least-squares methods, which is simply a truncated version of the eigenfunction expansion obtained using separation of variables, for this separable partial differential equation using the orthonormal eigenfunctions of the differential operator as the basis functions. Clearly, this will not be the case in more general problems.

13.4 Projection Methods for Discrete Systems

Here, the Galerkin projection framework considered for continuous systems in the previous section is adapted for discrete systems. As you will see, the same procedure is followed except that inner products of functions are replaced with inner products of vectors. We consider three scenarios in order to develop a strong intuition for the approach as well as to see how it applies for linear and nonlinear differential equations.

13.4.1 System of Linear Algebraic Equations

Let us begin by reconsidering a problem from Chapter 1 – determining the solution of a system of linear algebraic equations of the form

$$\mathbf{A}\hat{\mathbf{u}} = \mathbf{b}, \tag{13.31}$$

where the coefficient matrix $\mathbf{A}$ is $N \times N$, and the solution and right-hand-side vectors, $\hat{\mathbf{u}}$ and $\mathbf{b}$, respectively, are $N \times 1$ vectors. We know that the solution can be expressed exactly in terms of a linear combination of N $N \times 1$ linearly independent *basis vectors*. However, let us approximate the solution using $R \leq N$ basis vectors as follows:

$$\mathbf{u} \approx \sum_{n=1}^{R} a_n \boldsymbol{\phi}_n = a_1 \boldsymbol{\phi}_1 + a_2 \boldsymbol{\phi}_2 + \cdots + a_n \boldsymbol{\phi}_n + \cdots + a_R \boldsymbol{\phi}_R = \boldsymbol{\Phi}\mathbf{a}, \tag{13.32}$$

where the $N \times R$ matrix $\boldsymbol{\Phi}$ has the basis vectors $\boldsymbol{\phi}_n$ as its columns, and the coefficient vector is

$$\mathbf{a} = \begin{bmatrix} a_1 & a_2 & \cdots & a_n & \cdots & a_R \end{bmatrix}^T. \tag{13.33}$$

The residual is defined by

$$\mathbf{r} = \mathbf{b} - \mathbf{A}\mathbf{u}. \tag{13.34}$$

Substituting (13.32) gives

$$\mathbf{r} = \mathbf{b} - \mathbf{A}\boldsymbol{\Phi}\mathbf{a}. \tag{13.35}$$

Defining the *weight vectors* $\mathbf{w}_i$, $i = 1, \ldots, R$, which are the columns of the $N \times R$ matrix $\mathbf{W}$, and taking the orthogonal projection of the residual onto these weight vectors yields

$$\langle \mathbf{r}, \mathbf{w}_i \rangle = 0, \quad i = 1, \ldots, R, \tag{13.36}$$

which is equivalent to

$$\mathbf{W}^T \mathbf{r} = \mathbf{0}. \tag{13.37}$$

Substituting the residual from (13.35) leads to

$$\mathbf{W}^T \mathbf{b} - \mathbf{W}^T \mathbf{A} \boldsymbol{\Phi} \mathbf{a} = \mathbf{0};$$

thus,

$$\mathbf{W}^T \mathbf{A} \boldsymbol{\Phi} \mathbf{a} = \mathbf{W}^T \mathbf{b}.$$

If the $R \times R$ matrix $\mathbf{W}^T \mathbf{A} \boldsymbol{\Phi}$ is not singular, then the coefficients we seek are given by

$$\mathbf{a} = \left(\mathbf{W}^T \mathbf{A} \boldsymbol{\Phi} \right)^{-1} \mathbf{W}^T \mathbf{b}. \tag{13.38}$$

Thus, we can obtain the solution for the coefficients in the linear combination (13.32) when $R \leq N$. The system (13.38) is a reduced-order model of the system of equations (13.31). Observe that the additional calculations required to evaluate the ROM (13.38) as compared to simply evaluating $\mathbf{u} = \mathbf{A}^{-1}\mathbf{b}$ only makes sense if $R \ll N$.

For the case when $R = N$, for which the number of basis vectors corresponds to the size of the original problem, we should obtain the exact solution $\mathbf{u} = \hat{\mathbf{u}}$. In this case, $\mathbf{A}$, $\boldsymbol{\Phi}$, and $\mathbf{W}$ are all square and $N \times N$. If the coefficient matrix $\mathbf{A}$ is invertible, such that rank($\mathbf{A}$) $= N$, then evaluating the inverse in (13.38) yields

$$\mathbf{a} = \boldsymbol{\Phi}^{-1} \mathbf{A}^{-1} \left(\mathbf{W}^T \right)^{-1} \mathbf{W}^T \mathbf{b},$$

which simplifies to

$$\mathbf{a} = \boldsymbol{\Phi}^{-1} \mathbf{A}^{-1} \mathbf{b}. \tag{13.39}$$

Therefore, the solution (13.32) with $R = N$ of the system of linear algebraic equations (13.31) is then

$$\mathbf{u} = \boldsymbol{\Phi} \mathbf{a} = \boldsymbol{\Phi} \boldsymbol{\Phi}^{-1} \mathbf{A}^{-1} \mathbf{b} = \mathbf{A}^{-1} \mathbf{b}. \tag{13.40}$$

Consequently, orthogonal projection via the method of weighted residuals applied to a system of linear algebraic equations is perfectly consistent with using the inverse of the coefficient matrix to obtain the solution of a nonsingular system. Note that no assumptions have been made about the weight functions $\mathbf{W}$; it has only been assumed that the basis functions, that is, the columns of $\boldsymbol{\Phi}$, are linearly independent.

In Section 2.3.2, we considered the special case when the coefficient matrix of the linear system of algebraic equations is real and symmetric. It was shown that a solution could be written as a linear combination of the orthonormal eigenvectors of $\mathbf{A}$. Let us now see what happens to the preceding analysis if these orthonormal eigenvectors are used as the basis vectors, in which case $\boldsymbol{\Phi} = \mathbf{Q}$ is an orthogonal matrix whose columns are the orthonormal eigenvectors of real, symmetric $\mathbf{A}$.

As before, the solution of $\mathbf{Au} = \mathbf{b}$ is expressed in the form

$$\mathbf{u} = \mathbf{Qa},$$

and the residual is

$$\mathbf{r} = \mathbf{b} - \mathbf{Au} = \mathbf{b} - \mathbf{AQa}.$$

Premultiplying by $\mathbf{W}^T$, and recalling that $\mathbf{W}^T\mathbf{r} = \mathbf{0}$, this now gives

$$\mathbf{W}^T\mathbf{AQa} = \mathbf{W}^T\mathbf{b}.$$

Applying the Galerkin method, for which $\mathbf{W} = \mathbf{Q}$, this yields

$$\mathbf{Q}^T\mathbf{AQa} = \mathbf{Q}^T\mathbf{b}.$$

As shown at the end of Section 2.3.3, this leads to the result that

$$\mathbf{a} = \begin{bmatrix} \frac{\langle \mathbf{q}_1, \mathbf{b} \rangle}{\lambda_1} & \frac{\langle \mathbf{q}_2, \mathbf{b} \rangle}{\lambda_2} & \cdots & \frac{\langle \mathbf{q}_N, \mathbf{b} \rangle}{\lambda_N} \end{bmatrix},$$

in which the solution is written in terms of the eigenvalues λ_n and orthonormal eigenvectors $\mathbf{q}_n$ of the real, symmetric coefficient matrix $\mathbf{A}$.

13.4.2 System of First-Order, Linear Ordinary Differential Equations

Next we return to a problem that occupied us in Section 2.6 – determining the solution of a system of N first-order, linear ordinary differential equations of the form

$$\frac{d\hat{\mathbf{u}}}{dt} = \mathbf{A}\hat{\mathbf{u}} \tag{13.41}$$

along with initial conditions. The solution vector $\hat{\mathbf{u}}(t)$ is $N \times 1$, and the coefficient matrix $\mathbf{A}$ is $N \times N$. For now, we consider the case when the system is autonomous, in which case $\mathbf{A}$ is comprised only of constants, and homogeneous. Suppose that the solution $\hat{\mathbf{u}}(t)$ of this system of ordinary differential equations can be approximated in terms of a linear combination of R $N \times 1$ linearly independent *basis vectors* $\boldsymbol{\phi}_n, n = 1, \ldots, R$ according to

$$\mathbf{u}(t) \approx \sum_{n=1}^{R} a_n(t)\boldsymbol{\phi}_n = a_1(t)\boldsymbol{\phi}_1 + a_2(t)\boldsymbol{\phi}_2 + \cdots + a_n(t)\boldsymbol{\phi}_n + \cdots + a_R(t)\boldsymbol{\phi}_R = \boldsymbol{\Phi}\mathbf{a}(t), \tag{13.42}$$

where the $N \times R$ matrix $\boldsymbol{\Phi}$ has the basis vectors $\boldsymbol{\phi}_n$ as its columns, and the time-dependent coefficients comprise the vector

$$\mathbf{a}(t) = \begin{bmatrix} a_1(t) & a_2(t) & \cdots & a_n(t) & \cdots & a_R(t) \end{bmatrix}^T.$$

In spectral numerical methods and proper-orthogonal decomposition, the basis vectors $\boldsymbol{\phi}_n, n = 1, \ldots, R$ are specified or predetermined, and the coefficients $a_n(t), n = 1, \ldots, R$ are determined using Galerkin projection. From Section 2.6, we know that the $a_n(t)$ coefficients are exponential functions containing the eigenvalues if the basis vectors are taken to be the eigenvectors of the coefficient matrix $\mathbf{A}$. Here, however, let us consider the more general case where the basis vectors are simply linearly independent, but otherwise unspecified, at this point.

The central question is how do we determine the coefficients $a_n(t), n = 1, \ldots, R$ for a given set of basis vectors $\boldsymbol{\phi}_n, n = 1, \ldots, R$, which are the columns of $\boldsymbol{\Phi}$? Substituting the expression for the solution (13.42) into the system of ordinary differential equations (13.41) gives

$$\frac{d}{dt}\left[\boldsymbol{\Phi}\mathbf{a}(t)\right] = \mathbf{A}\boldsymbol{\Phi}\mathbf{a}(t),$$

or

$$\boldsymbol{\Phi}\dot{\mathbf{a}}(t) = \mathbf{A}\boldsymbol{\Phi}\mathbf{a}(t). \tag{13.43}$$

As in the continuous case, let us define the *residual vector* by

$$\mathbf{r}(t) = \dot{\mathbf{u}} - \mathbf{A}\mathbf{u} = \boldsymbol{\Phi}\dot{\mathbf{a}}(t) - \mathbf{A}\boldsymbol{\Phi}\mathbf{a}(t). \tag{13.44}$$

In addition to the basis and residual vectors, we also define a set of *weight vectors* $\mathbf{w}_i, i = 1, \ldots, R$ that form the columns of the $N \times R$ matrix $\mathbf{W}$. We seek to determine the coefficients $a_n(t), n = 1, \ldots, R$ for which the projection of the residual vector on each of these weight vectors is zero. In other words, the inner product of the residual vector with each of the weight vectors must vanish, such that they are orthogonal according to

$$\langle \mathbf{r}(t), \mathbf{w}_i \rangle = 0, \quad i = 1, \ldots, R,$$

which is equivalent to specifying that

$$\mathbf{W}^T\mathbf{r}(t) = \mathbf{0}. \tag{13.45}$$

Because of how it is constructed, this is called the *method of weighted residuals*, as discussed in Section 13.2.4.

Now let us return to the definition of the residual given in (13.44). Substituting into (13.45) gives

$$\mathbf{W}^T\boldsymbol{\Phi}\dot{\mathbf{a}}(t) - \mathbf{W}^T\mathbf{A}\boldsymbol{\Phi}\mathbf{a}(t) = \mathbf{0},$$

or

$$\mathbf{W}^T\boldsymbol{\Phi}\dot{\mathbf{a}}(t) = \mathbf{W}^T\mathbf{A}\boldsymbol{\Phi}\mathbf{a}(t).$$

If the $R \times R$ matrix $\mathbf{W}^T\boldsymbol{\Phi}$ is not singular, then this can be written in the general form

$$\dot{\mathbf{a}}(t) = \left(\mathbf{W}^T\boldsymbol{\Phi}\right)^{-1}\mathbf{W}^T\mathbf{A}\boldsymbol{\Phi}\mathbf{a}(t) \tag{13.46}$$

subject to the initial conditions. For a given set of basis and weight vectors, this provides a system of ordinary-differential equations to solve for the corresponding time-dependent coefficients $a_n(t), n = 1, \ldots, R$. This is the ROM for the system of first-order equations (13.41). Again, solving (13.46) for $\mathbf{a}(t)$ as compared to (13.41) for $\hat{\mathbf{u}}(t)$ only makes sense if $R \ll N$ in the ROM.

For the case when $R = N$, for which the number of basis vectors corresponds to the size of the original problem, we should obtain the exact solution $\mathbf{u} = \hat{\mathbf{u}}$. In this case, $\mathbf{\Phi}$ and $\mathbf{W}$ are square and $N \times N$. Because $\left(\mathbf{W}^T\mathbf{\Phi}\right)^{-1} = \mathbf{\Phi}^{-1}\left(\mathbf{W}^T\right)^{-1}$ in this case, it follows that (13.46) can be simplified to

$$\dot{\mathbf{a}}(t) = \mathbf{\Phi}^{-1}\mathbf{A}\mathbf{\Phi}\mathbf{a}(t), \tag{13.47}$$

where the matrix of weight vectors $\mathbf{W}$ has been eliminated. Consequently, even in this method of weighted residuals sense, we can see that if the basis vectors $\phi_n, n = 1, \ldots, N$ that form the columns of $\mathbf{\Phi}$ are taken to be the eigenvectors of the coefficient matrix $\mathbf{A}$, then this is the same as solving $\dot{\mathbf{u}} = \mathbf{A}\mathbf{u}$ using diagonalization, for which $\mathbf{D} = \mathbf{\Phi}^{-1}\mathbf{A}\mathbf{\Phi}$ is diagonal and uncouples the equations for $a_n(t), n = 1, \ldots, N$. In this case, $\mathbf{\Phi}$ is the modal matrix $\mathbf{U}$, and $\mathbf{a}(t) = \mathbf{v}(t)$, where

$$a_n(t) = v_n(t) = e^{\lambda_n t}, \quad n = 1, \ldots N.$$

While this appears to simply provide an alternative justification for the diagonalization process to solve systems of first-order, linear ordinary differential equations, let us remind ourselves that in the general form given by (13.46), the time-dependent coefficients $\mathbf{a}(t)$ could be determined using *any* set of basis vectors $\mathbf{\Phi}$ and weight vectors $\mathbf{W}$. Therefore, one can think of Galerkin projection in the linear discrete case as being a generalization of the standard diagonalization procedure.

13.4.3 System of First-Order, Nonlinear Differential Equations

Let us go through the same procedure as before for a more general scenario that will allow us to extend these ideas to nonlinear ordinary and partial differential equations. To do so, consider the generalization of the system of differential equations (13.41) in the form

$$\frac{d\hat{\mathbf{u}}}{dt} = \mathbf{f}\left[\hat{\mathbf{u}}(t), t\right], \tag{13.48}$$

with the required initial conditions, where now $\mathbf{f}$ could be any linear or nonlinear, autonomous or nonautonomous, homogeneous or nonhomogeneous function of the solution vector and time. We could even allow for spatial derivatives to appear in $\mathbf{f}$ to accommodate certain partial differential equations as well.

As before, we substitute (13.42) for the approximate solution vector written in terms of our set of basis vectors, which now gives

$$\mathbf{\Phi}\dot{\mathbf{a}}(t) = \mathbf{f}\left[\mathbf{\Phi}\mathbf{a}(t), t\right]. \tag{13.49}$$

As a result, now the residual vector is given by

$$\mathbf{r}(t) = \dot{\mathbf{u}} - \mathbf{f}\left[\mathbf{u}(t), t\right] = \mathbf{\Phi}\dot{\mathbf{a}}(t) - \mathbf{f}\left[\mathbf{\Phi}\mathbf{a}(t), t\right]. \tag{13.50}$$

Projecting the residual onto each of the weight vectors $\mathbf{w}_i, i = 1, \ldots, R$ by premultiplying the transpose of the weight matrix $\mathbf{W}^T$ on both sides of this equation, and applying (13.45), gives

$$\mathbf{W}^T \boldsymbol{\Phi}\dot{\mathbf{a}}(t) = \mathbf{W}^T \mathbf{f}\left[\boldsymbol{\Phi}\mathbf{a}(t), t\right].$$

Once again, if the $R \times R$ matrix $\mathbf{W}^T \boldsymbol{\Phi}$ is not singular, then this can be written in the general form

$$\dot{\mathbf{a}}(t) = \left(\mathbf{W}^T \boldsymbol{\Phi}\right)^{-1} \mathbf{W}^T \mathbf{f}\left[\boldsymbol{\Phi}\mathbf{a}(t), t\right] \tag{13.51}$$

subject to the initial conditions. As with (13.46), for a given set of basis and weight vectors, this provides a system of ordinary-differential equations, which now may be nonlinear, to solve for the corresponding time-dependent coefficients $a_n(t), n = 1, \ldots, R$. This system of equations is the ROM of the original system.

Now consider the case when $R = N$, for which the number of basis vectors corresponds to the size of the original problem. In this case, $\boldsymbol{\Phi}$ and $\mathbf{W}$ are square and $N \times N$. Because $\left(\mathbf{W}^T \boldsymbol{\Phi}\right)^{-1} = \boldsymbol{\Phi}^{-1} \left(\mathbf{W}^T\right)^{-1}$ if both $\boldsymbol{\Phi}$ and $\mathbf{W}$ are invertible, it follows that (13.51) can be simplified to

$$\dot{\mathbf{a}}(t) = \boldsymbol{\Phi}^{-1} \mathbf{f}\left[\boldsymbol{\Phi}\mathbf{a}(t), t\right], \tag{13.52}$$

where again the matrix of weight vectors $\mathbf{W}$ has been eliminated.

REMARKS:

1. *Equations (13.46) and* (13.47) *are simply special cases of the more general forms* (13.51) *and* (13.52), *respectively, when* $\mathbf{f}\left[\mathbf{u}(t), t\right] = \mathbf{A}\mathbf{u}$ *for a linear system.*
2. *In either formulation, we can express the original system of equations in terms of a projection matrix* $\mathbf{P}$ *as follows. Because* $\mathbf{u}(t) = \boldsymbol{\Phi}\mathbf{a}(t)$ *from expression* (13.42), *differentiating gives the expression* $\dot{\mathbf{u}}(t) = \boldsymbol{\Phi}\dot{\mathbf{a}}(t)$. *Substituting* (13.51) *for* $\dot{\mathbf{a}}(t)$ *gives the form*

$$\dot{\mathbf{u}}(t) = \mathbf{P}\mathbf{f}\left[\mathbf{u}(t), t\right], \tag{13.53}$$

   where the $N \times N$ *projection matrix is*

$$\mathbf{P} = \boldsymbol{\Phi} \left(\mathbf{W}^T \boldsymbol{\Phi}\right)^{-1} \mathbf{W}^T. \tag{13.54}$$

   This applies for orthogonal or nonorthogonal projection schemes.
3. *Including the continuous case considered in the previous section, our three scenarios span first-order, linear ordinary differential equations all the way to nonlinear partial differential equations.*

13.5 Galerkin Projection and Reduced-Order Modeling for Discrete Systems

Reduced-order models of the original systems result when $R < N$ and are given by (13.38), (13.46), and (13.51) for the coefficients in each of the three scenarios considered. The solution of the ROM provides an approximation to the $N \times 1$ solution vector $\hat{\mathbf{u}}(t)$ by using a reduced number of basis vectors in (13.32) or (13.42) to form the ROM of the solution. Although the basis vectors are still $N \times 1$, we only have R

Table 13.1 Sizes of vectors and matrices in Galerkin projection.

Name	Symbol	Size
Solution vector	$\mathbf{u}(t)$	$N \times 1$
Right-hand-side function	$\mathbf{f}[\mathbf{u}(t),(t)]$	$N \times 1$
Coefficient matrix	$\mathbf{A}$	$N \times N$
Vector of time-dependent coefficients	$\mathbf{a}(t)$	$R \times 1$
Basis vectors	$\phi_n, n = 1, \ldots, R$	$N \times 1$
Matrix of basis vectors	$\mathbf{\Phi}$	$N \times R$
Residual vector	$\mathbf{r}(t)$	$N \times 1$
Weight vectors	$\mathbf{w}_i, i = 1, \ldots, R$	$N \times 1$
Matrix of weight vectors	$\mathbf{W}$	$N \times R$
	$\mathbf{W}^T\mathbf{\Phi}$	$R \times R$
Projection matrix	$\mathbf{P}$	$N \times N$

of them and matrix $\mathbf{\Phi}$ is, therefore, $N \times R$, and $\mathbf{a}(t)$ is $R \times 1$, but $\mathbf{u}(t) = \mathbf{\Phi}\mathbf{a}(t)$ is still $N \times 1$. The size of each vector and matrix is given in Table 13.1.

Because the matrices of basis vectors $\mathbf{\Phi}$ and weight vectors $\mathbf{W}$ are no longer square, the final step in each derivation that eliminates the matrix of weight vectors $\mathbf{W}$ producing (13.39), (13.47), and (13.52) is no longer possible. Therefore, we need some way to relate the matrix of weight vectors $\mathbf{W}$ to the matrix of basis vectors $\mathbf{\Phi}$. By far the most common approach is to specify the weight vectors to be the same as the basis vectors, in which case $\mathbf{W} = \mathbf{\Phi}$. This projection procedure is called *Galerkin projection*, for which (13.51) becomes

$$\dot{\mathbf{a}}(t) = \left(\mathbf{\Phi}^T\mathbf{\Phi}\right)^{-1}\mathbf{\Phi}^T\mathbf{f}\left[\mathbf{\Phi}\mathbf{a}(t),t\right] = \mathbf{\Phi}^+\mathbf{f}\left[\mathbf{\Phi}\mathbf{a}(t),t\right], \tag{13.55}$$

where $\mathbf{\Phi}^+ = \left(\mathbf{\Phi}^T\mathbf{\Phi}\right)^{-1}\mathbf{\Phi}^T$ is the pseudo-inverse of $\mathbf{\Phi}$ (see Section 2.8). In the special case when $\mathbf{f}\left[\mathbf{u}(t),t\right] = \mathbf{A}\mathbf{u}$, (13.46) becomes

$$\dot{\mathbf{a}}(t) = \left(\mathbf{\Phi}^T\mathbf{\Phi}\right)^{-1}\mathbf{\Phi}^T\mathbf{A}\mathbf{\Phi}\mathbf{a}(t) = \mathbf{\Phi}^+\mathbf{A}\mathbf{\Phi}\mathbf{a}(t). \tag{13.56}$$

Note that in both cases the square $R \times R$ matrix $\left(\mathbf{\Phi}^T\mathbf{\Phi}\right)^{-1}$ is invertible even though $\mathbf{\Phi}$ is $N \times R$. Observe the resemblance of (13.56) to the diagonalization approach used to solve systems of first-order differential equations in Chapter 2.

For a given set of basis vectors that comprise matrix $\mathbf{\Phi}$, (13.55) [or (13.56)] represents a system of $R < N$ equations to solve for the $a_n(t), n = 1, \ldots, R$ time-dependent coefficients in the reduced-order expansion for the solution (13.42). It represents a ROM for the full system of equations as it provides an approximation for its solution vector $\mathbf{u}(t)$ with fewer than N modes. It then remains to determine the $N \times 1$ basis vectors $\phi_n, n = 1, \ldots, R$ that form the columns of the $N \times R$ matrix $\mathbf{\Phi}$. If possible, we desire a method that produces an *optimal* set of R basis vectors in order to provide the best approximation with the fewest number of basis vectors possible. This is accomplished using proper-orthogonal decomposition in the next section.

REMARKS:

1. *Galerkin projection can be applied using any set of linearly independent basis vectors or functions. As we have seen, there are certain advantages if the basis vectors or functions are also orthogonal.*
2. *Because* $\mathbf{W} = \mathbf{\Phi}$ *for Galerkin projection, the projection matrix* (13.54) *becomes*

$$\mathbf{P} = \mathbf{\Phi}\left(\mathbf{\Phi}^T\mathbf{\Phi}\right)^{-1}\mathbf{\Phi}^T = \mathbf{\Phi}\mathbf{\Phi}^+.$$

If, in addition, the basis vectors $\boldsymbol{\phi}_n, n = 1, \ldots, R$, *that comprise* $\mathbf{\Phi}$ *are orthonormal, then* $\mathbf{\Phi}^T\mathbf{\Phi} = \mathbf{I}$, *and the projection matrix simplifies to*

$$\mathbf{P} = \mathbf{\Phi}\mathbf{\Phi}^T,$$

which is symmetric and positive semidefinite.

3. *If the weight vectors are chosen to be different from the basis vectors, that is,* $\mathbf{W}$ *and* $\mathbf{\Phi}$ *are not the same, then it is called* Petrov–Galerkin projection.

Example 13.3 A reduced-order model is sought for the system of first-order, linear ordinary differential equations

$$\dot{\mathbf{u}}(t) = \mathbf{A}\mathbf{u}(t),$$

where

$$\mathbf{A} = \begin{bmatrix} 1 & 0 & 2 & 0 & 0 & 1 \\ 0 & 2 & 0 & 3 & 0 & 0 \\ 0 & 0 & 3 & 0 & 1 & 0 \\ 0 & 3 & 0 & 4 & 0 & 7 \\ 0 & 0 & 1 & 0 & 5 & 0 \\ 1 & 0 & 0 & 2 & 0 & 6 \end{bmatrix}.$$

Use Galerkin projection with the eigenvectors of the matrix $\mathbf{A}$ being the basis vectors in the ROM.

Solution

The eigenvalues of $\mathbf{A}$ from largest to smallest by magnitude are

$$\lambda_1 = 9.4337, \lambda_2 = 5.4142, \lambda_3 = 3.7558, \lambda_4 = 2.5858, \lambda_5 = -1.0645, \lambda_6 = 0.87495.$$

The corresponding eigenvectors, which we take as the basis vectors, are the columns of

$$\mathbf{\Phi} = \begin{bmatrix} -0.057794 & 0.17305 & 0.13 & 0.57706 & 0.10227 & -0.88487 \\ -0.32606 & -0.08013 & -0.79792 & 0.47794 & -0.68003 & -0.42370 \\ 0. & 0.37399 & 0. & 0.56939 & 0. & 0. \\ -0.8079 & -0.09119 & -0.46700 & 0.093323 & 0.69464 & 0.15889 \\ 0. & 0.90288 & 0. & -0.23585 & 0. & 0. \\ -0.48742 & 0.015917 & 0.35826 & -0.22369 & -0.21113 & 0.11065 \end{bmatrix}.$$

We choose $R < N = 6$ to be the number of basis vectors to use in our ROM, where we take the eigenvectors corresponding to the largest R eigenvalues (by magnitude).

Thus, $\mathbf{\Phi}$ is an $N \times R$ matrix with its columns being the R corresponding basis vectors. In Galerkin projection, the matrix $\mathbf{W}$ of weight vectors equals $\mathbf{\Phi}$. Let us then define the coefficient matrix in the ROM (13.46) to be

$$\mathbf{B} = \left(\mathbf{\Phi}^T \mathbf{\Phi}\right)^{-1} \mathbf{\Phi}^T \mathbf{A} \mathbf{\Phi}.$$

For the case with $R = 2$:

$$\mathbf{B} = \begin{bmatrix} 9.4337 & 0 \\ 0 & 5.4142 \end{bmatrix}.$$

For the case with $R = 3$:

$$\mathbf{B} = \begin{bmatrix} 9.4337 & 0 & 0 \\ 0 & 5.4142 & 0 \\ 0 & 0 & 3.7558 \end{bmatrix}.$$

For the case with $R = 4$:

$$\mathbf{B} = \begin{bmatrix} 9.4337 & 0 & 0 & 0 \\ 0 & 5.4142 & 0 & 0 \\ 0 & 0 & 3.7558 & 0 \\ 0 & 0 & 0 & 2.5858 \end{bmatrix}.$$

For the case with $R = 5$:

$$\mathbf{B} = \begin{bmatrix} 9.4337 & 0 & 0 & 0 & 0 \\ 0 & 5.4142 & 0 & 0 & 0 \\ 0 & 0 & 3.7558 & 0 & 0 \\ 0 & 0 & 0 & 2.5858 & 0 \\ 0 & 0 & 0 & 0 & -1.0645 \end{bmatrix}.$$

In each case, the ROM consists of an uncoupled system of R equations consisting of the largest R eigenvalues along the diagonal. As in the continuous case (Example 13.1), increasing R does not alter the lower-order coefficients. This is because the eigenvectors of $\mathbf{A}$ have been used as the basis vectors for the Galerkin projection.

REMARKS:

1. *When $R = N$, Galerkin projection reduces to the usual diagonalization procedure in which the coefficient matrix for the trial function coefficients becomes fully diagonalized, where $\mathbf{\Phi}$ becomes the modal matrix of $\mathbf{A}$, and $\mathbf{B}$ is the diagonal matrix with the eigenvalues of $\mathbf{A}$ along the diagonal.*
2. *As for the continuous case, we see that truncating the linear combination of basis vectors to include only the R eigenvectors having the largest eigenvalues indeed produces a valid ROM in the sense of Galerkin projection as imagined in the introduction of this chapter.*
3. *In the general case, when the basis vectors are not chosen to be the eigenvectors of $\mathbf{A}$, the $R \times R$ system of equations would be a fully coupled system.*

13.6 Proper-Orthogonal Decomposition (POD) for Continuous Data

Galerkin projection takes any basis vectors (for discrete data) or functions (for continuous data) and determines the best coefficients in the trial vector or function to represent the solution of the governing differential equation(s). In the continuous case, the ability of Galerkin projection to produce effective ROM or spectral numerical solutions hinges on the choice of basis functions. The quality of the ROM is determined by how well the basis functions represent the underlying behavior of the system. However, Galerkin projection does not address how the basis functions are to be obtained or which are "best." Although the original differential equation(s) governing the system's behavior is at the heart of the Galerkin projection procedure for determining a reduced-order model, the following methods for determining the optimal set of basis functions for a specific data set do not take into account the governing differential equation(s). That is, such methods are *data driven*, not *model driven*.

All of the techniques addressed thus far are predicated on the fact that we can express a function, which may be the solution of a differential equation, for example, in the form (13.14). Given a set of linearly independent basis functions $\phi_n(x), n = 1, \dots, N$, spectral numerical methods and Galerkin projection provide a means to obtain approximate solutions or ROM of the system by determining the coefficients $a_n, n = 1, \dots, N$ to complete the linear combination. The key to these methods is that the basis functions have already been specified or otherwise preselected. This raises the question as to how can these basis functions be determined such that they are optimal for the experimental or numerical data set under consideration? This is commonly accomplished using *proper-orthogonal decomposition* (POD) and its variants as outlined in this section for continuous data and the next section for discrete data.

Recall that in spectral numerical methods, the basis functions are chosen such that they satisfy the boundary conditions, and they are also typically selected to be mutually orthogonal for their ease of integration. In ROM contexts, however, there are a limited number of modes that comprise the low-dimensional model of the system. In order for these modes to provide as much model reduction as possible and still yield a faithful representation of the system, the R modes must be chosen in some "optimal" sense from data for that particular system. Such data could arise experimentally or numerically, and the ability to accommodate most any data set from any source is a primary reason for its popularity. The ultimate objective is to represent the state of a system in a low-dimensional form while retaining the essential features of the original large-dimensional data set.

In what sense are POD basis functions, or modes, optimal? Optimal basis functions would be expected to contain some of the characteristics of the underlying system; therefore, they should be obtained from data obtained from the system itself. This would be expected to allow for representing the data faithfully with the fewest number of modes. Formally, this is accomplished by determining the basis functions (POD modes) for which the maximum amount of information, such as energy, is contained in the smallest number of modes.

Keep in mind that there are two main objectives of POD: first to extract the dominant features from the data set, and second for use in further analysis, such as the ROM discussed in the previous section. To facilitate these objectives, there are several desired properties of such a decomposition. Because of their ease of use, orthonormal basis functions are generally desirable. As alluded to previously, a representation that is as faithful as possible to the original data with the fewest number of basis functions is sought. Finally, the POD analysis should be completely independent of the source of the data; for example, no knowledge of the governing equations is necessary. To accomplish these objectives, a set of optimal basis functions are sought that allow us to represent a high-dimensional system with a low-dimensional one containing most of the relevant dynamics of the system.

We first consider the case when the data are in the form of a single continuous data function $u(x)$. In this case, its POD modes are determined using variational methods as discussed in this section. Proper-orthogonal decomposition also goes by the names *principal-component analysis* (PCA) in the statistics community and *Karhunen–Loève decomposition* in the field of pattern recognition.

13.6.1 Single Continuous Data Function

Recall that in projection methods (and spectral numerical methods), we seek to represent the continuous data set $u(x)$, which may be an approximation to the solution of the differential equation $\mathcal{L}\hat{u}(x) = f(x)$, as a linear combination of a set of basis functions $\phi_n(x)$ in the form

$$u(x) = \phi_0(x) + \sum_{n=1}^{N} a_n\phi_n(x) = \phi_0(x) + a_1\phi_1(x) + \cdots + a_n\phi_n(x) + \cdots + a_N\phi_N(x), \tag{13.57}$$

where $\phi_0(x)$ is chosen to satisfy any nonhomogeneous boundary conditions. Our objective in POD is to determine the basis functions $\phi_n(x)$, $i = 1, \ldots, N$ that are optimal for representing the continuous data $u(x)$ for the particular system under consideration in the form (13.57). These POD modes are effective in identifying dominant spatial features in a system, and Galerkin projection can then be used to determine the coefficients in the trial function (13.57), that is, how much of each particular mode is relevant for a given system. It is important to realize that POD analysis can be applied to any set of data, including from a nonlinear system or one for which no mathematical model exists.

Keeping in mind how the basis functions will be used in the projection process, the objective is to minimize the sum of the differences between the known data function $u(x)$ and the projection of the data function onto each of the sought-after basis functions in a mean square sense. This suggests that we minimize

$$\begin{aligned} J[\phi_1(x), \ldots, \phi_N(x)] &= \sum_{n=1}^{N} \left\| u(x) - \langle u(x), \phi_n(x) \rangle \phi_n(x) \right\|^2 \\ &= \sum_{n=1}^{N} \int_{x_0}^{x_1} \left[u(x) - \langle u(x), \phi_n(x) \rangle \phi_n(x) \right]^2 dx. \end{aligned} \tag{13.58}$$

Let us minimize the functional (13.58) subject to the constraint that the norm of each resulting basis function is unity, such that

$$\|\phi_n(x)\|^2 = \langle \phi_n(x), \phi_n(x) \rangle = \int_{x_0}^{x_1} \phi_n^2(x) dx = 1, \quad n = 1, \ldots, N. \tag{13.59}$$

This constraint also prevents trivial solutions for the basis functions when minimizing the objective functional. Keep in mind that the objective is to determine the basis functions $\phi_n(x), n = 1, \ldots, N$ that minimize J for a given data function $u(x)$. Expanding the integrand gives

$$J[\phi_1, \ldots, \phi_N] = \sum_{n=1}^{N} \int_{x_0}^{x_1} \left[u^2 - 2 \langle u, \phi_n \rangle u \phi_n + \left(\langle u, \phi_n \rangle \phi_n \right)^2 \right] dx,$$

which can be written

$$J[\phi_1, \ldots, \phi_N] = \sum_{n=1}^{N} \left[\langle u, u \rangle - 2 \langle u, \phi_n \rangle^2 + \langle u, \phi_n \rangle^2 \|\phi_n\|^2 \right].$$

Because the norm of the basis functions is unity, simplifying further reveals that we seek to minimize the functional

$$\begin{aligned} J[\phi_1(x), \ldots, \phi_N(x)] &= \sum_{n=1}^{N} \left[\langle u(x), u(x) \rangle - \langle u(x), \phi_n(x) \rangle^2 \right] \\ &= \sum_{n=1}^{N} \left[\int_{x_0}^{x_1} u^2(x) dx - \left(\int_{x_0}^{x_1} u(x) \phi_n(x) dx \right)^2 \right]. \end{aligned} \tag{13.60}$$

The first term only involves the given function $u(x)$; therefore, it does not vary or influence the minimization process. Because the second term is squared, it is nonnegative for all $u(x)$ and $\phi_n(x)$, and minimizing (13.60) would be equivalent to maximizing $\sum_{n=1}^{N} \langle u(x), \phi_n(x) \rangle^2$ owing to the minus sign. Consequently, minimizing the functional (13.60) can be regarded as determining the optimal basis functions that maximize the average projection of $u(x)$ onto each of the basis functions $\phi_n(x)$.

Incorporating the constraint (13.59) using Lagrange multipliers, the augmented functional to be minimized is

$$\tilde{J}[\phi_1, \ldots, \phi_N] = \sum_{n=1}^{N} \left[\langle u, u \rangle - \langle u, \phi_n \rangle^2 + \lambda_n \left(\langle \phi_n, \phi_n \rangle - 1 \right) \right],$$

where $\lambda_n, n = 1, \dots, N$ are the Lagrange multipliers enforcing the constraints. Recalling that the basis functions $\phi_n(x), n = 1, \dots, N$ vary, but that the data function $u(x)$ does not, taking the variation and setting equal to zero, that is, $\delta \tilde{J} = 0$, gives

$$\sum_{n=1}^{N} \left[-2 \langle u, \phi_n \rangle \delta \left(\langle u, \phi_n \rangle \right) + \lambda_n \delta \langle \phi_n, \phi_n \rangle \right] = 0.$$

Observe that

$$\delta \left(\langle u, \phi_n \rangle \right) = \langle u, \delta\phi_n \rangle, \quad \delta \langle \phi_n, \phi_n \rangle = 2 \langle \phi_n, \delta\phi_n \rangle;$$

therefore,

$$\sum_{n=1}^{N} \left[- \langle u, \phi_n \rangle \langle u, \delta\phi_n \rangle + \lambda_n \langle \phi_n, \delta\phi_n \rangle \right] = 0,$$

or because the inner product is just a constant value and can be taken inside another inner product (integral), this may be rewritten as

$$\sum_{n=1}^{N} \left[- \langle \langle u, \phi_n \rangle u, \delta\phi_n \rangle + \langle \lambda_n \phi_n, \delta\phi_n \rangle \right] = 0.$$

Combining terms

$$\sum_{n=1}^{N} \langle -u \langle u, \phi_n \rangle + \lambda_n \phi_n, \delta\phi_n \rangle = 0.$$

In order for each term in the sum to vanish, the coefficients of $\delta\phi_n$ must each vanish. Consequently,

$$u \langle u, \phi_n \rangle = \lambda_n \phi_n, \quad n = 1, \dots, N.$$

If we define the integral operator

$$\mathcal{R}\phi_n(x) = u(x) \langle u(x), \phi_n(x) \rangle = u(x) \int_{x_0}^{x_1} u(x)\phi_n(x) dx, \quad n = 1, \dots, N, \tag{13.61}$$

then we have the eigenproblem

$$\mathcal{R}\phi_n = \lambda_n \phi_n, \quad n = 1, \dots, N. \tag{13.62}$$

It is proven in Holmes et al. (2012) that this integral operator is self-adjoint. Therefore, solution of the integral equation gives the eigenvalues, which are real and nonnegative, and the resulting basis functions are mutually orthogonal and normalized to length one (owing to the constraint). It is important to observe that the orthogonality of the POD modes is a natural consequence of the optimization procedure, not a property that is imposed a priori. Equation (13.62) is an eigenproblem to be evaluated for the basis functions $\phi_n(x)$, for which the eigenvalues are the Lagrange multipliers.

Summarizing, minimization of the functional (13.58) subject to the constraint (13.59) is equivalent to solving the eigenproblem (13.62), where the eigenfunctions are the optimal basis functions that we seek; these are the POD modes. The eigenvalues, and their associated eigenfunctions, are typically ordered from largest to smallest, as the eigenfunctions having the largest eigenvalues are most dominant in the original data function $u(x)$. It is because the eigenfunctions are obtained from experimental (or numerical) data that they are often called *empirical eigenfunctions* or *empirical basis functions*. In many physical applications, such as fluid mechanics, each eigenvalue λ_n represents the average energy contained in the corresponding POD mode $\phi_n(x)$. Therefore, the POD modes with the largest eigenvalues represent the most energetic dynamic features in the system (recall that the eigenvalues are arranged in decreasing order). The POD modes are optimal in the sense that for N modes obtained from a linear, self-adjoint operator, no other set of N basis functions contains more average energy in an L_2-norm sense.

For a given data function $u(x)$ and orthonormal basis functions $\phi_n(x)$, the coefficients in the trial function (13.57) are obtained from the inner products

$$a_n = \left\langle u(x), \phi_n(x) \right\rangle, \quad n = 1, \dots, N; \tag{13.63}$$

that is, by projecting the data function onto each of the basis functions. Now that we know that the basis functions that arise from minimizing the objective functional (13.58) subject to the constraint (13.59) are orthonormal, in which case the coefficients in the trial function are simply (13.63), we see that our original objective functional (13.58) can be written in the form

$$J[\phi_1(x), \dots, \phi_N(x)] = \sum_{n=1}^{N} \left\| u(x) - a_n \phi_n(x) \right\|^2. \tag{13.64}$$

Thus, the basis functions are such that the difference between the data function and each of the terms in the trial function (including the coefficients) are minimized in a least-squares sense. This makes sense in light of our overall objective to determine the basis functions that best represent (optimally span) the data function in the form of the expansion (13.57).

Because the POD basis functions are optimal, we can represent the most important information contained in the original data set with the first $R < N$ POD modes. In typical applications of POD analysis, $R = O(10)$ is often sufficient. This provides the model reduction that we seek, in which the original large data set can be represented by a relatively small number of optimally chosen POD modes.

Interestingly, the projection methods used to obtain the ROM utilize the governing equation of the system, while the POD approach to determining the basis functions does not. That is, POD is a *data-driven* method. While this is advantageous for data sets obtained from systems for which a mathematical model is not available, in a Galerkin projection context, where a mathematical model is known, this model does not come into play in formation of the POD basis functions. Consequently, the primary advantage of POD analysis is also its primary shortcoming. Because the POD modes are determined directly from the experimental or numerical data, they do not take into account any information about the mathematical model of the system that may have produced the data. While this means that no model is necessary in order to obtain the POD modes for a data set, even when such a model does exist, it is not taken into account when obtaining the POD modes (it is taken into account in the Galerkin projection process, however).

Because the optimal basis functions are determined from the actual data set, they provide the best representation of the original data with the fewest POD modes for the set of parameters used to obtain the data. However, this means that they are problem – and parameter – dependent, requiring us to obtain a new set of basis functions each time the data set changes, whether from consideration of a different dynamical system or a different set of data from the same system.

REMARKS:

1. *Here, the data function has been taken to be a single function for simplicity of discussion, whereas in practice it is more likely to be an ensemble of data for different parameters or "snapshots" in time (see the next section).*
2. *POD analysis is very general owing to its data-driven framework, in which the data function can arise from any experimental or numerical data set. In addition, it is very powerful and convenient owing to its production of orthonormal POD modes for use in projection methods.*
3. *POD modes are those orthonormal eigenfunctions that allow for a reconstruction that is closest to the original data set in a least-squares sense. Once again, the orthogonality of the POD modes is a natural consequence of the optimization procedure, not a property that is imposed a priori.*
4. *Variants and extensions of POD analysis, including the method of snapshots for transient data, balanced truncation, dynamic-mode decomposition, Koopman operators, resolvent-mode analysis, et cetera, are available to address some of its shortcomings in various situations for which POD is not suitable. These are discussed in subsequent sections.*
5. *In the more common discrete case, where the data function $u(x)$ becomes a vector, the self-adjoint operator $\mathcal{R}$ is replaced by a real, symmetric matrix, and the eigenproblem becomes algebraic, producing orthonormal eigenvectors for the POD modes. These orthonormal eigenvectors are the principal components (axes) of the data giving rise to the nomenclature principal-component analysis. This is considered in Section 13.7.*

13.6.2 Unsteady Data and the Method of Snapshots

Let us now suppose that the data are available for a range of the spatial coordinate x and time t, such that $u = u(x,t)$. The most natural extension of the POD technique outlined in the previous section to the unsteady case would be to simply treat time as an additional independent variable along with the spatial coordinate. Then the trial function would be of the form

$$u(x,t) = \phi_0(x) + \sum_{n=1}^{N} a_n \phi_n(x,t), \tag{13.65}$$

such that the basis functions are now functions of both space and time. These basis functions would then be sought such that the functional

$$J[\phi] = \int_0^{t_f} \int_{x_0}^{x_1} \left[u(x,t) - \langle u(x,t), \phi_n(x,t) \rangle \phi_n(x,t) \right]^2 dx dt, \tag{13.66}$$

which is now an integral over space and time, is to be minimized. This approach, which is most faithful to the original formulation of POD by Lumley (1967), leads to a very large eigenproblem to be solved for the basis functions.

To avoid solving an extremely large eigenproblem for the basis functions, the usual method for dealing with time-evolving data is to use snapshots, which are data sets over the entire spatial domain $u(x)$ taken at a set of discrete times. This is called the *method of snapshots* and was introduced by Sirovich (1987). While the spatial data remain continuous (for now), the temporal data are discrete in time. For example, if the data set contains M snapshots, then the functional (13.60) would become

$$J[\phi_n] = \sum_{n=1}^{N} \left\{ \frac{1}{M} \sum_{j=1}^{M} \int_{x_0}^{x_1} \left[u_j(x) - \langle u_j(x), \phi_n(x) \rangle \phi_n(x) \right]^2 dx \right\}, \tag{13.67}$$

which is equivalent to maximizing the time-averaged functional

$$J[\phi_n] = \sum_{n=1}^{N} \left\{ \frac{1}{M} \sum_{j=1}^{M} \langle u_j(x), \phi_n(x) \rangle^2 \right\} = \sum_{n=1}^{N} \left\{ \frac{1}{M} \sum_{j=1}^{M} \left[\int_{x_0}^{x_1} u_j(x) \phi_n(x) dx \right]^2 \right\}. \tag{13.68}$$

This analysis still results in N spatial basis functions; however, they are optimized over the set of M snapshots in time.

While POD with snapshots leads to a more manageable eigenproblem to solve for the basis functions, it may not account as faithfully for the temporal evolution of dynamic features in the system's behavior. In particular, the method is not able to detect distinct coherent structures that may exist and evolve from one time snapshot to another. Such dynamic features are essentially averaged across the time interval via the snapshot data as the sequential ordering of the snapshots is lost. As discussed in Section 13.8.2, Koopman-mode decomposition, and its discrete form dynamic-mode decomposition, are designed to mitigate this weakness. In addition, Towne et al. (2018) outline a method called *spectral POD* that restores the sequential temporal evolution of the snapshot data in order to identify dynamically significant coherent features in the solution. In fact, these two methods are closely related to each other.

13.7 Proper-Orthogonal Decomposition (POD) for Discrete Data

Rather than the data being in the form of a continuous function, it is more likely that it is a discrete vector or matrix from experimental measurements or computational simulations. This is the case considered in this section. We start with POD applied to a single discrete data vector to fix ideas and illustrate similarities with the continuous case considered in the last section followed by the general case for large data sets using matrix methods, where its true power comes to the fore.

13.7.1 Single Discrete Data Vector

In the discrete case, we first imagine that the data set consists of a single vector of discrete values. Given the $N \times 1$ data vector $\mathbf{u}$, and taking our cues from (13.57) for the continuous case, we seek the optimal basis vectors $\boldsymbol{\phi}_n$ for approximating the data vector according to

$$\mathbf{u} = \sum_{n=1}^{N} a_n \boldsymbol{\phi}_n = a_1 \boldsymbol{\phi}_1 + a_2 \boldsymbol{\phi}_2 + \cdots a_n \boldsymbol{\phi}_n + \cdots + a_N \boldsymbol{\phi}_N = \boldsymbol{\Phi}\mathbf{a}, \tag{13.69}$$

where $\boldsymbol{\Phi}$ is $N \times N$ and contains the basis vectors as its columns. We could imagine that this data vector is the solution of some system of linear algebraic equations $\mathbf{Au} = \mathbf{b}$, for example. If only $R < N$ of the terms are retained in the linear combination (13.69), then $\boldsymbol{\Phi}$ would be $N \times R$, and $\boldsymbol{\Phi}\mathbf{a}$ would approximate the original data vector $\mathbf{u}$.

Our objective in POD is to determine the basis vectors $\boldsymbol{\phi}_n, n = 1, \ldots, N$ that are optimal for representing the discrete data $\mathbf{u}$ for the particular system under consideration in the form (13.69). These POD modes are effective in identifying dominant spatial features in a system, and Galerkin projection can then be used to determine the coefficients in the trial function (13.69), that is, how much of each particular mode is relevant for a given system. It is important to realize that POD analysis can be applied to any set of data, including from a nonlinear system or one for which no mathematical model exists.

Keeping in mind how the basis vectors will be used in the projection process, the objective is to minimize the sum of the differences between the known data vector $\mathbf{u}$ and the projection of the data vector onto each of the sought-after basis vectors in a mean square sense. This suggests that we minimize

$$J[\boldsymbol{\phi}_1, \ldots, \boldsymbol{\phi}_N] = \sum_{n=1}^{N} \left\| \mathbf{u} - \langle \mathbf{u}, \boldsymbol{\phi}_n \rangle \boldsymbol{\phi}_n \right\|^2. \tag{13.70}$$

The basis vectors are sought that minimize this objective function subject to the constraint that the basis vectors are normalized according to

$$\|\boldsymbol{\phi}_n\|^2 = \langle \boldsymbol{\phi}_n, \boldsymbol{\phi}_n \rangle = 1, \quad n = 1, \ldots, N. \tag{13.71}$$

This constraint also prevents trivial solutions for the basis vectors. Writing the norm in (13.70) in terms of an inner product gives

$$J = \sum_{n=1}^{N} \left\langle \mathbf{u} - \left\langle \mathbf{u}, \boldsymbol{\phi}_n \right\rangle \boldsymbol{\phi}_n, \mathbf{u} - \left\langle \mathbf{u}, \boldsymbol{\phi}_n \right\rangle \boldsymbol{\phi}_n \right\rangle.$$

Alternatively, the inner product can be written

$$J = \sum_{n=1}^{N} \left(\mathbf{u} - \left\langle \mathbf{u}, \boldsymbol{\phi}_n \right\rangle \boldsymbol{\phi}_n\right)^T \left(\mathbf{u} - \left\langle \mathbf{u}, \boldsymbol{\phi}_n \right\rangle \boldsymbol{\phi}_n\right),$$

or

$$J = \sum_{n=1}^{N} \left(\mathbf{u}^T - \left\langle \mathbf{u}, \boldsymbol{\phi}_n \right\rangle \boldsymbol{\phi}_n^T\right) \left(\mathbf{u} - \left\langle \mathbf{u}, \boldsymbol{\phi}_n \right\rangle \boldsymbol{\phi}_n\right).$$

Multiplying out yields

$$J = \sum_{n=1}^{N} \left(\mathbf{u}^T \mathbf{u} - \left\langle \mathbf{u}, \boldsymbol{\phi}_n \right\rangle \mathbf{u}^T \boldsymbol{\phi}_n - \left\langle \mathbf{u}, \boldsymbol{\phi}_n \right\rangle \boldsymbol{\phi}_n^T \mathbf{u} + \left\langle \mathbf{u}, \boldsymbol{\phi}_n \right\rangle^2 \boldsymbol{\phi}_n^T \boldsymbol{\phi}_n\right).$$

Note that $\left\langle \mathbf{u}, \boldsymbol{\phi}_n \right\rangle = \mathbf{u}^T \boldsymbol{\phi}_n = \boldsymbol{\phi}_n^T \mathbf{u}$ and $\boldsymbol{\phi}_n^T \boldsymbol{\phi}_n = \left\langle \boldsymbol{\phi}_n, \boldsymbol{\phi}_n \right\rangle = \|\boldsymbol{\phi}_n\|^2 = 1$. Therefore, our objective function simplifies to

$$J[\boldsymbol{\phi}_1, \ldots, \boldsymbol{\phi}_N] = \sum_{n=1}^{N} \left(\langle \mathbf{u}, \mathbf{u} \rangle - \left\langle \mathbf{u}, \boldsymbol{\phi}_n \right\rangle^2\right), \tag{13.72}$$

where we note the similarity to (13.60) for the continuous case taking into account the differences in inner products in the discrete and continuous cases. Augmenting this function with the constraint (13.71) yields

$$\tilde{J} = \sum_{n=1}^{N} \left[\langle \mathbf{u}, \mathbf{u} \rangle - \left\langle \mathbf{u}, \boldsymbol{\phi}_n \right\rangle^2 + \lambda_n \left(\left\langle \boldsymbol{\phi}_n, \boldsymbol{\phi}_n \right\rangle - 1\right)\right]. \tag{13.73}$$

In order to determine the basis vectors that minimize the augmented objective function, we differentiate with respect to each unknown basis vector $\boldsymbol{\phi}_n$ and set equal to zero. Recall that the data vector $\mathbf{u}$ is known, so the first term in the augmented function does not influence the minimization process. Recalling that

$$\frac{\partial}{\partial \boldsymbol{\phi}_n} \left\langle \mathbf{u}, \boldsymbol{\phi}_n \right\rangle = \mathbf{u}, \quad \frac{\partial}{\partial \boldsymbol{\phi}_n} \left\langle \boldsymbol{\phi}_n, \boldsymbol{\phi}_n \right\rangle = 2\boldsymbol{\phi}_n,$$

differentiating the augmented function (13.73) with respect to each of the basis vectors gives the N equations

$$-2\left\langle \mathbf{u}, \boldsymbol{\phi}_n \right\rangle \mathbf{u} + 2\lambda_n \boldsymbol{\phi}_n = \mathbf{0}, \quad n = 1, \ldots, N.$$

This result simplifies to

$$\mathbf{u}\left\langle \mathbf{u}, \boldsymbol{\phi}_n \right\rangle = \lambda_n \boldsymbol{\phi}_n, \quad n = 1, \ldots, N. \tag{13.74}$$

Defining the operator

$$\mathcal{R}\boldsymbol{\phi}_n = \mathbf{u}\langle\mathbf{u}, \boldsymbol{\phi}_n\rangle, \tag{13.75}$$

we have the algebraic eigenproblem

$$\mathcal{R}\boldsymbol{\phi}_n = \lambda_n \boldsymbol{\phi}_n. \tag{13.76}$$

Let us further consider the operator

$$\mathcal{R}\boldsymbol{\phi}_n = \mathbf{u}\langle\mathbf{u}, \boldsymbol{\phi}_n\rangle = \mathbf{u}\mathbf{u}^T \boldsymbol{\phi}_n,$$

where it is noted that the outer product $\mathbf{u}\mathbf{u}^T$ produces the $N \times N$ symmetric matrix

$$\mathcal{R} = \mathbf{u}\mathbf{u}^T. \tag{13.77}$$

Because $\mathcal{R}$ is real and symmetric, its N eigenvalues are real, and the corresponding $N \times 1$ eigenvectors, which are the basis vectors or POD modes, are mutually orthogonal. In addition, $\mathcal{R}$ is positive semidefinite; therefore, its eigenvalues are nonnegative. As in the continuous case, it is important to observe that the orthogonality of the POD modes is a natural consequence of the optimization procedure, not a property that is imposed a priori. Equation (13.76) is an eigenproblem to be evaluated for the basis vectors $\boldsymbol{\phi}_n$, for which the eigenvalues are the Lagrange multipliers.

Summarizing, minimization of the objective function (13.70) subject to the constraint (13.71) is equivalent to solving the eigenproblem (13.76), where the eigenvectors are the optimal basis vectors that we seek; these are the POD modes. As in the continuous case, the eigenvalues, and their associated eigenvectors, are typically ordered from largest to smallest as the eigenvectors having the largest eigenvalues are most dominant in the original data function $\mathbf{u}$.

Analogous to the continuous case, for a given data vector $\mathbf{u}$ and orthonormal basis vectors $\boldsymbol{\phi}_n$, the coefficients in the trial function (13.69) are obtained from the inner products

$$a_n = \langle\mathbf{u}, \boldsymbol{\phi}_n\rangle, \quad n = 1, \ldots, N,$$

and our objective function (13.70) can be written in the form

$$J[\boldsymbol{\phi}_1, \ldots, \boldsymbol{\phi}_N] = \sum_{n=1}^{N} \left\|\mathbf{u} - a_n \boldsymbol{\phi}_n\right\|^2. \tag{13.78}$$

Thus, the basis functions are such that the difference between the data function and each of the terms in the trial vector (including the coefficients) are minimized in a least-squares sense, which makes sense in light of our overall objective to determine the basis vectors that best represent (optimally span) the data vector in the form of the expansion (13.69).

POD modes are those orthonormal eigenvectors that allow for a reconstruction that is closest to the original data set in a least-squares sense. The orthonormal eigenvectors of $\mathcal{R}$ are the principal components of the data, giving rise to the nomenclature principal-component analysis.

REMARKS:

1. *Observe that the resulting basis vectors are mutually orthogonal for any data vector no matter its origin. This is why it is called* proper-orthogonal decomposition. *We once again emphasize that the operator being self-adjoint and the basis vectors orthogonal is a natural consequence of the choice of objective function; they are not assumptions imposed on the analysis.*
2. *Comparing this result for the discrete case to the continuous case is quite revealing. Although differential calculus has been used here for the discrete case, and variational calculus was used for the continuous case, we arrive at the same result in both cases with proper definitions of inner products and norms. As the reader should appreciate by now, this is no coincidence when linear algebraic and differential operators are utilized.*
3. *Recall that we have considered the simplified case in which the data vector is a single sequence of data. The more general case is considered in the next section, which provides the basis for the vast majority of POD analyses in practice. As we will see in the next section, $\mathcal{R}$ is (nearly) a covariance matrix for the data vector.*

Example 13.4 As in Example 2.5, the solution of a system of equations $\mathbf{Au} = \mathbf{b}$ is sought, where

$$\mathbf{A} = \begin{bmatrix} -2 & 1 & 1 & 1 & 2 \\ 1 & 2 & 1 & -1 & 1 \\ 1 & 1 & -2 & 1 & 1 \\ 1 & -1 & 1 & 2 & 1 \\ 2 & 1 & 1 & 1 & -2 \end{bmatrix}, \quad \mathbf{b} = \begin{bmatrix} -1 \\ 2 \\ 1 \\ -2 \\ 1 \end{bmatrix}.$$

Determine the solution $\mathbf{u}$ in terms of a linear combination of the optimal basis vectors obtained using POD.

Solution

For later comparison, the solution of the system of linear algebraic equations is given by

$$\mathbf{u} = \mathbf{A}^{-1}\mathbf{b} = \begin{bmatrix} 0.321429 \\ 0.809524 \\ -0.285714 \\ -0.52381 \\ -0.178571 \end{bmatrix}. \tag{13.79}$$

The symmetric matrix $\mathcal{R}$ is the outer product of $\mathbf{u}$ with itself, which is

$$\mathcal{R} = \mathbf{u}\mathbf{u}^T = \begin{bmatrix} 0.103316 & 0.260204 & -0.0918367 & -0.168367 & -0.057398 \\ 0.260204 & 0.655329 & -0.231293 & -0.424036 & -0.144558 \\ -0.0918367 & -0.231293 & 0.0816327 & 0.14966 & 0.0510204 \\ -0.168367 & -0.424036 & 0.14966 & 0.274376 & 0.0935374 \\ -0.057398 & -0.144558 & 0.0510204 & 0.0935374 & 0.0318878 \end{bmatrix}.$$

The orthogonal matrix with its columns being the eigenvectors of real symmetric $\mathcal{R}$ is given by

$$\boldsymbol{\Phi} = \begin{bmatrix} 0.300185 & 0.0507728 & 0.717397 & 0.619831 & 0.0919885 \\ 0.756022 & 0.127872 & -0.208037 & -0.0459654 & -0.605547 \\ -0.266831 & -0.0451314 & -0.526575 & 0.775125 & -0.2206 \\ -0.489191 & -0.0827409 & 0.405932 & -0.113482 & -0.75907 \\ -0.16677 & 0.985996 & 0. & 0. & 0. \end{bmatrix}.$$

The optimal basis vectors ϕ_n are the columns of $\boldsymbol{\Phi}$.

The solution of the system of linear algebraic equations $\mathbf{Au} = \mathbf{b}$ is then constructed using the columns of $\boldsymbol{\Phi}$ as the basis vectors ϕ_n in a Galerkin projection according to (13.39) and (13.40) as follows:

$$\mathbf{u} = \boldsymbol{\Phi}\mathbf{a} = \boldsymbol{\Phi}\boldsymbol{\Phi}^{-1}\mathbf{A}^{-1}\mathbf{b} = \mathbf{A}^{-1}\mathbf{b} = \begin{bmatrix} 0.321429 \\ 0.809524 \\ -0.285714 \\ -0.52381 \\ -0.178571 \end{bmatrix}.$$

Apparently, this simply gets us back to the original solution of the system of linear algebraic equations. However, we now have a route to a ROM of the system. To see this, take a look at the coefficients in the linear combination

$$\mathbf{u} = a_1\phi_1 + a_2\phi_2 + a_3\phi_3 + a_4\phi_4 + a_5\phi_5,$$

which are given by

$$\mathbf{a} = \boldsymbol{\Phi}^{-1}\mathbf{A}^{-1}\mathbf{b} = \begin{bmatrix} 1.07077 \\ 5.551 \times 10^{-17} \\ 3.989 \times 10^{-17} \\ 4.510 \times 10^{-17} \\ 1.110 \times 10^{-16} \end{bmatrix}.$$

All but one of the coefficients in the linear combination of basis vectors are zero to machine precision. Consequently, it is only necessary to use the first basis vector, that is, the first column of $\boldsymbol{\Phi}$ (with the nonzero coefficient), to reconstruct an accurate representation of the solution. This gives the approximate solution

$$\mathbf{u} \approx a_1\phi_1 = 1.07077 \begin{bmatrix} 0.300185 \\ 0.756022 \\ -0.266831 \\ -0.489191 \\ -0.16677 \end{bmatrix} = \begin{bmatrix} 0.321429 \\ 0.809524 \\ -0.285714 \\ -0.52381 \\ -0.17857 \end{bmatrix}.$$

As you can see, even using only one basis vector in the ROM leads to an approximation to the solution that is accurate to five significant figures. This clearly leads to a much improved ROM of the solution as compared to that in Example 2.5, in which all five basis vectors were required for a good reconstruction of the solution.

13.7.2 Matrix Approach for Large Data Sets

Introduction

In order to ease discussion, we have thus far limited our considerations to situations where the data function or vector is a single sequence of data. This has allowed us to overlook some of the more nuanced aspects and terminology of POD analysis. The real power of POD analysis, however, comes to the fore when we deal with very large data sets that contain multiple instances of data. For instance, we may have measurements of multiple data points in time or multiple measurements for a large sample of people, objects, et cetera. Therefore, our $N \times 1$ data vector $\mathbf{u}$ becomes an $N \times M$ data, or measurement, matrix $\mathbf{X}$.[8] M is the number of samples, occurrences, trials, or objects, while N is the number of data elements, or measurements, per sample. N represents the dimensionality of the data set. Thus, the data matrix $\mathbf{X}$ is formed by stacking the M data vectors $\mathbf{u}_m, m = 1, \ldots, M$ as its columns, as follows:

$$\mathbf{X} = \begin{bmatrix} \vdots & \vdots & & \vdots & & \vdots \\ \mathbf{u}_1 & \mathbf{u}_2 & \cdots & \mathbf{u}_m & \cdots & \mathbf{u}_M \\ \vdots & \vdots & & \vdots & & \vdots \end{bmatrix}.$$

We have emphasized from the beginning that vectors and matrices provide a convenient means of tabulating data. We have also seen how each such representation corresponds to a particular vector basis. It stands to reason that some bases are better than others for representing a particular data set. In fact, we might define an optimal basis as that which encapsulates the maximum amount of information in the minimum number of dimensions, or modes. In other words, the most efficient means of representing a particular data set would be the one that provides the closest approximation to the original data with the fewest number of data elements in the reduced set of data.

For example, imagine that we have three-dimensional measurements of a system that only exhibits two-dimensional behavior, such as projectile motion. In a general coordinate system, we require all three dimensions to characterize the data set. If we could determine the two-dimensional plane in which the projectile motion occurs, however, we could fully represent the measured data in terms of these two coordinates, thereby providing a more efficient and physically meaningful representation of the measurements.

[8] Note the departure from our usual convention, where M is the number of rows and N is the number of columns, so that N indicates the dimensionality of the data.

We seek a general framework that allows us to perform similar representations involving much larger data sets for which there is no known underlying correlation in the data. This is provided by POD, which quantifies the level of approximation using the variability, defined by the variance, that is contained within the data set in the form of the covariance matrix. POD then determines the directions in the data that exhibit the most variation by projecting the data matrix onto the eigenvectors of the covariance matrix that correspond to the largest eigenvalues of the covariance matrix.

We are getting ahead of ourselves; consider a simple example to illustrate. Let us use data from the 2015–2016 Chicago Bulls basketball team. For each of the 15 players, we record their (1) years as a pro, (2) height in feet, and (3) weight in pounds. Thus, we have $M = 15$ samples of $N = 3$ dimensional data. Note that each sample corresponds to an individual player, and $N = 3$ measurements are recorded for each player. Stacking each sample as a column in the 3×15 data, or measurement, matrix gives[9]

$$\mathbf{X} = \begin{bmatrix} 1. & 7. & 4. & \cdots & 0. & 6. & 2. \\ 6.75 & 6. & 6.58333 & \cdots & 6.91667 & 6.25 & 6.58333 \\ 250. & 161. & 220. & \cdots & 230. & 190. & 200. \end{bmatrix}.$$

This original data set could be plotted as $M = 15$ points – one point for each player – on an $N = 3$ dimensional coordinate system, where each coordinate corresponds to the years as a pro, height, or weight. While representing the complete data set in this way allows for easy extraction and interpretation of specific data elements, it does not necessarily correspond to the most efficient manner of representing the data as a whole or draw attention to correlations in the data. While efficiency may not seem terribly important when considering a data set with only $N = 3$ measurements for $M = 15$ samples, it is not uncommon for data sets in science and engineering applications to consist of thousands or millions of data elements, giving rise to the modern moniker "big data." The most efficient means of representing a particular data set would be the one that provides the closest approximation to the original data set with the fewest number of data elements in the reduced set of data. An increasingly popular technique for accomplishing this data reduction is POD, which also goes by the name principal-component analysis (PCA) in the statistics community.

The mathematical basis for POD and PCA data-reduction techniques is singular-value decomposition (SVD), which can be used to take a high-dimensional data set and reduce it to a lower dimensional data set that provides an optimal approximation to the original data. See Sections 2.8 and 11.8, where this was illustrated in the context of image compression and linear regression. It essentially makes items that are similar more similar and items that are dissimilar more so through a change in basis vectors. SVD can be thought of as a means of determining the coordinate system with respect to which the data exhibit the most variation along the coordinate directions. In other words, it transforms a system having correlated variables into the one in which the variables are least correlated, thereby exposing the strongest relationships –

[9] For brevity, we only include the first three and last three columns in the 3×15 matrix.

correlations – within the data. More simply, we recognize that the basis with respect to which we collect the data – usually for convenience of data collection – may not be the same basis with respect to which the data are optimally represented.

For our Chicago Bulls roster example, we have 15 samples of three data elements; however, some of the data may exhibit interdependencies that can be taken advantage of in its representation. For example, we would expect that there is a correlation between a player's height and weight; can we take advantage of such a correlation to provide a reduced-order representation of the data? Formally, this is facilitated through a change of basis from the original to the new representation of the data.

Obviously, modeling 100% of the variability would require using 100% of the data's full information, that is, its dimensionality. One of the advantages of this SVD-based data-reduction approach is that one can decide explicitly the trade-off between the level of data reduction and the accuracy of the data reconstruction, where retaining more data will result in a more accurate data reconstruction.

POD and SVD

In order to formulate the POD problem for a large $N \times M$ data matrix $\mathbf{X}$, let us begin with that for the single $N \times 1$ data vector considered in the last section. Recall that the data vector was written as a linear combination of the basis vectors according to (13.69), where the basis vectors are the eigenvectors of the $N \times N$ symmetric matrix $\mathcal{R} = \mathbf{u}\mathbf{u}^T$ according to (13.76) and (13.77).

Note that rather than solving the eigenproblem (13.76) for the real symmetric matrix $\mathcal{R} = \mathbf{u}\mathbf{u}^T$, one could equivalently obtain the SVD for the original data vector $\mathbf{u}$. As discussed in Section 6.5.4, this has numerical advantages owing to better conditioning. More importantly, this reveals the path to formulating the POD optimization problem for large data matrices.

For the single-data vector case, the linear combination (13.69) can be written in terms of the SVD according to

$$\mathbf{u} = \mathbf{\Phi}\mathbf{a} = \mathbf{U}\mathbf{\Sigma}\mathbf{V}^T, \tag{13.80}$$

where recall that $\mathbf{U}$ and $\mathbf{V}$ are orthogonal matrices, and $\mathbf{\Sigma}$ is a diagonal matrix with the singular values of $\mathbf{u}$ along its diagonal. In this case, the basis vectors are given by $\mathbf{\Phi} = \mathbf{U}$, and the coefficients in the linear combination are $\mathbf{a} = \mathbf{\Sigma}\mathbf{V}^T$. That is,

$$\mathbf{U} = \begin{bmatrix} \vdots & \vdots & & \vdots & & \vdots \\ \phi_1 & \phi_2 & \cdots & \phi_n & \cdots & \phi_N \\ \vdots & \vdots & & \vdots & & \vdots \end{bmatrix}.$$

Because the data vector is $N \times 1$, the matrix $\mathbf{U}$ of left singular vectors is $N \times N$; the diagonal matrix of singular values $\mathbf{\Sigma}$ is $N \times 1$, which is the same size as $\mathbf{u}$; and the matrix $\mathbf{V}$ of right singular vectors is 1×1. For the $N \times M$ data matrix $\mathbf{X}$, this is now

$$\mathbf{X} = \mathbf{U}\mathbf{\Sigma}\mathbf{V}^T = \sum_{n=1}^{M} \sigma_m \boldsymbol{\phi}_m \mathbf{v}_m^T, \tag{13.81}$$

where the matrix $\mathbf{U}$ of left singular vectors remains $N \times N$, but the diagonal matrix of singular values $\mathbf{\Sigma}$ is now $N \times M$, and the matrix $\mathbf{V}$ of right singular vectors is $M \times M$. Similar to the eigenproblem (13.76) with (13.77), we now have the eigenproblem

$$\frac{1}{M}\mathbf{X}\mathbf{X}^T \phi_n = \lambda_n \phi_n.$$

Consequently, the eigenvectors of $\mathbf{X}\mathbf{X}^T$ are the left singular vectors of the data matrix $\mathbf{X}$, and the eigenvalues are related to the singular values according to $\sigma_n^2 = M\lambda_n$.

Observe that the objective function (13.70) for the case with a single data vector can be written in the form

$$J[\phi_1, \dots, \phi_N] = \sum_{n=1}^{N} \|\mathbf{u} - \phi_n \phi_n^T \mathbf{u}\|_2^2.$$

The data now consist of an $N \times M$ matrix $\mathbf{X}$, where M is the number of samples (players), and N is the number of measurements per sample (years as pro, height, and weight). Therefore, the analogous form of the POD objective function corresponding to the previous SVD approach is given by

$$J[\mathbf{\Phi}] = \|\mathbf{X} - \mathbf{\Phi}\mathbf{\Phi}^T\mathbf{X}\|_F^2 = \sum_{n=1}^{N} \|\mathbf{u}_n - \mathbf{\Phi}\mathbf{\Phi}^T \mathbf{u}_n\|_2^2,$$

where we note use of the Frobenius norm in place of the summation. We have used the property that the square of the Frobenius norm is the sum of the squared L_2 norms of the columns of a matrix. The matrix $\mathbf{\Phi}$ contains the orthogonal basis vectors ϕ_n as its columns, and $\mathbf{u}_n$ is the nth column of the data matrix. As before, the matrix of basis functions is then the left singular vectors of the data matrix, that is, $\mathbf{\Phi} = \mathbf{U}$, in the SVD of the data matrix. Alternatively, if the number of samples M is much less than the number of data points N per sample, then it may be advantageous to solve the eigenproblem for $\mathbf{X}^T\mathbf{X}$, which is only $M \times M$ instead of the full SVD for $\mathbf{X}$, which is $N \times M$.

Covariance Matrix

When dealing with various collections of data, we need some means of characterizing the nature of that data using quantitative measures. In statistics, we have the mean and standard deviation for characterizing multiple samples of the same measurement. For example, we could calculate the mean and standard deviation of the heights for the Chicago Bulls roster, where the standard deviation is a measure of the spread in the data. Another measure of the spread in data is called the *variance*, which is simply the square of the standard deviation. When we have multiple measurements, we would also like to have a way to characterize how closely related they may be to one another. This is accomplished using the *covariance*, which is a generalization of the variance to multiple dimensions. Recall that we expect that the players' heights and weights would be somewhat correlated to one another. How can we quantify such a correlation? Thinking all the way back to Chapter 1, recall that the inner product of two vectors tells how closely related the vectors are. If the inner product is zero, then they are orthogonal and completely uncorrelated, and the larger the inner product, the more closely they are correlated. For example, if the vectors each have length one, that is, they are normalized, then an inner product of unity indicates that the vectors are parallel and fully correlated. In statistics, these inner products are called the *covariance*, and the matrix of all possible inner products is the *covariance matrix*.

Just as we must subtract off the mean of a set of data containing a single measurement before computing its standard deviation, we must subtract off the means of each measurement (years as a pro, height, and weight) before computing the covariances. Let us call this the *mean-adjusted measurement matrix* $\mathbf{A}$

$$\mathbf{A} = \mathbf{X} - \begin{bmatrix} \vdots & \vdots & & \vdots \\ \hat{\mathbf{x}} & \hat{\mathbf{x}} & \cdots & \hat{\mathbf{x}} \\ \vdots & \vdots & & \vdots \end{bmatrix} = \begin{bmatrix} \vdots & \vdots & & \vdots \\ \mathbf{x}_1 - \hat{\mathbf{x}} & \mathbf{x}_2 - \hat{\mathbf{x}} & \cdots & \mathbf{x}_M - \hat{\mathbf{x}} \\ \vdots & \vdots & & \vdots \end{bmatrix},$$

where the vector $\hat{\mathbf{x}}$ contains the mean of each row, that is, measurement. For the Chicago Bulls data, the mean-adjusted measurement matrix is

$$\mathbf{A} = \begin{bmatrix} -3.6 & 2.4 & -0.6 & \cdots & -4.6 & 1.4 & -2.6 \\ 0.1055 & -0.6444 & -0.06111 & \cdots & 0.2722 & -0.3944 & -0.06111 \\ 31.06 & -57.93 & 1.066 & \cdots & 11.06 & -28.93 & -18.93 \end{bmatrix}.$$

The $N \times N$ covariance matrix $\mathbf{C}$ for the mean-adjusted measurement matrix $\mathbf{A}$ is defined by

$$\mathbf{C} = \frac{1}{M-1}\mathbf{A}\mathbf{A}^T, \tag{13.82}$$

where recall that M is the number of samples in the data matrix. For the Chicago Bulls data, the $N \times N$ covariance matrix is

$$\mathbf{C} = \begin{bmatrix} 19.6857 & 0.05 & -2.45714 \\ 0.05 & 0.0768519 & 6.93294 \\ -2.45714 & 6.93294 & 860.638 \end{bmatrix}.$$

Note that $\mathbf{C}$ is symmetric and positive definite, which is always the case for the product of any matrix $\mathbf{A}$ and its transpose (see Section 2.4), and observe how it is related to the matrix $\mathcal{R}$ in the case for a single data vector.[10] Along the main diagonal, each element is the inner product of the corresponding row with itself. Therefore, each of the main diagonal elements is the variance, that is, spread around the mean, of that quantity. For example, $C[1,1] = 19.69$ is the inner product of the years as a pro with itself, $C[2,2] = 0.077$ is the inner product of the player's height with itself, and $C[3,3] = 860.6$ is the inner product of the player's weight with itself. Thus, we see that there is a great deal of variance in the weights of the players and very little variance in their heights. The total variance is the sum of the variances along the main

[10] $\mathcal{R}$ would be a covariance matrix if we had adjusted the data vector using its mean.

diagonal, which is the trace of the covariance matrix $\mathbf{C}$. In this case, the total variance is 880.401.

The off-diagonal elements of $\mathbf{C}$ are the covariances (inner products) of the quantities corresponding to the respective rows and columns. Therefore, $C[i,j] = C[j,i]$ is the covariance between the ith and jth quantities (rows) in the data matrix. A covariance of zero means that the two corresponding quantities are uncorrelated, and a large covariance indicates that the two quantities are highly correlated. For example, $C[1,2] = C[2,1] = 0.05$ is small, indicating that the players' heights and years as a pro are not highly correlated. $C[1,3] = C[3,1] = -2.46$ is also rather small, indicating that the players' years as a pro and their weights are also uncorrelated. On the other hand, $C[2,3] = C[3,2] = 6.93$ is larger, indicating that the players' heights and weights are more highly correlated. These correlations make intuitive sense as taller players generally weigh more, and neither of these physical attributes has much to do with how long a player has been in the league.

Note the similarity between the covariance matrix $\mathbf{C}$ as defined here and the Gram matrix $\mathbf{G}'$ used in Section 1.7.1 to check for the linear independence of a set of vectors. This is no coincidence as both involve inner products of vectors that indicate how closely aligned they are.

The singular values of the mean-adjusted measurement matrix $\mathbf{A}$ are the square roots of the eigenvalues of $\mathbf{A}\mathbf{A}^T$, which are

$$\sigma_1 = 109.772, \quad \sigma_2 = 16.5983, \quad \sigma_3 = 0.539031,$$

and the eigenvectors are the left singular vectors in the SVD of $\mathbf{A}$, which form the columns of the matrix $\mathbf{U}$. Thus, the eigenvectors of $\mathbf{A}\mathbf{A}^T$, which are the left singular vectors of $\mathbf{A}$, are the basis vectors $\boldsymbol{\phi}_n, n = 1, \ldots, N$ that form the orthonormal POD modes. Recall that the singular values are ordered from largest to smallest, that is, from the one with the highest correlation to the one with the least in terms of the bases for the new coordinate system. The sum of the singular values is $\sigma_1 + \sigma_2 + \sigma_3 = 126.909$. Therefore, the first mode contains $\sigma_1/(\sigma_1 + \sigma_2 + \sigma_3) = 86.5\%$, the second mode contains $\sigma_2/(\sigma_1 + \sigma_2 + \sigma_3) = 13.1\%$, and the third mode contains only $\sigma_3/(\sigma_1 + \sigma_2 + \sigma_3) = 0.4\%$ of the information. Therefore, keeping only the first mode would encompass 86.5% of the variability in the data, and keeping the first two modes would provide 99.6%.

As stated, the columns of $\mathbf{U}$ are the eigenvectors $\boldsymbol{\phi}$ of $\mathbf{A}\mathbf{A}^T$ and are called the left singular vectors of $\mathbf{A}$; these are the principal axes or directions of the data. The first vector $\boldsymbol{\phi}_1$ points in the direction having the largest variance (spread) in the data, and the magnitude of this spread is given by the corresponding singular value; the second vector $\boldsymbol{\phi}_2$ points in the direction having the second largest variance, and so forth. This new coordinate system (basis) highlights the directions in which the spreads in the data (variances) are most prominent, in which case the data are least correlated with respect to these coordinate directions.

The projections of the data on each of the principal axes, that is, the product of the original (mean-adjusted) measurement matrix $\mathbf{A}$ with each of the vectors $\boldsymbol{\phi}$, are called the principal components of the data, with each singular value having its corresponding principal component. Because the singular values are in order from largest to smallest, their corresponding principal directions provide a ranked list of the directions along which the variances are greatest (and covariances smallest). In other words, the left singular vectors having the largest singular values are the dominant principal components of the data matrix.

The columns of $\mathbf{V}$ are the eigenvectors of $\mathbf{A}^T\mathbf{A}$ and are called the right singular vectors of $\mathbf{A}$. In the present example, $\mathbf{V}$ is a 15×15 matrix. Recall from our discussion of SVD in Section 2.8 that the nonzero eigenvalues of $\mathbf{A}^T\mathbf{A}$, which is $M \times M$, and $\mathbf{A}\mathbf{A}^T$, which is $N \times N$, are the same and that the eigenvectors are related via (2.47). Therefore, it is prudent to solve the eigenproblem corresponding to the smaller of the two products if $N \ll M$ or $M \ll N$. In practice, the left and right singular vectors along with the singular values are computed directly using an SVD of the mean-adjusted data matrix $\mathbf{A}$. This is generally faster than working with the matrix $\mathbf{A}\mathbf{A}^T$ as it avoids the need to take the transpose and perform matrix multiplication before solving the eigenproblem. In addition, SVD is robust to round-off errors that may arise during the computations.

REMARKS:

1. *SVD is a technique for transforming a system having correlated variables into the one in which the variables are least correlated, thereby exposing the strongest relationships (correlations) within the data. It essentially makes items that are similar more similar and items that are dissimilar more so.*
2. *SVD can be thought of as a means of determining the coordinate system with respect to which the data exhibit the most variation along the coordinate directions. This can form the basis for data-reduction techniques, such as POD, whereby a high-dimensional data set is reduced to a lower-dimensional data set that provides an optimal approximation to the original data. Thus, POD (and other data-reduction techniques) is intimately connected with SVD.*
3. *If the data matrix is instead formed with M rows and N columns, for which the data are stacked as rows, rather than columns, then the covariance matrix involves $\mathbf{A}^T\mathbf{A}$, rather than $\mathbf{A}\mathbf{A}^T$, and the roles of the orthogonal matrices $\mathbf{U}$ and $\mathbf{V}$ switch in the SVD as compared to that described here.*

Proper-Orthogonal Decomposition

POD determines an optimal projection of a high-dimensional data set onto a lower-dimensional subspace allowing for a large data set to be represented in a more compact fashion but without losing too much essential information about the data. It is optimal in the sense that for a given number of dimensions, which is less than the dimensionality of the original data set, the low-dimensional basis contains the maximum variance for the given data set. Alternatively, it can be viewed as the least squares, or L_2 norm, reconstruction error being minimized.

Let us think in more detail about what this implies. In particular, we ask what the covariance matrix should look like if the data are represented in terms of an "optimal" basis. If the data were such that there was no correlation between any of the sets of measurements, then their covariances would all be zero, and the covariance matrix would be diagonal with respect to the new transformed coordinate system. This is accomplished via diagonalization of the covariance matrix, resulting in a diagonal matrix for which all of the covariances – off-diagonal elements – are zero with respect to the new coordinate system. We could then argue that the coordinate system (basis) for which this is true is the optimal basis for representing the particular data set.

Let us consider diagonalization of the covariance matrix. If $\mathbf{U}$ is the orthogonal modal matrix containing the eigenvectors of the symmetric covariance matrix $\mathbf{C} = \frac{1}{M-1}\mathbf{A}\mathbf{A}^T$, then the diagonalized covariance matrix would be

$$\mathbf{D} = \mathbf{U}^T\mathbf{C}\mathbf{U}.$$

The diagonalizing modal matrix $\mathbf{U}$ is then the orthogonal matrix of left singular vectors of the mean-adjusted measurement matrix $\mathbf{A}$. Using the SVD of $\mathbf{A}$, and taking advantage of $\mathbf{U}$ and $\mathbf{V}$ being orthogonal, the diagonal matrix is then as follows:

$$\begin{aligned}
\mathbf{D} &= \mathbf{U}^T\mathbf{C}\mathbf{U} \\
&= \frac{1}{M-1}\mathbf{U}^T\mathbf{A}\mathbf{A}^T\mathbf{U} \\
&= \frac{1}{M-1}\mathbf{U}^T(\mathbf{U}\boldsymbol{\Sigma}\mathbf{V}^T)(\mathbf{U}\boldsymbol{\Sigma}\mathbf{V}^T)^T\mathbf{U} \\
&= \frac{1}{M-1}\mathbf{U}^T\mathbf{U}\boldsymbol{\Sigma}\mathbf{V}^T\mathbf{V}\boldsymbol{\Sigma}^T\mathbf{U}^T\mathbf{U} \\
&= \frac{1}{M-1}\boldsymbol{\Sigma}\boldsymbol{\Sigma}^T
\end{aligned}$$

$$\mathbf{D} = \frac{1}{M-1}\begin{bmatrix} \sigma_1^2 & 0 & \cdots & 0 \\ 0 & \sigma_2^2 & \cdots & 0 \\ \vdots & \vdots & \ddots & \\ 0 & 0 & & \sigma_N^2 \end{bmatrix}.$$

Given the eigen-decomposition σ_n^2 and $\mathbf{U}$ of the product $\mathbf{A}\mathbf{A}^T$, each column vector of $\mathbf{U}$, that is, the left singular vectors, is a *principal component* of the covariance matrix, and they are ordered by magnitude of the corresponding eigenvalue σ_n^2, where σ_n are the singular values of $\mathbf{A}$.

While we may regard this diagonalization of the covariance matrix as optimal, is it always possible? Because the covariance matrix, which is essentially the product of $\mathbf{A}$ and its transpose, is symmetric and positive definite, it can be diagonalized using the orthogonal eigenvectors of their product. Recall that the eigenvalues of $\mathbf{A}\mathbf{A}^T$ are the squares of the singular values of the original matrix $\mathbf{A}$. Therefore, the most efficient technique for obtaining the required eigenvalues and eigenvectors is via an SVD of the matrix $\mathbf{A}$, which is faster than computing the eigen-decomposition of the covariance matrix. Recall that $\mathbf{A}$ is the matrix of original measurement data with the means subtracted off of each measurement.

A full reconstruction of the mean-adjusted measurement matrix can then be obtained from the SVD as follows:

$$\mathbf{A} = \mathbf{U}\mathbf{\Sigma}\mathbf{V}^T = \sum_{n=1}^{N} \sigma_n \boldsymbol{\phi}_n \mathbf{v}_n^T,$$

where we note the use of the outer product of the vectors $\boldsymbol{\phi}_n$ and $\mathbf{v}_n$ as derived in (2.50). The vectors $\boldsymbol{\phi}_n, n = 1, \ldots, N$, which are the columns of $\mathbf{U}$, are the orthogonal POD modes. A low-dimensional representation can then be obtained by noting that each successive singular value is smaller in magnitude. Therefore, a reconstruction that includes only the first $R < N$ POD modes, or principal components, should provide a very good "low-dimensional" representation of the data according to

$$\mathbf{A} \approx \sum_{n=1}^{R} \sigma_n \boldsymbol{\phi}_n \mathbf{v}_n^T. \tag{13.83}$$

The number of modes $R < N$ is chosen such that

$$\sum_{n=1}^{R} \sigma_n \approx \sum_{n=1}^{N} \sigma_n.$$

That is, we only neglect terms having small singular values. See Figure 2.12 in Example 2.14 for an illustration of this. Because the third singular value in the Chicago Bulls data is very small, we would expect that a reconstruction using (13.83) that only includes the first two POD modes should provide a very good "low-dimensional" representation of the data. For $R = 2$, this leads to the approximate mean-adjusted measurement matrix

$$\mathbf{A} = \begin{bmatrix} -3.600 & 2.399 & -0.6002 & \cdots & -4.599 & 1.399 & -2.599 \\ 0.2377 & -0.4587 & 0.006468 & \cdots & 0.07294 & -0.2284 & -0.1619 \\ 31.06 & -57.93 & 1.066 & \cdots & 11.06 & -28.93 & -18.93 \end{bmatrix}.$$

As you can see, the approximate reconstruction of $\mathbf{A}$ is very close to the original despite only using two of the three modes in its reconstruction. While the dimensionality of the Chicago Bulls data set has only been reduced from three- to two-dimensional, the real power of POD analysis arises for very large data sets that are amenable to large reductions in their dimensionality owing to significant correlations in the data.

Take note of how we progressed from the variational underpinnings for the continuous case (Section 13.6), where POD was posed as a constrained optimization problem, through the directly analogous discrete case using differential calculus, to the full-blown case based on SVD for very large discrete data sets. It is hoped that this long, seemingly circuitous, path to SVD-based POD provides a clear articulation of the fundamental mathematics underlying POD, thereby providing a clear understanding of the technique as well as a platform for further improvements. For example, it clearly shows that the orthogonality of the POD modes in both the continuous and discrete cases is a natural outgrowth of the optimization procedure, not an assumption that must be made. That is, it is not a choice that we make for convenience, as is the case in other contexts, but a necessary consequence of the fact that the covariance matrix is symmetric and positive semidefinite. It also enables us to see the direct connections between POD analysis and regression analysis using least-squares methods as illustrated in Section 11.3.

REMARKS:

1. *Normally, we do not know ahead of time how many modes will be required to obtain a satisfactory reconstruction of a given data set. However, once we have the SVD, it is straightforward to determine the number of modes required to achieve a desired level of accuracy in the reconstruction by eliminating those modes having small singular values.*
2. *Recall that POD is not changing the data in any way; it is simply providing another, more optimal, way of looking at it that facilitates the model reduction.*
3. *As with the continuous case, time-evolving systems are typically handled such that the samples are measurements in time, commonly called "snapshots." For data of the form $u(x,t)$ at discrete values of $x_i, i = 1, \dots, N$ and $t_j, j = 1, \dots, M$, the data would be arranged such that at each spatial location, they are placed as the rows of the matrix, and the temporal data (snapshots) would be placed in successive columns. That is, each column of the data matrix is the spatial data at a particular time, as follows:*

$$\mathbf{X} = \begin{bmatrix} u(x_1,t_1) & u(x_1,t_2) & \cdots & u(x_1,t_j) & \cdots & u(x_1,t_M) \\ u(x_2,t_1) & u(x_2,t_2) & \cdots & u(x_2,t_j) & \cdots & u(x_2,t_M) \\ \vdots & \vdots & \ddots & \vdots & & \vdots \\ u(x_i,t_1) & u(x_i,t_2) & \cdots & u(x_i,t_j) & \cdots & u(x_i,t_M) \\ \vdots & \vdots & & \vdots & \ddots & \vdots \\ u(x_N,t_1) & u(x_N,t_2) & \cdots & u(x_N,t_j) & \cdots & u(x_N,t_M) \end{bmatrix}.$$

In this case, N is the number of spatial points at which the data are measured, and M is the number of snapshots in time. If more than one dependent variable is measured at each spatial location, these are stacked as additional rows in the data matrix.

4. *POD is used to identify dominant features and structures in a wide variety of physical phenomena, as the basis for Galerkin projection to form ROM for analysis and control, in image and signal processing, and in pattern recognition of statistical data.*
5. *Galerkin projection of a set of governing equations onto POD modes that are obtained purely from experimental or numerical data, with no knowledge of the underlying governing equations, can sometimes lead to anomalous behavior. This*

is particularly true for nonnormal systems, in which the eigenvectors are not orthogonal, allowing for transient growth behavior that is very different than the long-time asymptotic behavior. This is addressed via a series of extensions and alternative methods outlined in the next section.

13.8 Extensions and Alternatives to POD

POD is a data-driven optimization technique that seeks the optimal basis vectors or functions for a particular data set. Its data-driven framework is its primary advantage as it can be applied to any data set in a wide variety of application areas in engineering, the sciences, economics, and beyond. However, it is not always the best approach to extract dominant features from various types of data sets, nor are POD modes always optimal for use in Galerkin projection or ROM. This can be the case for nonnormal (see Section 2.4) or nonlinear systems, for example. In addition, POD does not take advantage of any mathematical model that may be available for the system, that is, it is not model driven. In areas such as fluid mechanics, therefore, where the governing equations are well known, POD does not take advantage of these models. As a result, there is room in such fields for additional enhancements and alternatives to traditional POD analysis. The following methods are currently in use in fluid mechanics and are active areas of research. We keep the discussion brief, but readers may find natural applications in other areas as well.

Owing to its inherent nonlinearity and large dimensionality, fluid dynamics has proven to be a fertile area for further developments in ROM. One advantage in fluid dynamics is that the governing equations are well known, and this knowledge can be used to enhance the data-driven nature of traditional POD methods. Note that the power and elegance of the methods presented here are a product of their foundations in linear theory. Therefore, their optimality only can be established formally in such contexts. Because of their inherently linear nature, there are a number of interrelationships between these methods that allow one to clearly see the similarities and differences as applied to different types of problems. This leads to a solid foundation for further efforts in developing ROM techniques. However, it must be kept in mind that nonlinear systems, and nonlinear methods for obtaining basis functions, may prove advantageous in certain contexts despite their loss of beneficial properties, such as orthogonality.

In Towne et al. (2018), it is shown how all of the methods mentioned here can be unified using what is referred to as *spectral proper-orthogonal decomposition* (SPOD), which actually traces its roots back to the original formulation of POD in which no distinction was made between space and time. That is, they were treated the same, with the spatial and temporal behavior both being included in the optimization procedure. The POD framework as usually articulated, including in Section 13.7 here, is referred to as *space-only POD*, as the POD modes are spatial modes and do not inherently take temporal evolution of the system into account. They go on to show how dynamic-mode decomposition (DMD) and the related Koopman analysis are essentially spectral POD with the temporal modes being emphasized. Resolvent-mode analysis is also unified with the spectral POD framework by showing that resolvent modes are essentially approximations of the spectral POD modes.

It is important to recognize that Galerkin projection is neutral with regard to the linearity or nonlinearity of the system and its mathematical model; it applies equally well to any discrete or continuous system. For this reason, it is found to be at the center of a wide variety of numerical and ROM methods. POD and its kin, on the other hand, are deeply imbued with linear systems theory. While this is the source of their mathematical beauty and practical efficiency, it is also the source of their greatest limitation.

For a general review of POD and some of its more popular variants, see Benner et al. (2015). For summaries that emphasize applications to fluid mechanics, see Rowley and Dawson (2017), Mezić (2013), and Towne et al. (2018).

13.8.1 Balanced Truncation and Balanced POD

While POD modes optimally span the vector or function space of the data set used to obtain them, they are not necessarily optimal for use in Galerkin projection. Balanced truncation addresses this by working in a linear input–output framework. Whereas POD is based on maximizing the energy accounted for in each successive mode, *balanced truncation* approaches the problem from a philosophically different point of view based on the controllability and observability of a system. The most controllable states of a system are those that are most susceptible to the inputs via forcing or control, while the most observable states are the ones that lead to the largest outputs. Inputs arise from external influences on the system, whereas outputs are the quantities that are sought in order to identify the state of the system, that is, observations are the measurements taken of the system.

The motivation for balanced truncation arises out of control theory and is based on balancing the controllability and observability of the system (see section 10.4.3 of Cassel 2013). The idea is that the most useful ROM is that which contains both the most controllable modes and the most observable modes. This is the balancing aspect; the truncation comes in when the least controllable and observable states are neglected. This requires a technique that produces two sets of modes, one corresponding to the inputs and one corresponding to the outputs of the system. The controllability modes are obtained as in POD, while the observability modes are computed from the adjoint of the state equation. The resulting direct and adjoint modes are biorthogonal with one another. Note that whereas POD is executed on any data set, and it is not necessary to have a model of the system, balanced truncation requires evaluation of adjoints; therefore, a model of the system is necessary. The forward direct state and backward adjoint systems must be linear, and the input data matrix is comprised of simulation data from solving the state and adjoint systems.

The controllability and observability of a system are quantified by Gram matrices that define the extent of the controllability and observability of the various states. The objective of balanced truncation is to determine the transformation that diagonalizes these Gramians, thereby balancing the controllable and observable modes, such that the most controllable modes are also the most observable ones. The balancing transformation is determined using SVD, which then allows one to easily truncate the modes that have small singular values. The modes that are most observable are of interest because they represent those that are most susceptible to influence dynamically within the system.

Balanced truncation is computationally more expensive than POD as it involves calculation of the forward state and the backward adjoint solution. Moreover, it is limited to linear systems, whereas POD applies to any data set regardless of its origin. This also means that balanced truncation is not suitable for purely experimental data. In order to reduce the computational cost of balanced truncation, *balanced POD* has been developed. It is an approximation of balanced truncation that is more computationally efficient by incorporating the method of snapshots from POD and using the adjoint modes for the projection. It involves approximating the very large Gramians using data from simulations, which greatly reduces the size of the Gramians for use in the SVD calculation. Balanced POD is somewhat of a hybrid data- and model-driven approach in that the input to the algorithm is data, but the data are obtained directly from a model (and its adjoint) of the system. Interestingly, Ma et al. (2011) have shown that balanced POD is mathematically equivalent to the eigensystem realization algorithm (ERA) popular in system identification as discussed in Section 13.9. Because the ERA is a purely data-driven method, there is potential for mitigating some of the disadvantages of balanced POD, as no solutions of the adjoint system are required for ERA.

Balanced POD is typically more effective than traditional POD for nonnormal systems that exhibit transient growth (see Section 5.3.3). This is because the POD modes that contain the most energy are not necessarily those that are most dynamically significant in a nonnormal system. In other words, the POD modes may not include those small modes that are most susceptible to transient growth in finite time. This is clearly illustrated by a series of examples in chapter 5 of Holmes et al. (2012). As with balanced truncation, each of the two sets of modes that result from the balanced POD procedure are not orthogonal; however, these balancing modes and adjoint modes are biorthogonal. See Holmes et al. (2012) for more details on balanced truncation and balanced POD.

13.8.2 Koopman Operator and Dynamic-Mode Decomposition

When POD is applied to time-dependent data using the method of snapshots, the temporal behavior of the system is simply treated as a series of unrelated data samples. Therefore, it cannot be expected to faithfully capture the time evolution of coherent structures contained within the system. Instead, the temporal snapshots are incorporated into formation of the spatial basis vectors or functions. This is overcome through use of the *Koopman operator*, which specifies how a system is transformed from one snapshot in time to the next. In addition, the Koopman operator allows for treatment of nonlinear systems.

We have seen that systems governed by linear differential equations can only exhibit a limited range of behaviors, namely some combination of oscillatory and/or exponential growth or decay in time. Therefore, the form of the solution is known – in terms of exponentials – and the task is to determine the constants in the exponentials. These constants are eigenvalues, which in general can be complex, of the coefficient matrix that describes the state-space form of the system. The eigenvalues give the temporal behavior, and the eigenvectors give the spatial modes that are subject to the temporal evolution. The Koopman modes, which are the eigenfunctions of the Koopman operator, contain information about the oscillation frequency and exponential growth or decay rate in the dynamical system. They are not necessarily orthogonal. Koopman modes are similar to global stability modes as they are spatial modes that evolve at a single frequency.

For nonlinear systems, it has been shown by Mezić (2005) that the finite-dimensional, nonlinear state-space formulation of the discrete problem can be converted to an equivalent infinite-dimensional, linear Koopman operator. The eigenfunctions of the Koopman operator give a nonlinear coordinate transformation in which the system is linear. In this way, the nonlinearity is traded for infinite dimensionality. While linear Koopman operators can capture certain types of nonlinear behavior, it cannot represent all such dynamics. In particular, deterministic chaos, as discussed in Section 5.6, cannot be captured using linear Koopman modes.

DMD is the most common approach to determining discrete approximations of the continuous Koopman modes. This connection between the Koopman operator and DMD has been established by Rowley (2005) and applies when the DMD modes that approximate the Koopman modes have linearly independent snapshots.

As suggested by the name, DMD focuses on capturing the time-evolving aspects of a system. Each DMD mode contains a single oscillation frequency and exponential growth (or decay) rate in a manner similar to modal stability analysis (see chapter 9 of Cassel 2013). In fact, DMD analysis is equivalent to a modal analysis of a system using Fourier analysis in certain situations. Note, however, that the ability to separate distinct Fourier modes depends on the linearity of the system. In a DMD context, then, the time-dependent coefficients in our previous methods would be of the form

$$a_n(t) = g_n e^{\lambda_n t},$$

where the eigenvalues λ_n may be complex, and g_n are constants.

The input to a DMD analysis is two data matrices obtained from the same system that contain a series of data measurement pairs that are separated by a fixed time interval Δt. In other words, a set of quantities are measured at a series of time steps that contain information about the evolving dynamics of the system. The data are placed in two matrices that are offset by one time step. The DMD analysis then identifies common spatial structures, the DMD modes, along with their growth or decay rates, which are the eigenvalues. It uses an Arnoldi-type algorithm to obtain approximations of the eigenvalues and eigenvectors (see Section 6.5.3). The DMD modes are the eigenvectors of the product of the pseudo-inverse of one of the data matrices with the other data matrix.

Say we have the two data matrices

$$\mathbf{X} = \begin{bmatrix} \mathbf{x}(t_1) & \mathbf{x}(t_2) & \cdots & \mathbf{x}(t_m) & \cdots & \mathbf{x}(t_M) \end{bmatrix},$$

and

$$\mathbf{Y} = \begin{bmatrix} \mathbf{x}(t_1 + \Delta t) & \mathbf{x}(t_2 + \Delta t) & \cdots & \mathbf{x}(t_m + \Delta t) & \cdots & \mathbf{x}(t_M + \Delta t) \end{bmatrix},$$

where the $\mathbf{x}$ vectors contain the discrete spatial data at each time, that is, snapshot. It is assumed that there is a linear transformation $\mathbf{K}$ that maps one snapshot to another in the form

$$\mathbf{x}(t_m + \Delta t) = \mathbf{K}\mathbf{x}(t_m).$$

The matrix $\mathbf{K}$ is a finite-dimensional approximation of the Koopman operator, which is infinite dimensional. The DMD eigenvalues and modes come from an eigen-decomposition of

$$\mathbf{A} = \mathbf{Y}\mathbf{X}^+,$$

where the superscript $+$ indicates the pseudo-inverse (see Section 2.8).

As with POD, DMD is a purely data-driven method. However, the DMD modes are not guaranteed to be orthogonal, nor are they naturally ordered from most to least significant as in POD. Because it assumes linearity of the underlying system, DMD analysis often performs poorly if the data are obtained from a nonlinear system, which can only be compensated for by including data at finer and finer time intervals. In addition, note that whereas the DMD analysis is applied directly to the data, POD analysis is applied to the covariance matrix, a statistical representation of the original data.

REMARKS:

1. *Because DMD modes are generally not orthogonal, as opposed to providing basis vectors or functions for a Galerkin projection–based ROM, DMD is primarily used to identify important time-evolving features in a solution in order to better understand its dynamics.*
2. *Koopman and DMD analysis are more closely aligned with spectral Fourier analysis than POD, which explains their relationship to global mode stability analysis. Specifically, the DMD modes of a linear system supply the global stability modes of the system.*
3. *DMD was introduced by Schmid (2010). See Mezić (2013) for a review of the applications of methods based on Koopman operators in fluid mechanics. Also, see Hemati et al. (2017) for an example of current research in DMD algorithms. Finally, see Kutz et al. (2016) for more on DMD.*

13.8.3 Resolvent Analysis

We motivate and define the resolvent matrix or operator by considering a non-homogeneous, first-order, linear system of ordinary differential equations. Given a constant matrix $\mathbf{A}$, we have

$$\dot{\mathbf{u}}(t) = \mathbf{A}\mathbf{u}(t) + \mathbf{f}(t), \tag{13.84}$$

where $\mathbf{u}(t)$ is the state vector of outputs, and $\mathbf{f}(t)$ are inputs to the system. For a linear ordinary differential equation with constant coefficients, we know that the homogeneous solutions are of the form (see Section 2.6.1)

$$\mathbf{u}_H(t) = \mathbf{u}_n e^{\lambda_n t}. \tag{13.85}$$

Given that

$$\dot{\mathbf{u}}_H(t) = \lambda_n \mathbf{u}_n e^{\lambda_n t},$$

the homogeneous solution (with $\mathbf{f} = \mathbf{0}$) has λ_n and $\mathbf{u}_n$ that satisfy the eigenproblem

$$\mathbf{A}\mathbf{u}_n = \lambda_n \mathbf{u}_n. \tag{13.86}$$

Let us consider the case with harmonic forcing at frequency ω according to

$$\mathbf{f}(t) = \hat{\mathbf{f}} e^{\mathrm{i}\omega t}, \tag{13.87}$$

where $\hat{\mathbf{f}}$ is a constant vector. Then the particular solution is of the form

$$\mathbf{u}_P(t) = \hat{\mathbf{u}} e^{\mathrm{i}\omega t}, \tag{13.88}$$

where $\hat{\mathbf{u}}$ is a constant vector. Then

$$\dot{\mathbf{u}}_P(t) = \mathrm{i}\omega \hat{\mathbf{u}} e^{\mathrm{i}\omega t}.$$

Substituting the particular solution (13.88) and the harmonic forcing (13.87) into (13.84) yields

$$\mathrm{i}\omega \hat{\mathbf{u}} e^{\mathrm{i}\omega t} = \mathbf{A}\hat{\mathbf{u}} e^{\mathrm{i}\omega t} + \hat{\mathbf{f}} e^{\mathrm{i}\omega t},$$

or

$$(\mathrm{i}\omega \mathbf{I} - \mathbf{A})\, \hat{\mathbf{u}} = \hat{\mathbf{f}}.$$

Thus, the vector in the particular solution (13.88) becomes

$$\hat{\mathbf{u}} = (\mathrm{i}\omega \mathbf{I} - \mathbf{A})^{-1} \hat{\mathbf{f}}. \tag{13.89}$$

The general solution of the system of linear ordinary differential equations (13.84) is then

$$\mathbf{u}(t) = \mathbf{u}_H(t) + \mathbf{u}_P(t) = \mathbf{u}_n e^{\lambda_n t} + (\mathrm{i}\omega \mathbf{I} - \mathbf{A})^{-1} \hat{\mathbf{f}} e^{\mathrm{i}\omega t}. \tag{13.90}$$

The *resolvent matrix* is defined by

$$\mathbf{R}_\omega = (\mathrm{i}\omega \mathbf{I} - \mathbf{A})^{-1}.$$

It is a linear operator that relates the input $\hat{\mathbf{f}}$ to a system to its corresponding output $\hat{\mathbf{u}}$ according to (13.89). Observe that the resolvent matrix does not exist, that is, $(\mathrm{i}\omega \mathbf{I} - \mathbf{A})$ is singular, when $\mathrm{i}\omega = \lambda$ is an eigenvalue of $\mathbf{A}$. When $\mathrm{i}\omega \neq \lambda$, the resolvent matrix determines how the inputs $\mathbf{f}(t)$ are related to the outputs $\mathbf{u}(t)$. The inputs could be in the form of any forcing functions, corresponding to actuation, for example, and the outputs could be any observable quantities.

Resolvent analysis consists of determining the maximum amplification (gain) in the output of the system for each frequency ω of input in a manner similar to transient growth stability analysis in Section 5.3.3. Given an SVD of the resolvent matrix $\mathbf{R}_\omega$, the largest singular value equals the square of the maximum gain, and the right and left singular vectors provide the corresponding input (forcing) and output (state) vectors, respectively.

REMARKS:

1. *For nonlinear systems, the nonlinear governing equation is rewritten as a linear equation with the nonlinear terms being included in the forcing function* $\mathbf{f}(t)$.
2. *Note the correspondence between global stability analysis* ($i\omega = -ic$) *and harmonic forcing.*
3. *The reason for this collection of extensions and alternatives to standard POD is largely due to the fact that we are attempting to model data from systems that are often highly nonlinear with fundamentally linear methods. As a result, there is limited guidance as to which method is most appropriate or provides the best results for the data from a given system.*

13.9 System Identification

Reduced-order modeling as presented in this chapter depends on knowledge of a mathematical model of the system under consideration in order to project the equation(s) onto the POD modes to produce the reduced-order model. While the POD modes are generated directly using data from the system itself, that is, they are data driven, the governing equations are required to perform the Galerkin projection. What if no such system model is available? Can a model for the system be generated solely from data produced by that system? This is the subject of *system identification*.

Typically, system identification consists of choosing a model structure, such as the linear state-space model considered throughout this chapter, but where the coefficients and model parameters within the mathematical model are to be determined using an optimization procedure. In its simplest form, system identification is an extension of least-squares methods. For example, polynomial regression is an example of system identification in which the form of the polynomial is specified beforehand, and data obtained from the system are used to determine the coefficients in the polynomial trial function using least-squares methods. This produces an algebraic model.

Recall that in state estimation, we use data and a system model to determine the optimal state of the system. Rather than using data and a system model to estimate the state, however, the goal of system identification is to use the data to determine the system model itself. A popular method for performing this is called the *eigensystem realization algorithm* (ERA), which is used to determine the matrices in a linear state-space model of a system. By determining these model parameters, a relationship between the inputs to the system and the corresponding outputs that they produce is identified. The ERA is closely related to balanced POD discussed in Section 13.8.1 but without the need to calculate the adjoint equations. Of course, if the model is indeed linear, then such a state-space approach would be appropriate. If it is nonlinear, then more advanced system identification methods would need to be employed. For example, DMD introduced in the previous section can be used for this purpose.

In general, we prefer a mathematical model of a system that is obtained from first principles for analysis and prediction. As with other reduced-order models, however, a simple model obtained using system identification methods may be sufficient for use in closed-loop control settings, where the accuracy of the model can be sacrificed for computational speed as the control must be computed on-the-fly during the system's operation. System identification is a broad and growing topic of research. Because it largely relies on statistical methods and – increasingly – machine learning techniques, which are beyond our scope, we must leave it to others. For recent treatments of system identification methods, see Brunton et al. (2016) and Duriez et al. (2017).

13.10 Epilogue

Observe how in this chapter, we have progressed from the inverse variational problem to Galerkin projection and formulated the proper-orthogonal decomposition optimization problem. POD modes are the optimal basis vectors or functions from a data set for use in Galerkin projection to formulate a reduced-order model of the discrete or continuous system from which the data arose. In this way, we have taken maximum advantage of both variational and matrix methods to formulate a comprehensive framework using projection-based methods to better understand and solve complex linear and nonlinear dynamical systems. This framework is illustrated in Figure 13.1, which summarizes the various sources of basis functions considered throughout the text and the applications of Galerkin projection in numerical methods and optimization and control.

We encountered Fourier modes, Chebyshev polynomials, and Legendre polynomials in Chapter 3. These functions are solutions of self-adjoint, that is, Sturm–Liouville, differential equations and provide convenient orthogonal basis functions for use in spectral numerical methods as discussed in Chapter 7. Linear and quadratic shape functions are commonly used as basis functions in Rayleigh–Ritz and finite-element methods as introduced in Chapter 7 of the present text and chapter 3 of Cassel (2013). In this chapter, we have focused on data-driven methods in which the basis functions are obtained directly from experimental or numerical data. In particular, we focused on feature-extraction techniques, such as POD, that provide the optimal basis functions or vectors for the data from a particular system. The data, along with a mathematical model of the system, can then be used to obtain a reduced-order model of the system for use in optimization and control settings, for example. In addition, preselected polynomial basis functions are the basis for least-squares regression as described in Chapter 11.

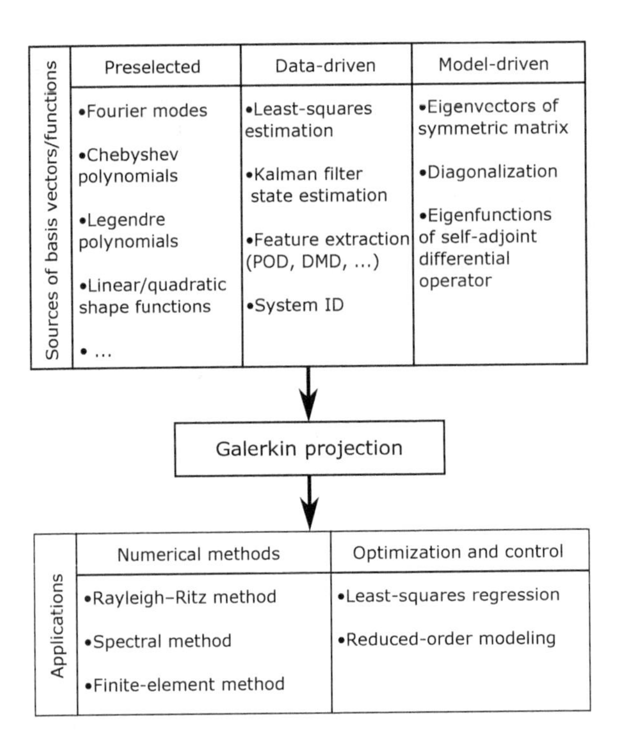

Figure 13.1 Summary of projection-based methods.

As we have seen throughout the text, the selection or determination of basis vectors or functions plays a significant role in a wide variety of solution techniques, numerical methods, and dynamical system diagnostics. They may be chosen for their mathematical properties or determined from actual data based on an optimization principle. In either case, Galerkin projection provides the link between the basis vectors or functions and the various applications.

References

Alvarez-Ramirez, J., Solis-Duan, J., and Puebla, H. (2005). Control of the Lorenz system: destroying the homoclinic orbits. *Physics Letters A* **338**, 128–140.

Aref, H. and Balachandar, S. (2018). *A first course in computational fluid dynamics*, Cambridge University Press.

Arnoldi, W. E. (1951). The principle of minimized iterations in the solution of the matrix eigenvalue problem. *Quarterly of Applied Mathematics* **9**, 17–29.

Asmar, N. H. (2005). *Partial differential equations with Fourier series and boundary value problems*, 2*nd* edition, Pearson Prentice Hall.

Benner, P., Gugercin, S., and Willcox, K. (2015). A survey of projection-based model reduction methods for parametric dynamical systems. *SIAM Review* **57**(4), 483–531.

Borrelli, F., Bemporad, A., and Morari, M. (2017). *Predictive control for linear and hybrid systems*, Cambridge University Press.

Boyd, S. and Vandenberghe, L. (2018). *Introduction to applied linear algebra: vectors, matrices, and least squares*, Cambridge University Press.

Briggs, W. C., Henson, V. E., and McCormick, S. F. (2000). *A multigrid tutorial*, 2*nd* edition, SIAM.

Brunton, S. L., Proctor, J. L., and Kutz, J. N. (2016). Discovering governing equations from data by sparse identification of nonlinear dynamical systems. *Proceedings of the National Academy of Sciences* **113**(15), 3932–3937.

Calafiore, G. C. and El Ghaoui, L. (2014). *Optimization models*, Cambridge University Press.

Canuto, C., Hussain, M., Quarteroni, A., and Zang, T. A. (1988). *Spectral methods in fluid dynamics*, Spring-Verlag.

Cassel, K. W. (2013). *Variational methods with applications in science and engineering*, Cambridge University Press.

Charru, F. (2011). *Hydrodynamic instabilities*, Cambridge University Press.

Chung, T. J. (2010). *Computational fluid dynamics*, Cambridge University Press.

Demmel, J. W. (1997). *Applied numerical linear algebra*, SIAM.

Drazen, P. G. (1992). *Nonlinear systems*, Cambridge University Press.

Drazin, P. G. (2002). *Introduction to hydrodynamic stability*, Cambridge University Press.

Duriez, T., Brunton, S., and Noack, B. R. (2017). *Machine learning control – taming nonlinear dynamics and turbulence*, Springer.

E, W. and Liu, J. G. (1996). Essentially compact schemes for unsteady viscous incompressible flows. *Journal of Computational Physics* **126**, 122–138.

Farrell, B. F. and Ioannou, P. J. (1996). Generalized stability theory. Part 1: autonomous operators. *Journal of the Atmospheric Sciences* **53**, 2025–2040.

Fasshauer, G. F. (2007). *Meshfree approximation methods with MATLAB*, World Scientific Publishing.

Ferziger, J. H., Peric, M., and Street, R. L. (2020). *Computational methods for fluid dynamics*, 4th edition, Springer-Verlag.

Fletcher, C. A. J. (1984). *Computational Galerkin methods*, Springer-Verlag.

Fletcher, C. A. J. (1991a). *Computational techniques for fluid dynamics 1: fundamental and general techniques*, Springer-Verlag.

Fletcher, C. A. J. (1991b). *Computational techniques for fluid dynamics 2: specific techniques for different flow categories*, Springer-Verlag.

Fletcher, R. and Reeves, C. M. (1964). Function minimization by conjugate gradients. *Computer Journal* **7**, 149–154.

Gilat, A. and Subramaniam, V. (2014). *Numerical methods for engineers and scientists: an introduction with applications using MATLAB*, 3*rd* edition, John Wiley and Sons.

Golub, G. H. and Van Loan, C. F. (2013). *Matrix computations*, 4*th* edition, Johns Hopkins University Press.

Greenberg, M. D. (1998). *Advanced engineering mathematics*, 2*nd* edition, Prentice Hall.

Gustafsson, B. (2008). *High order difference methods for time dependent PDE*, Springer-Verlag.

Hemati, M. S., Rowley, C. W., Deem, E. A., and Cattafesta, L. N. (2017). De-biasing the dynamic mode decomposition for applied Koopman spectral analysis of noisy datasets. *Theoretical and Computational Fluid Dynamics* **31**, 349–368.

Higham, N. J. (2002). *Accuracy and stability of numerical algorithms*, SIAM.

Hildebrand, F. B. (1976). *Advanced calculus for applications*, Prentice Hall.

Holmes, P., Lumley, J. L., Berkooz, G., and Rowley, C. W. (2012). *Turbulence, coherent structures, dynamical systems and symmetry*, 2*nd* edition, Cambridge University Press.

Horn, R. A. and Johnson, C. R. (2013). *Matrix analysis*, 2*nd* edition, Cambridge University Press.

Incropera, F. P., DeWitt, D. P., Bergman, T. L., and Lavine, A. S. (2007). *Fundamentals of heat and mass transfer*, 6*th* edition, John Wiley and Sons.

Jeffrey, A. (2002). *Advanced engineering mathematics*, Harcourt/Academic Press.

Kalman, R. E. (1960). A new approach to linear filtering and prediction problems. *Transactions of the ASME, Series D, Journal of Basic Engineering* **82**, 35–45.

Kreyszig, E. (2011). *Advanced engineering mathematics*, 10*th* edition, John Wiley and Sons.

Kutz, J. N., Brunton, S. L., Brunton, B. W., and Proctor, J. L. (2016). *Dynamic mode decomposition – data-driven modeling of complex systems*, SIAM.

Lay, D. C., Lay, S. R., and McDonald, J. J. (2016). *Linear algebra and its applications*, 5*th* edition, Pearson Education Limited.

Lele, S. K. (1992). Compact finite difference schemes with spectral-like resolution. *Journal of Computational Physics* **103**, 16–42.

Lorenz, E. N. (1963). Deterministic nonperiodic flow. *Journal of Atmospheric Sciences* **20**, 130–141.

Lumley, J. L. (1967). The structure of inhomogeneous turbulence, In A. M. Yaglom and V. I. Tatarski, editors, *Atmospheric turbulence and wave propagation*, 166–178. Nauka, Moscow.

Ma, Z., Ahuja, S., and Rowley, C. W. (2011). Reduced order models for control of fluids using the eigensystem realization algorithm. *Theoretical and Computational Fluid Dynamics* **25**, 233–247.

Mancho, A. M., Wiggins, S., Curbelo, J., and Mendoza, C. (2013). Lagrangian descriptors: a method for revealing phase space structures of general time dependent dynamical systems. *Communications in Nonlinear Science and Numerical Simulation* **18**, 3530–3557.

Meckes, E. S. and Meckes, M. W. (2018). *Linear algebra*, Cambridge University Press.

Meiss, J. D. (2007). *Differential dynamical systems*, SIAM.

Messac, A. (2015). *Optimization in practice with MATLAB*, Cambridge University Press.

Mezić, I. (2005). Spectral properties of dynamical systems, model reduction and decompositions. *Nonlinear Dynamics* **41**, 309–325.

Mezić, I. (2013). Analysis of fluid flows via spectral properties of the Koopman operator. *Annual Review of Fluid Mechanics* **45**, 357–378.

Miller, G. (2014). *Numerical analysis for engineers and scientists*, Cambridge University Press.

Moin, P. (2010). *Fundamentals of engineering numerical analysis*, 2*nd* edition, Cambridge University Press.

Morton, K. W. and Mayers, D. F. (1994). *Numerical solution of partial differential equations*, Cambridge University Press.

Nair, S. (2009). *Introduction to continuum mechanics*, Cambridge University Press.

Nayar, N. and Ortega, J. M. (1993). Computation of selected eigenvalues of generalized eigenvalue problems. *Journal of Computational Physics* **108**, 8–14.

O'Neil, P. V. (2012). *Advanced engineering mathematics*, 7*th* edition, Cengage Learning.

Pikovsky, A. and Politi, A. (2016). *Lyapunov exponents: a tool to explore complex dynamics*, Cambridge University Press.

Press, W. H., Teukolsky, S. A., Vetterling, W. T., and Flannery, B. P. (2007). *Numerical recipes – the art of scientific computing*, 3rd edition, Cambridge University Press.

Radke, R. (1996). A MATLAB implementation of the implicitly restarted Arnoldi method for solving large-scale eigenvalue problems, MS Thesis, Rice University.

Rao, S. S. (2002). *Applied numerical methods for engineers and scientists*, Prentice Hall.

Reddy, J. N. (2013). *An introduction to continuum mechanics*, 2*nd* edition, Cambridge University Press.

Rowley, C. W. (2005). Model reduction for fluids using balanced proper orthogonal decomposition. *International Journal of Bifurcation and Chaos* **15**(3), 997–1013.

Rowley, C. W. and Dawson, S. M. (2017). Model reduction for flow analysis and control. *Annual Review of Fluid Mechanics* **49**, 387–417.

Saad, Y. (2003). *Iterative methods for sparse linear systems*, SIAM.

Saad, Y. and Schultz, M. H. (1986). A generalized minimum residual algorithm for solving nonsymmetric linear systems. *SIAM Journal of Scientific and Statistical Computing* **7**, 856–869.

Saltzman, B. (1962). Finite amplitude free convection as an initial value problem – I, *Journal of the Atmospheric Sciences*, **19**, 329–341.

Schmid, P. J. (2010). Dynamic mode decomposition of numerical and experimental data. *Journal of Fluid Mechanics* **656**, 5–28.

Sirovich, L. (1987). Turbulence and the dynamics of coherent structures, Parts I–III. *Quarterly of Applied Mathematics* **XLV**(3), 561–582.

Strang, G. (2006). *Linear algebra and its applications*, 4*th* edition, Cengage Learning.

Strang, G. (2016). *Introduction to linear algebra*, 5*th* edition, Wellesley Cambridge Press.

Tannehill, J. C., Anderson, D. A., and Pletcher, R. H. (1997). *Computational fluid mechanics and heat transfer*, Taylor and Francis.

Thomas, J. L., Diskin, B. and Brandt, A. T. (2003). Textbook multigrid efficiency for fluid simulations. *Annual Review of Fluid Mechanics* **35**, 317–340.

Tolstykh, A. I. (1994). *High accuracy non-centered compact difference schemes for fluid dynamics applications*, World Scientific Publishing.

Towne, A., Schmidt, O., and Colonius, T. (2018). Spectral proper-orthogonal decomposition and its relationship to dynamic mode decomposition and resolvent analysis. *Journal of Fluid Mechanics* **847**, 821–867.

Trefethen, L. N. and Bau, D. B. (1997). *Numerical linear algebra*, SIAM.

Trefethen, L. N. and Embree, M. (2005). *Spectra and pseudospectra: the behavior of nonnormal matrices and operators*, Princeton University Press.

Vanka, S. P. (1986). Block-implicit multigrid solution of Navier–Stokes equations in primitive variables. *Journal of Computational Physics* **65**, 138–158.

Viswanath, D. (2003). Symbolic dynamics and periodic orbits of the Lorenz attractor. *Nonlinearity* **16**, 1035–1056.

Viswanath, D. (2007). Recurrent motions within plane Couette turbulence. *Journal of Fluid Mechanics* **580**, 339–358.

Yariv, A. and Yeh, P. (2007). *Photonics: optical electronics in modern communications*, 6th edition, Oxford University Press.

Index

CPSIA information can be obtained
at www.ICGtesting.com
Printed in the USA
LVHW102043250221
679955LV00011B/413